Chemistry of Aluminium, Gallium, Indium and Thallium

Chemistry of Aluminium, Gallium, Indium and Thallium

Edited by

A.J. DOWNS
Inorganic Chemistry Laboratory
University of Oxford

BLACKIE ACADEMIC & PROFESSIONAL
An Imprint of Chapman & Hall
London · Glasgow · New York · Tokyo · Melbourne · Madras

Published by
Blackie Academic & Professional, an imprint of Chapman & Hall,
Wester Cleddens Road, Bishopbriggs, Glasgow G64 2NZ

Chapman & Hall, 2–6 Boundary Row, London SE1 8HN, UK

Blackie Academic & Professional, Wester Cleddens Road, Bishopbriggs,
Glasgow G64 2NZ, UK

Chapman & Hall Inc., 29 West 35th Street, New York NY 10001, USA

Chapman & Hall Japan, Thomson Publishing Japan, Hirakawacho
Nemoto Building, 6F, 1-7-11 Hirakawa-cho, Chiyoda-ku, Tokyo 102, Japan

DA Book (Aust.) Pty Ltd, 648 Whitehorse Road, Mitcham 3132, Victoria,
Australia

Chapman & Hall India, R. Seshadri, 32 Second Main Road, CIT East,
Madras 600 035, India

First edition 1993

© Chapman & Hall, 1993

Typeset in 10/12pt Times by EJS Chemical Composition, Bath

Printed in Great Britain by St Edmundsbury Press Ltd,
Bury St Edmunds, Suffolk

ISBN   0 7514 0103 X   *Po 4665*

A catalogue record for this book is available from the British Library

Library of Congress Cataloging-in-Publication data

Chemistry of aluminium, gallium, indium and thallium / edited by A.J.
    Downs. -- 1st ed.
        p.   cm.
    Includes bibliographical references and index.
    ISBN 0-7514-0103-X (acid-free)
    1. Aluminum.   2. Gallium.   3. Indium.   4. Thallium.
    I. Downs, A.J. (Anthony John), 1936–
QD181.A4C48   1993
546'.67--dc20   92-47139
                                                         CIP

Preface

Boron has all the best tunes. That may well be the first impression of the Group 13 elements. The chemical literature fosters the impression not only in the primary journals, but also in a steady outflow of books focussing more or less closely on boron and its compounds. The same preoccupation with boron is apparent in the coverage received by the Group 13 elements in the comprehensive and regularly updated volume of the Gmelin Handbook. Yet such an imbalance cannot be explained by any inherent lack of variety, interest or consequence in the heavier elements. Aluminium is the most abundant metal in the earth's crust; in the industrialised world the metal is second only to iron in its usage, and its compounds can justifiably be said to touch our lives daily — to the potential detriment of those and other lives, some would argue. From being chemical curios, gallium and indium have now gained considerably prominence as sources of compound semiconductors like gallium arsenide and indium antimonide. Nor is there any want of incident in the chemistries of the heavier Group 13 elements. In their redox, coordination and structural properties, there is to be found music indeed, notable not always for its harmony but invariably for its richness and variety.

This book seeks to redress the balance with a definitive, wide-ranging and up-to-date review of the chemistry of the Group 13 metals aluminium, gallium, indium and thallium. It takes as one of its chief reference points the excellent account, also entitled *The Chemistry of Aluminium, Gallium, Indium and Thallium*, written by Wade and Banister some twenty years ago. The aim has been to treat important advances featuring the Group 13 metals and their compounds, particularly in the past two decades, but to do this not as a catalogue nor in a piecemeal fashion, but within a framework designed to place new facts, developments and applications in the context of more general patterns of physical and chemical behaviour. The chapters have been written by members of an international team of authors, each selected as an expert, with practising experience in industrial or academic research, in the particular field under review.

After an outline, in chapter 1, of some of the distinctive features of the elements and their compounds, chapter 2 treats the elements themselves with due emphasis on the metallurgical and commercial aspects. The inorganic derivatives are then described in chapter 3, a substantial account divided mainly according to the oxidation state of the Group 13 metal. There follow two chapters reviewing solid compounds with extended structures: chapter 4 concerning oxide and hydroxide derivatives (dominated by the aluminium compounds) and chapter 5 concerning III–V compounds. The organometallic compounds are described in chapter 6, with a separate review, in chapter 7, detailing the roles of these and other Group 13 compounds in organic synthesis. The theme of chapter 8 is coordination and solution chemistry. This leads, naturally enough, to accounts (i) of environmental aspects, with particular reference to biochemical factors (chapter 9) and (ii) of the analytical chemistry of the elements (chapter 10).

A book of this scale cannot pretend to be comprehensive, even when it makes reference to some 2000 original papers, books and review articles, some published as recently as late 1992. We are well aware that some compounds and some topics have had relatively short shrift. Such is the case, for example, with the burgeoning topics of zeolites and aluminophosphates, themselves the subject of numerous specialist reviews, and readers will look in vain for information on materials like cements where aluminium is typically a secondary — but influential — component. We are aware too of the overlap that exists between some of the chapters. Some overlap is inescapable in an edited volume like this. While it may mean that our coverage is not everywhere as efficient as it might be, we dare to hope that the different perspectives, as well as the cross-linking between chapters, will actually add to, and not diminish, the book's appeal.

In aiming for a clear, concise and structured treatment, we have tried also to maintain an accessible style designed to give a text that will be found engaging and readable, by non-specialist no less than specialist readers. We see the book therefore not just as a source-book on the Group 13 metals, but as a monograph that can be read with profit by scientists in various walks of life. We hope that it will be found useful, not only by researchers and other experienced workers, but also by senior undergraduate students and their teachers. It is directed mainly at chemists and materials scientists, but includes sections that will surely court the interests of biochemists, physicists, chemical engineers and environmental and industrial scientists.

A.J.D.

Contents

6 Organometallic compounds: synthesis and properties   322
K.B. STAROWIEYSKI

7 Organic transformations mediated by Group 13 metal compounds   372
J.A. MILLER

8 The coordination and solution chemistry of aluminium, gallium, indium and thallium **430**

D.G. TUCK

9 The elements in the environment **474**

R.B. MARTIN

10 Analytical methods

H. ONISHI

Contributors

Dr A.J. Downs — Inorganic Chemistry Laboratory, University of Oxford, South Parks Road, Oxford OX1 3QR, UK

Professor I.J. Polmear — Department of Materials Engineering, Monash University, Clayton, Victoria 3168, Australia

Dr M.J. Taylor — Department of Chemistry, University of Auckland, Private Bag, Auckland, New Zealand

Dr P.J. Brothers — Department of Chemistry, University of Auckland, Private Bag, Auckland, New Zealand

Dr K.A. Evans — Development Director, Alcan Chemicals Ltd, Chalfont Park, Gerrards Cross, Buckinghamshire SL9 0QB, UK

Dr I.R. Grant — MCP Wafer Technology Ltd, 34 Maryland Road, Tongwell, Milton Keynes MK15 8HJ, UK

Dr K.B. Starowieyski — Department of Chemistry, Technical University (Politechnika), Noakowskiego 3, 00-664 Warsaw, Poland

Dr J.A. Miller — Exxon Chemical Company, Basic Chemicals Group, PO Box 4900, Baytown, Texas 77522-4900, USA

Professor D.G. Tuck — Department of Chemistry, University of Windsor, 401 Sunset Avenue, Windsor, Ontario N9B 3P4, Canada

Professor R.B. Martin — Chemistry Department, University of Virginia, McCormick Road, Charlottesville, VA 22903, USA

Professor H. Onishi — Department of Chemistry, University of Tsukuba, Tsukuba-shi, Ibaraki, Japan

1 Chemistry of the Group 13 metals: some themes and variations

A.J. DOWNS

> A sweet disorder in the dress
> Kindles in clothes a wantonness:
> A lawn about the shoulders thrown
> Into a fine distraction...
> A careless shoe-string, in whose tie
> I see a wild civility:
> Do more bewitch me, than when Art
> Is too precise in every part.
>
> Herrick: *Delight in Disorder*

1.1 Introduction: some characteristics of the Group 13 elements

Few families of elements in the Periodic Table are more given to 'wild civility' than the members of Group 13. The five elements in question, boron, aluminium, gallium, indium and thallium, each have atoms whose electronic ground state finds the three valence electrons with the configuration ns^2np^1. Herein is the root cause of the many similarities pervading the chemistries of the elements — witness, for example, the dominance of the +3 oxidation state and the acceptor properties which typify the resulting derivatives. Similarities may abound, but it is the *diversity* or *non-conformity* of behaviour that tends more often to catch the eye.[1-6] For the Group is strong in chemical individualism, exemplifying well the infinite variety of chemical personality among the elements which remains one of the most remarkable phenomena of the universe.

Most eccentric of the elements is unquestionably boron which is a non-metal unique in many ways and often seeming to have more in common with its horizontal neighbour carbon and its diagonal neighbour silicon than with its vertical neighbour aluminium. As revealed by the numerical properties summarised in Table 1.1 and illustrated in part in Figure 1.1, boron stands apart from the other members of Group 13 in the relatively tight binding of its valence electrons to the nucleus (signalled by the first three ionisation energies which are all substantially larger than the corresponding parameters for the other Group 13 elements) and the small size of the atom in combination. These factors, coupled with an electronegativity similar to that of a number of other important atoms (e.g. C and H), lead to an extensive and unusual type of covalent (molecular) chemistry in which multi-centre bonding is perhaps the most distinctive feature. There are, moreover, two other properties of the boron atom which have a significant

Table 1.1 Some properties of the Group 13 elements boron, aluminium, gallium, indium and thallium.

Property	B
(i) Properties of the isolated atom[a–d]	
Atomic number	5
Naturally occurring isotopes[a,e]	^{10}B (19.82%), ^{11}B (80.18%)
Relative atomic mass (^{12}C $=$ 12.0000)[a,f]	10.811
Ground-state electron configuration (term)	[He]$2s^2 2p^1$ ($^2P_{1/2}$)
Spin-own-orbit coupling constant (A/cm^{-1})[b,g]	10.17
Ionisation energies (kJ mol^{-1})[a,b,g]	
$\quad$ M $\rightarrow$ M$^+$	800.6
$\quad$ M$^+$ $\rightarrow$ M^{2+}	2427
$\quad$ M^{2+} $\rightarrow$ M^{3+}	3660
$\quad$ M^{3+} $\rightarrow$ M^{4+}	25025
Promotion energy, $ns^2 np^1$ ($^2P_{1/2}$) $\rightarrow$ $ns^1 np^2$ ($^4P_{1/2}$) (kJ mol^{-1})[g]	345
Main lines in atomic spectrum: wavelength (nm) (species)[a,b,h]	208.891 (I)
	208.957 (I)
	249.667 (I)
	249.773 (I) (AA)
	345.129 (I)
	1166.004 (I)
	1166.247 (I)
Electron affinity [M(g) $\rightarrow$ M$^-$(g)] (kJ mol^{-1})[a,c,i,j]	26.7
Effective nuclear charge, Z^{*a}	
$\quad$ Slater value	2.60
$\quad$ Clementi value	2.42
$\quad$ Froese–Fischer value	2.27
Static average electric dipole polarisability for ground-state atom (Å^3)[b]	3.03
Thermal neutron capture cross-section (barns)[a]	^{10}B 3837, ^{11}B 0.005
Nuclei accessible to NMR experiments (nuclear spin, I)[a,e]	[^{10}B (3)], ^{11}B (3/2)]
(ii) Properties of the bound atom	
Electronegativity, χ	
$\quad$ Pauling scale[a]	2.04
$\quad$ Allred scale[a]	2.01
$\quad$ Sanderson scales[k]	1.88
	1.53[B^I]
	2.28[B^{III}]
$\quad$ Pearson scale (eV)[a,l]	4.29
Atomic (metallic) radius (Å)[a,c,d]	0.80–0.90
Single-bond covalent radius (Å)[a,c,d]	0.88
van der Waals radius (Å)[a,c,d]	2.08
Covalent bonds in trivalent derivatives:	
$\quad$ length (Å) (mean bond enthalpy (kJ mol^{-1}))[a–d,m–r]	
$\quad$ M–M	1.75 (293)
$\quad$ M–H [coordination number]	1.19 (381) [4]
$\quad$ M–F	1.31 (646) [3]
$\quad$ M–Cl	1.74 (444) [3]
$\quad$ M–O	1.37 (520) [3]
$\quad$ M–C	1.58 (365) [3]^{bb}
Covalent bonds in univalent derivatives: length (Å) (D_0^0 (kJ mol^{-1}))[cc]	
$\quad$ M–H	1.232 (330)
$\quad$ M–F	1.263 (754)
$\quad$ M–Cl	1.716 (531)
$\quad$ M–Br	1.888 (433)
$\quad$ M–I	2.131 (c. 365)^{m,dd}

Al	Ga	In	Tl
13	31	49	81
^{27}Al (100%)	^{69}Ga (60.108%),	^{113}In (4.33%),	^{203}Tl (29.524%),
	^{71}Ga (39.892%)	^{115}In (95.67%)	^{205}Tl (70.476%)
26.981539	69.723	114.82	204.3833
[Ne]$3s^23p^1$ ($^2P_{1/2}$)	[Ar]$3d^{10}4s^24p^1$ ($^2P_{1/2}$)	[Kr]$4d^{10}5s^25p^1$ ($^2P_{1/2}$)	[Xe]$4f^{14}5d^{10}6s^26p^1$ ($^2P_{1/2}$)
74.69	550.79	1475.07	5195.1
577.4	578.8	558.3	589.3
1816.6	1979	1820.6	1971.0
2744.6	2963	2704	2878
11577	6175	5210	(4900)
347	454	418	541
308.215 (I)	287.424 (I) (AA)	303.936 (I) (AA)	276.787 (I) (AA)
309.271 (I) (AA)	294.364 (I)	325.609 (I)	291.832 (I)
309.281 (I) (AA)	403.299 (I)	325.856 (I)	**351.924 (I)**
394.401 (I)	417.204 (I)	410.176 (I)	352.943 (I)
396.152 (I)	**639.656 (I)**	**451.131 (I)**	377.572 (I)
	641.344 (I)		535.046 (I)
42.5	29	c. 29	c. 20
3.50	5.00	5.00	5.00
4.07	6.22	8.47	12.25
3.64	6.72	9.66	13.50
6.8	8.12	10.2	7.6
0.233	2.9	194	3.4
^{27}Al (5/2)	[^{69}Ga (3/2)], ^{71}Ga (3/2)	[^{113}In (9/2)],	[^{203}Tl (1/2)],
		^{115}In (9/2)	^{205}Tl (1/2)
1.61	1.81	1.78	1.62[Tl^I], 2.04[Tl^{III}]
1.47	1.82	1.49	1.44
1.54	2.10	1.88	1.96
0.84[Al^I]	0.86[Ga^I]	0.71[In^I]	0.99[Tl^I]
1.71[Al^{III}]	2.42[Ga^{III}]	2.14[In^{III}]	2.25[Tl^{III}]
3.23	3.2	3.1	3.2
1.431	1.22–1.40	1.62–1.68	1.704 (α-form)
1.25	1.25	1.50	1.55
2.05	1.90	1.90	2.00
2.86 (188)	2.44 (113)	3.25 (100)	3.41 (c. 63)
c. 1.50 (285)[s]	c. 1.50 (272) [4][t]	? (238)	? (192)
1.63 (589) [3][u]	1.71 (602) [3][u]	2.05 (c. 525) [6][v]	2.28 (460) [8][w]
2.07 (425) [3][x]	2.10 (363) [4]	2.26 (327) [3][y]	2.42 (c. 250) [4][z]
1.74 (585) [4][aa]	1.82 (c. 430) [4][aa]	2.15 (c. 360) [6][aa]	2.24 (?) [6][aa]
1.96 (274) [3][bb]	1.97 (245) [3][bb]	2.16 (162) [3][bb]	2.21 (125) [3][bb]
1.648 (<295)	1.663 (<274)	1.838 (239)	1.870 (190)
1.654 (665)	1.774 (577)	1.985 (507)	2.084 (441)
2.130 (494)	2.202 (475)	2.401 (428)	2.485 (369)
2.295 (427)	2.352 (416)	2.543 (385)	2.618 (330)
2.537 (364)	2.575 (335)	2.754 (331)	2.814 (266)

Table 1.1 (*Contd.*)

Property	B
X-ray diffraction: mass absorption coefficient, μ/ϱ $(\text{cm}^2\,\text{g}^{-1})^a$	
Cu K_α	2.39
Mo K_α	0.392

(iii) Properties of the element

Property	B
Crystal structure (cell dimensions (Å)), space group^{a,d,q,r,ee}	α-**B** *tetragonal* ($a = 8.740, c = 5.06$), $P4_2/nnm$ β-**B** *tetragonal* ($a = 10.161, c = 14.283$), $P4_1$ or $P4_3$ α-**B** *rhombohedral* ($a = 5.057$, $\alpha = 58.06°$), $R\bar{3}m$ β-**B** *rhombohedral* ($a = 10.145$, $\alpha = 65.20°$), $R\bar{3}m$, $R32$, $R3m$ Other polymorphs reported and partially characterised, e.g. *cubic*, *rhombic*, *monoclinic* and *hexagonal forms*
Melting point $(\text{K})^{a-c,m,r,ff}$	2453
Boiling point $(\text{K})^{a-c,m,r,ff}$	4050
$\Delta_{\text{fusion}}H$ at m.p. $(\text{kJ mol}^{-1})^{a-c,m,r,ff}$	22.2
$\Delta_{\text{vap}}H$ at 298.15 K $(\text{kJ mol}^{-1})^{a-c,m,r,ff}$	577.8
$\Delta H^{\ominus}$ atomisation at 298.15 K $(\text{kJ mol}^{-1})^{a,b,m}$	565
Standard entropy, $S^{\ominus}$, at 298.15 K $(\text{J K}^{-1}\text{mol}^{-1})^{a,b,m}$	5.90
Density $(\text{kg m}^{-3})^{a,c,r,ff}$	2340 (β-rhomb), 293 K
Electrical resistivity $(\Omega\,\text{m})^a$	18 000, 273 K
Thermal conductivity at 300 K $(\text{W m}^{-1}\text{K}^{-1})^{a,b}$	27.0

(iv) Properties of the cations and redox behaviour

Property	B
Ionic radius for six-fold coordination (Å)	
M^{3+} ^{aa}	(0.27)
M^+	–
$\Delta_f H^{\ominus}[M^{3+}(\text{g})]$ $(\text{kJ mol}^{-1})^{b,m}$	7468.8
$\Delta_f H^{\ominus}[M^+(\text{g})]$ $(\text{kJ mol}^{-1})^{b,m}$	1369.6
Thermodynamic properties for aqueous species (kJ mol^{-1})	
$\Delta_f G^{\ominus}[M^{3+}(\text{aq})]$ std. state, $m = 1^{b,m}$	$-$ 967.7 [B(OH)$_3$]
$\Delta_f H^{\ominus}[M^{3+}(\text{aq})]$ std. state, $m = 1^{b,m}$	$-$ 1072.8 [B(OH)$_3$]
$\Delta_{\text{hydration}}G^{\ominus}[M^{3+}(\text{g})]$, single ion^{n,hh}	n.a.
$\Delta_{\text{hydration}}H^{\ominus}[M^{3+}(\text{g})]$, single ion^{n,hh}	n.a.
Standard reduction potentials, $E^{\ominus}$ $(\text{V})^{a,b,ii}$	
$M^{\text{III}}(\text{aq}) + 3e^- \rightarrow M(\text{s}), a_{H^+} = 1$	-0.889
$M^{\text{III}}(\text{aq}) + 3e^- \rightarrow M(\text{s}), a_{OH^-} = 1$	-1.811 [B(OH)$_4$]$^-$
$M^{\text{I}}(\text{aq}) + e^- \rightarrow M(\text{s}), a_{H^+} = 1$	–
$M^{\text{III}}(\text{aq}) + 2e^- \rightarrow M^{\text{I}}(\text{aq}), a_{H^+} = 1$	–

Al	Ga	In	Tl
48.6	67.9	243	224
5.16	60.1	29.3	119
f.c.c. (a = 4.04959), *Fm3m*	α-**Ga** *orthorhombic* (a = 4.5186, b = 7.6570, c = 4.5258), *Cmca* **β-Ga** *monoclinic* (a = 2.766, b = 8.053, c = 3.332, β = 92.03°), *C2/c* **γ-Ga** *rhombic* (a = 10.593, b = 13.523, c = 5.203), *Cmcm* **δ-Ga** *rhombohedral* (a = 7.729, α = 72.03°), $R\bar{3}m$ $T(\gamma \rightarrow \alpha)$ 238 K High pressure forms **Ga(II)** *b.c.c.* (a = 5.951), $I\bar{4}3d$ **Ga(III)** *b.c. tetragonal* (a = 2.808, c = 4.458), *I4/mm*	*Face-centred tetragonal* (a = 3.2530, c = 4.9455), *I4/mmm*	α-**Tl** *hexagonal* (a = 3.456, c = 5.525), $P6_3/mmc$ **β-Tl** *cubic* (a = 3.882), *Im3m* **γ-Tl** *f.c.c.* (a = 4.851), *Fm3m* $T(\alpha \rightarrow \beta)$ 503 K
933.25	302.93	429.55	577
2793	2693	2343	1746
10.67	5.59	3.27	4.31
321.9	285.0	242.8	180.9
330.0	277.0	243.3	182.2
28.30	40.88	57.82	64.18
2698, 293 K	5907, 293 K	7310, 298 K	11850, 293 K
2.6548×10^{-8}, 293 K	27×10^{-8}, 273 K[gg]	8.37×10^{-8}, 293 K	18.0×10^{-8}, 273 K
237	40.6	81.6	46.1
0.535	0.620	0.800	0.885
(*c*. 1.00)	1.13[a]	1.32[a,c]	1.50[aa]
5484.0	5816	5345.3	5639.2
910.09	861.9	807.8	777.73
-491.5	-159	-97.9	$+214.6$
-538.4	-212	-105	$+196.6$
-4540	-4550	-4020	-4000
-4680	-4690	-4110	-4110
-1.677	-0.549	-0.3382	$+0.741$
-2.328 [Al(OH)$_4$]$^-$	-1.326 [Ga(OH)$_4$]$^-$	-1.007 [In(OH)$_4$]$^-$	-0.163 (Tl$_2$O$_3$)
($+0.3$)	(-0.32)	(-0.126)	-0.336
(-2.7)	(-0.66)	(-0.444)	$+1.280$

Table 1.1 (*Contd.*)

Property	B
(v) **Environmental properties**[a,jj]	
Abundances	
Sun (number of atoms per 10^{12} of H)	<125
Meteorite (number of atoms normalised to solar values for abundant and involatile elements)	1000
Earth (p.p.b.)	400
Continental crust (p.p.b.)	10^4
Ocean (10^{-12} mol kg^{-1} water)	4.1×10^8
Residence time in ocean (years) (oxidation state)	10^7 (III)
Geological data[a,d]	
Chief ores and sources	Borax and kernite, $Na_2[B_4O_5(OH)_4] \cdot xH_2O$; colemanite, $Ca_2[B_3O_4(OH)_3]_2 \cdot 2H_2O$
World production (tonnes year^{-1})	1×10^6 (B_2O_3)
Reserves (tonnes)	270×10^6 (as B_2O_3)
Biological role[a,ll]	Essential to plants; toxic in excess
Human body (mg kg^{-1} dry mass)[a,d,jj]	100
Daily dietary intake (mg)[a]	1–3
Toxic intake (mg)	4000
Lethal intake (mg)	?

(*a*) J. Emsley, *The Elements*, 2nd edn., Clarendon Press, Oxford, 1991.

(*b*) D.R. Lide, editor-in-chief, *Handbook of Chemistry and Physics*, 73rd edn., CRC Press, Boca Raton, FL, 1992–1993.

(*c*) J.A. Dean, ed., *Lange's Handbook of Chemistry*, 13th edn., McGraw-Hill, New York, 1985.

(*d*) Ref. 4.

(*e*) C.M. Lederer and V.S. Shirley, *Table of Isotopes*, 7th edn., Wiley, New York, 1978; J.R. De Laeter, K.G. Heumann and K.J.R. Rosman, *J. Phys. Chem. Ref. Data*, 1991, **20**, 1327. Nuclei enclosed in square brackets are seldom used for NMR studies.

(*f*) IUPAC, Inorganic Chemistry Division, Commission on Atomic Weights and Isotopic Abundances, *Pure Appl. Chem.*, 1991, **63**, 975, 991.

(*g*) C.E. Moore, *Atomic Energy Levels*, Vols. I–III, Circular of the National Bureau of Standards 467, U.S. Government Printing Office, Washington, DC, 1949–1958; C.E. Moore, *Ionization Potentials and Ionization Limits Derived from the Analyses of Optical Spectra*, Nat. Stand. Ref. Data Ser., National Bureau of Standards, No. 34, US Government Printing Office, Washington, DC, 1970.

(*h*) I denotes lines arising from the neutral atom. Stronger lines only are listed, with the strongest shown in bold type. Lines with application in atomic absorption spectroscopy are indicated by 'AA'.

(*i*) Ref. 2.

(*j*) H. Hotop and W.C. Lineberger, *J. Phys. Chem. Ref. Data*, 1985, **14**, 731.

(*k*) R.T. Sanderson, *Inorganic Chemistry*, Reinhold, New York, 1967; *Inorg. Chem.*, 1986, **25**, 1856.

(*l*) R.G. Pearson, *Inorg. Chem.*, 1988, **27**, 734.

(*m*) *Selected Values of Chemical Thermodynamic Properties*, National Bureau of Standards Technical Notes 270–1 to 270–8, US Government Printing Office, Washington, DC, 1965 onwards; *JANAF Thermochemical Tables*, 2nd edn., NSRDS–NBS 37, US Government Printing Office, Washington, DC, 1971; supplements, *J. Phys. Chem. Ref. Data*, 1974, **3**, 311; 1978, **7**, 793.

Al	Ga	In	Tl
3.3×10^6	630	45	8
3.4×10^6	1.6×10^3	8	8
1.5×10^7	2×10^3	4	8
8.3×10^7	1.8×10^4	50	400
10^4	400	1	68
150 (III)	10^4 (III)	Unknown (III)	10^4 (I)
Bauxite, found as boehmite and diaspore, AlO(OH), and gibbsite and hydrargillite, $Al(OH)_3$	Occurs up to 1% in other minerals; recovered as a by-product of zinc- and aluminium-refining	Occurs up to 1% in zinc and lead sulfide ores; obtained as a by-product of zinc- and lead-smelting	Rare; dispersed in potash, feldspar and pollucite; by-product of zinc- and lead-smelting and H_2SO_4 manufacture
18×10^6 $(Al)^{kk}$	60–80 $(Ga)^{kk}$	c. 100 $(In)^{kk}$	15 $(Tl)^{kk}$
6×10^9	–	>1500	–
Accumulates in the body from daily intake; may be implicated in Alzheimer's disease and dialysis encephalopathies	None; stimulatory	None; stimulatory; teratogenic	None; teratogenic
100	?	?	?
2.45	Not known but low	Not known but low	0.0015
5000	Not known but low	30	Not known
?	?	>200	600

(n) Ref. 128.

(o) J.E. Huheey, *Inorganic Chemistry*, 3rd edn., Harper International SI Edition, Cambridge, MA, 1983.

(p) M.E. O'Neill and K. Wade, in *Comprehensive Organometallic Chemistry*, eds. G. Wilkinson, F.G.A. Stone and E.W. Abel, Pergamon Press, Oxford, 1982, Vol. 1, p. 1.

(q) Ref. 30.

(r) See Refs. 8, 10 and 19.

(s) See, for example, J.L. Atwood, F.R. Bennett, C. Jones, G.A. Koutsantonis, C.L. Raston and K.D. Robinson, *J. Chem. Soc., Chem. Commun.*, 1992, 541.

(t) Ref. 22.

(u) A.N. Utkin, G.V. Girichev, N.I. Giricheva and S.V. Khaustov, *Zh. Strukt. Khim.*, 1986, **27**, 43.

(v) R. Hoppe and D. Kissel, *J. Fluorine Chem.*, 1984, **24**, 327.

(w) Ref. 151.

(x) V.P. Spiridonov, A.G. Gershikov, E.Z. Zasorin, N.I. Popenko, A.A. Ivanov and L. I. Ermolayeva, *High Temp. Sci.*, 1981, **14**, 285.

(y) V.M. Petrov, N.I. Giricheva, G.V. Girichev, V.A. Titov and T.P. Chusova, *Zh. Strukt. Khim.*, 1990, **31**, 46.

(z) J. Glaser, *Acta Chem. Scand., Ser. A*, 1980, **34**, 75.

(aa) R.D. Shannon, *Acta Crystallogr.*, 1976, **A32**, 751.

(bb) T. Fjeldberg, A. Haaland, R. Seip, Q. Shen and J. Weidlein, *Acta Chem. Scand., Ser. A*, 1982, **36**, 495.

(cc) Ref. 164.

Table footnote continued overleaf

(*dd*) J.A. Coxon and S. Naxakis, *J. Mol. Spectrosc.*, 1987, **121**, 453.

(*ee*) *Landolt–Börnstein Numerical Data and Functional Relationships in Science and Technology*, New Series, Group III: Crystal and Solid State Physics, *Structure Data of Elements and Intermetallic Phases*, Springer-Verlag, Berlin, Vol. 6, 1971; Vol. 14a, 1988.

(*ff*) E.A. Brandes, ed., *Smithells Metals Reference Book*, 6th edn., Butterworth, London, 1983.

(*gg*) Value varies significantly with the crystallographic axis.

(*hh*) W.E. Dasent, *Inorganic Energetics: an Introduction*, 2nd edn., Cambridge University Press, Cambridge, 1982.

(*ii*) Ref. 110.

(*jj*) P.A. Cox, *The Elements: their Origin, Abundance and Distribution*, Oxford University Press, Oxford, 1989.

(*kk*) See chapter 2.

(*ll*) J.J.R. Fraústo da Silva and R.J.P. Williams, *The Biological Chemistry of the Elements: the Inorganic Chemistry of Life*, Clarendon Press, Oxford, 1991.

bearing on its distinctive chemistry. In the first place, it is the atom with the *smallest ns → np* promotion energy, and the relative closeness in energy of the valence *s* and *p* orbitals favours a major contribution from the 2*s* orbital to the bonding of boron compounds. Because of this and because of the inherent strengths of the bonds which boron forms, the univalent state is rarely encountered, being confined to 'high-temperature' molecules like BF and BCl.[7–9] Secondly, the 2*p* orbitals of boron share to some extent with those of carbon, the ability to engage in relatively efficient π-type interactions, providing a mechanism for supplementing the bonding to electron-rich centres. Such interactions subscribe presumably to the relative stability of planar 3-coordinate environments for the boron atom, as in the trihalides, boric acid, amidoboranes containing the unit $\overset{\diagdown}{\underset{\diagup}{B}}{-}\overset{\diagup}{\underset{\diagdown}{N}}$, and derivatives of borazine, $H_3B_3N_3H_3$, formally analogous to benzene.[1–6,8–10] These special properties of boron, acting together but in varying degrees, give rise to the following extraordinary phenomena.

(i) *The element* itself is not a metal but a semiconductor with several hard and refractory allotropic forms characterised by unique and elaborate structures based on the B_{12} icosahedron. Indeed, this unit is something of a leitmotif in boron chemistry.

(ii) *Boron hydrides and related compounds* (e.g. carbaboranes and metal-substituted derivatives and boron subhalides) are typically discrete molecular species remarkable for their stoichiometries, structures and configurations.[8–18] To account for the properties of these polyboron cluster compounds it has been necessary substantially to revise and extend earlier theories of covalent bonding. By contrast, there is only one binary hydride of aluminium known to be stable under normal conditions, and that is a polymeric solid, the α-form being isostructural with AlF_3 and featuring 6-coordinated aluminium atoms.[19–21] Very recent studies leading to the identification and characterisation of the thermally fragile but discrete molecules $H_2Ga(\mu\text{-}H)_2GaH_2$,[22a] $H_2Ga(\mu\text{-}H)_2BH_2$[22b] and $H_2GaB_3H_8$[22c] suggest that, if anything, gallium comes closer than aluminium to shadowing the behaviour of boron in this respect.

(iii) *Metal borides* too find few parallels elsewhere in Group 13.[4,8,10,11,13,23] With an extraordinary range of stoichiometry — from (at least) M_5B to MB_{66} where M is the metal atom — they reflect the small size of the boron atom, as well as the propensity of such atoms to form branched and unbranched chains, planar networks and three-dimensional arrays. The resulting polyboron frameworks enjoy high intrinsic stability while acting as hosts to M atoms in proportions determined by the size and electronic properties of M.

These and many other features combine to make the chemistry of boron peculiarly diverse and complex. There are good grounds therefore for separating boron from the other Group 13 elements and considering it within its own domain. Just such an approach has been followed in numerous books,[8–18] ranging from the comprehensive and regularly updated volumes of the Gmelin Handbook,[8] through reviews of specific types of compound,[9,12–14,16] to accounts seeking to unify and explain certain aspects of boron chemistry within a wider context.[16–18]

All the other members of the Group are metals. Certainly they show a closer kinship to one another than they do to boron, but their properties are far from uniform and frequently vary in an irregular fashion ultimately reflecting the discontinuous build-up of the Periodic Table. Symptomatic of the irregularities are the ionisation energies of the atoms which, unlike those of the corresponding metals of Groups 1, 2 and 3, vary in a discontinuous way as a function of atomic number (see Table 1.1 and Figure 1.1). The energies reflect as a primary influence the electron core of the atom which changes considerably: for B and Al it is simply the [He] and [Ne] core, respectively; for Ga and In it is the preceding noble-gas core supplemented by the $3d^{10}$ and $4d^{10}$ shells, respectively; and for Tl it is [Xe] plus the $4f^{14}$ and $5d^{10}$ shells. Accordingly, the fourth ionisation energies of Ga, In and Tl are typically less than half those of B and Al. Although the +4 oxidation state does not feature in any known chemical compound of the Group 13 elements, and would surely be a prodigious oxidising agent, it may not be wholly out of reach to In and Tl, a view receiving some support from claims that a mercury(III) complex can be synthesised at low temperatures by electrochemical means.[24] In relation to Al, the additional nuclear charge of Ga is shielded somewhat imperfectly by the $3d^{10}$ shell. This causes the valence electrons of Ga to be more tightly held than might be expected on the basis of simple extrapolation from B and Al; it also discriminates significantly between the more penetrating $4s$ and the less penetrating $4p$ electrons, as revealed by the ionisation energies. Thus, the $4p$ electron of Ga

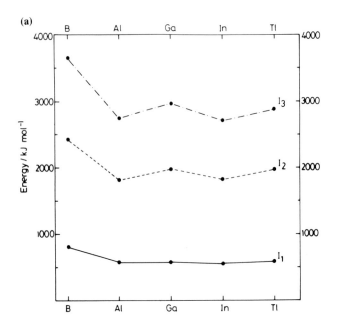

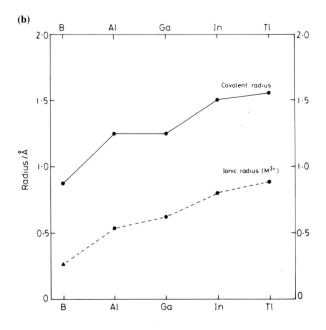

Figure 1.1 (a), (b)

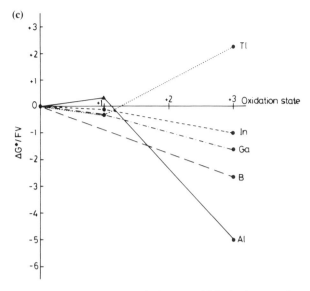

Figure 1.1 Some properties of the Group 13 elements: (a) ionisation energies of the valence electrons; (b) single-bond covalent radii and radii of the M^{3+} ions in six-fold coordination; and (c) oxidation state diagram for aqueous solutions with $a_{H^+} = 1$ and $T = 298\,K$.

is bound not less but marginally more tightly than the $3p$ electron of Al, whereas the valence s electrons experience binding energies nearly 10% greater in the heavier atom. Although indium reverts to type to some extent, with reduced binding energies for both the $5s$ and $5p$ electrons, the intervention of an *extra* complement of imperfectly screening electrons — in this case the $4f^{14}$ shell — strengthens the hold of the Tl nucleus on its valence electrons, but the $6s$ disproportionately more than the $6p$. While variations in the make-up and shielding of the electron core are surely important, the increased nuclear charge of the heavier atoms must cause electron speeds near the nucleus to become comparable with that of light, with the result that both the mass and binding energy of the electron increase and its orbital radius suffers a corresponding decrease. It has been argued that these relativistic effects are at the root of some of the most conspicuous anomalies displayed by the later elements in the Periodic Table.[25] On the evidence of recent theoretical treatments involving Group 13 derivatives, however, such effects make a significant, but not the dominant, contribution to the pattern of binding energies affecting the valence electrons.[26]

The energies of the valence electrons can be correlated, at least in part, with the increase in chemical reactivity of the elements at moderate temperatures and the emergence of relatively well defined cationic chemistry, features distinguishing the heavier members of Group 13 from boron. Yet it is the redox chemistry that highlights the irregular pattern of these energies. Although $+3$ persists as the characteristic oxidation state for all the members of the Group, the $+1$ state gains in importance as the atomic number increases. Thus, aluminium(I) is little more abundant than boron(I), being restricted mostly to high-temperature molecules or transient species produced electrochemically, although suitable polar media have been shown to support the metastable monohalides AlF and AlCl at low temperatures;[27] gallium(I) and, even more, indium(I) play significant roles, although both are powerful reducing agents susceptible to disproportionation under normal (aqueous) conditions; whereas thallium(I) is, by some margin, preferred to thallium(III) which is a powerful oxidising agent under normal conditions. The situation is illustrated by the standard reduction potentials listed in Table 1.1 and by the oxidation state diagram included in Figure 1.1. This emergence of an oxidation state corresponding formally to the loss of only the single np valence electron has for many years been associated with the *inert pair effect*,[1,28,29] referring to the resistance of the ns^2 electrons to be lost or to participate in covalent bond formation. The same effect is a prominent feature of other p-block elements in the succeeding Groups 14–18; it is even foreshadowed in the unusually low reactivity of elemental mercury. The term 'inert pair' as applied to the ns^2 electrons may be a useful label, but its implication of core-like behaviour is unduly simplistic. Such an assumption is certainly difficult to reconcile with the rich and varied stereochemistry distinguishing the Group 13 metals in the $+1$ oxidation state, where the role of the two extra valence electrons is far from easy to predict $(q.v.)$.[4,19,30-34] Just what factors ultimately determine the relative energy balance between the $+1$ and $+3$ state will be addressed more fully in section 1.4.1 and also in chapter 3.

All the elements, including boron, also form at least some compounds in which the oxidation state of the Group 13 atom is formally $+2$. However, authentic derivatives of the divalent state, which are monomeric and paramagnetic, are normally short-lived under conventional conditions; they can be detected and characterised only by measurements involving very short temporal resolution (e.g. pulse radiolysis) or by trapping in a suitable solid matrix, usually at low temperatures. Such species include Tl^{2+} (solv) (which may be an intermediate in certain redox reactions of the couple Tl^{3+}/Tl^+),[35] AlO[36] and $AlCl_2$.[37] On the other hand, divalent derivatives may achieve long-term stability in one of two ways: (i) through dimerisation with the formation of a metal–metal bond, as in $[M_2X_6]^{2-}$ (M = Ga or In; X = Cl, Br or I) or $Ga_2^{4+}(S^{2-})_2$; or (ii) through disproportionation to give a mixed valence derivative, e.g. $M^I[M^{III}X_4]$ (M = Ga or In; X = Cl, Br or I).[38,39] The balance between these two options is a delicate one, and the different elements show a relatively erratic pattern of behaviour. Thus, aluminium is not given to forming mixed valence compounds and there are only a few examples of systems with Al–Al bonds; on the other hand, thallium enters into the formation of mixed valence compounds but whether it will afford Tl_2^{4+} (isoelectronic with Hg_2^{2+}), in one form or another, remains a moot point.[40]

Acidity, in the broadest sense of electrophilicity, is a common theme of the characteristic tripositive oxidation state. It is much diminished in derivatives of the unipositive state which may even betray some nucleophilic character, as in the oxidative addition of HCl to the AlCl[41] or GaCl[42] molecule at low temperatures or of an alkyl halide to InBr or InI at ambient temperatures.[43] The natural acidity of the M^{III} centre leads typically to the coordination of electron-rich species (neutral molecules or anions) to form complexes in which the metal usually enjoys 4-, 5- or 6-fold co-ordination.[19,32,33] Such complexes are noteworthy on two grounds. In the first place, their relative stabilities imply that boron(III) and aluminium(III) are class 'a' acceptors; this is in keeping with their designation as 'hard' acids or 'oxophiles', a trait reflected in the high thermodynamic stabilities of their oxide derivatives, and by their occurrence as such in nature. With increasing atomic number and polarizability, however, the metal atoms become progressively 'softer' as acceptor centres, with a correspondingly enhanced preference for donors like S^{2-} and I^-. Accordingly the heavier elements tend to be found in nature not as oxides but in sulfide minerals. The second material point is that the complexes also vary widely and idiosyncratically in stoichiometry, structure and other properties. Elaboration of these points is deferred to section 1.4.2, but chapter 8 gives a fuller account of the coordination chemistry of the Group 13 metals.

Still the peculiarities of constitution and structure of the complexes give the merest hint of the diversity and anomalies that mark the structural chemistries of the Group 13 metals at large. Herein it might justifiably be argued, 'confusion now hath made his masterpiece!' Several factors contribute to the confusion. Firstly, polymorphism is relatively common-place — witness, for example, the different forms of the oxides and hydroxides of aluminium and gallium and TlI.[4,30] With no single, dominant influence evidently at work, the structure adopted by any given system typically reflects a compromise between the demands of bulk coulombic forces, dispersion and other secondary interactions, and covalent bonding; it is liable also to be a sensitive function of the external conditions of temperature and pressure. Secondly, isotropic structures with more-or-less close-packed atoms, and exemplified by the wurtzite or zinc blende structures of III–V compounds and the CsCl structure favoured by TlX (X = Cl, Br or I), are the exceptions more than the rule. Accordingly there are few situations in which the simple ionic model is likely to be a good approximation. Instead, the structures abound in the anisotropic effects of covalent bonding, with the metal atoms often bridged by ligands to generate layer structures, extended chains, or, ultimately, discrete molecular units. Thirdly, there is the conundrum of whether the lone pair of electrons in the valence shell of Ga^I, In^I or Tl^I has a stereochemical role to play.[30,31,44] Some structures imply an active role through the quintessential feature of a non-centrosymmetric site for the metal atom, as in Tl_2O, there being typically several short bonds lying all to one side of the metal and relatively long secondary bonds on the other side. Yet there are other structures, like that of TlCl, which find the metal at the centre of a more-or-less regular co-ordination polyhedron, and so appear to deny the lone pair any stereo-chemical influence. The structural theme will recur, not only later in this chapter (see section 1.4.3), but also in chapters 3, 4, 6 and 8. For a foretaste of the structural vagaries to come, we need look no further than the solid elements themselves. The position, as summarised in Table 1.1, is that only aluminium and indium exist in unique forms and that only aluminium and thallium (in its α-form) assume under normal conditions the close-packed structures typical of metals. Indium has an unusual structure which involves a slightly distorted version of cubic close-packing, each atom having four neighbours at $3.25\,\text{Å}$ and eight more at $3.38\,\text{Å}$. Gallium is the most remarkable of all. It is little less prolific than boron in the range of poly-morphs it forms, but even stranger is the structure of the normal form of the metal, which is akin to that of iodine with the atoms being associated in *pairs*. It is notable too for its low melting point (29.78°C), melting being accompanied by *contraction*, and there is evidence that the Ga_2 units persist in the liquid. The extraordinary properties of gallium and the relatively large interatomic distances in indium and thallium all seem to suggest that binding in these post-transition metals derives mainly from the delocalisation of the single p electron, with the ion cores approximating to M^+ more closely than to M^{3+}. Here, then, is evidence of a lone pair that is relatively core-like or 'inert', while retaining the potential to affect the structure of the system.

1.2   History and occurrence

1.2.1   History

Of the Group 13 elements, only boron and aluminium have histories extending back more than 150 years.[45,46] Aluminium derives its name from alum,[47] the double sulfate $KAl(SO_4)_2 \cdot 12H_2O$ which was used medicinally as an astringent in ancient Greece and Rome (Latin *alumen*, bitter salt), as a fireproofing agent for wood, and also as a mordant in dyeing. The records left to us seem to suggest that the different varieties of *alumen* were more or less impure mixtures of aluminium and iron(II) sulfates. It is not known when the salt now called alum was discovered, but commercial plants for the extraction of the salt were certainly in existence in the fourteenth and fifteenth centuries. In about 1760, Baron recognised that alum contained not only an alkali metal but also a second metal. All known methods of reduction to isolate the metal having failed, Lavoisier was led to observe in 1782 that the metal must have a stronger affinity for oxygen than any known reducing agent. Davy was no more successful than his predecessors in isolating the metal but proposed the name 'alumium' and then 'aluminum'; this version has persisted in North America up to the present day[48] but a third version, namely 'aluminium', was soon adopted and is now internationally sanctioned. It was not until 1825 that the Danish scientist Oersted succeeded in preparing the metal, albeit in an impure form, by the action of dilute potassium amalgam on aluminium chloride. His method was improved 2 years later by Wöhler who used metallic potassium, instead of its amalgam, to reduce aluminium chloride. Still the isolation of the pure metal was delayed until 1854 when the French chemist Deville, having perfected a process for the manufacture of sodium, used this metal to reduce aluminium chloride. In the same year he and Bunsen, working independently, also made the significant advance of generating the metal by electrolysis of the molten complex salt $Na[AlCl_4]$. So precious was the metal at this time that it was exhibited next to the crown jewels at the Paris Exposition of 1855 and the Emperor Louis Napoleon III used aluminium cutlery on state occasions. The transformation from precious curio to one of the cornerstones of the modern industrialised world is discussed in chapter 2.

All the other Group 13 metals came to light in the period 1860–1880,[45,46] and owe their discovery — and, in two cases, their names — to spectroscopic analysis of the light emitted by the incandescent vapours created by the action of a flame or an electric spark on appropriate metal compounds. Thallium was the first to be identified in this way, being discovered independently by Crookes[49] and Lamy[50] in 1861–1862. Thus, Crookes' examination of some seleniferous residues from a sulfuric acid plant with the aid of the spectroscope led to the following report:[49] 'suddenly a bright-green line flashed into view and quickly disappeared'. With an imagery suggestive of Keats:

'Then felt I like some watcher of the skies
When a new planet swims into his ken'

Crookes was led to conclude that his 'new planet' was a new element present in the residues he had been examining. The green line in the spectrum resembled the colour of spring vegetation and led to the naming of the element as 'thallium' (from the Greek *thallos*, a budding shoot or twig).[47] Lamy too observed the characteristic green line of the atomic emission spectrum and he was certainly the first to prepare the element in any quantity.[50] Closer acquaintance with the chemistry of thallium, founded on these and subsequent studies, revealed similarities to so many other elements (the alkali metals, silver, mercury and lead, for example) that Dumas was prompted to christen it the 'duck-billed platypus of the metals'.[51]

Hard on the heels of thallium came indium which was first identified in 1863 by Reich and Richter working at the Freiberg School of Mines.[52] Spectrographic examination of a crude zinc chloride liquor derived from samples of zinc blendes disclosed a brilliant indigo blue line and a second fainter blue line which had not been observed before. Reich and Richter were able to isolate the oxide of the new element and to reduce it to the metal by heating in a stream of hydrogen or coal gas. They were responsible too for coining the name 'indium' in token of the element's distinctive flame coloration (Latin *indicum*, indigo).[47]

Gallium remained a missing link up to 1870 when Mendeleev predicted its existence under the name of eka-aluminium.[53] Its discovery came, again through the agency of the spectroscope, in 1875. The French chemist Lecoq de Boisbaudran,[54] who was guided at the time by an independent theory of his own and had been searching for the missing element for some years, examined the deposit formed on zinc from an acid solution of a zinc blende ore. This he observed gave two new violet lines in an oxyhydrogen flame or a spark. Within a month he had isolated more than a gram of gallium metal from several hundred kilograms of the crude ore; the steps involved dissolution in acid, deposition with zinc, precipitation of hydroxides (including that of gallium), co-precipitation with zinc sulfide, repeated precipitation of the hydroxide, dissolution in caustic potash, and finally electrolysis. The element was named 'gallium'; the common belief is that this was in honour of France (Latin *Gallia*, Gaul).[47] That the physical and chemical properties displayed by the element matched so closely those predicted by Mendeleev was a conspicuous triumph for the Periodic Law, and did much to establish its widespread acceptance in the scientific community.[53,54]

1.2.2 Occurrence

The natural abundances and distributions of the Group 13 elements, as summarised in Table 1.1, reflect the workings of several influences.[55] Chief among these is the atomic number. Atoms of low atomic number, like Li, Be and B, are in short supply because of the way in which they are sidestepped in nucleosynthesis. There is also a dearth of atoms of high atomic number, the cosmic abundances of the metallic members of Group 13 decreasing in the order $Al \gg Ga > In \gtrsim Tl$, in keeping with the progressive decline in nuclear stability of atoms with atomic number >26. All the members of the Group have odd atomic numbers and their nuclei are therefore less stable than those of the adjacent members of the carbon group with even atomic numbers; again, the abundances are broadly consistent with this pattern. In addition, there are four chemical facets with a major bearing on the distribution of the elements as we find them on earth. Firstly, they vary in the volatilities of the forms in which they are thought to have occurred in the solar gas, from involatile aluminium oxides which condense at temperatures above 1400 K, and are therefore among the 'early condensates', through gallium (condensation temperature 1075 K) to relatively volatile indium and thallium (condensation temperatures *c.* 450 K).[55] As a result aluminium is the only Group 13 element to have experienced efficient condensation and to have a terrestrial abundance in keeping with its cosmic abundance. The ratios of the terrestrial to the cosmic abundances (B $0.3:1$, Al $1.3:1$, Ga $0.17:1$, In $0.03:1$ and Tl $0.04:1$)[55,56] imply that partial volatilisation has led to appreciable depletion of the earth in the other elements. Secondly, the rare elements Ga, In and Tl have atomic sizes compatible with those of more common elements, e.g. Al, Zn and Pb, which they can partially replace in minerals. They are therefore rather evenly spread over the earth's surface occurring widely, albeit in low concentrations, in various minerals. By contrast, boron is by virtue of its size and/or charge incompatible with common elements, causing its solubility in major minerals to be small; for this reason boron tended to remain in the silicate melt of the magma and to be concentrated in the rocks which solidified last. Thirdly, as already noted in the preceding section, the ligating preferences of the elements change as a function of atomic number. Chemical fractionation following condensation has led therefore to segregation of the oxophilic members, boron and aluminium, which are lithophiles, either partially or completely from the more chalcophilic members, gallium, indium and thallium. Finally, since the relevant oxido and sulfido derivatives have relatively low solubilities in water at or near pH 7, the metals normally enter into the hydrosphere only at relatively low concentrations.[57] As surface waters decrease in pH, however, so they offer an increased potential to leach aluminium, in particular, from the lithosphere, with what may be disastrous consequences to marine life.[57] These and other issues touching on the circulation of the elements in the environment at large, as well as the impact on living organisms, are elaborated in chapter 9.

Aluminium is notable for being the most abundant metal in the earth's crust (of which it forms 8.3% by weight); it is exceeded in abundance only by oxygen (45.5%) and silicon (25.7%), and is approached only by iron (6.2%) and calcium (4.6%).[4,19,55] It is a major constituent of many common igneous minerals including feldspars and micas. Weathering of these in temperate climates gives clay minerals such as kaolinite [$Al_2(OH)_4Si_2O_5$], montmorillonite and vermiculite. Under conditions of extreme weathering, silicon is leached out as well as the alkali metals and an impure mixture of hydrated oxides (so-called bauxite or laterite) is formed. Bauxite, the normal commercial source of aluminium metal, has a composition which varies considerably from place to place. In the tropics, the aluminium is mainly in the form of gibbsite or hydrargillite [$Al(OH)_3$], whereas in temperate countries, it occurs mainly as boehmite or diaspore ($AlOOH$) (see chapter 2). Aluminium also occurs in many well known, though rarer, minerals such as cryolite (Na_3AlF_6), spinel ($MgAl_2O_4$), garnet [Ca_3Al_2-($SiO_4)_3$], beryl ($Be_3Al_2Si_6O_{18}$) and turquoise [$Al_2(OH)_3PO_4 \cdot H_2O/Cu$]. The anhydrous oxide Al_2O_3 is found as corundum, and impure forms of this are notable as gemstones, e.g. ruby (Al_2O_3/Cr), sapphire (Al_2O_3/Fe) and oriental emerald.

Gallium, indium and thallium are very much less abundant than aluminium and tend to be found at low concentrations in sulfide minerals, rather than as oxides, although gallium is also associated with aluminium in bauxite.[4,19,55] Gallium is, nevertheless, about twice as abundant as boron, although it is more difficult to extract because, unlike boron, it is widely distributed and has not accumulated to give any ore in which it is a major constituent. Instead it always occurs as a minor ingredient in association either with zinc or germanium, its horizontal neighbours, or with aluminium, its vertical neighbour in the Periodic Table. The highest concentrations (0.1–1%) occur in the mineral germanite, a complex sulfide of zinc, copper, germanium and arsenic, but this is too rare to be useful as a primary source. Concentrations in zinc blende, bauxite or coal are typically lower by about two orders of magnitude. Zinc blende was formerly an important source of gallium, but has now been supplanted by bauxite whence the metal is obtained as a by-product of the aluminium industry (see chapter 2).

Indium, like gallium, does not form any minerals of its own. Instead it is widely distributed in minute amounts in many minerals, usually but not exclusively being concentrated in sulfide deposits. Its content in these ores is often related to that of tin. Zinc, which it also resembles in size and other properties, is an important carrier, and zinc blendes afford the principal commercial source of the metal.[4,19] Unlike germanium, another element shadowing zinc, indium is generally found in blendes that are geologically old. It is found to a lesser extent in association with copper and iron sulfides. With a terrestrial abundance only about 10^{-3} that of gallium,[55] it is a rare element, its total contribution to the earth's crust being comparable with that of silver. Few minerals contain more than 0.1%, although pegmatite dikes in western Utah are reported to contain up to 2.8% indium.

Thallium too is widely distributed, but generally in very low concentrations, with a terrestrial abundance not very different from that of indium.[4,19,55,58] Yet the geochemistry of the two elements is quite different, mainly because of the different sizes and charges of the naturally occurring ions, with $r(In^{3+}) < r(Tl^{+})$; accordingly Tl^{+} shows a much closer affinity to the ions Rb^{+} and Pb^{2+} which are comparable in size. Thus, both thallium and rubidium were concentrated in the late crystallisations of magmatic potassium minerals such as feldspars and micas. To this extent, it is a rare lithophile. However, it is also a chalcophile occurring not only in oxide, but also in sulfide minerals where, for example, it replaces lead to a small extent in minerals like galena, PbS. There exist some thallium-containing minerals, including crookesite, $(Cu, Tl, Ag)_2Se$ (17% Tl), lorandite, $TlAsS_2$ (59% Tl), urbaite, $TlAs_2SbS_5$ (30% Tl) and hutchinsonite, $(Tl, Cu, Ag)_2S \cdot PbS \cdot 2As_2S_3$, but the deposits are so small as to be of no commercial importance. In practice, thallium is extracted from sulfide minerals, being concentrated in the flue dusts which accumulate during roasting of sulfide ores, notably in the manufacture of sulfuric acid (see chapter 2).

1.3 Physical properties of metal atoms

The physical properties of the metal atoms are of two kinds: (i) those which can be identified more or less precisely with the *isolated* atoms, and (ii) those which can be defined or determined for the atom only when it is in the combined state. In practice, category (i) includes the nuclear properties of the different isotopes (Table 1.2) and those properties, determined mostly by spectroscopic methods, associated with the different energy states of the atoms and derived ions (Table 1.1); category (ii) includes properties such as atomic and ionic radii, bond lengths, bond energies and electronegativity (Table 1.1).

1.3.1 Nuclear properties: isotopes

All the Group 13 metals having odd atomic numbers, there are few stable isotopes.[55] Natural aluminium consists exclusively of the isotope ^{27}Al; natural gallium, indium and thallium each consist of just two isotopes, as listed in Tables 1.1 and 1.2. Therefore, there is not much scope for exploiting isotopic effects in spectroscopic studies of natural compounds, although the presence of two stable isotopes facilitates the identification of ions in the mass spectra of gallium-, indium- or thallium-containing species, and, under appropriate conditions, distinct isotopic features have been observed in the infrared spectra of simple molecules like $GaCl$,[42] $HGaCl_2$,[42] $OGaF$[59] and Ga_2H_6.[60] The stable nuclei vary widely, not only in their nuclear spin properties, but also in their cross-sections for the capture of thermal neutrons (see Table 1.1). Thus, aluminium is noteworthy for having a cross-section of only 0.233 barn, which, together with a short-lived irradiation product (^{28}Al), makes it an attractive material for containment within nuclear reactors. On the other hand, with a cross-section nearly 10^3 larger than this, indium is relatively opaque to neutrons and its alloys have been used in the control bars of such reactors.

1.3.1.1 NMR studies.

All the naturally occurring nuclei, being magnetically active, have lent themselves in varying degrees to NMR measurements.[61,62] The relevant properties, together with the corresponding details for ^{10}B and the much more extensively studied ^{11}B nuclei, are summarised in Table 1.3. Thallium stands apart in having two spin-½ nuclei, whereas all the other nuclei are quadrupolar. The properties are such that the ease of observation follows the order $Tl > Al > Ga \sim In$, the preferred nuclei being ^{205}Tl (because of its greater abundance), ^{71}Ga (because of its superior receptivity and width factor and despite its adverse abundance), and ^{115}In (because of its superior receptivity and natural abundance). ^{205}Tl is indeed the most receptive heavy metal nucleus with I ½, its receptivity being 30 times that of ^{119}Sn, its nearest rival. On the other hand, the quadrupolar

Table 1.2 Properties of the better characterised isotopes of the Group 13 metals.[a]

Nuclide	Atomic mass[b]	Source[c]	Half-life	Decay mode[d]	Nuclear spin, I; nuclear magnetic moment[e]
^{22}Al	22.079370	^{24}Mg(^{3}He, p4n)	70 ms	β^+, p	4+
^{23}Al	23.007265	^{24}Mg(p, 2n)	0.47 s	β^+, p	
^{24m}Al		^{24}Mg(p, n)	0.129 s	IT, β^+	1+
^{24}Al	23.999941	^{24}Mg(p, n)	2.07 s	β^+	4+
^{25}Al	24.990429	^{24}Mg(p,γ); ^{25}Mg(p, n)	7.17 s	β^+	5/2+; 3.646
^{26m}Al		^{23}Na(α, n); ^{26}Mg(p, n)	6.345 s	β^+	0
^{26}Al	**25.986892**	**^{26}Mg(p, n); ^{25}Mg(d, n);** ^{28}Si(d, α)	**7.1 × 10^5 y**	**β^+, EC**	**0**
^{27}Al	26.981539	**Naturally occurring**	**Stable**	–	**5/2+; +3.64151**
^{28}Al	27.981910	^{27}Al(n, γ); daughter ^{28}Mg	2.25 m	β^-	3+; 3.24
^{29}Al	28.980446	^{26}Mg(α, p)	6.5 m	β^-	5/2+
^{30}Al	29.982940	^{30}Si(n, p)	3.68 s	β^-	3+
^{31}Al	30.983800	^{18}O(^{18}O, αp); U(p)	0.64 s	β^-	
^{32}Al	31.9880	In(p)	33 ms	β^-	1+
^{34}Al	33.9965	Be or C(^{40}Ar)	0.05 s	β^-	
^{62}Ga	61.944178	^{64}Zn(p, 3n)	0.116 s	β^+, EC	0
^{63}Ga	62.939140	^{63}Cu(α, 4n); Ni(^{6}Li)	32 s	β^+, EC	
^{64}Ga	63.936836	^{63}Cu(α, 3n); ^{64}Zn(p, n); ^{64}Zn(d, 2n)	2.63 m	β^+, EC	0
^{65}Ga	64.932738	^{63}Cu(α, 2n); ^{64}Zn(d, n); ^{64}Zn(p, γ)	15.2 m	β^+, EC	3/2−
^{66}Ga	65.931590	^{63}Cu(α, n)	9.5 h	β^+, EC	0
^{67}Ga	**66.928204**	**^{66}Zn(d, n);** ^{65}Cu(α, 2n)	**3.260 days**	EC	**3/2−; +1.8507**
^{68}Ga	67.927981	**Daughter ^{68}Ge**	**1.130 h**	**β^+, EC**	**1+; 0.01175**
^{69}Ga	**68.925580**	**Naturally occurring**	**Stable**	–	**3/2−; +2.01659**
^{70}Ga	69.926028	^{69}Ga (n, γ)	21.1 m	β^-, EC	1+
^{71}Ga	**70.924700**	**Naturally occurring**	**Stable**	–	**3/2−; +2.56227**
^{72}Ga	**71.926365**	**^{71}Ga(n, γ)**	**14.10 h**	**β^-**	**3−; −0.13224**
^{73}Ga	72.925169	^{73}Ge(n, p); ^{76}Ge(d, αn)	4.87 h	β^-	3/2−
^{74m}Ga		^{74}Ge(n, p)	10 s	IT	1+
^{74}Ga	73.926940	^{76}Ge(d, α); ^{74}Ge(n, p)	8.1 m	β^-	3−
^{75}Ga	74.926499	^{76}Ge(n, pm); ^{76}Ge(γ, p)	2.10 m	β^-	3/2−
^{76}Ga	75.928670	^{76}Ge(n, p)	29 s	β^-	3−
^{77}Ga	76.928700	Fission	13.0 s	β^-	
^{78}Ga	77.931760	Fission	5.09 s	β^-	3+

Nuclide	Atomic mass	Production	Half-life	Decay mode	J^π; μ
^{79}Ga	78.932530	Fission	2.85 s	β^-	
^{80}Ga	79.936250	Fission	1.68 s	β^-	
^{81}Ga	80.937750	Fission	1.22 s	β^-	
^{82}Ga	81.94269	Fission	0.607 s	β^-	
^{83}Ga		Fission	0.310 s	β^-	
^{102}In	101.92440	Cu(^{40}Ca); ^{92}Mo(^{16}O)	23 s	EC	(5)
^{103}In	102.920110	^{92}Mo(^{16}O or ^{14}N)	1.1 m	β^+, EC	9/2+
^{104m}In			16 s	IT	
^{104}In	103.918440	^{92}Mo(^{16}O); ^{96}Ru(^{12}C)	1.84 m	β^+, EC	5+; +4.44
^{105m}In		^{90}Zr(^{19}F); ^{106}Cd(p, 2n)	43 s	IT	1/2−
^{105}In	104.914558	^{106}Cd(p or ^{3}He)	5.1 m	β^+, EC	9/2+; +5.675
^{106m}In		^{106}Cd(p, n)	5.3 m	**β^+, EC**	**3+**
^{106}In	105.913490	^{106}Cd(p, n); ^{92}Mo(^{16}O)	6.2 m	β^+, EC	7+; +4.92
^{107m}In		^{106}Cd(d, n); Ag(^{3}He)	51 s	IT	1/2−
^{107}In	106.910284	^{106}Cd(d, n); ^{106}Cd(p, γ)	32.4 m	β^+, EC	9/2+; +5.59
^{108m}In		^{107}Ag(α, 3n); ^{108}Cd(p)	57 m	β^+, EC	6+
^{108}In	107.909678	^{107}Ag(α, 3n); ^{108}Cd(p)	40 m	β^+, EC	3+; +4.56
^{109m}In		Daughter ^{109}Sn	1.3 m	IT	1/2−
^{109}In	108.907133	^{107}Ag(α, 2n)	4.2 h	β^+, EC	9/2+; +5.54
^{110m}In		^{107}Ag(α, n); ^{109}Ag(α, 3n)	4.9 h	EC	7+; +4.72
^{110}In	109.907230	Daughter ^{110}Sn; ^{107}Ag(α, n); ^{109}Ag(α, 3n)	1.15 h	β^+, EC	2+; +4.37
^{111m}In		^{109}Ag(α, 2n); ^{111}Cd(p, n)	7.7 m	IT	1/2−; +5.53
^{111}In	**110.905109**	**^{109}Ag(α, 2n); ^{111}Cd(p, n)**	**2.8049 days**	**EC**	**9/2+; +5.50**
^{112m}In		^{109}Ag(α, n)	20.8 m	IT	4+
^{112}In	111.905536	^{109}Ag(α, n); ^{112}Cd(p, n)	14.4 m	β^+, EC, β^-	1+; +2.82
^{113m}In		**Daughter ^{113}Sn**	**1.658 h**	**IT**	**1/2−; −0.210**
^{113}In	**112.904061**	**Naturally occurring**	**Stable**	–	**9/2+; +5.529**
^{114m}In		**^{113}In(n, γ)**	**49.51 days**	**IT, EC**	**5+; +4.7**
^{114}In	113.904916	Daughter ^{114m}In	1.198 m	β^-, EC	1+; +2.82
^{115m}In		Daughter ^{115}Cd; ^{115}In(n, p or α)	4.486 h	IT, β^-	1/2−; −0.255
^{115}In	**114.903880**	**Naturally occurring**	**4.4×10^{14} y**	**β^-**	**9/2+; +5.541**
^{116m2}In		^{115}In(n, γ)	2.16 s	IT	8−
^{116m1}In		^{115}In(n, γ)	54.1 m	β^-	5+; +4.3
^{116}In	115.905264	^{115}In(n, γ)	14.1 s	β^-	1+; 2.788
^{117m}In		^{116}Cd(n, γ); daughter ^{117}Cd	1.94 h	β^-, IT	1/2−; 0.25
^{117}In	116.904517	Daughter ^{117}Cd	44 m	β^-	9/2+; +5.52

Table 1.2 (*Contd.*)

Nuclide	Atomic mass[b]	Source[c]	Half-life	Decay mode[d]	Nuclear spin, I; nuclear magnetic moment[e]
^{118m2}In		^{118}Sn(n, p); ^{121}Sb(n, α)	8.5 s	IT, β^-	(8−)
^{118m1}In	117.906120	^{118}Sn(n, p); ^{119}Sn(γ, p)	4.40 m	β^-	5+
^{118}In		Daughter ^{118}Cd; ^{118}Sn(n, p)	5.0 s	β^-	1+
^{119m}In		^{120}Sn(γ, p); fission	17.9 m	β^-, IT	1/2−
^{119}In	118.905819	^{120}Sn(γ, p); fission	2.3 m	β^-	9/2+; +5.52
^{120m}In		^{120}Sn(n, p); fission	47 s	β^-	8−
^{120}In	119.907890	^{120}Sn(n, p); ^{123}Sb(n, α)	3.1 s	β^-	(1+); +4.30
^{121m}In		^{122}Sn(γ, p); fission	3.8 m	β^-, IT	1/2−
^{121}In	120.907847	^{122}Sn(γ, p); fission	23 s	β^-	9/2+; +5.50
^{122m}In		Fission; ^{124}Sn(d, α)	10 s	β^-	8−
^{122}In	121.910280	^{122}Sn(n, p); daughter ^{122}Cd	1.5 s	β^-	(1+)
^{123m}In		^{124}Sn(γ, p); fission	47 s	β^-	(1/2−); −0.40
^{123}In	122.910450	Fission	6.0 s	β^-	(9/2+); +5.49
^{124m}In		Fission	3.4 s	β^-	8−; +3.89
^{124}In	123.912980	^{124}Sn(n, p); fission	3.18 s	β^-	3+; +4.04
^{125m}In		Fission	12.2 s	β^-	1/2−; −0.43
^{125}In	124.913670	Fission	2.33 s	β^-	9/2+; +5.50
^{126m}In		Fission	1.53 s		3+; +4.03
^{126}In	125.916470	Fission	1.63 s	β^-	8−; +4.06
^{127m}In		Fission	3.73 s	β^-	(1/2−)
^{127}In	126.917320	Fission	1.14 s	β^-	(9/2+); +5.52
^{128m}In		Fission	0.7 s	β^-	(8−)
^{128}In	127.920560	Fission	0.80 s	β^-	3+
^{129m}In		Fission	1.23 s	β^-, n	1/2−
^{129}In	128.921600	Fission	0.63 s	β^-	9/2+
^{130m2}In		Fission	0.53 s	β^-	5+
^{130m1}In		Fission	0.51 s	β^-	10−
^{130}In	129.924870	Fission	0.29 s	β^-	1−
^{131m2}In		Fission	0.3 s	β^-	(21/2+)
^{131m1}In		Fission	0.35 s	β^-	(1/2−)

Nuclide	Atomic mass	Production	Half-life	Decay mode	J^π; μ
^{131}In	130.926410	Fission	0.28 s	β^-	(9/2+)
^{132}In	131.93214	Fission	0.20 s	β^-	(7−)
^{133}In		Fission	0.18 s	β^- (n)	
^{184m}Tl	183.981670		11 s	β^+, EC, α	
^{185}Tl		^{180}W(^{14}N)	1.8 s	IT, α	(9/2−)
^{186m}Tl		W(^{14}N)	4 s	IT	
^{186}Tl	185.978510	^{182}W(^{14}N); ^{197}Au(^{3}He)	28 s	β^+, EC	
^{187m}Tl		^{159}Tb(^{32}S); ^{182}W(^{14}N)	15.6 s	IT	(9/2+)
^{187}Tl	186.976240	^{180}W(^{14}N)	50 s	β^+, EC	1/2+
^{188m}Tl		^{181}Ta(^{16}O)	1.18 m	β^+, EC	(7+)
^{188}Tl	187.975880	^{181}Ta(^{16}O)	1.2 m	β^+, EC	(2−)
^{189m}Tl		^{181}Ta(^{16}O); Pb(p)	1.4 m	β^+, EC	(9/2−); +3.878
^{189}Tl	188.980780	^{181}Ta(^{16}O)	2.3 m	β^+, EC	(1/2+)
^{190m}Tl		^{181}Ta(^{16}O); Pb(p)	3.7 m	β^+, EC	(7+); +0.495
^{190}Tl	189.973490	Daughter ^{190}Pb; Pb(p)	2.6 m	β^+, EC	(2−)
^{191m}Tl		^{182}W(^{14}N, 5n); Hg(p)	5.2 m	β^+, EC	(9/2+); +3.903
^{192m}Tl		^{181}Ta(^{16}O)	10.8 m	β^+, EC	(7+); +0.518
^{192}Tl	191.972120	Pb(p); ^{181}Ta(^{16}O, 4n); U(p)	9.6 m	β^+, EC	(2−)
^{193m}Tl		^{181}Ta(^{16}O, 4n); ^{185}Re(^{12}C, 4n)	2.1 m	IT	(9/2−); +3.948
^{193}Tl	192.970520	^{184}W(^{14}N, 5n); Hg(p)	22 m	β^+, EC	(1/2+); +1.591
^{194m}Tl		Hg(p); Pb(p)	32.8 m	β^+, EC	(7+); +0.540
^{194}Tl	193.970920	Hg(p); daughter ^{194}Pb	34 m	β^+, EC	2−; 0.14
^{195m}Tl		Daughter ^{195}Pb; ^{187}Re(^{12}C, 4n)	3.6 s	IT	9/2−
^{195}Tl	194.969630	Hg(p); ^{196}Hg(d, 3n)	1.16 h	EC, β^+	1/2+; +1.58
^{196m}Tl		Hg(p); ^{197}Au(α, 5n)	1.41 h	β^+, EC	(7+)
^{196}Tl	195.970460	Daughter ^{196}Pb; Hg(p); ^{197}Au(α, 5n)	1.84 h	β^+, EC	2−; 0.07
^{197m}Tl		Daughter ^{197m}Pb; ^{197}Au(α, 4n)	0.54 s	IT, β^+, EC	9/2−
^{197}Tl	196.969498	^{197}Au(α, 4n); ^{198}Hg(d, 3n)	2.83 h	β^+, EC	1/2+; +1.58
^{198m}Tl		^{197}Au(α, 3n); ^{198}Hg(d, 2n)	1.87 h	β^+, EC, IT	7+; +0.64
^{198}Tl	197.940460	Daughter ^{198}Pb; ^{197}Au(α, 3n); Hg(d)	5.3 h	EC, β^+	2−; 0.00
^{199}Tl	198.969870	^{197}Au(α, 2n); ^{199}Hg(d, 2n)	7.4 h	EC	1/2−; +1.60
^{200}Tl	199.970934	Hg(d); ^{197}Au(α, n); daughter ^{200}Pb	1.087 days	EC	2−; 0.04
^{201}Tl	**200.970794**	**Daughter ^{201}Pb**	**3.038 days**	EC	**1/2+; +1.605**
^{202}Tl	201.972085	^{202}Hg(d, 2n); ^{201}Hg(d, n); ^{203}Tl(d, t)	12.23 days	EC	2−; 0.06
^{203}Tl	**202.972320**	**Naturally occurring**	**Stable**	−	**1/2+; +1.622258**
^{204}Tl		^{203}Tl(n, γ)	**3.78 y**	**β^-, EC**	**2−; 0.09**

Table 1.2 (*Contd.*)

Nuclide	Atomic mass[b]	Source[c]	Half-life	Decay mode[d]	Nuclear spin, I; nuclear magnetic moment[e]
205Tl	**204.974401**	**Naturally occurring**	**Stable**	–	**1/2+; +1.638215**
206mTl		204Hg(α, pn); 204Hg(7Li)	3.76 m	IT	12–
206Tl	205.976084	205Tl(n, γ); daughter 210mBi	4.20 m	β^-	0
207mTl		208Pb(t, α)	1.3 s	IT	11/2–
207Tl	206.977404	Descendant 227Ac	4.77 m	β^-	1/2+; +1.88
208Tl	**207.981988**	**Natural source; descendant 228Th**	**3.053 m**	**β^-**	**(5+)**
209Tl	208.985334	Descendant 233U, 229Th, 225Ac	2.2 m	β^-	(1/2+)
210Tl	209.990056	Descendant 226Ra	1.30 m	β^-	(5+)

(*a*) Sources of data: D.R. Lide, editor-in-chief, *Handbook of Chemistry and Physics*, 73rd edn., CRC Press, Boca Raton, FL, 1992–1993; C.M. Lederer and V.S. Shirley, *Table of Isotopes*, 7th edn., Wiley, New York, 1978; IUPAC, Inorganic Chemistry Division, Commission on Atomic Weights and Isotopic Abundances, *Pure Appl. Chem.*, 1986, **58**, 1677; 1991, **63**, 975, 991; P. Raghavan, *At. Data Nucl. Data Tables*, 1989, **42**, 189; E. Browne and R.B. Firestone, *Table of Radioactive Isotopes*, ed. V.S. Shirley, Wiley–Interscience, New York, 1986; J. Emsley, *The Elements*, 2nd edn., Clarendon Press, Oxford, 1991. Key isotopes are given in bold type.
(*b*) Atomic mass relative to 12C = 12.0000.
(*c*) Sources of artificially produced nuclei given in the form *target (projectile, outgoing particle/radiation)*. p. proton; n, neutron; d, deuteron; t, triton.
(*d*) p, proton emission; IT, isomeric transition; EC, orbital electron capture; n, neutron emission; α, α-particle emission.
(*e*) Units: I $h/2\pi$; nuclear magnetic moment in nuclear magneton units.

properties of Al, Ga and In can be a source of significant information and not necessarily a detriment to the experiment, as is commonly assumed. The nuclei benefit not only from their relatively high sensitivities to detection, but also from large ranges of chemical shift. As a result, they are all potentially sensitive probes for chemical studies, particularly relating to the chemical and physical nature of the metal atom environment.

Aluminium.[61,62] ^{27}Al is now one of the ten nuclei most frequently given to NMR studies, a position it owes largely to the advances in high-resolution magic-angle-spinning (MAS) measurements[62,63] which have recently given access to solid, insoluble aluminium-bearing materials. Its one drawback is its quadrupole moment which causes the resonance linewidth to vary markedly with the electric field gradient at the nucleus and so with the symmetry of the nuclear environment. At one extreme we find a minimum linewidth of 2.0–2.5 Hz (for $[Al(OH_2)_6]^{3+}$ in dilute aqueous solution at pH 1): at the other, we find linewidths for solution species of up to 60 kHz likely to defeat detection of the signal by normal means. The chemical shifts, with a known span of about 300 ppm, appear to be determined mainly by the number and nature of the ligands bound to the aluminium centre. With relatively few exceptions, the shifts fall into three main regions: (a) δ +150 ppm and more to high frequency of the reference $[Al(OH_2)_6]^{3+}$ (aq), alkyl-aluminium compounds; (b) δ +40 to +140 ppm, tetrahedrally coordinated aluminium derivatives; and (c) δ −46 to +40 ppm, octahedrally coordinated aluminium derivatives. Within a given family of compounds, the shift moves generally to high frequency as the number of coordinated ligands decreases, thus providing a useful, but not definitive, index to the coordination number.[62,64] Coordination of an aluminium centre by a chelate ligand follows the precedent set by boron in similar circumstances, with a chemical shift which varies appreciably with the number of atoms making up the chelate ring. Because relaxation of ^{27}Al is usually fast, spin–spin coupling to other nuclei is not commonly observed. Coupling constants J_{AlX} *have* been measured, directly or indirectly, for a variety of compounds in which the ^{27}Al nuclei enjoy relatively long relaxation times (e.g. X = ^{1}H, ^{2}H, ^{11}B, ^{13}C, ^{14}N, ^{19}F, ^{31}P, ^{37}Cl and ^{81}Br);[61,62] although the expected trend $^1J > ^2J > ^3J$ appears to hold, the database is as yet too restricted to justify any detailed analysis. Still more sparse are reliable quantitative results for the relaxation times affecting ^{27}Al.

In practice, therefore, it is mainly through the chemical shifts and linewidths that ^{27}Al NMR spectra have been turned to account for the characterisation of aluminium compounds. In the liquid phase, moreover, the exchange of coordinated ligands between aluminium centres is often so sluggish as to admit the observation of distinct ^{27}Al resonances corresponding to the components making up a mixture of aluminium-containing species. Slow exchange is certainly the rule in ionic solutions, and

Table 1.3 NMR properties of nuclei of the Group 13 atoms.[a]

Property	[^{10}B]	^{11}B	^{27}Al
Natural abundance (%)	19.82	80.18	100
Nuclear spin ($h/2\pi$)	3+	3/2−	5/2+
Magnetic moment (μ/μ_N)	+1.8006	+2.6886	+3.64151
Relative sensitivity ($^1H = 1.00$)	1.99×10^{-2}	0.17	0.21
Receptivity ($^{13}C = 1.00$)	22.1	754	1.17×10^3
Magnetogyric ratio (10^7 rad T^{-1} s^{-1})	2.8740	8.5794	6.9704
Quadrupole moment (10^{-30} m^2)	8.5	4.1	15
Frequency ($^1H = 100$ MHz; 2.3488 T) (MHz)	10.746	32.084	26.057
Width factor (Al = 1)[b]	0.20	0.31	1.00
Normal reference		Et$_2$O · BF$_3$	[Al(OH$_2$)$_6$]$^{3+}$
Approximate range of chemical shifts (ppm)		200	300

(a) Sources of data: D.R. Lide, editor-in-chief, *Handbook of Chemistry and Physics*, 73rd edn., CRC Press, Boca Raton, FL, 1992–1993; J. Emsley, *The Elements*, 2nd edn., Clarendon Press, Oxford, 1991; R.K. Harris and B.E. Mann, eds., *NMR and the Periodic Table*, Academic Press, London, 1978; C. Brevard and P. Granger, *Handbook of High Resolution Multinuclear NMR*, Wiley, New York, 1981; J. Mason, ed., *Multinuclear NMR*, Plenum Press, New York, 1987. Nuclei enclosed in square brackets are seldom used for NMR studies.

(b) Width factor = $Q^2(2I + 3)/I^2(2I - 1)$ and is the nuclear contribution to quadrupole relaxation, normalised to Al = 1.

^{27}Al NMR measurements have contributed much to our knowledge of solutions of aluminium salts in polar solvents, with regard to solvation, ion-pairing, hydrolysis and polymerisation, and complexation.[61,62] Such studies played a major part, for example, in the characterisation of the tridecameric cation [AlO$_4$Al$_{12}$(OH)$_{24}$(OH$_2$)$_{12}$]$^{7+}$ (with the well known Keggin ion structure) in highly concentrated basic solutions of aluminium salts.[65] Although the ^{27}Al resonance due to a complexed aluminium centre is often too broad to be detected, the relatively sharp signal due to [Al(OH$_2$)$_6$]$^{3+}$ can be used to determine the stoichiometry of the complex, such as that formed with thymulin.[66] In favourable cases, equilibrium constants can be determined and kinetic parameters derived from the temperature-dependence of ^{27}Al resonances, although ligand-based spin-½ nuclei are usually more amenable to studies of this sort. Highly reactive organo- and hydrido-derivatives of aluminium have also been interrogated through their ^{27}Al NMR spectra; how the spectra relate to the coordination sphere of the aluminium is well illustrated by the case histories of some tetrahydro-aluminate complexes of the heavier transition metals which appear to feature five-coordinated aluminium in units of the type (μ-H)$_2$AlH(μ-H)$_2$.[67]

In the explosion of research activity which has been triggered by the development of magic-angle-spinning and related methods, ^{27}Al has played a conspicuous part. Rapid rotation of a solid powder specimen of an aluminium compound about an axis inclined at the 'magic angle' of 54.736° [cos^{-1} (1/$\sqrt{3}$)] to the magnetic field direction can minimise the contributions

$[^{69}Ga]$	^{71}Ga	$[^{113}In]$	^{115}In	^{203}Tl	^{205}Tl
60.108	39.892	4.33	95.67	29.524	70.476
3/2−	3/2−	9/2+	9/2+	1/2+	1/2+
+2.01659	+2.56227	+5.529	+5.541	+1.622258	+1.638215
6.91×10^{-2}	0.14	0.34	0.34	0.18	0.19
237	319	83.8	1.890×10^3	289	769
6.420	8.158	5.8493	5.8618	15.3078	15.4584
17.8	11.2	80	81	–	–
24.003	30.495	21.866	21.914	57.149	57.708
5.85	2.32	6.6	6.7	–	–
	$[Ga(OH_2)_6]^{3+}$		$[In(OH_2)_6]^{3+}$		$TlNO_3(aq)$
	1400		1100		>5500

of the dipolar, first-order quadrupole and chemical shift anisotropy terms to the linewidth of the ^{27}Al resonance. A second-order quadrupole effect remains, although its magnitude is substantially smaller for the spinning than for the static sample and decreases at high magnetic fields; this results in a supplementary shift in the resonance frequency of an unsymmetrically sited ^{27}Al atom, leading to problems in setting up an accurate and universally applicable scale of chemical shifts.[62] With the additional stratagems of double-rotation (DOR) and dynamic-angle-spinning (DAS),[68a] even this effect can be averaged out, thus enabling the environments of ^{27}Al sites to be scrutinised in unprecedented detail.[68b] The feasibility of cross-polarisation *to* and *from* ^{27}Al nuclei has even been verified experimentally in an alumino-phosphate molecular sieve;[68c] this bodes well for the future use of hetero-nuclear correlations to unravel local microstructure in solids. Under typical MAS conditions, the ^{27}Al resonances of solid samples are narrowed to the point where it is possible to distinguish not only between aluminium sites of different coordination number, but also, in many cases, between sites that differ not in their primary but in their secondary coordination sphere.[62,63] ^{27}Al MAS measurements thus complement similar measurements involving the spin-½ nucleus ^{29}Si as a means of determining the ordering within aluminosilicate frameworks, although the ^{27}Al chemical shifts are less sensitive than their ^{29}Si counterparts to the occupants of the second and more distant coordination shells. On the other hand, ^{27}Al has the advantage of its quadrupole moment, giving linewidths that can be used to assess the electric field gradient at the nuclear site. Moreover, the 100% isotopic abundance of ^{27}Al and its very short spin-lattice relaxation time impart high sensitivity to ^{27}Al MAS measurements, even traces of aluminium being detectable in this way, and the aluminium content of a sample can be deter-mined quantitatively, for example by comparing absolute resonance intensities before and after the addition of a known amount of a suitable aluminium reference compound. What makes the NMR technique important is that it can be applied to polycrystalline and amorphous samples that can be characterised only partially or not at all by conventional diffraction methods. MAS NMR experiments involving ^{27}Al have focused particularly on (i) zeolites, made important by their remarkable sorptive and catalytic properties, (ii) the sol-gels which act as precursors to the zeolites and other ceramic materials, (iii) clays and related layer silicates,[69] (iv) glasses, (v) cements, and (vi) a variety of other oxo- and hydroxo-aluminium compounds including alumina, aluminium hydroxides and metal aluminates.[70] Typical of the results which have thus come to light is the identification of five-coordinated aluminium sites in barium aluminium glycolate,[71a] the mineral andalusite[71a] and the aluminophosphate molecular sieve precursor AlPO$_4$-21.[71b] Polyoxyaluminium clusters even larger than $[AlO_4Al_{12}(OH)_{24}(OH_2)_{12}]^{7+}$ have been characterised in the solid state and the ^{27}Al MAS NMR spectrum of the sulfate salt formed by one of these is consistent with the heavy-atom formulation $Al_{24}O_{72}$;[72] the spectrum also witnesses the thermal transformation of the cluster to a material containing a high proportion of five-coordinated aluminium sites. One of the most important applications of such measurements has been to explore the way in which aluminium can be removed from the frameworks of zeolites, e.g. by hydrothermal treatment, and to gain insights into the chemical status of the extra-framework aluminium species.[61,62,73] The dealumination process has a critical bearing on the thermal stability and catalytic activity of the product; indeed, so-called 'ultrastable' zeolites prepared in this way now occupy a commanding position in the petroleum industry by virtue of their efficiency in activating hydrocarbons to a wide range of reactions, e.g. cracking, hydrocracking, oxidation and isomerisation. Through ^{27}Al NMR studies of static samples, it has even been possible to investigate the nature of catalyst surfaces and to shed some light on the dynamic events occurring at these surfaces.[74]

Gallium and indium.[61]   Much less widely used in NMR experiments are gallium and indium nuclei, partly because the linewidths are typically much greater, partly because the chemistries of these elements have attracted much less attention, by comparison with aluminium. The resolution of individual resonances due to different metal sites is impaired by the broader signals, although the problem is offset to some extent by the increased dispersion of the chemical shifts characterising gallium and indium. Correlations between the chemical shifts of analogous ionic species of the three elements Al, Ga and In suggest that In is about 1.5 times more responsive to change in chemical environment than is Ga, which is in turn 5.7 times more responsive than is Al. Such correlations imply that ^{71}Ga follows ^{27}Al in the pattern of its chemical shifts, with coordination number and the nature of the ligands as the primary influences, although there are some differences of detail in the way the two nuclei respond to certain ligands. Relatively few results are available for ^{115}In, but there is a linear correlation between the ^{71}Ga and ^{115}In chemical shifts exhibited by tetrahalo anions of the type $[MX_nY_{4-n}]^-$ (M = Ga or In; X, Y = halogen; n = 0–3). ^{71}Ga NMR studies have been brought to bear on the base hydrolysis of gallium(III) and mixtures of gallium(III) and aluminium(III);[75] the results indicate that the hydrolysis products include the polyoxocations $[GaO_4\text{-}Ga_{12}(OH)_{24}(OH_2)_{12}]^{7+}$ and $[GaO_4Al_{12}(OH)_{24}(OH_2)_{12}]^{7+}$ which are presumed to be isostructural with the corresponding Al_{13} species.[75b] The heteronuclear polyoxocation $[GaO_4Al_{12}(OH)_{24}(OH_2)_{12}]^{7+}$ has even been characterised by the ^{71}Ga NMR MAS spectrum displayed by its solid sulfate derivative.[75c] The ^{71}Ga NMR spectra of the compounds having the compositions $GaCl_2$ and $GaBr_2$, whether as melts or in solution, each show two resonances, one relatively sharp corresponding to Ga(I) and the other broad corresponding to Ga(III); only the Ga(I) resonance can be observed in the ^{69}Ga and ^{71}Ga spectra of the solids.[76] By contrast, the ^{115}In NMR spectrum of the molten compound with the composition $InCl_2$ gives little support to the formulation $In^I[In^{III}Cl_4]$.[77] Only rarely have the results for liquid samples afforded estimates of spin-spin coupling constants or reliable quantitative details regarding the relaxation behaviours of gallium and indium nuclei.

Thallium.[61] The lack of a quadrupole moment, unique in Group 13, means that the relaxation times of ^{203}Tl and ^{205}Tl are much longer than those characteristic of the lighter nuclei, and leads to quite different operational techniques resembling more closely those appropriate to ^{13}C, although the relaxation times are not particularly long. Not only does thallium have a high sensitivity, it can in principle be observed in all possible chemical environments, irrespective of symmetry, shows a chemical shift highly responsive to its chemical state, and gives access also to spin-spin coupling constants (with values up to 15 000 Hz[78]) as a regular part of its NMR database. There are indeed few metal nuclei better adapted to the NMR experiment than ^{205}Tl, and such experiments have contributed significantly to our knowledge of thallium chemistry. So sensitive are thallium nuclei to their environment, the NMR parameters for a given thallium compound are likely to vary widely as a function of temperature, solvent, concentration and counter-ion. This property has been exploited in biological studies where Tl^+ has been used as a probe to emulate the functions of alkali-metal ions, and particularly Na^+ and K^+.[61,79] Thus, the interaction of Tl^+ with a variety of anti-biotics results in profound changes in chemical shift and relaxation behaviour; the shifts span a range of almost 1000 ppm, indicating very different modes of bonding to the various ionophores.[61] The same approach is exemplified by studies (i) of the monovalent cation binding sites of bovine plasma activated protein C (APC)[80a] and (ii) of the equilibrium binding of such cations by gramicidin A in dimyristoylphosphatidylcholine vesicles.[80b] The behaviour of Tl^+ on complexation, e.g. by chloride,[81a] EDTA and various macrocyclic ligands,[81b] or in non-aqueous media has also been investigated. Another ion showing NMR properties unusually responsive to its surroundings is Me_2Tl^+, the behaviour of which has been the focus of much attention. In a somewhat different vein, anionic clusters incorporating thallium in company with other Main Group elements, e.g. $[Sn_{8-x}Pb_xTl]^{n-}$ ($x = 1$–4, n probably 5), have been successfully characterised by their NMR spectra.[82] Elsewhere there is important information to be gained about the halide complexes which thallium(III) forms in aqueous solution, concerning, for example, stability constants, geometries and rates of ligand exchange.[83] Thallium relaxation processes in these and other solutions have invited numerous studies; with spin-rotation and chemical shift anisotropy mechanisms believed to be instrumental, these processes are usually very efficient, and particularly in the presence of oxygen or other paramagnetic species. Nor has the action been confined to conventional solutions, for solids or melts of thallium metal, alloys and salts have also been examined, and several investigators have reported the use of thallium NMR to study Tl^+ ions adsorbed on zeolites.[61]

1.3.1.2   NQR and ESR studies.   The very quadrupolar properties which are apt to limit the practical usefulness of the nuclei ^{27}Al, ^{69}Ga, ^{71}Ga, ^{113}In and ^{115}In in conventional NMR experiments make them highly eligible for NQR experiments which afford primarily a means of investigating the interaction of such nuclei with the electric fields to which they are exposed by their environments.[84] Through the frequencies, splitting, linewidths and relaxation rates, and the temperature-dependence of one or more of these parameters, NQR spectra can be highly enlightening, particularly in dealing with solid materials that do not give useful single crystals. The main information which they hold relates, firstly, to the number of chemically distinct sites occupied by the quadrupolar nuclei in the sample and, secondly, to the local environment of the nuclei in each of these sites. Hence it may be possible to estimate charge distributions and to assess the nature of the interaction between the quadrupolar nucleus and the atoms in its coordination sphere. Although the quadrupolar Group 13 atoms have featured less often in such measurements than, say, the heavier halogen atoms, they have still attracted attention, particularly in the form of their halide complexes, e.g. $M^+GaCl_4^-$ and $M^+Ga_2Cl_7^-$ (M = alkali metal or Ga),[85a] salts of the $Al_2Br_7^-$ anion,[85b] $[R_4N]_2Ga_2Br_6$ (R = Me or Et),[85c] and various adducts with neutral donor molecules.[85d] Solid oxide derivatives lending themselves to NQR measurements include the mineral Muscovite mica[86a] and the mixed indium antimony oxide $In_3Sb_5O_{12}$.[86b] Salient points of many of these studies include the identification and characterisation of solid phase changes.

Another important domain where the nuclear spins play a crucial role is that of ESR spectroscopy. Interactions between the electron and nuclear spin magnetic moments lead to the appearance of hyperfine structure in the ESR spectrum of a paramagnetic derivative of one of the Group 13 metals. An individual nucleus with nuclear spin I should give rise in a first order spectrum to a multiplet composed of $2I + 1$ equally spaced lines: the multiplicity thus gives an explicit measure of the identity and number of such nuclei that the species contains and the spacing may well hold the key to the disposition of ligands about the nucleus, to the distribution of the electron spin density and to the nature of the bonding. The technique is well adapted to the study of matrix-isolated species[87] and the response of the spectrum to changes of conditions then provides a means of monitoring intramolecular and intermolecular properties, including aggregation and matrix inter-actions, and the orientational or motional behaviour of the paramagnetic centre within the matrix host. Hence, for example, neutral Al and Ga atoms have been trapped and characterised,[88a] as have the dipositive ions Ga^{2+} and Tl^{2+} [35b] or derivatives of such ions, $viz.$ AlH^+ [88b] and AlF^+.[88c] The carbonyls $M(CO)_2$ (M = Al, Ga or In), formed on co-condensation of the appropriate metal atoms with carbon monoxide, have also been identified and their magnetic parameters analysed to deduce that they are bent, planar π radicals of C_{2v} symmetry.[89] Other fugitive organometallic derivatives, produced and characterised in a similar fashion and typically with relatively weak binding of the metal, include $(\eta^2-C_2H_4)M$,[90a] aluminocyclopentane (I),[90a] aluminocyclopentene (II),[90b] and $(C_6H_6)M$ (M = Al or Ga).[90c] On the

(I)   (II)

other hand, the paramagnetic compounds $(Bu^tNCHCHNBu^t)_2M$ (M = Al or Ga) contain not the divalent but the trivalent metal bound to one singly and one doubly reduced ligand.[91] Less ambiguous, though, is the 16-line ESR spectrum displayed by the matrix formed by co-condensing aluminium vapours with an excess of an inert hydrocarbon;[92] it would be difficult to identify this signal with anything other than the Al_3 cluster.

1.3.1.3 Radionuclides. The better characterised radionuclides of aluminium, gallium, indium and thallium are listed, with some of their properties and typical methods of nucleosynthesis, in Table 1.2; the more important of these are highlighted by bold characters. The number of such nuclides with half-lives in excess of 10 min are: Al 1, Ga 7, In 17 and Tl 15. Indium is noteworthy on two counts: firstly, its more abundant naturally occurring isotope (^{115}In) is a very long-lived radionuclide; secondly, it forms a multitude of isobaric nuclei. Indium isotopes of high mass number are formed in the fission of actinide nuclei, but only as transient intermediates. Several of the heavier thallium isotopes are formed in the natural radioactive decay chains that originate in thorium and uranium. Of more practical use for the production of radionuclides of the elements is the bombardment of stable isotopes of various elements with neutrons (yielding, for example, ^{28}Al, ^{72}Ga, ^{114m}In, ^{116m}In or ^{204}Tl), with protons, deuterons or heavier projectiles (e.g. ^{14}N, ^{16}O or ^{20}Ne), or with high-energy photons. Some nuclei are conveniently reached as daughters of accessible radionuclides of other elements; such is the case, for instance, with ^{68}Ga (from ^{68}Ge),[93] ^{113m}In (from ^{113}Sn), and ^{201}Tl (from ^{201}Pb).[94] Access can be gained to samples enriched in the following radionuclides either currently or in the recent past, although short-lived nuclides like ^{68}Ga and ^{113m}In have to be produced on site: ^{26}Al (e.g. as AlCl$_3$ in 0.5 M HCl), ^{66}Ga, ^{67}Ga (e.g. as GaCl$_3$ in aqueous HCl), ^{68}Ga (available from ^{68}Ge and HCl, for example[93]), ^{72}Ga, ^{111}In (e.g. as InCl$_3$ in aqueous HCl), ^{113m}In (available from ^{113}Sn as [SnCl$_6$]$^{2-}$ in 6 M HCl, for example), ^{114}In, ^{114m}In, ^{201}Tl (daughter of ^{201}Pb[94]), ^{204}Tl (e.g. as TlNO$_3$ or Tl$_2$SO$_4$ in aqueous solution), ^{206}Tl, ^{207}Tl, ^{208}Tl and ^{210}Tl. The nuclear reactions induced by neutron irradiation in isotopically natural samples form the basis of neutron activation analysis, a technique that has found considerable use for the quantitative determination of the Group 13 metals, typically at trace levels (see chapter 10).

Radionuclides of the Group 13 metals have not found extensive use in general chemical studies, but radio-gallium,[95] -indium[95] and -thallium[96] have, through their diagnostic and therapeutic applications — real or potential — awoken considerable interest in medical circles. There follows a brief survey, element by element, giving more specific details about some of the more noteworthy aspects.

Aluminium.   The long-lived radionuclide ^{26}Al was used, together with ^{24}Na and ^{18}F, as a tracer in transport experiments designed to elucidate the identities of the ions carrying the current in the electrolysis of alumina dissolved in cryolite[97] (see chapter 2). Hence it was established that nearly all the current is carried by Na$^+$ ions, but about 1% is carried by an anion containing aluminium and fluorine in the proportions 1:2 (AlOF$_2^-$ or AlO$_2$F$_2^{3-}$ perhaps). Bombardment of atmospheric constituents by galactic cosmic rays produces both ^{26}Al and another long-lived radionuclide ^{10}Be (half-life 1.5×10^6 years); these are removed from the atmosphere chiefly by rain and the portion falling on the ocean eventually reaches the oceanic floor. Because aluminium and beryllium have similar geochemical properties, the isotopic ratio ^{26}Al/^{10}Be has been suggested as a means of 'dating' various marine reservoirs.[98] Manganese nodules dredged from the ocean bed have been analysed in this way, for example, giving results supporting the hypothesis that they have accumulated only very slowly. The effects of weathering on chondrites have also been investigated by reference to the [26]Al content.[99] Such measurements have been hampered in the past by the difficulties inherent in determining [26]Al, but these have been overcome, to some extent at least, by the deployment of γ–γ coincidence or accelerator mass spectrometry.[98] The truly ubiquitous character of aluminium is emphasised by recent measurements made on the γ-rays emanating from the centre of the galaxy;[100] prominent among the emissions are those centred near 1809 keV having as their source [26]Al.

Gallium and indium.[95]   Gallium and indium each have not one but two isotopes that lend themselves well to the detection methods of nuclear medicine. These are ^{67}Ga, ^{68}Ga, ^{111}In and [113m]In with half-lives ranging from 1.13 h to 3.26 days; the longer-lived ^{67}Ga and ^{111}In are cyclotron-produced, usually by proton reactions involving [67]Zn and [111]Cd, respectively, and the short-lived ^{68}Ga and [113m]In are produced on site from appropriate radio-nuclide generators. Radiopharmaceuticals incorporating the positron-emitting ^{68}Ga nuclide are of particular interest because their distribution *in vivo* can be quantified to give the true distribution of activity in the source. The design of pharmaceuticals labelled with radio-gallium or -indium has to take account of the fact that both Ga[III] and In[III] form very stable chelates with the plasma protein transferrin, although the problem is mitigated to some extent by the sluggishness of the approach to equilibrium with gallium or indium complexes having stability constants in excess of 10^{20}. One of the major goals of research in this area has therefore been the development of strongly binding bifunctional chelating agents. The first of these was 1-(*p*-benzenediazonium)-ethylenediamine-*N,N,N',N'*-tetra-acetic acid; others tried subsequently, or offering good prospects, include 8-hydroxy-quinoline, tropolone,[101a] 3-hydroxy-4-pyridinones (**III**),[101b] *N,N'*-bis-(2-hydroxybenzyl)-ethylenediamine-*N,N'*-di-acetic acid (**IV**) and its derivatives,[101c] and bisaminoethanethiol derivatives, e.g. **V**.[101d]

Interest in the radiopharmaceutical possibilities of gallium and indium

R = H, Me, OMe or NO$_2$

(III)   (IV)   (V)

was first engendered by the discovery in the 1960s, by Edwards and Hayes,[102] that ^{67}Ga administered as the citrate localised in soft tumour tissue. The citrate is now widely used in oncological medicine for tumour detection, although the mechanism of uptake of ^{67}Ga by tumour cells is far from clear. The use of ^{67}Ga and ^{68}Ga has been limited in practice to the detection of soft tissue tumours and inflammatory lesions. Development of lipophilic ^{68}Ga tracers for perfusion imaging of the brain and heart has not been successful, but the gallium(III) complex of **V** is reported to be highly promising as a myocardial perfusion imaging agent for positron emission tomography.[101d] Major uses of ^{111}In in medicine include not only tumour scanning but also imaging or scanning of bone, bone marrow, thrombus, lymph node, heart, spleen and abscesses and in cisternography.[95] Perhaps the most significant application in recent years has been in the labelling of blood cells. It has been shown, for example, that mixing of the 8-hydroxy-quinoline complex with cells separated from plasma causes the indium to become firmly bound inside the cell; there is evidence that the lipophilic chelate diffuses inside the cell and that the indium exchanges at intracellular binding sites. As the indium is attached *inside* the blood cell, a stable label results for reinjection into a patient because the cell membrane denies plasma transferrin access to the labelled protein. Labelled platelets, labelled white cells and labelled lymphocytes have all been studied extensively. Labelled platelets have been shown for a series of normal patients to behave in the same manner as unlabelled platelets, and in patients with thrombosis or atherosclerosis to localise at the site of the lesions. Hence various methods have been devised for *in vivo* imaging of thrombi and vascular lesions.[95, 103] The kinetics of uptake of the labelled cells are such that ^{113m}In is too short-lived for many applications of this sort, but ^{113m}In-labelled chelate complexes can be used to study many organs of the human body in a non-invasive manner, e.g. the brain and renal functions.

The two successive γ-rays emitted from the bound ^{111}In atom can also be utilised in quite different fields in the perturbed γ–γ angular correlation method (PAC). ^{111}In has thus been exploited as a probe for hyperfine interactions, with particular reference to solids and their surfaces. Representative studies include the characterisation of copper (100) surfaces[104a] and of lattice defects in cold-worked copper,[104b] identification of the phases and their interrelation in the Cu–In–S system,[104c] examination of defect-acceptor pairs in doped germanium samples,[104d] and the diffusion behaviour of helium in metals.[104e] The method has also been applied to studies of the molecular dynamics of cellular macromolecules such as DNA and DNA moieties.[104f] Other radioisotopes of indium have found applications in areas as diverse as the structural defects in heavy-ion implanted GaAs (^{112}In),[105a] quality control of the surfaces of materials (^{113m}In)[105b] and the distribution of indium trace impurities in organometallic compounds ($^{114+114m}$In).[105c]

Thallium.   The relatively long-lived radionuclide ^{204}Tl has seen service in numerous chemical studies. Among these are the determination of dissolved oxygen in water[106a] through the reaction

$$4Tl(s) + O_2 + 2H_2O \rightarrow 4Tl^+ + 4OH^- \qquad (1.1)$$

studies of the rate of exchange between Tl(I) and Tl(III),[106b] and determination of the solubilities of thallium crown ether complexes in organic solvents.[106c] More striking, however, is the short-lived nuclide ^{201}Tl which has gained some prominence in recent years as a basis for myocardial scintigraphy in the evaluation of coronary diseases.[96,107a] The X-ray emission induced by a ^{201}Tl implant has also been used to image the thyroid gland and determine the iodine distribution therein.[107b]

1.3.2   Properties of the atoms

Table 1.1 presents for the free atoms of the Group 13 elements details of relative atomic mass, spectroscopic properties, $ns \rightarrow np$ promotion energy, ionisation potentials, electron affinity, electric dipole polarizability and thermodynamic properties. It also includes information about the following parameters more germane to the bound atoms: electronegativity, metallic, covalent, van der Waals and ionic radii, representative bond lengths and bond energies, and thermodynamic properties of derivatives of the elements. Some of the more salient aspects of these properties have already been pointed out in section 1.1.

The separation of the components of the doublet that forms the ground state of each atom implies the following values (in cm^{-1}) for the spin-own-orbit coupling constants, A: B 10.17, Al 74.69, Ga 550.79, In 1475.07, Tl 5195.1.[108] Corresponding to this sequence, whereas the spin-orbit coupling in boron and aluminium is adequately described by the Russell–Saunders scheme, the large magnetic interaction in thallium is compatible less with simple L,S- than with j,j-coupling. The energy gap between the valence s- and p-orbitals is indicated by the promotion energy for the process $ns^2np^1\ (^2P_{1/2}) \rightarrow ns^1np^2\ (^4P_{1/2})$ which varies as follows: B 28 805 cm^{-1} (345 kJ), Al 29 020 cm^{-1} (347 kJ), Ga 37 972 cm^{-1} (454 kJ), In 34 978 cm^{-1} (418 kJ), Tl 45 220 cm^{-1} (541 kJ).[108] Hence it follows that gallium and, above all, thallium are the atoms least well adapted to exploit fully the valence ns as well as the valence np orbitals for their bonding interactions. At energies about 32 000–36 500 cm^{-1} (380–440 kJ) above the np orbitals come the vacant nd orbitals.[108] Although these have the potential to fulfil an acceptor role, facilitating the coordination of the Group 13 metal atom by more than four donor atoms, all the recent calculations seem agreed that this role is only a secondary, supporting one.[109] From gallium onwards there is also the possibility that the $(n-1)d^{10}$ shell may make some contribution to the bonding, beyond merely enhancing the polarisability of the atom. It is not possible to gauge the energy separation between the tightly bound $(n-1)d$ electrons and the valence ns or np electrons for the neutral atom, but the transition $(n-1)d^{10}\ (^1S) \rightarrow (n-1)d^9ns^1\ (^1D)$ can be observed for the M^{3+} ions at the following energies (in cm^{-1}): Ga^{3+} 155 810, In^{3+} 138 764, Tl^{3+} 96 727. The energies for the corresponding transitions of adjacent isoelectronic ions, also in cm^{-1}, are: Cu^+ 26 265, Ag^+ 46 046, Au^+ 29 621; Zn^{2+} 83 509, Cd^{2+} 88 872, Hg^{2+} 61 086. The heaviest member of each series stands out for having a relatively small $(n-1)d-ns$ separation, but thallium(III) is decidedly less well placed than gold(I) or mercury(II) to take advantage of any admixture of $(n-1)d$ states. Hence it is not altogether surprising that thallium(III) does not show the same tendency as gold(I) and mercury(II) to occur in linear environments.[19,30-33,58] That linear coordination is encountered, however, notably in the $(CH_3)_2Tl^+$ cation, isoelectronic with $(CH_3)_2Hg$, gives some grounds for believing that the $5d$ electrons are not wholly lost to the core.

For a given stage of ionisation of the valence shell, the ionisation potentials vary irregularly in the sequence $B \gg Al \lesssim Ga > In < Tl$. The first ionisation potentials of the Group 13 metals are considerably larger than those of the preceding alkali metals. The contrast is particularly pronounced for the three heaviest metals; on the other hand, the first ionisation potentials of these same metals are considerably smaller than those of the preceding members of Group 11, viz. Cu, Ag and Au. With atomisation energies intermediate between those of the corresponding Group 1 and Group 11 metals, Ga, In and Tl are therefore much less predisposed to oxidation to univalent derivatives than are K, Rb and Cs, but better predisposed to such oxidation than are Cu, Ag and Au. At 5139, 5521, 5083 and 5438 kJ, the total energy input for the formation of Al^{3+}, Ga^{3+}, In^{3+} and Tl^{3+}, respectively, is well in excess of the investment needed for the formation of a Group 3 cation (Sc^{3+} 4255, Y^{3+} 3777, La^{3+} 3455 kJ). As a result the Group 13 metals are that much less electropositive than the Group 3 metals, being significantly less potent reducing agents. There is a further difference: the relatively smooth decrease in total ionisation energy with atomic number characterising the Group 3 metals gives way in Group 13 to energies which show little overall decrease and two appreciable *increases*. The chemical energy recouped through solvation or lattice-formation, determined to a first approximation by the size of the ion ($q.v.$), tends in both cases to become less exoergic as the atomic number increases. The greater the atomic number, therefore, the more noble the Group 13 metal becomes (witness, for example, the standard reduction potentials in Table 1.1). By contrast, the energetics of ion-formation and ion-coordination are such that the heavier Group 3 metals are, if anything, more electropositive than the lighter ones (e.g. $E^\ominus$ for the aqueous couple $M^{3+}/M(s)$ is -2.09, -2.38 and -2.38 V for M = Sc, Y and La, respectively[110]). Although, in relation to the metal in its standard state, *both* the $+1$ and the $+3$ oxidation states become less favoured for the heavier Group 13 metals, the decline in stability is more acute for + 3 than it is for + 1, for reasons yet to be analysed (see section 1.4.1 and chapter 3). From being strongly reducing and liable to undergo disproportionation, the + 1 state thus gains in stability at the expense of + 3, until it emerges as the dominant condition of thallium.

While electronegativity cannot be assigned an absolute value irrespective of the chemical environment of the atom, application of different criteria yields coefficients with relative magnitudes typically in the order B > Al < Ga > In < Tl. In fact, the coefficients for the metallic members do not vary greatly from element to element. In keeping with estimates of the effective nuclear charge (see Table 1.1), the metals are superior to the members of Groups 1–3, but comparable with the transition metals, with regard to electronegativity. As might be expected, there is a significant diminution in electronegativity when the oxidation state changes from + 3 to + 1. On the evidence of thermochemical and spectroscopic measurements, the energy of the bond formed by a Group 13 atom M in a particular oxidation state and with a given substituent X appears usually to follow the order B–X > Al–X > Ga–X > In–X > Tl–X. Increasing the atomic number of the ligand atom X within a particular Group, while keeping M constant, seems always to evoke a decrease in the bond energy, e.g. M–F > M–Cl > M–Br > M–I, but the degree of discrimination varies both with the nature of M and with its oxidation state. Here we note that the spread of values, say from M–F to M–I, decreases appreciably, not only from M = B or Al to M = Tl but also, for the heavier metals, when the tripositive gives way to the unipositive state. That M^I should thus be a softer acceptor than M^{III} is a sign of a general phenomenon, and one which has vital implications for the relative stabilities of the two oxidation states (see sections 1.4.1 and 1.4.2).

Much more extensive and, usually, more precise than bond energies are the interatomic distances which offer, in practice, the main index to the interactions and spatial properties of the Group 13 atoms. The distances have provided the input for estimates of the various single-atom or single-ion radii. Even by the standards of the highly simplified models which these radii embody, the values show greater variance than usual, reflecting the reality of distances that are only roughly additive and of metal atom environments that are not always isotropic. The shortcomings of the estimates are a symptom of the general problem that the metal atoms in the combined state are neither 'fish nor fowl' in their adherence to either a simple ionic or a simple covalent model of bonding. If no undue weight can be accorded the absolute values, the covalent and ionic radii for the M^{III} species take the following order: $B < Al \lesssim Ga < In < Tl$. The increase in bulk with atomic number is less pronounced than for the corresponding species in Groups 1–3 — indeed aluminium and gallium are remarkable less for their difference than for their similarity in size — but the overall trend reflects chemical interactions which wane in strength with increasing atomic number, in keeping with the usual pattern of Main Group chemistry. The length of a specific bond appears to be influenced less by the geometry of the unit in which it occurs than by the coordination number of the metal atom and the charge distribution at the metal and substituent centres. There is little sign of multiple bonding outside high-temperature molecules such as AlCl or GaP. It is true that the molecule $Al[N(SiMe_3)_2]_3$, which features planar AlN_3 and $AlNSi_2$ moieties, has Al–N bonds measuring only 1.78 Å (cf. distances of 1.90–1.97 Å for 4-coordinated aluminium derivatives[111]), but the revelation that the $AlNSi_2$ planes are canted at an angle of 50° to the AlN_3 plane makes it improbable that there is appreciable π-type electron delocalisation over the whole $Al(NSi_2)_3$ skeleton.[112a] The corresponding gallium and indium compounds are isostructural with M–N distances of 1.86 and 2.06 Å, respectively, in each case close to the values estimated for a normal σ-bond.[112b] Relatively short Al–N bonds displayed by the Al–NEt unit of the cyclic molecule $Cl_2AlN(Et)C_2H_4NMe_2$ (1.77 Å),[112c] by two of the $Al–N=CBu^t_2$ units of $LiAl(N=CBu^t_2)_4$ (1.78 Å),[112d] and the Al–NMe–Si portion of the tricyclic species $[ClAl(NMeSiMe_2)_2NMe]_2$ (1.80 Å)[112e] may signal π-bonding, but the evidence is hardly compelling. The sterically crowded phosphido-gallium compounds VI^{113a} and VII^{113b} both contain formally unsaturated tri-coordinated gallium centres with relatively short Ga–P distances (2.27 and 2.30 Å, respectively), but the Ga_3P_3 ring of VII is non-planar and the stereochemistry at the phosphorus atoms of both molecules is tellingly pyramidal. The stratagem of inhibiting aggregation by steric crowding has not, however, been exploited to any marked degree, and Group 13 offers few analogies to the stable compounds affording the heavier Group 14 and Group 15 elements the opportunity of multiple bonding.[114]

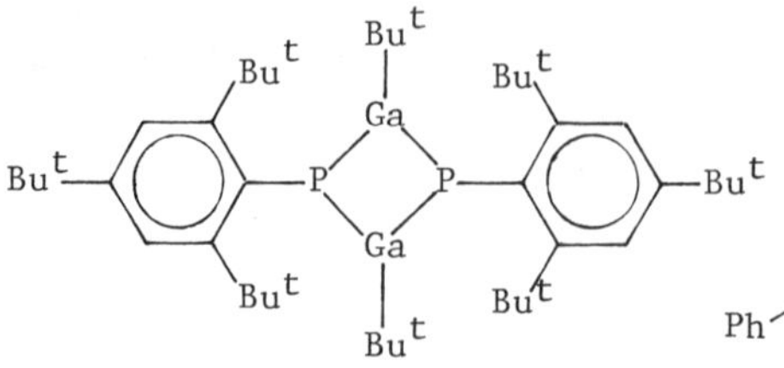

(VI)

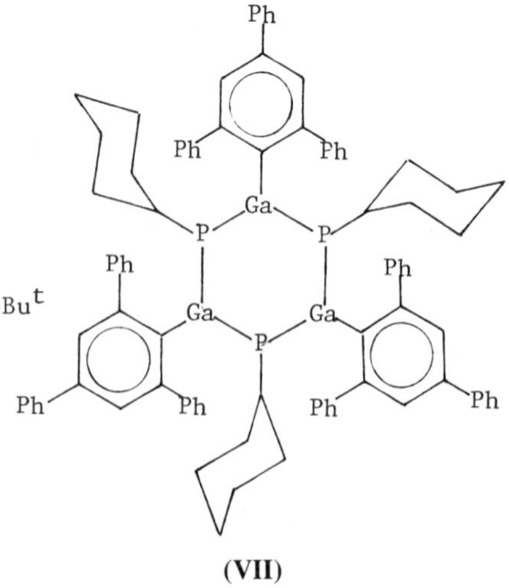

(VII)

Information about the chemical reactions of Group 13 metal atoms is still comparatively sparse. However, co-condensation of the vapours of the metal and a potential reagent, with or without a suitable diluent, followed by spectroscopic interrogation of the solid deposit, has shed light on a little of this chemistry. Laser ablation of the solid element has considerable promise as an expedient source of the metal atoms for such studies.[115] Reaction pathways thus identified include aggregation (e.g. with the formation of Al_3[92]) and simple addition to each of the following molecules without bond-cleavage, albeit with varying degrees of electron-transfer: O_2 (to form superoxo- or peroxo- derivatives),[116] O_3,[117] SiH_4,[118] NH_3,[119] H_2O,[120] CO,[89] C_2H_2,[121] C_2H_4,[90a] $CH_2=CH \cdot CH=CH_2$[90b] and benzene.[90c] Cleavage of the reagent molecule with oxidative addition to the metal atom typically ensues; in some cases this requires only thermal activation, in others it calls for photolytic activation. For example, ground-state Al atoms react spontaneously with H_2O,[120] NH_3[119] and CH_4[122] to give Al^{II} hydrides of the type HAlX (X = OH, NH_2 and CH_3, respectively), whereas photolytic assistance is typically needed to engineer the corresponding reactions of the heavier Group 13 atoms. Oxidative addition to give M^{II} species is also the reaction channel favoured by the reaction mixtures $Al + Cl_2$[37] and $M + CH_3Br$ (M = Al, Ga or In).[123] Atomic aluminium or gallium inserts into the H–H bond of dihydrogen to yield both MH and HMH (M = Al or Ga), but only when it is in the excited ^2D state and not in its ^2P ground state;[124] by contrast, the dimer Ga_2 reacts spontaneously with H_2 to form $Ga(\mu-H)_2Ga$. Such experiments have seldom been scaled up, but co-condensation of aluminium or gallium vapour with $Bu^tN=CH \cdot CH=NBu^t$ (Bu^tdab) does afford a synthetic route to the appropriate diazadiene complex $M(Bu^tdab)_2$ (M = Al or Ga; see also section 1.3.1.2).[125]

1.4 Chemical properties of the elements and their compounds

From the properties of the atoms we turn to the chemistry of the elements and their compounds by highlighting just four aspects, not by any means of equal weight, but each contributing to the chemical personality of this particular group. Firstly, there is the question of the relative stabilities of the different oxidation states. There follows a brief commentary on the coordination chemistry of the metals. Thirdly, the structural chemistry of the metals and, finally, metal–metal bonding complete the survey.

1.4.1 Choice of oxidation state: +1, +2 or +3?

How it is that the heavier members of Groups 13 and 14 come to develop a valency two less than the normal valency (N) has provoked no little controversy. The classical rationalisation of an 'inert pair' of tightly bound valence s electrons, first articulated by Grimm and Sommerfeld[28] and popularised by Sidgwick,[1,29] has been challenged by Drago.[126] Contrary to the simple notion that the outer s electrons become more stable for the heavier elements, Drago pointed out that the energy required to remove them from the isolated atom is no greater for these than it is for the elements in the middle of the Group; for example, the sum of the ionisation potentials I_2 and I_3 for Ga, In and Tl is 4942, 4525 and 4849 kJ mol^{-1}, respectively. Drago went on to conclude, through analysis of the spectroscopic and thermodynamic data then available, that the factor primarily responsible for the increasing stability of the $(N-2)$-valent chloride (MCl_{N-2}) relative to the N-valent one (MCl_N) with the descent of Groups 13 and 14 is the fall in bond energy in MCl_N. A similar verdict has been delivered very recently, on the basis of *ab initio* quantum mechanical calculations taking in binary hydride and halide molecules of the Group 13 elements.[26] Drago contended that the fall in bond energy is due not so much to a rise in s–p promotion energy, as the inert-pair hypothesis would suggest, as to a fall in the *intrinsic* bond energy with increasing atomic number.[126] Here, however, his arguments seem less secure.[127] For one thing, they were based on an estimated s–p promotion energy which was substantially wide of the mark for thallium; for another, they involved the questionable assumption that the intrinsic bond energy is unaffected by any 'inertness' of the valence s electrons. Nevertheless, Drago's views have been widely adopted in the past two decades.[2,4]

The standard reduction potential $E^{\ominus}$ for the aqueous couple M^{3+}/M^+ is an important quantity in any analysis of the relative stabilities of the M^{III} and M^I states in Group 13. Some of the figures for this potential (enclosed in brackets in Table 1.1) are only estimates, but the accuracy is probably sufficient to throw light on the energy balance between M^{III} and M^I. The potential is related to the standard free energy change of the reaction

$$M^+(aq) + 2H^+(aq) \rightleftarrows M^{3+}(aq) + H_2(g) \tag{1.2}$$

by the equation $\Delta_{1,2}G^{\ominus} = 2FE^{\ominus}$, and the problem of how $\Delta_{1,2}G^{\ominus}$ varies from metal to metal may be restated by means of the thermodynamic cycle shown in Figure 1.2(a). $\Delta_{1,2}G^{\ominus}$ is then related to the enthalpy of ionisation of M^+ to M^{3+}, i, and the free energies of hydration, $\Delta_{hydration}G$, of the relevant ions by equation (1.3).

$$\Delta_{1,2}G^{\ominus} = i + \Delta_{hydration}G^{\ominus}[M^{3+}(g)] - \Delta_{hydration}G^{\ominus}[M^+(g)] + \text{constant} \tag{1.3}$$

Values of i and $\Delta_{hydration}G^{\ominus}[M^{3+}(g)]$ are available from the data included in Table 1.1; $\Delta_{hydration}G^{\ominus}[M^+(g)]$ must be estimated empirically by means of the Born equation and the radius of the cation.[128] The relative contribution of each term on the right-hand side of (1.3) to the variations in $\Delta_{1,2}G^{\ominus}$ may now be assessed by reference to Figure 1.3(a). Hence it is evident that the pattern of $\Delta_{1,2}G^{\ominus}$ is determined mainly by the combined effects of i and

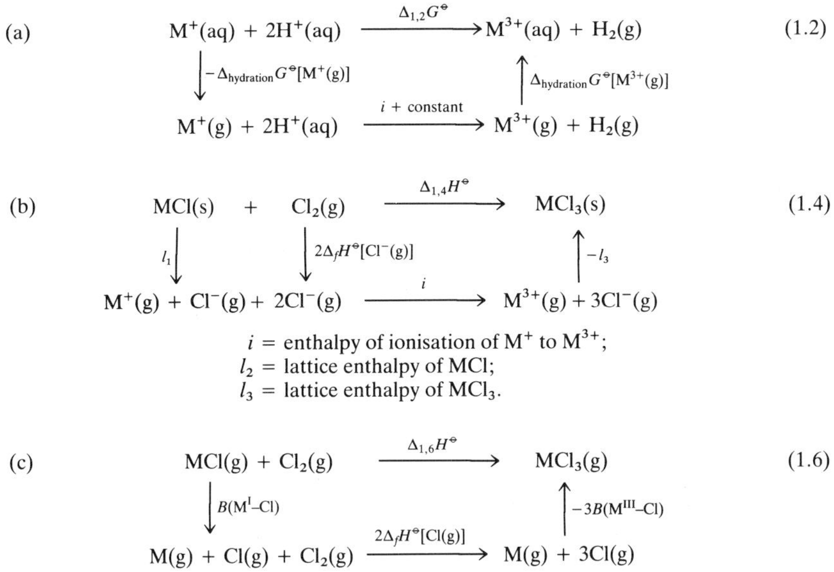

(a)
$$M^+(aq) + 2H^+(aq) \xrightarrow{\Delta_{1,2}G^\ominus} M^{3+}(aq) + H_2(g) \qquad (1.2)$$

$$\downarrow -\Delta_{\text{hydration}}G^\ominus[M^+(g)] \qquad\qquad \uparrow \Delta_{\text{hydration}}G^\ominus[M^{3+}(g)]$$

$$M^+(g) + 2H^+(aq) \xrightarrow{i + \text{constant}} M^{3+}(g) + H_2(g)$$

(b)
$$MCl(s) + Cl_2(g) \xrightarrow{\Delta_{1,4}H^\ominus} MCl_3(s) \qquad (1.4)$$

$$l_1 \downarrow \qquad \downarrow 2\Delta_f H^\ominus[Cl^-(g)] \qquad \uparrow -l_3$$

$$M^+(g) + Cl^-(g) + 2Cl^-(g) \xrightarrow{i} M^{3+}(g) + 3Cl^-(g)$$

i = enthalpy of ionisation of M^+ to M^{3+};
l_2 = lattice enthalpy of MCl;
l_3 = lattice enthalpy of MCl_3.

(c)
$$MCl(g) + Cl_2(g) \xrightarrow{\Delta_{1,6}H^\ominus} MCl_3(g) \qquad (1.6)$$

$$\downarrow B(M^{\text{I}}-Cl) \qquad\qquad \uparrow -3B(M^{\text{III}}-Cl)$$

$$M(g) + Cl(g) + Cl_2(g) \xrightarrow{2\Delta_f H^\ominus[Cl(g)]} M(g) + 3Cl(g)$$

Figure 1.2 Thermodynamic cycles for the processes (a) $M^+(aq) + 2H^+(aq) \rightleftharpoons M^{3+}(aq) + H_2(g)$, (b) $MCl(s) + Cl_2(g) \rightleftharpoons MCl_3(s)$, and (c) $MCl(g) + Cl_2(g) \rightleftharpoons MCl_3(g)$.

$\Delta_{\text{hydration}}G^\ominus[M^{3+}(g)]$. Although the overall decrease in magnitude of the hydration energy accounts in large measure for the reduced stability of In^{III} and Tl^{III}, the sawtooth variation in i has a marked modulating effect which causes Ga^{I} and Tl^{I} to be significantly weaker reducing agents than would otherwise be predicted.

Reaction (1.4) affords an alternative means of assessing the relative stabilities of M^{III} and M^{I}.

$$MCl(s) + Cl_2(g) \rightleftharpoons MCl_3(s) \qquad (1.4)$$

In this case, changes in $\Delta_{1.4}G^\ominus$ are determined mainly by the way the standard enthalpy change, $\Delta_{1.4}H^\ominus$, varies with the nature of M, and this in turn reflects the influences of i and the lattice enthalpies of MCl (l_1) and MCl_3 (l_3), as indicated by the cycle depicted in Figure 1.2(b) and expressed in equation (1.5).

$$\Delta_{1.4}H^\ominus = i + l_3 - l_1 + \text{constant} \qquad (1.5)$$

Values of $\Delta_{1.4}H^\ominus$, i, l_1 and l_3 can be calculated from the standard enthalpies of formation of the appropriate solid chlorides;[129] these are well determined in some cases but AlCl and GaCl require that l_1 be estimated empirically and $\Delta_f H^\ominus$ then deduced accordingly.[130] As revealed by the details plotted in Figure 1.3(b), the trend in $\Delta_{1.4}H^\ominus$ owes most to the variables i and

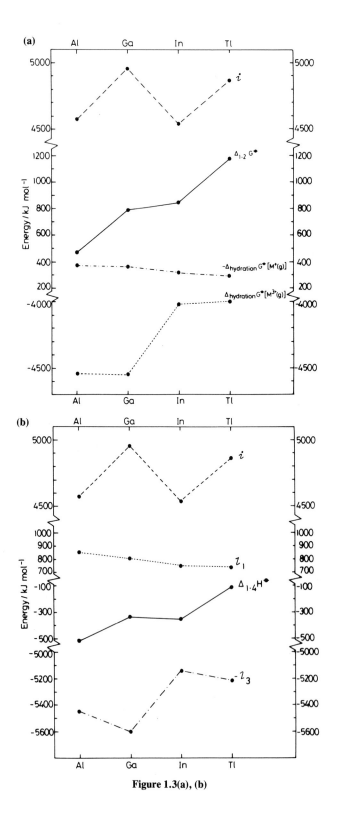

Figure 1.3(a), (b)

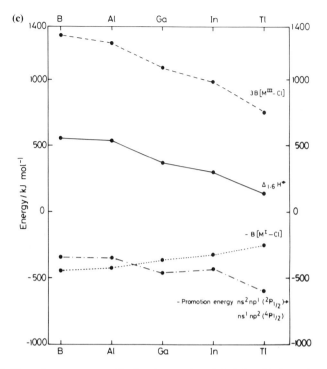

Figure 1.3 Plots of energy terms affecting the reactions 1.2, 1.4 and 1.6 as a function of the Group 13 element. (a) **Reaction 1.2**: variations of $\Delta_{1.2}G^{\ominus}$, i, $-\Delta_{\text{hydration}}G^{\ominus}[M^{+}(g)]$ and $\Delta_{\text{hydration}}G^{\ominus}[M^{3+}(g)]$. (b) **Reaction 1.4**: variations of $\Delta_{1.4}H^{\ominus}$, i and the lattice enthalpies l_{1} and $-l_{3}$. (c) **Reaction 1.6**: variations of $\Delta_{1.6}H^{\ominus}$, the bond enthalpy terms $-B[M^{I}\text{–}Cl]$ and $3B[M^{III}\text{–}Cl]$, and $-P$, where P is the promotion energy for the transition $ns^{2}np^{1}$ ($^{2}P_{1/2}$) → $ns^{1}np^{2}(^{4}P_{1/2})$.

the lattice enthalpy term l_{3}, which follows a pattern similar to that of $\Delta_{\text{hydration}}G^{\ominus}[M^{3+}(g)]$ in its dependence on the metal M. Indeed, the effect of i is more pronounced in this case, countering that of l_{3} to make $GaCl_{3}$ and $TlCl_{3}$ less, and not more, stable than $InCl_{3}$.

Both of these analyses, which emphasise the roles of ionisation and coordination (whether by water or by chloride anions), are predicated on the assumption that the ionic model has some validity in its interpretation of the species M^{I} and M^{III}. While the model does not give a good account of these species, it is probably no less applicable here than to the chemistry of first-row transition metals, where it has been turned to some account.[128] If the making and breaking of covalent bonds are judged to be the dominant principles of Group 13 chemistry, reaction (1.6) and the thermodynamic cycle of Figure 1.2(c) are more apt as a basis for analysis.

$$MCl(g) + Cl_{2}(g) \rightleftarrows MCl_{3}(g) \tag{1.6}$$

The entropy change accompanying (1.6) varies but little from metal to metal, and variations in the standard free energy change are therefore determined principally by the enthalpy change, $\Delta_{1.6}H^{\ominus}$, which is in turn a function of the bond enthalpy $B(M^I-Cl)$ and the mean bond enthalpy $B(M^{III}-Cl)$, in accordance with equation (1.7).

$$\Delta_{1.6}H^{\ominus} = B(M^I-Cl) - 3B(M^{III}-Cl) + \text{constant} \qquad (1.7)$$

Plots of the relevant variables, as reproduced in Figure 1.3(c), confirm that $\Delta_{1.6}H^{\ominus}$ emulates $3B(M^{III}-Cl)$ closely in its dependence on the atomic number of M. Whereas $B(M^I-Cl)$ varies relatively smoothly, it is noteworthy that $3B(M^{III}-Cl)$ responds not only more sharply but also in a decidedly stepwise fashion. That the s–p promotion energy of M should follow much the same pattern as $3B(M^{III}-Cl)$ makes it hard to resist the circumstantial evidence that the difference in energy between the valence s and p orbitals carries significant weight among the factors determining the strength of the $M^{III}-Cl$ bond.

Whereas the unipositive oxidation state is an established fact of life in Group 13, there are no authenticated cases of mononuclear derivatives of the dipositive state which are long-lived under normal conditions. If we assume that the hypothetical dichloride is composed of the dipositive ions M^{2+}, some idea of its standard heat of formation can be gained either by estimating a lattice enthalpy on the basis of a likely value for the radius of the cation or by equating the lattice enthalpy with that of a neighbouring well characterised compound, e.g. $GeCl_2$, $SnCl_2$ or $PbCl_2$.[128] The results of such calculations are included in the oxidation-state diagram of Figure 1.4. Hence it is clear that all the dichlorides are highly susceptible to disproportionation in accordance with the equations

$$MCl_2(s) \rightleftarrows 1/3M(s) + 2/3MCl_3(s) \qquad (1.8)$$

$$MCl_2(s) \rightleftarrows 1/2MCl(s) + 1/2MCl_3(s) \qquad (1.9)$$

With M = Al, Ga, In or Tl, the respective enthalpy changes are -197, -190, -233 or $-232\,\text{kJ mol}^{-1}$ for equation (1.8) and -174, -197, -237 or $-282\,\text{kJ mol}^{-1}$ for equation (1.9). In practice, disproportionation to M^I and M^{III}, to the accompaniment of halide transfer, accounts for the properties of the known dihalides of the Group 13 metals which are most aptly formulated as mixed valence compounds. In addition, there is a second type of reaction to which the M^{2+} radical ion is made vulnerable by the vacancy in its relatively stable valence s orbital; this involves dimerisation, $viz.$

$$2M^{2+} \rightleftarrows [M-M]^{4+} \qquad (1.10)$$

By contrast with Hg_2^{2+}, however, the concentration of positive charge on the product restricts its formation to conditions in which it is coordinated by electron-rich ligands, as in the species $[M_2X_6]^{2-}$ and $Ga_2Cl_4 \cdot 2\text{dioxane}$.

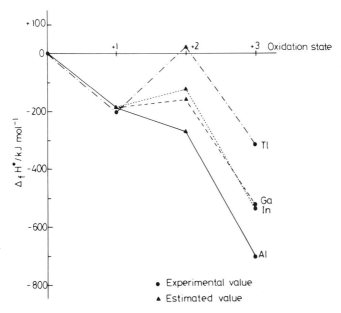

Figure 1.4 Oxidation state diagram giving experimental or estimated values for the standard heats of formation of solid chlorides of the Group 13 metals of the types MCl, MCl_2 and MCl_3.

Whether disproportionation or dimerisation takes precedence and whether kinetically stable monomeric M^{II} derivatives can be made viable are issues reviewed recently by Tuck,[38] with the conclusions to be described in chapter 3. Suffice it to say here that normal derivatives of the open-shell ion M^{2+} are likely always to be at an energetic disadvantage with respect to disproportionation. Part of this disadvantage comes from the extra 'size' that the single s electron imparts to the M^{2+} ion (compare Zn^{2+} with Ga^{2+}, for example), at the expense of lattice or solvation energy terms; part comes from the extra energy needed to ionise the second electron not from the p but from the more tightly bound s shell.

In Group 13, as in Group 14,[127] the inference to be drawn is that the stability of the $N-2$ compared with the normal valence state of N is dominated by no single factor. The binding energy of the M^{III} species is undoubtedly important, but there is a second, interrelated influence at work, namely the relatively tight binding of the valence s electrons. It is not that the outer s electrons necessarily become harder to remove with increasing atomic number, rather that *they do not become easier to remove to the extent that might have been expected* (to judge, for example, by the precedent set by the Group 3 elements). The same combination of binding energies and tightly held s electrons appears also to deny the stable existence of simple monomeric M^{II} species. Hence there would seem to be substance to the classical notion of the 'inert pair'. It would be wrong, however, to think of the valence s electrons as being completely inactive; they can affect the stereochemistry of lower-valent compounds, and contribute to the bonding in the higher-valent ones. Even so, they are inactive enough to have an effect on valency and, in the words of Hopkins and Nelson,[127] 'thus may be fairly described as a "reluctant" pair, if not a positively "inert" one'.

1.4.2   Coordination chemistry of the metals[32,33]

The acidity of the Group 13 metals in the tripositive state finds expression in the more limited Brønsted sense, for example, in the properties of the hydrated cations $[M(OH_2)_6]^{3+}$ which all act in some degree as proton sources in aqueous solution. However, with pK values corresponding to the first step in the hydrolysis reaction of 4.99, 2.60, 3.9 and 0.6 for M = Al, Ga, In and Tl,[32,33,131] respectively, the pattern of acid strengths is by no means uniform. The properties of the aquated Al^{3+} ion are of particular concern on several grounds: firstly, there is the key role played in the environment through the leaching of aluminium from silicates by acid rain; secondly, aluminium is now known to be toxic to aquatic life; and thirdly, there is the question of whether aluminium is implicated as a toxic factor in a number of human diseases (e.g. Alzheimer's disease, the dialysis encephalopathy syndrome, etc.).[132] These issues will be taken up more fully in chapter 9. Compounds based on the +3 oxidation state are typically electron acceptors, and the addition products they form with electron-rich species make for a rich and variegated coordination chemistry, as will become clear in chapter 8. Aluminium trichloride is, for example, an unusually potent Lewis acid, a property which underlies its widespread use in promoting electrophilic processes like Friedel–Crafts' reactions;[4,19,133] it has recently been employed in this way to mediate the reaction between fullerene or polychloro-fullerenes, $C_{60}Cl_n$, and an aromatic hydrocarbon to give polyaryl-fullerenes.[134] The tris(trifluoromethanesulfonate) derivatives of boron, aluminium and gallium have also been shown to function efficiently as Friedel–Crafts' catalysts.[135] In solid catalysts too Al^{III} is an important agent in the creation of active, acidic sites; outstanding among such catalysts are the zeolites and natural and synthetic clays. It has been shown that the natural acidity of some metal compounds and sulfonic acid resins can be enhanced by treatment with $AlCl_3$ to give materials capable, for instance, of activating paraffins to isomerisation.[136]

Attempts to assess relative Lewis acidities have served only to emphasise the variety of factors contributing to the efficiency of coordination and the strength of the coordinate link.[4,19,137,138] There is no unique series of acceptor strength. With respect to a relatively hard O- or N-atom base, the acceptor ability of MX_3 compounds varies, for a given substituent X, typically in the order M = Al > Ga > In, with the position of Tl somewhat indeterminate. With respect to a relatively soft base like Me_2S, however, the order is more likely to be $M = Ga > Al > B$. The substituents X have a more predictable effect and, with respect to all the bases studied so far, the acceptor strength for a given M diminishes in the sequence $MX_3 > MPh_3 > MMe_3$ (where $M = B$, Al, Ga or In, and X = halogen). According to the data included in Table 1.4, the monohalide complexes, $[MX]^{2+}$, have stability constants which vary under aqueous conditions in the order $X = F > Cl \gtrsim Br \gtrsim I$ when $M = Al$, Ga or In, *but* in the order $X = F < Cl < Br < I$ when $M = Tl$.[131] The picture which thus emerges is that B^{III} and Al^{III} are significantly harder acceptors than the post-transition metal species Ga^{III}, In^{III} and Tl^{III}, with the consequences already alluded to in section 1.1. Differences in acceptor properties can be exploited in displacement reactions such as

$$Me_3N \cdot GaH_3 + BF_3 + Me_2S \rightarrow Me_2S \cdot GaH_3 + Me_3N \cdot BF_3 \quad \text{(ref. 139a)}$$
(1.11)

$$Al(BH_4)_3 + CO \rightarrow 1/x\,[HAl(BH_4)_2]_x + H_3B \cdot CO \quad \text{(ref. 139b)} \quad (1.12)$$

The change from hard to relatively soft acceptor action on the part of M^{III} species as the atomic number of M increases can be attributed to the enhanced contribution made by electron delocalisation (as opposed to coulombic forces) to the coordinate link.[137,140] It is a consequence of the condition that the heavier Group 13 atoms are relatively more susceptible to short-range orbital perturbations, in keeping, for example, with their greater polarisabilities, than are the lighter ones. The same principles hold presumably for Tl^I which, while being a much weaker acid, shares the class 'b' or soft character of Tl^{III} (see Table 1.4). In the absence of comparable quantitative information, however, the coordination properties of the other unipositive cations remain largely *terra incognita*.

Whatever the relative acid strengths may suggest, and contrary to popular

Table 1.4 Some stability constants, $-\log_{10} K_1$, of $1:1$ complexes formed by the ions of the Group 13 metals in aqueous solution at 298 K.[a]

Ligand[b]	Metal ion				
	Al^{3+}	Ga^{3+}	In^{3+}	Tl^{3+}	Tl^+
OH^-	9.01	11.4	10.1	13.4	0.79
F^-	7.0	5.9	4.6	c. 3	0.10
Cl^-	<0	c. 0.01	2.36	7.72	0.49
Br^-	<0	c. −0.10	2.08	9.7	0.91
I^-	<0	c. −0.2	1.64	c. 14	c. 1.2
SCN^-	0.42	2.04	3.15	–	0.58
$EDTA^{4-}$	16.5	21.0	24.9	35.3	6.41

(a) Values taken from Ref. 131 and corrected, where possible, to zero ionic strength.
(b) $EDTA^{4-}$, $[(O_2CCH_2)_2NC_2H_4N(CH_2CO_2)_2]^{4-}$.

misconceptions, it does not follow that any member of Group 13 departs from the normal order of thermochemical bond strengths, that is, $B(M–F) > B(M–Cl) > B(M–Br) > B(M–I)$ and $B(M–O) > B(M–S)$, etc. The crucial point is that the *spread* of these energies, as in the range $B(M–F)$ to $B(M–I)$, is much smaller for $M = Tl$ than it is for $M = Al$ (see section 1.3.2). In a real world where there is competition for different acid and base centres, as in aqueous solution, the final balance is that which minimises the free energy of the system *as a whole*. Changes in the ligating preferences of the cation for halide ions in solution must be understood to reflect not only the properties of the cation but also those of the solvent. Thus, the stability constants of Table 1.4 make it clear that water molecules compete with the Group 13 cations for coordination to the halide anions, relatively successfully in the face of the aquated Ga^{3+} ion, much less so in the face of the aquated Tl^{3+} ion. This principle of competition underlies the outcome of numerous meta-thesis, precipitation and complexation reactions, e.g.

$$TlOH + C_5H_6 \xrightarrow{\text{aq. soln.}} (\eta^5-C_5H_5)Tl \downarrow + H_2O \qquad (1.13)$$

$$FClO_3(g) + AlCl_3(s) \rightarrow ClClO_3(g) + AlCl_2F(s) \quad \text{(ref. 141)} \quad (1.14)$$

Complexes of Group 13 metal compounds come with a miscellany of stoichiometries and structures.[32,33,142] Thus, derivatives of trivalent compounds may involve discrete mononuclear species, usually with 4-, 5- or 6-fold coordination of the metal centre. They may feature dissociation into two or more ionic moieties, e.g. $[cis\text{-}InI_2(OSMe_2)_4]^+[InI_4]^-$. They may incorporate discrete units with two or more metal atoms linked *via* bridging ligands, as with $[Cl_3Al(\mu-Cl)AlCl_3]^-$ and $[Cl_3Tl(\mu-Cl)_3TlCl_3]^{3-}$. In yet another variation, they may consist of polymeric networks of metal atoms, again linked *via* appropriate bridging ligands, as in $KAlF_4$.[143] Of the sundry factors determining the stoichiometry and structure of such complexes, the following appear to be most significant: (a) the electronic properties and size of the M^{3+} cation, (b) the acidity of the metal compound, (c) the properties of the substituents in this compound, and (d) the coordinating properties of the ligand, including its size, shape and denticity. Complexes of the metals in lower oxidation states are much less plentiful, but are still noteworthy (i) for the incidence of metal–metal bonding, as in $[X_3M–MX_3]^{2-}$ ($M = Ga$ or In; $X = Cl$, Br or I) and $Ga_2X_4(dioxane)_2$ ($X = Cl$ or Br), and (ii) for the proclivity of M^I centres to be partnered by soft ligands like C_6H_6 (as in $[(\eta^6-C_6H_6)_2Ga]^+[GaX_4]^-$ with the structure **VIII**[144a]), [2.2]paracyclophane (as in $[([2.2]paracyclophane)Ga]^+[GaBr_4]^-$ with the structure **IX**[144b]), and thiourea (tu) (as in $TlX \cdot 4tu$ where $X = Cl$, Br or I[144c]). High coordination numbers are common for the M^I centres, but now the picture is apt to be clouded by uncertainties about where primary interactions give way to secondary interactions. Such questions call, however, for a wider perspective on the structural chemistry of Group 13 metal compounds.

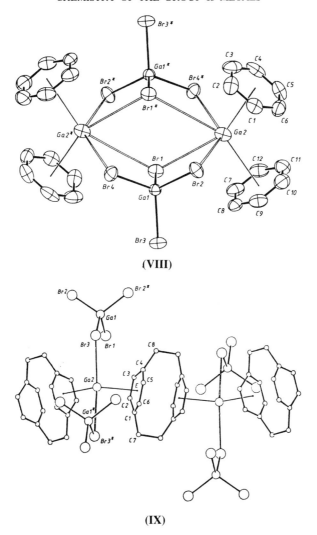

(VIII)

(IX)

1.4.3 Structural aspects

1.4.3.1 Trivalent derivatives. The liaison with an appropriate number of 1-, 2- or 3-electron ligands affords discrete molecules in which the Group 13 metal atom attains the oxidation state $+3$ with a coordination number varying from 3 to 1. Such species are relatively scarce, however, being confined mostly to the regime of high-temperature vapours, although some of them, such as the OAlX (X = F or Cl)[145] and mononuclear trihalide molecules, are amenable to characterisation by matrix-isolation methods.[87] On the other hand, 1-electron ligands X which are sufficiently bulky (e.g. $X = N(SiMe_3)_2$,[112a] $As(mesityl)_2$,[146] $2,4,6\text{-}Bu_3^tC_6H_2S$[147] or 2,4,6-

Table 1.5 Examples of coordination geometries exhibited by trivalent Group 13 metal compounds.

Coordination number	Coordination geometry	Examples[a]
1	Diatomic	$ME(g)$ (M = Al, Ga, In or Tl; E = N, P, As, Sb or Bi)
2	Linear array	Me_2Tl^+; $OAlX$ (X = F or Cl)[b]
3	Trigonal planar array	$MX_3(g)$ (M = Al, Ga or In; X = halogen); $Al(mesityl)_3$; $MMe_3(g)$ (M = Al, Ga, In or Tl); $In[Co(CO)_4]_3$; $Al[N(SiMe_3)_2]_3$;[c] H_2GaCl;[d] $HMCl_2$ (M = Al or Ga);[d] $M[S(2,4,6\text{-}Bu_3^tC_6H_2)]_3$ (M = Al or Ga);[e] $Ga[Se(2,4,6\text{-}Bu_3^tC_6H_2)]_3$;[f] $[(2,4,6\text{-}Ph_3C_6H_2)GaP(cyclo\text{-}C_6H_{11})]_3$ (VII);[g] $Ga[As(mesityl)_2]_3$[h]
	Pyramid	$Tl[Mo(CO)_3(C_5H_5)]_3$[i]
4	Tetrahedron	$[MX_4]^-$ and M_2X_6 (M = Al, Ga, In or Tl; X = halogen); Al_2Me_6; Ga_2H_6;[j] $Al_2H_6, 2NMe_3$;[k] $[Me_2NAlH_2]_3$; AlO_4 units in aluminosilicates and $AlPO_4$; $GaPO_4$; $LiGaO_2$; $ME(s)$ (M = Al, Ga or In; E = N, P, As or Sb); $MOCl$ (M = Al or Ga); α- and γ-Ga_2S_3; $[RNAlR']_n$ (R,R' = H or organic group); $[Bu^tGaS]_4$;[m] $NaAl(SiMe_3)_4$
	ψ-Trigonal bipyramid	$[(Me_3SiCH_2)_2TlCl]_2$[n]
5	Trigonal bipyramid	$AlX_3(NMe_3)_2$ (X = H or Cl); $(Me_3P)_3H_3W(\mu\text{-}H)_2Al(H)(\mu\text{-}H)WH_3(PMe_3)_3$;[o] $AlCl_3(morpholine)_2$;[p] $[In(NCS)_5]^{2-}$; $InCl_3(PPh_3)_2$; $Me_2GaCl(phen)$; AlO_5 units in aluminates and zeolites; GaO_5 in $CuGaInO_4$[q] and $In_{1.2}Ga_{0.8}MgO_4$;[r] $In(dithizonate)_3$; $TlCl_3L_2$ (L = 4-pyridinecarbonitrile N-oxide); $GaCl_3 \cdot$ 1,4-dioxane; $[H_2B(pz)_2]_2GaCl$ (pz = pyrazolyl ring);[s] $TlBr_3 \cdot$ 1,4-dioxane; $TlX_3 \cdot 4H_2O$ (X = Cl or Br)
	Distorted trigonal bipyramid	$[Me_2Ga(tropolonate)]_2$;[t] $[Al_2(1,2\text{-ethanediolate})_4]^{2-}$[u]
	Square pyramid	$MMe_3(s)$ (M = In or Tl); $[InCl_5]^{2-}$; $EtAl(sal_2en)$; $p\text{-tolyl}Tl(S_2CNEt_2)$; $MeAl(BH_4)_2$; $MeGa(O_2CMe)_2$; $GaCl(L)$;[v] $InX(porph)$ (X = Cl, Me or MeOSO; porph = substituted porphyrinate ligand); $MCl(phthalocyaninate)$ (M = Al or Ga)[w]
6	Octahedron	$[M(OH)_6]^{3+}$, MF_6^{3-}, $M_2(SO_4)_3$, $M(acac)_3$, $[M(C_2O_4)_3]^{3-}$ and $M(8\text{-quinolinate})_3$ (M = Al, Ga, In or Tl); $\alpha\text{-}Al_2O_3$ and -Ga_2O_3; $In(OH)_3$; AlO_6 units in aluminosilicates and aluminophosphates; $M'MO_2$ (M' = alkali metal) and MPO_4 (M = Ga or In); $MCl_3(s)$ (M = Al, In or Tl); $InOX$ (X = Cl, Br or I); $AlH_3(s)$; AlH_6^{3-}; $MF_3(s)$ (M = Ga or In); $GaSbO_4$ (statistical rutile structure)
	Distorted octahedron	In_2O_3 and Tl_2O_3 (C-M_2O_3 structure); InO_6 in $CuGaInO_4$;[q] $[ON(\mu\text{-}O)_2Al(ONO)_4]^{2-}$; $Al(1,3\text{-}Ph_2N_3)_3 \cdot$ toluene;[x] $Tl(O_2CC_6F_5)(O_2CC_6F_5)(OPPh_3)]_2$;[y] $[In_2(1,2\text{-dithiooxalate})_5]^{4-}$;[z] $[AlCl_2(12\text{-crown-4})]^+$;[aa] $Me_3In \cdot [Pr^iNCH_2]_3$[bb]
7	Trigonal prism	$Al(BH_4)_3$; $M(S_2CNR_2)_3$ (M = Ga or In); form of $GaO(OH)$
	Pentagonal bipyramid	$[AlCl_2(benzo\text{-}15\text{-crown-5})]^+$; $[InF_4(OH_2)]_n^{n-}$; $[In(DAPSC)(OH_2)_2]^{3+}$;[cc] $[In_2(C_2O_4)_3(OH_2)_4] \cdot 2H_2O$; $InCl(1,4,7\text{-triazacyclononanetriacetate-H}) \cdot 0.64H_2O$;[dd] $Rb_2In_3F_{11}$,[ee] $Al(BH_4)_3 \cdot NMe_3$
	Monocapped trigonal prism	Tl_6TeO_{12}; $\beta\text{-}Al_2S_3$
8	Cube	$TlOF$; $NaTlF_4$
	Square antiprism	$NH_4[In(C_2O_4)_2(OH_2)_2]$; $TlSX$ (X = Cl, Br or I)

Dodecahedron	$In(O_2CMe)_3(phen)$
Bicapped trigonal prism	$TlF_3(s)^{ff}$
Hexagonal bipyramid	$[Me_2Tl(dibenzo\text{-}18\text{-}crown\text{-}6)]^{+gg}$
Irregular	$Tl(O_2CMe)_3$; $Tl(O_2CMe)_3 \cdot H_2O$; $Tl[Tl^{III}(O_2CMe)_4]$
Tricapped trigonal prism	$Tl(NO_3)_3 \cdot 3H_2O$
9	

(a) Abbreviations: phen = 1,10-phenanthroline; sal_2en = bis(salicylaldehyde)ethylenediimine; DAPSC = 2,6-diacetylpyridine–disemicarbazone. Entries without superscript letters are taken from the following sources: Refs. 2, 4, 6, 19, 30, 32 and 33.

(b) Ref. 145.
(c) Ref. 112a.
(d) Ref. 42.
(e) Ref. 147.
(f) Ref. 148.
(g) Ref. 113b.
(h) Ref. 146.
(i) Ref. 150.
(j) Ref. 22a.
(k) Ref. 157.
(l) Ref. 111.
(m) M.B. Power and A.R. Barron, J. Chem. Soc., Chem. Commun., 1991, 1315.
(n) Ref. 153.
(o) A.R. Barron, D. Lyons, G. Wilkinson, M. Motevalli, A.J. Howes and M.B. Hursthouse, J. Chem. Soc., Dalton Trans., 1986, 279.
(p) G. Müller and C. Krüger, Acta Crystallogr., 1984, C40, 628.
(q) Ref. 161.
(r) J. Barbier, J. Solid State Chem., 1989, 82, 115.
(s) D.L. Reger, S.J. Knox and L. Lebioda, Inorg. Chem., 1989, 28, 3092.
(t) Ref. 158.
(u) M.C. Cruickshank and L.S. Dent Glasser, Acta Crystallogr., 1985, C41, 1014.
(v) H_2L is the compound V; Ref. 101d.
(w) K.J. Wynne, Inorg. Chem., 1984, 23, 4658.
(x) J.T. Leman, A.R. Barron, J.W. Ziller and R.M. Kren, Polyhedron, 1989, 8, 1909.
(y) Ref. 159.
(z) L. Golič, N. Bulc and W. Dietzsch, Inorg. Chem., 1982, 21, 3560.
(aa) Ref. 155.
(bb) Ref. 156.
(cc) J. Davis and G.J. Palenik, Inorg. Chim. Acta, 1985, 99, L51.
(dd) A.S. Craig, I.M. Helps, D. Parker, H. Adams, N.A. Bailey, M.G. Williams, J.M.A. Smith and G. Ferguson, Polyhedron, 1989, 8, 2481.
(ee) Ref. 162.
(ff) Ref. 151.
(gg) Ref. 154.

$Bu^t_3C_6H_2Se^{148}$) or ill-adapted to bridge two or more metal atoms (e.g. $X = Me$) are capable of producing simple mononuclear species of the type MX_3 which are long-lived under normal conditions. The directly co-ordinated ligand atoms are found at the vertices of a trigonal array which conforms almost invariably to the expectations of simple VSEPR[149] or electrostatic arguments by being coplanar with the metal atom. It is therefore a matter of some surprise that the thallium complex $Tl[Mo(CO)_3(\eta^5-C_5H_5)]_3$ should favour a decidedly pyramidal geometry at thallium, as well as a considerable spread in the Tl–Mo bond distances;[150] this may mean that there is considerable transfer of electron density to the thallium so that the compound is far from being a conventional Tl^{III} derivative.

These rather special cases apart, the prevailing pattern is for trivalent derivatives of the Group 13 metals to adopt structures in which the metal atom achieves a coordination number higher than three, either by taking up additional ligands or by establishing supplementary links to the existing ligands (resulting, for example, in chelation of a single metal centre or bridging of two or more such centres). Table 1.5 lists in order of increasing coordination number the sort of stereochemistries that the resulting structures impose on the metal atom, together with representative examples. Pre-eminent among the modes of coordination are those where the metal makes contact with either four or six nearest neighbours approximating more or less closely to a regular tetrahedron or octahedron, respectively. On the other hand, five-fold coordination is quite common and coordination numbers greater than six are not out of the ordinary, particularly in the context of indium and thallium compounds.

The coordination number and geometry assumed by the metal atom depend only partly on the properties of the M^{III} centre. Unquestionably the size of this centre is a significant factor and the greater bulk of indium and thallium, compared with aluminium and gallium, must facilitate the achievement of environments with 7-, 8- or 9-fold coordination of the metal, as in TlF_3,[151] $NH_4[In(C_2O_4)_2(OH_2)_2]^{33}$ and $Tl(NO_3)_3 \cdot 3H_2O$.[33] Size alone is not enough, however. If it were, aluminium and gallium, which have very similar radii, would be expected to be almost identical in their structural preferences. In fact, while having much in common with aluminium, gallium is more prone to enter 4-coordinate environments. This point is illustrated by the structures of the solid trichlorides (contrast the octahedral coordination of Al in the layer structure of $AlCl_3$ with the tetrahedral coordination of Ga in solid $GaCl_3$ which is built of Ga_2Cl_6 molecules), and by the finding that $MgGa_2O_4$ is not a 'normal' spinel like $MgAl_2O_4$ but an 'inverse' spinel, with the Ga atoms distributed over octahedral *and* tetrahedral sites.[30] Covalent interactions leading to the transfer of charge from the ligands to the metal atom, in accordance with the Electroneutrality Principle, vary in the part they play but can seldom, if ever, be overlooked. In this respect, the energy separation between the valence ns and np orbitals of the metal is pertinent, with a large separation tending to militate against optimum binding ($q.v.$) as well as the attainment of high coordination numbers. The occupied $(n-1)d$ orbitals may also have a contribution to make, directly or indirectly. If so, thallium is best placed to take advantage of them ($q.v.$), and we look for evidence of the linear coordination which is such a distinctive feature in mercury(II) chemistry.[2,4,30,31,152] Only in derivatives of linear R_2Tl^+ cations (R = organic group), however, does Tl^{III} appear to emulate its isoelectronic neighbour Hg^{II}; the bonds to carbon are invariably supplemented by weaker ones to the anions, and/or additional ligands in some cases, so that the actual coordination number of the metal is at least 4. The result may be an unusual geometry, as with (a) $[(Me_3SiCH_2)_2 TlCl]_2$ (**X**)[153] where the Tl atom is four-coordinate having a coordination geometry based on a distorted trigonal bipyramid with one equatorial position vacant, and (b) $[Me_2Tl(dibenzo-18-crown-6)]^+$ (**XI**) isolated as its 2,4,6-trinitrophenolate salt, and where the linear Me_2Tl moiety is threaded through the eye of the crown ether to produce an 8-coordinate array in the

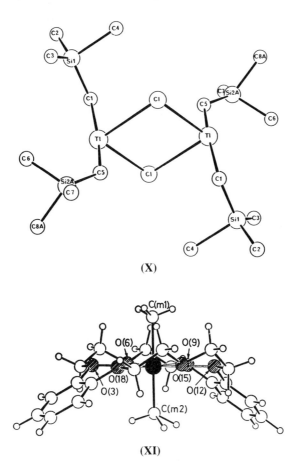

(**X**)

(**XI**)

form of a hexagonal bipyramid.[154] Presumably the $(n-1)d^{10} \rightarrow$ $(n-1)d^9 ns^1$ promotion energy becomes unprofitably large when the Tl^{III} centre is coordinated by ligands harder than organic groups. Secondary interactions involving, for example, dispersion forces are sometimes critical to the selection of coordination environments, and the enhanced polarisabilities of the post-transition metals Ga, In and Tl may well be another structure-determining influence, and one favouring high co-ordination numbers. Where alternative structures differ but little in energy, the contribution of dispersion, symmetry or crystal-packing factors may be decisive. This may be the reason why the $[InCl_5]^{2-}$ anion differs from its isoelectronic relatives $[SnCl_5]^-$ and $SbCl_5$ in approximating the form not of a trigonal bipyramid but of a square pyramid, for example in its tetraethylammonium salt (XII).[33] Trivalent metal centres are unlikely, however, to participate in dispersion interactions to the extent that the corresponding univalent metal centres do (q.v.).

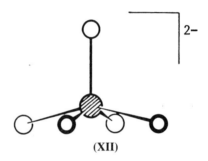

(XII)

In many situations, it is the ligands or the host lattice that dictates the mode of coordination of the metal. Simple ligands have the capacity to regulate the coordination of the metal through their steric properties (as noted above), the formal charges they carry, and their donor characters (reflecting the influences of electronegativity, polarisability and frontier orbitals). A polydentate ligand in particular is likely to generate a metal environment with a design owing much more to the properties of the ligand than to those of the metal. If differences in direct bonding interactions are small, stabilisation of a particular arrangement is then determined by the optimum combination of (a) non-bonding repulsions in the primary co-ordination sphere, (b) coulombic repulsions, and (c) the geometrical constraints due to the stereochemistry and pliancy of the ligand. For example, the irregular 6-coordinate environments of the complexes $[AlCl_2(12\text{-crown-}4)]^+$ [155] and $Me_3In \cdot [Pr^iNCH_2]_3$[156] reflect the conflicting demands of the monodentate and polydentate ligands, with the $AlCl_2$ and $InMe_3$ units standing proud of the plane formed by the donor atoms of the polydentate ligand. Multidentate ligands affect coordination geometry not only according to their shape and whether they are rigid or flexible, but also according to their 'bite' (that is, the preferred separation of the donor atoms). Ligands like BH_4^-, NO_3^-, $MeCO_2^-$ and $R_2NCS_2^-$ with unusually short bites are always prone to give what may appear to be unusual geometries. In reality, such a ligand containing two chemically equivalent donor atoms much closer together than the sum of the non-bonding radii tends to interact *via* multi-centre bonding through *both* atoms so that the mean position of the pair of atoms lies roughly at the vertex of a conventional coordination polyhedron. Thus, $Al(BH_4)_3$, with bidentate BH_4 groups, and several dithizonates of the type $M(S_2CNR_2)_3$ opt not for octahedral but for trigonal prismatic coordination of the metal atom; molecules like $MeAl(BH_4)_2$ and $MeGa(O_2CMe)_2$ favour 5-coordinated environments which approximate more closely to a square pyramid than to a trigonal bipyramid; and the achievement of a high coordination number is often associated with the presence of such ligands, as exemplified by the cases of $Al(BH_4)_3 \cdot NMe_3$, $Tl(NO_3)_3 \cdot 3H_2O$, $Tl(O_2CMe)_3$ and $In(O_2CMe)_3 \cdot$ phen (phen = 1,10-phenanthroline).[33] That ligands are commonly called upon to bridge Group 13 metal atoms in discrete oligomers or in extended 1-, 2- or 3-dimensional networks imposes a further constraint on the environment of the metal, leading typically to deviations from regular tetrahedral, trigonal bipyramidal or octahedral geometries or, on occasion, to an uncharacteristic setting. Table 1.5 includes numerous illustrations. Molecular species are exemplified by the trihalide dimers M_2X_6, $[Me_2NAlH_2]_3$,[111] $Al_2H_6 \cdot 2L$ (L = NMe_3, NMe_2CH_2Ph or $MeNCH_2CH_2CH=CHCH_2$),[157] $[Me_3In]_4$, $[Me_2Ga(tropolonate)]_2$[158] and $[(C_6F_5)_2Tl(O_2CC_6F_5)(OPPh_3)]_2$.[159] Extended chains are represented by the structures of solid AlI_3, the $[AlF_5]_n^{2n-}$ anion (formed by the linking of AlF_6 octahedra through *trans* fluorines), the $[InF_4(OH_2)]_n^{n-}$ anion (isolated as its hydrazinium(1+) salt and where InF_6O pentagonal bipyramids share edges),[30] and $In(C_5H_5)_3$ (with the structure **XIII**, each metal atom being linked by σ-bonds to two terminal and two bridging cyclopentadienyl groups).[30] Layer structures are found in solid MCl_3 (M = Al, In or Tl), MOCl (M = Al or Ga, involving layers of tetrahedral MO_3Cl groups), InOX (X = Cl, Br or I, involving layers of octahedral InO_3X_3 groups), the recently reported macroanion

(XIII)

Table 1.6 Examples of coordination geometries exhibited by univalent Group 13 metal compounds.

Coordination number	Coordination geometry[a]	Examples[b]
1	Diatomic	$MX(g)$ (M = Al, Ga, In or Tl; X = H, F, Cl, Br or I)
2	Bent	$Al(\mu\text{-}F)_2Al$;[c] $Ga(\mu\text{-}H)_2Ga$;[d] $[2,4,6\text{-}(CF_3)_3C_6H_2OM]_2$ (M = In[e] or Tl[f])
3	Trigonal pyramid	$[MeOTl]_4$; $[Ph_3SiOTl]_4$;[g] $\{Tl_2(OSiMe_2)_2O\}_2]_n$;[g] Tl_2O; Tl_2S; M^ITlO (M^I = alkali metal),[h] $Tl(\mu\text{-}OBu^t)_3Sn$;[i] Tl_3BO_3; Tl_6TeO_6; Tl_2HPO_4; Tl_3PO_4; $Tl(L\text{-ascorbate})$;[j] $[TlSBu^t]_8$;[k] $[TlSCH_2Ph]_n$;[k] $[Tl_7(SPh)_6]^+$;[k] $[TlI]^+PF_6^-$[l]
4	ψ-Trigonal bipyramid	$[Tl_5(SPh)_6]^-$;[k] $[TlSBu^t]_8$;[k] $[Tl_2Sn(OEt)_6]_n$[m]
	Square pyramid	$Tl(L\text{-ascorbate})$[j]
5	Pentagonal-based pyramid	$(\eta^5\text{-}C_5H_5)M$ (M = In or Tl);[n] $Tl(salicylate)(phen)$
	Distorted square pyramid	$Tl(hfac)$; $[Bu^i_2NCS_2Tl]_2$
6	Octahedron	$TlH_2PO_4 \cdot H_3PO_4$
	Distorted octahedron	β-TlF;[o] α-InCl;[p] $[Pr^n_2NCS_2Tl]_2$; $Tl_4[SiE_4]$ (E = S or Se);[q] $Tl_6[Ge_2Te_6]$[r]
7	Distorted trigonal prism	(benzotriazolato)Tl;[s] $[Et_2NCS_2Tl]_2$
	Monocapped octahedron	$Tl_3F(CO_3)$; TlH_2PO_4; $Tl^I[Tl^{III}(O_2CMe)_4]$[h]
	Monocapped trigonal prism	β-TlI; β-InCl, InBr;[t] InI
8	Cube	TlCl; TlBr; α-TlI; $TlNO_3$-III
	Distorted cube	$Tl^I[Tl^{III}Se_2]$;[u] $TlNO_3 \cdot 4SC(NH_2)_2$
	Dodecahedron	$[Tl(1,4\text{-dioxane})]^+[TlBr_4]^-$; $Ga^+[Ga^{III}Cl_4]^-$;[v] α-$Ga^+[Ga^{III}Br_4]^-$;[v] $In^+[In^{III}X_4]^-$ (X = Br[t] or I[v])
	Bicapped trigonal prism	$Ga^+[Ga^{III}I_4]^-$;[v] $Ga_2^+[Ga_2I_6]^{2-}$[v]
	Distorted square antiprism	$In^I[In^{III}Te_2]$
	5 + 3	$[(\eta^5\text{-}C_5Me_5)Al]_4$;[w] $Tl_2[Mo_7O_{22}]$[h]
9	Tricapped trigonal prism	$TlCd_3$;[x] red-Tl(picrate);[y] β-$Ga^+[Ga^{III}Br_4]^-$;[v] $Tl_3Ga_9O_2S_{13}$[z]
	5 + 4	$[(\eta^5\text{-}C_5Me_5)In]_6$;[aa] $Tl[Zn(SO_4)Cl]$[h]
	6 + 3	$Tl_2[Cu(SO_3)_2]$;[h] $Tl_3[Al_{13}S_{21}]$[bb]
10	3 + 4 + 3	$TlVO_3$[h]
	Irregular	$TlBrO_4$[cc]
11	3 + 5 + 3	$Tl_3[Al_{13}S_{21}]$[bb]
	Dodecahedron	$Tl_3Ga_9O_2S_{13}$[z]
12	3 + 6 + 3	$Tl_3[Al_{13}S_{21}]$;[bb] $TlIO_3$[dd]
	Cubic close-packed	$TlMnF_3$
	Distorted hexagonal prism	$Tl_{0.3}WO_3$[h]
14	Structure IX	$[2.2]ParacyclophaneGa^+[Ga^{III}Br_4]$[ee]
16	Structure VIII	$[(\eta^6\text{-}C_6H_6)_2Ga]^+[Ga^{III}X_4]^-$ (X = Cl or Br)[ff]

(a) Some of the descriptions refer to points on the surface of a sphere arranged on parallel planes intersecting the sphere.
(b) Abbreviations: phen = 1,10-phenanthroline; hfac = hexafluoroacetylacetonate. Entries without superscript letters are taken from the following sources: Refs. 2, 4, 19, 30, 32 and 33.

(c) R. Ahlrichs, L. Zhengyan and H. Schnöckel, *Z. Anorg. Allg. Chem.*, 1984, **519**, 155.
(d) Ref. 124.
(e) Ref. 173a.
(f) Ref. 173b.
(g) Ref. 165.
(h) Ref. 44.
(i) M. Veith and R. Rösler, *Angew. Chem., Int. Ed. Engl.*, 1982, **21**, 858.
(j) Ref. 167.
(k) Ref. 166.
(l) L = N,N',N''-trimethyl-1,4,7-triazacyclononane: K. Wieghardt, M. Kleine-Boymann, B. Nuber and J. Weiss, *Inorg. Chem.*, 1986, **25**, 1309.
(m) M.J. Hampden-Smith, D.E. Smith and E.N. Duesler, *Inorg. Chem.*, 1989, **28**, 3399.
(n) Ref. 184.
(o) Ref. 168.
(p) Ref. 169.
(q) G. Eulenberger, *Acta Crystallogr.*, 1986, **C42**, 528.
(r) G. Eulenberger, *J. Solid State Chem.*, 1984, **55**, 306.
(s) J. Reedijk, G. Roelofsen, A.R. Siedle and A.L. Spek, *Inorg. Chem.*, 1979, **18**, 1947.
(t) Ref. 170b.
(u) Ref. 58, p. 104.
(v) Ref. 170a.
(w) Ref. 174a.
(x) H.W. Zandbergen, G.C. Verschoor and D.J.W. Ijdo, *Acta Crystallogr.*, 1979, **B35**, 1425.
(y) F.H. Herbstein, M. Kapon and S. Wielinski, *Acta Crystallogr.*, 1977, **B33**, 649.
(z) S. Jaulmes, M. Julien-Pouzol, J. Dugué, P. Laruelle and M. Guittard, *Acta Crystallogr.*, 1986, **C42**, 1111.
(aa) Ref. 174b.
(bb) Ref. 172a.
(cc) Ref. 171.
(dd) Ref. 172b.
(ee) Ref. 144b.
(ff) Ref. 144a.

$[Al_3P_4O_{16}]_n^{3n-}$ isolated with interlamellar $[H_3N(CH_2)_4NH_3]^{2+}$ cations,[160] and $CuGaInO_4$ (where In has a position between cubically packed O layers with trigonally compressed octahedral coordination and Ga shares with Cu positions near to hexagonally packed O layers with trigonal bipyramidal coordination).[161] Further cross-linking to produce 3-dimensional networks may give not only more-or-less regular coordination polyhedra, but also distorted or unusual ones, as is the case, for example, with the C-M_2O_3 structures favoured by In_2O_3 and Tl_2O_3,[30] with TlF_3,[151] and with the complex fluoroindate $Rb_2In_3F_{11}$ (formed from parallel sheets of edge- and corner-sharing InF_7 pentagonal bipyramids joined together by infinite and parallel chains of corner-sharing InF_6 octahedra).[162]

Table 1.5 does not include any examples of homonuclear metal–metal bonds which are the normal hallmark of the +2 oxidation state. In fact, the known structures of species like $M_2X_6^{2-}$ (**XIV**: M = Ga or In; X = Cl, Br or

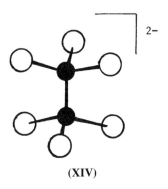

(**XIV**)

I), $Ga_2X_4 \cdot 2L$ (X = Cl or Br; L = 1,4-dioxane or pyridine) and GaE (Figure 1.5(a): E = S, Se or Te) are characterised by near-tetrahedral coordination of the metal atoms, whereas coordination of the M_2 unit (M = Al, Ga or In) by four bulky $CH(SiMe_3)_2$ groups gives rise to trigonal planar coordination at each M centre.[163] These properties emphasise that, but for the presence of a homonuclear metal–metal in place of a heteronuclear metal–ligand bond, there is no fundamental difference between M^{II} and M^{III} species. In this respect, M^{II} and M^{III} bear the same relationship to each other as do Hg^I and Hg^{II} in mercury chemistry.[2,4,30,152]

1.4.3.2 Univalent derivatives. Simple diatomic molecules containing a Group 13 metal atom bound to a 1-electron ligand like H or a halogen are the stuff of high-temperature vapours or low-temperature matrices.[87,164] Under normal conditions the metal strives to achieve coordination numbers of 2 or more. As revealed by the representative examples cited in Table 1.6, the coordination numbers now span a range so wide as to defy the laying down of rules about what constitutes 'normal behaviour'. The most common coordination numbers are 3, 6 and 8, although examples of 5- and 7-fold coordination are also quite numerous. As noted previously, it is often difficult to differentiate clearly between primary and secondary metal–ligand contacts and therefore to specify in simple, unequivocal terms the coordination geometry of the metal atom; in some cases the geometry is just not amenable to descriptions based on any of the normal polyhedra.

In addition to the variety of behaviour and the problems of defining the sphere of influence of the metal, there are other peculiarities that catch the eye. Firstly, there is a distinct lack of symmetry about many of the environments in which the metal is found, particularly when the coordination number is 8 or less. Thus, the structures often put the metal at the apex of a pyramid with all the near neighbours making up the base. The resulting geometry may be a trigonal pyramid as in the cubane-like molecules [ROTl]$_4$ (**XV**: R = Me or Ph$_3$Si[165]), the thiolate clusters [Tl$_7$(SPh)$_6$]$^+$ and [TlSBu^t]$_8$,[166]

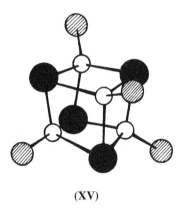

(**XV**)

the binary compounds Tl$_2$O and Tl$_2$S (which have layer structures of the anti-CdI$_2$ type, see Figure 1.5(b))[30] and various oxysalts of thallium(I). Much less common is a square-based pyramid, as in one form of Tl(L-ascorbate),[167] or a pentagonal-based pyramid, as in (η^5–C$_5$H$_5$)M (M = In or Tl).[30] These highly eccentric arrangements, in which the ligands are all confined to the same coordination hemisphere, give way, as the coordination number increases to 6 or 7, to arrangements where the extra ligands are located on the other side of the metal atom while still implying a strongly aspherical M$^+$ ion. The position is well illustrated by the 6-coordinate structures of β-TlF and α-InCl and the 7-coordinate structure of β-TlI, β-InCl, InBr and InI.[19,30,31,168,169] β-TlF has a modified rock-salt structure containing two independent metal ions each surrounded by a distorted octahedron of halide ions of which two (*cis*) are much more distant than the others (Figure 1.5(c)). By contrast, the yellow (β) form of TlI is made up of slices of the rock-salt structure which are displaced relative to one another, with the

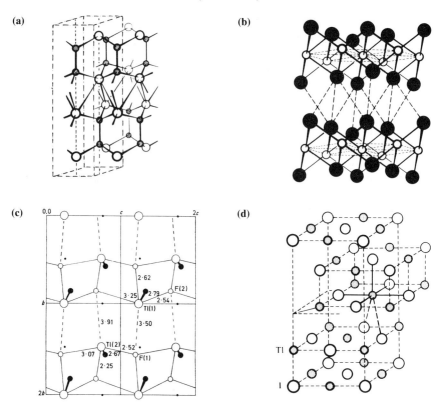

Figure 1.5 Structures of some compounds of the Group 13 metals in lower oxidation states. (a) GaS: hatched circles are Ga, open circles are S atoms (reproduced with permission from Ref. 30). (b) Anti-CdI$_2$-type structure of Tl$_2$O. (c) Structure of β-TlF shown in projection down the a-axis: large circles are Tl atoms at $x = 0.25$, small circles are F atoms: open at $x = 0.25$, and infilled at $x = 0.75$ and -0.25; dashed lines indicate long interactions in the b direction; Tl atoms at $x = 0.75$ (whose interactions are not shown) are indicated by dots. For clarity, Tl(1) and F(1) are shown in the cell $(y + 1, z + 1)$, Tl(2) is shown in the cell $(y + 1)$, and F(2) is shown in the cell $(z + 1)$ (reproduced with permission from Ref. 168). (d) Yellow (β) form of TlI (reproduced with permission from Ref. 30).

result that the metal ion has five nearest neighbours at the vertices of a square-based pyramid and then two nearly equidistant next-nearest neighbours to produce what approximates to a monocapped trigonal prism (Figure 1.5(d)). As the coordination number and the average metal–ligand distances increase still further, so it appears that the metal is more likely to be enclosed by a relatively uniform distribution of ligands. Thus, 8-coordinate arrays may take the form of a regular cube (e.g. TlCl, TlBr, α-TlI and TlNO$_3$-III) or a dodecahedron [e.g. Ga$^+$[Ga^{III}Cl$_4$]$^-$,[170a] α-Ga$^+$[Ga^{III}Br$_4$]$^{-}$[170a] and In$^+$[In^{III}X$_4$]$^-$ (X = Br[170b] or I[170a])], as well as less regular forms like a distorted cube or square antiprism, or a bicapped trigonal prism. Instances of metal atoms in 9-, 10-, 11- or 12-coordinate environments can also be found. Occasionally these environments are irregular, as with the distorted hexagonal prism which is the setting of thallium in the tungsten bronze $Tl_{0.3}WO_3$[44] or the 10-coordinate arrangement in $TlBrO_4$;[171] more often there is a relatively uniform filling of the space around the metal centre, even if the arrangements do not always conform to familiar topology (e.g. the environments of Tl in $TlVO_3$ (10-coordinate),[44] $Tl_3[Al_{13}S_{21}]$ (11-coordinate),[172a] $TlIO_3$ (12-coordinate)[172b] and $TlMnF_3$ (perovskite structure with cubic close-packing of Tl and F)[30]).

The nature of the ligands is obviously crucial. As with trivalent derivatives, bulky ligands can enforce uncharacteristic coordination geometries at the metal atom. An example is provided by the dimeric phenolates $[2,4,6-(CF_3)_3C_6H_2OM]_2$ (M = In[173a] or Tl[173b]) which have as their nucleus 4-membered $O(\mu-M)_2O$ rings with 2-coordinate M atoms. Ligands with particularly small bites and the ability to engage in multi-centre bonding with the metal are likely also to make their mark, notably in creating metal environments of high coordination number. This is a conspicuous property in the structures of the organometallic species $[(\eta^5-C_5Me_5)Al]_4$,[174a] $[(\eta^5-C_5Me_5)In]_6$,[174b] $[([2.2]paracyclophane)Ga]^+[GaBr_4]^-$ (**IX**)[144b] and $[(\eta^6-C_6H_6)_2Ga]^+[GaX_4]^-$ (X = Cl or Br) (**VIII**)[144a] where the formal coordination numbers of the M^I atom are 8, 9, 14 and 16, respectively.

The aspherical form which the M^I centre commonly betrays must be associated with a non-spherical charge distribution of the valence-shell lone pair. It is usually supposed that mixing of ns and np orbitals, made possible by the reduced symmetry of the crystal field, causes the lone pair to assume some p-character and, with it, the sort of directional properties that are manifest in a simple molecule like NH_3. The electrostatic repulsion between this unsymmetrically distributed electron pair and the ligands would then account qualitatively for the deviations from cubic symmetry which characterise so many derivatives of the Group 13 metals in the +1 state. The behaviour can be explained in terms of the unusually high polarisability which the ns^2 long-pair imparts to the M^+ ion.[31,169] The importance of cation, as opposed to anion, polarisation is strongly urged not only by the layer structures favoured by some compounds (e.g. Tl_2O and Tl_2S) but also by the circumstantial evidence that distortion is driven by short-range forces which are more likely to prevail at short than at long mean metal–ligand distances.

The variety of structures displayed by univalent compounds cannot be easily rationalised, despite the various explanations it has attracted, ranging in sophistication from VSEPR to relativistic molecular orbital calculations.[25,31,44,169,175,176] Neither VSEPR, with its requirement that lone-pair electrons demand angular space around the central atom, nor relativity explains the gamut of structural behaviour characterising these and isoelectronic derivatives formed by the heavier members of Groups 14–16. Of the simpler approaches, one offering some insight into the problem is based on the second-order Jahn–Teller model which treats nuclear displacements as perturbations and uses perturbation theory to determine what effect the displacements have on the electronic energy of the molecular aggregate.[175–177] For displacement from an equilibrium geometry in the mode Q_i the change in electronic energy, correct to second order, is given by

$$\Delta E(Q_i) = \frac{1}{2}[V_{00}^{ii} + 2\sum_k{}' |V_{0k}^i|^2/(E_0 - E_k)]\delta Q_i^2 \qquad (1.15)$$

The first term is the contribution to ΔE arising from the nuclear displacement in the presence of the *undisturbed* charge density ϱ_{00}; the second term represents the contribution to ΔE arising from the *change* in ϱ caused by the nuclear displacement, that is

$$V_{0k}^i = \int(\partial V_{ne}/\partial Q_i)\varrho_{0k}\, d\tau \qquad (1.16)$$

The transition density, ϱ_{0k}, is the change in ϱ_{00} following 'mixing-in' with the ground-state function, ψ_0, the singly excited state function, ψ_k. Only when ϱ_{00} relaxes by varying in a way that does not correspond to rigid following of the nuclei does the change affect the electronic energy. Equation (1.16) implies, however, that the integral V_{0k}^i vanishes unless ϱ_{0k} and Q_i contain the same irreducible representation. According to equation (1.15), moreover, the energy response to relaxation in ϱ_{00} depends on a series of terms each with an inverse dependence on the energy gap $(E_0 - E_k)$. The largest of these terms is clearly the one associated with the first excited electronic state (required also to have the same spin multiplicity as the ground state of the unperturbed system). Most treatments focus on this term and take the relaxation in ϱ_{00} for any mode Q_i to approximate to a single ϱ_{0k} which may be expressed roughly as a product of the HOMO and LUMO of a single determinantal wave function ψ_0. The model then provides a mechanism for determining which, if any, of the normal modes of the undistorted aggregate causes the HOMO and LUMO to mix together with a resultant stabilisation of the HOMO. The essential prerequisite is that the vibration shall have an irreducible representation common to the transition density $\psi_0 \psi_k$ corresponding to HOMO → LUMO excitation.

Figure 1.6 shows a simplified molecular orbital diagram for the σ-framework of a fragment formed by a unipositive Group 13 metal cation M^+ surrounded by a regular octahedron of X^{z-} anions. The normal modes of the MX_6 fragment span the representation $a_{1g} + e_g + t_{1u}$ (stretching) and $t_{1u} + t_{2g} + t_{2u}$ (bending). Application of the principles outlined above gives a transition density of species t_{1u} $(2a_{1g} \rightarrow 2t_{1u})$ and so establishes the underlying susceptibility of the system to distortion. The action of t_{1u} deformations may result in distortion along a 3-fold, a 2-fold or a 4-fold rotation axis of the octahedron to give geometries belonging to the point group C_{3v}, C_{2v} or C_{4v}, respectively, as illustrated in Figure 1.7.[31,169,178] The trigonal pyramidal coordination of Tl^I in Tl_2O (and other compounds) would correspond to an

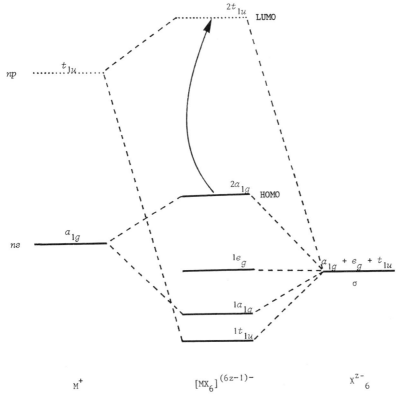

Figure 1.6 Schematic MO energy-level scheme for an isolated regular octahedral complex formed by a univalent cation of a Group 13 metal M with σ-bonding ligands X^{x-}, indicating the electronic excitation most likely to lead to stabilisation with accompanying distortion. Solid lines denote occupied, dotted lines unoccupied orbitals.

extreme case of the C_{3v} geometry, the distorted octahedron of β-TlF approximates to the C_{2v} model, and the metal environment in β-TlI shows an obvious kinship to the C_{4v} model. A common feature of all these distortions is that the inversion centre of the octahedron is lost, thereby permitting the mixing of valence ns and np functions of the metal atom to give a lone pair that appears to be stereochemically active. In each case, the LUMO (t_{1u}) develops a component that is totally symmetric under the lower symmetry of the distorted geometry and therefore capable of interacting with the HOMO to an extent inversely dependent on the HOMO-LUMO energy gap. This opens up a larger HOMO-LUMO gap, as shown for the case of a trigonal distortion in Figure 1.8. If the driving force is large enough, the regular environment must be unstable and a static distortion results. Whether a regular or a distorted structure prevails depends on the HOMO-LUMO gap

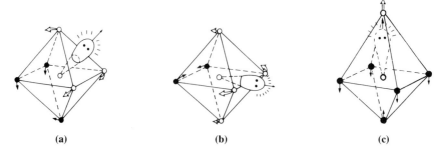

Figure 1.7 Possible modes of distortion of an isolated regular octahedral complex formed by a univalent Group 13 metal cation resulting from the action of t_{1u} deformations, including a portrayal of the influence exerted by the lone pair. (a) Distortion along a 3-fold axis of the octahedron giving a unit with C_{3v} symmetry; (b) distortion along a 2-fold axis of the octahedron giving a unit with C_{2v} symmetry; and (c) distortion along a 4-fold axis of the octahedron giving a unit with C_{4v} symmetry. Based on illustrations reproduced with permission from Ref. 178.

in the unperturbed system: the smaller the gap, the greater is the incentive for distortion. For a relatively electropositive atom like In or Tl, the gap is likely to follow the ns–np separation quite closely. On this basis, distorted structures implying stereochemical activity of the lone pair are more likely to be formed by In^I than by Tl^I. There is insufficient information rigorously to test this prognosis, but the irregular structures of InCl and InBr contrast tellingly with the regular structures of TlCl and TlBr. The other major consideration is the electronegativity or basicity of the ligand. Here the effect is less easy to predict since the character of the HOMO is likely to change markedly according to whether the metal ns or the ligand orbitals lie deeper in energy, and the role of other orbitals (not included in Figure 1.6) may need to be considered.[176] There are empirical grounds, however, for believing that the tendency to distort increases as the ligands become more basic.[44] Not to be overlooked, moreover, is the part played by ligand–ligand repulsions in tempering distortion.

The treatment described in the preceding paragraphs is strictly concerned with discrete molecular units. Similar frontier orbital arguments can be applied to extended solids that appear to parallel the corresponding molecular systems in their behaviour, the function of the distortion now being to enlarge the band gap of the assembly.[175,176] The result may be to create a ferroelectric material, as in $TlGaSe_2$.[179] In such solids, however, the nature of the distortions may be revealed only at low temperatures, for, unless the distortions at different metal sites are ordered, the structure as a whole will behave at normal temperatures as if it were regular.[31,180] It follows that even superficially regular structures, like that of TlCl, may actually conceal dynamic averaging of different possible orientations of metal-centred coordination polyhedra which are inherently of lower than cubic symmetry.

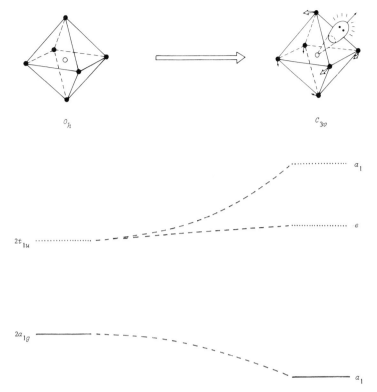

O_h C_{3v}

a_1

e

$2t_{1u}$

$2a_{1g}$

a_1

Figure 1.8 Correlation diagram for the frontier orbitals of a regular octahedral and trigonally distorted octahedral complex of a univalent Group 13 metal cation. Solid lines denote occupied, dotted lines unoccupied orbitals.

1.4.4 Compounds with metal–metal bonds[30,32,33,38,39,150]

Outside the solid elements and their alloys, homonuclear bonding contributes less to the chemistries of the heavier Group 13 elements than it does, say, to the chemistries of boron and adjacent Group 14 elements. On the evidence of the mean bond enthalpies included in Table 1.1, the M–M bond is amongst the weakest to be formed by a Group 13 metal M and grows ever more tenuous as the atomic number of M increases. The spectroscopic properties of discrete M_2 molecules in the gas phase reveal a similar pattern, with the following properties for the $^3\Sigma_g^-$ ground state (corresponding to the simple MO bonding scheme $(\sigma_g ns)^2(\sigma_u^* ns)^2(\pi_u np)^2$): B_2 $r_e = 1.590$ Å, $D_0^0 = 291$ kJ mol^{-1}; Al$_2$ $r_e = 2.466$ Å, $D_0^0 = 150$ kJ mol^{-1}; Ga$_2$ $D_0^0 = 135$ kJ mol^{-1}; In$_2$ $D_0^0 = 97$ kJ mol^{-1}; Tl$_2$ $D_0^0 < 87$ kJ mol^{-1}.[164] To judge by the case of boron, there is little difference in the length and energy of the M–M bond in M_2 and in the tetrahalide molecule X_2M–MX$_2$.[38] Whereas compounds containing B–B bonds are typically subject to kinetic stabilisation, reaction being opposed by substantial activation barriers, similar derivatives of the heavier Group 13 elements are much more vulnerable to attack and their survival demands altogether more rigorous control of the steric and electronic properties of the associated ligands. Only in the past two decades has significant progress been made with the preparation and characterisation of compounds containing Al–Al, Ga–Ga or In–In bonds, and which are long-lived under normal conditions. Whether Tl–Tl bonding is ever likely to amount to a significant force is still open to doubt.[179,181]

Many compounds contain two or more Group 13 atoms constrained by the action of bridging ligands to be separated by distances well within the sum of their van der Waals radii. It is notoriously difficult, however, to infer the significance of metal–metal interactions in such compounds by distance criteria alone.[181b] For example, the $Ga\cdots Ga$ distances in the molecules $H_2Ga(\mu\text{-H})_2GaH_2$[22a] and $Me_2Ga(\mu\text{-H})_2GaMe_2$[182] are, at 2.58 and 2.61 Å, respectively, comparable not only with twice the covalent radius of tetra-hedrally coordinated gallium, but also with the Ga–Ga distances ranging from 2.47 to 3.07 Å in the different forms of elemental gallium. Combined with the relatively small amplitude of vibration of the $Ga\cdots Ga$ vector, the distances appear to make a persuasive case for direct interaction between the two metal atoms, although most theoretical treatments of these and related molecules contend that no such bond path exists.[183] Only when the metal is in a formal oxidation state less than +3 does there appear to be scope for significant *direct* interaction between the bridged metal atoms. Even then there is no clear consensus about its extent, for example, in compounds like $InCl$,[169] $(\eta^5\text{-}C_5H_5)M$ (M = In or Tl),[184] $[TlOR]_4$ (R = Et or Pr^n),[185] $\{[\eta^5\text{-}(C_6H_5CH_2)_5C_5]Tl\}_2$[181a] and $Tl^IGa^{III}Se_2$.[179]

What cannot be contested, however, is that homonuclear metal–metal bonding unsupported by any bridging ligands is central to the stabilisation of authentic derivatives of the +2 oxidation state. Neutral molecules of the type $X_2M\text{–}MX_2$ involving tri-coordinated Group 13 metal atoms, M, and 1-electron ligands, X, have proved highly elusive, presumably because of the facility of intramolecular ligand transfer leading to disproportionation (see section 1.4.1). This process can be inhibited, however, either (a) by making X both bulky and deficient in the means to bridge metal centres or (b) by coordinative saturation of the M centres, with an appropriate increase in coordinate number to 4 (or sometimes higher), through the attachment of additional ligands. The working out of the first strategy is demonstrated by the successful preparation of the compounds $M_2[CH(SiMe_3)_2]_4$ with M–M bond distances of 2.660, 2.541 and 2.828 Å for M = Al,[163a] Ga[163b] and In,[163c] respectively. More familiar, however, are the results of exploiting the second strategy. Thus, Ga–Ga or In–In bonding is now a well established fact of life in ethane-like molecules of the type $LX_2M\text{–}MX_2L$ (cf. **XIV**) formally generated by the dimerisation of MX_2L radicals (X = Cl, Br or I; and L = neutral or anionic ligand), with M–M bond lengths of 2.39–2.42

(M = Ga) and c. 2.78 Å (M = In).[32,33,38] No less prominent in chalcogenides like GaS (see Figure 1.5(a)), such bonding acts to hold together puckered layers of Ga and S atoms (with $r(Ga–Ga) = 2.45$ Å) in a structure related to that of wurtzite (ZnS), distortion of which to accommodate the extra electron per MS unit may be regarded as another manifestation of the second-order Jahn–Teller theorem.[30,175c,186] Neither aluminium nor thallium is known to enter into the formation of metal–metal bonded systems analogous to these. Even with gallium and indium there are quite stringent coordination requirements to be met if stable metal–metal bonds are to be sustained under these conditions, as exemplified by the cases of InS and InTe.[30] Thus, InS may be described as being built of puckered 3-connected layers similar to those in GeS, except that the layers are translated relative to their positions in GeS so that there are close In⋯In contacts between the layers (2.80 Å). By contrast, InTe shows the alternative pattern of behaviour open to compounds of this sort, with a structure having no close In⋯In contacts but two widely disparate metal sites intimating the formulation $In^{I}In^{III}Te_2$.

Oxidation states in the range 0 to +1 furnish a handful of well characterised examples of homonuclear clusters incorporating more than two metal atoms. At one extreme, the species are highly reactive 'naked' clusters like Al_3,[92] which can be maintained at appreciable concentrations and studied only by the stratagems of cooled molecular beams or low-temperature matrices.[87] At the other extreme, the conjunction of the univalent metal and a bulky ligand with minimal bridging facility may yield a molecular cluster which is stable and amenable to study under normal conditions. Such is the case, for example, with the aluminium(I) compound $[(\eta^5\text{-}C_5Me_5)Al]_4$;[174a] this consists of a tetrahedral Al_4 core coordinated radially through the Al atoms to the $\eta^5\text{-}C_5Me_5$ ligands (**XVI**). The corresponding indium compound is a hexamer, the hub of which is an In_6 octahedron.[174b] The average M–M distance in the aluminium compound (2.77 Å) is intermediate between those in $Al_2[CH(SiMe_3)_2]_4$ (2.66 Å) and metallic

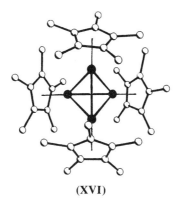

(**XVI**)

aluminium (2.86 Å): the corresponding distance in the indium compound (3.95 Å) is comparable with the longer of the two distances in metallic indium (3.36 and 3.94 Å). Cationic indium clusters have also come to light in metal-rich chalcogenides. For example, the compound $Na_{24}In_5O_{15}$ has been shown to include In_5^{6+} clusters in the form of indium-centred tetrahedra with $r(In-In) = 2.74$ Å,[187a] and $In_{11}Mo_{40}O_{62}$ contains linear In_5^{7+} and In_6^{8+} ions, with $r(In-In) = 2.62-2.68$ Å, occupying channels between the anionic clusters.[187b] Slightly angular In_3^{5+} units, with the In–In–In angle = 157.8° and $r(In-In) = c.$ 2.77 Å, act to cross-link the endless interlocking chains of five-membered In–Se rings which make up the crystal structure of In_4Se_3.[188]

The Group 13 metals are not well adapted to the formation of homoatomic Zintl ions.[189] Unlike the members of Groups 14–16, they lack the valence electrons needed for optimum bonding in clusters of this sort, and which can be achieved otherwise only at the expense of a large accumulated negative charge. Nevertheless, persistent and unusual aggregation of Group 13 metal atoms is found in some intermetallic phases, e.g. Li_9Al_4 and Li_2Ga which contain zigzag chains of Al and Ga atoms, respectively, and Li_3M_2 (M = Al, Ga or In) which contains puckered layers of M atoms.[190] Isolated tetrahedral M_4 and triangular M_3 clusters (M = Al or Ga) characterise the intermetallic compounds Ba_8Ga_7 and Sr_8M_7.[191] Even more remarkable are the three-dimensional networks formed by the interlinking of polyhedral gallium clusters found in gallium-rich binary derivatives of the alkali metals. Compositions like K_3Ga_{13},[192a] $RbGa_3$,[192b] KGa_3,[192c] $RbGa_7$,[192d] Na_7Ga_{13}[192e] and $Na_{22}Ga_{39}$[192f] exhibit complex structures containing large, interconnected and usually empty Ga_n polyhedra—icosahedra, dodecahedra, octadecahedra, etc. Still more intricate structures emerge for ternary phases like $Li_3Na_5Ga_{19.56}$,[193a] $K_4Na_{13}Ga_{49.57}$,[193b] $K_3Li_9Ga_{28.83}$[193c] and $Rb_{0.60}Na_{6.25}Ga_{20.02}$.[193d] Here gallium is strongly reminiscent of boron in its behaviour.[194] Electron transfer from the alkali metal to the gallium network seems to have occurred, but apparent fractional occupancy of both gallium and cation positions, as well as disorder, complicates the understanding and interpretation of the bonding requirements of many of these structures. Clearcut and successful analysis *has* been possible in some cases,[190,194,195] but has not generally been submitted to experimental test. Much less is known about alkali metal–indium phases, but the properties are by no means foreshadowed by those of their gallium counterparts. Only very recently has the Zintl-type phase K_8In_{11} come to light, and with it the first sighting of the isolated and unprecedented In_{11}^{7-} cluster.[196] With an electron count well below the $2n + 2$ skeletal minimum prescribed by Wade's rules, this species takes the form of a compressed pentacapped trigonal prism, the In–In separations ranging from 2.96 to 3.28 Å. Gallium and indium do resemble each other in that $K_{22}In_{39}$ consists, like $Na_{22}Ga_{39}$, of a network of heavily interbonded In_{12} and In_{15} clusters.[197] A novel network structure is found, however, in the well crystallized phase $Na_7In_{11.8}$:[197] it consists of interbonded *closo*-In_{16} icosioctahedra, *nido*-In_{11} icosahedra and In atoms. By contrast, the isostructural phases Na_2In and Na_2Tl contain relatively well isolated metal clusters, in this case tetrahedral M_4^{8-} with $r(M–M) = 3.07–3.15$ and $3.18–3.30 \text{ Å}$ for $M = In$[196] and Tl,[198] respectively, and inviting comparisons with isoelectronic species like Sn_4^{4-} (in KSn) and Sb_4.

The limitations of the Group 13 metals as sources of homoatomic Zintl ions can be alleviated by combination with their more electron-rich neighbours to produce heteroatomic ions. Practical implementation of this approach has led, for example, to the successful synthesis and characterisation of a compound containing the polyanions $TlSn_8^{3-}$ and $TlSn_9^{3-}$,[199a] the first with a tricapped trigonal-prismatic (**XVII**) and the second with a bicapped square-antiprismatic geometry (**XVIII**); Tl takes up a capping position in each case. Prepared and isolated in a similar fashion is the cyclic $Tl_2Te_2^{2-}$ anion, which has a butterfly shape, the Tl atoms being positioned along the fold of the wings (**XIX**).[199b] In addition, the new Zintl compound $Ca_{14}GaAs_{11}$ has a structure composed of isolated Ga-centred $GaAs_4$ tetrahedra that are separated by linear As_3 moieties, As and Ca atoms.[200]

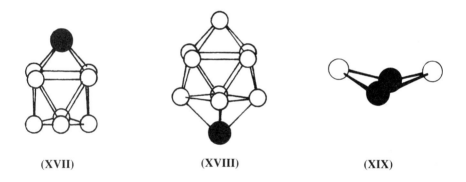

(XVII)   **(XVIII)**   **(XIX)**

Heteronuclear metal–metal or metal–metalloid bonds can also be sustained by the Group 13 metals in more highly oxidised states, provided that appropriate steric and electronic control is exercised over the conditions of coordination. For example, mixed metal complexes of the types $InGaX_4 \cdot 2L$ and $[X_3GaInX_3]^{2-}$ (X = Cl or Br; L = pyridine, piperidine, piperazine, 1,4-dioxane, tetrahydropyran or tetrahydrofuran) have been described and appear, on the evidence of their Raman spectra, to contain Ga–In bonds.[201] Coordinative saturation of the Group 13 atom appears, in addition, to admit the stabilisation of M–Sn bonds in complexes of the form $Li[Me_3MSnMe_3]$ (M = Al, Ga, In or Tl).[202] Bonds to the heavier Group 15 elements (As, Sb and Bi) feature not only in the extended structures of the binary III–V compounds (see chapters 3 and 5), but also in discrete molecules like $Ga[As(mesityl)_2]_3$,[146] $[R_2GaAsBu^t_2]_2$ (R = Me, Et or Bu^n),[203] $(C_5Me_5)_2GaAs(SiMe_3)_2$,[203] $[Cl_2GaSbBu^t_2]_3$,[203] $[Me_2InAsMe_2]_3$[203] and

[ClIn(SbBu$^t_2$)$_2$]$_2$.[203] A Ga$_5$As$_7$ cluster (**XX**) provides the core of the organo-gallium compound [(PhAsH)(R$_2$Ga)(PhAs)$_6$(RGa)$_4$] (R = Me$_3$SiCH$_2$).[204] There exists too a variety of compounds in which gallium, indium or thallium is engaged in bonding with a transition-metal atom.[39,150,205] Examples include simple monomeric species with structures centred on a tri-coordinated Group 13 metal atom, e.g. In[Co(CO)$_4$]$_3$ and M[Mo(CO)$_3$-(C$_5$H$_5$)]$_3$ (M = In or Tl); in the aluminium complex Al[W(CO)$_3$-(C$_5$H$_5$)]$_3$ · 3thf, by contrast, the interaction is not between Al and W but

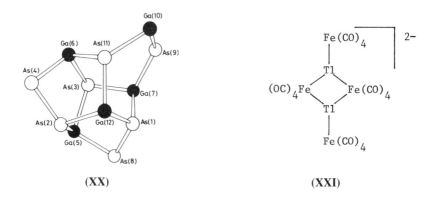

(**XX**)   (**XXI**)

between Al and carbonyl O atoms.[150] Several compounds have structures based on a central four-membered ring made up of alternating Group 13 and transition-metal atoms. Such is the case, for instance, with [M$'_2$(CO)$_8$-{μ-MM'(CO)$_5$}$_2$] (M = Ga or In; M' = Mn or Re), [{Fe(CO)$_4$}$_2$-{μ-InMn(CO)$_5$}$_2$] and [{Fe(CO)$_4$}$_2${μ-TlFe(CO)$_4$}$_2$]$^{2-}$ (**XXI**). Triangular GaRe$_2$ and tetrahedral GaRe$_3$ clusters form the nuclei of the compounds [Re$_2$(CO)$_4$(PPh$_3$)$_2$(μ-X)$_2${μ-GaRe(CO)$_4$(PPh$_3$)}] (X = Br or I) and [Re$_3$(CO)$_6$(PPh$_3$)$_3$(μ-Cl)$_3${μ_3-GaRe(CO)$_4$(PPh$_3$)}], respectively, and a tetrahedral TlPt$_3$ cluster is central to the cationic complex [TlPt$_3$(μ-CO)$_3$-(PCy$_3$)$_3$]$^+$ (Cy = cyclohexyl). Still more elaborate assemblies of metal atoms have been characterised in the anionic complexes [Tl$_2$Fe$_6$(CO)$_{24}$]$^{2-}$, [Tl$_4$Fe$_8$(CO)$_{30}$]$^{4-}$ and [Tl$_6$Fe$_{10}$(CO)$_{36}$]$^{6-}$, the bonding in which has excited considerable curiosity. For further details, the reader is directed to section 3.4.3 of chapter 3.

1.5   Organisation of subsequent chapters

In the following chapters aluminium, gallium, indium and thallium are to be treated, for the most part, side by side. This has been done deliberately in order to give a broader perspective, showing up both parallels and divergences in behaviour. There are places, as in chapters 2, 4 and 9, where

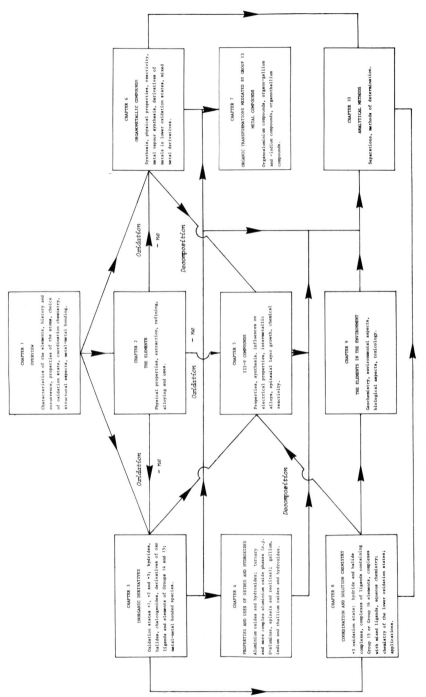

Figure 1.9 Plan of chapters and major topics.

CHAPTER 1

OVERVIEW

Characteristics of the elements, history and occurrence, properties of the atoms, choice of oxidation state, coordination chemistry, structural aspects, metal-metal bonding.

CHAPTER 6

ORGANOMETALLIC COMPOUNDS

Synthesis, physical properties, reactivity, metal vapour synthesis, derivatives of metals in lower oxidation states, mixed metal derivatives.

CHAPTER 7

ORGANIC TRANSFORMATIONS MEDIATED BY GROUP 13 METAL COMPOUNDS

Organoaluminium compounds, organo-gallium and -indium compounds, organothallium compounds.

CHAPTER 10

ANALYTICAL METHODS

Separations, methods of determination.

CHAPTER 2

THE ELEMENTS

Physical properties, extraction, refining, alloying and uses.

CHAPTER 5

III-V COMPOUNDS

Properties, synthesis, influences on electrical properties, intermetallic alloys, epitaxial layer growth, chemical reactivity.

CHAPTER 9

THE ELEMENTS IN THE ENVIRONMENT

Geochemistry, environmental aspects, biological aspects, toxicology.

CHAPTER 3

INORGANIC DERIVATIVES

Oxidation states +1, +2 and +3; hydrides, halides, chalcogenides, derivatives of oxo ligands and elements of Groups 14 and 15; metal-metal bonded species.

CHAPTER 4

PROPERTIES AND USES OF OXIDES AND HYDROXIDES

Aluminium oxides and hydroxides; ternary and more complex aluminium oxide phases (e.g. β-aluminas, spinels and zeolites); gallium, indium and thallium oxides and hydroxides.

CHAPTER 8

COORDINATION AND SOLUTION CHEMISTRY

+3 oxidation state: hydride and halide complexes, complexes of ligands containing Group 15 or Group 16 elements, complexes with mixed ligands, aqueous chemistry; chemistry of the lower oxidation states; applications.

Oxidation

- ne

Oxidation

- ne

Oxidation

- ne

Decomposition

Decomposition

Addition of donor species

aluminium and its compounds will inevitably dominate the scene by virtue of their abundance, their technological importance, their potential impact on the biosphere, and the much richer treasury of information that has accumulated in this area. The arrangement and main interconnections of the topics are outlined schematically in Figure 1.9.

The next chapter treats the elements themselves, with due emphasis on the metallurgical and commercial aspects. The inorganic derivatives of the elements are then described in chapter 3; this is a substantial survey divided into two main parts, one dealing with the +3 oxidation state and the other with the +1 and +2 oxidation states. There follow two chapters reviewing solid compounds with extended structures, chapter 4 concerning oxide derivatives (dominated by alumina and aluminates), and chapter 5 III–V compounds. The organometallic compounds are described in chapter 6, followed by a separate chapter devoted to the roles of these and other Group 13 metal compounds in organic synthesis. The theme of chapter 8 is co-ordination and solution chemistry. This links to, and is often inseparable from, some of the topics treated in chapter 3. It also leads naturally enough to accounts of environmental behaviour, with particular reference to biochemical factors (chapter 9), and of separation and analytical procedures appropriate to the elements (chapter 10). So "mighty things from small beginnings grow".

References

1. N.V. Sidgwick, *The Chemical Elements and their Compounds*, Clarendon Press, Oxford, 1950, Vol. I, p. 334 *et seq.*
2. F.A. Cotton and G. Wilkinson, *Advanced Inorganic Chemistry*, 5th edn, Wiley-Interscience, New York, 1988, p. 162 *et seq.*
3. A.G. Sharpe, *Inorganic Chemistry*, 3rd edn, Longman, London, 1992, p. 254 *et seq.*
4. N.N. Greenwood and A. Earnshaw, *Chemistry of the Elements*, Pergamon Press, Oxford, 1984, p. 155 *et seq.*
5. D.F. Shriver, P.W. Atkins and C.H. Langford, *Inorganic Chemistry*, Oxford University Press, Oxford, 1990, p. 331 *et seq.*
6. A.G. Massey, *Main Group Chemistry*, Ellis Horwood, Chichester, UK, 1990, p. 175 *et seq.*
7. A.G. Massey, *Adv. Inorg. Chem. Radiochem.*, 1983, **26**, 1.
8. *Gmelin Handbook of Inorganic Chemistry*, 8th edn, *Boron Compounds*, Syst. No. 13, Parts 1–20, 1974–1979; 1st Supplement, Vols. 1–3, 1980–1981; 2nd Supplement, Vols. 1–2, 1982–1983; 3rd Supplement, Vols. 1–4, 1987–1988; 4th Supplement, Vols. 3a and 4, 1991; Springer-Verlag, Berlin.
9. *Supplement to Mellor's Comprehensive Treatise on Inorganic and Theoretical Chemistry*, Vol. V, *Boron*. Part A: *Boron–Oxygen Compounds*; Part BI: *Boron–Hydrogen Compounds*; Longman, London, 1980–1981.
10. N.N. Greenwood and B.S. Thomas, *The Chemistry of Boron*, Pergamon Press, Oxford, 1975; also as Chapter 11 in *Comprehensive Inorganic Chemistry*, eds. J.C. Bailar, Jr., H.J. Eméleus, R. Nyholm and A.F. Trotman-Dickenson, Pergamon Press, Oxford, 1973, Vol. 1.
11. E.L. Muetterties, ed., *The Chemistry of Boron and its Compounds*, Wiley, New York, 1967.
12. E.L. Muetterties, ed., *Boron Hydride Chemistry*, Academic Press, New York, 1975.

13. J.F. Liebman, A. Greenberg and R.E. Williams, eds., *Advances in Boron and the Boranes*, VCH, Weinheim, Germany, 1988.
14. R.N. Grimes, ed., *Metal Interactions with Boron Clusters*, Plenum Press, New York, 1982.
15. S. Heřmánek, ed., *Boron Chemistry*, World Scientific, Singapore, 1987.
16. G.A. Olah, K. Wade and R.E. Williams, eds., *Electron Deficient Boron and Carbon Clusters*, Wiley, New York, 1991.
17. D.M.P. Mingos and D.J. Wales, *Introduction to Cluster Chemistry*, Prentice Hall, London, 1990.
18. C.E. Housecroft, *Boranes and Metalloboranes*, Ellis Horwood, Chichester, UK, 1990.
19. K. Wade and A.J. Banister, *The Chemistry of Aluminium, Gallium, Indium and Thallium*, Pergamon Press, Oxford, 1975; also as Chapter 12 in *Comprehensive Inorganic Chemistry*, eds. J.C. Bailar, Jr., H.J. Eméleus, R. Nyholm and A.F. Trotman-Dickenson, Pergamon Press, Oxford, 1973, Vol. 1.
20. E. Wiberg and E. Amberger, *Hydrides of the Elements of Main Groups I–IV*, Elsevier, Amsterdam, 1971.
21. F.M. Brower, N.E. Matzek, P.F. Reigler, H.W. Rinn, C.B. Roberts, D.L. Schmidt, J.A. Snover and K. Terada, *J. Am. Chem. Soc.*, 1976, **98**, 2450; J.W. Turley and H.W. Rinn, *Inorg. Chem.*, 1969, **8**, 18.
22. (a) A.J. Downs, M.J. Goode and C.R. Pulham, *J. Am. Chem. Soc.*, 1989, **111**, 1936. C.R. Pulham, A.J. Downs, M.J. Goode, D.W.H. Rankin and H.E. Robertson, *J. Am. Chem. Soc.*, 1991, **113**, 5149; (b) C.R. Pulham, P.T. Brain, A.J. Downs, D.W.H. Rankin and H.E. Robertson, *J. Chem. Soc., Chem. Commun.*, 1990, 177; (c) C.R. Pulham, A.J. Downs, D.W.H. Rankin and H.E. Robertson, *J. Chem. Soc., Chem. Commun.*, 1990, 1520; *J. Chem. Soc., Dalton Trans.*, 1992, 1509.
23. B. Post, in *Boron, Metallo-Boron Compounds and Boranes*, ed. R.M. Adams, Interscience, New York, 1964, p. 301; N.N. Greenwood, R.V. Parish and P. Thornton, *Q. Rev. Chem. Soc.*, 1966, **20**, 441.
24. R.L. Deming, A.L. Allred, A.R. Dahl, A.W. Herlinger and M.O. Kestner, *J. Am. Chem. Soc.*, 1976, **98**, 4132.
25. K.S. Pitzer, *Acc. Chem. Res.*, 1979, **12**, 271; P. Pyykkö and J.-P. Desclaux, *Acc. Chem. Res.*, 1979, **12**, 276; P. Pyykkö, *Chem. Rev.*, 1988, **88**, 563.
26. P. Schwerdtfeger, G.A. Heath, M. Dolg and M.A. Bennett, *J. Am. Chem. Soc.*, 1992, **114**, 7518.
27. W.N. Rowlands, A.D. Willson, P.L. Timms, B. Mile, J.H.B. Chenier, J.A. Howard and H.A. Joly, *Inorg. Chim. Acta*, 1991, **189**, 189; M. Tacke and H. Schnöckel, *Inorg. Chem.*, 1989, **28**, 2895.
28. H.G. Grimm and A. Sommerfeld, *Z. Phys.*, 1926, **36**, 36.
29. N.V. Sidgwick, *The Electronic Theory of Valency*, Oxford University Press, Oxford, 1927; *Ann. Rep. Prog. Chem.*, 1933, **30**, 120.
30. A.F. Wells, *Structural Inorganic Chemistry*, 5th edn., Clarendon Press, Oxford, 1984.
31. J.D. Dunitz and L.E. Orgel, *Adv. Inorg. Chem. Radiochem.*, 1960, **2**, 1.
32. M.J. Taylor, in *Comprehensive Coordination Chemistry*, eds. G. Wilkinson, R.D. Gillard and J.A. McCleverty, Pergamon Press, Oxford, 1987, Vol. 3, p. 105.
33. D.G. Tuck, in *Comprehensive Coordination Chemistry*, eds. G. Wilkinson, R.D. Gillard and J.A. McCleverty, Pergamon Press, Oxford, 1987, Vol. 3, p. 153.
34. G. Davidson, *Coord. Chem. Rev.*, 1979, **30**, 74; 1981, **34**, 74; 1982, **40**, 97; 1983, **49**, 117; 1984, **56**, 113; 1985, **66**, 119; 1986, **75**, 128; 1988, **85**, 121; 1990, **102**, 146.
35. (a) B. Cercek, M. Ebert and A.J. Swallow, *J. Chem. Soc. A*, 1966, 612; (b) M.C.R. Symons and J.K. Yandell, *J. Chem. Soc. A*, 1971, 760; (c) C.E. Burchill and G.G. Hickling, *Can. J. Chem.*, 1970, **48**, 2466; (d) C.E. Burchill and W.H. Wolodarsky, *Can. J. Chem.*, 1970, **48**, 2955; (e) D.R. Stranks and J.K. Yandell, *J. Phys. Chem.*, 1969, **73**, 840; (f) B. Falcinella, P.D. Felgate and G.S. Laurence, *J. Chem. Soc., Dalton Trans.*, 1974, 1367.
36. S.J. Bares, M. Haak and J.W. Nibler, *J. Chem. Phys.*, 1985, **82**, 670.
37. G.A. Olah, O. Farooq, S.M.F. Farnia, M.R. Bruce, F.L. Clouet, P.R. Morton, G.K.S. Prakash, R.C. Stevens, R. Bau, K. Lammertsma, S. Suzer and L. Andrews, *J. Am. Chem. Soc.*, 1988, **110**, 3231.
38. D.G. Tuck, *Polyhedron*, 1990, **9**, 377.

74   CHEMISTRY OF ALUMINIUM, GALLIUM, INDIUM AND THALLIUM

39. M.J. Taylor, *Metal-to-Metal Bonded States of the Main Group Elements*, Academic Press, London, 1975, Chapter 3.
40. The reduced oxomolybdate phase $Tl_{0.8}Sn_{0.6}Mo_7O_{11}$ contains pairs of Tl atoms separated by only 2.84 Å, a distance consistent with the presence of a Tl–Tl bond; there is circumstantial evidence that this unit approximates to Tl_2^{4+}, being stabilised by the oligomeric transition metal clusters. R. Dronskowski and A. Simon, *Angew. Chem., Int. Ed. Engl.*, 1989, **28**, 758.
41. Hg. Schnöckel, *J. Mol. Struct.*, 1978, **50**, 275.
42. R. Köppe, M. Tacke and H. Schnöckel, *Z. Anorg. Allg. Chem.*, 1991, **605**, 35. R. Köppe and H. Schnöckel, *J. Chem. Soc., Dalton Trans.*, 1992, 3393.
43. D.G. Tuck, in *Comprehensive Organometallic Chemistry*, eds. G. Wilkinson, F.G.A. Stone and E.W. Abel, Pergamon Press, Oxford, 1982, Vol. 1, p. 702.
44. I.D. Brown and R. Faggiani, *Acta Crystallogr.*, 1980, **B36**, 1802.
45. J.W. Mellor, *A Comprehensive Treatise on Inorganic and Theoretical Chemistry*, Longman, Green and Co., London, 1924, Vol. V.
46. M.E. Weeks and H.M. Leicester, *Discovery of the Elements*, 7th edn., published by the Journal of Chemical Education, Easton, PA, 1968.
47. D.W. Ball, *J. Chem. Educ.*, 1985, **62**, 787; J.G. Stark and H.G. Wallace, *Educ. Chem.*, 1970, **7**, 152.
48. G.J. Leigh, ed., *Nomenclature of Inorganic Chemistry, Recommendations 1990*, International Union of Pure and Applied Chemistry, Blackwell Scientific, Oxford, 1990.
49. W. Crookes, *Chem. News*, 1861, **3**, 193.
50. A. Lamy, *Compt. Rend.*, 1862, **54**, 1255.
51. J.B.A. Dumas, J. Pelouze and H. St. C. Deville, *Compt. Rend.*, 1862, **55**, 866.
52. F. Reich and T. Richter, *Chem. News*, 1863, **8**, 236, 280; *J. prakt. Chem.*, 1863, **89**, 441; 1863, **90**, 179; 1864, **92**, 480.
53. D.I. Mendeléev, *J. Russ. Phys. Chem. Soc.*, 1869, **1**, 60; 1871, **3**, 47; *Liebig's Ann. Suppl.*, 1872, **8**, 133; *Compt. Rend.*, 1875, **81**, 969.
54. Lecoq de Boisbaudran, *Compt. Rend.*, 1875, **81**, 493, 1100; 1876, **82**, 163, 1036; 1876, **83**, 611, 824, 1044; *Chem. News*, 1877, **35**, 148, 157, 167; *Ann. Chim. Phys.*, 1877, **10**, 100.
55. P.A. Cox, *The Elements: their Origin, Abundance and Distribution*, Oxford University Press, Oxford, 1989; A.E. Ringwood, *Origin of the Earth and Moon*, Springer-Verlag, New York, 1979.
56. The values are normalised to make the abundance ratio of silicon equal to one.
57. H.J.M. Bowen, *Environmental Chemistry of the Elements*, Academic Press, London, 1979.
58. A.G. Lee, *The Chemistry of Thallium*, Elsevier, London, 1971.
59. Hg. Schnöckel and H.J. Göcke, *J. Mol. Struct.*, 1978, **50**, 281.
60. C.R. Pulham, A.J. Downs and I.M. Mills, unpublished results.
61. J.F. Hinton and R.W. Briggs, in *NMR and the Periodic Table*, eds. R.K. Harris and B.E. Mann, Academic Press, London, 1978, p. 279.
62. J.W. Akitt, *Prog. NMR Spectrosc.*, 1989, **21**, 1.
63. C.A. Fyfe, J.M. Thomas, J. Klinowski and G.C. Gobbi, *Angew. Chem., Int. Ed. Engl.*, 1983, **22**, 259; J. Klinowski, *Prog. NMR Spectrosc.*, 1984, **16**, 237; R.J. Kirkpatrick, K.A. Smith, S. Schramm, G. Turner and W.-H. Yang, *Annu. Rev. Earth Planet. Sci.*, 1985, **13**, 29; J.M. Thomas and J. Klinowski, *Adv. Catal.*, 1985, **33**, 199; E. Hallas, B. Schnabel and M. Haehnert, *Silikattechnik*, 1986, **37**, 279; G. Engelhardt, U. Lohse, M. Magi and E. Lippmaa, *Stud. Surf. Sci. Catal.*, 1984, **18**, 23.
64. R. Benn, A. Rufińska, H. Lehmkuhl, E. Janssen and C. Krüger, *Angew. Chem., Int. Ed. Engl.*, 1983, **22**, 779.
65. R. Bertram, W. Gessner, D. Müller, H. Görz and S. Schönherr, *Z. Anorg. Allg. Chem.*, 1985, **525**, 14.
66. J.-P. Laussac, P. Lefrancier, M. Dardenne, J.-F. Bach, M. Marraud and M.-T. Cung, *Inorg. Chem.*, 1988, **27**, 4094.
67. A.R. Barron and G. Wilkinson, *J. Chem. Soc., Dalton Trans.*, 1986, 287.
68. (a) A. Llor and J. Virlet, *Chem. Phys. Lett.*, 1988, **152**, 248; A. Samoson, E. Lippmaa and A. Pines, *Mol. Phys.*, 1988, **65**, 1013; B.F. Chmelka, K.T. Mueller, A. Pines, J. Stebbins, Y. Wu and J.W. Zwanziger, *Nature (London)*, 1989, **339**, 42; K.T. Mueller, B.Q. Sun, G.C. Chingas, J.W. Zwanziger, T. Terao and A. Pines, *J. Magn. Reson.*, 1990, **86**, 470;

(b) Y. Wu, B.F. Chmelka, A. Pines, M.E. Davis, P.J. Grobet and P.A. Jacobs, *Nature (London)*, 1990, **346**, 550; J. Rocha, W. Kolodziejski, H. He and J. Klinowski, *J. Am. Chem. Soc.*, 1992, **114**, 4884; (c) C.A. Fyfe, H. Grondey, K.T. Mueller, K.C. Wong-Moon and T. Markus, *J. Am. Chem. Soc.*, 1992, **114**, 5876.

69. H.D. Morris, S. Bank and P.D. Ellis, *J. Phys. Chem.*, 1990, **94**, 3121 and refs. cited therein.

70. See, for example, J.J. van der Klink, W.S. Veeman and H. Schmid, *J. Phys. Chem.*, 1991, **95**, 1508.

71. (a) L.B. Alemany and G.W. Kirker, *J. Am. Chem. Soc.*, 1986, **108**, 6158; M.C. Cruickshank, L.S. Dent Glasser, S.A.I. Barri and I.J.F. Poplett, *J. Chem. Soc., Chem. Commun.*, 1986, 23; (b) R. Jelinek, B.F. Chmelka, Y. Wu, P.J. Grandinetti, A. Pines, P.J. Barrie and J. Klinowski, *J. Am. Chem. Soc.*, 1991, **113**, 4097.

72. L.F. Nazar, G. Fu and A.D. Bain, *J. Chem. Soc., Chem. Commun.*, 1992, 251.

73. See, for example, J. Rocha and J. Klinowski, *J. Chem. Soc., Chem. Commun.*, 1991, 1121; A. Madani, A. Aznar, J. Sanz and J.M. Serratosa, *J. Phys. Chem.*, 1990, **94**, 760.

74. B.A. Huggins and P.D. Ellis, *J. Am. Chem. Soc.*, 1992, **114**, 2098.

75. (a) J.W. Akitt and D. Kettle, *Magn. Reson. Chem.*, 1989, **27**, 377; (b) S.M. Bradley, R.A. Kydd and R. Yamdagni, *J. Chem. Soc., Dalton Trans.*, 1990, 413, 2653; *Magn. Reson. Chem.*, 1990, **28**, 746; (c) S.M. Bradley, R.A. Kydd and C.A. Fyfe, *Inorg. Chem.*, 1992, **31**, 1181.

76. H. Schmidbaur, T. Zafiropoulos, W. Bublak, P. Burkert and F.H. Köhler, *Z. Naturforsch.*, 1986, **41a**, 315.

77. C. Margheritis, *Z. Naturforsch.*, 1984, **39a**, 1112.

78. J. Blixt, B. Györi and J. Glaser, *J. Am. Chem. Soc.*, 1989, **111**, 7784.

79. J.P. Manners, K.G. Morallee and R.J.P. Williams, *J. Chem. Soc., Chem. Commun.*, 1970, 965; J.J. Dechter and J.I. Zink, *J. Chem. Soc., Chem. Commun.*, 1974, 96.

80. (a) K.A.W. Hill, S.A. Steiner and F.J. Castellino, *J. Biol. Chem.*, 1987, **262**, 7098; (b) D.C. Shangu, J.F. Hinton, R.E. Koeppe, II and F.S. Millett, *Biochemistry*, 1986, **25**, 6103.

81. (a) J. Glaser, U. Henriksson and T. Klason, *Acta Chem. Scand., Ser. A*, 1986, **40**, 344. (b) See, for example, Y.-C. Lee, J. Allison and A.I. Popov, *Polyhedron*, 1985, **4**, 441.

82. W.L. Wilson, R.W. Rudolph, L.L. Lohr, R.C. Taylor and P. Pyykkö, *Inorg. Chem.*, 1986, **25**, 1535.

83. B.N. Figgis, *Trans. Faraday Soc.*, 1959, **55**, 1075; J. Glaser and U. Henriksson, *J. Am. Chem. Soc.*, 1981, **103**, 6642; *Acta Chem. Scand., Ser. A*, 1985, **39**, 355.

84. A.L. Porte, *Annu. Rep. Prog. Chem., Sect. C, Phys. Chem.*, 1983, **80**, 149; K.V.S. Rama Rao and S. Ramaprabhu, *Phys. Stat. Sol.*, 1986, **93a**, 17; J.A.S. Smith, *Chem. Soc. Rev.*, 1986, **15**, 225; J.A.S. Smith, *Z. Naturforsch.*, 1986, **41a**, 453; Yu. A. Buslaev, E.A. Kravčenko and L. Kolditz, *Coord. Chem. Rev.*, 1987, **82**, 3.

85. (a) T. Deeg and A. Weiss, *Ber. Bunsenges. Phys. Chem.*, 1975, **79**, 497; (b) K. Yamada and T. Okuda, *J. Phys. Chem.*, 1985, **89**, 4269; K. Yamada, T. Okuda and S. Ichiba, *Bull. Chem. Soc. Jpn.*, 1987, **60**, 4197; (c) T. Okuda, N. Yoshida, M. Hiura, H. Ishihara, K. Yamada and H. Negita, *J. Mol. Struct.*, 1982, **96**, 169; H. Ishihara, K. Yamada and T. Okuda, *Bull. Chem. Soc. Jpn.*, 1986, **59**, 3969; (d) see, for example, K. Yamada, T. Okuda and H. Negita, *Z. Naturforsch.*, 1986, **41a**, 230; T. Okuda, K. Yamada, H. Ishihara and S. Ichiba, *Z. Naturforsch.*, 1987, **42b**, 835; T. Okuda, M. Sato, H. Hamamoto, H. Ishihara, K. Yamada and S. Ichiba, *Inorg. Chem.*, 1988, **27**, 3656; L.A. Popkova, E.N. Gur'yanova, V.I. Muromtsev and A.P. Zhukov, *J. Gen. Chem. USSR*, 1988, **58**, 2225; 1989, **59**, 25.

86. (a) S. Sengupta, S. Rhadakrishna and R.A. Marino, *Z. Naturforsch.*, 1986, **41a**, 341; (b) W. Zapart, M.B. Zapart, A.P. Zhukov, V.I. Popolitov and L.A. Shuvalov, *Phys. Lett. A*, 1987, **121**, 248.

87. M.J. Almond and A.J. Downs, *Spectroscopy of Matrix Isolated Species*, in *Advances in Spectroscopy*, eds. R.J.H. Clark and R.E. Hester, Wiley, Chichester, UK, 1989, Vol. 17.

88. (a) J.H. Ammeter and D.C. Schlosnagle, *J. Chem. Phys.*, 1973, **59**, 4784; (b) L.B. Knight, Jr., R.L. Martin and E.R. Davidson, *J. Chem. Phys.*, 1979, **71**, 3991; (c) L.B. Knight, Jr., E. Earl, A.R. Ligon, D.P. Cobranchi, J.R. Woodward, J.M. Bostick, E.R. Davidson and D. Feller, *J. Am. Chem. Soc.*, 1986, **108**, 5065.

89. P.H. Kasai and P.M. Jones, *J. Am. Chem. Soc.*, 1984, **106**, 8018; *J. Phys. Chem.*, 1985,

89, 2019; J.H.B. Chenier, C.A. Hampson, J.A. Howard, B. Mile and R. Sutcliffe, *J. Phys. Chem.*, 1986, **90**, 1524; J.A. Howard, R. Sutcliffe, C.A. Hampson and B. Mile, *J. Phys. Chem.*, 1986, **90**, 4268; W.G. Hatton, N.P. Hacker and P.H. Kasai, *J. Phys. Chem.*, 1989, **93**, 1328.

90. (a) J.H.B. Chenier, J.A. Howard and B. Mile, *J. Am. Chem. Soc.*, 1987, **109**, 4109; J.A. Howard, B. Mile, J.S. Tse and H. Morris, *J. Chem. Soc., Faraday Trans. 1*, 1987, **83**, 3701; P.M. Jones and P.H. Kasai, *J. Phys. Chem.*, 1988, **92**, 1060; (b) J.H.B. Chenier, J.A. Howard, J.S. Tse and B. Mile, *J. Am. Chem. Soc.*, 1985, **107**, 7290; (c) J.A. Howard, H.A. Joly, B. Mile and R. Sutcliffe, *J. Phys. Chem.*, 1991, **95**, 6819.

91. W. Kaim and W. Matheis, *J. Chem. Soc., Chem. Commun.*, 1991, 597.

92. J.A. Howard, R. Sutcliffe, J.S. Tse, H. Dahmane and B. Mile, *J. Phys. Chem.*, 1985, **89**, 3595.

93. K.D. McElvany, K.T. Hopkins and M.J. Welch, *Int. J. Appl. Radiat. Isot.*, 1984, **35**, 521; C. Loc'h, B. Maziere and D. Comar, *J. Nucl. Med.*, 1980, **21**, 171; B. Maziere, C. Loc'h and D. Comar, *J. Radioanal. Chem.*, 1983, **76**, 295.

94. H.B. Hupf, S.D. Tischer and F. Al-Watban, *Nucl. Instrum. Methods Phys. Res.*, 1985, **B10/11**, 967; R.M. Lambrecht, M. Sajjad, R.H. Syed and W. Meyer, *Nucl. Instrum. Methods Phys. Res.*, 1989, **A282**, 296.

95. M.J. Welch and S. Moerlein, in *Inorganic Chemistry in Biology and Medicine*, ed. A.E. Martell, ACS Symposium Series 140, American Chemical Society, Washington, DC, 1980, p. 121; M.K. Karimeddini and R.P. Spencer, *Prog. Clin. Cancer*, 1982, **8**, 181; M.A. Green, M.J. Welch, C.J. Mathias, K.A.A. Fox, R.M. Knabb and J.C. Huffman, *J. Nucl. Med.*, 1985, **26**, 170; M.F. Tsan and U. Scheffel, *J. Nucl. Med.*, 1986, **27**, 1215; M.A. Green and M.J. Welch, *Nucl. Med. Biol.*, 1989, **16**, 435; C.J. Mathias, Y. Sun, M.J. Welch, M.A. Green, J.A. Thomas, K.R. Wade and A.E. Martell, *Nucl. Med. Biol.*, 1988, **15**, 69; D.A. Moore, M.J. Welch, K.R. Wade, A.E. Martell and R.J. Motekaitis, *J. Labelled Compd. Radiopharm.*, 1989, **26**, 362; F.C. Hunt, *Nucl. Med. Biol.*, 1988, **15**, 659.

96. P. Robinson, *Eur. Heart J.*, 1984, **5** (Suppl. A), 65; A.S. Iskandrian and A.H. Hakki, *Am. Heart J.*, 1985, **109**, 113; R. Schmoliner and R. Dudczak, *Dev. Nucl. Med.*, 1985, **8**, 225; F.L. Datz, *J. Nucl. Med. Technol.*, 1988, **16**, 119; S. Yasui and M. Meguro, *Pharma Med.*, 1989, **7**, 91.

97. W.B. Frank and L.M. Foster, *J. Phys. Chem.*, 1957, **61**, 1531.

98. F. Guichard, J.-L. Reyss and Y. Yokoyama, *Nature (London)*, 1978, **272**, 155; D. Bourles, G.M. Raisbeck, F. Yiou, J.M. Loiseaux, M. Lieuvin, J. Klein and R. Middleton, *Nucl. Instrum. Methods Phys. Res.*, 1984, **B5**, 365; G.M. Raisbeck and F. Yiou, *Nucl. Instrum. Methods Phys. Res.*, 1984, **B5**, 91.

99. G.F. Herzog and P.J. Cressy, Jr., *Meteoritics*, 1976, **11**, 59.

100. See, for example, N. Prantzos, *AIP Conf. Proc.*, 1991, **232**, 129, and other articles in the same issue.

101. (a) L. Mortelmans, A. Verbruggen, S. Malbrain, M.J. Heynen, C. De Bakker, M. Boogaerts and M. De Roo, *Eur. J. Nucl. Med.*, 1988, **14**, 159; (b) Z. Zhang, S.J. Rettig and C. Orvig, *Inorg. Chem.*, 1991, **30**, 509; (c) R.J. Motekaitis, Y. Sun, A.E. Martell and M.J. Welch, *Inorg. Chem.*, 1991, **30**, 2737; (d) L.C. Francesconi, B.-L. Liu, J.J. Billings, P.J. Carroll, G. Graczyk and H.F. Kung, *J. Chem. Soc., Chem. Commun.*, 1991, 94.

102. C.L. Edwards and R.L. Hayes, *J. Nucl. Med.*, 1969, **10**, 103.

103. See, for example, S.C. Srivastava and G.E. Meinken, in *Radiolabeled Monoclonal Antibodies for Imaging and Therapy*, ed. S.C. Srivastava, *NATO ASI Ser. A*, 1988, **152**, 817.

104. (a) T. Klas, J. Voigt, W. Keppner, R. Wesche and G. Schatz, *Phys. Rev. Lett.*, 1986, **57**, 1068; (b) M. Deicher, G. Gruebel, W. Reiner and T. Wichert, *Mater. Sci. Forum*, 1987, **15–18** (*Vacancies Interstitials Met. Alloys*, Pt. 2), 635; (c) M. Bruessler, H. Metzner, K.D. Husemann and H.J. Lewerenz, *Hyperfine Interact.*, 1990, **60**, 805; (d) U. Feuser, R. Vianden and A.F. Pasquevich, *Hyperfine Interact.*, 1990, **60**, 829; (e) T. Wichert, *Radiat. Eff.*, 1983, **78**, 177; (f) E.G. Sideris, C.A. Kalfas and N. Katsaros, *Inorg. Chim. Acta*, 1986, **123**, 1.

105. (a) S. Winter, S. Blaesser, H. Hofsaess, S. Jahn, G. Lindner, U. Wahl and E. Recknagel, *Nucl. Instrum. Methods Phys. Res.*, 1990, **B48**, 211; (b) R. Otto, P. Hecht and H.G.

Koennecke, *Isotopenpraxis*, 1985, **21**, 49; (c) T.V. Sharova, M.R. Leonov and I.R. Feshchenko, *Zh. Priklad. Khim. (Leningrad)*, 1990, **63**, 1884.

106. (a) H.G. Richter and A.S. Gillespie, Jr., *Anal. Chem.*, 1962, **34**, 1116; (b) R.D. Cannon, *Electron Transfer Reactions*, Butterworth, London, 1980, p. 87; (c) J.E. Hardcastle, T.A. Jordan, I. Alam and L.R. Caswell, *J. Radioanal. Nucl. Chem., Letters*, 1984, **87**, 259.

107. (a) A.E. Mays, Jr. and F.R. Cobb, *J. Clin. Invest.*, 1984, **73**, 1359; P.A. Siffring, N.C. Gupta, S.M. Mohiuddin, D.J. Esterbrooks, D.E. Hilleman, S.C. Cheng, M.H. Sketch and P.F. Mathis, *Radiology (Easton, Pa)*, 1989, **173**, 769; (b) R. Amano, N. Tonami, A. Ando, T. Hiraki and K. Hisada, *Int. J. Appl. Radiat. Isot.*, 1984, **35**, 123; R. Amano, A. Ando, T. Hiraki and N. Tonami, *Radioisotopes*, 1986, **35**, 266.

108. C.E. Moore, *Atomic Energy Levels*, Vols. I–III, Circular of the National Bureau of Standards 467, U.S. Government Printing Office, Washington, DC, 1949–1958; C.E. Moore, *Ionization Potentials and Ionization Limits Derived from the Analyses of Optical Spectra*, Nat. Stand. Ref. Data Ser., National Bureau of Standards, No. 34, U.S. Government Printing Office, Washington, DC, 1970; D.R. Lide, editor-in-chief, *Handbook of Chemistry and Physics*, 73rd edn., CRC Press, Boca Raton, FL, 1992–1993.

109. See, for example, A.E. Reed and F. Weinhold, *J. Am. Chem. Soc.*, 1986, **108**, 3586; W. Kutzelnigg, *Angew. Chem., Int. Ed. Engl.*, 1984, **23**, 272; A.E. Reed and P.v.R. Schleyer, *J. Am. Chem. Soc.*, 1990, **112**, 1434; E. Magnusson, *J. Am. Chem. Soc.*, 1990, **112**, 7940.

110. S.G. Bratsch, *J. Phys. Chem. Ref. Data*, 1989, **18**, 1.

111. A.J. Downs, D. Duckworth, J.C. Machell and C.R. Pulham, *Polyhedron*, 1992, **11**, 1295.

112. (a) G.M. Sheldrick and W.S. Sheldrick, *J. Chem. Soc. A*, 1969, 2279; (b) P.G. Eller, D.C. Bradley, M.B. Hursthouse and D.W. Meek, *Coord. Chem. Rev.*, 1977, **24**, 12; (c) M.J. Zaworotko and J.L. Atwood, *Inorg. Chem.*, 1980, **19**, 268; (d) H.M.M. Shearer, R. Snaith, J.D. Sowerby and K. Wade, *J. Chem. Soc., Chem. Commun.*, 1971, 1275; (e) U. Wannagat, T. Blumenthal, D.J. Brauer and H. Bürger, *J. Organomet. Chem.*, 1983, **249**, 33.

113. (a) D.A. Atwood, A.H. Cowley, R.A. Jones and M.A. Mardones, *J. Am. Chem. Soc.*, 1991, **113**, 7050; (b) H. Hope, D.C. Pestana and P.P. Power, *Angew. Chem., Int. Ed. Engl.*, 1991, **30**, 691.

114. A.H. Cowley and N.C. Norman, *Prog. Inorg. Chem.*, 1986, **34**, 1.

115. T.R. Burkholder and L. Andrews, *J. Chem. Phys.*, 1991, **95**, 8697; L. Andrews, private communication.

116. S.M. Sonchik, L. Andrews and K.D. Carlson, *J. Phys. Chem.*, 1983, **87**, 2004; M.J. Zehe, D.A. Lynch, Jr., B.J. Kelsall and K.D. Carlson, *J. Phys. Chem.*, 1979, **83**, 656; B.J. Kelsall and K.D. Carlson, *J. Phys. Chem.*, 1980, **84**, 951.

117. S.M. Sonchik, L. Andrews and K.D. Carlson, *J. Phys. Chem.*, 1984, **88**, 5269.

118. M.A. Lefcourt and G.A. Ozin, *J. Phys. Chem.*, 1991, **95**, 2616, 2623.

119. J.A. Howard, H.A. Joly, P.P. Edwards, R.J. Singer and D.E. Logan, *J. Am. Chem. Soc.*, 1992, **114**, 474.

120. M.A. Douglas, R.H. Hauge and J.L. Margrave, *J. Chem. Soc., Faraday Trans. 1*, 1983, **79**, 1533.

121. P.H. Kasai, D. McLeod, Jr., and T. Watanabe, *J. Am. Chem. Soc.*, 1977, **99**, 3521.

122. K.J. Klabunde and Y. Tanaka, *J. Am. Chem. Soc.*, 1983, **105**, 3544; J.M. Parnis and G.A. Ozin, *J. Am. Chem. Soc.*, 1986, **108**, 1699.

123. Y. Tanaka, S.C. Davis and K.J. Klabunde, *J. Am. Chem. Soc.*, 1982, **104**, 1013.

124. J.M. Parnis and G.A. Ozin, *J. Phys. Chem.*, 1989, **93**, 1215, 1220; Z.L. Xiao, R.H. Hauge and J.L. Margrave, *Inorg. Chem.*, 1993, **32**, 642.

125. F.G.N. Cloke, G.R. Hanson, M.J. Henderson, P.B. Hitchcock and C.L. Raston, *J. Chem. Soc., Chem. Commun.*, 1989, 1002; F.G.N. Cloke, C.I. Dalby, M.J. Henderson, P.B. Hitchcock, C.H.L. Kennard, R.N. Lamb and C.L. Raston, *J. Chem. Soc., Chem. Commun.*, 1990, 1394.

126. R.S. Drago, *J. Phys. Chem.*, 1958, **62**, 353.

127. K.G.G. Hopkins and P.G. Nelson, *J. Chem. Soc., Dalton Trans.*, 1984, 1393.

128. D.A. Johnson, *Some Thermodynamic Aspects of Inorganic Chemistry*, 2nd edn., Cambridge University Press, Cambridge, 1982.

129. D.R. Lide, editor-in-chief, *Handbook of Chemistry and Physics*, 73rd edn., CRC Press,

Boca Raton, FL, 1992–1993; *Selected Values of Chemical Thermodynamic Properties*, National Bureau of Standards Technical Notes 270-1 to 270-8, U.S. Government Printing Office, Washington, DC, 1965 onwards.

130. T.C. Waddington, *Adv. Inorg. Chem. Radiochem.*, 1959, **1**, 214.
131. R.M. Smith and A.E. Martell, *Critical Stability Constants*, Plenum Press, New York, 1974–1989, Vols. 1–6.
132. S.J.A. Fatemi, F.H.A. Kadir, D.J. Williamson and G.R. Moore, *Adv. Inorg. Chem.*, 1991, **36**, 409; Articles in *Met. Ions Biol. Syst.*, 1988, **24**; D.R. Crapper McLachlan and B.J. Farnell, *Neurol. Neurobiol.*, 1985, **15**, 69; J.M. Candy, J. Klinowski, R.H. Perry, E.K. Perry, A. Fairbairn, A.E. Oakley, T.A. Carpenter, J.R. Atack, G. Blessed and J.A. Edwardson, *Lancet*, 1986, **i**, 354; T.P.A. Kruck, W. Kalow and D.R. Crapper McLachlan, *J. Chromatogr.*, 1985, **341**, 123.
133. G.A. Olah, ed., *Friedel–Crafts and Related Reactions*, Interscience, New York, 1963–1965, Vols. 1–4; G.A. Olah, *Friedel–Crafts Chemistry*, Wiley-Interscience, New York, 1973.
134. G.A. Olah, I. Bucsi, C. Lambert, R. Aniszfeld, N.J. Trivedi, D.K. Sensharma and G.K. Surya Prakash, *J. Am. Chem. Soc.*, 1991, **113**, 9385, 9387.
135. G.A. Olah, O. Farooq, S.M.F. Farnia and J.A. Olah, *J. Am. Chem. Soc.*, 1988, **110**, 2560.
136. G.A. Olah, G.K. Surya Prakash and J. Sommer, *Superacids*, Wiley, New York, 1985, p. 53; E.E. Getty and R.S. Drago, *Inorg. Chem.*, 1990, **29**, 1186.
137. J.E. Huheey, *Inorganic Chemistry*, 3rd edn., Harper International SI Edition, Cambridge, MA, 1983, p. 286.
138. W.B. Jensen, *Chem. Rev.*, 1978, **78**, 1; W.B. Jensen, *The Lewis Acid-Base Concepts: an Overview*, Wiley, New York, 1980.
139. (a) N.N. Greenwood, A. Storr and M.G.H. Wallbridge, *Inorg. Chem.*, 1963, **2**, 1036; (b) C.R. Pulham, L.A. Jones and A.J. Downs, unpublished results.
140. See, for example, R.S. Drago, D.C. Ferris and N. Wong, *J. Am. Chem. Soc.*, 1990, **112**, 8953; T.K. Ghanty and S.K. Ghosh, *Inorg. Chem.*, 1992, **31**, 1951.
141. H.S.P. Müller and H. Willner, *Inorg. Chem.*, 1992, **31**, 2527.
142. A.J. Carty and D.G. Tuck, *Prog. Inorg. Chem.*, 1975, **19**, 243; M.F. Lappert, ed., *MTP International Review of Science, Inorganic Chemistry Series 2*, Butterworth, London, 1975, Vol. 1, pp. 165, 219, 257, 281 and 311.
143. J. Nouet, J. Pannetier and J.L. Fourquet, *Acta Crystallogr.*, 1981, **B37**, 32.
144. (a) M. Uson-Finkenzeller, W. Bublak, B. Huber, G. Müller and H. Schmidbaur, *Z. Naturforsch.*, 1986, **41b**, 346; (b) H. Schmidbaur, W. Bublak, B. Huber and G. Müller, *Organometallics*, 1986, **5**, 1647; (c) J.C.A. Boeyens and F.H. Herbstein, *Inorg. Chem.*, 1967, **6**, 1408.
145. Hg. Schnöckel, *J. Mol. Struct.*, 1978, **50**, 267.
146. C.G. Pitt, K.T. Higa, A.T. McPhail and R.L. Wells, *Inorg. Chem.*, 1986, **25**, 2483.
147. K. Ruhlandt-Senge and P.P. Power, *Inorg. Chem.*, 1991, **30**, 2633.
148. K. Ruhlandt-Senge and P.P. Power, *Inorg. Chem.*, 1991, **30**, 3683.
149. R.J. Gillespie, *Chem. Soc. Rev.*, 1992, **21**, 59.
150. N.A. Compton, R.J. Errington and N.C. Norman, *Adv. Organomet. Chem.*, 1990, **31**, 91.
151. Ch. Hebecker, *Z. Anorg. Allg. Chem.*, 1972, **393**, 223.
152. C.A. McAuliffe, ed., *The Chemistry of Mercury*, Macmillan, London, UK, 1977.
153. F. Brady, K. Henrick, R.W. Matthews and D.G. Gillies, *J. Organomet. Chem.*, 1980, **193**, 21.
154. K. Henrick, R.W. Matthews, B.L. Podejma and P.A. Tasker, *J. Chem. Soc., Chem. Commun.*, 1982, 118.
155. J.L. Atwood, H. Elgamal, G.H. Robinson, S.G. Bott, J.A. Weeks and W.E. Hunter, *J. Inclusion Phenom.*, 1984, **2**, 367.
156. D.C. Bradley, D.M. Frigo, I.S. Harding, M.B. Hursthouse and M. Motevalli, *J. Chem. Soc., Chem. Commun.*, 1992, 577.
157. J.L. Atwood, F.R. Bennett, F.M. Elms, C. Jones, C.L. Raston and K.D. Robinson, *J. Am. Chem. Soc.*, 1991, **113**, 8183.
158. I. Waller, T. Halder, W. Schwarz and J. Weidlein, *J. Organomet. Chem.*, 1982, **232**, 99.

159. K. Henrick, M. McPartlin, G.B. Deacon and R.J. Phillips, *J. Organomet. Chem.*, 1981, **204**, 287.

160. R.H. Jones, J.M. Thomas, R. Xu, Q. Huo, A.K. Cheetham and A.V. Powell, *J. Chem. Soc., Chem. Commun.*, 1991, 1266.

161. A. Roesler and D. Reinen, *Z. Anorg. Allg. Chem.*, 1981, **479**, 119.

162. J.-C. Champarnaud-Mesjard and B. Frit, *Acta Crystallogr.*, 1978, **B34**, 736.

163. (a) W. Uhl, *Z. Naturforsch.*, 1988, **43b**, 1113; (b) W. Uhl, M. Layh and T. Hildenbrand, *J. Organomet. Chem.*, 1989, **364**, 289; (c) W. Uhl, M. Layh and W. Hiller, *J. Organomet. Chem.*, 1989, **368**, 139.

164. K.P. Huber and G. Herzberg, *Molecular Spectra and Molecular Structure. IV. Constants of Diatomic Molecules*, van Nostrand Reinhold, New York, 1979.

165. S. Harvey, M.F. Lappert, C.L. Raston, B.W. Skelton, G. Srivastava and A.H. White, *J. Chem. Soc., Chem. Commun.*, 1988, 1216.

166. B. Krebs and A. Brömmelhaus, *Z. Anorg. Allg. Chem.*, 1991, **595**, 167.

167. D.L. Hughes, *J. Chem. Soc., Dalton Trans.*, 1973, 2209.

168. N.W. Alcock and H.D.B. Jenkins, *J. Chem. Soc., Dalton Trans.*, 1974, 1907.

169. C.P.J.M. van der Vorst and W.J.A. Maaskant, *J. Solid State Chem.*, 1980, **34**, 301.

170. (a) H. Schmidbaur, R. Nowak, W. Bublak, P. Burkert, B. Huber and G. Müller, *Z. Naturforsch.*, 1987, **42b**, 553; W. Hönle, A. Simon and G. Gerlach, *Z. Naturforsch.*, 1987, **42b**, 546; H.P. Beck, *Z. Naturforsch.*, 1984, **39b**, 310; G. Gerlach, W. Hönle and A. Simon, *Z. Anorg. Allg. Chem.*, 1982, **486**, 7; (b) T. Staffel and G. Meyer, *Z. Anorg. Allg. Chem.*, 1987, **552**, 113.

171. J.C. Gallucci, R.E. Gerkin and W.J. Reppart, *Acta Crystallogr.*, 1989, **C45**, 701.

172. (a) B. Krebs and H. Greiwing, *Acta Chem. Scand.*, 1991, **45**, 833; (b) J.G. Bergman and J.S. Wood, *Acta Crystallogr.*, 1987, **C43**, 1831.

173. (a) M. Scholz, M. Noltemeyer and H.W. Roesky, *Angew. Chem., Int. Ed. Engl.*, 1989, **28**, 1383; (b) H.W. Roesky, M. Scholz, M. Noltemeyer and F.T. Edelmann, *Inorg. Chem.*, 1989, **28**, 3829.

174. (a) C. Dohmeier, C. Robl, M. Tacke and H. Schnöckel, *Angew. Chem., Int. Ed. Engl.*, 1991, **30**, 564; (b) O.T. Beachley, Jr., M.R. Churchill, J.C. Fettinger, J.C. Pazik and L. Victoriano, *J. Am. Chem. Soc.*, 1986, **108**, 4666.

175. (a) J.K. Burdett, *Molecular Shapes*, Wiley, New York, 1980; (b) J.K. Burdett and J.-H. Lin, *Acta Crystallogr.*, 1981, **B37**, 2123; (c) J.K. Burdett, *Adv. Chem. Phys.*, 1982, **49**, 47; (d) T.A. Albright, J.K. Burdett and M.-H. Whangbo, *Orbital Interactions in Chemistry*, Wiley, New York, 1985.

176. R.A. Wheeler and P.N.V. Pavan Kumar, *J. Am. Chem. Soc.*, 1992, **114**, 4776.

177. R.F.W. Bader, *Mol. Phys.*, 1960, **3**, 137; *Can. J. Chem.*, 1962, **40**, 1164; L.S. Bartell, *J. Chem. Educ.*, 1968, **45**, 754; R.G. Pearson, *J. Mol. Struct. (THEOCHEM)*, 1983, **103**, 25; R.F.W. Bader and P.J. MacDougall, *J. Am. Chem. Soc.*, 1985, **107**, 6788; P.J. MacDougall, *Inorg. Chem.*, 1986, **25**, 4400.

178. L.S. Bartell and R.M. Gavin, Jr., *J. Chem. Phys.*, 1968, **48**, 2466.

179. K.A. Yee and T.A. Albright, *J. Am. Chem. Soc.*, 1991, **113**, 6474.

180. O. Knop, A. Linden, B.R. Vincent, S.C. Choi, T.S. Cameron and R.J. Boyd, *Can. J. Chem.*, 1989, **67**, 1984.

181. (a) P. Schwerdtfeger, *Inorg. Chem.*, 1991, **30**, 1660; (b) C. Janiak and R. Hoffmann, *J. Am. Chem. Soc.*, 1990, **112**, 5924.

182. P.L. Baxter, A.J. Downs, M.J. Goode, D.W.H. Rankin and H.E. Robertson, *J. Chem. Soc., Dalton Trans.*, 1990, 2873.

183. See, for example, K. Lammertsma and J. Leszczyński, *J. Phys. Chem.*, 1990, **94**, 2806; R.F.W. Bader and D.A. Legare, *Can. J. Chem.*, 1992, **70**, 657.

184. E. Frasson, F. Menegus and C. Panattoni, *Nature (London)*, 1963, **199**, 1087.

185. V.A. Maroni and T.G. Spiro, *Inorg. Chem.*, 1968, **7**, 193.

186. J.K. Burdett, *J. Am. Chem. Soc.*, 1980, **102**, 450.

187. (a) G. Wagner and R. Hoppe, *J. Less-Common Met.*, 1986, **116**, 129; (b) Hj. Mattausch, A. Simon and E.-M. Peters, *Inorg. Chem.*, 1986, **25**, 3428.

188. J.H.C. Hogg, H.H. Sutherland and D.J. Williams, *Acta Crystallogr.*, 1973, **B29**, 1590.

189. J.D. Corbett, *Chem. Rev.*, 1985, **85**, 383.

190. H. Schäfer and B. Eisenmann, *Rev. Inorg. Chem.*, 1981, **3**, 29; H. Schäfer, *J. Solid State Chem.*, 1985, **57**, 97.

191. M.L. Fornasini, *Acta Crystallogr.*, 1983, **C39**, 943.
192. (a) C. Belin, *Acta Crystallogr.*, 1980, **B36**, 1339; (b) R.G. Ling and C. Belin, *Z. Anorg. Allg. Chem.*, 1981, **480**, 181; (c) C. Belin and R.G. Ling, *C. R. Acad. Sci., Ser. II*, 1982, **294**, 1083; (d) C. Belin, *Acta Crystallogr.*, 1981, **B37**, 2060; (e) U. Frank-Cordier, G. Cordier and H. Schäfer, *Z. Naturforsch.*, 1982, **37b**, 119, 127; (f) R.G. Ling and C. Belin, *Acta Crystallogr.*, 1982, **B38**, 1101.
193. (a) M. Charbonnel and C. Belin, *Nouv. J. Chim.*, 1984, **8**, 595; (b) C. Belin and M. Charbonnel, *J. Solid State Chem.*, 1986, **64**, 57; (c) C. Belin, *J. Solid State Chem.*, 1983, **50**, 225; (d) M. Charbonnel and C. Belin, *J. Solid State Chem.*, 1987, **67**, 210.
194. C. Belin and R.G. Ling, *J. Solid State Chem.*, 1983, **48**, 40.
195. J.K. Burdett and E. Canadell, *J. Am. Chem. Soc.*, 1990, **112**, 7207; *Inorg. Chem.*, 1991, **30**, 1991.
196. S.C. Sevov and J.D. Corbett, *Inorg. Chem.*, 1991, **30**, 4875.
197. S.C. Sevov and J.D. Corbett, *Inorg. Chem.*, 1992, **31**, 1895.
198. D.A. Hansen and J.F. Smith, *Acta Crystallogr.*, 1967, **22**, 836.
199. (a) R.C. Burns and J.D. Corbett, *J. Am. Chem. Soc.*, 1982, **104**, 2804; (b) R.C. Burns and J.D. Corbett, *J. Am. Chem. Soc.*, 1981, **103**, 2627.
200. S.M. Kauzlarich, M.M. Thomas, D.A. Odink and M.M. Olmstead, *J. Am. Chem. Soc.*, 1991, **113**, 7205; R.F. Gallup, C.Y. Fong and S.M. Kauzlarich, *Inorg. Chem.*, 1992, **31**, 115.
201. I. Sinclair and I.J. Worrall, *Inorg. Nucl. Chem. Lett.*, 1981, **17**, 279.
202. A.T. Weibel and J.P. Oliver, *J. Am. Chem. Soc.*, 1972, **94**, 8590; *J. Organomet. Chem.*, 1973, **57**, 313.
203. A.H. Cowley and R.A. Jones, *Angew. Chem., Int. Ed. Engl.*, 1989, **28**, 1208.
204. R.L. Wells, A.P. Purdy, A.T. McPhail and C.G. Pitt, *J. Chem. Soc., Chem. Commun.*, 1986, 487.
205. K.H. Whitmire, *J. Coord. Chem.*, 1988, **17**, 95.

2 The elements

I.J. POLMEAR

2.1 Characteristics and physical properties

Aluminium, gallium, indium and thallium all have a silvery lustre and are comparatively soft metals but these are virtually their only common features. Aluminium is the most abundant metal in the earth's crust whereas the other three, although not rare elements, occur only in fractions of one part per million (ppm).[1] Aluminium is widely distributed in clays and shales, but is also present in high concentrations in bauxite minerals. The other three are only thinly dispersed, usually as oxides or sulfides, and are recovered as by-products during the extraction and refining of other metals such as zinc. Current annual world production of aluminium is around 18 million tonnes which makes it the second most used metal after iron. On the other hand, total production of the other three metals, combined, probably does not exceed 200 tonnes. Aluminium is the base for a wide range of industrial cast and wrought alloys. This contrasts with gallium which has only one important application, namely as a source of semiconductor materials, and the other metals which find some use as alloying additions in a limited range of speciality products.

Physical properties of the four elements are compared in Table 2.1. Particular properties that account for the ubiquitous use of aluminium and its alloys are as follows.

1. *Light weight.* Aluminium (density $2.7\,\mathrm{g\,cm^{-3}}$) is second only to magnesium as a lightweight structural metal.
2. *Electrical conductivity.* The conductivity of electrical conductor (EC) grades of aluminium ($>99.6\%$ Al) average about 62% that of the International Annealed Copper Standard (IACS) and, because of its lower density, aluminium will conduct twice as much electricity as an equivalent weight of copper.
3. *Thermal conductivity.* The thermal conductivity of aluminium is 56% that of copper and the second highest of the commonly used metals.
4. *Crystal structure.* Aluminium has a face centred cubic structure which facilitates plastic deformation and enables the metal and most of its alloys to be readily fabricated into plate, rod, sheet, foil, etc.

In addition, aluminium and its alloys are non-toxic, or so it is generally held, and are covered by a tenacious oxide film that protects against corrosion in many environments. The metal and its alloys melt at relatively low temperatures compared, for example, with steel and copper alloys and many compositions display good casting properties.

Table 2.1 Physical properties of the elements.[2-4,a]

Property	Aluminium	Gallium	Indium	Thallium	
Atomic number	13	31	49	81	
Atomic volume (cm^3 mol^{-1})	9.995	11.8	15.71	17.25	
Relative atomic mass (^{12}C = 12.0000)	26.9815	69.72	114.82	204.38	
Boiling point (°C)	2520	2420	2070	1473 ±10	
Melting point (°C)	660.10	29.78	156.4	303.9	
Crystal structure	f.c.c.	α-Form, orthrhombic[a]	Tetragonal	<230°C α, c.p.h.	>230°C β, b.c.c.
a (Å)	4.0496	4.519	3.253	3.456	3.882
b (Å)		7.657			
c (Å)		4.526		5.525	
Closest distance of approach of atoms (Å)	2.862	2.449	4.9455	3.408	
Density at 20°C (g cm^{-3})	2.698	5.907	3.253	11.85	
Elastic modulus at 20°C (GPa)	69	9.8	7.31	7.9	
Electrical resistivity at 20°C (nΩ m)	26.5	174	10.6	180 (0°C)	
Enthalpy at 25°C (J mol^{-1})	4540	5569	83.7	6828	
Entropy at 25°C (J K^{-1} mol^{-1})	28.30	40.88	6602	64.18	
Heat of fusion (kJ kg^{-1})	395.4	80.17	57.82	21.1	
Heat of evaporation at b.p. (kJ kg^{-1})	10780	3880	28.5	813.2	
Specific heat at 25°C (J K^{-1} kg^{-1})	903	371	2024	129	
Electrochemical equivalent (valence 3) (μg)	93.2	241	233	706	
Magnetic susceptibility (volumetric) (m.k.s.)	7.88×10^{-9}	2.71×10^{-4}	396	-3.1×10^{-6}	
Thermal conductivity at 0°C (W m^{-1} K^{-1})	237	33.5 (29.8°C)	7.0×10^{-6}	46.1	
Thermal expansion, linear coefficient at 20°C ($\times 10^6$ K^{-1})	23.6	18.0	81.6	28.0	
Volume change on freezing (%)	−6.5	+3.2	24.8	−3.2	

(a) See also Table 1.1.

Apart from having properties quite distinct from aluminium, the other three elements differ notably one from another. Gallium shares with mercury and rubidium the property of being liquid at ambient temperatures and has one of the largest differences (2390°C) between the melting and boiling temperatures to be displayed by any element. Gallium also, rather unusually, expands rather than contracts on freezing. Each of the three elements has a different crystal structure and gallium and thallium have several allotropic forms, the α and β forms of thallium transforming at 230°C. Densities vary from 5.9 to 11.85 $g\,cm^{-3}$.

Natural aluminium consists entirely of ^{27}Al whereas each of the other three elements has two naturally occurring isotopes.[5] Gallium exists as ^{69}Ga (60.1%) and ^{71}Ga (39.9%), indium as ^{113}In (4.33%) and ^{115}In (95.67%) and thallium as ^{203}Tl (29.52%) and ^{205}Tl (70.48%). All four elements have a number of radioactive nuclides most of which have very short half-lives (see chapter 1). Exceptions are ^{26}Al (7.1×10^5 years) and ^{204}Tl (3.78 years).

2.2 Extraction and refining of aluminium

As mentioned earlier, aluminium is the most abundant metal in the earth's crust (8%) and only oxygen and silicon are present in larger amounts. Because of its high chemical reactivity, aluminium is never found in nature as the metal; it always occurs in its oxidized form and most commonly as aluminates and aluminosilicates in clays and shales. Within these materials, it occurs as the free oxide alumina (Al_2O_3) combined with water or as derivatives of this oxide.

The most important and virtually the only current commercial source of aluminium is bauxite, which is the generic name given to minerals that usually contain 40–60% hydrated alumina together with compounds such as iron oxides, silica and titania, as well as a range of minor impurities including gallium. The name originates from the district Les Baux in Provence, France, where the mineral was first mined in the first half of the 19th century. Bauxite is formed by the surface leaching of aluminium-bearing rocks such as granite and basalt, usually under tropical conditions and the largest known reserves are in Australia, Jamaica, Guyana and Brazil where the deposits are amenable to open-cut, strip mining.

Production of aluminium from bauxite involves two distinct processes which are often operated at quite different locations. First it is necessary to separate the alumina from the bauxite, which is carried out almost exclusively by the Bayer process, after which the metal is extracted by electrolysis of the oxide once it has been dissolved in molten cryolite (Na_3AlF_6).

2.2.1   Bayer process for alumina production[6,7]

The Bayer process was developed and patented by Karl Josef Bayer in Austria in 1888 and essentially involves digesting crushed bauxite in strong sodium hydroxide solutions at temperatures up to 240°C. Most of the alumina is dissolved leaving an insoluble residue known as 'red mud' which mainly comprises iron oxides and silica and is removed by filtration. The particular concentration of sodium hydroxide as well as the temperature and pressure of the operation are optimised according to the nature of the bauxite ore, notably the respective proportions of the different forms of alumina (α, β or γ). This first stage of the Bayer process can be expressed by the equation

$$Al_2O_3 \cdot xH_2O + 2NaOH \rightarrow 2NaAlO_2 + (x+1)H_2O \qquad (2.1)$$

Subsequently, in the second stage, conditions are adjusted so that the reaction is reversed. This is referred to as the decomposition stage:

$$2NaAlO_2 + 2H_2O \rightarrow 2NaOH + Al_2O_3 \cdot 3H_2O \qquad (2.2)$$

The reaction is achieved by cooling the liquor and seeding with crystals of the trihydrate, $Al_2O_3 \cdot 3H_2O$, to promote precipitation of this compound as fine particles rather than in a gelatinous form. Decomposition is commonly carried out at around 50°C in slowly stirred vessels and may require up to 30 h to complete. The $Al_2O_3 \cdot 3H_2O$ is removed and washed and the sodium hydroxide liquor is recycled back to the digestors.

Alumina is then produced by calcining the trihydrate in rotary kilns or, more recently, fluidised beds. Calcination occurs in two stages with most of the water of crystallisation being removed in the temperature range 400–600°C. This produces alumina in the more chemically active γ form which further heating to temperatures as high as 1200°C converts partly or completely to relatively inert α-alumina. Each form has different physical characteristics and individual aluminium smelters may specify differing mixtures of α- and γ-alumina. Currently, Australia produces some 35% of the world's alumina and other major suppliers are Jamaica, Guyana and Brazil.

Invariably alumina produced by the Bayer process has small amounts of impurities such as SiO_2 and Fe_2O_3 but these seldom present problems in subsequent refining. Another impurity is gallium which is present in all bauxites at levels up to 0.01 wt% and is mostly dissolved during digestion. An average large refinery would have a 'throughput' of as much as 500 tonnes of gallium annually. Although the concentration of gallium is very low (e.g. $0.5 \, g \, l^{-1}$), this is the richest commercial source of gallium, as discussed later.

2.2.2   Alternative sources of alumina

The amphoteric nature of aluminium provides the opportunity to use acid as well as alkaline processes to recover alumina. As one example, some attention has been given to acid extraction of alumina from kaolinite which is both a widely distributed clay mineral and a major constituent of the ash in coal. The mineral alunite, $K_2SO_4 \cdot Al_2(SO_4)_3 \cdot 4Al(OH)_3$, is processed commercially in the CIS in plants located in regions remote from sources of bauxite. However, alumina produced from these or other alternative sources is 1.5–2.5 times more costly than that from the Bayer process.

2.2.3 Production of aluminium by the Hall–Héroult process[6–10]

The first commercial production of aluminium occurred in France in 1855 when H. Sainte-Claire Deville reduced aluminium chloride with sodium. As is so often the case, the prospect of military applications of this new metal led to government support because Napoleon the Third foresaw its potential for use as lightweight body armour. Comparatively little aluminium was produced, however, until independent discoveries in 1886 by Hall in the United States and Héroult in France led to the development of an economic method for its electrolytic extraction. Hall–Héroult technology remains the basis for production today and the emergence of aluminium as a commercial metal has relied on the availability of bulk quantities of relatively cheap electrical power. The cost of aluminium which had been more than US$1000 a kilogram in 1855 was reduced to less than $US1 in 1888. In recent times it has varied between $US1 and $US2 per kilogram.

Alumina has a high melting point (2040°C) and is a poor conductor of electricity. The key to the successful production of aluminium lies in dissolving the oxide in molten cryolite (Na_3AlF_6) and a typical electrolyte contains 80–90% of this compound and 2–8% of alumina, together with additives such as AlF_3 and CaF_2. Cryolite was first obtained from relatively inaccessible sources in Greenland but is now made synthetically.

An electrolytic reduction cell (known as a pot) consists essentially of baked carbon anodes, that are consumed and require regular replacement, the molten cryolite-alumina electrolyte, a pool of liquid aluminium, a carbon-lined container to hold the metal and electrolyte, and a gas collection system to prevent fumes from the cell escaping into the atmosphere (Figure 2.1). There are also alumina feeders that are operated intermittently under some form of automatic control. A typical modern cell is operated at around 950°C and takes 250 kA at an anode current density around $0.7\,A\,cm^{-2}$. The anode and cathode are separated by 4–5 cm and there is a voltage drop of approximately 4.5 V across each cell. The cell is operated so that the carbon side-linings are protected with a layer of frozen cryolite and the upper surface of the bath is covered with a crust of alumina. Some 1800 kg of aluminium is produced daily, the metal being syphoned out regularly and cast into ingots, and alumina is replenished as required. About 150 cells are connected in series to make up a potline (Figure 2.2).

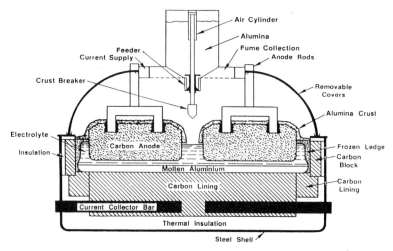

Figure 2.1 Hall–Héroult electrolytic cell for producing aluminium.[6]

Figure 2.2 Potline of electrolytic cells for producing aluminium (courtesy Comalco Limited).

The exact mechanism for the electrolytic reaction in the cell is uncertain but it is probable that the current-carrying ions are Na^+, AlF_4^-, AlF_6^{3-} and one or more ternary complex ions such as $AlOF_3^{2-}$. At the cathode it is considered that the fluroaluminate anions are discharged via a charge transfer at the cathode interface to produce aluminium metal and F^- ions while, at the anode, the oxofluoroaluminate ions dissociate to liberate oxygen which forms CO_2. The overall reaction can be written as follows:

$$2Al_2O_3 + 3C \rightarrow 4Al + 3CO_2 \qquad (2.3)$$

Commonly some 3.5–4 tonnes of bauxite are needed to produce 2 tonnes of alumina which, in turn, yield 1 tonne of aluminium. Significant quantities of other materials, such as 0.4 tonne of carbon, are also consumed. However, the most critical factor is the consumption of electricity which, despite continual refinements to the process, still amounts to 13 000 to 15 000 kWh for each tonne of aluminium produced from alumina. These values compare with the theoretical thermodynamic requirement of approximately 6500 kWh per tonne.

Of the total voltage drop of 4.5 V across a modern cell, only 1.2 V represents the decomposition potential or free energy of the reaction associated with the formation of molten aluminium at the cathode. The largest component of the voltage drop arises from the electrical resistance of the electrolyte in the space between the electrodes and this amounts to around 1.7 V, or 35–40% of the total. Efficiency can be increased if the anode–cathode distance is reduced and this aspect has been the focus of recent changes in cell design. One modification that shows promise is to coat the cathode with titanium diboride which has the property of being readily wetted by molten aluminium.[10] This results in the formation of a thinner, more stable film of aluminium that can be drained away into a central sump if a sloped cathode is used. Reductions in spacing from the normal 4–6 cm down to 1–2 cm have been claimed permitting a decrease of 1–1.5 V in cell voltage. Overall predictions on cell performance by the year 2000 suggest that electrical consumption will be reduced to an average of 12 500 kWh per tonne of aluminium produced.

2.2.4 Alternative methods for producing aluminium

Because of the large disparity between the theoretical and actual requirements for electrical energy to produce each tonne of aluminium, it is to be expected that alternative methods of production would have been investigated. One example has been a chloride-based smelting process developed by Alcoa[11] which commenced operation in the United States in 1976 with an initial capacity of 13 500 tonnes of aluminium per year and the potential to achieve a 30% cost saving. This process also used alumina as a starting material which was combined with chlorine in a reactor to produce $AlCl_3$.

This chloride served as an electrolyte in a closed cell to produce aluminium and chlorine, the latter being recycled back into the reactor. The process had the advantage of being continuous but the provision of materials of construction that could resist attack by chlorine over long periods of time proved to be difficult. This factor, together with improvements in efficiencies in conventional electrolysis, led to the process being discontinued in 1985.

Several companies have been investigating carbothermic methods for producing aluminium. One process involves mixing aluminium ore with coking coal to form briquettes that are then reduced in stages in a type of blast furnace operating at temperatures ranging from 500 to 2100°C. The molten metal product comprises aluminium combined with iron and silicon which is scrubbed and absorbed by a spray of molten lead at the bottom of the furnace. Since aluminium and lead are immiscible, the lighter aluminium rises to the surface where it can be skimmed off. Further purification of the aluminium is required. Although overall cost savings have been predicted, no commercially viable process has so far eventuated.

2.3 Production of gallium, indium and thallium

2.3.1 Production of gallium

Gallium exists in nature to roughly the same extent as lead (5–15 ppm) but it is widely distributed and present only in low concentrations. It tends to be associated with aluminium and is thus present in bauxite and most aluminosilicate minerals, especially clays, but again at levels of only a few parts per million. Gallium is also present in some zinc ores such as sphalerite and is commonly found in coal fly ash and phosphate flue gases.[4,12–15] The richest single source is the rare mineral germanite which is a complex zinc-copper-arsenic-germanium sulfide containing 0.1–0.8% gallium.

The major commercial source of gallium is the Bayer process for the extraction of alumina from bauxite (see section 2.2.1). Gallium collects in the sodium aluminate liquor and, because this liquor is recycled, the gallium content increases until it reaches an equilibrium value of 100–125 ppm. When gallium is required, a batch of liquor is taken and a concentrate of sodium gallate is prepared. Formerly this was carried out in one of two ways.

1. The Beja process in which the liquor is treated with CO_2 in several controlled stages to remove most of the Al_2O_3 and Na_2CO_3 and the remaining gallium concentrate is then dissolved in sodium hydroxide.
2. The De la Bretèque process in which the liquor is electrolysed directly using a continuously agitated mercury or sodium amalgam cathode into which the gallium is absorbed. The gallium is then leached from the cathode using hot sodium hydroxide.

Both these processes are becoming obsolete and are being replaced by solvent extraction in which the gallium is removed from the liquor by reaction with organic chelating agents.

An impure form of gallium is extracted from the sodium gallate solutions by electrolysis at 40–60°C using stainless steel electrodes. Gallium is deposited as a liquid at the bottom of the vessel and, after washing in hydrochloric or nitric acids, it can be filtered through porous ceramic or glass plates to produce 99.9–99.99% pure metal. Many applications of gallium, particularly in electronics, require much higher purity (e.g. >99.9999%). This is achieved by operations such as heating under vacuum to remove volatile elements, electrochemical purification (anodic dissolution and cathodic deposition) or fractional crystallisation.

Gallium is also occasionally recovered as a by-product during zinc production from sphalerite ore. In this case, the sequence of operations is as follows.

1. Zinc sulfate solution obtained from leaching the ore in sulfuric acid is purified by neutralising the excess acid; this precipitates out an 'iron mud' that may contain up to 0.07% gallium.
2. The iron mud is leached with sodium hydroxide to dissolve both gallium and the associated aluminium after which the hydroxides are precipitated by neutralising with acid.
3. These hydroxides are leached with hydrochloric acid to obtain a solution containing gallium and some aluminium.
4. Gallium trichloride is extracted with ether, which is then distilled off, and the gallium trichloride is dissolved in concentrated sodium hydroxide.
5. Gallium metal is electrodeposited from the sodium gallate solution and purified by one of the methods described above.

The cost of gallium depends on purity and may be as much as US$3000 per kg for 99.99999 grade.[3] Annual world production is difficult to estimate but may be in the range 60–80 tonnes.

2.3.2 Production of indium[4,14,15]

Indium is estimated to have an abundance in the earth's crust of 0.05 ppm, which is similar to that of silver, but again it is widely dispersed in concentrations of 0.001% or less. As with other comparatively rare metals, indium becomes more concentrated in by-products arising during the extraction of major metals.

Indium is most commonly associated with zinc in minerals such as sphalerite and commercial production arises almost exclusively from metal residues, slags and flue dusts formed in the smelting of zinc ores. Recovery of indium can be achieved by several methods.

1. Leaching of a suitable by-product in sulfuric or hydrochloric acid and purification of the leach solution by using indium strips to give a sponge of crude indium on zinc or aluminium plates. For solutions with low concentrations of indium, solvent extraction with *bis*(2-ethylhexyl)-phosphoric acid or tributyl phosphate can be employed.
2. Indium phosphate can be precipitated from slightly acidic solutions. The phosphate can then be converted to oxide by further leaching in sodium hydroxide and the oxide can be readily reduced to indium metal.
3. In zinc retort smelting, indium is distilled together with the zinc and concentrates in the molten zinc-lead metal at the bottom during the first stage evaporation and reflux purification of the zinc. The indium can be separated as a high-grade slag from which it can be recovered by leaching and sponging on zinc or aluminium plates, as mentioned above.

Sponge indium is usually from 99.0 to 99.5% pure and requires upgrading for most uses, notably for those in the semiconductor industry. Refining techniques involve soluble-anode electrolysis supplemented by other techniques if required. World production was estimated to be close to 100 tonnes per annum in 1989 when the cost was approximately US$250 per kg.[2]

2.3.3 Production of thallium[11-13]

Thallium is not particularly rare, with an estimated concentration in the earth's crust of 0.4 ppm. It occurs as chemical compounds in four rare minerals of which the most common are crookesite, $(CuTlAg)_2Se$, which is found mainly in Sweden, and lorandite, $Tl_2S \cdot 2As_2S_3$, which occurs in Greece and in Wyoming in the United States. The other two are urbaite, $TlAs_2SbS_5$, and hutchinsonite, $(Tl, Cu, As)_2S \cdot PbS \cdot 2As_2S_3$. However, thallium is recovered commercially as a by-product in the roasting of zinc blende (ZnS) and lead sulfide ores used for the production of sulfuric acid. The thallium collects in flue dust, usually in the form of oxide, Tl_2O, or sulfate, Tl_2SO_4, together with other volatile by-products including elements like cadmium, tellurium, selenium, indium and germanium. The thallium content is low and further enrichment is necessary. If the thallium compounds are soluble, i.e. oxides or sulfates, direct leaching with hot water or dilute acid will separate them from other insoluble metal compounds. If not, then the thallium compounds are made more soluble by oxidising roasts, sulfatising, or by treatment with alkali. The dissolved thallium sulfate is further purified by repeated precipitation as thallium(I) chloride and finally dissolved in sulfuric acid. Thallium is then extracted by electrolysis using platinum or stainless steel electrodes.

Thallium is available in several grades of purity, the price for 99.9% pure metal being around US$80 per kg in recent years. Annual production is estimated to be between 5 and 15 tonnes.[3]

2.4 Alloying of aluminium

2.4.1 Alloying elements

Although most metals will alloy with aluminium, comparatively few have sufficient solid solubility to serve as major additions (Table 2.2).[17,18] Only zinc, magnesium, lithium and silver (all greater than 10 at.%) and copper and silicon have significant solubilities. Of these elements, lithium, which is reactive and presents problems during melting and casting, is a relatively new addition to a class of lower density, higher stiffness aerospace alloys. Silver is too expensive to use in quantity but does induce interesting microstructural changes in certain alloys if present as a minor addition (e.g. 0.1 at.%).[19] There are, however, several other elements with solubilities below 1 at.% which can confer significant improvements to properties. Examples are the transition metals chromium, manganese and zirconium which form intermetallic compounds such as Al_3Zr and Al_6Mn that control grain size and provide some dispersion hardening. Iron and silicon are normally present as impurities and form intermetallic compounds such as Al_3Fe and $Al(Fe,Mn,Si)$. With the exception of hydrogen which can cause

Table 2.2 Solid solubility of elements in aluminium.

Element	Temperature (°C)	Maximum solid solubility[a,b]	
		wt%	at.%
Cadmium	649	0.4	0.09
Cobalt	657	<0.02	<0.01
Copper	548	5.65	2.48
Chromium	661	0.77	0.40
Germanium	424	7.2	2.7
Iron	655	0.05	0.025
Lithium	600	4.2	14.5
Magnesium	450	17.4	18.5
Manganese	658	1.82	0.90
Nickel	640	0.04	0.02
Silicon	577	1.65	1.59
Silver	566	55.6	23.8
Tin	228	~0.06	~0.01
Titanium	665	~1.3	~0.74
Vanadium	661	~0.4	~0.21
Zinc	443	83.1	67.0
Zirconium	660.5	0.28	0.08

(a) Maximum solid solubility occurs at eutectic temperatures for all elements except chromium, titanium, vanadium, zinc and zirconium for which it occurs at peritectic temperatures.

(b) Solid solubility at 20°C is estimated to be approximately 2 wt% for magnesium and zinc, 0.1 to 0.2 wt% for germanium, lithium and silver and below 0.1% for all other elements.

serious porosity in castings, elemental gases have no detectable solubility in either liquid or solid aluminium.

2.4.2 Classification of aluminium alloys[17,20]

As a world average, some 80% of aluminium is converted into wrought products and the remaining 20% is used for castings although proportions differ from country to country. Wrought products include rolled plate, sheet and foil which commonly make up 50% of wrought materials, extrusions (bar, rod and tube), forgings, wire and a limited range of products made from powders. Most are fabricated from direct chill (DC) semi-continuously cast ingots such as the rolling slab shown in Figure 2.3. Cast products are produced mainly by pressure die casting (usually around 70%), permanent mould and sand casting.

The nomenclature used to identify aluminium alloys varies in different countries. However, for wrought alloys most have accepted the four digit International Alloy Designation System (IADS) which is shown in Table 2.3. The first digit indicates the alloy group. In each case the principal alloying element is shown but compositions usually contain other elements. For example, 2xxx series alloys commonly have present magnesium and

Figure 2.3 Direct chill (DC) cast aluminium alloy ingots for rolling into plate, sheet or foil (courtesy Australian Aluminium Development Council).

Table 2.3 Aluminium alloy designation systems.

Alloy type	Four-digit designation
Wrought alloys	
99.00% (min.) aluminium	1xxx
Copper	2xxx
Manganese	3xxx
Silicon	4xxx
Magnesium	5xxx
Magnesium and silicon	6xxx
Zinc	7xxx
Others	8xxx
Casting alloys	Three-digit designation
99.00% (min.) aluminium	1$xx.x$
Copper	2$xx.x$
Silicon with added copper and/or magnesium	3$xx.x$
Silicon	4$xx.x$
Magnesium	5$xx.x$
Zinc	7$xx.x$
Tin	8$xx.x$
Others	9$xx.x$

manganese in addition to copper, whereas the 7xxx series alloys invariably contain magnesium and often copper in addition to zinc. The last two digits signify individual alloys, except for the 1xxx series in which they indicate purity of the aluminium. Finally, the second digit indicates modifications to the original composition. Principal applications of the various classes of alloys are included in Table 2.4.

No internationally accepted method has been adopted for identifying aluminium casting alloys. However, the Aluminum Association of the United States uses a system which has some similarity to that employed for wrought alloys and this is included in Table 2.3. The first digit again signifies the alloy group but the second and third digits now identify the aluminium alloy (or purity for the 1$xx.x$ series). The last digit which is preceded by a decimal point indicates product form (e.g. 0 for casting, 1 for ingot).

Aluminium alloys can be divided into non-heat treatable and heat treatable series, depending on whether or not they respond to precipitation hardening (see section 2.4.4).

2.4.3 Non-heat treatable alloys

Wrought compositions that do not respond to strengthening by heat treatment mainly comprise the various grades of commercial purity aluminium (1xxx series) as well as those alloys with manganese (3xxx) or magnesium (5xxx) as the major additions. Approximately 95% of all aluminium alloy rolled products are made from these groups. Mechanical

Table 2.4 Nominal compositions[a] and applications of selected wrought aluminium alloys.

IADS designation	Cu	Mg	Zn	Mn	Cr	Si	Other	Typical applications
Non-heat treatable alloys								
1145							Al 99.45 (min)	Sheet, foil, packaging
1199							Al 99.99 (min)	Electrical and electronic foil
3003				1.2				Sheet, plate, beverage containers,
3004		1.0		1.2				cooking utensils
5005		0.8						Sheet, electrical conductor wire
5083		4.5		0.7		0.15		Marine, pressure vessels, truck bodies
Heat treatable alloys								
2014	4.4	0.5		0.8		0.8		Aircraft structures
2024	4.4	1.5		0.6				Aircraft structures
2618	2.2	1.5				0.18	1.1 Fe, 1.1 Ni	Aircraft structures
6061	0.3	1.0		0.2		0.6		Marine, welded structures
6063		0.7				0.4		Sheet, pipes, architectural sections
7005		1.4	4.5		0.13		0.14 Zr	Welded structures
7075	1.6	2.5	5.6		0.23			Aircraft structures
7050	2.3	2.3	6.2				0.12 Zr	Aircraft structures

(a) Compositions in wt%.

strength is achieved primarily through combinations of:

1. solution hardening by dissolved elements such as magnesium;
2. dispersion hardening by relatively insoluble particles of intermetallic phases which aluminium readily forms with transition metals such as iron, manganese, chromium and zirconium; and
3. work hardening due to dislocation multiplication and interaction during fabrication. Examples of alloy compositions, tensile properties and applications are shown in Table 2.4.

Silicon is the principal addition to most aluminium-base casting alloys because of the high fluidity of the Al–Si eutectic. Hard silicon particles form as one component of the binary eutectic and they contribute to strengthening of non-heat treatable casting alloys (Table 2.5). However, these silicon particles may be coarse and so adversely affect ductility and toughness. Refinement may be achieved by a process called modification which involves the addition of small amounts of sodium (e.g. 0.01 wt%) or strontium (e.g. 0.04 wt%) to the melt.[21] Some Al–Mg casting alloys are also not amenable to heat treatment (e.g. 514.0).

2.4.4 Heat treatable alloys

As mentioned earlier, heat treatable aluminium alloys are those that undergo precipitation (age) hardening. This process may lead to the formation of one or more finely dispersed precipitates (e.g. Figure 2.4) which cause hardening by impeding dislocation motion and delaying the onset of plastic deformation. For an alloy to be amenable to age hardening, the solubility of one or more of the alloying elements must decrease with decreasing temperature in the manner shown in the hypothetical phase diagram in Figure 2.5. Heat treatment normally involves the following stages:

1. solution treatment at a relatively high temperature within the single phase region, e.g. T_s in Figure 2.5, to dissolve the alloying elements;
2. rapid cooling or quenching of the alloy (e.g. composition C), usually to room temperature, to obtain a supersaturated solid solution (SSSS) of these elements in aluminium; and
3. controlled decomposition of the SSSS to form finely dispersed precipitates by ageing for convenient times at one or, sometimes, two intermediate temperatures (T_A).

The complete decomposition of the SSSS is usually a complex process that may involve several metastable stages and the ageing conditions are usually chosen to obtain a critical dispersion of precipitates that causes maximum hardening (and strengthening). A typical ageing treatment may be 12 h at 170°C. The precipitates are usually very fine (<0.1 μm) and densities may

Table 2.5 Nominal compositions[a] and typical uses of selected cast aluminium alloys.

US Aluminum Association No.	Casting process[b]	Si	Cu	Mg	Other	Typical uses
208.0	S	3.0	4.0			Manifolds, valve bodies
238.0	PM	4.0	10.0	0.3		Soleplates domestic irons
242.0	S, PM		4.0	0.25	2.0 Ni	Pistons
295.0	S	1.1	4.5			Housings, crankcases
308.0	PM	5.5	4.5			General purpose, weldable castings
332.0	PM	12.0	1.0	1.0		Pistons, pulleys
356.0	S, PM	7.0		0.3	2.0 Ni	Cylinder heads, wheels
390.0	D	17.0	4.5	0.55		Cylinder blocks, compressors
413.0	D	12.0				Thin-walled castings
514.0	S			4.0		Dairy fittings, cooking utensils

(a) Compositions in weight percent.
(b) S, sand casting; PM, permanent mould (gravity-die) casting; D, pressure die casting.

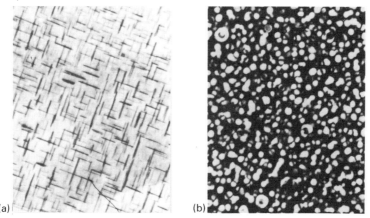

Figure 2.4 Precipitates in age-hardened aluminium alloy. (a) Plates of θ' precipitate in an Al–Cu alloy ($\times 25\,000$). (b) Spherical precipitate of δ' phase in Al–Li alloy ($\times 100\,000$).

be as high as $10^{17}\,\mathrm{cm}^{-3}$. Alloys that show a high response may have tensile yield strengths exceeding 500 MPa which compare with around 10 MPa for pure, unalloyed aluminium.

The common wrought alloys that respond to age hardening are covered by the three series 2*xxx* (Al–Cu, Al–Cu–Mg), 6*xxx* (Al–Mg–Si) and 7*xxx* (Al–Zn–Mg, Al–Zn–Mg–Cu) (Table 2.3). They can be divided into two groups: those that develop medium strength (e.g. with a yield strength of 300 MPa), such as 6061 and 7016 (Table 2.4), and the higher strength alloys

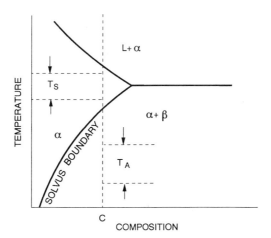

Figure 2.5 Section of binary phase diagram for an alloy system capable of responding to age hardening. For alloy composition C, T_s represents the temperature range for solution heat treatment and T_A represents the temperature range for an ageing treatment.

(e.g. with a yield strength of 500 MPa) that have been developed particularly for aerospace use (e.g. 7075). In general, the high-strength alloys have complex compositions and require more careful processing. Fabrication is mostly carried out at elevated temperatures and grains and dispersions of intermetallic particles (e.g. $Al_6(Mn, Fe)$ and $Al_{20}Cu_2Mn_3$) become aligned in the direction of working. Sheet materials, in particular, also may develop textures in which certain crystallographic planes adopt preferred orientations. All these effects cause the mechanical properties of most wrought products to be anisotropic with minima in directions transverse to working directions.

Cast Al–Si alloys show some response to age hardening if small additions of magnesium are made through precipitation of the phase Mg_2Si (e.g. alloy 356.0, Table 2.5). Alloys based on the systems Al–Cu (e.g. 242.0), Al–Cu–Si (e.g. 238.0) and Al–Zn–Mg may also be heat treated in this way.

2.4.5   Corrosion behaviour[20]

Aluminium is an active metal that oxidises readily under the influence of the high free energy of the reaction whenever the necessary conditions for oxidation prevail. Nevertheless, the fact that a stable oxide is formed generally leads to protection of the metal and most of its alloys in aqueous solutions within the pH range 4.5–8.5. Growth of this natural oxide film is self-limiting and the thickness (maximum ~10 nm) depends on conditions and exposure time. Thicker surface oxide films are more protective and these can be developed by various chemical and electrochemical treatments. One example is anodising in which an alloy or component is made the anode in an electrolyte such as an aqueous solution containing dilute sulfuric acid. Strong acids and alkalis usually cause rapid attack of aluminium and its alloys, with exceptions being concentrated nitric acid, glacial acetic acid and aqueous ammonia.

The electrode potential of aluminium with respect to other metals becomes particularly important when considering galvanic effects arising from dissimilar metal contact. Table 2.6 compares electrode potentials with respect to the 0.1 M calomel electrode ($Hg–HgCl_2$, 0.1 M KCl) for various metals and alloys immersed in a solution of 1 M NaCl and 0.1 M H_2O_2. The table suggests that sacrificial attack of aluminium and its alloys will occur when they are in contact with most metals in a corrosive environment. However, it should be noted that, because of polarisation effects, electrode potentials serve only as a guide to the possibility of galvanic corrosion. For example, contact between aluminium and stainless steels usually results in less electrolytic attack than might be expected from the large difference in electrode potentials, whereas contact with copper causes severe galvanic corrosion of aluminium even though this difference is smaller.

As mentioned earlier, alloying elements may be present as solid solutions

Table 2.6 Electrode potentials of various metals and alloys with respect to the 0.1 M calomel electrode in aqueous solution containing 53 g dm^{-3} NaCl and 3 g dm^{-3} H$_2$O$_2$ at 25°C.[17]

Metal or alloy		Potential (V)
Magnesium		−1.73
Zinc		−1.10
Alclad 6061, Alclad 7075		−0.99
5083		−0.87
Aluminium (99.95%)	aluminium	−0.85
3004	alloys[a]	−0.84
1100, 3003, 6063, 6061, Alclad 2024		−0.83
2014		−0.69
Cadmium		−0.82
Mild steel		−0.58
Lead		−0.55
Tin		−0.49
Copper		−0.20
Stainless steel		−0.09
Nickel		−0.07
Chromium		−0.49 to +0.18

(a) Compositions corresponding to the numbers are given in Table 2.4.

in aluminium, or as a variety of microconstituents comprising the element itself (e.g. silicon), intermetallic compounds (e.g. Al$_2$CuMg and Al$_7$Cr$_2$Fe) or as fine precipitates (e.g. MgZn$_2$). In general, a solid solution is the most corrosion-resistant form in which an alloy can exist. Magnesium dissolved in aluminium renders it more anodic although dilute alloys, e.g. Al–4.5%Mg, retain a relatively high resistance to corrosion, particularly to sea water and alkaline solutions. Silicon and zinc in solid solution in aluminium have only minor effects on corrosion-resistance, although zinc does cause a significant increase in electrode potential. Iron and silicon occur as impurities and may form compounds most of which are cathodic with respect to aluminium. These compounds can result in non-uniform corrosive attack at localised areas on an alloy surface. Pitting and preferential attack at grain boundaries are examples of this form of attack with exfoliation (or layer) corrosion of the surface of components having a marked directionality of grain structure being an extreme example (Figure 2.6). Copper reduces the corrosion-resistance of aluminium more than any other common alloying element mainly because of its presence in various microconstituents. Nevertheless, it should be noted that, when added in small amounts (0.05–0.2%) so that it remains in solid solution, corrosion of aluminium and its alloys tends to become more general and pitting attack is reduced. Thus, although under corrosive conditions the overall weight loss is greater, perforation by pitting is retarded.

It should also be noted that the corrosion-resistance of sheet alloys used particularly in the aircraft industry is normally improved by roll-cladding the

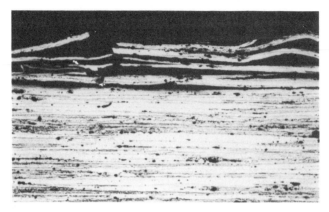

Figure 2.6 Microsection showing exfoliation (layer) corrosion at the surface of an extrusion of a 2*xxx* series aluminium alloy (× 100). Note elongation of grains in the extrusion direction.

surfaces with either high purity aluminium or an Al–1%Zn alloy in order to provide protection. This is achieved by attaching cladding plates to each side of freshly scalped and cleaned ingots at the first rolling pass. Good bonding is obtained and each clad surface is normally 5% of the final thickness of the so-called Alclad composite sheet (Table 2.6).

2.4.6   New developments in aluminium alloys

2.4.6.1   Wrought lithium-containing alloys.[20–23] Lithium has a high solubility in aluminium with a maximum of approximately 4 wt% (14.5 at.%) at 610°C which reduces to less than 1% at room temperature. This is significant because binary and more complex alloys respond to age hardening. Moreover, because of the low density of lithium (0.54 g cm^{-3}), the density of an aluminium alloy is reduced by 3% for each 1% of added lithium. Lithium is also unique amongst the more soluble elements in that it causes a marked increase in Young's modulus of aluminium (6% for each 1% added). Because of all these features, lithium-containing alloys are being developed as a new generation of low-density, high-stiffness materials for possible use in the structures of future aircraft and helicopters. Lithium, which must be of high purity to avoid grain boundary embrittlement of the alloys by impurities such as sodium, is expensive and melting must be done in a protective atmosphere. As a consequence, the current lithium-containing alloys are two to three times more costly than existing aluminium-base materials but the prospect of weight savings still improves the competitive position of aluminium alloys with regard to non-metallic composites. Examples of the compositions of some of these alloys are 2090 (Al–2.7%Cu–2.3%Li–0.12%Zr), 8090 (Al–1.3%Cu–2.5%Li–0.9%Mg–0.12%Zr) and 2091 (Al–2.1%Cu–2.0%Li–1.5%Mg–0.10%Zr).

2.4.6.2 Rapid solidification processing.[20,24,25] During the last decade, much attention has been given to the technique of rapid solidification processing (RSP) to produce an entirely new range of experimental aluminium alloys having mechanical and corrosion properties superior to those obtainable by conventional metallurgical practices. Special interest has centred on creep resistance at elevated temperatures because these materials have the potential to be competitive with titanium alloys up to temperatures of 300°C, or higher, which compare with around 125°C for existing wrought, age hardenable alloys fabricated from cast ingots.

Techniques have been developed which allow cooling rates from the melt of 10^4–10^7°C s^{-1}, resulting in fine, stable microstructures in either sprayed powders or thin ribbon which is produced by squirting molten alloys on to an internally water-cooled, rotating wheel. What is also significant is that normally sparingly soluble elements such as iron, molybdenum and vanadium, which have low diffusivities in aluminium (coefficients of 10^{-11}–10^{-13} cm^2 s^{-1} compared with 10^{-5} cm^2 s^{-1} for solutes such as copper and magnesium), are retained in supersaturated solid solution. Subsequently these elements can form dispersions of fine particles such as Al_6Fe which are thermally stable and resist coarsening at elevated temperatures.

Gas atomization is a well known industrial process and methods of producing strip, for which the highest cooling rates are achieved, have been scaled up to yield commercial quantities. Powders or pulverised ribbons are compacted in vacuum, sealed in cans and fabricated, usually by extrusion.

RSP may lead to the formation of amorphous (glassy) structures in certain ferrous and non-ferrous alloys that are normally crystalline, leading to quite distinctive mechanical, physical and chemical properties. The first amorphous, single phase RSP aluminium alloys were observed in 1981 with the Al–Fe–B and Al–Co–B systems.[26] However, these alloys were very brittle. Subsequently improved ductility was achieved with Al–Ni–Si and Al–Ni–Ge alloys[27] and, more recently, with a range of aluminium alloys containing lanthanide and transition metal elements. Examples are Al–8%La–5%Ni and Al–8%Y–5%Ni–2%Co.[28] Small samples of these alloys have shown tensile strengths of up to 1250 MPa which is more than double that observed with conventionally produced high strength aluminium alloys. These new alloys also display relatively high thermal stability and some must be heated to temperatures around 420°C before the metastable, amorphous structure becomes crystalline.

2.4.6.3 Metal matrix composites.[20,29,30] Composites such as fibre-glass have relatively low values of Young's modulus and are restricted in their elevated temperature use because of their polymeric matrices. Replacing such matrices with metals improves these properties and special attention has been given to the incorporation of fine ceramic fibres or particulates in aluminium (e.g. Figure 2.7). Most commonly used are fibres of SiC or Al_2O_3 and several methods have been developed to incorporate them as uniform dispersions in aluminium or its alloys such as 6061. These include:

1. *stir casting* in which the ceramic particles are stirred into the melt before casting;
2. *squeeze casting* in which the molten alloy is forced under pressure into a three-dimensional, porous 'preform' containing the ceramic particles separated by a suitable binder;
3. *spray casting* in which the matrix alloy is atomised and the reinforcement is introduced into the spray so that both components are deposited together on a cold substrate; and
4. *powder metallurgy* in which alloy powder and the ceramic reinforcement are mixed and subsequently compacted in a die by a combination of pressure and temperature.

Once a metal matrix composite has been produced, it can be remelted, cast and fabricated by normal practices. As one example, Alcan Aluminium Ltd has opened a plant in Canada which produces a composite under the trade name Duralcan and has an annual capacity to produce 11 000 tonnes. Prices currently range from US$6.5 to 9 per kg.

Specific moduli (Young's modulus divided by density) of conventional high strength aluminium alloys, most steels and titanium alloys are close to

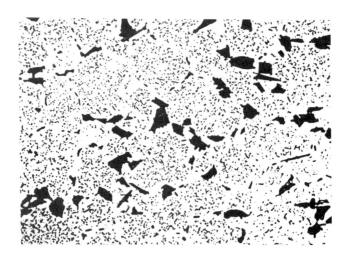

Figure 2.7 Microsection of a metal matrix composite containing silicon carbide particles (courtesy J. White) (×400).

24–26 units. Values for aluminium alloy metal matrix composites range from 32 to 38 units, indicating an improvement of up to around 50%. Applications for which aluminium alloy metal matrix composites are being evaluated include automotive components such as connecting rods and pistons, aerospace structures, sporting goods and bicycle frames.

2.5   Uses of the elements

2.5.1   Aluminium and its alloys

Examples of applications of wrought and cast alloys are included in Tables 2.4 and 2.5 and this discussion is confined to a consideration of general trends. In most countries aluminium is used in five major areas: (i) building and construction, (ii) containers and packaging, (iii) transportation, (iv) electrical conductors and (v) machinery and equipment. The total amounts and patterns of usage vary widely from country to country depending on the levels of industrialisation and economic growth. As examples, the *per capita* consumption of aluminium in the United States in the mid 1980s was 27 kg which compares with 18 kg for Japan and Australia, 10 kg for Britain, 0.4 kg for India and an overall world average of 5 kg. More recent figures show that containers and packaging were the major outlet in the United States (28% of all sales) and Australia (34%); elsewhere the major outlets varied from building and construction in Japan (41%), transportation in the former West Germany (30%), to electrical conductors in India (52%). The actual distribution of recent shipments of all products made from aluminium alloys in the United States (total of 7.6 million tonnes in 1990) and Japan (total of 2.3 million tonnes in 1991) are shown in Figure 2.8.

The use of aluminium alloys to lower the weight of motor cars is expected to be a future growth area in many countries in which they are manufactured because of anticipated legislative requirements to reduce pollution by means of greater fuel economies. In this regard, each 10% reduction in the average weight of a motor car is considered to result in a 5–6% decrease in fuel consumption. As an example, the average aluminium content in cars produced in the United States rose from about 8 kg in 1947 to 37 kg in 1973 and 67 kg in 1987.[20] The current figure is around 90 kg.

Aluminium is an attractive metal to recycle because the remelting of scrap requires only some 5% of the energy needed to produce the same weight of primary aluminium from bauxite. Currently the proportion of secondary (scrap) to primary aluminium is 25–30% which is well below that recovered with steel and copper. However, aluminium alloys do present a special problem because they cannot be refined. Remelting thus tends to downgrade the alloys so that they tend to be used for foundry castings which,

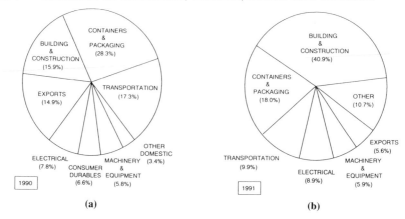

Figure 2.8 Distribution of aluminium alloy products in (a) the United States (1990) and (b) Japan (1991). Examples of applications in the various sectors are as follows:

Containers and packaging: beverage cans, foil, collapsible tubes

Transportation: automotive engine components, aircraft and space vehicles, boats, railway trucks

Building and construction: window frames, architectural panels, mobile homes, roofing and guttering, bridges

Electrical: busbars, overhead cables, capacitors

Consumer durables: kitchen utensils, refrigerators, furniture

Machinery and equipment: heat exchangers, chemical equipment, tanks, printing plates.

in turn, make up only some 20% of the market for aluminium and thus are limited in the amount of wrought alloy scrap they can absorb. More than 60% of scrap originates from fabricating processes so that some recycling into wrought products is possible by retaining known scrap within a closed circuit. A notable example is provided by beverage cans. Some 8.8×10^{10} cans were produced in the United States in 1990 and the amount recovered by recycling (64%) made up some 10% of the total consumption of aluminium in that country.[31]

As mentioned earlier, a large amount of electricity is required to reduce alumina to aluminium. Aluminium has therefore been proposed as a potential 'energy bank', providing this energy can be released in a controlled way by electrochemical conversion to aluminium hydroxide. A particularly attractive prospect for making this conversion is to couple an aluminium anode through an aqueous electrolyte to an electrode which has access to an inexhaustible supply of the cathode reactant (oxygen) from air.[32] In such an arrangement, the theoretical specific energy (Wh/g) of aluminium is 8.1; only lithium anodes have a higher value and these are both costly and difficult to fabricate.

The design of the aluminium/air battery is conceptually simple and the electrolyte may be either a neutral chloride (e.g. NaCl) or an alkali (e.g.

NaOH). The net reaction is

$$4Al + 6H_2O + 3O_2 \rightarrow 4Al(OH)_3 \qquad (2.4)$$

Refuelling is also an easy process as it would involve only the regular addition of water and the removal of the solid reaction products, together with the occasional replacement of the metal anodes. Major reasons for delays in the commercial exploitation of the system have been difficulties in manufacturing cheap and reliable cathodes and in finding a convenient method for separating and removing the aluminium hydroxide reaction product.

It is also interesting to note that another barrier to overcome is the adherent oxide film which inteferes with the controlled dissolution of the aluminium anode. One solution has been to use high-purity aluminium in which small amounts (e.g. 200 ppm) of so-called activating elements, such as gallium and indium, are dissolved. During electrolysis, these elements are deposited on the surface of the anode and thereby promote localised galvanic dissolution of the aluminium.

Two major applications are being promoted for the aluminium/air battery. One is as a storage system for power supply grids and the other is for use in electric road vehicles. For this latter purpose, specific energy yields of around 4 Wh/g have been achieved and it has been claimed that such a battery is capable of providing power to drive a normal-sized motor car some 400 km between stops to replenish the electrolyte, and 2000 km before the aluminium anodes need replacing.

2.5.2 Uses of gallium[2,33,34]

Rather unusually, gallium has only one major application and this as a source of compound semiconductors. Such devices were first developed using the Group 14 element germanium but are now more than 90% based on silicon because this element is plentiful and cheaper to produce. Group 14 elements do not, however, possess the most outstanding electronic properties and for some applications compounds made by combining elements from Groups 13(III) and 15(V) offer superior performance. Of these, the most important is gallium arsenide (GaAs), although other, more complex gallium-containing compounds also exhibit semiconducting properties; examples are GaAsP, GaAsAl and InGaAsP (see chapter 5).

Gallium arsenide semiconductors have the following advantages over silicon:

1. There is a narrower gap between the electron conduction and valence bands so that less energy is required to induce conduction. This allows more compact integrated circuits to be assembled without the danger of

overheating. Moreover, gallium arsenide semiconductors can operate at higher temperatures than those made from silicon.

2. Electron mobility is some six times greater, a feature which has special significance for fast electronic switching in applications such as computing.

3. As with most other compound semiconductors, the energy evolved when electrons move out of the conduction band is usually in the form of light waves having frequencies in the visible and infrared ranges. As a consequence, devices can be constructed to exploit this two-way conversion between light waves and electronic currents and include light-emitting diodes (LEDs) that were first used for displays on calculators and digital watches, laser diodes and photodetectors. The opportunity to link electronic components to optical communications systems has revolutionised telecommunications and has led to the development of the opto-electronic industry.

4. They are less susceptible to radiation damage and for this reason find application in space probes and satellites.

Gallium arsenide is more difficult than silicon to produce as bulk crystals that are pure and free of defects. Accordingly they have been used more commonly for thin film devices that are produced by techniques such as molecular beam epitaxy. Gallium arsenide semiconductors are also more expensive than those made of silicon and, as a consequence, they tend to be used in niche markets rather than general circuitry. Nevertheless, the market for gallium arsenide is expected to reach US$4 billion by 1992.

Other applications of gallium are of minor significance. These include single crystal garnets (compounds of mixed M_2O_3 metal oxides) for which gallium gadolinium garnet is used as a substrate for one type of bubble memory device. Gallium can be used as a substitute for mercury in high-temperature thermometers, and as a coating for mirrors.

Although not of practical importance, it may be noted that liquid gallium wets most surfaces. It is also remarkable for its ability to penetrate the atomic lattices of other metals and this mobility is particularly evident with aluminium and its alloys resulting in so-called liquid metal embrittlement at grain boundaries.

2.5.3   Uses of indium[2,35]

Although indium was discovered in 1863, no use was made of this metal for many years and the world supply was measured in grams until well into the twentieth century. The first reported commercial application was as a minor addition to gold-based dental alloys in which indium served as a scavenger for oxygen. Hardness and ductility were also improved, as was resistance to discoloration, and indium is still added to some of these alloys.

Major current uses of indium are as follows:

1. *Semiconductors*. Indium combines with Group 15(V) elements such as antimony and phosphorus to produce compounds that exhibit semi-conductor characteristics (see chapter 5). One example is InSb which has been used for infrared detectors in military applications. However, its use is limited because it has to be cooled to liquid nitrogen temperatures in order to achieve optimal performance.

2. *Fusible alloys*. Indium is a component of a number of alloys which have very low melting points. One example is Bi–22%Pb–18%In–11%Sn–8.5%Cd which melts at 48°C. These alloys are used in temperature overload devices such as fuses and plugs for sprinkler systems.

3. *Speciality solders*. A number of solders based on indium are available with melting points as low as 93°C. Examples are In–42%Sn–14%Cd and In–15%Pb–5%Ag. Indium is also a minor addition to solders based on lead (e.g. Pb–5%In–5%Ag), bismuth (e.g. Bi–26%In–17%Sn) and tin (e.g. Sn–26%Pb–20%In).

4. *Seals*. Indium metal and a range of indium alloys such as In–48%Sn and In–5%Ag will wet glass, quartz and a range of ceramics. For this reason they serve as seals for joints in equipment, particularly in vacuum systems where their low vapour pressure is a special advantage.

5. *Speciality bearings*. Indium may be applied as an electroplated or sprayed layer to lead-based bearings and then allowed to diffuse into the surface by heating; this leads to improved hardness, corrosion-resistance and anti-seizure properties. Such bearings are expensive and are used only in high performance engines.

6. *Nuclear control rods*. Rods of composition Ag–15%In–5%Cd were developed in the 1950s and have a high capture cross-section for neutrons. They have been used in the majority of pressurised water reactors since that time.

7. *Conductive films*. Because of the high transparency of films of the compounds In_2O_3 and $(InSn)_2O_3$ to visible light, they are used on glass as conductive patterns for liquid crystal displays (LCDs) and as demister strips on motor car windscreens.

8. *Sodium vapour lamps*. A major application for indium in Europe is in the manufacture of low-pressure sodium vapour lamps that are commonly used for outdoor lighting. The indium is applied on the inside of the glass cylinder that forms the outer envelope of the lamp. This coating reflects infrared waves emitted by the lamp while, at the same time, transmitting the visible light which permits the lamp to operate at a higher temperature, thereby raising its efficiency.

9. *Solar cells*. The compounds InP and $CuInSe_2$ are relatively efficient at converting sunlight into electricity and are under active investigation for this purpose.

10. *Jewellery*. Indium is used in alloys with gold and one composition, Au–20%Ag–5%In has a distinctive colour referred to as 'green gold'.

2.5.4 Uses of thallium[4,36]

There are no known uses for thallium metal or for alloys in which thallium is the principal alloying element. Thallium is, however, a minor addition in a number of materials although concern with its toxicity has led to its replacement in some areas where substitute metals are available. Examples of applications are as follows:

1. *Low melting point glasses* which are fluid at 125–150°C. These glasses may also contain sulfur, arsenic and selenium and are claimed to be more durable and insoluble in water than ordinary glasses.
2. *Bearing alloys* such as Pb–15%Sb–8%Tl–5%Sn which are superior to other lead-based alloys. Ag–Tl alloys, which are also available, have been used for special bearings where low coefficients of friction and good resistance to acid attack are required.
3. *A Pb–20%Tl alloy* which shows negligible solubility in mineral acids and has been used for containers.
4. *An alloy Pb–20%Sn–10%Tl* which has been used for the production of insoluble anodes for the electrolytic deposition of copper.
5. *A Hg–8.5%Tl alloy* that has a freezing point of −60°C, which is more than 20° lower than that of mercury itself, and has application in thermometers, switches and other instruments used in polar regions.
6. *Photocells* which depend on the fact that the electrical conductivity of thallium sulfide changes on exposure to light.
7. *Discriminator circuits* for scintillators used in gamma-ray spectrometry which are made from alkali halide crystals containing thallium. These crystals separate the slowly decaying pulses of protons produced by fast neutrons from electron pulses arising from gamma ray absorption. One example is sodium iodide activated with 0.1–0.2% thallium.

Table 2.7 Family of thallium-containing ceramic superconductors.[37–39]

Family	General formula	Specific compounds	T_c (K)
Tl–Ca–Ba–Cu–O	$TlCa_{n-1}Ba_2Cu_nO_{3+2n}$	$TlBa_2Cu_nO_5$	c. 60
		$TlCa_1Ba_2Cu_2O_7$	85
		$TlCa_2Ba_2Cu_3O_9$	117
	$Tl_2Ca_{n-1}Ba_2Cu_nO_{4+2n}$	$Tl_2Ba_2CuO_6$	81
		$Tl_2Ca_1Ba_2Cu_2O_8$	110
		$Tl_2Ca_2Ba_2Cu_3O_{10}$	125

An unexpected use of thallium has arisen following the achievement of superconductivity in ceramic compounds of the general formula $La_{2-x}Sr_xCuO_4$. Subsequently studies were made of yttrium-based compounds because it was known that T_c, the critical temperature at which superconductivity is achieved on cooling, is dependent on pressure. Chu and his colleagues[36] sought to introduce a 'chemical' pressure by replacing the large lanthanum ion with the smaller ion of yttrium and were successful in raising T_c to 93 K. This has been a very important development because the ceramic is superconducting if cooled with liquid nitrogen rather than requiring more expensive helium. More recently, work with compounds containing thallium has raised T_c to 125 K (Table 2.7).[37-39] Work now is focused more on producing these ceramics in forms (e.g. thin films and wires) suitable for commercial applications and on achieving high current densities without destroying superconducting behaviour.

References

1. R.J. Stanner, *Am. Sci.*, 1976, **64**, 258.
2. *Metals Handbook*, 10th edn, American Society for Metals International, Metals Park, OH, 1990, Vol. 2, p. 1114.
3. D.R. Lide, editor-in-chief, *Handbook of Chemistry and Physics*, 73rd edn., CRC Press, Boca Raton, FL, 1992–1993.
4. K. Wade and A.J. Banister, *The Chemistry of Aluminium, Gallium, Indium and Thallium*, Pergamon Press, Oxford, 1975.
5. C.M. Lederer and V.S. Shirley. *Table of Isotopes*, 7th edn., Wiley, New York, 1978.
6. K. Gjotheim and B.J. Welch, *Aluminium Smelter Technology*, Aluminium-Verlag, Düsseldorf, 1980.
7. A.R. Burkin, ed., *Production of Aluminium and Alumina, Critical Reports in Applied Chemistry*, Vol. 20, Wiley, Chichester, UK, 1987.
8. E.W. Dewing, *Can. Inst. Min. Met.*, 1991, **30**, 153.
9. P.D. Stobart, ed., *Centenary of the Hall and Héroult Processes 1886–1986*, International Primary Aluminium Institute, London, 1986.
10. L.G. Boxall, A.V. Cooke and H.W. Hayden, *J. Metals*, 1984, **36** (No. 11), 35.
11. A.S. Russell, *Proc. 7th Int. Light Metals Congress*, Aluminium-Verlag, Düsseldorf, 1981, p. 40.
12. Reference 2, p. 739.
13. M.B. Bever, ed., *Encyclopedia of Materials Science and Engineering*, Pergamon Press, Oxford, 1986, Vol. 3, p. 1895.
14. M. Grayson, ed., *Kirk-Othmer Encyclopedia of Chemical Technology*, 3rd edn., Wiley, New York, 1983, Vol. 22, p. 835.
15. Reference 2, p. 750.
16. L. Sanderson, *Can. Min. J.*, 1944, **65**, 624.
17. K.R. Van Horn, ed., *Aluminium. Vol. 1: Properties, Physical Metallurgy and Phase Diagrams*, American Society for Metals, Metals Park, OH, 1967.
18. L.F. Mondolfo, *Aluminium Alloys: Structure and Properties*, Butterworth, London, 1976.
19. I.J. Polmear, *Mater. Sci. Forum*, 1987, **13, 14**, 195.
20. I.J. Polmear, *Light Alloys: Metallurgy of the Light Metals*, 2nd edn, Edward Arnold, London, 1989.
21. J.W. Martin, *Annu. Rev. Mater. Sci.*, 1988, **18**, 101.
22. E.J. Lavernia and T.S. Srivatsan, *J. Mater. Sci.*, 1990, **25**, 1137.
23. A.A. Gokale and T.R. Ramachandran, *Indian J. Technol*, 1990, **28**, 235.

24. T.R. Anantharaman and C. Suryanaryana, *Rapidly Solidified Metals*, Trans. Tech., Switzerland, 1987, p. 143.
25. C.M. Adam and R.E. Lewis, in *Rapidly Solidified Crystalline Alloys*, eds. S.K. Das, B.H. Kear and C.M. Adam, AIME, Warrendale, PA, 1985, p. 157.
26. A. Inoue, A. Kitamura and T. Matsumoto, *J. Mater. Sci.*, 1981, **16**, 1895.
27. A. Inoue, M. Yamamoto, H.M. Kimura and T. Matsumoto, *J. Mater. Sci., Lett.,* 1987, **6**, 194.
28. A. Inoue, N. Matsumoto and T. Matsumoto, *Mater. Trans. Jpn. Inst. Met.,* 1990, **31**, 493.
29. T.C. Willis, *Met. Mater.,* 1988, **4**, 485.
30. P. Rohatgi, *Adv. Mater. and Processes*, 1990, **137** (No. 2), 39.
31. D.D. Roeber and R.E. Sanders, Jr., *Proc. Int. Conf. on Recent Advances on Science and Engineering of Light Metals*, Japan Institute of Light Metals, Tokyo, 1991, p. 747.
32. N. Fitzpatrick and G.M. Scamans, *New Scientist*, 1986, No. 1517, 17 July.
33. *Gallium and Gallium Arsenide: Supply, Technology and Uses*, U.S. Bureau of Mines, Report IC9208, 1988.
34. S. Jones, G.B. Thomas and B. Wiltshire, *Met. Mater.*, 1986, **2**, 353.
35. E.F. Milner and C.E.T. White in *Kirk-Othmer Encyclopedia of Chemical Technology*, 3rd edn., ed. M. Grayson, Wiley, New York, 1981, Vol. 13, p. 207.
36. M.K. Wu, J.R. Ashburn, C.J. Torng, P.H. Hor, R.L. Meng, L. Gao, Z.J. Huang, Y.Q. Wang and C.W. Chu, *Phys. Rev. Lett.*, 1987, **58**, 908.
37. Z.Z Sheng and A.M. Hermann, *Nature (London)*, 1988, **332**, 55.
38. Z.H. Kang, R.T. Kampwirth and K.E. Gray, *Phys. Lett. A*, 1988, **131**, 208.
39. S.S.P. Parkin, U.Y. Lee, E.M. Engler, A.I. Nazzal, T.C. Huang, G. Gorman, R. Savoy and R.B. Beyers, *Phys. Rev. Lett.*, 1989, **60**, 2539.

3 Inorganic derivatives of the elements

M.J. TAYLOR and P.J. BROTHERS

3.1 Introduction

The chemistries of aluminium, gallium, indium and thallium have much in common with each other, but also display some striking differences. Features in common include the obvious ones, such as the group oxidation state of $+3$, their existence in water as the solvated $[M(OH_2)_6]^{3+}$ cations which are moderately strong acids, and the tendency to exhibit a variety of coordination numbers, up to four with certain ligands and up to six with others. Properties where the members of Group 13 show divergent behaviour include some with a clear trend, for example the tendency for the $+1$ oxidation state, with its distinctive chemistry, to expand in influence and become predominant for thallium. In other cases, however, as noted in chapter 1, the behaviour appears erratic. The assortment of mixed-valence compounds is a good example. Another erratic property is the ability to form compounds with metal-to-metal bonds.

The Group 13 elements, with the ns^2np^1 outer shell configuration, possess fewer valence electrons than valence orbitals. All of the valence shell electrons are formally used in bonding in the $+3$ (M^{III}) state. The $+1$ (M^I) state is attained by loss of the sole valence p electron, leaving the s^2 electron pair which is usually regarded as non-bonding, although the structures of some Tl^+ and In^+ compounds call this assumption into question (see chapter 1). M^{2+} species are normally transitory and need to be isolated from further reaction when they are mononuclear with a single valence-shell electron. In the form of the M_2^{4+} dimer, they are the basis of the formal $+2$ (M^{II}) state which has a significant place in the chemistry of the group. Given that the mercury(I) state — based on the dinuclear $[Hg-Hg]^{2+}$ species — is of long standing, it is perhaps surprising that the adjacent element Tl does not furnish thallium(II) compounds containing the $[Tl-Tl]^{4+}$ entity isoelectronic with the Hg_2^{2+} ion. Indeed, thallium is now the sole member of Group 13 for which homonuclear element-to-element bonds are not well authenticated.

On questions of structure and bonding in Group 13, a good deal remains to be learned. Descriptions of bonding in simple ionic terms are seldom, if ever, adequate and the covalent description frequently has to come to terms with situations where the compound is unsaturated, the valence shell is incomplete, and the structure exhibits special features in consequence. Bonding by means of bridging ligands emerges as a strong theme in the inorganic chemistry of Group 13. It can be seen not only in the tendency of MX_3 compounds to dimerise to M_2X_6, but also in their marked inclination to form polymers. In the extent to which Group 13 atoms — Al and Ga especially — can participate in the building of complicated lattice-type structures, the group is second only to Group 14, although the possibilities which this affords for purposeful synthetic chemistry have only lately been recognised. The expanding knowledge of compounds in which an Al, Ga, In or Tl centre bears ligands of several different types is another strong theme of this chapter.

The ready availability of chemical reagents of high purity now makes a major contribution to inorganic chemical research. Aluminium metal is available as sheet, turnings or powder. Familiar reagent forms of aluminium are the acetate, the lactate, various alkoxides, the halides, and hydrated forms of the nitrate, perchlorate and sulfate. Gallium, which melts at 28°C, can be dispersed by vigorous shaking in warm water, then cooled, to provide a powdered form which dissolves in acids more efficiently than does the bulk solid or the liquid metal. Commercially available gallium compounds include GaAs, GaS, Ga_2O_3, the trichloride and tribromide, and the hydrated nitrate and sulfate. The chloride Ga_2Cl_4 can also be purchased. Indium is a soft metal and cuts into shavings for chemical use. Its readily available compounds include the anhydrous trihalides and the hydrate $InCl_3 \cdot 4H_2O$, the acetate, and hydrates of the nitrate, perchlorate and sulfate. Two indium(I) compounds are marketed, namely the monochloride and cyclopentadienylindium(I). Thallium is available as the metal, the oxide Tl_2O_3, and as Tl^I and Tl^{III} compounds, including both the acetates, the chlorides TlCl and $TlCl_3 \cdot 4H_2O$, thallium(I) nitrate and thallium(III) perchlorate, $Tl(ClO_4)_3 \cdot 6H_2O$. Extra care is needed in working with thallium to take account of the high toxicity of the metal and all its chemical forms. Table 3.1 gives a list of the more commonly available reagent compounds of Al, Ga, In and Tl, including some organometallic compounds. The volatility and ease of purification of the organometallic compounds by vacuum-line techniques are useful attributes for synthetic work, and ones which have been exploited in the production of III–V semiconductor compounds (see chapter 5). In other situations a metal alkoxide, one of the halides, or the metal itself will usually be the preferred starting material. Where aqueous solutions are required, one of the oxyanion salts will be employed, preferably the perchlorate, if the effects of complexation are to be minimised.

As noted in chapter 1, some fundamental atomic properties underlie the chemical behaviour of the Group 13 elements. Figure 3.1 compares the ionisation energies with those of the neighbouring elements of Groups 12 and 14. The trends in successive ionisation energies are irregular, revealing the influence of the $3d^{10}$ electron shell in the case of gallium, and of the $4f^{14}$ shell for thallium. One effect of the greater energy needed to ionise the gallium atom to Ga^{3+}, compared with that to convert aluminium to Al^{3+}, is to cause gallium to resemble boron rather than aluminium in some of its chemistry. The tendency for gallium, like boron, to remain four-coordinate

Table 3.1 Group 13 metal research-grade chemicals available commercially.

Aluminium compounds	Gallium compounds	Indium compounds	Thallium compounds
$\alpha\text{-Al}_2\text{O}_3$	Ga_2O_3	In_2O_3	Tl_2O_3
$\gamma\text{-Al}_2\text{O}_3$	GaS	In_2S_3	TlF
$\alpha\text{-Al(OH)}_3$	$\text{GaF}_3 \cdot 3\text{H}_2\text{O}$	InF_3	TlCl
Al_4C_3	Ga_2Cl_4	$\text{InF}_3 \cdot 3\text{H}_2\text{O}$	$\text{TlCl}_3 \cdot 4\text{H}_2\text{O}$
AlBO_3	GaCl_3	InCl	TlBr
AlF_3	GaBr_3	InCl_3	TlI
$\text{AlF}_3 \cdot 3\text{H}_2\text{O}$	$\text{Ga(NO}_3)_3 \cdot x\text{H}_2\text{O}$	$\text{InCl}_3 \cdot 4\text{H}_2\text{O}$	TlPF_6
AlCl_3	$\text{Ga}_2(\text{SO}_4)_3 \cdot x\text{H}_2\text{O}$	InBr_3	TlNO_3
$\text{AlCl}_3 \cdot 6\text{H}_2\text{O}$	$\text{Ga(acac)}_3{}^a$	InI_3	Tl_2SO_4
AlBr_3	Me_3Ga	$\text{In(NO}_3)_3 \cdot x\text{H}_2\text{O}$	$\text{Tl(ClO}_4)_3 \cdot 6\text{H}_2\text{O}$
AlI_3	Et_3Ga	$\text{In}_2(\text{SO}_4)_3$	TlO_2CCH_3
$\text{Al(NO}_3)_3 \cdot 9\text{H}_2\text{O}$		$\text{In}_2(\text{SO}_4)_3 \cdot \text{H}_2\text{O}$	$\text{Tl(O}_2\text{CCH}_3)_3 \cdot x\text{H}_2\text{O}$
$\text{Al}_2(\text{SO}_4)_3 \cdot 18\text{H}_2\text{O}$		$\text{In(ClO}_4)_3 \cdot 8\text{H}_2\text{O}$	$\text{Tl(O}_2\text{CCF}_3)_3$
$\text{AlNH}_4(\text{SO}_4)_2 \cdot 12\text{H}_2\text{O}$		$\text{In(O}_2\text{CCH}_3)_3$	TlOEt
$\text{Al(ClO}_4)_3 \cdot 9\text{H}_2\text{O}$		$\text{In(O}_2\text{CCF}_3)_3$	Tlacac^a
$\text{Al(OPr}^i)_3$		$\text{In(acac)}_3{}^a$	CpTl^a
$\text{Al(acac)}_3{}^a$		Me_3In	
Me_3Al		Et_3In	
Et_3Al		CpIn^a	
Et_2AlCl			
$\text{Et}_3\text{Al}_2\text{Cl}_3$			
$\text{Me}_3\text{N} \cdot \text{AlH}_3$			

(a) acac = acetylacetonate; Cp = $\eta^5\text{-C}_5\text{H}_5$.

in situations where aluminium favours six-fold coordination is a case in point. There are other similarities between gallium and boron, with signs that the chemistry of gallium hydrides may be capable of expansion, if not to match that of the polyboranes, at least to surpass that of aluminium hydrides.

The stabilities of the oxidation states of the aluminium sub-group are reflected in their redox potentials which are given in Table 3.2 for the +3 and +1 states. The factors which influence the relative stabilities of the +1 and +3 states and the significance of the 'inert-pair' effect have recently been addressed by computational methods. This subject, which was introduced in chapter 1, is elaborated in section 3.3.1. Compounds of thallium(III) tend to be powerful oxidising agents and the reduction potential of the aqueous couple $\text{Tl}^{3+}/\text{Tl}^+$ is strongly positive. Of the M^+ ions only Tl^+ exists in aqueous solutions. It is instructive to combine the $E^\ominus$ values for the M^{3+} and M^+ ions to determine $\Delta G^\ominus$ for the hypothetical reaction of the M^{3+} ion and the metal:

$$\text{M}^{3+}(\text{aq}) + 2\text{M}(\text{s}) \rightleftharpoons 3\text{M}^+(\text{aq}) \tag{3.1}$$

The $\Delta G^\ominus$ values in Table 3.2 show that the reaction is highly unfavourable for aluminium, and that the equilibrium also lies well to the left-hand side for

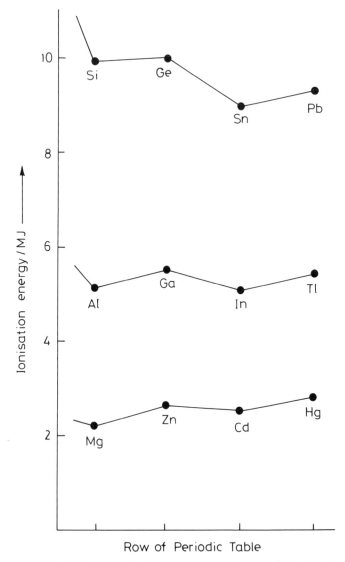

Figure 3.1 Total ionisation energy for the gaseous process $M \rightarrow M^{n+} + ne^-$ for the Group 13 and neighbouring elements.

gallium and indium. However, another factor needs to be taken into account, namely the existence of the Ga–Ga bonded species Ga_2^{4+} which can be considered to arise from the reaction (3.2):

$$Ga^+ (aq) + Ga^{3+}(aq) \rightleftharpoons Ga_2^{4+}(aq) \qquad (3.2)$$

Unpredictable situations such as this add spice to the pursuit of unfamiliar

Table 3.2 Some properties of the Group 13 metals.

Property	Al	Ga	In	Tl
Electron configuration	$[Ne]3s^2p^1$	$[Ar]3d^{10}4s^2p^1$	$[Kr]4d^{10}5s^2p^1$	$[Xe]4f^{14}5d^{10}6s^2p^1$
Electronegativity[a]	1.47	1.82	1.49	1.44
$E^{\ominus}$ (V)[b] of M^{3+}(aq) $+3e^- \rightleftharpoons$ M(st)	−1.677	−0.549	−0.338	+0.741
$E^{\ominus}$ (V)[b] of M^{3+}(aq) $+2e^- \rightleftharpoons$ M^+(aq)	(−2.7)	(−0.66)	−0.444	+1.280
$E^{\ominus}$ (V)[b] of M^+(aq) $+e^- \rightleftharpoons$ M(st)	(+0.3)	(−0.32)	−0.126	−0.336
$\Delta G^{\ominus}$ (kJ mol^{-1}) of M^{3+}(aq) $+2$M(st) $\rightleftharpoons$ $3M^+$(aq)	+570	+66	+61	−312

(a) Allred scale (see Table 1.1).
(b) J. Emsley, *The Elements*, 2nd edn., Clarendon Press, Oxford, 1991; D.R. Lide (editor-in-chief), *Handbook of Chemistry and Physics*, 73rd edn., CRC Press, Boca Raton, FL, 1992–1993; S.G. Bratsch, *J. Phys. Chem. Ref. Data*, 1989, **18**, 1.

valencies. Redox behaviour in non-aqueous systems does not necessarily follow the same pattern, and fused salt media tend to favour the existence of species containing the metal in low oxidation states. It is under these conditions that much of the knowledge of Ga^I, Ga^{II}, In^I and In^{II} systems described in section 3.3 has been acquired.

Much precise information is available from the determination of structures of Group 13 compounds by X-ray or electron diffraction. To facilitate comparisons with bond lengths and other dimensions in the present account, Table 3.3 gives ionic and covalent radii of Al, Ga, In and Tl. For all except Tl, the determination of M–M bond distances in the formal M^{II} compounds provides an additional measurement of the covalent radius, i.e. it is half the M–M bond length; these data, included in Table 3.3, are seen to be reasonably close to the values from other sources. The additional inclusion of the single-bond covalent radii of atoms E to yield bond lengths M–E for the elements of Groups 14–17 in Table 3.3(b) provides a set of data against which particular compounds can be assessed. Table 3.4 draws attention to the difference between bridging and terminal bond lengths which is a characteristic feature of many Group 13 compounds.

This chapter is in several parts: section 3.2 deals with the +3 state and

Table 3.3 (a) Ionic, covalent and metallic radii of the Group 13 metals.

Property	Al	Ga	In	Tl
Ionic radius of M^{3+} for four-fold coordination $(Å)^a$	0.40	0.47	0.62	0.75
Ionic radius of M^{3+} for six-fold coordination $(Å)^a$	0.535	0.62	0.80	0.885
Ionic radius of M^+ for six-fold coordination (Å)	(c. 1.0)	1.13^b	$1.32^{b,c}$	1.50^a
Single-bond covalent radius of M $(Å)^{a,c,d}$	1.25	1.25	1.50	1.55
One-half the length of the M–M single bond $(Å)^e$	1.32	1.20	1.38	–
Metallic radius of M $(Å)^f$	1.43	1.22	1.625	1.705

Table 3.3 (b) Single-bond covalent radii (Å) of atoms of Groups 13–17.^{a,c,d}

B	0.88	Al	1.25	Ga	1.25	In	1.50	Tl	1.55
C	0.79	Si	1.15	Ge	1.20	Sn	1.40		
N	0.73	P	1.09	As	1.20	Sb	1.41		
O	0.72	S	1.03	Se	1.16	Te	1.35		
F	0.74	Cl	1.01	Br	1.15	I	1.33		

(a) R.D. Shannon, *Acta Crystallogr.*, 1976, **A32**, 751.
(b) J. Emsley, *The Elements*, 2nd edn., Clarendon Press, Oxford, 1991.
(c) J.A. Dean, ed., *Lange's Handbook of Chemistry*, 13th edn., McGraw-Hill, New York, 1985.
(d) N.N. Greenwood and A. Earnshaw, *Chemistry of the Elements*, Pergamon Press, Oxford, 1984.
(e) Data in section 3.3.
(f) See Table 1.1.

Table 3.4 Terminal and bridging Al–X bond distances $(d\ (\text{Å}))^{a,b}$ and calculated bond indices $(I)^{a,c}$ for species $Al_2X_6.^d$

Parameter	X							
	Cl	Br	OH	OBu^t	NH$_2$	N=CH$_2$	CH$_3$	H
$d(\text{Al–X})_t$	2.06	2.21	1.70	1.69	1.86	1.79	1.95	1.56
$d(\text{Al–X})_{br}$	2.21	2.33	1.84	1.82	1.97	1.93	2.12	1.72
Ratio $d_t/d_{br}^{\ e}$	0.93	0.95	0.92	0.93	0.94	0.93	0.92	0.91
$I(\text{Al–X})_t$	0.983				0.847	0.841	0.856	0.955
$I(\text{Al–X})_{br}$	0.538				0.438	0.411	0.413	0.496
Ratio $I_t/I_{br}^{\ f}$	1.83				1.93	2.05	2.07	1.93

(a) S.J. Bryan, W. Clegg, R. Snaith, K. Wade and E.H. Wong, *J. Chem. Soc., Chem. Commun.*, 1987, 1223; D.R. Armstrong, P.G. Perkins and J.J. Stewart, *J. Chem. Soc. A*, 1971, 3674; 1973, 627.

(b) Additional data from Ref. 2 and section 3.2.

(c) Bond indices express the electron density per bonded atom.

(d) Computed structures in the cases X = H, OH or NH$_2$.

(e) Mean $d_t/d_{br} = 0.93$, implying that the terminal Al–X bond distance is some 7% shorter than the bridging Al–X bond distance.

(f) Mean $I_t/I_{br} = 1.96$. The bond indices of the terminal Al–X bonds are roughly twice those of the bridging bonds, regardless of whether the species Al$_2$X$_6$ is formally electron-precise or electron-deficient (X = H or CH$_3$). Since bond indices correlate well with bond enthalpies, this implies that the terminal bonds in species Al$_2$X$_6$ are about twice as strong as the bridging bonds.

section 3.3 with the lower oxidation states. Some chemistry of the formal low oxidation states, e.g. +2 as in species based on the Ga_2^{4+} nucleus or non-integral values as in the catenated triatomic species In_3^{5+}, is associated with the presence of homonuclear M–M bonds and such compounds are included in section 3.3. Other facets of metal-to-metal bonding involving the Group 13 elements, *viz.* intermetallic compounds, Zintl phases, and combinations of Al, Ga, In or Tl with the transition metals in the form of carbonyl and related complexes, are examined in section 3.4.

For classification, the formal oxidation state is preferable to the valency, although the latter is sometimes used, being sanctioned by custom and generally unambiguous when applied to the +1 or +3 states. The Lewis acidity characteristic of unsaturated compounds results in the formation of the coordination compounds which are a major feature of Group 13 chemistry, to be examined further in chapter 8. In the present chapter, we do not attempt to distinguish between covalent and dative bonds, although the physicochemical properties of the two can sometimes be quite different. This distinction has been examined recently by Haaland[1] using many Group 13 examples.

The present treatment of the inorganic chemistry of aluminium, gallium, indium and thallium is organised on a ligand-by-ligand basis. Group 13 was examined in this way by Wade and Banister with a literature coverage up to about 1970.[2] A review on gallium[3] and monographs on the chemistry of gallium[4] and thallium[5] also appeared in the period 1963–1971. Aspects of the inorganic chemistry of Al, Ga, In and Tl were examined in 1984[6] and also feature in the comprehensive accounts of their coordination chemistry compiled in 1985.[7] Unreferenced information will usually be from one or other of these sources. The treatise on structural inorganic chemistry by Wells[8] describes the important structural types which are adopted in well known crystalline solids.

Among the many factors which spur the progress of research into the chemistry of the Group 13 metals are the technological importance of aluminium and its compounds, and the special significance of gallium and indium to the electronics industry. Some of the uses of the inorganic compounds and the underlying chemical developments are identified in the present chapter. The account thus provides a detailed, comparative examination of the state of the inorganic chemistry of aluminium, gallium, indium and thallium, with clear links to the topics which are dealt with elsewhere, for example in chapter 4 on the solid oxides, chapter 5 on III–V semiconductors, chapter 6 on organometallic compounds and chapter 8 on coordination and solution chemistry.

3.2   Derivatives of the trivalent state

3.2.1   Hydrides

Aluminium hydride (alane) was first reported as an impure solid $[AlH_3]_x$ and subsequently prepared as amine or ether adducts. A convenient synthesis of the etherate was developed in 1947 using the newly discovered lithium tetrahydroaluminate which reacts with $AlCl_3$ in ether:[9]

$$3LiAlH_4 + AlCl_3 \xrightarrow[-LiCl]{Et_2O} 4AlH_3 \cdot xEt_2O \qquad (3.3)$$

The ether solution is unstable and soon deposits a solid, $AlH_3 \cdot 0.3Et_2O$,[10,11] from which the remaining diethyl ether is not easily removed. An alternative procedure which provides a stable solution of alane in tetrahydrofuran (thf) is to add just enough sulfuric acid to a thf solution of $LiAlH_4$ to precipitate Li_2SO_4 completely.[12] The technique to prepare ether-free alane, AlH_3, by crystallisation of the ether solution in the presence of benzene was published in 1976,[13] having been classified under US Air Force contracts at the time of its development in the early 1960s.

Several solid phases of alane can be prepared. The crystal structure of one of these phases was deduced from X-ray and neutron diffraction results using samples with a crystallite size too small for single crystal work.[14] In this structure, the Al has octahedral coordination which resembles that in the essentially ionic structure of the fluoride, AlF_3. The Al–H distance, bridging between two Al atoms, is 1.715 Å. This distance is much longer than the covalent bond length of 1.55 Å found in $LiAlH_4$, but less than in $Al(BH_4)_3$. Here the Al–H distance is 2.1 Å in the solid state and 1.80 Å in gaseous $Al(BH_4)_3$.[15]

Alane itself, its adducts with ethers or amines,[11] and hydrido complexes of the types $MAlH_4$ and M_3AlH_6 — particularly $LiAlH_4$ — are powerful and selective reducing and hydrogenating agents for both organic and inorganic compounds.[16,17] Other specifically formulated reducing agents can be prepared by alcoholysis of tetrahydroaluminates, or by ligand exchange between $[Al(OR)_4]^-$ and $[AlH_4]^-$ ions.[18] An important example is $Na[AlH_2(OC_2H_4OMe)_2]$ which is soluble in aromatic hydrocarbons and safer to handle in large quantities than $LiAlH_4$.[19] Mixed alkoxyhydrides $AlH_n(OPr^i)_{3-n}$ are produced by redistribution reactions of $Al(OPr^i)_3$ with $AlH_3 \cdot thf$.[20]

Alane derivatives and complexes, such as the tertiary amine adducts of AlH_3, are of interest as hydride sources for hydroalumination of unsaturated substrates,[21] in preparing the hydrides of other metals, and as precursors for chemical vapour deposition of aluminium metal in thin film technology.[22,23] The trimethylamine adduct, $H_3Al \cdot NMe_3$, is monomeric in the vapour but dimeric in the solid state,[24] and is capable of adding a second molecule of NMe_3 to form $H_3Al(NMe_3)_2$, notable as a compound in which aluminium is five-coordinate.[25] The experimental and computed geometries of the $Al_2(\mu\text{-}H)_2$ core for $[H_3AlNR_3]_2$ ($R_3 = H_3$ or Me_2CH_2Ph; Figure 3.2) indicate that there is minimal direct $Al \cdots Al$ interaction (in contrast to the situation in the dimers of $AlMe_3$ and $HAlMe_2$ for which this is believed to be of some importance). In the present case, the bridge is unsymmetrical with $Al\text{-}\mu\text{-}H$

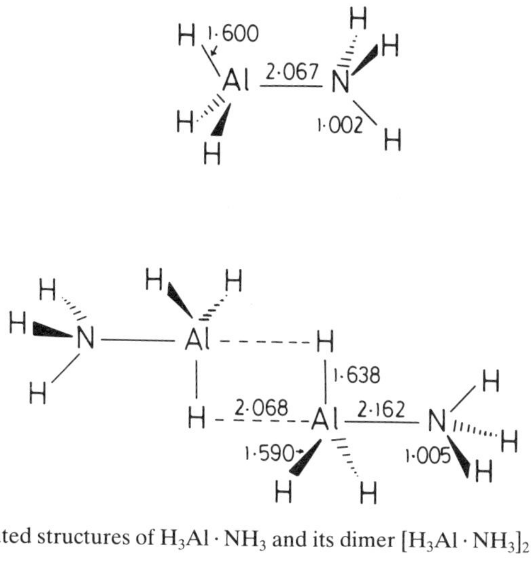

Figure 3.2 Computed structures of $H_3Al \cdot NH_3$ and its dimer $[H_3Al \cdot NH_3]_2$ (distances in Å).[24]

distances of $c.$ 1.60 and 2.05 Å (compare the Al–μ-H distance of 1.72 Å in the AlH$_3$ crystal structure). The structures, being of the type LH$_2$Al(μ-H)$_2$-AlH$_2$L, can be viewed as derivatives of the elusive dialane Al$_2$H$_6$, the existence of which in the gas phase finds some support.[26] Polydentate tertiary amines may give polymeric species,[27] or products incorporating cationic as well as anionic aluminium hydride moieties.[28]

A recent development of aluminium hydride chemistry has been the preparation of a variety of aluminohydride complexes of transition metals[29,30] in which a notable feature is a double hydrogen-bridged moiety, (μ-H)$_2$Al, involved in the bonding of pairs of Al atoms or in the attachment of Al to the transition metal centre. As already noted, bridging hydrides are recognised as a structural unit of some alane adducts with unidentate tertiary amines.

A possible approach to the formation of metal hydrides akin to the boron hydrides is one based on the properties of small clusters, M$_n$ ($n = 2$–6). Computed heats of formation of aluminium hydrides (up to Al$_6$H$_{10}$) from Al$_n$ clusters plus hydrogen suggest this route to be thermodynamically feasible.[31] The molecules AlH$_3$ and Al$_2$H$_6$ have been detected by mass spectrometry when aluminium evaporates into a hydrogen atmosphere at 1100–1200°C.[32] The alane radical anion, [AlH$_3$]$^-$, and the radical complex [(Me$_3$N)$_2$AlH$_2$] have been generated in γ-radiolysis experiments.[33,34] However, and despite the fact that solid alane can be prepared by appropriate techniques, the uncoordinated molecular aluminium hydrides continue to prove elusive.

For gallium, Ga atoms, the dimer Ga$_2$ and the trimer Ga$_3$ have been shown to react with H$_2$ in matrices to form low-valent species GaH, GaH$_2$ and Ga$_2$H$_2$ (section 3.3.2). The GaH$_3$ and InH$_3$ molecules were reported as transient vapour species in 1965,[32] and some thallium hydride species have been claimed.[5]

Attempts to prepare gallium hydride have a chequered history, and the successful synthesis of the binary hydride, gallane [GaH$_3$]$_n$, dates only from 1989.[35,36] The precursor is monochlorogallane, [H$_2$GaCl]$_2$, a compound conveniently prepared by the reaction of gallium(III) chloride with an excess of trimethylsilane (Scheme 3.1). This compound reacts in vacuo with freshly prepared lithium tetrahydrogallate near −30°C to give gallane in yields of 5–15% based on the equation

$$(1/2)[\text{H}_2\text{GaCl}]_2 + \text{LiGaH}_4 \rightarrow (1/n)[\text{GaH}_3]_n + \text{LiGaH}_3\text{Cl} \qquad (3.4)$$

All-glass apparatus is essential, and very careful fractional condensation at temperatures not exceeding −10°C is required to separate the product from unchanged [H$_2$GaCl]$_2$ and other chlorogallanes. Gallane condenses at low temperatures as a white solid which melts at $c.$ −50°C to a colourless, viscous liquid. Vapour-phase investigations by electron diffraction and infrared spectroscopy have established the structure of the diborane-like molecule,

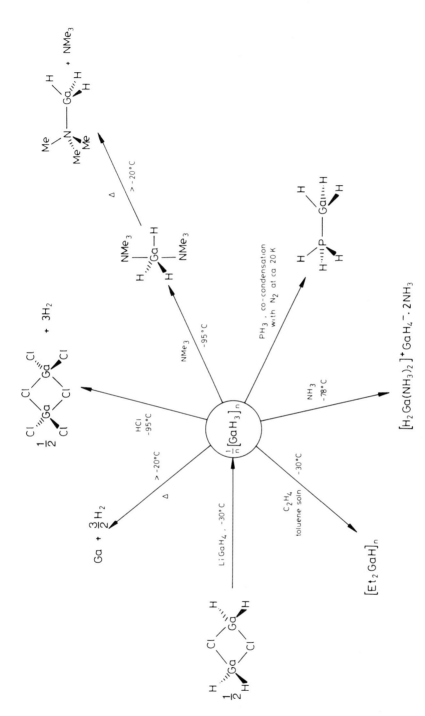

Scheme 3.1 Preparation and reactions of gallane.

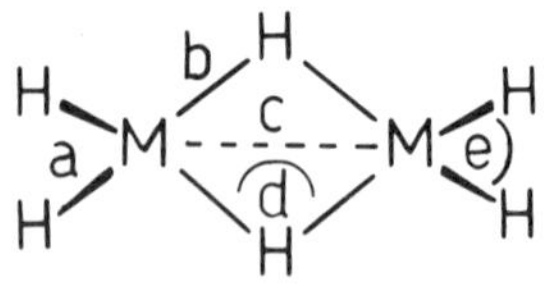

Dimensions

Molecule	Bond distance/Å		M·····M distance/Å	M–H$_{br}$–M/°	H$_t$–M–H$_t$/°	Reference
	a	b	c	d	e	
Al$_2$H$_6$ (calc.)	1.56	1.72	2.59	98	127	50
Ga$_2$H$_6$ (calc.)	1.56	1.77	2.65	97	130	50
Ga$_2$H$_6$ (expt.)	1.52	1.71	2.58	98	130	36

Figure 3.3 Molecular structures of Al$_2$H$_6$ and Ga$_2$H$_6$.

H$_2$Ga(μ-H)$_2$GaH$_2$, shown in Figure 3.3. The dimensions are Ga$\cdots$Ga = 2.580, Ga–H$_{ter}$ = 1.52, Ga–H$_{br}$ = 1.71 Å, and the angle Ga–H$_{br}$–Ga = 98°.[36] The Ga$\cdots$Ga distance of 2.58 Å is considerably *less* than the Al$\cdots$Al distance of *c.* 2.9 Å in the tertiary amine-stabilised dialane molecules discussed above, a finding consistent with the presence of weaker, un-symmetrical, bridge bonds in the dialane derivatives.

In the case of gallane, there is spectroscopic evidence of an oligomer [GaH$_3$]$_n$ with $n > 2$ in the solid state and in solutions at low temperature, but Ga$_2$H$_6$ appears to be the principal component of the vapour. Its reactions, summarised in Scheme 3.1, appear mostly to parallel those of diborane. It is noteworthy that symmetrical cleavage of the Ga(μ-H)$_2$Ga bridges occurs with NMe$_3$ to give the molecular species H$_3$Ga(NMe$_3$)$_n$ ($n = 1$ or 2), whereas NH$_3$ causes unsymmetrical cleavage with the formation of [H$_2$Ga(NH$_3$)$_4$]$^+$ [GaH$_4$]$^-$.

Gallane is potentially important as a means of vapour transport of gallium at low temperatures and as an intermediate in thermolysis reactions involved in the chemical vapour deposition of gallium-bearing films. The gallane adducts are also of interest in this context, and a side effect of the synthesis of gallane has been to stimulate fresh research in this area. Gallane-amine adducts, like those of alane with NMe$_3$, are important as source materials for the epitaxial growth of high-purity metal nitride films on semiconductor silicon or gallium arsenide substrates; consequently these and other volatile gallane derivatives have become the subject of considerable investigation.[37,38] Gallane-phosphine, H$_3$Ga·PH$_3$, is now well characterised by matrix-isolation experiments.[36] Gallane-arsine, H$_3$Ga·AsH$_3$, predicted to be a stable adduct with a binding energy of 66 kJ mol^{-1},[39] offers the prospect of replacing the mixtures of trimethyl-gallane and arsine traditionally used to prepare gallium arsenide films by CVD techniques, especially since the elimination of carbon which may be incorporated into the films represents an important research goal.

Ab initio molecular orbital studies can now contribute strongly to developments in synthetic as well as structural chemistry, and this is well illustrated by recent work on the equilibrium molecular structures, vibrational properties and binding energies of the electron-deficient molecules Al_2H_6 and Ga_2H_6.[40–43] These calculations show that the dimers are stable relative to the AlH_3 or GaH_3 monomers, with binding energies of 132 and 98 kJ mol^{-1}, respectively. The results support the $bis(\mu$-hydrido) structure akin to that of diborane (dimerisation energy, 155 kJ mol^{-1}) in keeping with the infrared spectra and electron-diffraction results for Ga_2H_6. Figure 3.3 compares the calculated and experimental dimensions of Ga_2H_6. Similar approaches have been used to examine the nature of bonding in the mixed diborane-type molecules $AlBH_6$, $BGaH_6$ and $AlGaH_6$. The calculated order of the binding energies with respect to the monomers is $AlBH_6 > B_2H_6 > Al_2H_6 > BGaH_6 > AlGaH_6 > Ga_2H_6$.[44,45] For species in which alternative bridging ligands are present, for example chlorogallanes, quantum mechanical methods predict that the dimers are likely to be chlorine-bridged.[46,47] This is borne out by experiment for the dimers $[HGaCl_2]_2$ and $[H_2GaCl]_2$.[36] The monomers $HGaCl_2$ and H_2GaCl have also been characterised;[48,49] they are formed from GaCl in matrix reactions and are convincingly identified by their vibrational spectra.

$$GaCl + HCl \xrightarrow[\text{Ar matrix}]{hv} HGaCl_2 \tag{3.5}$$

$$GaCl + H_2 \xrightarrow[\text{Ar matrix}]{hv} H_2GaCl \tag{3.6}$$

Trialane Al_3H_9 and trigallane Ga_3H_9 have also been studied by computational methods.[50] The energetically favoured structures are planar ones with six-membered $[MH]_3$ rings and D_{3h} symmetry (Figure 3.4, structure **A**), but these are predicted to be only slightly more stable than the acyclic structures with a pentacoordinate central atom (structure **B**). The calculations show that Al has a stronger tendency than Ga to form hypervalent structures, and this may go some way to explain the failure so far to isolate dialane Al_2H_6. There is also experimental evidence that Al prefers to exist in structures of higher coordination number. As will be discussed in the context of borohydrides, gallium forms $GaH(BH_4)_2$ with the five-coordinate structure **B**, whereas in similar reactions aluminium forms $Al(BH_4)_3$ in which the central Al atom is six-coordinate. *Ab initio* molecular orbital calculations for $Al(BH_4)_3$ support the prismatic model[51] found by electron-diffraction studies.[15] The calculations predict a similar, predominantly covalent structure for aluminium aluminohydride, $Al(AlH_4)_3$, but suggest that $B(AlH_4)_3$ is unstable and not likely to be observed.

Development of the chemistry of indium and thallium hydrides to match that of the corresponding aluminium and gallium compounds appears somewhat improbable. The compounds $LiInH_4$ and $LiTlH_4$ are known but decompose rapidly at 0°C, whereas the similar decompositions of $LiAlH_4$

(a)

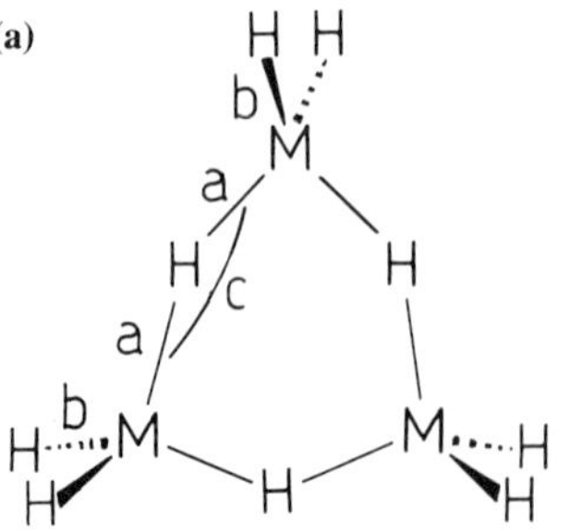

(b)

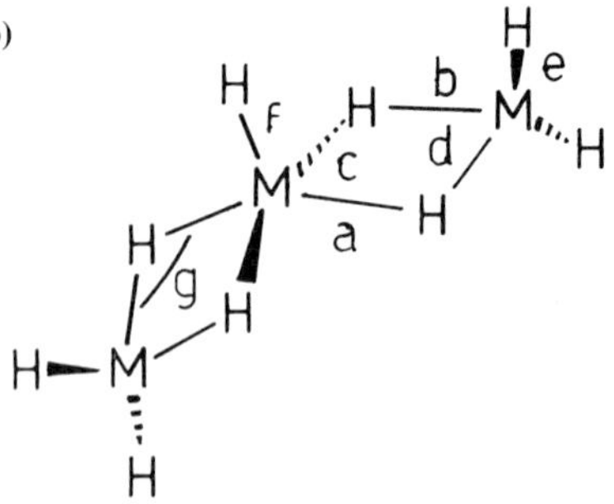

| (a) | Bond distances/Å | | Bridge angle/° |
	a	b	c
Al$_3$H$_9$	1.70	1.57	146
Ga$_3$H$_9$	1.73	1.56	142

| (b) | Bond distances/Å | | | | | | Bridge angle/° |
	a	b	c	d	e	f	g
Al$_3$H$_9$	1.82	1.76	1.70	1.68	1.57	1.56	98
Ga$_3$H$_9$	2.09	1.95	1.65	1.65	1.56	1.55	95

Figure 3.4 Computed structures of Al$_3$H$_9$ and Ga$_3$H$_9$.[50]

and LiGaH$_4$ take place at c. 100 and 50°C, respectively. Complex hydrides In(AlH$_4$)$_3$, Tl(AlH$_4$)$_3$ and Tl(GaH$_4$)$_3$ can be obtained from InCl$_3$ or TlCl$_3$ and LiMH$_4$ (M = Al or Ga) in ether at very low temperature but are extremely unstable substances.[2,7] Despite some claims,[52] it is unlikely that any binary hydrides of the type [InH$_3$]$_x$ or [TlH$_3$]$_x$ have been synthesised.

Until recently, the only well characterised compounds with In–H bonds were a few alkylindates, e.g. Na[InEt$_2$H$_2$] and M[InR$_3$H] (M = Na or K; R = Me, Et or Me$_3$SiCH$_2$). Use of the large tris-(trimethylsilyl)methyl (Tsi) group has enabled the synthesis of an alkylhydroindate sufficiently stable to be investigated in the solid state and in solution.[53] The crystal structure of [Li(thf)$_2$][Tsi$_2$In$_2$H$_5$] reveals a complex anion [(Tsi)H$_2$In(μ-H)InH$_2$(Tsi)]$^-$, the In atoms of which are linked to the Li$^+$ ion via two of the H atoms. Although the H atoms were not located by the X-ray study, their presence is confirmed by IR and NMR spectra. The structure, with its single hydrogen bridge, has affinities with that of Na[Me$_3$Al(μ-H)AlMe$_3$],[54] and similar hydrogen bridges have been postulated in other aluminium compounds, e.g. [Me$_2$AlH]$_n$.

The borohydrides of aluminium and gallium present another aspect of the hydride chemistry of the Group 13 metals, namely that of metallaboranes.[55] Aluminium tris(tetrahydroborate), Al(BH$_4$)$_3$, known since the 1950s, is a colourless liquid, m.p. -64, b.p. $45°C$, which can be prepared by the reaction of B_2H_6 and AlH_3 in ether or, more conveniently, by heating $AlCl_3$ with $NaBH_4$ in $1:3$ molar proportions.[7] The molecular structure of gaseous $Al(BH_4)_3$ is close to D_{3h} in symmetry.[15] Each BH_4 group is bonded to Al through two H-bridge bonds and the line through these H atoms is perpendicular to the AlB_3 plane, or almost so: bond distances are $Al–H = 1.80$, $B–H_{br} = 1.28$ and $B–H_{ter} = 1.20\,\text{Å}$. The volatile metal tetrahydroborates form an intriguing class of electron-deficient compounds. In this context, $Al(BH_4)_3$ has been studied widely by physical methods including vibrational, photoelectron and NMR spectroscopy.[56–60] Products of lower borane/alane ratio, $HAl(BH_4)_2$ and H_2AlBH_4, have been described. Their structures may be polymeric and akin to that of alane or ionic, e.g. tending towards $[AlH_2]^+[BH_4]^-$. In ether and thf solution,[60,61] the NMR spectra indicate equilibria of the type

$$AlH_3 + nBH_3 \rightleftharpoons AlH_{3-n}(BH_4)_n \qquad (3.7)$$

(with all species coordinated by thf). Further reactions involving disproportionation of the hydridoaluminium tetrahydroborate species were also detected. Study of the closely related system $LiBH_4/Al(BH_4)_3/thf$[60] reveals equilibria involving $Al(BH_4)_3 \cdot thf$ and the complex ions $[Al(BH_4)_4]^-$ and $[AlH(BH_4)_3]^-$. The formation of adducts like $Al(BH_4)_3 \cdot thf$ is a characteristic feature of aluminium *tris*(tetrahydroborate) chemistry.

Thermal decomposition of $Al(BH_4)_3$ at $70°C$ eliminates diborane to leave a hydrido derivative, believed to be $HAl(BH_4)_2$. A clean synthesis of $HAl(BH_4)_2$ is by treatment of $Al(BH_4)_3$ with CO which removes one mole of borane as $H_3B \cdot CO$.[62] The structure of this liquid appears to involve dimeric units $(BH_4)_2Al(\mu\text{-}H)_2Al(BH_4)_2$, with Al–H–Al bridges and BH_4 groups attached in a bidentate fashion to each Al atom. In contrast, the recently discovered gallaboranes, $HGa(BH_4)_2$[63] and H_2GaBH_4[64] are volatile and favour monomeric structures in the vapour phase, e.g. $H_2Ga(\mu\text{-}H)_2BH_2$. NMR spectra of the latter molecule in solution at low temperature are characteristic of a 'rigid' terminal GaH_2 and a 'fluxional' BH_4 unit, but also reveal a second component, probably $[H_2GaBH_4]_2$, and possibly with a structure based on a non-planar 8-membered ring composed of alternating GaH_2 and BH_2 groups linked *via* single hydrogen bridges.

The reactions of aluminium *tris*(tetrahydroborate) with volatile boron hydrides generate further aluminaborane compounds of remarkably high thermal stability. For example, the reaction with B_2H_6 in benzene solution at $100°C$ yields AlB_4H_{11}, and that with B_5H_9 yields AlB_5H_{12}.[65] These aluminaboranes have formulae matching those of the *arachno* boron hydrides B_5H_{11} and B_6H_{12}. Whether they have similar structures is doubtful; we return to this point later in describing recent work on the gallaboranes. Chloro- and alkyl-substituted alanes, and alane adducts such as $Me_3N \cdot AlH_3$, react with borane anions, e.g. $B_3H_8^-$, or with polyboranes, e.g.

$B_{10}H_{14}$, to form metallaborane derivatives.[66,67] A gallium representative, $Me_2GaB_3H_8$, was also obtained and both it and $Me_2AlB_3H_8$ were shown to undergo intramolecular ligand-exchange processes that are rapid on the NMR timescale at room temperature.[66,68]

The chemistry of gallaboranes, like that of gallane, claims particular attention through the reactions with N, P, As or Sb bases as possible routes to contaminant-free III–V semiconductors. The simplest molecule H_2Ga $(\mu\text{-}H)_2BH_2$ decomposes at temperatures above $-35°C$ into gallium, diborane and hydrogen. With ammonia it yields $[H_2Ga(NH_3)_2]^+BH_4^-$ as a white solid stable at room temperature.[64] This adduct shares the ionic formulation of the ammonia adduct of diborane, formally $BH_3 \cdot NH_3$, for which the structure $[H_2B(NH_3)_2][BH_4]$ was recognised by Schultz and Parry in 1958.[69] The postulated existence of the GaH_2^+ cation as the ammoniate receives some support from the recent X-ray characterisation of a structurally related organogallium complex, $[Me_2Ga(NH_2Bu^t)_2]^+$ in which Ga has the expected tetrahedral coordination, although with highly distorted geometry.[70] As noted already, cationic aluminium hydride derivatives are certainly known.

The novel hydride *arachno*-2-gallatetraborane(10), $H_2GaB_3H_8$, has been prepared by metathesis involving monochlorogallane and the $B_3H_8^-$ ion:[71]

$$\frac{1}{2}[H_2GaCl]_2 + Bu^n_4NB_3H_8 \xrightarrow{\text{neat reagents, } -30°C} H_2GaB_3H_8 + Bu^n_4NCl \quad (3.8)$$

The structure of this gallaborane, determined in the gas phase by electron diffraction, displays a terminal GaH_2 group linked to the B_3H_8 framework by

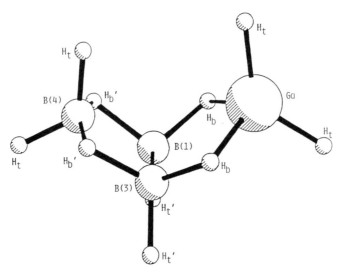

Figure 3.5 The molecular structure of $H_2GaB_3H_8$.[71]

a pair of Ga(μ-H)B bonds to complete a tetraborane(10)-like structure (see Figure 3.5). Some NMR studies performed on $H_2GaB_3H_8$ show that, unlike $Me_2GaB_3H_8$ mentioned earlier, it does not appear to undergo rapid exchange of hydrogens, at least in the range -80 to $0°C$. With ammonia, $H_2GaB_3H_8$ yields a white solid stable at room temperature, which on the evidence of its IR spectrum is probably $[H_2Ga(NH_3)_2]^+[B_3H_8]^-$. The reactions of monochlorogallane which give rise to H_2GaBH_4, $H_2GaB_3H_8$ and other gallane derivatives are summarised in Scheme 3.2.

Old records of indium or thallium borohydrides cite the preparation of an unstable adduct $In(BH_4)_3 \cdot thf$ by the reaction of trimethylindium with diborane in thf solution, and mention ill-characterised products derived from the reactions between $InCl_3$ or $TlCl_3$ and $LiBH_4$ in ether. In contrast, the thallium(I) compound $TlBH_4$ is thermally stable. Dimethylindium

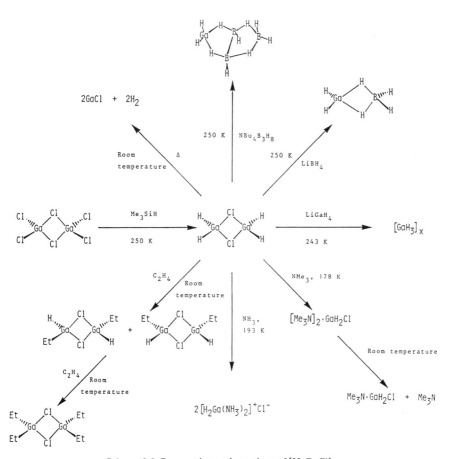

Scheme 3.2 Preparation and reactions of $[H_2GaCl]_2$.

tetrahydroborate, Me_2InBH_4, can be vaporised without decomposition and is now under investigation.[72] Reports of the reactions of trimethylindium and trimethylthallium with $B_{10}H_{14}$ describe the preparation of $MeInB_{10}H_{12}$ and the ionic complexes $[Me_2M]^+[Me_2MB_{10}H_{12}]^-$ (M = In or Tl).[73] X-Ray examination of the thallium-containing anion, crystallised as the Ph_3PMe^+ salt, shows the Me_2Tl group attached by η^4-coordination to a face of the decaborate ligand.[74] The Tl-bond distances range from 2.51 to 2.77 Å. It is interesting to compare this structure with the metallacarboranes, particularly those of the $C_2B_9H_{11}^{2-}$ system, discussed in section 3.2.6.2.

3.2.2 Binary halides

3.2.2.1 Fluorides.

Group 13 fluorides, MF_3, are colourless, crystalline solids with high melting points (AlF_3 sublimes at 1270°C). The MF_3 solid structures are based on a distorted six-coordinate layer lattice and are predominantly ionic in character. Preparation is by the action of HF, F_2 or other fluorinating agent on the heated oxide, M_2O_3. GaF_3 (and InF_3 similarly) is obtained as colourless needles by the thermal decomposition of $GaF_3 \cdot 3NH_3$ or $(NH_4)_3GaF_6$ at 600°C in an inert atmosphere.[75] AlF_3, GaF_3 and InF_3 are insoluble in, and impervious to, water. TlF_3 on the other hand is hydrolysed to form $Tl(OH)_3$. The enthalpies of formation of the trifluorides show that they become less stable as the atomic number of the metal increases ($\Delta_f H^\ominus$ in kJ mol^{-1}: Al, -1504; Ga, -1160; In, -900; Tl, -600).

In addition to $AlF_3 \cdot 3H_2O$, a highly water-soluble salt $AlF_3 \cdot 9H_2O$ can be prepared from Al powder and aqueous HF.[76,77] Solutions of the latter were investigated by ^{19}F NMR spectroscopy which finds signals attributable to the Al–F complexes $[AlF(OH_2)_5]^{2+}$, $[AlF_2(OH_2)_4]^+$, $[AlF_3(OH_2)_3]$ and $[AlF_4(aq)]^-$. The last of these presumably has two water molecules in the primary coordination shell, although precedent for a possible drop in the coordination number in changing from cationic to anionic species can be found in the behaviour of gallium in aqueous chloride and bromide solutions. In non-aqueous solution, aluminium chloro/fluoro and iodo/fluoro complexes have been identified by ^{27}Al NMR,[78] and the signal at δ_{Al} 47 ppm in fluoride-rich CH_2Cl_2 solution is assigned to $[AlF_4]^-$. Whereas mixed halide complexes of the type $[AlX_nY_{4-n}]^-$ (X and Y = Cl, Br or I) occur in this solvent, the iodo/fluoro species rearrange to $[AlI_4]^-$ and $[AlF_4]^-$.

Solutions of gallium(III) and indium(III) fluorides in hydrofluoric acid contain aquo/fluoro complexes and yield hydrates $MF_3 \cdot 3H_2O$ on evaporation. The water of crystallisation is not easily removed. For example, $GaF_3 \cdot 3H_2O$ on heating gives $GaF_2(OH) \cdot xH_2O$. This is converted on exposure to ammonia into $GaF_3 \cdot 3NH_3$ which, as already mentioned, yields crystalline GaF_3 by heating under anhydrous conditions. There have been unsuccessful attempts to prepare pure AlF_3 by a similar route.

All the Group 13 metals form complex fluorides, including $NaMF_4$, Na_2MF_5, Na_3MF_6 and other, more complicated stoichiometries. The various types are based on discrete or linked MF_6 octahedra which form chains, sheets and three-dimensional structures through bridging fluoride ligands.[8] Regular octahedral $[AlF_6]^{3-}$ ions ($Al–F = 1.81$ Å; compare AlF_3 crystals where $Al–F$ distances of 1.70 and 1.89 Å are found) are present in $H_3AlF_6 \cdot 6H_2O$ (Figure 3.6), whilst $H_2AlF_5 \cdot 5H_2O$ has infinite, chain-like $[AlF_5]^{2-}$ polyanions which are fluoride-bridged through the *trans*-positions.[79] Discrete $[AlF_5(OH_2)]^{2-}$ ions ($Al–O = 1.94$ Å, mean $Al–F = 1.79$ Å) are found in the NH_4^+ salt.[80]

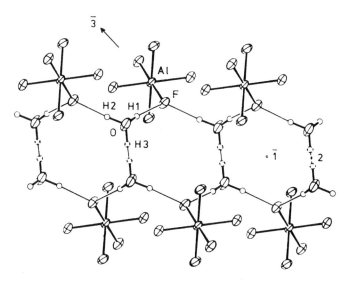

Figure 3.6 Part of a ribbon of hydrogen-bonded atoms in the structure of $H_3AlF_6 \cdot 6H_2O$. H atoms in half-occupied positions are shown as dotted circles. Reproduced with permission from D. Mootz, E.-J. Oellers and M. Wiebeke, *Acta Crystallogr.*, 1988, **C44**, 1334.

'Cryolite', Na_3AlF_6, which occurs naturally, and can be synthesised from $Al_2O_3 + HF + NaOH$, is used with alumina (Al_2O_3) and AlF_3 in the conducting melt from which aluminium is manufactured by electrolysis (see chapter 2). Because of its industrial interest, the structure of molten cryolite has been studied intensively for nearly a century. It is generally agreed that $[AlF_6]^{3-}$ anions, coming from the ionisation of Na_3AlF_6, are further dissociated into $[AlF_4]^-$ and $2F^-$. Re-investigation of $NaF–Na_3AlF_6$ molten mixtures at 1000°C by Raman spectroscopy provides good evidence for the existence of $[AlF_5]^{2-}$ ions, and it is now suggested that dissociation occurs by the following sequence:[81]

$$Na_3AlF_6 \rightleftharpoons 3Na^+ + [AlF_6]^{3-} \tag{3.9}$$

$$[AlF_6]^{3-} \rightleftharpoons [AlF_5]^{2-} + F^- \quad K \sim 0.29 \tag{3.10}$$

$$[AlF_5]^{2-} \rightleftharpoons [AlF_4]^- + F^- \quad K \sim 0.016 \tag{3.11}$$

Figure 3.7 shows how the mixture responds to an increase in the AlF_3 content. The vapour phase above such sodium fluoroaluminate melts produces solid particulates of $NaAlF_4$; solid $NaAlF_4$ is metastable at room temperature whereas the thermodynamically stable compounds are AlF_3 and $Na_5Al_3F_{14}$.[82]

The monomeric AlF_3 molecule exists in the vapour phase. It is trigonal planar with an aluminium-fluorine bond length of 1.63 Å.[83] *Ab initio* calculations,[84] giving Al–F = 1.620 Å, support this value. The GaF_3 molecule has also been identified as a vapour species, and the bond distance (Ga–F = 1.71 Å) measured by electron diffraction.[83] For both aluminium and gallium fluorides there is evidence from mass spectrometry and matrix-isolation studies[2] for the existence of the dimeric molecule, M_2F_6, in the vapour phase; such molecules assume major importance for the heavier halides of the Group 13 metals.

3.2.2.2 Chlorides, bromides and iodides. Accounts of Group 13 tend to give pride of place to the halides, and their properties are tabulated

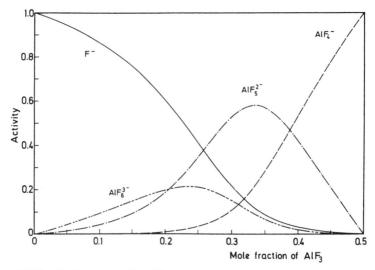

Figure 3.7 Distribution curves of the different species present in NaF–AlF_3 molten mixtures at 1000°C. Reproduced with permission from B. Gilbert and T. Materne, *Appl. Spectrosc.*, 1990, **44**, 2299.

elsewhere.[2] The trihalides MX_3 are well known compounds which in their structures and reactions exemplify many of the characteristic features of Group 13 metal chemistry. It is also within the halide chemistry of the Group that the lower oxidation states, $+1$ and $+2$, are best established (see section 3.3.3).

Aluminium(III) chloride is produced by the chlorination of Al_2O_3, or by burning the metal in a stream of chlorine. An older method, the action of chlorine on a strongly heated mixture of alumina and coke, gives an impure, yellowish product. It is usual to resublime the chloride under vacuum to produce the colourless solid before use. The reaction of dry HCl gas with pure Al at 230°C[85] is an alternative preparation.

Crystalline $AlCl_3$ forms a layer lattice in which the Al^{III} ions are six-coordinate. On melting (under pressure) or sublimation (at $c.$ 150°C), the compound assumes the molecular form Al_2Cl_6 which dissociates on further heating ($\Delta_{dissoc}H = 170\,kJ\,mol^{-1}$).[86]

Solid $AlCl_3$ fumes in moist air and reacts exothermically with water, alcohol and organic donor solvents. It is only slightly soluble in hydrocarbons or chlorinated solvents. Solutions of aluminium chloride in liquid SO_2 or $SOCl_2$ (a weak donor in which $AlCl_3$ adducts exist)[87] are convenient reagents which moderate, but do not block, the characteristic Lewis acid behaviour of $AlCl_3$. More powerful donors (like thf, CH_3CN, pyridine, etc.) dissolve aluminium chloride with complexation, either in molecular forms, $LAlCl_3$, L_2AlCl_3 or L_3AlCl_3, or to produce ionic species of which $[L_2AlCl_2]^+$, $[AlCl_4]^-$ and $[AlL_6]^{3+}$ are typical.[88–92] Solutions of $AlCl_3$ in methanol and higher alcohols contain $[AlCl_2(ROH)_4]^+$, $[AlCl_4]^-$ and $AlCl_3(ROH)_3$ according to the evidence of NMR spectra.[93]

Aluminium bromide and aluminium iodide are synthesised by direct reaction of the heated elements (or from HBr gas and Al) and are also purified by sublimation. $AlBr_3$ (m.p. 98°C) exists in the solid state as the dimeric molecular form, Al_2Br_6. In consequence, it has the ready solubility in hydrocarbons and non-polar solvents which $AlCl_3$ lacks. There are alkyl halide adducts of the type $Al_2Br_6 \cdot L$ and a benzene adduct $Al_2Br_6 \cdot C_6H_6$, m.p. 37°C. AlI_3 (m.p. 189°C) forms a chain structure, shown in Figure 3.8, with the repeating unit $-AlI_2(\mu\text{-}I)-$ so that Al is four-coordinate and there are single iodine-bridge linkages;[94] however, the liquid consists of molecular Al_2I_6.

The trichlorides and tribromides of gallium and indium are prepared as colourless, hygroscopic solids by heating the metal in a stream of the halogen, using nitrogen or argon as a diluent and carrier gas. Important precautions are to exclude air, moisture and organic materials, including tap-greases, and to ensure complete oxidation, since lower halides are a feature of these systems (see section 3.3.3). Crystalline $InCl_3$ and $InBr_3$ can also be synthesised by the anodic oxidation of the metal in contact with a solution of the elemental halogen in benzene/methanol.[95] The iodides, GaI_3

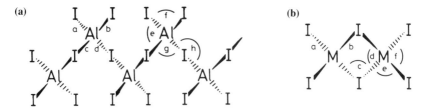

Bond distances in AlI$_3$: a = 2.46, b = 2.45, c = 2.70, d = 2.74 Å
Bond angles in AlI$_3$: e = 107–110, f = 119, g = 99, h = 119°

Dimensions in MI$_3$:	a/Å	b/Å	c/°	d/°	e/°	f/°
GaI$_3$	2.48	2.67	85	95	109 av	121
InI$_3$	2.64	2.84	86	94	108 av	125

Figure 3.8 Structural units in the crystal structures of (a) AlI$_3$ and (b) GaI$_3$ and InI$_3$.[94]

and InI$_3$, can be made by the reaction of the metal with I$_2$ in CS$_2$ or hexane under reflux.[96] InI$_3$ can be prepared without heating from In + I$_2$ in diethyl ether: vacuum distillation completely removes the ether leaving yellow, hygroscopic InI$_3$ crystals.[97]

In the solid state, the gallium halides consist of Ga$_2$X$_6$ molecules analogous to Al$_2$Br$_6$, and have low melting points (GaCl$_3$ 78, GaBr$_3$ 122, GaI$_3$ 211°C). Crystalline InCl$_3$ and InBr$_3$ have crystalline YCl$_3$-type structures[8] in which the metal is six-coordinate (InCl$_3$, m.p. 586°C; InBr$_3$, m.p. 436°C). Indium iodide is usually obtained as yellow crystals (m.p. 210°C): this is the β-form which consists of In$_2$I$_6$ molecules. It changes slowly into a red phase, α-InI$_3$, having the six-coordinate structure favoured by InBr$_3$[98] (cf. the change of AlCl$_3$ from six- to four-fold coordination of Al on melting). The dimensions of the Ga$_2$I$_6$ and In$_2$I$_6$ molecules are compared in Figure 3.8. Table 3.5 summarises the structures of Group 13 trihalides in the solid state.

The structures of the halogen-bridged dimeric M$_2$X$_6$ molecules of the Group 13 halides have been determined by electron diffraction,[2,7,8,99,100] and have been the subject of many spectroscopic investigations. There are available detailed vibrational analyses which confirm the D_{2h} symmetry and permit the comparison of force constants.[101] NQR spectra have been interpreted to show a higher asymmetry parameter for bridging than for terminal atoms.[102] Studies of the vapours have yielded the vibrational spectra[103] and dimensions[104] of the trigonal planar AlCl$_3$ and GaCl$_3$ molecules and have sought to investigate their bonding and to explore both the monomer/dimer equilibria and the dissociation to the monochloride. Matrix-isolation studies have provided further information about the monomers,[104,105] including vibrational frequencies for the AlCl$_n$Br$_{3-n}$ series.[106] For indium, the Raman spectra of the vapours over the molten

Table 3.5 The structures of Group 13 trihalides in the solid state.

$AlCl_3$	$AlBr_3$	AlI_3
6-coord. ionic[a]	dimeric Al_2Br_6	polymeric $[AlI_3]_x$
$GaCl_3$	$GaBr_3$	GaI_3
dimeric Ga_2Cl_6	dimeric Ga_2Br_6	dimeric Ga_2I_6
$InCl_3$	$InBr_3$	InI_3
6-coord. ionic[a]	6-coord. ionic[a]	6-coord. ionic (α-form)[a]
		dimeric In_2I_6 (β-form)
$TlCl_3$	$TlBr_3$	TlI_3
6-coord. ionic[a]	?	$Tl^+[I_3]^-$

(a) A slightly distorted version of the $CrCl_3$ layer structure based on cubic close-packing of halogens with M atoms in one-third of the octahedral holes.

trihalides are also characteristic of mixtures of InX_3 and In_2X_6 molecules.[107] The photoelectron spectra[108] and electron-diffraction patterns[99,104] of indium trihalide vapours indicate that the trichloride is mainly the monomer in the gas phase, while the dimer predominates in the tribromide and triiodide vapours.

The structures of organoaluminium and -gallium dihalides of the type RMX_2 (X = Cl or Br) are halogen-bridged and are shown by vibrational spectra to assume the *trans*-form of C_{2h} symmetry.[109] Cyclopentadienyl halides, e.g. $Ga(C_5Me_5)Cl_2$ and $Ga(C_5Me_5)_2Cl$, follow the same pattern,[110] and it is noteworthy that the rings exhibit η^1-coordination to Ga^{III}, in contrast to the η^5-coordination to Ga^I (section 3.2.3). Despite the presence of bulky ligands, the solid-state structures of Bu^tGaCl_2 and Cy_2GaCl ($Cy = C_6H_{11}$) also consist of Cl-bridged dimers,[111] although with Ga–Cl bonds which are distinctly longer than those of the Ga_2Cl_6 molecule. Like the trihalides, organogallium halides are Lewis acids and form adducts with bases, e.g. $Cy_2GaBr \cdot NH_2Ph$.[112] However, the reaction can take a different course which yields ionic compounds exemplified by $[Me_2Ga(NH_2Bu^t)_2]Br$ produced by the combination of Me_2GaBr with Bu^tNH_2 in diethyl ether solution.[70] The trend towards ionic compounds continues in the chemistry of organoindium and -thallium halides, as demonstrated by the fact that the solids $[Me_2In]Br$[113] and $[Me_2Tl]Cl$[114] have structures in which the cations $[Me_2M]^+$ (M = In or Tl), with a C–M–C bond angle of 180°, are an identifiable feature.

The vapours of AlX_3 and GaX_3 can effect chemical transport of the halides of other Main Group and transition metals, and this aspect of the Lewis acid behaviour of MX_3 species has received considerable attention, focused particularly on the structures, thermodynamic properties and prospective applications. The species concerned are molecular complexes which typically have $M(\mu$-X) bridges linking the metal centres and include a variety of cyclic oligomers,[86] some of which are illustrated in Figure 3.9.

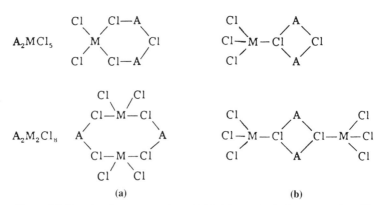

Figure 3.9 Postulated structures of metal chloride species in the vapour phase.[86]

Aluminium chloride in combination with NaCl provides a molten salt medium (m.p. 175°C) for studies of complexation and electrolytic processes under Lewis acidic, basic or neutral conditions as regulated by the $AlCl_3/NaCl$ ratio. A lower melting point (142°C) is achieved by using the $NaAlCl_4 + NaAlBr_4$ eutectic in which the mixed halide anions $[AlCl_{4-n}Br_n]^-$ are present.[115] The principal constituents of chloride melts are the ions M^+, $[AlCl_4]^-$ and $[Al_2Cl_7]^-$. Addition of gallium trichloride to a $KCl–AlCl_3$ melt generates the mixed-metal complex $[Cl_3Al(\mu\text{-}Cl)GaCl_3]^-$.[116] The analogous alkali chloride-gallium chloride melts contain $[GaCl_4]^-$, $[Ga_2Cl_7]^-$ and more highly polymerised complexes $[Ga_xCl_{3x+1}]^-$ $(x = 3, 4,...)$.[117] The structure of $[Al_2Cl_7]^-$ involves a single bridging chloride ligand between two $AlCl_3$ groups. This complex has been identified in several crystal structures and the counterparts $[Al_2Br_7]^-$ and $[Ga_2Cl_7]^-$ are also known;[106, 118-120] their dimensions are given in Table 3.6. However, the assumption that $[Al_2Cl_7]^-$ exists in solids with the composition $MCl \cdot 2AlCl_3$ (M = K, Rb or Cs) has been challenged, since the Raman and ^{27}Al NMR spectra of the solids correspond to a superposition of those of the $[AlCl_4]^-$ ion and $AlCl_3(s)$.[121]

The effect of adding fluoride ions to chloroaluminate melts has also been studied. In acidic melts (those with an $AlCl_3/NaCl$ molar ratio >1), the species $[Al_2Cl_6F]^-$ is formed, with a fluoride replacing a terminal chloride. In basic melts ($AlCl_3/NaCl$ molar ratio <1), fluoride replaces chloride in $[AlCl_4]^-$ and the species $[AlCl_3F]^-$, $[AlCl_2F_2]^-$, $[AlClF_3]^-$ and $[AlF_4]^-$ are observed.[122] Traces of oxide (or of water) in $AlCl_3/MCl$ melts generate oxochloroaluminates, e.g. $[AlOCl_2]^-$ and other forms solvated by $AlCl_3$ or $[AlCl_4]^-$, such as $[Al_2OCl_5]^-$, $[Al_2OCl_6]^{2-}$, $[Al_3OCl_8]^-$ and $[Al_4O_2Cl_{10}]^{2-}$.[123-126] Inorganic oxides are often deliberately treated with aluminium chloride to produce acid catalysts[127] in which species of this kind are an intrinsic feature. The $[Al_2OCl_6]^{2-}$ complex has been found adventitiously

Table 3.6 Geometry of halogen-bridged anions $[M_2X_7]^-$.

Compound[a]	Bridge angle M–X–M (°)	Torsion angles (°)[b]		Distances (Å)		
		τ_1	τ_2	$M\cdots M$	$M–X_{br}$	$M–X_t$
$[(C_6Me_6)_3Zr_3Cl_6][Al_2Cl_7]$	$\begin{cases} 112 \\ 116 \end{cases}$	2 35	−2 −35	3.75 3.83	2.28 2.26	2.08 2.08
$[(C_6H_6)_2Pd_2][Al_2Cl_7]_2$	116	1.3	−3.2	3.83	2.26	2.10
$[Te_4][Al_2Cl_7]_2$	111	13	−53	3.69	2.23	2.11
$K[Al_2Br_7]$	109	13	−59	3.91	2.40	2.28
$[NH_4][Al_2Br_7]$	108	31	−30	3.93	2.43	2.27
$K[Ga_2Cl_7]$	109	29	−35	3.74	2.30	2.14
$Ga[Ga_2Cl_7]$	109	30	−34	3.75	2.30	2.14
$[S_5N_5][Ga_2Cl_7]$	109	23	−30	3.74	2.31	2.13

(a) F. Stollmaier and U. Thewalt, *J. Organomet. Chem.*, 1981, **208**, 327; W. Frank, W. Hönle and A. Simon, *Z. Naturforsch.*, 1990, **45b**, 1 and refs. cited therein.

(b) Torsion angles are measured relative to the plane of the M–X–M bridge. In two cases the anion occurs in the eclipsed conformation; in the remainder, it is in the staggered form.

among the reaction products of a cyclosilazane with Al_2Cl_6. X-Ray analysis shows the oxo ligand to occupy the bridging position between $AlCl_3$ groups.[128] Figure 3.10 compares this structure with that of $[Al_4O_2Cl_{10}]^{2-}$ obtained as the Ag^+ salt during an attempted synthesis of $AgAlCl_4$.[126]

Ambient-temperature melts which can be used as reaction media and as electrolytes for aluminium-plating and in advanced batteries, e.g. for electric vehicles, have been developed using mixtures of $AlCl_3$ with *N*-butyl-pyridinium chloride or an alkylimidazolium chloride.[129–131] In addition to

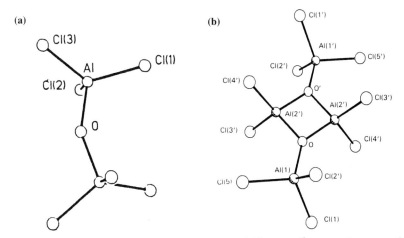

Figure 3.10 The structures of the oxochloroaluminates (a) $[Al_2OCl_6]^{2-}$ and (b) $[Al_4O_2Cl_{10}]^{2-}$. Reproduced with permission from D. Jentsch, P.G. Jones, E. Schwarzmann and G.M. Sheldrick, *Acta Crystallogr.*, 1983, **C39**, 1173.

molecular Al_2Cl_6, the now familiar anions $[AlCl_4]^-$, $[Al_2Cl_7]^-$ and $[Al_3Cl_{10}]^-$ have been identified in these systems.[132, 133] In the $AlCl_3$/PyHCl system there are four congruently melting compounds with the mole ratios $1:1, 1:2, 1:3$ and $2:3$,[134] but structural evidence is incomplete.

Aluminium chloride finds major laboratory and industrial use as a Friedel–Crafts catalyst for processes involving alkylation and acylation. Incipient carbocations are generated from halocarbon molecules of the type RCl or RCOCl simultaneously with the formation of $[AlCl_4]^-$ or $[Al_2Cl_7]^-$. $AlCl_3$ is also used to catalyse condensation, polymerisation and isomerisation; for example, it activates hydrocarbons to reactions leading to an enhanced octane-rating of motor fuels. Recently reported is the use of aluminium(III) iodide to promote reductive dehalogenation and aldol condensation of α-halocarbonyl compounds.[135] The value of chloroaluminate melts, including those which are liquid at room temperature, as solvent systems and electrolytes has already been discussed. Mixtures of $LiAlCl_4$ with $SOCl_2$ or SO_2Cl_2 are also used commercially in lithium batteries.

The halides $AlCl_3$, $GaCl_3$ and $InCl_3$, or less often the tribromides, are the convenient starting materials for the synthesis of many other Al, Ga or In compounds, e.g. alkoxides, alkylamides and organometallics. As powerful Lewis acids, members of the MX_3 series form numerous coordination compounds by interaction with one, two or three additional uncharged or anionic ligands possessing Group 15, 16 or 17 donor atoms (chapter 8). Recent research describes complexes with more than one other ligand besides the halogen, *viz.* $[InCl_3(Me_3PO)_2(MeOH)]$.[136] These studies also demonstrate just how subtle are the factors which control the nature of the products; for example, $[InCl_3(Me_2SO)_3]$ and $[InBr_3(Me_2SO)_3]$ have *fac*-octahedral structures but $[InCl_3(Me_3PO)_3]$ has the *mer*-configuration, while methyldiphenylphosphine oxide gives covalent $[InCl_3(Ph_2MePO)_3]$ but ionic $[InBr_2(Ph_2MePO)_4]^+[InBr_4]^-$.

With reactive organic halides, the acceptor capacity of the Group 13 metal halide MX_3 leads to the formation of $[MX_4]^-$ and a discrete carbocation is produced, e.g. $[Ph_3C]^+$ from Ph_3CCl, or $[C_3X_3]^+$ from C_3X_4 (X = Cl or Br).[137] Similar reactions with inorganic substrates promote the formation and stabilisation of non-metal cations, e.g. ICl_2^+, NS^+, PI_4^+ and $P_2I_5^+$,[138,139] and cationic clusters, e.g. Bi_8^{2+} and Te_4^{2+}.[140] With non-metal halides, the Group 13 halides are useful to bring about trans-halogenation reactions of the type $BF_3 + AlCl_3 \rightarrow AlF_3 + BCl_3$ in which the most electropositive element is united with the most electronegative, here Al with F.

The strongly oxidising character of the Tl^{III} state puts the halides of thallium on a different footing from those of the lighter Group 13 elements. Anhydrous $TlCl_3$, m.p. 155°C, which has a layer lattice, can be prepared by oxidising TlCl with Cl_2 under pressure, by thermal decomposition of the adduct $TlCl_3 \cdot NOCl$, or simply by dehydration of the hydrate $TlCl_3 \cdot 4H_2O$

using carbonyl chloride or thionyl chloride.[141] $TlBr_3 \cdot 4H_2O$ decomposes very readily, giving TlBr. Thallium(III) chloride and bromide solutions are produced by halogen oxidation of the appropriate thallium(I) halide suspended in the desired solvent. This method lends itself to the preparation of thallium(III) halide complexes, including those of the iodide, by simply adding the chosen ligand to the reaction mixture prior to the oxidation step. $CsTlI_4$, which contains the tetrahedral $[TlI_4]^-$ complex anion, is synthesised by heating TlI, CsI and I_2 at 450°C in a sealed tube.[142] The survival of iodide in contact with the oxidising Tl^{III} centre must be due to the stability of the complex ion, since TlI_3 itself is unstable with respect to $Tl^I[I_3]$. Solutions of $TlCl_3$ and $TlBr_3$ are powerful oxidising agents and find application as such, particularly in organic and organometallic chemistry.

Solutions of $AlCl_3$ in concentrated hydrochloric acid yield crystalline $AlCl_3 \cdot 6H_2O$, which contains the $[Al(OH_2)_6]^{3+}$ ion (section 3.2.3.2). Hexahydrates can also be crystallised from solutions of Al^{3+} in aqueous HBr or HI. The halides of Ga, In and Tl form a wide range of crystalline hydrates.[2,4,5] In these cases, but not that of Al^{3+}, there is ample evidence that the aqueous solutions contain halide complexes, generally of the six-coordinate type $[MX_{6-n}(OH_2)_n]^{n-3}$ and including cationic, anionic and uncharged species with halide ions and water molecules competing for places in the coordination sphere of M.[143-145]

Aluminium, gallium and indium trihalides in combination with the alkali halides, MX (M = Na, K or Rb; X = Cl or Br), form compounds of the kind $M^I[M^{III}X_4]$ and also in this category are the Group 13 complex halides wherein $M^I = Ga^I$ or In^I. Six different structure types are observed;[146] these contain $[MX_4]^-$ tetrahedra which are often slightly distorted in the solid state. The structural variety seems to be dictated mainly by the requirements of the counter-ions M^+ which display coordination numbers between six and twelve.

The addition of a large cation, typically R_4N^+ or Ph_4P^+, causes the precipitation of complex halide salts from aqueous solutions of gallium, indium or thallium halides. For gallium, only the $[GaX_4]^-$ complexes are obtained. Indium and thallium show greater variety in their coordination: depending on the choice of cation, they can give solids containing $[MX_4]^-$, $[MX_5]^{2-}$ (in either square pyramidal or trigonal bipyramidal form), $[MX_5(OH_2)]^{2-}$ and $[MX_6]^{3-}$ when X is Cl or Br, although only $[MI_4]^-$ in the iodide case.[145,147-154] Crystallisation of an aqueous solution of $TlCl_3 + CaCl_2$ yields $CaTlCl_5 \cdot 7H_2O$, shown by X-ray analysis to feature the $[Tl_2Cl_{10}]^{4-}$ ion, which has an edge-sharing bisoctahedral structure.[155] The cofacial bis-octahedral ions $[In_2Cl_9]^{3-}$ and $[Tl_2Cl_9]^{3-}$ are encountered in the compounds $Cs_3M_2Cl_9$. The discrete octahedral $[TlCl_6]^{3-}$ and $[TlBr_6]^{3-}$ ions occur in the solids $[Co(NH_3)_6]TlX_6$ which are precipitated when aqueous solutions of the thallium(III) halide and $[Co(NH_3)_6]X_3$ are combined.[5] The complex halide ions of the Group 13 metals are discussed at greater length in chapter 8.

3.2.3   Derivatives with bonds to oxygen

3.2.3.1   Oxides and hydroxides.   Aluminium oxide and hydroxide occur in many natural and synthetic forms, and their applications pervade the modern world. The mineral 'bauxite', crude $Al_2O_3 \cdot xH_2O$ of which the principal constituents are forms of $Al(OH)_3$ and $AlO(OH)$, is the aluminium source for the production of the metal (see chapter 2). Chapter 4 looks in detail at the properties and uses of oxide derivatives of the Group 13 metals. Here we review aspects of their preparation, chemical properties and structures in the context of the inorganic chemistry of the Group as a whole.

There are two principal forms of Al_2O_3 (alumina) which differ in the oxide lattice structures, and consequently in many of their properties. α-Al_2O_3 is the stable form at high temperature and exists as the metastable state under ambient conditions. The h.c.p. lattice of oxide ions has Al^{3+} in octahedral holes, giving AlO_6 coordination.[156] Pure α-Al_2O_3 results from the combustion of the metal or the calcination of aluminium compounds above 1000°C. It is made industrially by strongly heating $Al(OH)_3$, $AlO(OH)$, or the low-temperature phase, γ-Al_2O_3. This other form of alumina, γ-Al_2O_3, is produced via γ-$AlO(OH)$ by the dehydration, below 450°C, of the gelatinous hydroxide precipitated when ammonia is added to a solution of an aluminium salt. γ-Al_2O_3 has a defect-spinel ($MgAl_2O_4$) structure, comprising a c.c.p. oxide lattice with Al^{3+} occupying octahedral and some tetrahedral cation sites. The result is a less dense structure (densities in $g\,cm^{-3}$: γ-Al_2O_3, 3.4; α-Al_2O_3, 4.0).

The major uses of alumina are related to the properties of the polymorphs. α-Al_2O_3 (corundum) is chemically inert, insulating, refractory (m.p. 2045°C), and very hard. It is widely used in refractories, ceramics and abrasives. The crystalline form, when coloured by traces of other metal ions, provides some of the most highly prized gemstones, emerald, ruby and sapphire, for example. There are modifications of the hydroxide and the oxide hydroxide which are structurally related to α-Al_2O_3 and based on a lattice of h.c.p. O or OH constituent ions. These are the materials 'bayerite' [α-$Al(OH)_3$] and 'diaspore' [α-$AlO(OH)$]. α-$Al(OH)_3$ is produced by rapid precipitation on passing CO_2 gas into a cold alkaline aluminate solution. α-$AlO(OH)$, which is stable in the range 280–450°C, can be made by hydrothermal treatment of γ-$AlO(OH)$ in dilute aqueous NaOH at 380°C and 500 atm. Fibrous forms of alumina can now be produced which are impervious to chemical attack and withstand extended heating to 1400°C. Their many applications include the manufacture of lightweight composites of Mg or Al and other space-age materials.[6]

γ-Al_2O_3, the structurally related γ-$Al(OH)_3$ 'gibbsite', γ-$AlO(OH)$ 'boehmite' and other materials obtained by low-temperature dehydration of these less dense forms, are the basis of the 'activated aluminas' employed as absorbents, catalyst supports, chromatographic media and ion exchangers.

This is also the form of alumina favoured for medicinal use. The preparation of aluminium oxide and hydroxide for such applications requires careful control of conditions to regulate factors such as surface area, reactivity, pore size and the degree of hydration of the surface. ^{27}Al solid-state NMR[157] and infrared[158] measurements are particularly important for monitoring both the bulk and surface structural features of aluminas which may include hydroxyl groups in a variety of situations, as illustrated in Figure 3.11.

Another form of Al(OH)$_3$, 'nordstrandite', is made by ageing gelatinous aluminium hydroxide in a solution containing ethylene glycol. This gives a material intermediate in structure between the α- and γ-forms. Special purpose aluminas have been developed for use as fillers, synthetic fibres, etc. Aluminium oxide is also encountered as a film on the surface of the metal. The deliberate production of adherent oxide films on the metal by electrolytic means is the basis of the widely used anodising process.[159] The adsorbent properties of the oxide towards dyestuffs enable the films to be coloured for decorative effect.

Aluminium oxide is a constituent of many ternary and more complex oxide phases.[6] The practical importance of such materials is emphasised in chapter 4. Sodium-β-alumina has an interesting history, having been thought for a long time to be a form of Al$_2$O$_3$. However, sodium oxide is an essential component and the idealised formula is NaAl$_{11}$O$_{17}$ (i.e. Na$_2$O · 11Al$_2$O$_3$).[160] The Na$^+$ ions are situated between aluminium oxide blocks. They are free to diffuse within the structure and can readily be

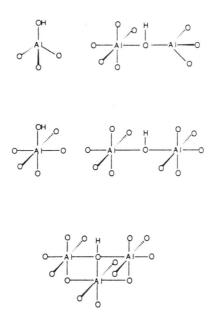

Figure 3.11 Five types of hydroxyl groups which may exist on the high surface area aluminas.[157]

exchanged with other cations from molten salts. Sodium-β-alumina has been thoroughly investigated as a solid-state electrolyte,[161] and this activity has led to the discovery of numerous other alkali-rich aluminates.[162]

The sodium aluminate of 1 : 1 composition, $NaAlO_2$ ($Na_2O \cdot Al_2O_3$), can be made by heating Al_2O_3 with sodium oxalate at 1000°C. The 1 : 1 compound of Al_2O_3 and MgO is the mineral spinel, $MgAl_2O_4$, an important prototype for many complex aluminates. In this structure Al^{3+} occupies octahedral sites in the cubic close-packed oxide lattice. As already noted, γ-Al_2O_3 is an example of a defect-spinel structure: not all the cation sites are occupied and the occupancy is divided between octahedral (AlO_6) and tetrahedral (AlO_4) sites.

Aluminate structures of increasing complexity can be built by the linking of AlO_4 tetrahedra. Na_5AlO_4 ($5Na_2O \cdot Al_2O_3$) contains discrete AlO_4 units with Al–O distances of 1.76 to 1.79 Å. $Na_{17}Al_5O_{16}$ ($17Na_2O \cdot 5Al_2O_3$) has chains of AlO_4 tetrahedra sharing corners, while $Na_7Al_3O_8$ ($7Na_2O \cdot 3Al_2O_3$) contains rings of six AlO_4 groups with further cross-links to build an infinite chain structure; a roughly linear relationship between the mean Al–O bond length and the Al–O–Al angle has been noticed.[163] In discriminating between structures with AlO_4, AlO_5 and AlO_6 groups, ^{27}Al solid-state NMR spectroscopy proves a valuable adjunct to, or substitute for, X-ray crystallography.[164]

Five polymorphs of gallium oxide have been discovered.[165] The stable form, β-Ga_2O_3, m.p. 1725°C, results when other forms of the oxide, or GaO(OH), are heated above 600°C, and can be prepared as crystals from the melt or hydrothermally above 300°C. In β-Ga_2O_3 the cubic close packing of oxide ions is distorted to produce edge-sharing GaO_6 octahedra and vertex-sharing GaO_4 tetrahedra.[166] The lower coordination number of half the Ga atoms causes the density of β-Ga_2O_3 to be c. 10% less than that of the corundum-like α-Ga_2O_3. The preference of Ga^{III} for four-fold coordination in the stable form of its oxide, despite the fact that it is marginally larger than Al^{III}, shows up in other structural contrasts, for example the well known difference between the solid trichlorides (Ga_2Cl_6 molecules, with tetrahedral Ga as against a chloride lattice with Al in octahedral sites).

The ambient temperature forms of In_2O_3 and Tl_2O_3 adopt a different structure from those of Al_2O_3 or Ga_2O_3, namely the type-C rare-earth sesquioxide structure.[8] This is related to fluorite (CaF_2), with vacancies in the oxide lattice such that the coordination of the metal atoms, although six-fold, is highly irregular. There is no obvious electronic reason why In_2O_3 or Tl_2O_3 (or the δ-form of Ga_2O_3) should assume this structure. In_2O_3 and Tl_2O_3 adopt the corundum structure of α-Al_2O_3 when heated under pressure above 1000 (In) or 600°C (Tl).[167] The physical properties of the stable forms of Al_2O_3, Ga_2O_3, In_2O_3 and Tl_2O_3 have been tabulated elsewhere.[2] Metal–oxygen distances in their structures[2,5,156,168] are compared in Table 3.7.

Gallium hydroxide is precipitated by the addition of ammonia to aqueous

Table 3.7 The coordination sphere and metal-oxygen distance in Group 13 metal oxides.

Phase	Coordination sphere(s)	M–O distance (Å)	Reference
α-Al_2O_3 (corundum)	AlO_6 octahedron	1.86 and 1.97	168
α-Ga_2O_3 (corundum)	GaO_6 octahedron	1.83	2
β-Ga_2O_3	GaO_4 tetrahedron GaO_6 octahedron	1.83 2.00	2
α-In_2O_3 (corundum)	InO_6 octahedron	2.07 and 2.27	156
In_2O_3 (C-type)	InO_6 (irregular)	2.18 (average)	2
Tl_2O_3 (C-type)	TlO_6 (irregular)	2.26 (average)	5

Ga^{3+} solutions. This gelatinous material forms crystalline GaO(OH) after dehydration at 100°C. Heating GaO(OH) in air to 450–550°C converts it into α-Ga_2O_3. In the case of indium, addition of ammonia in slight excess to aqueous In^{3+} solutions at 100°C, followed by ageing of the precipitate at this temperature, yields crystalline $In(OH)_3$. The conversion to InO(OH) requires heating to 250–400°C and high pressure.[169] The oxide, In_2O_3, can be obtained by heating the hydroxide (or indium nitrate) to constant weight in air. Strong heating *in vacuo* will remove oxygen, producing some In_2O and/or In metal. The hydroxides $Ga(OH)_3$ and $In(OH)_3$ are only very slightly soluble in water. Gallium hydroxide is amphoteric, dissolving in acids, and forming gallates with alkali, which are akin to aluminates. By contrast, indium hydroxide is, to all intents and purposes, a basic hydroxide. It reacts readily with acids to give indium salts, but indates are obtained only by using extremely alkaline conditions and are decomposed by water to regenerate the hydroxide.

Addition of hydroxide to an aqueous thallium(III) solution precipitates a brown solid approximating to the composition $Tl_2O_3 \cdot 1.5H_2O$. This is easily dehydrated to Tl_2O_3. Alternatively, an alkaline thallium(I) solution can be oxidised by hydrogen peroxide, followed by removal of water and other volatiles by heating to 450°C in an atmosphere of oxygen. The Tl^{III} oxide is insoluble in water but dissolves in acids to give Tl^{3+} salt solutions. Tl_2O_3 forms black crystals which decompose to Tl_2O and O_2 if heated, but survive to the melting point (716°C) under 1 atm of O_2. A mixed valence oxide, Tl_4O_3 ($3Tl_2O \cdot Tl_2O_3$) can be prepared by heating a mixture of Tl_2CO_3 and Tl_2O_3 in the proportions 3 : 1 at 450°C in an inert atmosphere. Thallates of the Tl^{III} state can be prepared by fusing together the oxides concerned, e.g. $NaTlO_2$ ($Na_2O \cdot Tl_2O_3$) and Li_3TlO_3 ($3Li_2O \cdot Tl_2O_3$).[5] The mixed oxide Tl_4O_3 is, in effect, a thallate of this kind. There are ternary oxides of the M_2O_3 perovskite type which contain Tl^{III}, e.g. $FeTlO_3$ and $CrTlO_3$. Interest in thallium oxide phases has been stimulated by the discovery that thallium is a constituent of some superconducting ceramics, e.g. $Tl_2Ba_2Ca_{m-1}Cu_mO_{2m+4}$ and $TlA_2Ca_{m-1}Cu_mO_{2m+3}$ (A = Sr or Ba).[170]

3.2.3.2 Hydrated cations.

The hydrated ions, $[M(OH_2)_6]^{3+}$, exist in acidic aqueous solutions and crystalline salts of Al, Ga, In and Tl. X-Ray studies of M^{3+} aqueous solutions (usually with ClO_4^- as the counter-ion) give consistent results for the M–O internuclear distances,[171] as shown in Table 3.8. Insight into cation hydration can also be gained from observations of the totally symmetric M–OH_2 stretching vibration in Raman spectra. For aqueous solutions of the M^{3+} ions the values of ν_{sym}(M–OH_2) are: Al 525, Ga 520, In 475 and Tl 450 cm^{-1}. The value for $[Al(OH_2)_6]^{3+}$ seems lower than expected in comparison with the others, and it has been suggested[172] that this arises from the mismatching of the small Al^{3+} ion with the size of the cavity created by the coordinated water molecules. Spectroscopic data are also available for the $[M(OH_2)_6]^{3+}$ ions in alum crystals[173] wherein the ν_{sym}(M–OH_2) values are: Al 542, Ga 537 and In 505 cm^{-1}. For $AlCl_3 \cdot 6H_2O$, which also contains the $[Al(OH_2)_6]^{3+}$ ion, ν_{sym}(M–OH_2) is 524 cm^{-1}.[174] Crystallographic analyses of alums and other hydrates have repeatedly demonstrated the octahedral coordination of the metal by six water molecules. The Al–OH_2 bond lengths are typically in the range 1.85–1.89 Å and hydrogen bonding to the anion is a common feature of such hydrates.[175]

The hexahydrated cations of the aluminium group are acidic:

$$[M(OH_2)_6]^{3+} + H_2O \rightleftharpoons [M(OH_2)_5(OH)]^{2+} + H_3O^+ \qquad (3.12)$$

The acid strength, measured in the presence of non-complexing anions, is least for aluminium and increases irregularly for the other members of the series as shown by the following pK_a values:[176] Al^{3+} 4.9, Ga^{3+} 2.6, In^{3+} 3.9, Tl^{3+} 0.6 (see Table 3.8 for a comparison with other data). The mononuclear cation $[Al(OH_2)_5(OH)]^{2+}$ has been detected in dilute solution by ^{27}Al NMR spectroscopy,[177] but polymerised hydrolysis products dominate the

Table 3.8 Some properties of the hydrated cations $[M(OH_2)_6]^{3+}$.

Species	M–O distance (Å)[a]	Totally symmetric mode (cm^{-1})[b]	pK_a^c
$[Al(OH_2)_6]^{3+}$	1.89	525	4.9
$[Ga(OH_2)_6]^{3+}$	1.90[d]	520	2.6
$[In(OH_2)_6]^{3+}$	2.16	475	3.9
$[Tl(OH_2)_6]^{3+}$	2.23	450	0.6

(a) Ref. 171.
(b) Ref. 172. Results refer to concentrated aqueous solutions, typically with M^{3+}: H_2O = 1:20.
(c) Ref. 176. Values given at zero ionic strength.
(d) Estimated value.

chemistry of Al^{3+} in aqueous solutions over a wide range of concentration and pH.[77,178] Knowledge of these, which is important in understanding the physiological behaviour of aluminium and its toxicity to life forms, has motivated considerable research into the chemical nature of aluminium(III) hydrolysis species.[179] The dimer, $[(H_2O)_4Al(\mu\text{-}OH)_2Al(OH_2)_4]^{4+}$, various trimers, and the terdecamer, $[AlO_4Al_{12}(OH)_{24}(OH_2)_{12}]^{7+}$, are especially well characterised.[178–182] The last species gives distinctive ^{27}Al NMR signals for the central AlO_4 moiety and the surrounding edge-linked AlO_6 units. Both the dimer and the Al_{13} complex, shown in Figure 3.12, have been identified in crystalline basic salts.[180] Control of the degree of hydrolysis, and hence the polymerisation, of aqueous Al^{III} solutions plays an important part in the use of hydrolysed aluminium solutions in waste water treatment,

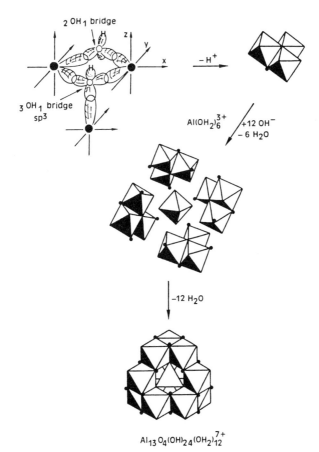

Figure 3.12 The aluminium terdecamer $[Al_{13}O_4(OH)_{24}(OH_2)_{12}]^{7+}$ found in the crystal structures of some basic aluminium salts.[183] Reproduced with permission from M. Henry, J.P. Jolivet and J. Livage, *Struct. Bonding (Berlin)*, 1992, **77**, 153.

e.g. for the trapping of particulate matter, and for the removal of fluoride as fluoroaluminate complexes from the effluent of aluminium manufacturing plants.[183-185] Production of the Al_{13}^{7+}-modified Keggin-type ion, alone or in combination with silicates, titanates, molybdates and other heteropolyions, is part of the developing technology of pillared layer catalyst systems. Techniques such as carefully controlled base hydrolysis of aqueous aluminium chloride solutions in the presence of alkylamines are employed.[186-188] Recent work has extended the investigations of pillaring to Al–Ga and Ga polyoxycations.[189]

As shown by the behaviour depicted in Figure 3.13, aqueous Ga^{3+} solutions tend to be more acidic than those of Al^{3+}. Their hydrolysis may involve the presence of $[Ga(OH)]^{2+}$ and $[Ga(OH)_2]^+$ species, according to

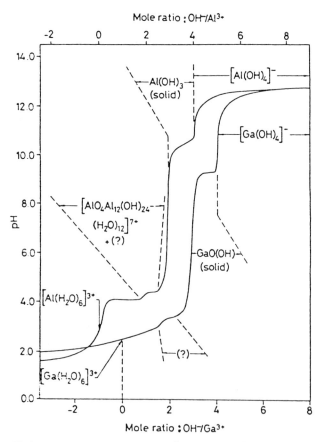

Figure 3.13 Hydrolytic behaviour of 0.2 mol dm^{-3} AlCl$_3$ and GaCl$_3$ aqueous solutions showing the speciation involved. Note that the OH/M scales for aluminium and gallium(III) are offset from one another by one unit. Reproduced with permission from S.M. Bradley, R.A. Kydd and R. Yamdagni, *J. Chem. Soc., Dalton Trans.*, 1990, 413.

potentiometric studies.[190,191] However, NMR investigations of the hydrolysis find no evidence of monomer (other than $[Ga(OH_2)_6]^{3+}$) and scant evidence of other species, but reveal the terdecamer, $[GaO_4Ga_{12}(OH)_{24}(OH_2)_{12}]^{7+}$, at pH 3–4.[191,192] This complex appears to be much less stable than the analogous aluminium species. NMR spectra show that $[GaO_4Al_{12}(OH)_{24}(OH_2)_{12}]^{7+}$ forms when combined Al^{III} and Ga^{III} aqueous solutions are hydrolysed. This proves to be more stable than either the Al_{13} or Ga_{13} species,[193] possibly because the somewhat larger Ga^{3+} ion provides a better fit into the structure of the polyoxycation. As these results would suggest, Ga^{III}, like Al^{III}, can also participate in the formation of hetero-polyanions of the Keggin type alongside molybdate or tungstate.[194]

Hydrolysis of Al^{III} or Ga^{III} solutions beyond an OH/M ratio of 2.5 : 1 produces a gel, followed at a ratio of 3 : 1 by the precipitation of $Al(OH)_3$ in the one case and $GaO(OH)$ in the other. These compounds are amphoteric and dissolve in the presence of alkali as well as in acids. Mononuclear aluminate, $[Al(OH)_4]^-$, or gallate, $[Ga(OH)_4]^-$, anions are the dominant species in NaOH solutions. At high concentrations, ^{27}Al NMR spectra can also detect species such as $[(HO)_3Al(\mu\text{-}O)Al(OH)_3]^{2-}$ related in compo-sition to the sodium aluminate hydrate phase which crystallises from the solution.[195]

The indium cation in non-complexing aqueous solutions, such as per-chloric acid, is $[In(OH_2)_6]^{3+}$. As in the gallium case, potentiometric investigations of the hydrolysis point to the presence of mononuclear $[In(OH)]^{2+}$ and $[In(OH)_2]^+$ species, as well as several polynuclear complexes.[196] The dimer $[In_2(OH)_2]^{4+}$, like the monomers, is expected to contain six-coordinate indium with additional H_2O ligands. The probable structure is $[(H_2O)_4In(\mu\text{-}OH)_2In(OH_2)_4]^{4+}$. A complex of higher nuclearity may be $[In_4(OH)_6]^{6+}$. A tetrameric structure built from InO_6 octahedra sharing corners with single hydroxo bridges between the indium atoms (Figure 3.14), is consistent with X-ray scattering data for In^{III} solutions.[197] The structure in question has been identified as the $[In_4(\mu\text{-}OH)_6]^{6+}$ cationic core of the crystalline hydrolysis product of (1,4,7-triazacyclononane)InBr$_3$ in alkaline solution.[198]

Base-induced hydrolysis of In^{III} solutions causes precipitation well before the OH/M ratio reaches 3.0. As in the Al and Ga systems, the outcome depends on the nature of the counter-anion and on the particular base, and is influenced by temperature and by ageing of the precipitate. If halide ions are present, they are likely to be incorporated in the precipitate, e.g. $In_2(OH)_3Cl_3$,[199] but in their absence $In(OH)_3$ is the eventual product. In contrast to $Al(OH)_3$ and $Ga(OH)_3$, the indium compound is essentially a basic hydroxide. A slight solubility in concentrated NaOH solution has been ascribed to the formation of $[InO(OH)_2]^-$,[200] but it seems more reasonable to formulate the solution species as four-coordinate $[In(OH)_4]^-$ or six-coordinate $[In(OH)_4(OH_2)_2]^-$.

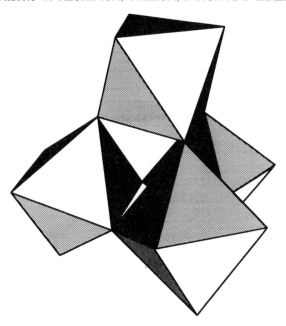

Figure 3.14 Model for the $[In_4(OH)_6]^{6+}$ complex consisting of four corner-sharing InO_6 octahedra with the indium atoms forming a regular tetrahedron.[197]

Thallium in the +3 state is strongly oxidising and is unstable in alkaline media. $[Tl(OH_2)_6]^{3+}$ exists in acidic solutions, and $Tl(ClO_4)_3$ at a concentration of $1\,mol\,dm^{-3}$ can be prepared by anodic oxidation of $TlClO_4$. The structure of the aquated cation is borne out by X-ray studies of the solution.[201] The crystalline solid $Tl(ClO_4)_3 \cdot 6H_2O$ also contains the exactly octahedral $[Tl(OH_2)_6]^{3+}$ ion with $Tl–OH_2$ bond lengths of $2.23\,\text{Å}$.[202] In contrast, the tetrahydrates $TlX_3 \cdot 4H_2O$ (X = Cl or Br) have crystal structures based on trigonal bipyramidal TlO_2X_3 units linked by hydrogen bonds through the remaining two water molecules.[203] Halide coordination persists in aqueous solutions of these compounds where a series of aquo-halogeno complexes occurs (section 3.2.2.2).

A feature of the aqueous solution chemistry of the Group 13 cations is that the extent to which ligands other than H_2O and OH^- enter the coordination sphere of the M^{3+} ions varies considerably between the members of the Group. Such coordination is least important for Al^{3+} solutions, except in the case of multiply charged anions, notably phosphate which forms a series of complexes identifiable by Raman and ^{31}P NMR spectroscopy.[204] For Ga^{3+} aqueous solutions there is similar evidence of complexing by phosphate, although the actual structures are still in doubt.[143] There is spectroscopic evidence of sulfate complexes in aqueous solutions of Al^{3+},[205] Ga^{3+} [143] and In^{3+};[199] the species concerned is probably $[M(SO_4)(OH_2)_4]^+$ but could also be $[M(SO_4)_2(OH_2)_2]^-$ (see section 3.2.3.5). Coordination of nitrate is significant in the case of In^{3+} solutions, and here the species appears to be $[In(NO_3)(OH_2)_5]^{2+}$ in which the nitrate ligand is monodentate.[206]

The replacement of the H_2O ligand in aqueous solutions by other species, including various uni-, bi- and multidentate complexing agents, shows an overall trend towards faster ligand substitution rates in the order $Al <$ $Ga \ll In$.[207] The hydrolysed species $[M(OH)(OH_2)_5]^{2+}$ react faster than the $[M(OH_2)_6]^{3+}$ ions by at least two orders of magnitude. The rate of exchange of water molecules of the aquated cation with those of the solvent is much faster for $[Ga(OH_2)_6]^{3+}$ than for $[Al(OH_2)_6]^{3+}$, according to ^{17}O NMR measurements of relaxation rates,[208] which imply dissociative activation in both cases. Dissociative or dissociative interchange mechanisms are now favoured for most ligand-exchange reactions of M^{III} complexes with unidentate ligands.[207]

3.2.3.3 Oxide halides.   Methods of preparation, structures and reactions of the Group 13 oxide fluorides were reviewed by Siegel in 1968,[209] and by Holloway and Laycock in 1983.[210] Solids of the formula MOF are known for all members of the Group. Siegel prepared AlOF from AlOBr using BrF_3 as the fluorinating agent, and obtained other fluorinated products, up to Al_2OF_4, from this reaction.[209] The addition of ammonia to solutions of aluminium sulfate containing various proportions of AlF_3 or the hydrolysis of AlF_3 at high temperature affords hydroxide fluorides, $Al(OH)_xF_{3-x}$, but does not provide a satisfactory route to the oxide fluoride. The solids with compositions between $Al(OH)_2F$ and $Al(OH)F_2$ have a common cubic structure in which F^- and OH^- are close-packed with Al^{3+} in octahedral holes. Gaseous AlOF is formed above 1200°C by the reaction

$$Al_2O_3(s) + AlF_3(g) \rightleftharpoons 3AlOF(g) \qquad (3.13)$$

which is reversed on cooling. Molecular AlOF has been obtained in a rare-gas matrix by the reaction of AlF with O_2 and shown to form the oxygen-bridged dimer $FAl(\mu-O)_2AlF$.[211] The gallium compound, GaOF, has been prepared from Ga_2O_3 and MnF_2 at 800°C, and by the action of fluorine on GaOI at room temperature.[209] The GaOF molecular species is formed in an argon matrix from GaF and oxygen atoms.[212] Indium oxide fluoride, InOF, has been prepared from either InOI or InOCl by the action of F_2 at 100°C,[213] and from In_2O_3 with InF_3 by a sealed-tube reaction at 900°C.[214] TlOF is readily prepared from Tl_2O_3 using either gaseous HF at 100°C or a 40% aqueous HF solution at room temperature.[210] X-Ray studies have revealed sixfold coordination of indium by oxygen and fluorine in crystalline InOF[215] and $In(OH)F_2$,[216] and eightfold coordination of thallium in TlOF.[217]

Whereas the oxide fluorides are inert materials, the other Group 13 oxide halides MOX (X = Cl, Br or I) tend to decompose under vacuum or on heating, and are hydrolysed by water or moist air. A recommended means of preparing the aluminium oxide halides[2] is by the reaction

$$3AlX_3 + As_2O_3 \xrightarrow{250°C} 3AlOX + 2AsX_3 \quad (X = Cl, Br \text{ or } I) \quad (3.14)$$

All three compounds AlOX decompose when heated under vacuum, with vaporisation of the halide AlX_3 and formation in stages of the solids Al_3O_4X, Al_5O_7X and Al_2O_3.[218] The compounds AlOX are hygroscopic and are hydrolysed to γ-AlO(OH). With ammonia they are said to form $AlO(NH_2)$ which decomposes at 120°C to Al_2O_3, AlN and NH_3.[2] Reactions with an elemental halogen proceed with replacement of a heavier by a lighter halide substituent; thus AlOBr reacts with chlorine at 100°C to form AlOCl, and AlOI can be converted to any of its congeners, including AlOF, by this means.

Gallium oxide iodide[209] and indium oxide iodide[213] are synthesised by heating the iodide MI_3 to 300°C in a stream of oxygen. Halogen displacement reactions can be used to prepare the compounds MOCl and MOBr (M = Ga or In) from MOI. No oxide iodide of thallium has been obtained. On the other hand, TlOCl has been prepared by the prolonged action of Cl_2O on TlCl, and TlOBr by the reaction of $TlBr_3$ with ozone.[5,219] Both compounds decompose readily.

3.2.3.4 *Alkoxides and aryloxides.*

Aluminium alkoxides, particularly the isopropoxide $Al(OPr^i)_3$, are convenient reagents from which to prepare other aluminium compounds, being readily available, inexpensive, and soluble in most organic solvents. They find a number of uses in synthetic organic chemistry, e.g. as catalysts in the oxidation of alcohols to aldehydes or ketones (Oppenauer oxidation) or the reverse process (Meerwin–Ponndorf–Verley reduction). The catalytic step is thought to depend on the ability of the unsaturated aluminium centre to attract carbonyl groups, leading to hydrogen transfer reactions within the coordination sphere.

Several methods are available for the synthesis of aluminium, gallium or indium alkoxides and aryloxides.[220–222]

1. Direct reaction of the metal with an alcohol or phenol under reflux:

$$M + 3ROH \rightarrow M(OR)_3 + 3/2H_2 \quad (3.15)$$

   This method is successful for the lower alcohols, in the presence of $HgCl_2$ or I_2 to initiate the reaction, and for phenol which requires no catalyst.
2. Reaction of MCl_3 with an alcohol ROH or with NaOR. This is liable to

generate mixtures of compounds of the type $MCl_{3-x}(OR)_x$. It can be driven to completion (a) by passing ammonia into the mixture to remove hydrogen chloride as NH_4Cl or (b) by using an excess of NaOR, with the alcohol as solvent from which NaCl is deposited.

3. From $Al(OR)_3$ ($R = Et$ or Pr^i) by alcoholysis or by transesterification:

$$Al(OR)_3 + nR'OH \rightleftharpoons Al(OR)_{3-n}(OR')_n + nROH \quad (3.16)$$

$$Al(OR)_3 + nAcOR' \rightleftharpoons Al(OR)_{3-n}(OR')_n + nAcOR \quad (3.17)$$

This method provides a way of preparing the higher alkoxides or aryl-oxides from the ethoxide or the isopropoxide, since the lower boiling alcohol or ester can be removed azeotropically with benzene to displace the equilibrium to the right.

4. From an organoaluminium compound R_3Al by reaction with an alcohol or phenol. This route can be used to obtain $Al(OR')_3$, but its main use is for the synthesis of organoaluminium alkoxides, R_2AlOR' and $RAl(OR')_2$.

5. From the amide $Al(NMe_2)_3$ by reaction with an excess of Bu^tOH in benzene.

A method for the direct electrochemical synthesis of metal alkoxides[223] by anodic oxidation of the metal in a conducting solution of a halide in ethanol or isopropanol gives polymeric products $AlO_x(OR)_yX_z$ when applied to aluminium. This proved, however, a satisfactory route to $Ga(OPr^i)_3$ and gave a 65% yield based on the current passed. A new method[224] for the preparation of monomeric alkoxides (e.g. $Ba(OR)_2$) employing $CH_3(OCH_2CH_2)_nOH$ ($n = 2$ or 3), which stabilises the product in part by chelation, may be applicable to the Group 13 metals, especially the heavier members.

The aluminium alkoxides are white powders, waxy solids or viscous liquids, which sublime or can be distilled by heating under reduced pressure. $Al(OMe)_3$ sublimes under vacuum at 240°C; $Al(OEt)_3$ melts at c. 130°C, b.p. 210°C/10 mm; $Al(OPr^i)_3$ melts at c. 118°C, b.p. 145°C/10 mm; $Al(OBu^s)_3$ is liquid, b.p. 204°C/30 mmHg. The trend in physical properties reflects the tendency of the $Al(OR)_3$ molecules to form associated structures. Association is most extensive for the methoxide which is a solid, but is hindered by the bulk of the ligand in the case of the higher alkyl derivatives. $Al(OBu^t)_3$ and $Al(OPr^i)(OBu^t)_2$ exist as dimers with terminal and bridging alkoxide ligands.[225] $Al(OPr^i)_3$, which is also dimeric in the vapour phase, condenses to a liquid which remains supercooled for a considerable time. Once solidified, the product has a melting point that tends to

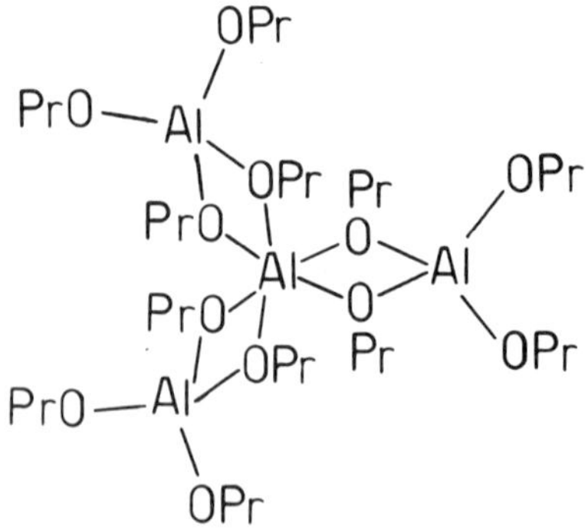

Figure 3.15 The 'Mitzubishi motif' structure of the Al(OPr^i)$_3$ tetramer.[226] Source: *Polyhedron*, 1991, **10**, 1639. Bond distances: Al–O$_{core}$, 1.92; Al–O$_{br}$, 1.80; Al–O$_{ter}$, 1.70 Å.

increase over time. The explanation of this 'ageing' phenomenon is that the product is trimeric when freshly distilled but slowly transforms to a tetra-meric structure in which the central Al atom achieves octahedral co-ordination *via* two bridging alkoxide ligands from each of three Al(OPr^i)$_4$ groups.[226] Figure 3.15 shows this structure.

Above 200°C at atmospheric pressure, the aluminium alkoxides decompose into mixed alkoxide-hydroxides and -oxides, ultimately to leave Al$_2$O$_3$.[227] Hydrolysis is rapid and causes the liquid alkoxides to form a skin on exposure to air. The action of water liberates the alcohol and forms white solid materials 'AlO(OR)' in which Al–O–Al bonds are a feature. The controlled hydrolysis of Al(OR)$_3$ mixtures dissolved in alcohols gives sols and gels which can be sintered into oxide ceramics.[228–230] Aluminosiloxanes, e.g. Al(OPr^i)$_{3-x}$(OSiMe$_3$)$_x$, made from aluminium isopropoxide and tri-methylsilyl acetate by transesterification, can also be used as precursors to gels and ceramics.[231] The siloxide groups are less susceptible to hydrolysis than are the alkoxide ones: very homogeneous gels are formed leading to Si–O–Al bonded ceramics with improved properties as refractory materials.

The effect of incorporating OSiMe$_3$ groups into an aluminium alkoxide is to favour smaller oligomers. Al(OSiMe$_3$)$_3$, m.p. 238°C, b.p. 155°C/1 mmHg, was first prepared more than 30 years ago[232] by the reaction of sodium trimethylsiloxide with AlCl$_3$, and shown on spectroscopic evidence to be a dimer, in contrast to the boron compound which is a volatile, mono-meric substance. Recently the bridged dimeric structure (Me$_3$SiO)$_2$-Al(μ-OSiMe$_3$)$_2$Al(OSiMe$_3$)$_2$ has been confirmed by X-ray crystal-lography.[233,234] A similar structure with tetrahedral coordination of aluminium has also been established for Al$_2$(OBu^t)$_6$. The central

$Al(\mu\text{-}O)_2Al$ moiety is planar with Al–O bond lengths of 1.82 Å. With lengths of 1.69 Å, the terminal Al–O bonds are considerably shorter than the sum of the relevant covalent radii.

The four-membered Al_2O_2 ring of alkoxide-bridged dimers is akin to that of the alumoxanes with Al–O–Al bonds which result from the reactions of organoaluminium compounds with water or species containing active oxygen.[235] In some cases, e.g. dimethylaluminium methoxide from the reaction of Me_3Al with MeOH, the product is a trimer, $[Me_2AlOMe]_3$, the structure of which contains a non-planar Al_3O_3 ring with the methoxide group acting as the bridging ligand.[236] Examples of the corresponding gallium and indium compounds are known. There are interesting parallels between the structural chemistry of polymeric alkoxides and alumoxanes[237] and that of the various oligomers with $(Al\text{-}N)_n$ frameworks which occur in systems with amido and imido ligands, to be discussed in section 3.2.5.

Structures with OR groups bridging aluminium to a second metal are formed by complex alkoxides such as $K[Al(OBu^t)_4]$, m.p. 164°C, and $Mg[Al(OPr^i)_4]_2$, m.p. 20°C. The compounds with other mixed systems, for instance the condensation products of $Al(OPr^i)_3$ with metal acetates, extend the range of properties of the metal alkoxides. Hydrolysis of $Mg[Al(OPr^i)_4]_2$ to spinel $MgAl_2O_4$ demonstrates the molecular precursor route to glasses and solid oxides which makes use of such products.[238] The formation of $NbAlO_4$ from amorphous materials prepared by the simultaneous hydrolysis of niobium and aluminium isopropoxides illustrates current techniques in this area.[239]

Compared with the Al compounds, the alkoxides of Ga, In and Tl^{III} have received scant attention. Gallium methoxide, $[Ga(OMe)_3]_x$, is a polymeric solid which decomposes without melting, but can be sublimed at 280°C/0.5 mmHg, whilst the ethoxide is crystalline and melts at 144°C. The liquids $[Ga(OPr^i)_3]_2$ (b.p. 120°C/1 mmHg) and $[Ga(OBu^t)_3]_2$ are dimeric according to NMR evidence.[225] The siloxide $Ga(OSiMe_3)_3$, being monomeric, contains a Ga atom with trigonal coordination, but the formation of a tetrahedral complex ion $[Ga(OSiMe_3)_4]^-$ is testimony to the unsaturation of the metal centre. In general, the Group 13 alkoxides are weak Lewis acids and give 1 : 1 adducts with ammonia, pyridine and other bases.

Gallium alkoxides, $Ga(OR)_3$, and chloride alkoxides, $GaCl_2(OR)$ and $GaCl(OR)_2$, have been obtained from $GaCl_3$ via the adducts $Ga_2Cl_6 \cdot ROH$ and $GaCl_3 \cdot ROH$.[204] Aryloxides have been prepared by the reaction of the phenol ArOH with Ga metal and shown to form pyridine adducts $Ga(OAr)_3 \cdot py$.[241] Similar preparative methods should be applicable to the indium compounds. However, the attempt to make $In(OPr^i)_3$ by reaction of anhydrous $InCl_3$ with 3 equiv. of $NaOPr^i$ in a benzene-isopropanol mixed solvent yielded the pentanuclear oxoalkoxide, $In_5O(OPr^i)_{13}$.[242] X-Ray analysis shows the presence of the μ_5-oxygen atom bonded in a square pyramidal configuration to five metal atoms which carry terminal OPr^i groups *trans* to the oxo atom. Variable temperature ^1H NMR studies reveal a greater lability of the bridging OPr^i groups within the indium cluster compared with the terminal ligands.[242] Thallium alkoxides have been little explored, and the way to thallium aryloxides would seem to be closed by the oxidation of phenols by the thallium(III) centre. A single alkoxide ligand can be attached to thallium by the reaction of the alcohol with thallium(III) acetate to produce $Tl(OAc)_2(OR)$ (R = Me, Et or Pr^i).[5]

The synthesis of volatile alkoxides or aryloxides of the Group 13 elements (like that of the corresponding thiolates and selenates) is of practical importance in the search for precursors to electronic materials (see chapter 5). It is also of some interest in the context of multiple bonding involving the heavier Main Group elements. Calculations and structural studies[243,244] support the view that π-type interactions account for the short M–O distances, e.g. Al–O = 1.71 Å, exhibited by compounds of the type Bu^t_2MOAr (M = Al or Ga; Ar = bulky aryl group). On the other hand, molecular orbital calculations on the model compound $Al_2(OH)_6$, performed in conjunction with the X-ray analysis of $Al_2(OBu^t)_6$, suggest that the extent of possible Al–O π-bonding is small and that the major cause of the short Al–O distances is the ionic contribution to the bonding.[233]

Halide alkoxides of the type $MX_{3-x}(OR)_x$ were mentioned earlier as possible products of the reaction between the metal halide MX_3 and the alcohol ROH. Their synthesis is a difficult problem which has recently received attention in the case of aluminium chloride alkoxides and their thf adducts.[245] The action of HCl or an acyl halide on $Al(OR)_3$ can be employed:[222]

$$Al(OR)_3 \xrightarrow{HCl} AlCl_{3-x}(OR)_x + xROH \qquad (3.18)$$

$$Al(OR)_3 + xCH_3COX \rightarrow AlX_x(OR)_{3-x} + xCH_3COOR \qquad (3.19)$$

Note, however, that the reaction of the tertiary butoxide, $Al(OBu^t)_3$, with CH_3COCl has been shown to follow a different route, with Bu^tCl and aluminium acetate as the chief products. The evidence suggests that the Al and Ga halide alkoxides may be di-, tri- or tetrameric, but that in all cases alkoxide ligands occupy the bridging sites.

3.2.3.5   Oxyanion salts and mixed oxides.

Salts of Group 15, 16 or 17 oxanions.   The combination of an aluminium group cation with an oxyanion of a Group 15, 16 or 17 element gives rise commonly to hydrated salts. As such, the compounds display *inter alia* the properties of the $[M(OH_2)_6]^{3+}$ ions, the characteristic behaviour of which includes the tendency to produce hydrolysed species, and to incorporate the oxyanion in the coordination sphere of the metal. These aspects of the aqua cations have been considered already in section 3.2.3.2. The oxyanions which form compounds with members of the aluminium sub-group are usually the more highly oxidised species, although Ga, In and Tl can form nitrites and NO_2^- complexes.[5,7,246] Sulfites, selenites and tellurites are mentioned in earlier literature[2] but have received little attention recently.

The usual nitrate of aluminium is $Al(NO_3)_3 \cdot 9H_2O$, m.p. 73°C, and which is very soluble in water and in ethanol. Hydrates with less water crystallise from nitric acid solutions. Heating to 135°C produces a basic nitrate, $Al(OH)_2NO_3 \cdot 1.5H_2O$,[247] and strong heating leads to amorphous alumina. Gallium forms $Ga(NO_3)_3 \cdot 9H_2O$, whereas indium yields $In(NO_3)_3 \cdot xH_2O$ ($x = 3$, 4 or 5) from aqueous nitric acid.[248,249] Thallium(III) nitrate crystallises as $Tl(NO_3)_3 \cdot 3H_2O$ which readily forms the oxide, turning brown in the process. The crystal structure of this nitrate shows both NO_3 and H_2O to be bonded to the metal;[250] by contrast, $Al(NO_3)_3 \cdot 9H_2O$ contains the $[Al(OH_2)_6]^{3+}$ ion and additional water of crystallisation. The preparation of the anhydrous nitrates $M(NO_3)_3$ (M = Al, Ga or In) demands special conditions and techniques which involve the reactions of N_2O_4, N_2O_5 or $ClNO_3$ with the metal or the halide MX_3.[7,251]

Aluminium sulfate is manufactured as $Al_2(SO_4)_3 \cdot 16H_2O$, which is the hydrate to crystallise from sulfuric acid solutions at ambient temperatures. The corresponding selenate is obtained from selenic acid solutions. Thermal decomposition of $Al_2(SO_4)_3 \cdot 16H_2O$ yields $Al_2(SO_4)_3$ at 450° and Al_2O_3 at temperatures above 600°C. Various basic salts and also acidic HSO_4^- derivatives are known.[2,7] Gallium sulfate forms $Ga_2(SO_4)_3 \cdot 18H_2O$ which loses water in stages to give $Ga_2(SO_4)_3$ at temperatures above 150°C. The indium and thallium sulfates occur with reduced levels of hydration as $In_2(SO_4)_3 \cdot H_2O$ and $In_2(SO_4)_3 \cdot 5H_2O$, and $Tl_2(SO_4)_3 \cdot 7H_2O$ (from dilute solution) or $Tl_2(SO_4)_3 \cdot 4H_2O$ (from concentrated sulfuric acid); heating gives the anhydrous salts and ultimately results in decomposition to the appropriate oxide. The structures of the Al, Ga, In and Tl sulfates are based on octahedral MO_6 coordination, irrespective of whether the ligands are H_2O, OH or SO_4 (generally bidentate). Examples include the crystalline hydrates, basic sulfates such as $Al_2(OH)_4(SO_4) \cdot 7H_2O$ (aluminite)[252] and $M_2(OH)_2(SO_4)_2 \cdot 4H_2O$ (M = In or Tl).[5,7] Acidic complexes of the type $HM(SO_4)_2 \cdot 4H_2O$ have also been described (M = Al, Ga or In)[253] and X-ray analysis shows that they must be formulated as $[H_5O_2]^+[M(OH_2)_2(SO_4)_2]^-$.

The aluminium group metals form a large number of double salts, the sulfates (both hydrated and anhydrous) being particularly numerous. The term 'alum' covers the crystalline double sulfates (and double selenates) analogous to $KAl(SO_4)_2 \cdot 12H_2O$ which exist for many combinations of M(I) and M(III) ions. There are also 'alum anhydrides' $M^IM^{III}(SO_4)_2$ and 'pseudoalums' in which a single M^{2+} cation replaces $2M^+$ of an ordinary alum, as well as the equivalent selenates.[2]

Salts of the M^{III} ions with the halogen oxyanions are highly hydrated and likely to contain the $[M(OH_2)_6]^{3+}$ cation in their crystals. Their thermal decomposition proceeds *via* basic salts.[254] Crystalline gallium(III) perchlorate, $Ga(ClO_4)_3 \cdot 6H_2O$, is easily made in high yield by dissolving the metal in concentrated perchloric acid,[255] and is the usual choice of compound from which to prepare aqueous Ga^{3+} solutions. The ease with which $HClO_4$ is able to dissolve gallium, in contrast to the slow attack of this metal by concentrated HCl and other acids, suggests that its role includes the further oxidising of the Ga^{II} species which are the first products when Ga dissolves in halogen acid solutions (see section 3.3.3). The perchlorates of In^{III} and Tl^{III} crystallise as $In(ClO_4)_3 \cdot 8H_2O$ and $Tl(ClO_4)_3 \cdot 6H_2O$. The perchlorate ion is normally assumed to be non-ligating towards M^{3+} ions. The high viscosity of concentrated aqueous solutions of aluminium, gallium or indium perchlorate is due to the susceptibility of the $[M(OH_2)_6]^{3+}$ cations to hydrolysis and polymerisation (see section 3.2.3.2). There has been some recent investigation of the anhydrous perchlorates, $M(ClO_4)_3$ for M = Al, Ga, In or Tl, and perchlorato complexes, $M^I_nAl(ClO_4)_{3+n}$ ($n = 1, 2$ or 3).[256] Fluorosulfates and chlorosulfates, $M(SO_3X)_3$ (M = Al, Ga or In; X = F or Cl) have been obtained: they are synthesised by the action of SO_2, SO_3, $S_2O_6F_2$ or HSO_3F on the chloride MCl_3 in moisture-free conditions,[7] and typically undergo hydrolysis to $M(SO_3OH)_3$.

Borates, carbonates, silicates, phosphates and arsenates. These compounds differ markedly from the hydrated salts just discussed and are best regarded as mixed oxides of the Group 13 metals and other oxides of Groups 13–15. Aluminium borates, $AlBO_3$ and others loosely designated $(Al_2O_3)_n \cdot B_2O_3$, amount to chemical variants of alumina,[257a] with some utility as catalysts, synthetic fibres, special purpose glasses, etc. The solid phases are thermally stable at temperatures up to *c.* 1000°C, above which B_2O_3 volatilises from the system.[257b] Borates $GaBO_3$ and $InBO_3$, isostructural with $CaCO_3$, have been obtained by heating Ga_2O_3 or In_2O_3 with boric acid.[2]

There are no simple carbonates of the aluminium group metals, only some basic salts in which carbonate is present, e.g. $NH_4[Ga(OH)_2CO_3]$. On the other hand, the aluminosilicates provide a huge array of minerals and synthetic materials of vast technological significance, e.g. clays, pillared clays and zeolites. Gallosilicate zeolites can be synthesised[258] and these offer advantages in some particular applications. Pillared clays incorporating either gallium polyoxycations[259] or mixed Ga–Al polyoxycations[260] intercalated between the layers of smectitic silicates have also been prepared and characterised; these are of especial interest in the field of shape-selective catalysis.[261] The inorganic chemistry of aluminosilicates depends on the structures formed by condensed SiO_4 units built up into rings, chains, layers and three-dimensional frameworks. A few Al orthosilicates contain discrete SiO_4 units; one is the garnet mineral $Ca_3Al_2(SiO_4)_3$ and another is Al_2SiO_5. The latter can be regarded alternatively as $Al^{3+}[AlSiO_5]^{3-}$ and is related to sheet silicates with $[Si_2O_5]^{2-}$ repeating units. This is an example of a situation common in aluminosilicate chemistry wherein silicon atoms are replaced by aluminium, and which has the effect of introducing an extra negative charge for each Si atom replaced. Clays and other minerals commonly have Al/Si ratios intermediate between limiting values of 0 and 0.33, with the extra negative charges balanced by Na, K, Mg, Ca or other cations. Minerals of the sodalite type have structures based on $(Al,Si)O_2$ frameworks with Na^+ or other positive ions in the interstices and also contain negative ions such as Cl^-, SO_4^{2-} and S^{2-}. Numerous forms have been made and hydrosodalites of general composition $Na_{6+x}(SiAlO_4)_6(OH)_x \cdot nH_2O$ can be synthesised from kaolinite and aqueous NaOH solution.[262] Sodalites attract attention as photochromic and ion-conducting materials with potential application as matrices for superconductor composites.

Anhydrous aluminium phosphate ($AlPO_4$) is isoelectronic with silicon dioxide (SiO_2) and is known in structures analogous to the three main polymorphs of silica (quartz, cristobalite and tridymite), including their α- and β-forms. The hydrated aluminium phosphates comprise an extensive range of materials and eighteen well-characterised compounds occur in the $AlPO_4–H_2O–H_3PO_4$ segment of the ternary system.[263] They can be classified as either acidic (Al/P < 1), containing aluminium ions and protonated phosphate groups, or neutral (Al/P = 1), containing Al^{3+}, OH^- and PO_4^{3-} ions. All the compounds exist as hydrates. The structures and uses of aluminium phosphate complexes have been reviewed recently.[263]

Hydrous aluminium phosphate, a gelatinous solid, $AlPO_4 \cdot xH_2O$, is formed when near-neutral Al^{3+} and phosphate solutions are combined. Heating the mixture, acidified with phosphoric acid, to 250°C in a sealed tube produces crystalline $AlPO_4 \cdot 2H_2O$. X-Ray study reveals cis-$AlO_4(OH_2)_2$ octahedra sharing vertices with PO_4 units.[264] An acid phosphate, $Al(H_2PO_4)_3$, can be obtained from solutions of alumina in H_3PO_4. Pyrolysis of these solids yields glassy materials with polyphosphate chain structures.

The highly viscous aluminium phosphate solutions tend to form polymeric aggregates, and equilibria are reached only very slowly. As a result, synthesis of a pure material requires rigorous adherence to prescribed conditions to avoid crystallisation of more than one compound. The presence of Na^+, K^+ or NH_4^+, added as hydroxides to control the pH, may lead to coprecipitation of impurities. Several of the higher hydrates can be converted by thermal treatment into compounds incorporating a reduced number of water molecules.

Given the difficulty of preparing pure compounds, structural methods, and especially X-ray crystallography, have been important agents of characterisation. The Al^{3+} ions are often octahedrally coordinated. It is usual for OH ligands and the non-protonated oxygens of phosphate groups to bridge two Al centres, while protonated P–OH groups and H_2O molecules are terminal. Bridging leads to the building of one-, two- and three-dimensional structures. For example, aluminium metaphosphate with the composition $Al(PO_3)_3$ contains AlO_6 octahedra, the phosphate part consisting of PO_4 units linked to produce $(PO_3)_n^{n-}$ infinite chains. Solid-state ^{27}Al NMR measurements (see chapter 1) have also been useful in determining the coordination geometry and identifying the ligands at aluminium.[265]

$AlPO_4$ is thermally stable, very insoluble and chemically inert. It has important uses as a filler and in inorganic binding and coating agents.[266] For these applications it is generally prepared by thermal dehydration of a hydrated phase. Aluminium phosphates are also significant in soil science and waste water treatment, and are used in medicine to regulate stomach acidity. The most significant recent development arises from the preparation of aluminophosphates by hydrothermal synthesis in the presence of an organic amine or quaternary ammonium salt to act as a template.[267] These are materials (designated APO-n) derived from the hydrous gel of phosphoric acid and aluminium hydroxide, and which have three-dimensional crystal frameworks consisting of alternate PO_4 and AlO_4 tetrahedral units. Some variants have aluminium in four-, five- and six-fold coordination with oxygen.[268] Entrapment of the template groups within the mineral framework leads to structures with microporous channels and cavities closely related to those of the zeolites, and which may share the surface activity and catalytic properties of this latter class. There is considerable impetus to investigations of microporous aluminophosphates and new variations on their synthesis are frequently reported.[269]

Gallium, indium and thallium phosphates precipitate from aqueous solutions as solids of the type $MPO_4 \cdot 2H_2O$ which are structurally similar to the corresponding Al compound. They lose water on heating to give the anhydrous MPO_4 solids. The gallium compound resembles $AlPO_4$ and features tetrahedral coordination. Crystalline $GaPO_4$ in quartz- and crystobalite-like modifications can be prepared by the solid-state reaction of

Ga_2O_3 with $NH_4H_2PO_4$ at 1000°C, followed by hydrothermal treatment under pressure in aqueous H_3PO_4 at 180°C.[270] $InPO_4$ and $TlPO_4$ (which are isostructural) display the metal in sixfold coordination which involves oxygen atoms from six different PO_4^{3-} groups.[5,271] The analogy between aluminophosphates and silicates extends to gallophosphates and offers scope for devising new materials which mimic the naturally occurring aluminosilicate minerals. Thus, a series of gallophosphates with framework structures enclosing alkylammonium ions can be synthesised from solutions of Ga^{3+} in phosphoric acid,[272] and a novel, clay-like gallophosphate with a layer structure has been described recently.[273]

Arsenates $MAsO_4 \cdot xH_2O$ are precipitated when solutions containing Ga^{3+} or In^{3+} and arsenate ions are neutralised and they can be dehydrated to the $MAsO_4$ solids. $GaAsO_4$ has one of the quartz-type SiO_2 structures. The antimonate, $GaSbO_4$, has a structure related to that of rutile in which Ga^{III} and Sb^V occupy at random the Ti^{IV} positions in the TiO_2 lattice.[8] Thallium forms $TlAsO_4$ and $TlAsO_4 \cdot 2H_2O$.[5] A few dihalophosphates have been described, e.g. $Al(PO_2F_2)_3$ and $Tl(PO_2Cl_2)_3$.[2] These can be prepared from the trihalides in non-aqueous media, but their chemistry has been little explored.

Carboxylates. Aluminium carboxylates are useful, low-priced reagents with applications in many fields. There are uses in the textile, paper and pharmaceutical industries which take advantage of the ability of Al^{3+} centres to bind oxygen ligands strongly. Some Al derivatives of the higher carboxylic acids are used as soaps in lubrication technology. A less welcome application finds them as constituents of the gelling agent in Napalm incendiary weapons.

The history of the preparation of aluminium carboxylates records considerable difficulty in attaining products with a carboxylate/Al ratio of more than $2:1$.[274] Special methods, avoiding the possibility of hydrolysis, needed to be developed before the production of tri-carboxylates could become routine. A wide range of 'basic carboxylates' can be prepared, including the types $Al(OH)(O_2CR)_2$ and $Al(OH)_2(O_2CR)$, and polynuclear species such as $Al_3O(OH)(O_2CR)_6$ obtained from the first type by partial dehydration. Diprotonic acids H_2B commonly give rise to basic salts $Al(OH)B$ and readily form anionic complexes, $[AlB_2]^-$ and $[AlB_3]^{3-}$, of which the *bis-* and *tris*(oxalato)aluminate ions are obvious examples.

The following methods enable the preparation of well-defined carboxylates of aluminium.[274,275]

1. Reaction of anhydrous aluminium chloride with a mixture of the carboxylic acid and acid anhydride:

$$AlCl_3 + 2RCOOH + (RCO)_2O \rightarrow Al(O_2CR)_3 + RCOCl + 2HCl$$

$$(3.20)$$

The product, $Al(O_2CR)_3$, crystallises from the reaction mixture.

2. Reaction of aluminium ethoxide or isopropoxide with the carboxylic acid or the acid anhydride:

$$Al(OR)_3 + 3(R'CO)_2O \rightarrow Al(O_2CR')_3 + 3R'COOR \quad (3.21)$$

where R = Et or Pr^i; R' = Me, Et, Pr or Bu;

$$Al(OPr^i)_3 + 3R'COOH \rightarrow Al(O_2CR')_3 + 3Pr^iOH \quad (3.22)$$

where R' = C_{15}, C_{17} or other higher alkyl group.

The reaction mixture represented in eq. (3.22) is heated under reflux with benzene or toluene and the liberated isopropanol is removed azeotropically. Normally the acid is present in excess; mixed aluminium alkoxide carboxylates can be synthesised by performing the reaction with a molar ratio $Al(OPr^i)_3/R'COOH = 1:1$ or $1:2$ to give $Al(OR)_2(O_2CR')$ or $Al(OR)(O_2CR')_2$, respectively. Controlled hydrolysis and heating to c. 150°C converts these compounds into polymeric oxide carboxylates, $[AlO(O_2CR)]_x$, which are hydrolytically stable and highly soluble in organic solvents.

A variation of method (1) is to heat $AlCl_3$ and the carboxylic acid under reflux in benzene until the evolution of HCl ceases. If a deficit of the acid is used, this method leads to mixed chloride carboxylates, $AlCl_{3-n}(O_2CR)_n$. Such mixed compounds can also be made by the action of acetyl chloride on the compounds $Al(OPr^i)_{3-n}(O_2CR)_n$ ($n = 1$ or 2).

Aluminium triacetate, $Al(O_2CMe)_3$, can be prepared by heating $AlCl_3$ or Al powder with a mixture of acetic acid and acetic anhydride. The basic acetates, $Al(OH)(O_2CMe)_2$ and $Al(OH)_2(O_2CMe)$, are obtained as white powders from aqueous aluminium acetate solution.[275] The diformate, $Al(OH)(O_2CH)_2 \cdot H_2O$, crystallises from concentrated solutions of freshly precipitated $Al(OH)_3$ in formic acid. With an excess of the acid, $Al(O_2CH)_3 \cdot 3H_2O$ crystallises.

The C_1, C_2 and C_3 carboxylates of aluminium, i.e. formates, acetates and propionates, are sparingly soluble in organic solvents. The carbon-rich carboxylates are reported to be associated in non-aqueous solutions. However, this may be the case only when traces of moisture have led to hydrolysis. All the carboxylates decompose when heated above 150–200°C, tending to form the oxide carboxylates $AlO(O_2CR)$, and ultimately Al_2O_3.

Carboxylates of Ga and In have received comparatively little attention. The preparative methods devised for Al carboxylates should work just as well for these elements; thus it is that crystalline $Ga(O_2CMe)_3$ has been prepared from Ga or $GaCl_3$ and acetic acid, the chloride acetates being obtained from $GaCl_3$ and a deficit of the acid.[241] Basic gallium citrate [from $Ga(OH)_3$ and citric acid] is a suitable form for administering the element during clinical trials which use the localisation of the ^{67}Ga radionuclide in soft tissue to study tumour growth.[276] Gallium tribenzoate and the phenylacetate $Ga(O_2CH_2Ph)_3$ have been prepared by prolonged reaction of the appropriate acid with metallic gallium.[241] The species $[Ga_3(\mu_3\text{-}O)\text{-}(\mu\text{-}O_2CPh)_6(4\text{-}Mepy)_3]^+$, a Main Group example of the class of oxo-centred trinuclear carboxylates, was obtained from a solution of $PhCO_2Na$ and Ga_2Cl_4 in 4-methylpyridine (4-Mepy).[277] The structure has a planar Ga_3O moiety in which the Ga–O bond distance is 1.87 Å. Each edge of the Ga_3O core is bridged by two benzoate ligands forming Ga–O bonds with an average length of 1.985 Å and the three axial positions are occupied by the 4-Mepy molecules. As Figure 3.16 shows, each gallium(III) centre thus acquires an octahedral coordination sphere. In^{III} formate and In^{III} acetate can be made by dissolving indium hydroxide in the minimum quantity of the hot acid. Electrolysis of indium in acetic acid leads to $In(O_2CMe)_2$ which is believed to be a mixed valence In^I/In^{III} compound.[278] The solids which crystallise from dilute solutions of gallium or indium hydroxide in carboxylic acids are apt to be the basic salts, as in the aluminium case. The oxalate systems, however, yield $M_2(O_4C_2)_3 \cdot xH_2O$ (M = Ga or In) which can be dehydrated as a first step before decomposition to the metal oxide occurs.[2]

Thallium(III) acetate is prepared from Tl_2O_3 and boiling glacial acetic acid; it crystallises on cooling as $Tl(O_2CMe)_3$, a moisture-sensitive solid. The structure of $Tl(O_2CMe)_3$ reveals the thallium atom coordinated by three chelating acetate groups. In $Tl(O_2CMe)_3 \cdot H_2O$, which forms when the anhydrous compound is exposed to the atmosphere, water molecules are also attached to the thallium atom which here displays eight-fold co-ordination.[279] Other Tl^{III} carboxylates can be generated from the acetate by metathesis reactions using the appropriate acid.[280] They are susceptible to photolysis, forming the Tl^I salts quantitatively:

$$Tl(O_2CR)_3 \xrightarrow{h\nu} TlO_2CR + CO_2 + \text{organic products} \qquad (3.23)$$

Thallium(III) carboxylates are useful reagents for a variety of organic and organometallic syntheses[5,281,282] (see chapter 7).

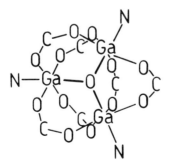

Figure 3.16 Coordination around the gallium atoms in the complex cation $[Ga_3(\mu_3 - O)\text{-}(\mu - O_2CPh)_6(4\text{-}Mepy)_3]^+$.[277] Source: *J. Am. Chem. Soc.*, 1992, **114**, 786.

Aluminium and the heavier Group 13 elements form trifluoroacetates, $M(O_2CCF_3)_3$. These can be prepared from the chloride MCl_3 and trifluoro-acetic acid or, in the case of Al and Ga, from a metal/Hg amalgam. The $M(O_2CCF_3)_3$ compounds are less susceptible to hydrolysis than are the tri-acetates. Thallium(III) trifluoroacetate is a particularly versatile Tl^{III} reagent. It is effective in metallating aromatic substrates to produce intermediates from which the $-Tl(O_2CCF_3)_2$ group can be displaced to introduce $-OH$, $-I$ and other substituents (see chapter 7).

β-Diketonates and allied derivatives. The *tris*(acetylacetonates) of aluminium, gallium and indium, $M(acac)_3$, have a lengthy history, having first been made by the action under reflux of acetylacetone on the hydroxide (formed by the progressive addition of ammonia to the aqueous metal nitrate solution).[283] They are crystalline compounds: $Al(acac)_3$, colourless, m.p. 194°C; $Ga(acac)_3$, yellow, m.p. 195°C; and $In(acac)_3$, orange, m.p. 187°C. All three can be sublimed at temperatures above 100°C and dissolve readily in benzene as monomers, but are insoluble in water and not easily hydrolysed.

The stability of the Group 13 acetylacetonates arises from the matching of the valency of three and the maximum coordination number of six. This point was nicely made 70 years ago in the early report.[283] 'The complete saturation of the principal and supplementary valencies is manifested in the acetylacetone derivatives of aluminium, gallium and indium by their monomeric molecular weights and by the absence of ammonia additive products of these compounds.'

Preparative methods for other Group 13 ketoenolates utilise the reactions of β-diketones or β-ketoesters with the freshly formed hydroxide $M(OH)_3$ (already mentioned) or with an alkoxide $M(OR)_3$, or with the halide MX_3, e.g. $AlCl_3$ in ether. Direct synthesis from the metal is possible; for Al or Ga an amalgam can be used. The technique of Al atom synthesis has also been shown to be effective.[284] $In(acac)_3$ has been prepared by electrochemical oxidation of In metal in a solution of acetylacetone (Hacac) in methanol.[285] These methods usually yield as products the *tris*-ketoenolates. If the ligand is added in the proportion 1 : 1 or 2 : 1 relative to the substrate $M(OR)_3$ or MX_3, the reaction may give products with residual alkoxide or halide ligands. Steps to introduce other alkoxide, aryloxide or siloxide groups by exchange reactions can then follow. Such products can be used to prepare *tris*-chelates in which the second chelating ligand is different from the first.

β-Diketonate and *β-ketoester* derivatives of Al, Ga and In are hydrolytically stable solids. They are slightly volatile, more so if the ligand contains fluorine atoms in place of hydrogen, but less so if aryl groups are present. The mixed β-diketone derivatives can form *cis/trans*-geometrical isomers or display optical activity in suitable cases. The rates and mechanisms of optical inversions have been investigated using NMR measurements in several instances.[286] X-Ray studies of compounds of the types $Al(OR)(acac)_2$ and $Al(OR)_2(acac)$ have revealed examples of mono-, di- and trimeric structures featuring 4- or 6-coordinate aluminium atoms;[222] the oligomeric systems are typically composed of four-membered Al_2O_2 rings formed by bridging alkoxide ligands.

Diffraction methods have confirmed the *tris*-chelated structures of $Al(acac)_3$ (acac = acetylacetonate) and the corresponding hexafluoro-acetylacetonate: each has an AlO_6 core with Al–O distances of 1.89 Å.[7] The *tris*(β-ketoenolates) of Al, Ga and In are stereochemically non-rigid molecules, although resolution of the enantiomers of $Al(acac)_3$ can be achieved by column chromatography at low temperature. Isomerisation of $M(acac)_3$ molecules (M = Al, Ga or In) and their analogues probably involves an intramolecular twist mechanism, although a dissociative mechanism, *via* a five-coordinate transition state in which one ligand is unidentate, receives support in the case of fluorinated derivatives.[287] The structures of $Al(acac)_3$, $Ga(acac)_3$ and $In(acac)_3$, their kinetic and thermodynamic parameters deduced from NMR and other results, and the trends in the vibrational spectra through the series point to a decrease in bond strength in the order Al–O > Ga–O > In–O.[7,207] Related molecules with MO_6 kernels include the tropolonates of Al, Ga, In and Tl, $M(O_2C_7H_5)_3$, and a number of complexes in which one or two of the chelating ligands have been replaced by other substituents.[7] Thallium in the +3 state forms $Me_2Tl(acac)$ but the simple acetylacetonate is a Tl^I compound, $Tl(acac)$.[5]

3.2.4   Derivatives with bonds to sulfur, selenium or tellurium

3.2.4.1   Chalcogenides.   The chalcogenides Al_2S_3 (white), Al_2Se_3 and Al_2Te_3 (grey) can be prepared by heating mixtures of the elements at c. 1000°C.[2,7,288] They react with moisture and are converted to aluminium hydroxide, releasing H_2Y (Y = S, Se or Te).[289] Crystal studies have established three forms of Al_2S_3. The polymorphs have structures based on hexagonal close-packing of sulfide. The Al atoms occupy tetrahedral sites in ordered and disordered fashion in the α- and β-forms, respectively, whereas the structure of γ-Al_2S_3 resembles that of γ-Al_2O_3 having Al in some octahedral sites (see Table 3.9).

Al^{III} participates in a number of ternary chalcogenides of the types M^IAlY_2 and $M^{II}Al_2Y_4$ (Y = S, Se or Te).[7] Ternary compounds of the type $M^{III}AlY_3$ are also formed by the heavier members of Group 13. $AlInS_3$ is an example: it has a structure closely related to that of In_2Se_3 but, as befits the different sizes of the atoms, Al enjoys tetrahedral and In fivefold co-ordination by S.[290] Studies of ternary systems based on aluminium sulfide are stimulated by the prospect of physical properties out of the ordinary, such as unusual conductivity or ion-transport capability useful in solid electrolytes.

Table 3.9 Aluminium, gallium and indium compounds with structures related to wurtzite (h.c.p. ZnS) and zinc blende (c.c.p. ZnS).

Anion lattice	Lattice-forming anion	Cation occupancy		Examples
		Tetrahedral sites	Octahedral sites	
Hexagonal close-packed (h.c.p.)	Oxide, O^{2-}		√	α-Al_2O_3
	Sulfide, S^{2-}	√		α-Al_2S_3, β-Ga_2S_3
	Sulfide, S^{2-}	√	√	γ-Al_2S_3
	Selenide, Se^{2-}	√		β-In_2Se_3
	Nitride, N^{3-}	√		AlN, GaN, InN
Cubic close-packed (c.c.p.)	Oxide, O^{2-}	√	√	γ-Al_2O_3
	Sulfide, S^{2-}	√		γ-Ga_2S_3, α-In_2S_3
	Selenide, Se^{2-}	√		Ga_2Se_3
	Telluride, Te^{2-}	√		Ga_2Te_3, β-In_2Te_3
	Phosphide, P^{3-}	√		AlP, GaP, InP
	Arsenide, As^{3-}	√		AlAs, GaAs, InAs
	Antimonide, Sb^{3-}	√		AlSb, GaSb, InSb

Phases with intriguing structures have been encountered; among these is $Tl_3Al_{13}S_{21}$ in which slightly distorted AlS_4 tetrahedra are connected via common corners to form a polymeric three-dimensional framework. The thallium atoms are located in tunnels running through the structure, and their coordination by 12, 11 or 9 sulfur atoms is rather irregular, betraying the stereochemical activity of the lone electron pair of Tl^+.[291]

For compounds of gallium, indium and thallium with sulfur, selenium and tellurium, the lower oxidation states of the metals come into play. The chemistry of these systems involves the M^I and/or M^{II}, as well as M^{III}, states; these aspects are treated in more detail in section 3.3.5. Table 3.9 summarises the structures of III–VI compounds of the type M_2Y_3 (M = Ga or In; Y = S, Se or Te) in comparison with the III–V compounds of the same metals.

Gallium(III) sulfide, Ga_2S_3, can be made from the elements, or by heating Ga in a stream of H_2S gas to 950°C. Prepared in this way, it is a white solid, m.p. 1095°C.[292] Ga_2S_3 dissolves in aqueous acids and decomposes slowly in moist air, evolving H_2S. The likely product is GaO(OH):

$$Ga_2S_3 + 4H_2O \rightarrow 2GaO(OH) + 3H_2S \qquad (3.24)$$

There are three polymorphs of Ga_2S_3. The crystal structures are related to those of ZnS with partial occupancy of the tetrahedral sites, thereby forming GaS_4 units. In yellow, crystalline α-Ga_2S_3 the sulfide lattice is of the wurtzite ZnS type and the Ga atoms have an ordered arrangement. Shrinkage of S_4 tetrahedra around the vacant sites leads to distortion from ideal packing.[293] In β- and γ-Ga_2S_3 the Ga atoms are disordered within sulfide lattices of wurtzite h.c.p.- and zinc blende c.c.p-types, respectively. The β-form is converted to the α-form by annealing at 1000°C. The selenide, γ-Ga_2Se_3, and telluride, γ-Ga_2Te_3, and mixed phases $Ga_2Y_xZ_{3-x}$, where Y and Z are different chalcogenides, can be synthesised by fusing mixtures of the elements in the appropriate proportions in sealed tubes. Ga_2S_2Te is notable for a structure containing GaS_3Te units which are linked to give parallel chains.[294]

The crystal chemistry of the indium chalcogenides is quite complicated.[295] The high-temperature stable forms of the In_2Y_3 compounds (Y = chalcogen) can be related to standard structures; for example, β-In_2S_3 is a defect spinel, while α-In_2S_3 has a structure like that of a cubic form of Al_2O_3. α-In_2Se_3 is related to wurtzite and β-In_2Te_3 to zinc blende.[296] In_2S_3 can be precipitated from near neutral aqueous In^{3+} solutions using H_2S or Na_2S. This gives the yellow form, α-In_2S_3. It is converted by heating above 300°C into red β-In_2S_3, m.p. 1100°C. The optical, photoelectric and other physical properties of the Ga and In chalcogenides have been studied in detail, particularly as photo- or semiconductor materials.[297] In this context, ternary phases assume major importance and $TlGaSe_2$ or $CuInSe_2$ are useful prototypes.[298,299] Many other $M^IM^{III}Y_2$-type thiogallates and thioindates and Se analogues of Tl^I, Cu^I and the alkali metals have been characterised. Ternary compounds with different compositions, e.g. $K_8M_4Y_{10} \cdot 14H_2O$ (M = Ga or In; Y = S or Se), are obtained from concentrated sulfide- or selenide-rich aqueous solutions as stable, colourless crystalline solids resistant to thermal and solvolytic decomposition. A notable feature of their structures is the presence of the adamantane-cage anions, $[M_4Y_{10}]^{8-}$ [300] shown in Figure 3.17. Even more remarkable is the hexameric anion $[Ga_6Se_{14}]^{10-}$ (Figure 3.17(d)) generated when GaSe and elemental Cs react to produce $Cs_{10}Ga_6Se_{14}$.[301] A chalcogenide of a different type is $[In_2Se_{21}]^{4-}$ in which each In^{III} centre is chelated by two Se_4^{2-} ligands and bound to a terminal Se atom of the Se_5^{2-} chain.[302] Hence this is in the nature of a polyselenide.

The range of semiconducting chalcogenides includes many gallium, indium and thallium thio-, seleno- and tellurometallates which are prepared from M_2S_3, M_2Se_3 and M_2Te_3, respectively, or from the elements, under anhydrous conditions. For example, gallium forms M^IGaY_2, $M^{II}Ga_2Y_4$ and $M^{III}GaY_3$ (Y = S, Se or Te), where M^I represents an alkali metal, M^{II} an alkaline earth and M^{III} a lanthanide. The products have different crystal structures. Thus M^IGaS_2 has the chalcopyrite ($CaFeS_2$) structure,[8] and $M^{II}Ga_2S_4$ has either the thiogallate structure typified by $ZnGa_2S_4$ (which contains $[SGa(\mu\text{-}S)_2GaS]^{2-}$ groups) or the spinel structure, as with $MgGa_2S_4$ matching the prototype $MgAl_2O_4$. Indium is found tetrahedrally coordinated, for example in $CuInSe_2$ which contains $[In_4Se_{10}]^{8-}$ groups,[303] or octahedrally coordinated as in $NaInSe_2$.[304]

There appears to be no simple thallium(III) sulfide Tl_2S_3. The sulfide TlS is not analogous to GaS (which has Ga^{II}–Ga^{II} bonds) but is a mixed valence compound, $Tl^ITl^{III}S_2$; the Tl^{III} atoms are at the centre of TlS_4 tetrahedra

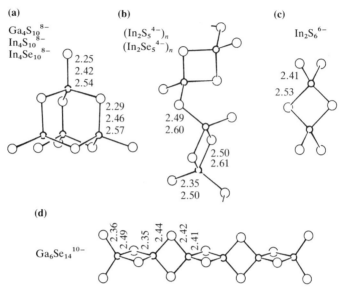

(a)

$Ga_4S_{10}^{8-}$
$In_4S_{10}^{8-}$
$In_4Se_{10}^{8-}$

2.25
2.42
2.54

2.29
2.46
2.57

(b)

$(In_2S_5^{4-})_n$
$(In_2Se_5^{4-})_n$

2.49
2.60

2.50
2.61

2.35
2.50

(c)

$In_2S_6^{6-}$

2.41
2.53

(d)

$Ga_6Se_{14}^{10-}$

2.36
2.49
2.35
2.44
2.42
2.41

Figure 3.17 The structures of some gallium and indium chalcogenide anions: adamantane-like $[Ga_4S_{10}]^{8-}$, $[Ga_4Se_{10}]^{8-}$, $[In_4S_{10}]^{8-}$ and $[In_4Se_{10}]^{8-}$ compared with $[In_2S_5^{4-}]_n$, $[In_2Se_5^{4-}]_n$, $[In_2S_6]^{6-}$ and $[Ga_6Se_{14}]^{10-}$, with bond lengths in Å.[300] Source: *Angew. Chem., Int. Ed. Engl.*, 1983, **22**, 113.

which share edges. Corner-sharing of $Tl^{III}S_4$ tetrahedra gives chains held together by Tl^I in the compound Tl_4S_3. Tl^{3+} ions and sulfide are incompatible in aqueous solutions for redox reasons, and the result of passing H_2S into a solution of a thallium(III) salt is to precipitate a mixture of Tl_2S and sulfur. Sulfur-rich thallium compounds can be made; these are thallium(I) polysulfides (see section 3.3.5). The Tl–Se and Tl–Te systems are rich in mixed-valence compounds, and the alleged phases Tl_2Se_3 and Tl_2Te_3 may not be authentic Tl^{III} materials since the structures give evidence of Tl–Tl interactions.

3.2.4.2 Chalcogenide halides.   The chalcogenide halides of Group 13 are compounds with the formula MYX (M = Al, Ga, In or Tl; Y = S, Se or Te; X = Cl, Br or I). All are known except the TlTeX members; thallium also forms chalcogenide halides of the Tl^I state. In addition, the gallium compounds $Ga_9S_8X_{11}$ (X = Cl or Br) are reported. The preparations and properties of the chalcogenide halides were reviewed in 1980,[305] with particular emphasis on their solid-state chemistry. The ionic character of the MYX solids increases, with a gain in chemical and thermal stability, to reach a maximum for indium. The stabilities of the thallium compounds are influenced by redox considerations. The usual synthesis of the Al, Ga and In compounds involves heating a mixture of the chalcogenide M_2Y_3 and excess

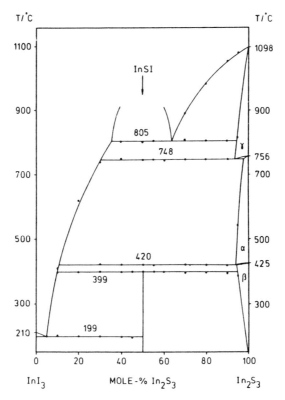

Figure 3.18 The phase diagram of InI_3–In_2S_3 mixtures showing the formation of InSI. Reproduced with permission from R. Kniep and W. Welzel, *Z. Naturforsch.*, 1985, **40b**, 29.

of the halide MX_3 in a sealed tube to 200–400°C for some days[306] (see, for example, the phase diagram in Figure 3.18). For the iodide, it is not necessary to use MI_3 since a mixture of the metal and iodine works just as well.

The aluminium and gallium compounds, MYX, are mostly colourless, hygroscopic solids (the telluride halides of Ga are yellow). Water vapour converts them to the hydroxide, and oxygen produces the oxide M_2O_3 at elevated temperatures. With ammonia, chalcogenide amide adducts, $MY(NH_2) \cdot NH_3$, are obtained. The indium chalcogenide halides are coloured and are more stable than their Al or Ga analogues. Only the sulfide halides decompose before melting and are moisture-sensitive. Strong acids cause decomposition, releasing the gas H_2Y. The action of alkali leads to the formation of $In(OH)_3$. The thallium(III) compounds TlYCl, TlYBr and TlYI (Y = S or Se) can be prepared by heating the corresponding Tl^I halide with sulfur or selenium.[5] They are insoluble solids with layer-lattice structures. Thermal decomposition is accompanied by oxidation of the chalcogenide component to S (or Se) and reduction to TlX. Heating the chlorides, TlYCl, to 500°C *in vacuo* produces the compounds Tl_4YCl_4, which are probably mixed valence derivatives, possibly with the structure $(Tl^+)_3$ $[TlYCl_4]^{3-}$ in which three-quarters of the thallium atoms have been reduced. The compounds produced by oxidising Tl_2S or Tl_2Se with the elemental halogen are not of the type TlYX. With bromine or iodine, they are formulated as Tl_2YX_4, but the halogen is readily lost on heating or to an organic solvent, and the structures have yet to be established.

The phase diagrams of most $MX_3-M_2Y_3$ systems have been investigated,[306] and the lattice constants of nearly all the MYX solids have been reported.[305] An analysis of the vibrational spectra of the aluminium sulfide halides and selenide halides shows that they form dimeric structures of the type $XAl(\mu-Y)_2AlX$ (Y = S or Se; X = halogen),[307] but other structural data are generally lacking.

3.2.4.3 Thiolates and related compounds.
Members of the aluminium group form thiolates, $M(SR)_3$, and halide thiolates, $MX_2(SR)$ and $MX(SR)_2$. There are selenium and tellurium equivalents. Several methods of preparation have been investigated.[308–311]

1. Reaction of the halide MX_3 (M = Al or Ga) with RYH (Y = S or Se) *via* the 1 : 1 adduct:

$$MX_3 \longrightarrow MX_3 \cdot RYH \longrightarrow MX_2(YR) \qquad (3.25)$$

2. Treatment of MX_3 in benzene with the thiolate of another metal, e.g. $Pb(SR)_2$:

$$2MX_3 + nPb(SR)_2 \rightarrow 2MX_{3-n}(SR)_n + nPbX_2 \qquad (3.26)$$

3. Reaction of GaX_3 with Me_3SiSR (R = Me, Et or Ph):

$$GaX_3 + Me_3SiSR \rightarrow GaX_2(SR) + Me_3SiX \qquad (3.27)$$

4. Interaction of $PhMI_2$ (M = Ga or In) with MeSSMe or PhSSPh, which yields the products MI_2SR by a redox reaction.[309]
5. The direct reaction between indium and Ph_2S_2 or Ph_2Se_2 in refluxing toluene:[310]

$$2In + 3Ph_2Y_2 \rightarrow 2In(YPh)_3 \quad (Y = S \text{ or } Se) \qquad (3.28)$$

6. Electrochemical synthesis by anodic oxidation of the metal (In or Tl) using an acetonitrile electrolyte containing the thiol RSH.[311]

Method 1 is impeded by the stability of the intermediates, particularly those of gallium,[312] which participate in the equilibrium

$$GaX_3 \cdot HYEt \rightleftharpoons GaX_2(YEt) + HX \quad (X = Cl, Br \text{ or } I; Y = S \text{ or } Se)$$
$$(3.29)$$

Attempts to remove the hydrogen halide HX with triethylamine were unsuccessful because products of the type $[Et_3NH][GaX_3(YH)]$ form. Gaseous HX can be eliminated by heating to 100°C, but this also causes other, unwanted decomposition. An interesting variation has employed the reaction of Ga_2X_4 with Me_2S_2. The iodide product has a thiol-bridged structure with a planar four-membered Ga_2S_2 ring in the dimer $[GaI_2$-$(SMe)]_2$.[313] Four-membered rings containing tetracoordinated elements of Group 13 are usually planar (exceptions are found in a few organoaluminium compounds), so this structure is of the expected type. In contrast, $[GaI_2(SPr^i)]_2$ (prepared from the methylthiolate by simple thiol exchange) has a folded ('butterfly') ring structure.[314] The fold angle achieved by bending the Ga_2S_2 ring on the Ga⋯Ga diagonal is 37°. A greater deviation from planarity is found in the ion $[Tl_2Te_2]^{2-}$, but this is a Tl^I species and not strictly comparable with the gallium example.

Iodo thiolates of the type $M_4I_4(SMe)_4S_2$ (M = Al or Ga)[313,315a] are formed by further reactions involving Me_2S_2; Figure 3.19 shows the Al compound which arises from $Ga[AlI_4]$.[315a] From Ga_2I_4 the first product is the S-bridged dimer, $[GaI_2(SMe)]_2$, already noted, and which then condenses to give the S-linked cage complex.[315a] Very large gallium-sulfido clusters $[Bu^tGaS]_4$, $[Bu^tGaS]_7$ and $[Bu^tGaS]_8$ have been characterised subsequently[315b] and an investigation of their chemistry has started.

Method 2 applied to gallium yields the moisture-sensitive products $GaX_{3-n}(SR)_n$ (X = Cl, Br or I; n = 1 or 2).[308] By using a thiolate with a bulky R group, namely LiSMes* (Mes* = $2,4,6$-$Bu^t_3C_6H_2$), the first monomeric, three-coordinate aluminium and gallium thiolates, $M(SMes^*)_3$, have been prepared in crystalline form.[316] Both have short M–S distances (average bond lengths: Al–S, 2.185; Ga–S, 2.205 Å), and the infrared spectra are also indicative of strong M–S bonding. Some π-interaction may occur but the failure of variable temperature ^1H NMR measurements to show restricted rotation about the M–S bond[316] weighs against this as a major influence.

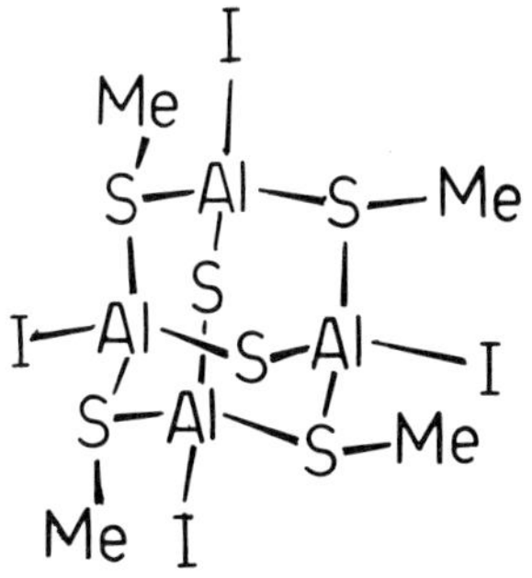

Figure 3.19. The molecular structure of $Al_4I_4(SMe)_4S_2$. Bond distances: Al–S, 2.18; Al–SMe, 2.30; Al–I, 2.50 Å. Source: *Inorg. Chim. Acta*, 1986, **120**, L23.

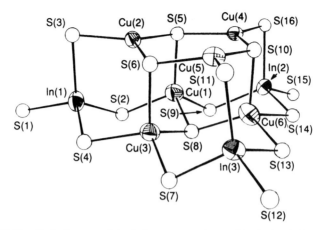

Figure 3.20 The $Cu_6In_3S_{16}$ core of the indium-copper cluster, $[Cu_6In_3(SEt)_{16}]^-$. Reproduced with permission from W. Hirpo, S. Dhingra and M.G. Kanatzidis, *J. Chem. Soc., Chem. Commun.*, 1992, 558.

In(SPh)$_3$ is obtained as an air-stable, white solid from InCl$_3$ and NaSPh in refluxing methanol.[317] It displays Lewis-acid character, giving rise to complexes of the types $[In(SPh)_4]^-$, $[InX(SPh)_3]^-$, In(SPh)$_3 \cdot$ L and In(SPh)$_3 \cdot$ L$_2$ (L and L$_2$ = N- or P-donors). The complex (Ph$_4$P)[In(SEt)$_4$] reacts with [Cu(NCCH$_3$)$_4$](PF$_6$) to give the cluster (Ph$_4$P)[Cu$_6$In$_3$(SEt)$_{16}$] with an adamantoid framework[318] (see Figure 3.20). Mixed metal thiolate and selenolate compounds of this type are potentially important as precursors to the photovoltaic materials CuInS$_2$ and CuInSe$_2$, respectively.

A feature of the direct routes to Group 13 thiolates and related selenium compounds (methods 5 and 6 above) is that various oxidation states are encountered, depending on the nature of the thiol or selenol and the experimental conditions. Anodic oxidation of the metal can give In^I, In^{II} or In^{III} derivatives, although the indium(III) products, In(ER)$_3$, are always formed if oxygen is present. The reaction with thallium anodes gives Tl^I products, e.g. TlSPh and TlSePh, and it is likely that the primary reaction of indium gives the In^I species, InER, followed by oxidative addition

$$InER + R_2E_2 \rightarrow In(ER)_3 \qquad (3.30)$$

Indium metal reacts with Ph$_2$Y$_2$ (Y = S or Se) in the presence of iodine to give compounds of the type InI(YPh)$_2$.[310] These are yellow, air-stable solids; the structures are unknown.

3.2.5   Derivatives with bonds to Group 15 atoms

3.2.5.1   Nitrides, phosphides, arsenides and antimonides.   The binary compounds of Group 13 with Group 15 elements include the combinations which are formally isoelectronic with silicon (AlP), germanium (GaAs), tin (InSb) and lead (TlBi) and range in properties and structural types from the covalent nitrides to intermetallic compounds as represented by TlSb and TlBi. Compounds of Al, Ga and In with N, P, As and Sb can be viewed as having properties intermediate between those of ionic and covalent materials; these are the well known III–V type semiconductors. Gallium arsenide, GaAs, is the most important technologically and is a key material in the electronics sphere (chapter 5). Excepting the nitrides, these compounds are manufactured in bulk by direct reaction of the pure elements at high temperature and under pressure when necessary, as coloured, chemically stable solids with relatively high melting points. The phosphides, arsenides and antimonides, MP, MAs and MSb, have cubic (zinc blende type) structures. The M^{III} atoms are regularly arranged in the tetrahedral sites of the lattice formed by the Group 15 atoms and the structure as a whole is diamond-like. The nitrides MN, with a greater degree of covalent character, adopt the hexagonal close-packed ZnS (wurtzite type) structure, as noted in Table 3.9.

Methods of synthesis for particular nitrides include:[319,320]

1. thermal decomposition at 900°C of ammonia adducts, e.g. $AlCl_3 \cdot NH_3$ to AlN or GaX_3/NH_3 (X = Cl, Br or I) to GaN;
2. reaction of Ga or Ga_2O_3 with ammonia gas at 600–1000°C to give GaN; and
3. pyrolysis in vacuum at 700°C of $(NH_4)_3[MF_6]$ to give MN (M = Ga or In).

Aluminium nitride is an attractive material as a substrate for electronic devices since it has high heat resistance and a large thermal conductivity. Unfortunately it reacts easily with atmospheric moisture to acquire a hydrated surface.[321] Gallium nitride has a higher chemical stability than AlN; it does not react readily with either water or acids, and withstands being heated in air to 1000°C.

Aluminium phosphide can be sublimed from a mixture of Al and Zn_3P_2 by heating at 950°C. On the small scale, AlP is prepared in the laboratory from aluminium powder and red phosphorus.[322] It is violently reactive with respect to dilute acids, and is also attacked by water, liberating phosphine, PH_3, and forming Al_2O_3.[323] Phosphides and arsenides of high purity can be prepared by the reactions of the metal halide or an organometal compound with $P(SiMe_3)_3$ or other phosphine or arsine derivative,[324–327] following the routes summarised by the following equations:

$$\left\{ \begin{array}{ll} MR_3 + HER'_2 \rightarrow R_2MER'_2 + RH & (3.31) \\ R_2MER'_2 \rightarrow ME + 2RR' & (3.32) \end{array} \right.$$

$$MR_3 + EH_3 \rightarrow ME + 3RH \qquad (3.33)$$

M = Al, Ga or In; E = P or As; R = Cl, Br, Ph or alkyl; R' = alkyl or silyl group.

Methods of preparation of III–V compounds by vapour-phase reduction or chemical decomposition of suitable compounds (CVD processes) have been much investigated to find means of laying down uniform surface layers and to achieve better control of purity of the products.[326, 327] Ways of producing gallium arsenide include the gas-phase reduction of GaCl by arsenic vapour, and the reaction of Me_3Ga with AsH_3 at elevated temperatures. Techniques have been developed recently for the preparation of InP epitaxial layers and exploited to produce other thin films.[328] The Group 13 phosphides, arsenides and antimonides are oxidised in air at temperatures above 400–500°C, InSb being the least susceptible to attack. Their tendency to undergo hydrolysis, although slight, requires the encapsulation of electronic devices.

The drive to improve the efficiency of semiconductor materials continues, and there is major interest in the preparation, properties and bonding of Lewis adducts of substituted gallanes and arsines as precursors to GaAs, and in Ga_xAs_y clusters which can be generated by laser vaporisation of the GaAs crystal, or by the gas-phase decomposition of gallium arsenic hydrides.[329–332] Recent molecular orbital calculations on Al_nP_n ($n = 1–3$) clusters deserve attention. They give an important lead in understanding structure and bonding which are intermediate between those of ionic clusters like Mg_nS_n and those of covalent species like Si_2, Si_4 and Si_6, and are being used as starting points to perform detailed investigations of the III–V clusters of heavier elements, e.g. GaAs.[333] There have also been recent advances in the experimental study of III–V clusters. A compound $Bu^t_6Ga_2P_4$ has been reported in which Bu^t_2Ga and Bu^t_3Ga moieties are attached to a P_4 unit[334] (see Figure 3.21). Another type of cluster is found in compounds of the type $Cs_6M_2E_4$ (M = Al or Ga; E = P or As).[335] There are terminal and bridging bonds in the complex ion $[EM(\mu\text{-}E)_2ME]^{6-}$ which conforms to D_{2h} symmetry. The M–E distances of the four-membered M_2E_2 rings correspond to single bonds, whereas the shorter terminal M–E distances correspond to a

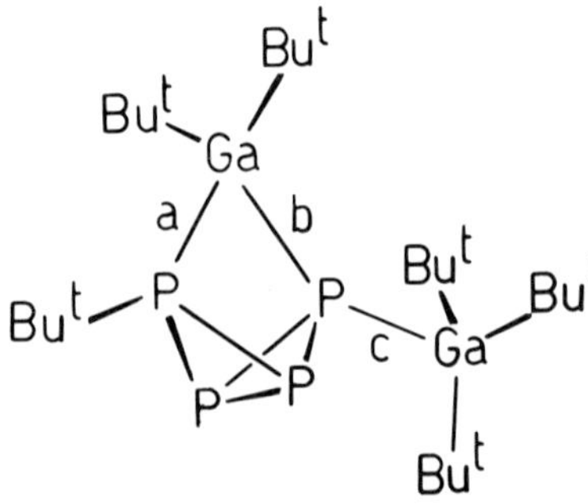

Figure 3.21 Structure of $Ga_2P_4Bu^t_6$ in the crystal.[334] Ga–P bond distances: a = 2.51; b = 2.48; c = 2.62; P–P = 2.19–2.21 Å. Source: *Angew. Chem., Int. Ed. Engl.*, 1991, **30**, 1353.

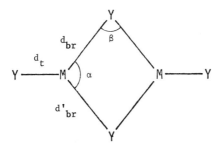

Species	Bond lengths/Å		Bond angles/°	
	d_t	$d_{br}{}^a$	α	β
$[Al_2P_4]^{6-}$	2.26	2.34	104	76
$[Ga_2P_4]^{6-}$	2.24	2.34	101	79
$[Al_2As_4]^{6-}$	2.35	2.45	105	75
$[Ga_2As_4]^{6-}$	2.36	2.46	102	78

^{a} d'_{br} is typically 0.01 Å longer than d_{br}.

Figure 3.22 Dimensions of the species $[M_2P_4]^{6-}$ and $[M_2As_4]^{6-}$.[335]

bond order of c. 1.5 (see Figure 3.22). Ring systems of this type feature strongly in the structures of amido-, phosphido- and related derivatives which are considered in the following sections 3.2.5.2 and 3.2.5.3. The compounds $Cs_6M_2E_4$, where M is a Group 13 metal and E is a Group 15 metalloid within a polyanion, are Zintl-phase materials of the I–III–V type and, as such, are treated in section 3.4.2 in the context of intermetallic compounds.

3.2.5.2 Amides, imides and related compounds. The extensive chemistry of aluminium-nitrogen compounds and their Ga and In counterparts includes that of amides, $M(NH_2)_3$, $MX(NH_2)_2$ and MX_2NH_2, and the corresponding species incorporating the organoamido substituents, –NHR and –NR₂. Also to be considered are imides and organoimido-derivatives containing the groups –NH– and –NR–, respectively, together with a few hydrazide and methyleneamide (ketimide) derivatives, e.g. $M(N{=}CR_2)_3$.

Some useful parallels can be drawn between these M–N systems and the M–O systems with –OH, –OR and –O– groups. To develop metal–nitrogen chemistry systematically on these lines would require the extensive study of liquid ammonia and amine solvent systems, an undertaking which has hardly been attempted except for the pioneering work of Wiberg's group.[2] Reactions of aluminium with alkali metals in liquid ammonia, or of NH_3 with $LiAlH_4$, lead to the amido-complexes mentioned above. Nitrogen-containing products result too from the action of ammonia on $Et_2O \cdot AlH_3$ at low temperature. All these materials readily lose hydrogen and NH_3 on warming to ambient temperatures. Further decomposition yields predominantly $Al_2(NH)_3$ at 150–180°C and AlN at 220°C.[336] The reactions of gallium compounds with ammonia appear to parallel those of aluminium compounds. The amide $Ga(NH_2)_3$ has been obtained from the reaction of NH_4Cl with $K[Ga(NH_2)_4]$ in liquid ammonia.

Dialkylamides are much more tractable than compounds with NH_2 groups and have received correspondingly more attention. Various methods can be used to synthesise the dialkylamides of aluminium, $Al(NR_2)_3$,[337–340] some of them being the counterparts of procedures used to prepare the alkoxides, $Al(OR)_3$.

1. Direct reaction of the metal with a secondary amine:

$$Al + 3R_2NH \rightarrow Al(NR_2)_3 + 3/2H_2 \qquad (3.34)$$

2. Reaction of a metal alkyl with Me_2NH:

$$Bu^t_3Al + 3Me_2NH \rightarrow Al(NMe_2)_3 + 3Bu^tH \qquad (3.35)$$

3. The action of a secondary amine on an AlH_3 derivative:

$$Me_3N \cdot AlH_3 + 3R_2NH \rightarrow Al(NR_2)_3 + Me_3N + 3H_2 \qquad (3.36)$$

4. Reaction of a metal halide MX_3 with a dialkylamide or disilylamide of an alkali metal (usually Li or Na):

$$AlCl_3 + 3LiNMe_2 \rightarrow Al(NMe_2)_3 + 3LiCl \qquad (3.37)$$

$$AlX_3 + 3LiN(SiMe_3)_2 \rightarrow Al[N(SiMe_3)_2]_3 + 3LiX \qquad (3.38)$$

Gallium, indium and thallium dialkylamides can also be synthesised by the above methods. Variations of the methods can be used to prepare Al and other Group 13 compounds of the type $MX_n(NR_2)_{3-n}$ (X = H, Cl, Br, I or an alkyl group; $n = 1$ or 2).[341] Transamination achieves the exchange of amido substituents which is controlled *inter alia* by the bulk and basicity of the alkyl or aryl substituents:

$$M(NR_2)_3 + xR'_2NH \rightarrow M(NR_2)_{3-x}(NR'_2)_x + (3-x)R_2NH \qquad (3.39)$$

For indium, the main preoccupation has been with the organometal compounds, $R_2InNR'_2$. An attempt to prepare $Me_2InN(SiMe_3)_2$ failed, but $Me_2TlN(SiMe_3)_2$ and $Tl[N(SiMe_3)_2]_3$ have been successfully synthesised.[342]

Organometal amides of Group 13 are obtained *via* the adducts of R_3M with the amine which then react by one of the following sequences:[343,344]

$$R_3M + HNR_2 \xrightarrow[-RH]{} 1/n[R_2M{-}NR_2]_n \qquad (3.40)$$

$$R_3M + H_2NR \xrightarrow[-RH]{} 1/n[R_2M{-}NHR]_n \xrightarrow[-RH]{} 1/n[RM{-}NR]_n \qquad (3.41)$$

Amides of the type $[R_2MNHR]_n$ and $[R_2MNR_2]_n$ are cyclic. On the other hand, imides of the type $[RMNR]_n$ are polycyclic.[345,346] There are subtle differences; for example, $[Me_2AlNMe_2]_2$ has a planar four-membered Al_2N_2 ring but $[Me_2AlNHMe]_3$ forms a six-membered Al_3N_3 ring which exists in either a chair or a 'skew-boat' conformation. The Al–N bond lengths (1.90–1.93 Å) are reasonably constant and show no obvious dependence on the ring size.[347] When the substituents on the amine are bulky or have electron-releasing properties, the amide products may be monomeric, e.g. $Me_2AlN(SiEt_3)_2$. However, no monomeric imides have ever been isolated. The polycyclic oligomers $[RAlNR]_n$ form an Al_nN_n skeleton which takes the form of a pseudo-cubic ($n = 4$), a hexagonal prismatic ($n = 6$), or a bridged- or fused-cage structure ($n = 5$, 7 or 8)[7,346] (see Figure 3.23).

The structural principles of organoaluminium amides and imides should hold good for the corresponding compounds of gallium, indium and thallium. The principles also apply to purely inorganic derivatives, specifically to the hydrides $[H_2MNR_2]_n$ and $[HMNR]_n$ and the halides $[X_2MNR_2]_n$ and $[XMNR]_n$, where these can be isolated.

The *tris*-dimethylamides, $M(NMe_2)_3$ (M = Al or Ga), have been prepared from MCl_3 and $LiNMe_2$ as colourless crystals and fully characterised in recent work.[348,349] The Al and Ga compounds are dimeric, whereas the boron compound $B(NMe_2)_3$ is a monomer. The dimers contain a planar, nearly square M_2N_2 core, with dimethylamide bridges between the metal

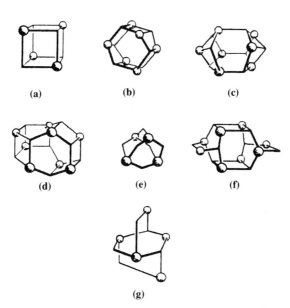

Figure 3.23 Cage structures formed by imidoalanes and related compounds. Reproduced with permission from M. Veith, *Chem. Rev.*, 1990, **90**, 9.

atoms. The M atoms have distorted tetrahedral coordination with N–M–N angles varying from 88 to 115° in the Al compound and 87 to 117° in the Ga compound. The terminal and bridging Al–N bond lengths average 1.80 and 1.965 Å, respectively, and the corresponding Ga–N bonds average 1.855 and 2.01 Å. Table 3.10 gives comparisons with the bond lengths of closely related compounds,[347–358] including other dimers, the trimer $[H_2AlNMe_2]_3$, and the silylamide derivatives $M[N(SiMe_3)_2]_3$ (M = Al, Ga or In) which are constrained to be monomeric by the bulky substituents. The bonding in these compounds has been discussed recently by Waggoner *et al.*[349] who note that the terminal $MNMe_2$ groups show a small but significant deviation from planarity. This is more pronounced in the case of gallium, perhaps because the more diffuse $4d$-orbitals of Ga do not participate in N→M π-type interactions to the same extent as do the $3d$-orbitals of Al.

While the unsaturation of the metal atom is the root cause of the oligomerisation of Group 13 amido derivatives, the factors which operate in particular instances are not fully understood. $[Cl_2AlNMe_2]_2$ is dimeric in the gas phase.[359] $[H_2AlNMe_2]_3$ has been found as a trimer in the solid phase,[347, 348, 360] whereas the analogous gallium compound is a dimer, $[H_2GaNMe_2]_2$, in both the gas[361] and solid phases. Yet the isomer of the gallium compound $[Me_2GaNH_2]_n$, formed by thermal decomposition of $Me_3Ga \cdot NH_3$, exists as a trimer with a 'skew-boat' $[GaN]_3$ ring.[362] The compound $[Cl_2GaN(H)SiMe_3]_n$ can exist in solution as either a dimer or a trimer (n = 2 or 3): the corresponding aluminium compound is a dimer[363]

Table 3.10 Structural features of some amido-aluminium derivatives.

Compound	Framework	Al–N bond distance (Å)[a]	Reference
$[H_2AlNMe_2]_3$	Al_3N_3 ring, chair	1.94 (av)	347, 348
cis-$[Me_2AlNHMe]_3$	Al_3N_3 ring, chair	1.94 (av)	350
trans-$[Me_2AlNHMe]_3$	Al_3N_3 ring, skew-boat	1.90 (av)	350
$[Me_2AlN(CH_2)_2]_3$	Al_3N_3 ring, skew-boat	1.91 (av)	351
$[Me_2AlNH_2]_3$	Al_3N_3 ring, skew-boat	1.94 (av)	352
$[Bu^t_2AlNH_2]_3$	Al_3N_3 ring, planar	1.95 (av)	352
$[Al(NMe_2)_3]_2$	Al_2N_2 ring, planar	1.97 (br), 1.81 (ter)	348, 349
$[HAl(NMe_2)_2]_2$	Al_2N_2 ring, planar	1.97 (br), 1.80 (br)	348, 349
$[Me_2AlNMe_2]_2$	Al_2N_2 ring, planar	1.96	350
$[X_2AlNMe_2]_2$ (X = Br or I)	Al_2N_2 ring, planar	1.94 (X = Br), 1.95 (X = I)	353
$[(Me_3Si)_2AlNH_2]_2$	Al_2N_2 ring, planar	1.96	354
$[Cl_2AlNHSiMe_3]_2$	Al_2N_2 ring, planar	1.91	355
$[Me_2AlNHSiR_3]_2$ (R = Et or Ph)	Al_2N_2 ring, planar	1.97	355
$[Al(N=CPh_2)_3]_2$	Al_2N_2 ring, planar	1.93 (br), 1.78 (ter)	356
$Cl_2AlNEtC_2H_4NMe_2$	$AlNC_2N$ chelate	1.77, 1.96	357
$Al[N(SiMe_3)_2]_3$	AlN_3 trigonal monomer	1.78	358

(a) br, bridging; ter, terminal; av, average.

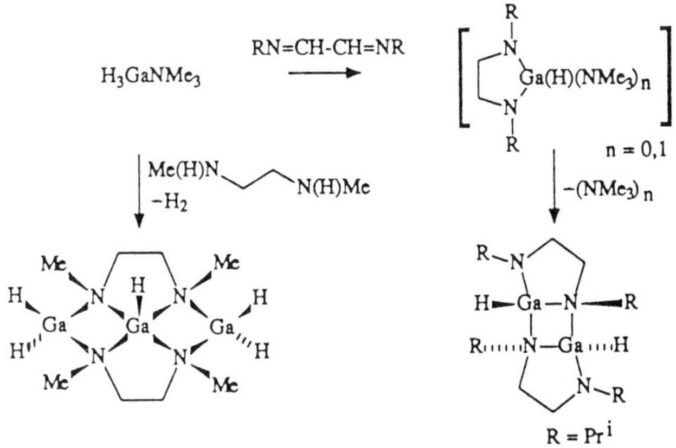

$$H_3GaNMe_3 \xrightarrow{RN=CH-CH=NR} \left[\begin{array}{c} R \\ | \\ N \\ \diagdown \\ Ga(H)(NMe_3)_n \\ \diagup \\ N \\ | \\ R \end{array} \right] \quad n = 0,1$$

Me(H)N⌢N(H)Me, −H₂ (downward)

−(NMe₃)ₙ (downward)

(left structure, Ga₃ oligomer with Me, H, N bridges)

(right structure)
$$R = Pr^i$$

Scheme 3.3 Routes to oligomeric gallium amide/hydride complexes from $H_3Ga \cdot NMe_3$.[365]

and the crystalline gallium compound is also made up of dimer molecules.[364] The amido groups bridge the metal atoms with their substituents adopting a mutually *trans*-configuration; the Ga_2N_2 ring is almost square, with Ga–N bonds of 1.964 and 1.974 Å, slightly shorter than the average (2.01 Å) of the bridging Ga–N bonds in $[Ga(NMe_2)_3]_2$.[349]

The structural chemistry of amidogallanes has recently taken a fresh turn with the synthesis of some remarkable complexes derived from reactions of $Me_3N \cdot GaH_3$ with a *bis*(secondary amine) or a *bis*(imine).[365] Scheme 3.3 shows the reactions involved. The structures of the novel products in the solid state have been established by X-ray crystallography and in solution by NMR measurements.

The chemical reactions of Group 13 alkylamido compounds have yet to be fully explored. The compounds can be sublimed *in vacuo* but tend to decompose when heated. Hydrolysis replaces $-NR_2$ by $-OH$, the ultimate products being hydroxides or oxides. Protonic reagents in general bring about M–N bond cleavage accompanied by the displacement of the amine R_2NH. This reaction can be used synthetically to convert alkylamides to alkoxides, for example:[366]

$$Al_2(NMe_2)_6 \xrightarrow[\text{toluene/room temperature}]{\text{excess Bu}^t\text{OH}} [Al_2(NMe_2)(OBu^t)_5] \xrightarrow[\text{toluene/reflux}]{\text{Bu}^t\text{OH}} Al_2(OBu^t)_6 \quad (3.42)$$

The last amido substituent proves remarkably persistent. The kinetic intermediate shown was inferred from NMR results, but isolated subsequently and characterised fully by X-ray diffraction, to reveal a dimeric structure with bridging NMe_2 and OBu^t groups, $(Bu^tO)_2Al(\mu\text{-}OBu^t)(\mu\text{-}NMe_2)Al(OBu^t)_2$.

Secondary amido derivatives, e.g. R_2MNHR', are much less well known than the tertiary analogues $R_2MNR'_2$. One reason is that they undergo ready intermolecular amine elimination to form products of the type $R_2MN(R)$ NMR_2. Another reason is that the NH proton may participate in alkane elimination, as in the conversion of an alkylamidoalane like $[Me_2AlNHPr^i]_2$ to an alkylimido-derivative like $[MeAlNPr^i]_4$. Lappert et al. have reported some success in preparing aluminium arylamides by the use of the bulky amine 2,4,6-tri-t-butylaniline $(ArNH_2)$,[367] gaining access to $[Me_2AlNHAr]_2$ by treatment of Me_3Al with $ArNH_2$. The same bulky aniline derivative forms an adduct with aluminium chloride, viz. $ArNH_2 \cdot AlCl_3$. Thermolysis of this adduct, which might have been expected to generate the anilido compound $Cl_2AlNHAr$ by loss of HCl, brings about a retro-Friedel–Crafts reaction to eliminate one of the ortho-Bu^t groups on the aryl ligand, 2-methylpropene being produced. On the other hand, an alternative route, using the reaction of $LiNHAr$ with $AlCl_3$, does indeed afford the bis-(anilido)aluminium chloride, $ClAl(NHAr)_2$.

Closely related to aluminium amides and imides are the hydrazides, $[R_2AlNHNR_2']_2$ and $[RAlNNR']_n$, prepared by reactions between a dialkylhydrazine, $R_2'NNH_2$, and an organoaluminium compound.[368] Other compounds with Al–N bonds include methyleneamides (ketimides) of the types $X_2AlN{=}CR_2$, $XAl(N{=}CR_2)_2$ and $Al(N{=}CR_2)_3$ (X = halogen or alkyl substituent). These can be prepared from an aluminium alkyl and an imide $R_2C{=}NH$ or nitrile $RC{\equiv}N$ (via the adduct $R_3Al \cdot L$, where L is the imide or nitrile), or from $AlCl_3$ and the lithium reagent $LiN{=}CR_2$.[369] The ketimides normally associate to form dimers containing four-membered Al_2N_2 rings in which the Al–N bond lengths (about 1.92 Å)[370] are similar to those in the amide and imide structures. A recent investigation of the structure and bonding of the tris(diphenylmethyleneamido)aluminium dimer, $[Al(N{=}CPh_2)_3]_2$, found Al–N distances (bridging 1.93, terminal 1.78 Å) and Al–N=C angles (bridging 130–133, terminal 148–175°) consistent with a bonding description in which each ligand functions as a source of three electrons, making the terminal Al–N bonds twice as strong as the bridging ones.[356] The study drew on molecular orbital bond index (MOBI) calculations[371] to identify this as a general characteristic of the species $X_2Al(\mu\text{-}X)_2AlX_2$, whether electron-precise (X = Cl, NR_2 or $N{=}CR_2$) or electron-deficient (X = Me or H).

3.2.5.3 Phosphido and other Group 15 analogues. Phosphine (PH_3) and arsine (AsH_3) adducts of unsaturated Group 13 compounds are readily prepared. Well characterised examples are $AlX_3 \cdot PH_3$ and $GaX_3 \cdot PH_3$ (X = Cl or Br).[372] Such adducts might be expected to yield phosphido products, $[X_2MPH_2]_n$ and $[XMPH]_n$, by the loss of HX on heating, but in practice these compounds are hard to characterise. Coordinatively saturated complexes are a simpler proposition, e.g. $Li[Al(PH_2)_4]$ or $Li[Al(AsH_2)_4]$,[373,374] and these become more stable when the ligands are dialkylphosphido or disilylphosphido groups, $-PR_2$, or their arsenic equivalents, $-AsR_2$.

Adducts of Group 13 compounds with organophosphine or organoarsine ligands, R_3P or R_3As, are commonplace. The preparation of unsaturated phosphido and arsenido compounds, $M(PR_2)_3$ or $M(AsR_2)_3$, is a more difficult proposition, and is hindered by the intractability of secondary phosphines or arsines, R_2EH ($E = P$ or As).[375] Lithium reagents can be employed, e.g. $LiPR_2$, and by this approach the white crystalline solid $[Cl_2AlPEt_2]_3$ was obtained from $LiPEt_2$ and $AlCl_3$ in ether:[376]

$$3AlCl_3 + 3LiPEt_2 \rightarrow [Cl_2AlPEt_2]_3 + 3LiCl \qquad (3.43)$$

Gallium diorganoarsenides, $Ga(AsR_2)_3$ ($R =$ mesityl or Me_3SiCH_2), and some products of the type $[Cl_{3-n}Ga(AsR_2)_n]_x$ have been made by the reaction of $GaCl_3$ in pentane with Me_3SiAsR_2.[377] R_2P- and R_2As- derivatives of the aluminium sub-group are chiefly the organometallic species R'_2MER_2 ($M =$ Al, Ga or In; $E = P$ or As) whose chemistry was pioneered by Coates et al.[378] Recent research has seen the synthesis of $Ga(PBu^t_2)_3$, $Ga(AsBu^t_2)_3$ and $Ga[As(mesityl)_2]_3$ by the low-temperature reaction of $GaCl_3$ with three equivalents of the appropriate lithium reagent, e.g. $LiEBu^t_2$ ($E = P$ or As).[379,380] These trisubstituted products are monomeric, with planar GaP_3 or $GaAs_3$ skeletal geometries. Treatment of $[(C_5Me_5)_2 GaCl]_2$ with $LiAs(SiMe_3)_2 \cdot 2thf$ in pentane solution affords monomeric $(C_5Me_5)_2GaAs(SiMe_3)_2$, with a Ga–As distance of 2.433 Å, just short of the sum of the relevant covalent radii (2.46 Å);[381] the product is converted by t-butanol to binary gallium arsenide, possibly by way of small GaAs clusters. Compounds of the type $R_2GaEBu^t_2$ have been found to be dimers; for example, $[Bu^n_2GaPBu^t_2]_2$ and $[Me_2GaAsBu^t_2]_2$ have structures with t-butyl-phosphido bridges to give a central Ga_2E_2 core which is essentially planar.[375] The same arrangement occurs in aluminium-arsenic dimers. Two examples with the Al_2As_2 core have been characterised by X-ray analysis, viz. $[Et_2AlAsBu^t_2]_2$[382] and $[Et_2AlAs(SiMe_3)_2]_2$.[383] The latter compound was isolated from the coupling reaction of Et_2AlCl with $LiAs(SiMe_3)_2$ at $-78°C$. The allied reaction between Et_2AlCl and $As(SiMe_3)_3$ failed to give dehalo-silylation products, yielding instead the adduct $Et_2(Cl)Al \cdot As(SiMe_3)_3$. With some relaxation of the steric demands at the Group 13 (M) and/or 15 (E) centres, so compounds of the type $R_2MER'_2$ tend to be trimers: this is the case, for example, with $[Bu^t_2GaPH_2]_3$, $[Me_2GaAsPr^i_2]_3$, $[Br_2GaAs (CH_2SiMe_3)_2]_3$, $[Me_2GaSbBu^t_2]_3$ and $[Cl_2GaSbBu^t_2]_3$, the conformation favoured by the six-membered M_3E_3 ring varying from compound to compound.[375]

The renewed interest in compounds with covalent bonds between the heavier Group 13 and Group 15 elements, spurred by the importance of III–V semiconductor materials, has seen testing for indium of the methods found to be successful for Al or Ga compounds. As already noted,[377] trimethylsilyl-substituted arsines react cleanly with gallium halides at low temperatures to yield products with Ga–As bonds and the corresponding trimethylsilyl halide. Similar methodology has been applied to the formation of In–P bonds. The reaction of $InCl_3$ with $P(SiMe_3)_3$ leads to an orange powder believed to be an oligomer $[Cl_2InP(SiMe_3)_2]_x$. From this Me_3SiCl can be eliminated by heating to yield polycrystalline indium phosphide, InP.[324] A different approach, using organoindium compounds with bulky substituents, e.g. $(Me_3CCH_2)_3In$ or $(Me_3SiCH_2)_3In$, to react with diphenylphosphine, Ph_2PH, and eliminate either neopentane or tetramethylsilane, gives initial products of the type R_2InPPh_2.[384] Interestingly, though, the expected 1:1 adducts, $R_3In \cdot P(H)Ph_2$, could not be isolated, although they are formed in the case of the organogallium compounds.

There are few molecular compounds with In–P or In–As bonds; again these are best known in organometal compounds of the type $R_2InER'_2$ (E = P or As).[375,378] X-Ray studies have established that the phosphides $(Me_3SiCH_2)_2InPPh_2$[375] and $C_5Me_5(Cl)InP(SiMe_3)_2$[385] are dimers with In–P distances of 2.62–2.65 Å, but that the arsenide $Me_2InAsMe_2$[375] is a trimer, two independent conformations of which (one with a planar and the other with a puckered six-membered In_3As_3 ring and In–As distances of 2.67–2.68 Å) coexist in the same asymmetric unit cell. The first stibido (R_2Sb) derivatives are of recent origin: the interaction of $InCl_3$ with $Bu^t_2SbSiMe_3$ affords the partially substituted compound $[ClIn(SbBu^t_2)_2]_2$. X-Ray analysis reveals that the solid comprises dimeric molecules and it is not Cl but the stibido groups which form the bridges.[386] In common with most of the M_2E_2 moieties encountered in Group 13, the In_2Sb_2 core is planar and nearly square. The bridging In^{III}–Sb bonds, for which there seem to be no prior measurements, have an average length of 2.865 Å, in agreement with the sum of the relevant covalent radii. Subsequent research has disclosed similar bond lengths (2.855 Å) in the trimer $[Me_2InSbBu^t_2]_3$, the central In_3Sb_3 ring of which assumes a 'skew-boat' conformation.[375] Table 3.11 is a compilation of structural data for phosphido and related As and Sb derivatives of the Group 13 metals; some representative structures are illustrated in Figure 3.24.

The process of ligand-elimination which must occur in the conversion of the adduct $R_3M \cdot ER'_3$ or the unsaturated compound $R_2MER'_2$ to the eventual product ME (anything from AlP to InSb) can in principle generate RMER' as an intermediate species. With M = Al and E = N, such species are the imidoalanes whose chemistry displays a rich array of oligomeric structures (section 3.2.5.2). Little of this kind has come to light in the chemistry of the compounds in which the heavier Group 13 and Group 15 elements are engaged, although a 30-year-old report that the decomposition of the 1:1 adduct of Me_3In with stibine, via a yellow polymer $[MeInSbH]_x$, yields indium antimonide, InSb, shows remarkable prescience of the

Table 3.11 Structural features of some neutral phosphido-, arsenido- and related derivatives of the Group 13 metals.

Compound	Framework	Core bond distance (Å)	Reference
Ga(PHAr)$_3$ $(Ar = 2,4,6\text{-}Bu^t_3C_6H_2)$	GaP$_3$ monomer	Ga–P, 2.32	380, 389
[Me$_2$GaPBu$^t_2$]$_2$	Ga$_2$P$_2$ ring	Ga–P, 2.47	380
[Bu$^n_2$GaPBu$^t_2$]$_2$	Ga$_2$P$_2$ ring	Ga–P, 2.48	380
[Bu$^t_2$GaP(H)C$_5$H$_9$]$_2$	Ga$_2$P$_2$ ring	Ga–P, 2.45	375
[Bu$^t_2$GaP(H)Bu^t]$_2$	Ga$_2$P$_2$ ring	Ga–P, 2.46	375
[Bu^{t}GaPAr]$_2$ $(Ar = 2,4,6\text{-}Bu^t_3C_6H_2)$	Ga$_2$P$_2$ ring	Ga–P, 2.27	389
[Bu$^t_2$GaPH$_2$]$_3$	Ga$_3$P$_3$ ring, planar	Ga–P, 2.44	375
[ArGaP(cyclohexyl)]$_3$ $(Ar = 2,4,6\text{-}Ph_3C_6H_2)$	Ga$_3$P$_3$ ring, puckered	Ga–P, 2.30	390
[Et$_2$AlAsBu$^t_2$]$_2$	Al$_2$As$_2$ ring	Al–As, 2.57	382
[Et$_2$AlAs(SiMe$_3$)$_2$]$_2$	Al$_2$As$_2$ ring	Al–As, 2.54	383
Ga(AsR$_2$)$_3$ $(R = mesityl)$	GaAs$_3$ monomer	Ga–As, 2.49	379
[R$_2$GaAsBu$^t_2$]$_2$ $(R = Me, Et$ or $Bu)$	Ga$_2$As$_2$ ring	Ga–As, 2.55	375, 380
[(Me$_3$SiCH$_2$)$_2$GaAsBu$^t_2$]$_2$	Ga$_2$As$_2$ ring	Ga–As, 2.59	375
[Ph$_2$GaAs(CH$_2$SiMe$_3$)$_2$]$_2$	Ga$_2$As$_2$ ring	Ga–As, 2.52	375
[BrGa{As(CH$_2$SiMe$_3$)$_2$}$_2$]$_2$	Ga$_2$As$_2$ ring	Ga–As, 2.52	375
[Ga{As(CH$_2$SiMe$_3$)$_2$}$_3$]$_2$	Ga$_2$As$_2$ ring	Ga–As, 2.56	375
[Me$_2$GaAsPr$^i_2$]$_3$	Ga$_3$As$_3$ ring, boat	Ga–As, 2.52	375
[Br$_2$GaAs(CH$_2$SiMe$_3$)$_2$]$_3$	Ga$_3$As$_3$ ring, skew-boat	Ga–As, 2.45	375
[ClGa(SbBu$^t_2$)$_2$]$_2$	Ga$_2$Sb$_2$ ring	Ga–Sb, 2.83	375
[Me$_2$GaSbBu$^t_2$]$_3$	Ga$_3$Sb$_3$ ring, skew-boat	Ga–Sb, 2.72	375
[Cl$_2$GaSbBu$^t_2$]$_3$	Ga$_3$Sb$_3$ ring, irregular boat	Ga–Sb, 2.66	375
[(Me$_3$SiCH$_2$)$_2$InPPh$_2$]$_2$	In$_2$P$_2$ ring	In–P, 2.65	375
[Me$_2$InAsMe$_2$]$_3$	In$_3$As$_3$ ring, $\begin{cases} \text{planar} \\ \text{puckered} \end{cases}$	In–As, $\begin{cases} 2.68 \\ 2.67 \end{cases}$	375
[ClIn(SbBu$^t_2$)$_2$]$_2$	In$_2$Sb$_2$ ring	In–Sb, 2.87	386
[Me$_2$InSbBu$^t_2$]$_3$	In$_3$Sb$_3$ ring, skew-boat	In–Sb, 2.86	375

potential importance of this route to semiconductor coatings.[387] Very recently, however, the cubane-like compounds [Bu^{i}AlPSiPh$_3$]$_4$ and [Bu^{t}Ga PSiPh$_3$]$_4$ have been prepared,[388] and by increasing still further the steric demands of the substituents it has been possible to isolate the dimer [Bu^{t}GaPAr]$_2$ (where Ar = 2,4,6-Bu$^t_3$C$_6$H$_2$) in accordance with the procedure summarised in equation (3.44):[389a]

$$[Bu^tGaCl_2]_2 + 2ArPHLi \xrightarrow[-78°C]{Et_2O} Bu^tGa(PHAr)_2$$

$$\xrightarrow[-H_2]{\Delta} Bu^tGa(\mu\text{-}PAr)_2GaBu^t \quad (3.44)$$

The crystal structure of the product reveals a central Ga$_2$P$_2$ rhombus with unusually short Ga–P distances [2.27 *vs.* 2.32 Å in the *bis*(phosphido)gallane Bu^{t}Ga(PHAr)$_2$]; while the dimensions, together with the yellow colour of

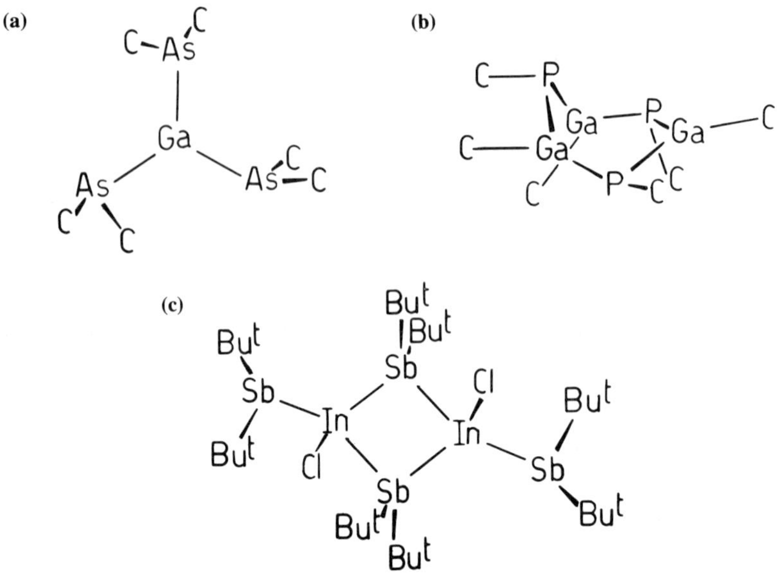

Figure 3.24 Types of M coordination in Group 13 – Group 15 compounds with representative bond lengths and bond angles. (a) *Tris*(dimesitylarsino)gallane:[379] Ga–As = 2.47–2.51 Å, As–Ga–As = 117–124°; the Ga atom is 0.15 Å out of the plane of the three As atoms. (b) [(2,4,6-Ph$_3$C$_6$H$_2$)GaP(*cyclo*-C$_6$H$_{11}$)]$_3$:[390] Ga–P = 2.28–2.34 Å, P–Ga–P = 115–121°. (c) [(Bu$^t_2$Sb)(Cl)In(μ-SbBu$^t_2$)]$_2$:[386] In–Sb = 2.80, 2.86, In–Cl = 2.41 Å, Sb–In–Sb = 85°.

the compound, are suggestive of p_π–p_π interactions, these cannot be extensive since the phosphorus atoms retain a pyramidal environment. On the other hand, the compound Bu$^t_2$GaPMes*(SiPh$_3$) (Mes* = 2,4,6-Bu$^t_3$–C$_6$H$_2$) turns out not only to be a monomer in the solid state, but also to have a flattened pyramidal geometry at phosphorus (with Ga–P = 2.295 Å); more strikingly still, solutions of the compound show clear NMR evidence of an appreciable barrier to rotation about the Ga–P bond.[389b]

Where the oligomers involve six-membered rings, e.g. [RMER']$_3$, they become akin to borazine, [HBNH]$_3$. This raises the possibility that the ring may display structural properties (such as planarity and shortened, equivalent ring bonds) which are indicative of quasi-aromatic character. Gallium is perhaps less likely than B or Al to produce a planar array, and indeed the structure of the first example of a Ga$_3$P$_3$ ring compound, [(2,4,6-Ph$_3$C$_6$H$_2$)GaP(*cyclo*-C$_6$H$_{11}$)]$_3$, bears this out.[390] Although the Ga centres have almost planar trigonal coordination, the geometry at the P atoms is pyramidal. The situation contrasts therefore with that of planar B$_3$N$_3$ or B$_3$P$_3$ rings, leading to the conclusion that the energy gained by forming Ga–P π-bonds is insufficient to overcome the inversion barrier at phosphorus.

Heteronuclear clusters with the cores Ga$_4$P$_5$ and Ga$_5$As$_7$ feature in the gallium phosphide [Ga$_4$(Trip)$_3${P(1-Ad)}$_4$P(H)(1-Ad)] (1-Ad = 1-adamantyl;

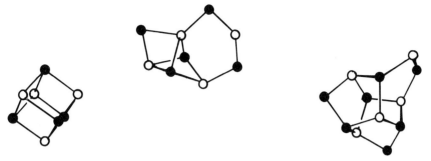

Figure 3.25 Cage structures present in some Al–P, Al–As, Ga–P and Ga–As compounds.[388,391,392] ○, Al or Ga; ● P or As.

Trip = 2,4,6-$Pr^i_3C_6H_2$)[391] and the gallium arsenide [(PhAsH)(R_2Ga) (PhAs)$_6$(RGa)$_4$] (R = CH_2SiMe_3),[392] respectively (see Figure 3.25). The phosphide was prepared by the reaction of Li_2P(1-Ad) with one equivalent of Cl_2GaTrip · thf, the arsenide by that of $PhAsH_2$ with (Me_3SiCH_2)$_3$Ga. These recent results suggest that the preparation and characterisation of cluster compounds involving the heavier members of Group 15 is likely to be another of the growth areas of Group 13 chemistry.

3.2.6 Derivatives with bonds to boron or Group 14 atoms

3.2.6.1 Borides, carbides and silicides.

Aluminium forms crystalline borides, among them AlB_2, AlB_4 and AlB_{12}, and ternary boron carbides, AlB_xC_y, in which Al atoms are incorporated in B (and C) network structures.[393–395] With the exception of AlB_2, the aluminium borides are high melting, very hard and extremely refractory. Beyond the reported preparation of GaB_{12},[396] borides of the heavier Group 13 elements are virtually uncharted.

Aluminium carbide, Al_4C_3, m.p. 2100°C, is made by heating the metal with carbon at temperatures above 1000°C in the absence of air (otherwise the product incorporates Al_2O_3 and AlN to give aluminium carbide oxide and carbide nitride compounds).[2] Al_4C_3 forms yellow crystals in which Al displays 4-, 5- and 6-fold coordination towards carbon, with Al–C distances ranging from 1.90 to 2.22 Å. It undergoes surface oxidation, burns in excess oxygen to Al_2O_3 and CO_2, and is hydrolysed by acids or boiling water, generating methane. Reduction can be achieved by a more electropositive metal; thus, calcium heated with Al_4C_3 forms CaC_2 and aluminium.

Reactions of Al_2O_3 with Al_4C_3 or Al_2O_3 with carbon at 1700–2200°C produce the solid phase Al_2OC (aluminium oxycarbide).[397,398] Single-crystal X-ray analysis[397] finds this solid to have a hexagonal lattice of the wurtzite type related to that of aluminium nitride, AlN (with which it is formally isoelectronic). A second product is Al_4O_4C (aluminium tetroxycarbide). This has a structure based on $Al(O_3C)$ tetrahedra which share corners and edges:[399] Al–O distances range from 1.71 to 1.87, and Al–C distances from 1.91 to 1.98 Å. The high-temperature reactions of Al_4C_3 with AlN in various proportions generate a series of aluminium carbonitrides, $Al_4C_3(AlN)_n$, of which the simplest is the $1:1$ product, Al_5C_3N. Crystallographic investigations show that the coordination of aluminium atoms is consistently tetrahedral throughout the series.[400]

Phase-mapping of the aluminium–silicon system locates a eutectic composition containing 12% Si, m.p. 577°C, but finds no silicide Al_4Si_3.[2] Metallurgy is concerned with numerous ternary and more complicated aluminium silicides, $M_xAl_ySi_z$, and analogous carbides (see chapter 2). Some of these phases can incorporate Ga, In or Tl, but there appear to be no gallium, indium or thallium carbides or silicides as such, possible Tl^I compounds excepted.

3.2.6.2 Metallacarboranes.   Carboranes, that is mixed hydrides of carbon and boron in which atoms of both these elements are incorporated in an electron-deficient molecular skeleton, can react with other Main Group elements or moieties to form heterocarboranes. When part of the carborane cage, the heteroatoms form bonds to C and/or B and acquire properties which are not typical of their usual chemical behaviour. The bonding principles and structural relationships of carboranes have been reviewed elsewhere.[402–406] Of particular interest here are the heterocarboranes containing a Group 13 moiety isolobal with BH.[407] Such a unit must be able to furnish two electrons and three suitably oriented orbitals, requirements met by AlR or GaR. Thus, the metallacarboranes $RMC_2B_9H_{11}$ (M = Al or Ga) and the carborane $C_2B_{10}H_{12}$ — a closed (*closo*) structure with the skeletal atoms occupying the corners of an icosahedron — are expected to have the same geometry. The synthesis, reactivity and structural characterisation of further Group 13 metallacarborane species has been described,[408] and this topic appears set to grow apace.

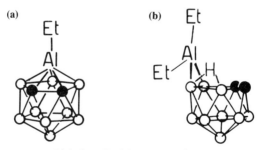

Figure 3.26 The structures of (a) *closo*-$EtAlC_2B_9H_{11}$ and (b) *nido*-$Et_2AlC_2B_9H_{12}$.[408, 410] ○ BH; ●, CH.

The first *closo*-aluminacarborane, $EtAlC_2B_9H_{11}$, was reported in 1968,[409] having been prepared by the reaction of $[C_2B_9H_{11}]^{2-}$ with $EtAlCl_2$ in tetrahydrofuran at $-50°C$. The structure, confirmed later by X-ray diffraction,[408,410] has the EtAl group situated almost symmetrically above the C_2B_3 face of the cage, as shown in Figure 3.26(a). The bond distances within the cage are $Al–C = 2.17$ and $Al–B = 2.14$ Å, whereas the $Al–C_2H_5$ bond measures 1.93 Å. An alternative synthesis of such metallacarboranes involves the initial formation of the *nido*-derivatives in which an R_2Al moiety bridges two boron atoms of the C_2B_3 face forming Al–B linkages of length 2.28 and 2.33 Å (see Figure 3.26(b)).[408,410] These then undergo cage closure with the elimination of one further molecule of alkane:[411]

$$C_2B_9H_{13} + R_3M \longrightarrow \textit{nido-}R_2MC_2B_9H_{12} + RH \qquad (3.44)$$

$$\textit{nido-}R_2MC_2B_9H_{12} \longrightarrow \textit{closo-}RMC_2B_9H_{11} + RH \qquad (3.45)$$

M = Al, R = Me or Et; M = Ga, R = Et. Aluminacarboranes of this type yield adducts with donors such as thf or diethyl ether, where the Al acts as a Lewis acid site. The reaction is thought to involve a return from the *closo* to the open (*nido*) form; this is certainly the case with an excess of PEt_3 which gives *endo*-10-{$EtAl(PEt_3)_2$}-7,8-$C_2B_9H_{11}$,[408] with the structure in Figure 3.27 in which the Al–B bond shown measures 2.13 Å. The use of stronger bases removes the apical aluminium group to yield the appropriate dianion, e.g. $[Me_2C_2B_9H_9]^{2-}$ from 2,3-Me_2-1-Et-1,2,3-$AlC_2B_9H_9$. This latter aluminacarborane is able to react with Main Group halides to replace the capping EtAl unit by MeB using $MeBBr_2$, by a germanium atom using GeI_2, or by tin using $SnCl_2$.[412]

Other heterocarboranes are of the sandwich (*commo*) type in which a metal atom — here aluminium — is situated between the faces of a pair of carborane cages. Figure 3.28 shows the structure of one such icosahedral *commo*-aluminacarborane, *viz* the $[Al(C_2B_9H_{11})_2]^-$ anion which was prepared as the Tl^+ salt from Tl_2 (*nido*-$C_2B_9H_{11}$) and Et_2AlCl.[408,413] Allied *commo*-aluminacarborane anions have been synthesised in which the inter-

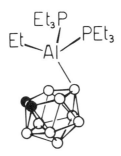

Figure 3.27 The structure of *endo*-$EtAl(PEt_3)_2C_2B_9H_{11}$.[408] ○, BH; ●, CH.

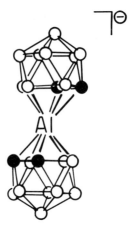

Figure 3.28 The structure of the *closo*-[Al(C$_2$B$_9$H$_{11}$)$_2$]$^-$ anion.[413] ○, BH; ●, CH.

linked carborane polyhedra are of the smaller AlC$_2$B$_4$, AlC$_2$B$_6$ and AlC$_2$B$_8$ varieties.[401] In these complexes, the carborane dianions act as η^2-ligands donating four electrons, *via* two carbon-based orbitals, to a tetrahedrally coordinated Al(III) centre. Figure 3.29(a) shows a representative structure for the [Al(*nido*-6,9-C$_2$B$_8$H$_{10}$)$_2$]$^-$ anion.[408] The structural relationship to EtAl(OEt$_2$)(*nido*-6,9-C$_2$B$_8$H$_{10}$),[408] shown in Figure 3.29(b), wherein the fourth position on Al is filled by the coordination of OEt$_2$, is obvious. Small cage aluminacarboranes of the AlC$_2$B$_4$ type are produced by the reaction of *nido*-2,3-Et$_2$-C$_2$B$_4$H$_6$ with the triethylamine adduct of alane, Et$_3$N · AlH$_3$.[414] By varying the conditions a *commo*- and two *nido*-products were isolated. In each of these the Et$_3$N base remained attached to aluminium.

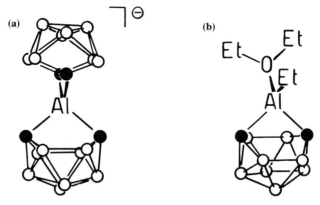

Figure 3.29 The structures of (a) the *nido*-[Al(C$_2$B$_8$H$_{10}$)$_2$]$^-$ anion and (b) the ether adduct *nido*-EtAl(OEt$_2$)C$_2$B$_8$H$_{10}$.[408] ○, BH; ●, CH.

In contrast to the progress reported for aluminacarboranes, there is little information on the carborane derivatives of the heavier Group 13 elements, and the literature contains few results to point the way ahead. Twelve-atom clusters incorporating Ga and the C_2B_9 fragment have already been mentioned. Addition of a toluene solution containing 2 molar equivalents of $GaCl_3$ to a suspension of 1 molar equivalent of $Tl_2[7,8-C_2B_9H_{11}]$ in toluene affords the *commo*-gallacarborane sandwich compound, $Tl[commo-3,3'-Ga(1,2-C_2B_9H_{11})_2]$.[408] The structure resembles that of the Al compound shown in Figure 3.28; intracage Ga–B distances range from 2.08 to 2.30 and Ga–C distances from 2.48 to 2.59 Å. Other carboranes incorporating either gallium or indium (M) are of the type *closo*-$MeMC_2B_4H_6$; these have been synthesised by the reaction of Me_3M with $C_2B_4H_8$ in the gas phase.[415] The compounds are thermally stable at temperatures up to 100°C. Stronger heating or the action of HCl regenerates the carborane. The structure of the gallium compound has been confirmed by X-ray crystallography. Having a closed seven-atom framework approximating to a pentagonal bipyramid, the structure has two distinctive features: (i) the position of the capping Ga atom is displaced slightly so that the Ga–B distances (2.11 and 2.22 Å) are shorter than the Ga–C distances (2.32 Å), and (ii) the Me–Ga axis is tilted by 20° with respect to the perpendicular to the equatorial plane (see Figure 3.30(a)). Extended Hückel calculations have been used to investigate the bonding in this system; hence it appears that the distortion enhances the

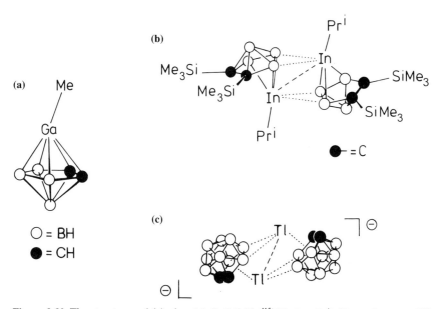

Figure 3.30 The structures of (a) *closo*-$MeGaC_2B_4H_6$,[415] (b) *closo*-$Pr^iIn[(Me_3Si)_2C_2B_4H_4]$[417] and (c) the *closo*-$[TlC_2B_9H_{11}]^-$ anion.[422]

skeletal bonding and is not necessarily a function of back-bonding involving the filled 3d-orbitals of the metal.[416] The *closo*-indacarborane, $Pr^iIn[(Me_3Si)_2C_2B_4H_4]$,[417] exists in the solid state as a dimer (see Figure 3.30(b)). The indium metal in each cage occupies an apical vertex of a trigonal bypyramid but is displaced above the C_2B_3 face in an η^3 fashion. The dimer has an In–In distance of 3.70 Å which leaves open the question of some direct bonding. The In–B distances are clearly in two categories; the intracage bond distances average 2.39 and the intercage distances average 3.04 Å.

The thallium(I) derivative $Tl_2C_2B_9H_{11}$, mentioned above as the reagent used to prepare $Tl[M(C_2B_9H_{11})_2]$ (M = Al or Ga), is formed when thallium(I) acetate reacts with the anion $[C_2B_9H_{12}]^-$ in aqueous alkaline solution.[418] Of the two Tl^I ions, one is associated with the anionic carborane cage and the other Tl^+ is the counter-ion. The crystal structure of the Ph_3PMe^+ salt of $[TlC_2B_9H_{11}]^-$ shows the Tl^I to occupy the apical position above the open face of the carborane fragment, with a slight slippage away from the carbon atoms. The shortest Tl–B and Tl–C distances (2.66–2.92 Å) are greater than expected on the basis of the covalent radii of the atoms involved, a finding taken to indicate an ion-pair formulation $[Tl^+C_2B_9H_{11}^{2-}]$.[419] However, the pale yellow colour of the ion derived from the colourless constituents Tl^+ and $[C_2B_9H_{11}]^{2-}$ implies some charge transfer. Be that as it may, the thallium is loosely bound to the carborane cage; hence, thallacarboranes are useful precursors to other metallacarboranes, just as thallium(I) cyclopentadienide is an expedient source of other metallocenes.

An alternative view of the bonding recognises the isolobal and isoelectronic relationship between carbollyl and cyclopentadienyl ligands, e.g. $[C_2B_9H_9Me_2]^{2-}$ and $[C_5Me_5]^-$.[420] Accordingly, Tl^I and other p-block elements in their low oxidation states are found either η^5-bonded to the planar face of the C_2B_9 ligand or in a slightly slipped (η^3) bonding situation. $Tl_2[C_2B_9H_9Me_2]$ is an example of the latter type;[421] its structure comprises Tl^+ ions and $[Tl(C_2B_9H_9Me_2)]^-$ counterions. The Tl–B distances of the metallacarborane anion are 2.68, 2.74 and 2.97 Å, whereas the Tl–C distances are 3.17 and 3.28 Å. The X-ray diffraction study also finds that there are sets of Tl^+ cations with a separation of only 3.67 Å, which is in the range calculated for weak bonding and so bears on the question of nonclassical Tl^I–Tl^I interactions. In another structural study,[422] the synthon $Tl_2C_2B_9H_{11}$[418] has been treated with $[PPN]^+Cl^-$ in acetonitrile to prepare $[PPN][closo-TlC_2B_9H_{11}]$ (PPN = bis(triphenylphosphoranylidene)ammonium). The separation of the Tl^I centres is 4.24 Å, and the structure shown in Figure 3.30(c), has a dimeric anion in which the carbollyl ligands occupy bridging positions between the two weakly interacting thallium atoms. The change of cation to $[Ph_3PMe]^+$ in the derivative described earlier[419] is sufficient to produce a structure in which the closest Tl–Tl distance is 7.97 Å, a result which seems to ensure that the final word has not yet been written on the subject of bonding in these intriguing compounds.

3.2.6.3 Silyl, germyl and stannyl derivatives. Few silyl or germyl derivatives of aluminium or its congeners are known. In comparison with the rich organometallic chemistry described in chapter 6, this area is under-developed, and one can only speculate on whether bonding between the elements concerned is unfavourable or whether it is simply a matter of experimental neglect. Early attempts to prepare *tris*(trimethylsilyl)-aluminium, $Al(SiMe_3)_3$, yielded only the cleavage products — mainly polysilanes and methylaluminium compounds — likely to have arisen from disproportionation, findings suggesting that the Al–Si bond is inherently weak. Alternatively, the unsaturation of the silicon centres may provide low-energy routes to decomposition. The synthesis of $Al(SiMe_3)_3$ was finally reported in 1977,[423] and that of the corresponding germyl, $Al(GeMe_3)_3$, in the following year.[424] The method involves the reaction of the mercury derivative with Al powder in thf and pentane at room temperature to give the thf adducts:

$$3Hg(EMe_3)_2 + 2Al \xrightarrow{\text{thf/pentane}} 2Al(EMe_3)_3 \cdot thf + 3Hg \quad (3.46)$$

E = Si or Ge. Both compounds $Al(EMe_3)_3 \cdot thf$ are colourless, crystalline solids which ignite spontaneously in air. The molecule of thf is not removed without decomposition, but this ligand can be replaced by another, e.g. $(Me_3Si)_3P$. The silyl decomposes at temperatures above 50°C to the expected polysilanes and methylaluminium compounds. The germyl derivative is decidedly more stable and melts without decomposition at 81°C. However, an attempt to prepare the analogous stannyl compounds using $Hg(SnMe_3)_2$ failed, the product being assumed to decompose below room temperature.

The organosilyl gallium compound, $Ga(SiMe_3)_3 \cdot thf$, cannot be prepared from Ga using the mercury reagent, but is obtained in good yield from $GaCl_3$ by the following method:[425]

$$GaCl_3 + 3Me_3SiCl + 6Li \xrightarrow{\text{thf, } -10°C} Ga(SiMe_3)_3 \cdot thf + 6LiCl \quad (3.47)$$

In this case, the adduct loses thf on vacuum sublimation to give the base-free product the vibrational spectrum of which confirms the planar structure of the $GaSi_3$ group, as expected for a monomer. A similar reaction of Me_3SiCl and lithium with Ga in the presence of ether yielded lithium *tetrakis*(trimethylsilyl)gallate coordinated by diethyl ether:

$$Ga + 4Me_3SiCl + 5Li \xrightarrow{\text{Hg/Et}_2O} Li[Ga(SiMe_3)_4] \cdot xEt_2O + 4LiCl \quad (3.48)$$

Solvent-free $Ga(SiMe_3)_3$, as produced by sublimation, ignites on exposure to air and decomposes at 50°C to gallium and hexamethyldisilane.

Interestingly, silyl derivatives of indium and thallium had been prepared before the aluminium and gallium compounds just discussed. The products were the base-free compounds, $M(SiMe_3)_3$ (M = In or Tl). Synthesis of *tris*-(trimethylsilyl)indium was accomplished by a route analogous to that leading to the Ga compound, but yielded the unsolvated product:[426]

$$InCl_3 + 3Me_3SiCl + 6Li \xrightarrow{\text{thf, } -10°C} In(SiMe_3)_3 + 6LiCl \quad (3.49)$$

The highly unstable compound $In(SiMe_3)_3$ forms greenish-yellow crystals which decompose at 0°C when exposed to light. As with the gallium compound, the Raman spectrum implies a monomeric structure derived from a planar MSi_3 skeleton. The vibrational frequencies were used to derive a stretching force constant of 1.25 mdyn $Å^{-1}$ for the In–Si bond. The more stable silyl derivatives of other fifth-period elements, $Sn(SiMe_3)_4$, $Sb(SiMe_3)_3$ and $Te(SiMe_3)_2$, have larger force constants, *viz.* 1.5–1.7 mdyn $Å^{-1}$, so that the small force constant in the indium case is consistent with the lability of the In–Si bond.

Tris(trimethylsilyl)thallium is conveniently prepared by the rapid reaction of the mercury reagent with trimethylthallium:[427]

$$3Hg(SiMe_3)_2 + 2Me_3Tl \rightarrow 2Tl(SiMe_3)_3 + 3Me_2Hg \quad (3.50)$$

It is remarkable for forming needle-shaped crystals which are blood-red at room temperature and yellow at $-196°C$, and, although highly unstable, it can be sublimed *in vacuo* at 40°C. ^1H NMR studies have shown that intermolecular exchange of Me_3Si groups occurs in CH_2Cl_2 or toluene solution.[428] The mechanism is not understood, but the effect of trimethylamine is to slow the exchange, thereby providing evidence for the formation of an adduct, $Me_3N \cdot Tl(SiMe_3)_3$. The base-free species is believed to be the monomer on the grounds of the NMR results and its Raman spectrum.[429] Various attempts to generate the *bis*(trimethylsilyl)thallium cation, $(Me_3Si)_2Tl^+$, analogous to Me_2Tl^+, proved unsuccessful. $Tl(SiMe_3)_3$ reacts with $CHCl_3$ to give $(Me_3Si)_2TlCl$ which itself reacts with HCl to form $TlCl_3$. The compound *bis*(trimethylsilyl)thallium chloride is a white solid for which a Cl-bridged, dimeric structure seems quite likely.

The foregoing compounds are highly sensitive to heat, light and oxygen. Greater kinetic stability is achieved by replacing Me_3Si by an even bulkier silyl ligand, e.g. $(Me_3Si)_3Si$. It is doubtful whether a Group 13 atom could accommodate three of these ligands and the sole products derived from the treatment of MCl_3 (M = Ga or In) with $LiSi(SiMe_3)_3 \cdot 3thf$ are the disubstituted compounds, $\{[(Me_3Si)_3Si]_2M(\mu\text{-}Cl)_2Li(thf)_2\}$.[430] Coordination at the M atom is tetrahedral but the angle Si–M–Si is unusually wide, *viz.* 138° at Ga and 140° at In, presumably as a result of steric interactions (see Figure 3.31). The average Ga–Si and In–Si bond lengths of 2.44 and 2.59 Å, respectively, are close to the sums of the relevant covalent radii (Ga–Si = 2.40; In–Si = 2.65 Å). In passing, we may note that an attempt to prepare the

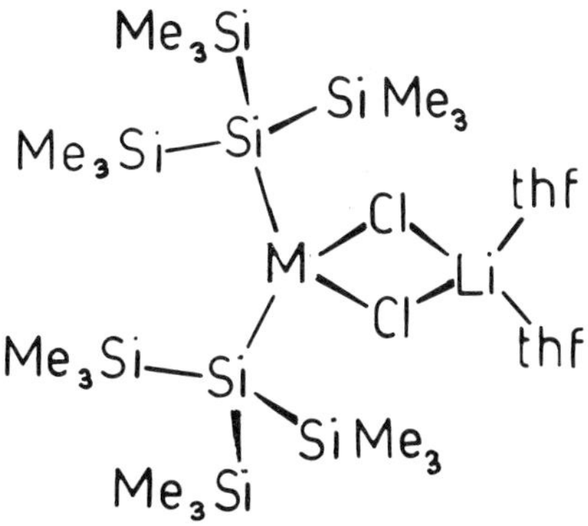

Figure 3.31 The structure of $[(Me_3Si)_3Si]_2M(\mu\text{-}Cl)_2Li(thf)_2$ (M = Ga or In).[430] Dimensions: Ga–Si = 2.37 Å, Si–Ga–Si = 138°; In–Si = 2.59 Å; Si–In–Si = 140°. Source: *J. Chem. Soc., Chem. Commun.*, 1986, 1776.

In^I derivative of this silyl ligand gave the In^{III} compound instead. Similar instances of ligand-induced disproportionation are frequent in low-valent indium chemistry (see section 3.3).

Most other examples of species with a silyl, germyl or stannyl ligand attached to Al, Ga, In or Tl are complex anions. Here the Group 13 atom is coordinatively saturated, a condition which allows simpler ligands, even hydride, to be present. An example is the anion $[H_3GeAlH_3]^-$ which can be made by the reaction of $KGeH_3$ with AlH_3, or GeH_4 with $LiAlH_4$, and is marginally stable in ether solvents at 25°C.[431] It is noteworthy that the Ge–Al compound is much less stable than the corresponding Ge–B compound which yields crystalline salts, e.g. $K[H_3GeBH_3]$. The Ge–Ga bond is perhaps more stable than the Ge–Al bond: the reagent $KGeH_3$ combines with Me_2GaCl in dimethoxyethane to give a solvated adduct from which germyldimethyl-gallane, Me_2GaGeH_3, has been prepared, admittedly in an impure condition.[432] This route has been used to prepare silyl as well as germyl derivatives:[433]

$$KSiH_3 + Bu^n_2AlCl \rightarrow K[Bu^n_2Al(SiH_3)Cl] \rightarrow Bu^n_2AlSiH_3 + KCl \quad (3.51)$$

$$KGeR_3 + Me_2GaCl \rightarrow K[Me_2Ga(GeR_3)Cl] \rightarrow Me_2GaGeR_3 + KCl \quad (3.52)$$

R = H or Ph. Alternatively, aryl-disilanes and -digermanes can be treated with $LiAlH_4$ to synthesise silyl- and germyl-aluminates, for example $Li[Ph_3GeAlH_3]$.[434] Trimethylstannyl derivatives of the form $Li[Me_3MSnMe_3]$ (M = Al, Ga, In or Tl) have been produced by the reaction of $LiSnMe_3$ with the corresponding alkyl, Me_3M,[435] or by cleavage of the tin–tin bond in Sn_2Me_6 by $LiMR_4$ (M = Al, Ga or Tl; R = H or Me).[436] The position of thallium is interesting: $LiTlMe_4$ is the most reactive member of the series towards Sn_2Me_6 and species with more than one stannyl group bound to Tl are formed, i.e. $Li[Me_nTl(SnMe)_{4-n}]$.

One other novel aluminium–tin compound deserves to be mentioned, although its original synthesis[437] has not been repeated. This is the complex *bis*(methylcyclopentadienyl)tin–aluminium chloride in which the Sn^{II} atom may act as a Lewis base, as in $(C_5H_5)_2Sn \cdot BF_3$. The characterisation of this intriguing compound, reported to be a viscous, golden oil, rests on spectroscopic evidence which should probably be re-examined.

In summary, silyl, germyl and stannyl derivatives of the Group 13 metals can be prepared. The methods introduce the Group 14 substituent R_3E as the halide, the mercury derivative or the dinuclear compound E_2R_6, in reactions with the Group 13 element as the metal, the halide or an organometallic compound. The products are typically highly unstable compounds. The oligomerisation which characterises the organometallics, e.g. Me_3Al to $Me_2Al(\mu\text{-}Me)_2AlMe_2$, has no counterpart with the silyl or germyl compounds, and the stability of the monomer $M(ER_3)_3$ tends to increase from Al to Tl. A different picture may emerge in the low-valent chemistry of the Group as recent calculations, discussed in section 3.3, indicate that M^I compounds of the type $AlSiR_3$ ($R = H$, Me or Bu^t) appear liable to form tetramers akin to Cp_4Al_4.

3.2.7   Pseudohalide derivatives

The pseudohalide ions, cyanide CN^-, cyanate NCO^-, thiocyanate NCS^- and azide N_3^-, show little tendency to interact with Al^{3+} ions in aqueous solution. Ga^{3+} and In^{3+}, on the other hand, form identifiable complexes with these ligands (chapter 8). That the pseudohalide is in competition with H_2O or OH^- under these conditions, however, influences the products which are isolable. The oxidising character of Tl^{3+} makes it incompatible with thiocyanate ligands, but there is evidence of azide complexes in aqueous solution.[2] Thallium(I) forms $TlNCS$ and TlN_3 as crystalline solids which resemble the corresponding potassium salts. Apparently the Tl^I compounds can be oxidised by the dihalogen X_2 to $Tl(SCN)X_2$ and $Tl(N_3)X_2$, respectively.[5]

The compounds $Ga(NCS)_3 \cdot 3H_2O$ and $In(NCS)_3 \cdot xH_2O$ can be prepared from aqueous sulfate solutions by the addition of $Ba(NCS)_2$ and removal of precipitated $BaSO_4$, followed by evaporation. The In compound can be dehydrated by heating *in vacuo* and yields complexes of the type $In(NCS)_3 \cdot L$ with donor ligands.[438,439] $In(NCS)_3$ can also be obtained from $InCl_3$ and $NaNCS$ in dry ethanol.[440] The infrared spectrum of the compound suggests that NCS bridges between In centres in an ambidentate fashion. The IR spectrum of $Ga(NCS)_3 \cdot 3H_2O$ indicates N-bound thiocyanate. Both the gallium and indium compounds are readily soluble in ethanol, acetone and ether, but are hydrolysed by hot water.

Some pseudohalide derivatives of the Group 13 elements have been prepared under non-aqueous conditions.[441] These may exist only in the form of complexes, e.g. $Al(NCS)_3$ as an etherate or an ammonia adduct.[2] Aluminium azide, $Al(N_3)_3$, can be made from NaN_3 and $AlBr_3$ in benzene, or by mixing ether solutions of AlH_3 and HN_3 at $-100°C$, and some of its chemistry, including the formation of complexes $[Al(N_3)_4]^-$ and $[Al(N_3)_5]^{2-}$, has been investigated. Al and Ga compounds of the types MX_2N_3 and $MX(N_3)_2$, where $M = Al$ or Ga and $X = Br$ or I, have been prepared by the reaction of MX_3 with the halogen azide XN_3 in benzene.[442] Aluminium cyanide is precipitated as $Al(CN)_3 \cdot OEt_2$ by adding HCN to ethereal AlH_3, and there is a report that the reaction of $Hg(CN)_2$ with AlH_3 in ether yields $AlH(CN)_2$.[2] Better known Al pseudohalides are the organoaluminium derivatives, R_2AlCN, R_2AlN_3 and $RAl(N_3)_2$, and there are Ga, In and Tl counterparts.[443] Some hydrazides, $R_2AlNHNR'_2$ and $[RAlN_2R']_x$, are mentioned in section 3.2.5.2 in the context of amido- or imido-aluminium compounds.

Gallium cyanide does not seem to have been made, but $In(CN)_3$ has been prepared in low yield by the action of cyanogen on indium oxyiodide, or from In or $In(OH)_3$ and HCN gas at $350°C$.[440] Both Ga^{III} and In^{III} form hexacyanoferrates which are precipitated from the aqueous M^{3+} solutions by adding the potassium salt of the iron(II) or iron(III) complex. The compounds are air-stable, white solids with low solubilities in water; indeed $Ga_4[Fe(CN)_6]_3$ can be employed for the gravimetric determination of gallium. Thallium ferrocyanide is a Tl^I compound which forms yellow crystals, $Tl_4Fe(CN)_6 \cdot 2H_2O$.

3.3   Derivatives of the lower valence states

3.3.1   Low oxidation states in Group 13

The $+3$ oxidation state prevails in much of the chemistry of the Group 13 metals, with the exception of thallium for which $+1$ is normally the more stable state. The $+1$ state is significant for indium and gallium, in diminishing order, but barely so for aluminium. Monovalent aluminium is almost entirely restricted to species which are transient under normal conditions; these are produced in high-temperature reactions or electrochemically, and may be observed by their spectra either in the gas phase or when trapped in a solid matrix at low temperatures. However, an important recent advance is the preparation of metastable AlF and AlCl in cooled solutions in polar solvents.[444,445]

The increasing sophistication of computational methods in chemistry sometimes allows theory to lead experiment. Molecular parameters have been calculated for a number of low-valent aluminium and other Group 13 compounds, serving either to confirm or to predict the properties of short-lived species. These are discussed in the following sections.

Although most of the monohalides of Ga and In can be readily prepared, their intractability and low solubility has limited their utility as precursors to other low-valent compounds. The preparation of organic solvent-soluble $(\eta^5\text{-}C_5H_5)In$ from solid InCl and LiC_5H_5 has helped to open up this area.[446] InX and GaX are strongly reducing species, unstable in air and water. However, indium halides can be dissolved in organic solvents containing suitable donor ligands, for example toluene/tetramethylethanediamine at 0°C,[447] and solutions of metastable GaCl in toluene/diethyl ether are also rich in synthetic promise.[448] The dependence of the disproportionation process $In^I \rightarrow In^0 + In^{II}$ or In^{III} on temperature, halide and donor ligand has been studied in some detail.[447]

The homogeneous disproportionation of In^I in aqueous solution has been difficult to study because in acidic solution the predominant process is oxidation of In^I by H^+, while under other conditions traces of indium metal are present and the predominant process is heterogeneous disproportionation. Kinetic and spectrophotometric studies on aqueous solutions of InBr have shown that slow homogeneous decomposition occurs during the induction period, with the deposition of colloidal indium metal. This step is faster in solutions containing H^+ or an excess of halide. The ensuing heterogeneous processes then accelerate the decomposition of In^I. The following reaction scheme (3.53) has been proposed, where step A is rate-limiting, while steps B and C (identified by pulse-radiolysis studies) are competing processes.[449]

$$In^I + H^+ \rightarrow In^{II} + \tfrac{1}{2}H_2 \tag{3.53A}$$

$$2In^{II} \rightarrow In^I + In^{III} \tag{3.53B}$$

$$In^I + In^{II} \rightarrow In^0 + In^{III} \tag{3.53C}$$

The disproportionation process is inhibited by nitrate ion, and indium(I) halides dissolved in aqueous nitrate solution are stable with respect to decomposition. Nitrate ion oxidises trace quantities of In^0 to In^I, thereby obviating the much more rapid heterogenous processes.[450]

The chemistry of Tl^I has much in common with that of K^+, Rb^+, Cs^+ and Ag^+. Compounds are predominantly ionic, and do not merit examination in detail. Information predating 1971 is accessible through the monograph by Lee.[5] An important recent application is the use of ^{205}Tl NMR measurements to probe biological processes in which Tl^+ replaces K^+ or Na^+ (see chapter 1). ^{205}Tl NMR studies have also correlated chemical shift with the stereochemical activity of the lone pair, and with the degree of covalency in Tl^I halides.[451] Structural data have been used to examine the degree of covalency or ionicity in Tl^I compounds, establishing that the effective ionic radius of Tl^I decreases as the covalency of the Tl–X bond increases.[452]

The predominance of the $+1$ state in thallium chemistry illustrates one of the important themes in Main Group chemistry, notably the increase in stability of the lower oxidation state, corresponding to the ion with the valence-electron configuration ns^2, with the descent of each group (see chapter 1). This so-called 'inert pair' effect has been examined in an *ab initio* theoretical treatment of the hydrides and fluorides of the Group 13 elements, MX and MX_3 (M = B, Al, Ga, In or Tl; X = H or F).[453] In 6th period elements the resistance of the $6s^2$ electrons to oxidation is often attributed to relativistic effects, and these were included in the calculations. The study found that low valencies in compounds of the heavy elements arise naturally as a consequence of the periodic trend towards lower M–X bond strength with increasing atomic number. Relativistic effects are significant but do not dominate this trend. Disproportionation is resisted by TlF, but is both favoured and assisted by relativity for monomeric TlH. With no discernible trend in the *ns*-orbital population between AlX_3 and TlX_3 (X = H or F), there is no computational evidence for a specially inert lone pair, for example in TlX_3. Although relativistic effects are important in 6th period elements, including thallium, it follows that the $6s$ electrons of thallium are no more inert than the valence *ns* electrons of lighter elements.[453] Whether or not the lone pair is stereochemically active is another matter; as noted in chapter 1, this is an issue pervading the structural chemistry of univalent derivatives.

Another theme in Main Group chemistry is that odd-electron oxidation states, such as the $+2$ state in Group 13, M^{II}, are rare. Indeed, monomeric, paramagnetic species containing M^{II} are known only as transients under normal conditions. Thermodynamic arguments indicate that the atomic states associated with an M^{II} ion are accessible, and that a typical M^{II}–X bond is thermodynamically stable. The ephemeral character of compounds such as $TlCl_2$ may then be a result of lattice-energy effects which promote decomposition or disproportionation.[454] The extra stability which the M^{III} state gains by complexation will also play a part.[455] Recently, the formally subvalent Ga^{II} compound $Ga(dbab)_2$ has been synthesised in 30% yield from the activated metal in the presence of 1,4-di-t-butyl-1,4-diaza-butadiene (dbab).[456] The same study reports a novel hydrometallation product, $(H_2Ga)_2(\mu\text{-}NBu^tCH_2)_2$, in which the two GaH_2 units are discrete and attached to a single, bridging diazabutadiene ligand. $Ga(dbab)_2$ has also been made by co-condensation of gallium vapour with dbab,[457] and similar methods have afforded the corresponding aluminium compound $Al(dbab)_2$.[458] On the evidence of the crystal structure of $Al(dbab)_2$ and the ESR spectrum of $Ga(dbab)_2$,[459] however, these compounds contain not the divalent but the trivalent metal bound to one singly and one doubly reduced ligand.

On the other hand, the M^{II} state is now well established for Al, Ga and In in dimeric, diamagnetic compounds containing metal–metal bonds.[455,460,461]

The metal-to-metal bonded states of Al, Ga, In and Tl have been surveyed earlier,[461] and recently the methods for the formation of homonuclear bonds between the Group 13 elements have been examined in detail.[460] Halide and chalcogenide derivatives of Ga and In form the bulk of this class, as discussed in sections 3.3.3 and 3.3.5. Examples containing Al are more limited, but may be found in sections 3.3.5 and 3.3.6. The existence of metal–metal bonded compounds of thallium remains controversial. This question is discussed in section 3.3.2 dealing with hydrides, as calculations on $[TlH]_2$ have been the focus of recent research.[462,463] The early report[464] of an organothallium compound with Tl–Tl bonds (perhaps the $[Tl_2Me_6]^{2-}$ ion) should not be entirely discounted, bearing in mind that some of the best evidence for Al–Al bonds has come from the organometallic area.

Both thermodynamic and kinetic factors govern the existence of the M^{II} dimers, and the balance between the M^{II} and M^{I},M^{III} states is delicate. The oft-quoted dihalides, MX_2, are actually mixed valence compounds better formulated as $M^{I}[M^{III}X_4]$. However, addition of a neutral or anionic donor ligand to such a compound produces the dimeric species $M^{II}_2X_4L_2$. The dihalides MX_2 also serve as a warning against the practice of assigning oxidation state simply on the basis of stoichiometry. Elsewhere we find some compounds containing the M^{II}–M^{II} moiety (e.g. GaSe, InSe and In_2Br_3), while other species of the same stoichiometry are M^{I},M^{III} mixed valence compounds (e.g. InTe and In_2Cl_3).

Since the stable oxidation states of the heavier Main Group elements are separated by two units, another general assumption is that redox reactions involving these elements are two-electron processes. This has been challenged in a study involving the oxidative addition of a quinone to InX (X = Cl, Br or I) to produce In^{III} catecholate species of the type $XIn(O_2C_6X_4)$.[465] Both In^{II} species and semiquinone (SQ˙) ligands have been observed as intermediates in this reaction, as summarised in Scheme 3.4.

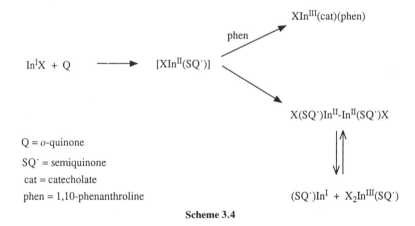

Scheme 3.4

Electron transfer in the intermediate $[XIn^{II}(SQ^{\cdot})]$ is promoted by the presence of a strong donor such as phenanthroline; otherwise dimerisation occurs to give the In^{II}–In^{II} species, leading to In^{I} and In^{III} semiquinone complexes. Evidence comes from ^{13}C NMR and EPR spectroscopy, from structural studies, and from the independent preparation of In^{I}, In^{II} and In^{III} semiquinone complexes.[465,466] Hence it appears that one-electron processes may be more prevalent in Main Group chemistry than was previously believed.

The account in *Comprehensive Inorganic Chemistry*[2] includes some discussion of lower valent derivatives of the Group 13 metals. Useful recent reviews dealing with aspects of low-valent Group 13 chemistry span the lower oxidation states of indium,[467] coordination chemistry,[7] monovalent Group 13 compounds,[468] the chemistry of In^{I} and In^{II} states,[469] metal–metal bonded compounds,[461] gallium and indium dihalides,[455] the chemistry of low-valent gallium derivatives,[470] π-complexes of Main Group elements including cyclopentadienyl derivatives of the Group 13 metals in the univalent state[446] and arene complexes of Ga^{I}, In^{I} and Tl^{I}.[471]

3.3.2   Low-valent hydrides

There are no well characterised monovalent or divalent hydrides in the condensed phase for the elements Al to Tl. AlH, GaH, InH and TlH have all been detected spectroscopically as short-lived species in gaseous M/H_2 mixtures (M = Al, Ga, In or Tl), although some form of activation to break the H–H bond is typically needed. For example, TlH has been detected spectroscopically when an electric discharge is passed through hydrogen between a copper anode and thallium cathode, or from the reaction of thallium metal with atomic hydrogen. Polymeric compounds $[InH]_n$ and $[TlH]_n$ have been reported as decomposition products of the corresponding trihydrides MH_3 (M = In or Tl), but are of doubtful authenticity. Cryogenic reactions of Al or Ga atoms or small clusters with H_2 in inert-gas matrices yield species like MH and MH_2 (M = Al or Ga),[472] and the photolytically induced matrix reaction of M atoms with CH_4 gives $HMCH_3$ (M = Al or Ga).[473,474] Group 13 metal atoms M form complexes with H_2O molecules on co-condensation with an excess of argon; these isomerise to molecules of the type HMOH spontaneously in the case where M = Al, only on photolysis in the cases where M = Ga, In or Tl;[475–477] further photolysis converts HMOH to MOH and MO. Co-condensation of Al atoms in their electronic ground state with NH_3 and an excess of adamantane at 77 K gives a matrix whose ESR spectrum attests to the formation of the paramagnetic species $Al(NH_3)_4$, $HAlNH_2$ and $Al(NH_3)_2$;[478] these are probably the initial products *en route* to $Al(NH_2)_3$ and dihydrogen.

The properties of all the diatomic hydrides deduced from the electronic

Table 3.12 Physical properties of the heteronuclear diatomic molecules MX (M = Al, Ga, In or Tl; X = H, F, Cl, Br or I) in their $^1\Sigma$ electronic ground states.[479]

Molecule	Dimensions, r_e (Å)	Vibrational properties		Dissociation energy, D_0 (kJ mol^{-1})
		Harmonic wavenumber, ω_e (cm^{-1})	Anharmonicity coefficient, $\omega_e x_e$ (cm^{-1})	
AlH	1.6478	1682.56	29.09	<295
AlF	1.654369	802.26	4.77	665
Al^{35}Cl	2.130113	481.30	1.95	494
Al^{79}Br	2.294807	378.0	1.28	427
AlI	2.537102	316.1	1.0	364
^{69}GaH	1.6630	1604.52	28.77	<274
^{69}GaF	1.774369	622.2	3.2	577
^{69}Ga^{35}Cl	2.201690	365.3	1.2	475
^{69}Ga^{81}Br	2.35248	263.0	0.81	416
^{69}GaI	2.57467	216.6	0.5	335
^{115}InH	1.8380	1476.04	25.61	239
^{115}InF	1.985396	535.35	2.64	507
^{115}In^{35}Cl	2.401169	317.4	1.01	428
^{115}In^{81}Br	2.54318	221.0	0.65	385
^{115}InI	2.75365	177.1	0.4	331
^{205}TlH	1.8702	1390.7	22.7	190
^{205}TlF	2.084438	477.3	2.3	441
^{205}Tl^{35}Cl	2.484826	283.75	0.818	369
^{205}Tl^{81}Br	2.618191	192.10	0.39	330
^{205}TlI	2.813676	(150)	–	266

spectra of the gaseous molecules are given, together with those of the corresponding halides, in Table 3.12.[479] The Group 13 hydrides MH and MH_3 have been investigated by extensive *ab initio* computations to evaluate trends in the M–H bond strengths and in the stability of the lower valent state.[453] The divalent species M_2H_4 (M = Al, Ga or In) and mixed metal compounds $AlGaH_4$ and $BGaH_4$ have also been the subjects of theoretical studies.[480,481] Adducts of AlH and AlH_2 with HCl and HF have also been examined in this way.[482]

The thallium(I) hydride dimer Tl_2H_2 has been used as a model compound for theoretical investigations probing the controversial existence of compounds with Tl–Tl bonds. This problem is related to the question of the stereochemical activity of the $6s^2$ lone pair in Tl^I compounds.[462] No compounds with a Tl–Tl separation less than twice the covalent radius of Tl (3.05 Å) and only one with a separation less than that in thallium metal (3.40 Å) have been reported. However, several compounds with Tl–Tl separations less than twice the van der Waals radius (4.00 Å) are known, generally as bridged complexes as shown in Figure 3.32(a). On the other hand, these short contacts may not indicate any substantial Tl–Tl interactions.[462] There is a limited number of examples of dimeric Tl complexes which do not contain bridging ligands, as in Figure 3.32(b). For example, the organometallic derivative $\{[\eta^5-C_5(CH_2Ph)_5]Tl\}_2$ contains the η^5-cyclopentadienyl ligands, L, in a *trans*-bent arrangement, L–Tl–Tl–L. The ligand angle β (L–Tl–Tl) is 131.8° and the Tl–Tl distance is 3.63 Å.[483]

Extended Hückel calculations on Tl_2 and Tl_2H_2 have been used to examine the Tl–Tl bonding interaction. The Tl–Tl overlap population in HTlTlH is sensitive to the ligand angle β. The linear arrangement is almost non-bonding, while the *trans*-bent arrangement shown in Figure 3.32(b) is strongly bonding, with a maximum close to 120°. The bridging geometry shown in Figure 3.32(a) is again non-bonding or anti-bonding for L = H, OMe or C_5H_5. This confirms the experimental conclusions that complexes with bridging ligands do not show any Tl–Tl interactions, while dimeric compounds with the *trans*-bent arrangement do involve such interactions despite the relatively long separation. These results also hold for the analogous In^I–In^I systems.[462] The argument has been extended to solid binary or ternary species with potential Tl–Tl or In–In interactions, for example TlSe, $TlMSe_2$ (M = Ga or In) and $TlCu_3Se_2$[298] falling outside the

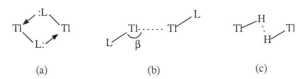

(a)   (b)   (c)

Figure 3.32 Bridged and metal–metal bonded structures for dimeric thallium(I) compounds.

scope of the immediate discussion. More rigorous *ab initio* configuration interaction (CI) calculations have been performed for TlH and Tl_2H_2.[463] The results agree qualitatively with those of the more simply based calculations in finding a minimum for the *trans*-bent structure with $\beta = 115.1°$. However, a second, deeper minimum is observed at $\beta = 38.75°$, indicating that a bridged structure (Figure 3.32(c)) might also be significant. The same study suggests that Tl_2H_2 may be a sufficient stable molecule to be observable by spectrocopic means used in conjunction with the matrix-isolation technique.[463] It should be noted that relativistic effects are also significant in determining the strength of the Tl–Tl bond.[463, 484]

3.3.3   Low-valent halides

3.3.3.1   Monohalides.   The monovalent halides MX are known for all four of the Group 13 metals M and all four of the long-lived halogens X, although their stability and chemistry varies greatly from metal to metal. The properties of the gaseous diatomic molecules, derived mainly from their electronic spectra, are given in Table 3.12.[479]

The aluminium species AlX are short-lived diatomics confined almost entirely to the gas phase and prepared by high-temperature reactions involving the elements or by anodic oxidation of aluminium metal at high current. At lower temperatures in the condensed phases, the AlX species disproportionate, typically to give a grey mixture of Al and AlX_3. AlCl was first detected in the gas phase through the reaction of Al with Cl_2 or HCl at 1000°C.[485] Spectroscopic studies of AlF,[486] AlCl[444,487] and AlBr[487] trapped in frozen-gas matrices have been reported. AlF, already known by the properties of the gaseous molecule,[479] was shown to dimerise to a four-membered ring $Al(\mu\text{-F})_2Al$.[486] The infrared spectrum of solid AlCl condensed at 77 K shows a broad absorption corresponding to $\nu(AlCl) = 320\,cm^{-1}$, in contrast to the value of $477\,cm^{-1}$ for the gaseous diatomic molecule.[444] The structures of the adducts of AlF and AlF_2 with HCl and HF have been examined in a theoretical study.[482]

The monovalent halides AlX in a new guise have recently proved useful in the preparation of aluminium organometallics. After co-condensation with unsaturated organic compounds at low temperature, AlCl reacts upon warming to give novel organometallic compounds. For example, with 2-butyne in pentane the dimeric dialuminacyclohexadiene (Figure 3.33(a)) is formed,[488] while with 2,3-dimethylbutadiene in toluene a beautiful hexameric species (Figure 3.33(b)) results.[489]

The most useful advance in monovalent aluminium chemistry has been the discovery that an ether adduct '$AlCl \cdot xEt_2O$' can be obtained in a metastable state in toluene/ether solution at low temperatures.[444] The ^{27}Al NMR spectrum indicates that the ether adduct may be dimeric, thus hinting

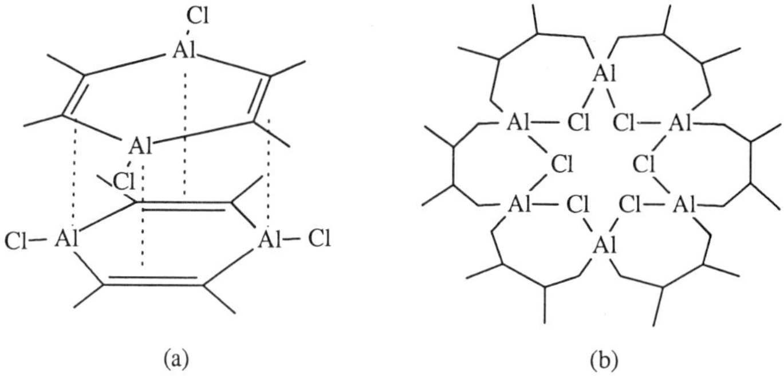

Figure 3.33 Products of the reactions of AlCl with organic substrates.[488,489]

at the possibility of Al^I–Al^I interaction akin to the Tl^I–Tl^I interaction discussed in the preceding section. $AlCl \cdot xEt_2O$ in toluene/ether reacts with $(\eta^5\text{-}C_5Me_5)_2Mg$ to give tetrahedral tetrameric $(\eta^5\text{-}C_5Me_5)_4Al_4$,[490] the first Al^I organometallic compound to be prepared and structurally characterised. Reactions of AlF with bipyridyl (bipy) in either tetrahydrofuran solution or an adamantane matrix at low temperature lead to paramagnetic species which give on evaporation a pyrophoric solid proposed to be AlF(bipy).[445] Given the importance of aluminium in technology, the chemistry of low-valent aluminium halides holds the prospect of yet more striking discoveries.

The gallium and indium monohalides, prepared from the elements or from the reaction of the metal M with MX_3 (M = Ga or In) or HgX_2, increase in stability as the halogen X changes from F to I, although all are readily oxidised by aqueous acid or air. GaF and InF are known only as the gaseous species prepared by the reaction of the metal M with MF_3 (M = Ga or In) at high temperature. The monohalides of Ga, In and Tl crystallise with mostly ionic M–X interactions in CsCl, distorted NaCl or α-TlI structural types (see chapter 1).[8] For example, the α-form of InCl crystallises in a distorted NaCl structure allowing some In–In interactions. The heavier monohalides disproportionate to the metal and trihalide in aqueous solution, although InI is the most stable under these conditions. They can be stabilised by co-ordination of the anion to a halide acceptor; for example, InCl adds to $AlCl_3$ to form $In^+[AlCl_4]^-$. This is similar to the case of gallium dichloride, i.e. $Ga^+[GaCl_4]^-$ (see section 3.3.3.2). Electronic states and spectroscopic constants have been calculated for GaCl and other Group 13 mono-halides.[491]

The vapour pressure above liquid InX (X = Cl, Br or I) has been measured over a range of elevated temperatures and the molar enthalpy of evaporation calculated. The monomers InX are the dominant vapour species, although dimers are also present in relatively low abundance.[492] For InBr the concentration of dimer In_2Br_2 in the vapour is approximately 14% at 680 and 4% at 810 K.[493]

Although the halides InX are useful as sources of In^I, they are insoluble, intractable materials which do not lend themselves readily to chemical synthesis. However, they have been shown to dissolve in toluene containing a donor ligand at temperatures below 0°C. For example, if InBr is dissolved in toluene/tetramethylethanediamine (TMEDA), then the adduct InBr · 3TMEDA is formed and the solution slowly precipitates solid InBr · 0.5TMEDA,[447] which has been identified spectroscopically. The corresponding species containing InCl and InI are less stable. The InX can thus be activated to oxidative addition in solutions containing alkyl halides. For example, InX dissolved in CH$_2$Cl$_2$/TMEDA reacts with EtX to give EtInX$_2$ · TMEDA (X = Br or I),[447] while InX in CH$_2$Cl$_2$/pyridine solution treated with PhEEPh gives InX(EPh)$_2$(py)$_2$ (X = Cl, Br or I; E = S or Se).[494] InX in a chemically reactive form can also be generated by electrochemical oxidation of In metal in CH$_2$X$_2$/CH$_3$CN solution (X = Cl, Br or I).[495]

In keeping with the increased stability of the monovalent state in the heaviest element of the Group, thallium(I) halides are air- and water-stable compounds which give neutral aqueous solutions and display physical and chemical properties generally similar to those of the corresponding alkali-metal and silver halides. There is of course one major difference from the alkali metals and silver: thallium salts are highly toxic. Thallium-doped Group 1 halide crystals show absorption and emission bands corresponding to species approximating to [TlX$_2$]$^-$ and [TlX$_4$]$^{3-}$ and are used as phosphors.[5] While TlF$_3$, TlCl$_3$ and TlBr$_3$ are thallium(III) halides, a consequence of the Tl^{3+}/Tl$^+$ and ½I$_2$/I$^-$ redox potentials is that TlI$_3$ is not a Tl^{III} halide but a Tl^I triiodide salt Tl$^+$I$_3^-$. The presence of asymmetric and slightly bent I$_3^-$ ions is confirmed by the crystal structure of the compound which is isotypic with RbI$_3$ and CsI$_3$.[496] TlI$_3$ in the presence of an excess of iodide ions forms the Tl^{III} complex [TlI$_4$]$^-$ in which the higher oxidation state is stabilised. Stability constants for species in the Tl$^+$/Cl$^-$ system have been determined by ^{205}Tl NMR spectroscopy. Titration of Tl$^+$ ions in aqueous KF solution with Cl$^-$ has led to the identification of the three aquo species Tl$^+$, TlCl and [TlCl$_2$]$^-$. The estimated stability constants agree well with those obtained previously by polarography or solubility methods.[497]

3.3.3.2 Halides of intermediate valency. Aluminium dihalides of the type AlX$_2$ (X = Cl or Br) have been investigated as the products of the reaction of AlX$_3$ with aluminium metal suspended in a dry hydrocarbon; AlCl$_2$ is also formed by the reaction of AlCl$_3$(g) with aluminium metal at low pressure.[498] AlCl$_2$ co-condensed with AlCl and AlCl$_3$ has been characterised by its infrared spectrum in a matrix study. There are signs that the AlX$_2$ species are paramagnetic, and that they are associated in the condensed state, possibly as dimers Al$_2$X$_4$ which feature both halogen bridging and Al–Al bonding, since they cannot assume the structure X$_2$Al–AlX$_2$ while at the same time remaining paramagnetic.

Whereas halides of intermediate valency are of little significance for aluminium and thallium, the gallium and indium dihalides with the formal stoichiometry GaX_2 and InX_2 (X = Cl, Br or I) form the cornerstone of a rich and developing area of chemistry in which authentic M^{II} compounds are now well established.[455] The gallium and indium dihalides MX_2, formed from the interaction of the metal M with MX_3, Hg_2X_2 or HgX_2, are more realistically formulated as the mixed valence compounds $M^I[M^{III}X_4]$. The exception is $InCl_2$, the existence of which is doubtful. The stoichiometries of all the known gallium and indium halides are given in Table 3.13, together with their ionic formulations.

Phase studies of the Ga and In binary systems M/MX_3 (X = Cl, Br or I) have identified for each metal a range of halides with stoichiometries intermediate between the limiting monovalent (MX) and trivalent (MX_3) states. Not all of these have been structurally characterised, and the formulae of some species reported in the early literature have subsequently been corrected. The intermediate halides can generally be prepared by heating M with MX_3 in the correct molar proportions. Phase equilibrium diagrams for eight of the twelve possible M/MX_3 systems (M = Ga, In or Tl) have been reviewed,[499] although the report omits some important results from the recent literature. The intermediate halides, which include the above-mentioned dihalides, may be generalised as mixed-valence species

Table 3.13 Binary compounds of gallium, indium or thallium with chlorine, bromine or iodine.[a]

Metal	Halogen					
	Chlorine		Bromine		Iodine	
Gallium		GaCl		GaBr		GaI
			Ga_2Br_3	$Ga^I_2[Ga^{II}_2Br_6]$	Ga_2I_3	$Ga^I_2[Ga^{II}_2I_6]$
	$GaCl_2$	$Ga^I[Ga^{III}Cl_4]$	$GaBr_2$	$Ga^I[Ga^{III}Br_4]$	GaI_2	$Ga^I[Ga^{III}I_4]$
	Ga_3Cl_7	$Ga^I[Ga^{III}_2Cl_7]$	Ga_3Br_7	$Ga^I[Ga^{III}_2Br_7]$	Ga_3I_7	$Ga^I[Ga^{III}_2I_7]$
	$GaCl_3$		$GaBr_3$		GaI_3	
Indium	InCl		InBr		InI	
	In_7Cl_9[b]	$In^I_6[In^{III}Cl_9]$				
	In_2Cl_3	$In^I_3[In^{III}Cl_6]$	In_2Br_3	$In^I_2[In^{II}_2Br_6]$		
			In_5Br_7	$In^I_3[In^{II}_2Br_6]Br$		
	In_5Cl_9	$In^I_3[In^{III}_2Cl_9]$				
			$InBr_2$	$In^I[In^{III}Br_4]$	InI_2	$In^I[In^{III}I_4]$
	$InCl_3$		$InBr_3$		InI_3	
Thallium	TlCl		TlBr		TlI	
	Tl_2Cl_3	$Tl^I_3[Tl^{III}Cl_6]$	Tl_2Br_3	$Tl^I_3[Tl^{III}Br_6]$		
	$TlCl_2$	$Tl^I[Tl^{III}Cl_4]$	$TlBr_2$	$Tl^I[Tl^{III}Br_4]$		
	$TlCl_3$		$TlBr_3$		TlI_3	$Tl^I[I_3]$

(a) References may be found in the text.
(b) Previously characterised as In_3Cl_4 or In_4Cl_5.

containing M^I cations with complex halide anions containing either M^{III} or dimeric $M^{II}-M^{II}$ units. In all the structures characterised to date, the complex anions feature four- or six-coordinated M atoms with M–M and M–X bond lengths consistent with the presence of predominantly covalent bonds. The packing of the complex anions at van der Waals distances determines the coordination geometry of the M^I cations which occupy interstices of variable coordination number ranging from 7 to 11, with much longer $M^I \cdots X$ distances. This behaviour is illustrated by the structural properties of the gallium dihalides, $Ga[GaX_4]$ (X = Cl, Br or I) itemised in Table 3.14. The structures of all compounds containing the $[GaBr_4]^-$ and $[InBr_4]^-$ ions have been reviewed.[500]

Three important halides have been identified for each of the three halogens X = Cl, Br and I in the system Ga/GaX_3, notably GaX_2, Ga_3X_7 and Ga_2X_3. Structural investigations have confirmed the $Ga^I[Ga^{III}X_4]$ formulation containing tetrahedral $[GaX_4]^-$ ions for $GaCl_2$,[501] $GaBr_2$ in both the α- and β-modifications[501,502] and GaI_2.[503] $GaCl_2$ and α-$GaBr_2$ are isotypic, as are β-$GaBr_2$ and GaI_2. Structural parameters are summarised in Table 3.14. The $\alpha \rightarrow \beta$ phase transition for $GaBr_2$ has been investigated by differential thermal analysis and X-ray powder measurements at different temperatures.[502] These studies showed considerable asymmetry in the environment of the Ga^+ ion in $Ga[GaCl_4]$ and α-$Ga[GaBr_4]$, a characteristic also signalled by the ^{69}Ga NMR spectra of the crystalline powders.[501] This may reflect the influence of the non-bonding electron pair of the Ga^+ ion.[503] The Ga^I,Ga^{III} mixed valence formulation is also found for Ga_3X_7, which is most aptly represented as $Ga^I[Ga^{III}_2X_7]$. The dinuclear Ga^{III} complex anion in $Ga[Ga_2Cl_7]$ contains tetrahedral Ga^{III} centres with a single chloro bridge (section 3.2.2.2).[120,504]

In contrast, halides with the composition Ga_2X_3 are Ga^I,Ga^{II} mixed valence salts $Ga^I_2[Ga^{II}_2X_6]$. The dimeric Ga^{II} anion is diamagnetic and contains a Ga–Ga bond.[503,505] Structural details for Ga^{II} metal–metal bonded complexes $M^+_2[Ga_2X_6]$ ($M^+ = Ga^+$, Li^+, NR_4^+ or PR_4^+) and $Ga_2X_4L_2$ (L = dioxane or pyridine) given in Table 3.15 show that the Ga–Ga bond is typically close to 2.4 Å in length, while the $[Ga_2X_6]^{2-}$ anion may adopt an eclipsed or a staggered, ethane-like conformation.[503,505-514] The reasons for

Table 3.14 Averaged Ga–X bond lengths for $Ga[GaX_4]$ (X = Cl, Br or I).

Compound	$r(Ga^{III}-X)$ (Å)	$r(Ga^I-X)$ (Å)	Coordination number of Ga^I	Reference
$Ga[GaCl_4]$	2.18	3.18	8	501
α-$Ga[GaBr_4]$	2.33	3.30	8	501
β-$Ga[GaBr_4]$	2.33	3.28	9	502
$Ga[GaI_4]$	2.56	3.53	8	503

Table 3.15 Structural data for compounds with Ga–Ga bonds.

Compound	r(Ga–Ga) (Å)	Geometry	Reference
$Li_2[Ga_2Cl_6]$	2.392	Eclipsed	506
$[Me_4N]_2[Ga_2Cl_6]$	2.390(2)	Staggered	507
$[Ph_3PH]_2[Ga_2Cl_6]$	2.407(1)	Staggered	508
$Ga_2Cl_4(dioxane)_2$	2.406(1)	Eclipsed (O opposite Cl)	509
$Ga_2Cl_4(pyridine)_2$	2.403(1)	*trans* staggered	510
$Ga_2[Ga_2Br_6]$	$\begin{cases} 2.427(4) \\ 2.439(6) \end{cases}$	Eclipsed	505
$Li_2[Ga_2Br_6]$	2.404	Eclipsed	511
$[Pr_4N]_2[Ga_2Br_6]$	2.419(5)	Staggered	512
$[Ph_3PH]_2[Ga_2Br_6]$	2.410(2)	Staggered	508
$Ga_2Br_4(dioxane)_2$	2.395(6)	Eclipsed (O opposite Br)	513
$Ga_2Br_4(pyridine)_2$	2.421(3)	*trans* staggered	514
$Ga_2[Ga_2I_6]$	2.387(5)	Staggered	503
$Li_2[Ga_2I_6]$	2.428	Eclipsed	506
$[Ph_3PH]_2[Ga_2I_6]$	2.414(5)	Staggered	508

the Ga^I, Ga^{III} formulation of some gallium halides versus the Ga^I, Ga^{II} formulation of others are not clear, and little enlightenment is to be found in the behaviour of the indium halide systems discussed below. Four-coordination is observed for Ga^{II}, In^{II} and Ga^{III} centres (with the exception of one In^{II} complex which incorporates a five-coordinated In atom), while In^{III} centres are characterised by four- or six-fold coordination, and the coordination sphere of the metal has a significant bearing on its oxidation state. The choice of halide is also important: whereas the Ga_2Br_3 solid phase is stable as $Ga_2^+[Ga_2Br_6]^{2-}$, the corresponding gallium-rich chloride phase is stable only in the melt and disproportionates to Ga and $GaCl_2$ on solidification.

NQR properties of several gallium halides have been collected. Thus, ^{79}Br and ^{127}I NQR resonances were observed for α-Ga_2Br_4, β-Ga_2Br_4 and Ga_2I_4 at various temperatures,[515] and phase transitions in $[NR_4]_2[Ga_2Br_6]$ (R = Me or Et) were also monitored by reference to the ^{69}Ga and ^{81}Br NQR spectra.[516] NQR spectra due to ^{35}Cl, ^{69}Ga, ^{81}Br and ^{127}I collected for complexes of Ga_2X_4 (X = Cl, Br or I) and GaX_3 (X = Cl or Br) with dioxane imply that electron transfer from oxygen to gallium is smaller in the Ga^{III} complexes,[517] inviting the inference that the coordinatively unsaturated (and so far unknown) molecule X_2Ga-GaX_2 should be a stronger Lewis acid than GaX_3.

One further aspect of the chemistry of low-valent gallium halides merits attention. This concerns the stability of the +2 oxidation state in solution. The Ga^{II} halide anions $[Ga_2X_6]^{2-}$ (X = Cl, Br or I) were originally detected

— and at first wrongly supposed to be Ga^I species — through their remarkable persistence in aqueous acid solutions prepared by dissolving the finely divided metal in the cold concentrated halogen acid, HX.[518,519] These solutions continue evolving hydrogen gas long after all the metal has dissolved and require heating to speed up the oxidation to the Ga^{III} state:

$$[Ga_2X_6]^{2-} + 2H^+ + 2X^- \rightarrow 2[GaX_4]^- + H_2 \qquad (3.54)$$

It may be remarked, parenthetically, that the reason for dissolving gallium in the cold when preparing its aqueous solutions is that any rise in temperature above 28°C causes the metal to melt and form a drop, thereby reducing the surface area to such an extent that dissolution becomes very slow. The $[Ga_2X_6]^{2-}$ ions can be precipitated as Me_4N^+ salts and separated from $Me_4N[GaX_4]$ by recrystallisation. They can also be isolated from the aqueous solution by ether extraction in the form of the complex acids $H_2Ga_2X_6$(aq) which undergo slow oxidation:

$$[Ga_2X_6]^{2-} + 2H^+ + 2Et_2O \xrightarrow{Et_2O} 2Et_2O \cdot GaX_3 + H_2 \qquad (3.55)$$

Other Ga^{II} complexes have been formed in side-reactions, e.g. $HGa_2Br_5 \cdot OEt_2$ and $Ga_2Br_4(OEt_2)_2$. The solids $(Me_4N)_2[Ga_2X_6]$ are stable indefinitely in air. They are sparingly soluble in water, which causes gradual hydrolysis accompanied by oxidation to $Ga(OH)_3$. The halide ligands of $[Ga_2X_6]^{2-}$ can be replaced by thiocyanate which coordinates through nitrogen to the Ga^{II} as it does to Ga^{III} centres.[461]

The systems $In/InCl_3$ and $In/InBr_3$ have been studied many times giving a range of often conflicting results, although recent work has helped to clarify the identities of the stable phases, the results being summarised in Table 3.13.[520-527] Binary compounds of other stoichiometries appearing in the older literature are questionable. The phase diagram for $InCl/InCl_3$ is shown in Figure 3.34.[520] Surprisingly, the chloro and bromo systems show some significant differences. The phase with the composition In_7Cl_9, previously characterised as In_3Cl_4[521] or In_4Cl_5, is best described as $In^I_6In^{III}Cl_9$. It shows a distorted NaCl-type structure, similar to that of the low-temperature form of InCl, containing In^I_3 triangles.[522] In_2Cl_3, In_2Br_3 and In_5Br_7 are mixed valence compounds, although In_2Cl_3 contains In^I, In^{III} while In_2Br_3 and In_5Br_7 contain In^I, In^{II}. In_2Cl_3 crystallises as $In^I_3[In^{III}Cl_6]$ with an octahedral In^{III} anion.[523] The ethane-like anion which appears in $In^I_2[In^{II}_2Br_6]$ (In_2Br_3)[521] and $In^I_3[In^{II}_2Br_6]Br$ (In_5Br_7)[527] contains an In–In bond and four-coordinated In^{II} centres.[521] Here we see the extra complication introduced by the tendency of anionic In^{III} chloro complexes to favour six-fold coordination, as in the $[InCl_6]^{3-}$ ion, although $[InBr_6]^{3-}$ is known, for example in aqueous In^{III}/HBr solutions. In_5Cl_9 also contains six-coordinated In^{III}, this time as the cofacial bisoctahedral anion $[In_2Cl_9]^{3-}$ in the mixed valence salt $In^I_3[In_2Cl_9]$,[520] yet has no counterpart in the Br system. The

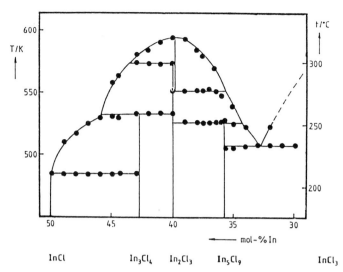

Figure 3.34 Phase diagram for the In/Cl system in the region 30–50 mol% In.[520] The species shown as In$_3$Cl$_4$ has subsequently been shown to be In$_7$Cl$_9$.[522] Reproduced with permission from G. Meyer and R. Blachnik, *Z. Anorg. Allg. Chem.*, 1983, **503**, 126.

dihalides InBr$_2$ and InI$_2$ are both formulated as In^I[In^{III}X$_4$], spectroscopic and structural studies having established that they contain four-coordinated, tetrahedral In^{III}.[521,525–527] The existence of InCl$_2$ has been debated in the literature but must now be regarded as doubtful,[7,527] in keeping with the observation that four-fold coordination is not favoured by In^{III} chloro complexes unless a bulky cation is present.

Although the intermediate indium halides are known on the evidence of structural and spectroscopic studies to be mixed valence compounds, a core photoelectron emission study of the indium chlorides InCl, In$_2$Cl$_3$, In$_5$Cl$_9$, InCl$_3$ and the disputed compound InCl$_2$ showed no split peaks corresponding to the presence of In^I and In^{III}. The internally referenced binding energy for In(3$d_{5/2}$)–Cl(2$p_{3/2}$) was essentially invariant for all the compounds studied, apparently reflecting the combined effects of changes of radius and Fermi level.[528]

Interest in mixed metal systems involving the low-valent halides of the Group 13 metals has been largely confined to phase studies. One example is the binary system InCl/SnCl$_2$ which has been re-examined; as a result the phase diagram has been corrected. It is a normal binary intersection of the In/Sn/Cl system only up to 50 mol% InCl. At higher concentrations of In^I the redox reaction In$^+$ + Sn^{2+} → In^{3+} + Sn^0 results in the deposition of metallic tin.[529]

The intermediate halides of thallium are less extensive, and Tl^{II} dimers are not a feature of this chemistry. Structural information concerning the

TlF/TlF$_3$ phase system is also lacking. Compounds with the composition TlX$_2$ and Tl$_2$X$_3$ (X = Cl or Br) are formulated as Tl^I[Tl^{III}X$_4$] and Tl$^I_3$[Tl^{III}X$_6$], respectively. The structure of α-Tl$_2$Cl$_3$ confirms the presence of octahedral [Tl^{III}Cl$_6$]$^{3-}$ groups (Tl^{III}–Cl = 2.50–2.65 Å), with 7-, 8- or 9-coordinated Tl^I ions experiencing longer Tl···Cl distances (3.06–3.83 Å).[530]

Remarkably, on reaction with a neutral or anionic donor ligand L, the gallium and indium dihalides M[MX$_4$] form true M^{II} complexes of the form LX$_2$M–MX$_2$L containing a metal–metal bond. For example, LCl$_2$Ga–GaCl$_2$L (L = dioxane) and $^-$Br$_3$In–InBr$_3^-$ are generated by the reactions of Ga[GaCl$_4$] with dioxane and of In[InBr$_4$] with [Bu$_4$N]Br in xylene, respectively, and both exhibit staggered ethane-like structures. Compounds of the type Ga$_2$X$_4$L$_2$ are known with N, O, S or P donor ligands,[510,531] and indium gives rise to similar compounds with N or P donor ligands.[522,533] As noted previously, the anions [M$^{II}_2$X$_6$]$^{2-}$ are found in the halides Ga$_2$X$_3$ (X = Cl, Br or I)[501–503] and In$_2$Br$_3$;[521] they have also been isolated as salts of the univalent cations R$_4$N$^+$ and R$_4$P$^+$.[455] The mixed metal complex anions, [X$_3$GaInX$_3$]$^{2-}$, with Ga–In bonds have been characterised too, as the Bu$^n_4$N$^+$ salts.[534] Most useful as a simple diagnostic test for the presence of a metal–metal bond in a species like [Ga$_2$X$_6$]$^{2-}$ is the Raman-active ν(Ga–Ga) mode (more correctly the in-phase Ga–Ga and Ga–X stretching mode) which ranges in energy from 235 cm^{-1} (X = Cl) to 122 cm^{-1} (X = I), or the corresponding ν(In–In) mode which falls in the range 150–120 cm^{-1}.[455,510,531] Earlier reports referring to the formation of products of the type [ML$_n$]$^+$ [MX$_4$]$^-$ from the reaction of M[MX$_4$] with neutral donor ligands L are incorrect,[510,531] except in the cases where L is an extremely poor donor such as an arene. For example, Ga[GaCl$_4$] dissolves in benzene forming [(η^6-C$_6$H$_6$)$_2$Ga]$^+$[GaCl$_4$]$^-$, which contains bis(η^6-benzene)gallium(I) cations weakly linked by [GaCl$_4$]$^-$ anions.[471] These are discussed further in section 3.3.6.2.

The remarkable structural and chemical variations found for the M$_2$X$_4$ species (M = Ga or In; X = Cl, Br or I), from mixed valent M^I[M^{III}X$_4$] compounds in the binary 'dihalides' MX$_2$ to the truly M^{II} metal–metal bonded fragments [X$_2$M–MX$_2$] found in the complexes [M$_2$X$_6$]$^{2-}$ or M$_2$X$_4$L$_2$ have received some attention.[455] An analysis based on spectroscopic data for M$_2$(g) and thermodynamic arguments for the disproportionation process

$$X_2MMX_2 \rightleftharpoons MX + MX_3 \rightleftharpoons M^+[MX_4]^- \qquad (3.56)$$

suggests that the relatively weak M–M bond is much less significant than the strong M–X bonds and the large lattice energy for the ionic species M$^+$[MX$_4$]$^-$. It is this last factor which may well determine the mixed valence composition of the binary 'dihalides' in the solid state,[455] since the process represented by equation (3.56) does not lie to the right in the gas phase and can be displaced to the left in molten salt phases.[465]

Again, the M–M bond strength is probably not a significant factor in rationalising the dominance of the M^{II} metal–metal bonded structures in the presence of a neutral or anionic donor ligand. Addition of negative charge to the $X_2M–MX_2$ fragment, in the form of a ligand L or X^-, may stabilise the system by reducing Coulombic repulsion between the positively charged metal centres. Thermodynamic arguments must address the effect of an added donor ligand on the disproportionation process shown above. Alternatively, the disproportionation of the metal–metal bonded species can be viewed as a halide transfer process, whereas the formation of the M–M bond formally involves oxidative insertion of M^I into the $M^{III}–X$ bond, e.g.

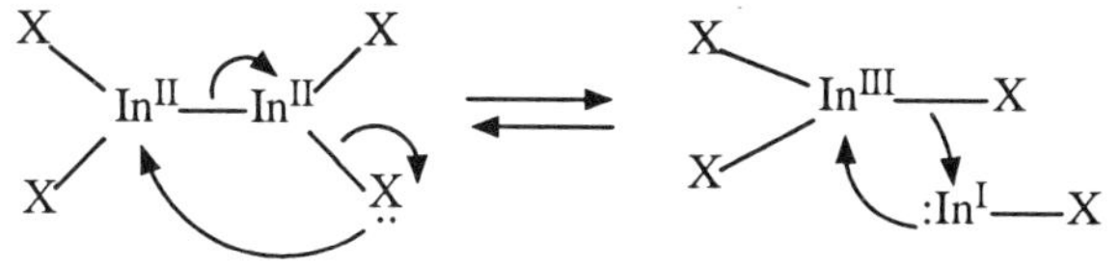

$$(3.57)$$

The presence of the donor ligands must inhibit the halide transfer that accompanies disproportionation of M_2X_4. A kinetic argument is that coordination of each of the metal centres in M_2X_4 by a ligand results in coordinative saturation, inhibiting the halide transfer. Greater stabilisation would be expected for gallium relative to indium because gallium favours lower coordination numbers, and there is experimental evidence to support this. Thus, the disproportionation

$$[M^{II}_2X_6]^{2-} \rightleftharpoons [M^IX_2]^- + [M^{III}X_4]^- \qquad (3.58)$$

has been observed to take place in solution for M = In (on the evidence of ^{115}In NMR measurements), but not for M = Ga. Ligands which are good donors (with N, O, S or P donor atoms) or poor leaving groups (amido[535] or organometallic[536] ligands) will also help to stabilise the metal–metal bond, and these desiderata appear to hold for the known M^{II} complexes.[455]

3.3.4 Low-valent oxides, hydroxides, alkoxides and oxyanion salts

The trivalent oxides M_2O_3 are known for all four metals, but stable, well-characterised monovalent oxides or hydroxides are important only for gallium and thallium. Gallium(I) oxide, Ga_2O, is an air-stable, dark-brown powder prepared from the high temperature reaction of Ga_2O_3 with Ga metal. It is strongly reducing, producing H_2S from H_2SO_4 and hydrogen from water. Metrical data for all four molecules of the type M_2O (M = Al, Ga, In or Tl) are derived from electron-diffraction studies of the vapours;[537] these indicate that the M–O–M bond angles fall in the range 140–150°. However, a very recent theoretical study of the oxides M_2O (M = Al or Ga) finds the linear structures to be global minima,[538] suggesting that this area is not yet fully understood. The same molecules have also been trapped in inert matrices at low temperatures and characterised by their infrared spectra.[539-543] The hydroxide molecules MOH have been detected not only in this way,[475,477,542] but also, in some cases, in flames.[2]

Thallium oxides and hydroxides include Tl_2O, TlO_2, Tl_4O_3, Tl_2O_3 and TlOH. Thallium(I) oxide, Tl_2O, is produced by heating TlOH or Tl_2CO_3 in the absence of air. The solid is hygroscopic, producing TlOH, TlOR or Tl^I salts in the reactions with water, alcohols or acids, respectively. The mixed oxidation state Tl^I,Tl^{III} oxide Tl_4O_3 is formulated as $3Tl_2O \cdot Tl_2O_3$. It displays a photoelectric effect, with an electrical conductivity which increases upon illumination. TlO_2, prepared by electrolysis of Tl_2SO_4 in the presence of oxalic acid, is described in the older literature as a peroxide, but more recent matrix-isolation studies indicate that a superoxide formulation $Tl^+O_2^-$ is more likely.[544] It is insoluble in water, dilute acid and dilute base, but reacts with dilute HCl to produce oxygen. TlOH is best prepared from thallium metal with aqueous ethanol in the presence of oxygen, and it resembles Group I hydroxides, giving strongly basic aqueous solutions. However, conductivity measurements on the solutions show that TlOH behaves as if it is highly associated, not as was at one time believed through simple ion-pairing, but perhaps by the formation of clusters like the tetramers $[TlOR]_4$ which are now well established for thallium(I) alkoxides.

The Tl^+ cation is a far less strong acid in aqueous solution than is Tl^{III}. Aquated Tl^I has an approximate pK_a of 13, compared with pK_{a1} of 1.1 and pK_{a2} of 1.5 for Tl^{III}. Many Tl^I salts, besides the halides and hydroxide, show signs of being associated in solution, but the reasons for this prove hard to fathom. The Tl^+ cation has a low enthalpy of hydration, and the precise nature of the hydrated ion in solution is not well established. Lee attributes the inability of Tl^I to form strong complexes to the shielding influence of the $6s^2$ electron pair.[5] As with the redox behaviour, there are obvious parallels with the chemistries of neighbouring elements, for example Pb^{II} vs. Pb^{IV}, or Sn^{II} vs. Sn^{IV}. Other aspects of the solution chemistry and possible complex formation of Tl^I are examined in chapter 8. Noteworthy too is the fact that some Tl^I salts, for example $TlNO_3$ and $TlClO_4$, are very soluble in liquid ammonia, where there is evidence of ion-association and solvation of the cation, as $[Tl(NH_3)_x]^+$.[545]

Ternary oxide phases involving each of the four Group 13 metals M combined with a second metal and oxygen form a large and important class of materials, as discussed in chapter 4, although few examples involve the lower oxidation states of M. A recent case of a low-valent ternary oxide is provided, however, by the binary phase system Tl_2O/MoO_3 in which six thallium(I) molybdates have been identified and characterised by their X-ray powder diagrams and IR spectra.[546] Another molybdate, $In_{11}Mo_{40}O_{62}$, is remarkable for the presence in channels between molybdate cluster anions of linear M–M bonded polycations, In_7^{5+} and In_6^{8+}.[547]

In addition to the univalent oxide molecules M_2O, several other low-valent oxides enjoying only transient existence under normal conditions have been examined spectroscopically either in the gaseous or matrix-isolated states. These include MO, M_2O_2,[548,549] M_4O_2,[540] peroxo- or superoxo- derivatives MO_2 and M_2O_4[550–553] and ozonides MO_3.[554] Mixed metal oxides, such as GaInO and $Ga_xIn_{4-x}O_2$, have also been characterised in this way.[540] The paramagnetic diatomic oxide AlO has a $^2\Sigma^+$ ground state and is characterised by the following properties:[479] $r_e = 1.6179$ Å, $\omega_e = 979.23$ cm^{-1}, $\omega_e x_e = 6.97$ cm^{-1} and $D_0 = 508$ kJ mol^{-1}. The dimer Al_2O_2 and several other aluminium oxide molecules have been examined by *ab initio* calculations in order to assess their molecular geometries, vibrational frequencies and relative stabilities. Species corresponding to calculated energy minima are Al_2O,[555,556] Al_2O_2,[556] AlO_2[557] and the planar hyperaluminium molecules Al_3O and Al_4O. The last of these is a square planar molecule with D_{4h} symmetry, which combines predominantly ionic Al–O bonding with substantial Al–Al interactions.[555] The reaction of gaseous Al^+ with alkyl halides and alcohols has been studied by ion cyclotron resonance spectroscopy. Dehydration of alcohols by Al^+ leads to the species $Al(OH_2)^+$.[558]

Thallium(I) alkoxides, $[TlOR]_4$, can be prepared from the action of the appropriate alcohols on thallium metal, TlOH or Tl_2O. The ethoxide, $[TlOEt]_4$, is commercially available and is useful as a precursor to other thallium(I) species. The alkoxides are tetrameric in solution and in the solid state, as evidenced by the partial crystal structure of $[TlOMe]_4$ which shows a tetrahedral arrangement of thallium atoms, presumed to be capped on each triangular face by a μ_3-OMe moiety.[8] The Tl_4O_4 cuboid arrangement is confirmed by the recent preparation and structural characterisation of $[Tl(\mu_3\text{-}OSiPh_3)]_4$ and $\{[Tl_2(OSiMe_2)_2O]_2\}_n$; these are derived from the reaction of thallium(I) ethoxide with Ph_3SiOH or $[OSiMe_2]_x$ (silicone grease), respectively. Each contains three-coordinated Tl^I centres in a Tl_4O_4 cuboid, the dimethylsiloxy derivative having $-Si(Me)_2OSi(Me)_2-$ links between the O atoms of adjacent cuboids forming a ladder polymer (Figure 3.35).[559]

The question of Tl–Tl interactions in the alkoxide tetramers has been examined by comparing the electronic absorption and emission spectra of $[TlOMe]_4$ and $TlNO_3$ in methanol. Such interactions might arise from *sp*

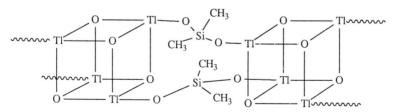

Figure 3.35 Polymeric $\{[Tl_2(OSiMe_2)_2O]_2\}_n$ containing $[Tl_4O_4]$ cuboids.[559]

orbital mixing in the highly symmetrical tetrahedral $[Tl^+]_4$ core. The absorption spectra of the two systems are similar, indicating that any ground state $Tl \cdots Tl$ interactions must be very weak. There is, however, a large shift to lower energy (red shift) in the emission band of the tetramer, when compared with Tl^+, and this has been interpreted in terms of a stronger metal–metal interaction in the *sp* excited state of the cluster.[560]

With sufficiently bulky substituents at oxygen, the tetrameric unit may give way to a dimeric one. Indeed, the lowest coordination number known for thallium(I) is found in the structurally characterised aryloxide dimer (Figure 3.36(a)) prepared from the action of the phenol $2,4,6-(CF_3)_3$ C_6H_2OH with thallium(I) ethoxide. The bulky aryl groups are oriented so as to be approximately perpendicular to the four-membered Tl_2O_2 ring.[561] Mixed metal alkoxides such as $Sn(\mu-OBu^t)_3M$ (Figure 3.36(b)) containing In^I or Tl^I as the second metal M are known.[562] The indium analogue of the thallium aryloxide, prepared from the same phenol with $(\eta^5-C_5H_5)In$ exemplifies the lowest coordination number for indium observed to date.[563] Other indium(I) species with aryloxide ligands are prepared by the electro-chemical oxidation of anodic indium in non-aqueous solutions of aromatic diols $Ar(OH)_2$, the air-sensitive products being of the type In^IOArOH with the hydroxy group presumed to be chelating as in Figure 3.36(c). The presence of the hydroxy group is confirmed by the infrared (ν_{OH}) and ^1H NMR spectra.[564] Oxidation of indium metal by 3,5-di(t-butyl)quinone (TBQ) in toluene gives an indium(I) semiquinone (TBSQ) complex which can be stabilised by coordination to 1,10-phenanthroline. The identity of the semiquinone moiety is confirmed by ESR spectroscopy.[565] A species identified spectroscopically and by elemental analysis as an indium(II) semiquinone dimer $InX_2(TBSQ)_2(py)_2$ was prepared from the reaction of TBQ with InX in toluene/pyridine solution (X = Cl or Br). If the reaction is carried out in toluene/TMEDA solution, an indium(III) product, $InX_2(TBSQ)(TMEDA)$, results (X = Cl, Br or I).[466]

The close relationship between Tl^+ and the Group 1 metal cations is evidenced by the range of salts or complexes of Tl^I with oxyanions that is not

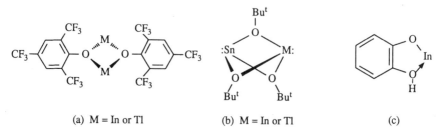

(a) M = In or Tl   (b) M = In or Tl   (c)

Figure 3.36 Indium(I) and thallium(I) complexes with bonds to oxygen: (a) $\{M[O-2,4,6-C_6H_2(CF_3)_3]\}_2$, M = In or Tl;[561, 563] (b) $Sn(\mu-OBu^t)_3M$, M = In or Tl;[562] and (c) $In(1,2-OC_6H_4OH)$.[564]

matched by the other Group 13 metals. These, including β-diketonates, salts of inorganic oxoacids and carboxylates, have been summarised in a recent review.[7] The structures of the solids show a range of geometries and coordination numbers ranging from 3 to 12 at the Tl^I centre; many are poly-nuclear species and some show evidence of a stereochemically active $6s^2$ lone pair (see chapter 1). Examples of compounds prepared and structurally characterised only recently include thallium(I) sulfite, Tl_2SO_3,[566] thallium(I) perbromate, $TlBrO_4$, which has the $BaSO_4$ structure,[567] and the iodate, $TlIO_3$.[568] The iodate resembles the corresponding potassium salt in having a particularly low solubility in water. In contrast, some thallium(I) carboxylates are extremely soluble in water. Concentrated solutions may have densities as high as $2-4\,g\,cm^{-3}$. Spectroscopic and X-ray diffraction studies of thallium(I) formate solutions indicate the presence of a tetramer, $Tl_4(O_2CH)_4$, at high concentrations (density $3.19\,g\,cm^{-3}$) which does not persist upon dilution.[569] Thallium(I) carboxylates have found important applications in organic synthesis; these will be elaborated in chapter 7. Compounds of thallium with sulfate, selenate, acetate or oxalate ions which have stoichiometries implying the presence of Tl^{II} are in fact Tl^I salts of Tl^{III} complex anions. For example, structural characterisation of $Tl(OAc)_2$ shows it to be the mixed valent salt $Tl^I[Tl^{III}(OAc)_4]$ with no evidence for a stereochemically active lone pair.[570] Diamagnetic crystals with the composition $In(OAc)_2$ formed by electrolysis of In metal in acetic acid have been shown by comparison of their infrared spectrum with that of $In(OAc)_3$ to be a mixed valent species, presumably incorporating In^I, In^{III} and acetate ions in the crystal lattice.[571] Similarly, the reaction of In metal with $In_2(SO_4)_3$ has been shown in spectroscopic and phase studies to give the salt $In^I[In^{III}(SO_4)_2]$.[572]

3.3.5   Low-valent sulfides, selenides and tellurides

The binary chalcogenides of aluminium are limited under normal conditions to the trivalent state in the compounds Al_2E_3 (E = S, Se or Te), with the exception of the recently reported compound Al_7Te_{10}.[573] The gaseous diatomic molecule AlS has been characterised by its electronic spectrum:[479] $r_e = 2.029\,Å$, $\omega_e = 617.12\,cm^{-1}$, $\omega_e x_e = 3.33\,cm^{-1}$ and $D_0 = 371\,kJ\,mol^{-1}$. Eight of the nine possible compounds M_2E_3 (M = Ga, In or Tl; E = S, Se or Te) are known, the apparent exception being Tl_2S_3. These compounds are discussed in section 3.2.4.1. A related compound is the indium mixed chalcogenide $In_{\sim2.01}$ $(S,Se,Te)_3$,[574] which is better described as $In^{III}_4SSe_2Te_3$ containing a small amount of In^I. The In^{III} centres are found in pre-dominantly tetrahedral sites with a small number of In^{III} and In^I centres in octahedral sites. In the light of these findings, it is possible that some ambiguities among indium chalcogenides are caused by partial replacement of In^{III} by In^I centres.[574]

Both Al^{III} and formally divalent Al^{II} centres are found in the aluminium telluride Al_7Te_{10}, prepared as red crystals from a melt of the elements.[573] The two important structural units are a double barrelane $[Te_4Al_4]_2$ linked by an Al^{II}–Al^{II} bond (2.60 Å) and a four-membered Al_2Te_2 ring (Figure 3.37). The compound, which can be described as a Zintl phase formulated as $\frac{1}{2}\{[Al^{III}]_{12}[Al^{II}-Al^{II}][Te^{-II}]_{20}\}$, exhibits diamagnetic and semiconducting behaviour.[573]

The M^I and M^{II} states, as well as the trivalent state M^{III}, are important in the binary chalcogenides of gallium, indium and thallium. In addition to the intrinsic interest in the wide variety of stoichiometries and structural types, these compounds are intriguing and potentially important because many are semiconductors, semi-metals, photoconductors or light emitters. Most of the compounds are prepared by the stoichiometric reaction of the elements in a sealed tube at high temperature. Ga_2S is formed in the high-temperature reduction of GaS with H_2 or Ga metal. All nine compounds of the type ME (M = Ga, In or Tl; E = S, Se or Te) are known, although they vary in formal composition. Phase studies show that, with sulfur, gallium forms Ga_2S, GaS, Ga_4S_5 and Ga_2S_3, indium forms In_5S_4, InS, In_6S_7 and In_2S_3, and thallium forms Tl_2S and TlS.[2,461,575] In addition, the compounds TlS_2, Tl_2S_2, Tl_2S_5 and Tl_2S_9 have also been described. The Group 13 -selenium and -tellurium systems are similarly rich in phases but the actual compounds are not always analogous to those in the sulfur systems. For example, The Ga/Se system contains Ga_3Se_2 while the In/Se system includes In_4Se_3 and In_6Se_7.[576]

The binary chalcogenides have in common with the binary halides the unorthodoxy that compounds of a given stoichiometry may vary from

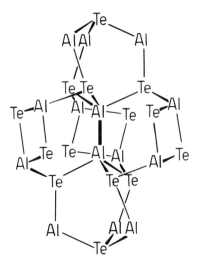

Figure 3.37 Important structural units in Al_7Te_{10}.[573]

species of genuinely intermediate oxidation state containing metal–metal bonds (as in the Ga_2^{4+} or In_3^{5+} fragments) to mixed valence M^I, M^{III} species. For example, GaS, GaSe, GaTe, red InS and InSe are layer-lattice M^{II} compounds containing M–M bonds, while InTe, TlS and TlSe are mixed valence $M^I[M^{III}X_2]$ species. In structural terms, the mixed valence compounds, like the mixed valence binary halides, show relatively short M^{III}–E distances, while M^I centres experience higher coordination numbers and much longer $M^I \cdots E$ distances. Important structural types are discussed below, a useful account of the area having been published quite recently.[6]

GaS, isotypic with GaSe, GaTe, InS and InSe, contains tetrahedral Ga^{II} centres each linked to three sulfurs and one gallium atom in metal–metal bonded Ga_2S_6 units forming a hexagonal layer structure (Figure 3.38(a));[8] the Ga–Ga bond lengths are 2.45 Å. In contrast, TlSe, like InTe and TlS, is a mixed valence compound $Tl^I[Tl^{III}Se_2]$. The $Tl^{III}Se_4$ tetrahedra share edges to form chains (Figure 3.38(b)) with the Tl^I lying between the chains and surrounded by a distorted cube of eight selenium atoms. Tl^{III}–Se and $Te^I \cdots Se$ distances are 2.68 and 3.42 Å, respectively.[5] The crystals are black with a metallic lustre, and show electrical conductivity along the axis of the chains.

Just how fine a balance there is between these two types of structure is demonstrated by the surprising discovery that, while GaSe and InSe are M^{II} compounds with metal–metal bonds, the mixed metal system $GaInSe_2$ shows the mixed valence structure $In^I[Ga^{III}Se_2]$, as does $GaInTe_2$.[577] Replacement of one gallium in the Ga_2^{4+} unit by indium with its more stable univalent state must be sufficient to induce disproportionation to the mixed valence form. Compounds of this type are an area of active research. For example, $TlGaS_2$, $TlGaSe_2$ and β-$TlInS_2$ are isostructural ferroelectric materials.[298]

Metal–metal bonded In_2^{4+} units (In–In = 2.74 and 2.76 Å) are found in In_6S_7 and In_6Se_7, described in ionic terms as $In^{III}_2\{[In_2]^{IV}\}_2(E^{-II})_7$ (E = S or Se).[576] In_5^{8+} units with a central In bonded to four other In atoms in a tetrahedral arrangement (In–In = 2.76 and 2.77 Å) occur in In_5S_4,[575] and In_3^{5+} moieties (In–In = 2.79 Å) occur in In_4Se_3 and In_4Te_3.[461,576] The latter compounds contain bent In_3 units linked through two selenium or tellurium atoms into a chain. The In^I cations and the third chalcogenide atom occupy spaces between the chains. The overall stoichiometry can be represented to

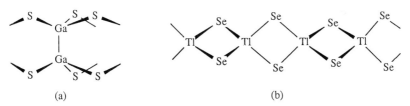

(a)   (b)

Figure 3.38 Coordination environment of (a) $(Ga_2)^{IV}$ in GaS[8] and (b) Tl^{III} in TlSe.[5]

a first approximation as $In^I[In_3]^V(E^{-II})_3$. Although Tl_4S_3 has the same formula, it corresponds more closely to $Tl^I_3[Tl^{III}S_3]$, showing the preference of the heavier metal Tl for the univalent state; the anions are made up of chains of corner-shared $Tl^{III}S_4$ tetrahedra held together by Tl^I ions (Tl^{III}–S = 2.54 Å; $Tl^I \cdots S = 2.90$–3.36 Å).[6] The sulfur-rich thallium compounds TlS_2, Tl_2S_5 and Tl_2S_9 are Tl^I polysulfides and are prepared by treating $TlNO_3$ with ammonium polysulfide solution.

Various ternary sulfur or selenium phases containing Tl^I in combination with other elements are known; one of these, $Tl_3Al_{13}S_{21}$, is discussed in section 3.2.4.1. Other examples are $Tl^I[M^{III}E_2]$ (M = Ga or In), Tl^IM^IE (M = Ag or Cu), $Tl^I_4[M^{IV}_2S_6]$ (M = Si or Ge) and $Tl^I_4[M^{IV}E_4]$ (M = Si, Ge or Sn). Compounds belonging to the last of these classes and which have been characterised recently include $Tl_4[SiS_4]$, which is isostructural with $Tl_4[GeS_4]$ and contains SiS_4^{4-} anions held together by Tl^+ cations in irregular six-fold coordination; $Tl_4[SiSe_4]$ consists of planar nets of Tl^+ cations enclosed between two parallel layers of $SiSe_4^{4-}$ tetrahedra with further Tl^I cations in the voids.[578]

The ternary phase Ga_3Te_3I (m.p. 692°C, congruent) is unique in containing the only mixed valent Ga^{II}/Ga^{III} combination to be identified to date. The compound forms one-dimensional chains which contain, in cross-section, Ga^{II}–Ga^{II} bonds (2.45 Å) in $Ga^{II}_4Te_2$ rings linked through Te atoms to $Ga^{III}Te_3I$ tetrahedra, giving an overall formula $[Ga^{II}_2Ga^{III}Te_3I]$ (Figure 3.39).[579] The following compounds with the same composition have also been identified: Ga_3Te_3Cl (m.p. 718°C, congruent), Ga_3Te_3Br (m.p. 733°C, congruent), In_3Te_3Br (m.p. 381°C, incongruent) and In_3Te_3I (m.p. 394°C, incongruent). The intense red crystals have high reflecting power and an acicular habit and are very easily cleaved along the fibres.

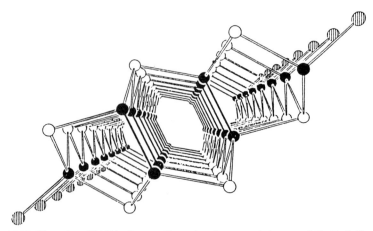

Figure 3.39 View along [0 0 1] in the one-dimensional structural element of Ga_3Te_3I; Ga atoms are black; I atoms are shaded; Te atoms are white.[579] Reproduced with permission from S. Paashaus and R. Kniep, *Angew. Chem., Int. Ed. Engl.*, 1986, **25**, 752.

Few species containing In^I in combination with sulfur anions are known. Electrochemical oxidation of anodic indium in non-aqueous solutions of a dithiol $HS(CH_2)_nSH$ ($n = 2$–6) gives $In^I[S(CH_2)_nSH]$.[580] Chelation through both the RS^- and RSH groups is presumed in structures similar to those of the In^I derivatives of aromatic diols which are formed in analogous circumstances (section 3.3.4).[564] Both classes of compounds undergo oxidative addition with I_2 or o-quinones to form In^{III} species; in addition, reversible CS_2 insertion occurs with the sulfur-containing species to form thioxanthate complexes.[580] The electrochemical oxidation of indium metal in CH_3CN solutions of alkyl or aryl thiols gives In^I, In^{II} or In^{III} thiolato products. The In^I compounds, InSR, are formed for R = Et or Bu^n, while In^{II} species $In_2(SR)_4$, formulated as metal–metal bonded dimers, occur when R = C_5H_{11} or naphthalide. Other arenethiols give $In(SR)_3$. The corresponding oxidation of Tl metal gives TlSR for all R groups.[581]

The Tl^I compounds with sulfur-based anions are predominantly ionic in character. However, the reaction of Tl^I salts with organic thiolates in non-aqueous solvents has produced a number of yellow polymeric or oligomeric thallium(I) thiolate species.[582] TlSPh consists of $[Tl_7(SPh)_6]^+$ and $[Tl_5(SPh)_6]^-$ clusters linked through bridging sulfur atoms; by contrast, $TlSBu^t$ contains $Tl_8(SBu^t)_8$ molecules with the thallium atoms in sites with trigonal pyramidal or ψ-trigonal bipyramidal coordination; and yet another option emerges with $TlSCH_2C_6H_5$ which forms a ladder-like arrangement of Tl_2S_2 four-membered rings linked through *trans* edges.[582]

3.3.6 Low-valent carbonyls and organometallic derivatives

In marked contrast to their prevalence in transition-metal chemistry, carbonyl derivatives of the early Main Group metals are rare and elusive species. Dicarbonyl molecules of the type $M(CO)_2$ have been generated by co-condensing M atoms (M = Al, Ga or In) with carbon monoxide molecules in an inert matrix at low temperatures; here they have been characterised by their IR and ESR spectra[583–590] (see chapter 6, section 6.3).

The tally of organometallic compounds containing Al, Ga, In or Tl in their lower oxidation states is small, yet it contains some unusual and exciting new bonding types. Although the major treatment of organo derivatives of the Group 13 metals is deferred to chapter 6, there are some features and some specific compounds which are illuminating in relation to the chemistry discussed in the preceding sections. The compounds include (i) M^I derivatives with M bearing η^5-cyclopentadienyl or η^6-arene ligands (systems which form a cornerstone of organo-transition-metal chemistry but are rare in the Main Groups) and (ii) M^{II} derivatives in which the feature of special interest is the presence of metal–metal bonds.

The $C_5H_5^-$ anion and its substituted analogues, e.g. $C_5Me_5^-$ and

$C_5(CH_2Ph)_5^-$, can be regarded, up to a point, as pseudohalide ions. However, the compounds they form with the Group 13 metals in the +1 oxidation state show great structural and chemical diversity, including some unique metal–metal bonded dimers and clusters. Whereas the aluminium(I) compound $(\eta^5\text{-}C_5Me_5)Al$ is tetrameric (see section 3.3.3.1),[490] the solid indium(I) compounds $[\eta^5\text{-}C_5H_2(SiMe_3)_3]In$, $[\eta^5\text{-}C_5(CH_2Ph)_5]In$ and $(\eta^5\text{-}C_5Me_5)In$ are monomeric, dimeric and hexameric, respectively.[446] The signs are that the metal–metal interactions in the clusters are typically quite weak.[591,592] Whereas $[\eta^5\text{-}C_5H_2(SiMe_3)_3]Tl$ is a monomer in the solid state, the *bis*-substituted complex $[\eta^5\text{-}C_5H_3(SiMe_3)_2]Tl$ is a hexamer but, unlike $[(\eta^5\text{-}C_5Me_5)In]_6$, it is made up not of a metal–metal bonded cluster but of a ring of bridging cyclopentadienyl ligands alternating with Tl atoms.[446] However, the extremely bulky pentabenzyl-substituted ligand does produce a metal–metal bonded dimer $[(\eta^5\text{-}C_5(CH_2Ph)_5]Tl$ analogous to the corresponding indium compound (see also section 3.3.2).

η^6-Arene species, formally isoelectronic with cyclopentadienyl derivatives, are noteworthy in providing unequivocal evidence for π-bonding interactions with M^I centres, where M = Ga, In or Tl. It had been known for 50 years that some compounds of Ga^I, In^I and Tl^I are soluble in aromatic solvents. A re-investigation of this chemistry since 1980, primarily by Schmidbaur and his group,[471] has revealed stable, isolable η^6-arene complexes with stabilities decreasing in the order $Ga^I > In^I > Tl^I$ (see, for example, chapter 1, p. 49 and chapter 6, section 6.6).

In addition to these derivatives of the univalent metal, recent research has also brought to light M^{II} dimers of the type $R_2M\text{–}MR_2$ [M = Al, Ga or In; R = $CH(SiMe_3)_2$].[536,593,594] The structures of all three compounds have been determined and reveal, rather surprisingly, that the metal–metal bond length is shortest for the gallium dimer (Al–Al 2.66, Ga–Ga 2.54 and In–In 2.83 Å). The reaction of Bu^i_2AlCl with potassium metal yields, in addition to the dimer $Bu^i_2Al\text{–}AlBu^i_2$, the remarkable cluster $K_2[Al_{12}Bu^i_{12}]$ containing a metal–metal-bonded Al_{12} icosahedron.[595] Each Al bears an alkyl group, and the $[Al_{12}Bu^i_{12}]^{2-}$ anion is analogous to $[B_{12}H_{12}]^{2-}$, thus being an example of a *closo* cluster which obeys Wade's rules.[596] The Al–Al distances are 2.68–2.70 Å, similar to the Al–Al bond length in $R_2Al\text{–}AlR_2$ (2.66 Å when R = $CH(SiMe_3)_2$)[594] and shorter than those in the $(\eta^5\text{-}C_5Me_5)_4Al_4$ cluster (2.77–2.78 Å).[490]

3.4   Intermetallic derivatives

3.4.1   Molecular species

The Group 13 diatomic molecules, M_2, are present at low concentration in the vapours of the metals or of strongly heated compounds, for example Al_2

Table 3.16 Dissociation energies of some Group 13 diatomic molecules in the gas phase.[a]

Group 13 molecules				
Molecule	Al_2	Ga_2	In_2	Tl_2
D_0° (kJ mol^{-1})	163	138	92	59
Group 13–Group 15 molecules				
Molecule	AlP	GaAs	InSb	TlBi
D_0° (kJ mol^{-1})	213	210	148	117
Group 13–coinage metal molecules				
Molecule	AlAg	AlAu	GaAu	InAu
D_0° (kJ mol^{-1})	186	322	210	208

(a) For sources of data see text.

above aluminium carbide, Ga_2 above gallium carbide, In_2 above indium antimonide or Tl_2 above thallium(I) fluoride. Table 3.16 gives the dissociation energies of the gaseous M_2 molecules which are derived from the mass spectra of these systems.[461] A pronounced drop in bond strength occurs in descending order of the Group from Al to Tl. Nevertheless, reasonably strong bonds can be ascribed to the pairing of the single p electrons of pairs of Group 13 atoms. The alkali metal dimers, in which the single s electrons are paired to form Na_2, K_2, Rb_2 or Cs_2 are much less strongly bound with dissociation energies in the range 75–40 kJ mol^{-1}. Included in Table 3.16 are the Group 13–15 molecules which are isoelectronic with the Group 14 dimers Si_2, Ge_2, Sn_2 and Pb_2. Again the dissociation energy indicates a moderately strong bond. Even stronger bonds are found in some of the molecules formed by Al, Ga or In with Ag or Au, examples of which are given in Table 3.16. AlAu is a particularly stable intermetallic molecule and the species $AlAu_2$ and Al_2Au have also been detected in the vapour phase of aluminium–gold mixtures.[597]

3.4.2 Intermetallic compounds

The metals Al, Ga, In and Tl are involved in a great variety of alloys, inter-metallic phases and compounds with other metals which may be either more or less electropositive than the Group 13 element concerned. Many of these are typical metallic solids, the structures of which are determined mainly by the principles of the close-packing of atoms. Neither ionic nor directed covalent bonding has much influence on the structures and properties of intermetallic compounds of this sort; the valence electrons can be assumed to occupy delocalised orbitals and high conductivity is a normal feature. Examples which involve transition and other Main Group metals will be cited in the following account.

Of greater interest from an inorganic perspective are compounds formed by the Group 13 metals, sometimes in association with a metal or metalloid of a later Group, in combination with one of the alkali or alkaline-earth metals. Characteristic features of the structures of these materials are polyanionic clusters. The bonding in such polyanions is predominantly covalent and the cations (Na^+, K^+, Rb^+, Ca^{2+}, etc.) are located within the interstices of the lattice. Of this nature are the Zintl-phase compounds.[598–601] Novel structures are involved and the properties, including colour, low electrical conductivity and the ability to dissolve in liquid ammonia or amines, distinguish the Zintl compounds from the ordinary run of metals and their alloys.

The method of preparing intermetallic phases of all types generally involves fusing a mixture of the appropriate metals in the correct stoichiometry under an inert atmosphere of nitrogen or argon, followed by careful cooling and annealing of the solid phase. X-Ray analysis is often necessary to confirm that the desired material has been produced.

Many metals form admixtures with aluminium and there are numerous commercially important aluminium alloys (see chapter 2, section 2.4). Lithium yields intermetallic compounds including Li_3Al, Li_9Al_4, Li_3Al_2 and $LiAl$, and magnesium yields $Mg_{17}Al_{12}$, Mg_9Al_{11}, $Mg_{23}Al_{30}$ and Mg_2Al_3.[2,600,602] The metals Na, K, Rb, Cs, Cd, In and Tl each have only a limited solubility in molten aluminium and do not give rise to binary compounds while Ga forms a simple eutectic system. Most other metals and metalloids readily alloy with Al; the 1970 account by Wade and Bannister[2] lists the structures of more than 150 intermetallic compounds of aluminium. To extend this list is beyond the present scope and specialist literature needs to be consulted for up-to-date information on any given system. For the most part, these involve typical intermetallic phases in that the compounds exist only in the solid state and do not dissolve in any solvent without chemical reaction. The structures are essentially metallic ones[8] in which questions of the efficient packing of metal atoms are usually paramount over any other bonding considerations. The bonding patterns among intermetallic compounds in general were discussed by Nesper,[602] and King[603] has treated the special case of quasicrystals containing Al_{12} icosahedra and larger complexes, e.g. $(Al,Cu)_{60}$. Quasicrystalline phases are a structural novelty encountered in Al–Li–Cu alloys whose study was initially stimulated by the need for very low density metals for aerospace purposes. Many complex binary and ternary aluminium-containing compounds are treated in the aforementioned reviews,[602,603] and the structures compared with those of prototype metal arrangements, e.g. $Mg_{17}Al_{12}$ and $Li_xMg_{17-x}Al_{12}$ with manganese, or Li_3CuAl_5 crystals with related alloys. Consideration is also given to the structural aspects of other aluminides and gallides in which nearly all features of chemical bonding that are known from boron compounds find a counterpart.

Discrete polyhedral clusters of aluminium are unknown, although there is a useful landmark in the shape of $K_2[Al_{12}Bu^i_{12}]$, mentioned in section 3.3.6. The structure of the $[Al_{12}Bu^i_{12}]^{2-}$ anion[595] is based on the Al_{12} icosahedron, and the formal removal of the alkyl groups from this species leaves Al_{12}^{14-} with precisely the number of skeletal electrons $(2n + 2)$ called for by Wade's rules.[596] Related examples of large polyanionic clusters of Ga and In will be encountered in the discussion of intermetallic compounds of gallium and indium below; however, there are also instances where the structures concerned must be assigned fewer than $2n + 2$ skeletal bonding electrons, so that fresh approaches are required for their understanding.[602,603]

Gallium forms intermetallic phases with the alkali metals in extraordinary variety; these are examined in the paragraphs below. Ga with magnesium forms Mg_5Ga_2, Mg_2Ga, $MgGa$, $MgGa_2$ and Mg_2Ga_5.[604] With calcium, strontium and barium there are compounds of the types MGa_2 and MGa_4, and others[600] such as Ca_5Ga_3, which is notable for a structure with equal numbers of Ga atoms and Ga_2 pairs per formula unit.[602] Interpenetration of gallium into aluminium is easily demonstrated; a line can be drawn on Al foil with solid Ga and after a few minutes the foil is weakened so that it tears along the mark. Solid solutions are formed in the system Ga–Al, and to a lesser extent in those of Ga with In or Sn. The low melting point of gallium is reflected in its alloy with indium (a eutectic with 25 at.% Ga, m.p. 16°C) and with tin (a eutectic with 92 at.% Ga, m.p. 15°C). Ga scarcely dissolves in solid Tl or Pb. Compound formation as such apparently plays no part in the systems of Ga with the metals of Group 13 or 14. With the transition metals (M) there are compounds of various stoichiometries, especially GaM and GaM_3. Gallium combines with the noble metals, forming alloys with platinum and gold.[2] There is a series of compounds of Ga with copper and silver constituting alloys which decrease in electrical conductivity and increase in brittleness at higher gallium content. Certain gallium alloys, e.g. GaM_3 (M = Zr, V or Nb), become superconducting at temperatures below 10–15 K and are used in the construction of high-field electromagnets.

With reference to the intermetallic phases of gallium with alkali metals, early work had indicated the formation of Li_2Ga, $LiGa$, Na_5Ga_8, $NaGa_4$, and some analogous compounds of K, Rb and Cs.[2,600,605] Belin and Ling et al.[606] then prepared and determined the crystal structures of seven further compounds: Li_3Ga_{14}, $Na_{22}Ga_{39}$, KGa_3, K_3Ga_{13}, $RbGa_3$, $RbGa_7$ and $CsGa_7$. A feature of these gallides is the presence of Ga clusters linked to each other within a three-dimensional network. The structures may also contain satellite Ga atoms attached to the polyhedra; in KGa_3, for example, there are Ga_8 dodecahedra connected via Ga atom spacers, and in MGa_7 (M = Rb or Cs) not only Ga_{12} icosahedra but also bonded pairs (Ga–Ga = 2.52 Å) of gallium atoms. Various attempts have been made by Belin and Ling,[606] Schäfer[607] and lately by King[603] to interpret these structures according to Wade's rules,[596] which specify $2n + 2$ skeletal electrons for an n-atomic cluster and to accommodate the new macroanions into the concept of Zintl phases at the junction between metallic and other types of bonding.

Subsequent investigations have explored some very complex non-stoichiometric, ternary phases, such as $Li_3Na_5Ga_{19.6}$, $Rb_{0.6}Na_{6.25}Ga_{20.02}$ and $K_2Li_9Ga_{28.83}$. The structures of these gallides contain large deltahedra of Ga atoms interspersed with the alkali cations. The deltahedra are linked to one another both directly and *via* extra atoms in the structure. Ga–Ga inter-atomic distances range from 2.47 to 2.90 Å. Within $Rb_{0.6}Na_{6.25}Ga_{20.02}$,[608] a pair of face-sharing icosahedra of stoichiometry Ga_{21}, shown in Figure 3.40(a), is vertex-linked to other such units as well as to single Ga_{12} icosahedra. Similar units are found in $Li_3Na_5Ga_{19.6}$[609] and $Li_3Na_5Ga_{28.83}$.[610] $K_4Na_{13}Ga_{49.57}$[611] has a curious triply fused isosahedron with the stoichiometry Ga_{28}. In these structures, and in some borides such as AlB_{12} which contain units of a similar sort, not all of the sites have an occupancy of 100%. Consequently the fused icosahedra may be Ga_{11} or Ga_{10} clusters of *nido-*, *arachno-* and *hypo-*types and present a further challenge to theory as well as to the experimentalist faced with their elucidation. An approach which examines complex gallides alongside borides and the allotropes of boron proves fruitful, and following these lines Burdett and Canadell[612,613] have extended Wade's rules to elucidate the electronic structure and bonding problems posed by the widespread occurrence of icosahedra and other large but often defective deltahedra.

Gallium forms other binary and ternary intermetallic compounds which contain Ga chains, clusters, fused clusters and cluster fragments. Isolated gallium polyhedra are unknown as yet in any alkali-metal compound, although Ga_3 and Ga_4 units occur in Ba_8Ga_7 where they display Ga–Ga distances of 2.63–2.75 Å, which resemble those of gallium metal.[614] According to the Zintl model, this phase can be formulated as $8Ba^{2+}(Ga_3^{9-})$-(Ga_4^{8-}) although this does not quite give the precise charge balance. The new Zintl compound $Ca_{14}GaAs_{11}$ has a structure composed of isolated Ga-centred $GaAs_4^{9-}$ tetrahedra that are separated by linear As_3 units, As and

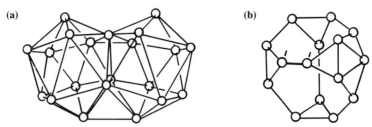

Figure 3.40 Large deltahedral gallium clusters: (a) Ga_{21} found in $Rb_{0.6}Na_{6.25}Ga_{20.02}$ which consists of a pair of Ga_{12} isosahedra with three atoms in common; (b) Ga_{15} found in Na_7Ga_{13} and $Na_{22}Ga_{39}$. In the latter case the o'–o' bond is unusually short, see text.

Ca.[615,616] A probable reason why isolated homonuclear clusters of gallium are rare is that the charge on such units would be too high (e.g. Ga_8^{10-} and Ga_{12}^{14-} for *closo* analogues of $B_8H_8^{2-}$ and $B_{12}H_{12}^{2-}$, respectively). Instead, intercluster bonding occurs to produce three-dimensional network structures. This formally converts the *s* electron pair on each Ga atom to a bonding role and lowers the cluster charge by one for each such *exo* bond. In the structure of $Na_{22}Ga_{39}$[617] there are eight vertex-linked icosahedra and four 'spacers' which consist of the 15-atom cluster shown in Figure 3.40(b). The closely related phase $Na_7Ga_{13}(Na_{21}Ga_{39})$[618] also contains the Ga_{15} cluster but there is a subtle distinction; in this instance the pair of Ga atoms shown primed in the figure are outwardly coordinated to other atoms, whereas in $Na_{22}Ga_{39}$ the atoms of this pair remain three-coordinate. In line with this difference, the Ga–Ga bond in Na_7Ga_{13} measures 2.54 Å (a normal value for gallide clusters) but is decidedly shorter at 2.43 Å in $Na_{22}Ga_{39}$. Burdett and Canadell[612] go so far as to suggest that it constitutes the first example of a homoatomic Ga=Ga double bond, but this is hard to justify in the wider context of gallium chemistry, given that the well-established covalent Ga–Ga single bond distance in dinuclear Ga^{II} (i.e. Ga_2^{4+}) derivatives is close to 2.40 Å (see section 3.3.3, Table 3.15). The metal itself exhibits pairs of atoms with a short Ga–Ga distance of 2.44 Å in α-Ga,[619] which persists in the molten metal.[620] However, 2.60–2.80 Å is the normal range of interatomic distances in other crystalline modifications of the metal. Noting this unique aspect of gallium, Pauling[621] long ago pointed out that the structure of α-Ga can be described as a metallic packing of Ga–Ga diatomic units. Whether this has any bearing on the readiness with which Ga_2^{4+} species can be prepared directly from the metal is a moot point.

The early literature records indium as forming a number of distinctive intermetallic phases.[2] With other Main Group metals, these include LiIn and others with higher lithium content; Na_2In, NaIn, Na_5In_8; Mg_2In, MgIn, $MgIn_2$; CaIn and $CdIn_3$.[2,600,605] Like the gallides, some of these may prove to be Zintl-type phases in which In is formally anionic. Indium forms low-melting alloys with tin (InSn, m.p. 117°C) and bismuth (InBi$_2$, m.p. 109, In_2Bi, m.p. 72°C). Of indium-transition-metal compounds, the majority are those of Ni, Pd, Pt, Cu, Ag and Au.[2]

Knowledge of the compositions and structures of binary alkali metal-indium phases is poor, although work by Corbett *et al.*[622] promises to rectify this situation. First indications suggest that close structural relationships between the systems formed by an alkali metal (A) with gallium or indium (M) may exist only for AM (NaTl-type), layered AM_3 and the cluster-containing $A_{22}M_{39}$ phases. Instead a novel and diverse indium cluster chemistry seems to be unfolding. The compound Na_2In (like Na_2Tl) contains isolated tetrahedral clusters, M_4^{8-}, which are isoelectronic with Sn_4^{4-} in KSn, or with the Sb_4 and P_4 molecules. Reactions of sodium and indium in proportions near 37.3 at.% Na at *c.* 500, 410 and then 250°C produce the well-crystallised phase $Na_7In_{11.8}$.[622] The network structure of this solid consists of interbonded *closo*-In_{16} icosioctahedra, *nido*-In_{11} icosahedra and In atoms. Extended Hückel calculations indicate that $2n + 4$ skeletal electrons are necessary for each of the units In_{16} and In_{11} with no non-bonding pairs on the latter, and give a good account of the observed properties. The crucial importance of structural analysis in systems of this complexity is emphasised by noting that the compound near $Na_7In_{11.8}$ in the Na–In phase diagram[623] had been assumed previously to be either Na_5In_8 (38.5 at.% Na) or Na_2In_3 (40%) on the basis of melting point data.

The potassium–indium system is different; here is found KIn_4 (an intermetallic compound of the $BaAl_4$ type),[624] a metallic phase K_8In_{11} (with isolated In_{11}^{7-} clusters and one electron per formula unit delocalised in a conduction band), and $K_{22}In_{39}$ which consists of a network of heavily inter-bonded In_{12} and In_{15} clusters.[625] The novel structure of the In_{11}^{7-} ion is depicted in Figure 3.41. The In–In distances range from 2.96 to 3.10 Å and compare well with 3.07–3.15 Å of the In_4^{8-} tetrahedra of Na_2In.[622] This structure represents a previously unknown cluster-bonding configuration of D_{3h} symmetry in which the skeletal electron count is well below the $2n + 2$ given by Wade's rules, being in this case $2n - 4$. The better known C_{2v} *closo* deltahedron with 11 vertices and 24 electrons (as for $B_{11}H_{11}^{2-}$) would require an unreasonable -13 charge with In_{11}. However, it is interesting to note that the In_{11}^{7-} cluster can be thought of as being derived from a *closo* In_9^{11-} (D_{3h}) ion by the addition of two In^+ ions in axial positions. The occurrence of isolated M_n clusters for In which are unknown for Al or Ga may owe something to the greater stability of the $+1$ oxidation state in the case of indium.

Thallium forms alloys with most other metals and a large number of intermetallic phases have been described, including at least five with lithium, four with sodium and three with potassium.[2,600,605] In Na_2Tl the

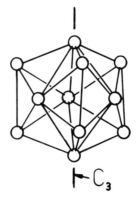

$\vdash C_3$

Figure 3.41 The discrete In_{11}^{7-} anion found in K_8In_{11}. The structure has D_{3h} symmetry and the C_3 axis is shown.

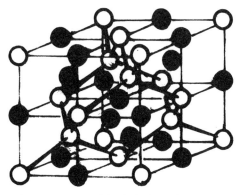

Figure 3.42 The crystal structure of NaTl (a Zintl phase with the B32 lattice). ●, Na; ○, Tl.

thallium atoms form isolated Tl_4^{8-} tetrahedra; the Tl–Tl distance is 3.22 Å and a bonding scheme can be constructed by mixing only the p orbital of thallium.[626] In NaTl[627] the thallium atoms are linked into a continuous network which is effectively a Tl_n^{n-} polymeric anion. Figure 3.42 shows the relationship of the NaTl structure to that of diamond on the one hand and to the primitive cubic lattice of CsCl on the other. The constituents (here formally Na^+ and Tl^-) are each surrounded in eight-fold coordination by four of their own kind and four of the opposite kind in a regular array.

The structure of NaTl is of the B32 lattice type which is formed by other Group 13 metals in combination with Li or Na. Table 3.17 gives the M–M distances which occur.[627] These separations are close to twice the atomic radii of the atoms concerned and are about 10% greater than the expected M–M distance for covalent bonding. However, the estimate for covalent M–M bonding applies to compounds in which the Group 13 element has a positive oxidation state and not the formal negative state which it assumes in a Zintl-phase compound. Other experimental data for the binary intermetallic compounds LiAl, LiGa, LiIn, NaIn and NaTl are cited in the review by Schmidt.[627] The same review also examines the electronic structures of

Table 3.17 M–M bond distances in some B32-type Zintl compounds.[a]

Compound	M–M bond	Bond distance (Å)
LiAl	Al–Al	2.76
LiGa	Ga–Ga	2.68
LiIn	In–In	2.94
NaIn	In–In	3.16
NaTl	Tl–Tl	3.24

(*a*) See Ref. 627.

some related ternary systems involving the Group 13 metals, e.g. $LiCd_{1-x}Tl_x$, $NaHg_{1-x}Tl_x$, $LiAg_{1-x}In_x$ and $LiCd_{1-x}In_x$. The lithium–cadmium–indium intermetallics are distinctively coloured with a hue which changes continuously from yellow to blue as a function of x. A common model can be applied to the bonding of all Zintl phases of B32 type; a feature of this is the metal-like charge distribution which encompasses the upper valence/conduction electronic states and allows the compounds to behave as semiconductors.

Zintl-phase compounds were first recognised among the derivatives of the metals and metalloids of Groups 14, 15 and 16 (i.e. in the germanium, arsenic and selenium sub-groups) with the highly electropositive metals of Groups 1 and 2, which transfer electrons to the partnering element. The finding of Zintl phases among the intermetallic compounds of Group 13, where the high degree of electron deficiency tends to inhibit the formation of anionic clusters,[628] thus represents an extension of the field, though not of the basic concept. With a few exceptions, such as Tl_4^{8-} tetrahedra in Na_2Tl, the diamond-type $(Tl^-)_\infty$ polyanion of NaTl and the graphite-type $(Ga^-)_\infty$ polyanion of $CaGa_2$,[599] most highly reduced cluster and network species involving the Group 13 metals are of recent discovery. Like the polycations of Main Group metals,[628] the polyanions (the Zintl anions) offer models for the stepwise oxidation or reduction of an element to its compound forms. Such clusters, which are an intermediate stage between an element and its isolated atoms or ions, test present understanding of electronic structures and bonding. There are close relationships between some of these species; for instance, Bi_9^{5+} is isoelectronic, although not isostructural, with Pb_9^{4-} and Sn_9^{4-}. The chemical utilisation of these clusters represents a further challenge and the scope for their application in semiconductors and other technological developments is immense.

To probe the limits of the Zintl concept a stage further, it is instructive to examine, if only briefly, some compounds in which a Group 13 metal combines with an element of Group 14, 15 or 16 to furnish the polyanionic structure. One such Zintl-phase material, $Ca_{14}GaAs_{11}$,[615,616] has already been mentioned. $K_3Ga_3As_4$[629] is another; it is a metallic grey solid, which, unlike some of the ternary Zintl phases containing chalcogenides, is insoluble in liquid ammonia or ethanediamine. The structure is composed of layers of K^+ ions and covalently bonded sheets $[Ga_3As_4^{3-}]_\infty$. Within the anionic sheets, each Ga atom is bonded to four As atoms at the corners of a distorted tetrahedron. The average Ga–As bond distance is 2.50, compared with 2.47–2.69 Å in several Ca–As–Ga compounds[615] and 2.45 Å in GaAs. Other ternary compounds contain, besides the alkali metal, Al and As, Al and Sb, or Ga and Sb.[629] Their reactions afford the possibility of solution-phase deposition of semiconductor materials; for example, combination of $K_3Ga_3As_4$ with $GaCl_3$ would produce a III–V phase and a water-soluble alkali-metal halide.

$$K_3Ga_3As_4 + GaCl_3 \rightarrow 4GaAs + 3KCl \qquad (3.59)$$

Such Zintl compounds are classified as being of the I–III–V or II–III–V types. In the idealised description of their bonding, the electropositive elements act merely as cation formers, releasing electrons to their more electronegative bonding partners. The latter elements then form bonds to satisfy the octet rule. However, this model sometimes fails, and in doing so signals the transition from one intermetallic bonding type to another. An example is provided by $Ba_7Ga_4Sb_9$,[630] for which it is impossible to find a conventional covalent network bearing a charge of -14 necessary to balance the $+14$ charges due to the seven Ba^{2+} cations present in the structure. Calculations reveal a half-filled σ^* band involved in the linking together of alternating four- and eight-membered Ga–Sb rings.

Some ternary phases involving Group 13 and Group 16 elements are of the kind $TlMY_2$, where M = Ga or In and Y = S, Se or Te,[298] and these have already been considered in section 3.2.4. The structures are characterised by Tl atoms running in chains, and there is good reason to suspect that the role of thallium goes beyond that of a passive Tl^+ cation and that Tl–Tl bonding has a part to play. In other Zintl compounds thallium is found within the polyanion. KTlTe is a case in point illustrating another important development, namely the use of a ligand to complex the alkali-metal cation (K^+) and so prevent electron transfer from the anion back to the cation in the solid state. The resulting compound (produced from KTlTe and 2,2,2-crypt in the presence of ethanediamine (en)) is isolated as dark brown crystals of (2,2,2-crypt-K^+)$_2Tl_2Te_2^{2-}$ · en.[631] The $Tl_2Te_2^{2-}$ anion adopts a butterfly shape with Tl atoms in the fold (see chapter 1, p. 69). The cross-ring Tl$\cdots$Tl distance of $3.60\,Å$ falls into the category for which the controversial Tl–Tl bonding interactions have been examined. Another compound dependent on this mode of stabilisation is (2,2,2-crypt-K^+)$_3$($TlSn_8^{3-}$,$TlSn_9^{3-}$)$_{1/2}$ · en obtained as red-brown crystals from the reaction of the alloy KTlSn with ethanediamine containing a stoichiometric quantity of the cryptand ligand.[632] Present in this structure are the polyhedral anions $TlSn_8^{3-}$ and $TlSn_9^{3-}$, the former a tricapped trigonal prism and the latter a bicapped square antiprism (see chapter 1, p. 69). In each case thallium occupies a capping position, with Tl–Sn bonds which measure 3.05–$3.07\,Å$. There is evidence from multinuclear NMR spectroscopy for the existence of $TlSn_8^{5-}$.[633] The species $TlSn_8^{5-}$ formally involves the substitution of a Tl atom for Sn in Sn_8^{4-} which is one of the original Zintl anions.[634,635] Molecular orbital calculations[632] have shown that p electrons are used for skeletal bonding in homo- and heteronuclear clusters of the Main Group elements. Any valence-shell s electrons are usually retained by the atoms as lone pairs and may have a profound effect on the structures adopted.

From the level of curiosities of marginal relevance to inorganic chemistry, intermetallic compounds of the Group 13 metals have progressed rapidly to

assume a position of considerable significance from which further developments on both the theoretical and the practical front are to be expected. It is
interesting to compare these compounds with the metal–metal-bonded
molecular species which are examined in the following section.

3.4.3 Group 13 transition-metal carbonyl derivatives

Gallium, indium and thallium form a number of well-authenticated compounds in which $M'(CO)_n$ ligands (M' = Mn, Fe or Co) are attached to the
Group 13 element.[461,636,637] The first compound of this type, *tris*(tetra-
carbonylcobalt)indium, $In[Co(CO)_4]_3$, was prepared in 1942 by the reaction
of the finely divided metals cobalt and indium with carbon monoxide under
200 atm pressure at 200°C.[638] Later single-crystal X-ray studies showed that
the compound consists of discrete molecules in which three $Co(CO)_4$ groups
are attached to indium in a trigonal planar arrangement.[639] The average
In–Co distance, 2.59 Å, is 0.15–0.20 Å less than might be expected from the
sum of the relevant covalent radii. The indium centre nevertheless retains
sufficient Lewis acidity to form the four-coordinate complex anion
$\{In[Co(CO)_4]_4\}^-$. There are other complex anions of the type $\{X_{4-n}In$
$[Co(CO)_4]_n\}^-$ (X = halogen; n = 1–3) in which the expected tetrahedral
coordination of indium has been confirmed by X-ray investigation.[640,641] In
the adduct $Cl_2In[Co(CO)_4] \cdot PPh_3$, triphenylphosphine provides the fourth
ligand in another tetrahedral array.[640] This compound is of particular
interest because it arises (admittedly in low yield) from the reaction between
$InCl_3$ and 1 equivalent of $K[Co(CO)_3(PPh_3)]$, showing that the Lewis-acidic
indium centre competes effectively with the cobalt atom for the phosphine
ligand. Somewhat unexpectedly, the In–Co bond length, at 2.54 Å, in
$Cl_2In[Co(CO)_4] \cdot PPh_3$ is even shorter than in $In[Co(CO)_4]_3$. In–Co bond
lengths of 2.60 to 2.70 Å are found in the four-coordinate complex anions
mentioned above.

Various compounds of the type $X_nM(M'L_x)_{3-n}$ (M = Ga or In; X = Cl
or Br; $M'L_x$ = $Co(CO)_4$ or $Mn(CO)_5$) have been reported.[637,642,643] They
are usually prepared by reactions of Ga_2X_4, GaX_3, InX or InX_3 with
$Na[Co(CO)_4]$, $Co_2(CO)_8$ or $Mn_2(CO)_{10}$, and can be rationalised as derivatives of trivalent gallium or indium in which the metal carbonyl ligand
functions as a pseudohalide group. Compounds containing both gallium or
indium and a transition metal hold promise as single-source precursors to
binary intermetallic phases useful to introduce ohmic contacts and Schottky
barriers in the formulation of III–V semiconductor devices, e.g. GaCo
or Ga_2Pt films.[644] During synthesis these compounds often retain a
molecule of the solvent (e.g. tetrahydrofuran, thf), which completes
the four-fold coordination of the Group 13 atom. The thf adduct of
(vinylgallium)tetracarbonyliron, $[(thf)(C_2H_3)GaFe(CO)_4]_2$, which occupies
a place between related zinc- and germanium-iron dimers with which it is

isoelectronic, has a structure based on a planar Ga_2Fe_2 ring in which each gallium atom has a pseudotetrahedral environment composed of the two iron atoms, a carbon atom of the vinyl group and the oxygen atom of the thf ligand.[645] $BrIn[Co(CO)_4]_2 \cdot thf$ occurs as yellow crystals which lose the solvent *in vacuo*, apparently forming the dimer $[(CO)_4Co]_2In(\mu\text{-}Br)_2In\text{-}[Co(CO)_4]_2$. There are evident parallels between the behaviour of these complexes and that of the parent InX_3 molecules which also form derivatives of the types In_2X_6, $[InX_4]^-$ and $InX_3 \cdot L$. The product of the reaction of InBr with $Co_2(CO)_8$ in benzene, i.e. in the absence of a solvent able to coordinate to the unsaturated indium centre, has the formula $Br_3In_3Co_4(CO)_{15}$. The structure, shown in Figure 3.43, comprises a six-membered ring of alternate In and Br atoms, with each indium bearing a $Co(CO)_4$ ligand and also bonded to the capping $Co(CO)_3$ group.[646]

The compounds $Ga[Mn(CO)_5]_3$ and $In[Mn(CO)_5]_3$ have been prepared by the direct reaction of the metal with $Mn_2(CO)_{10}$ heated in a sealed tube.[647] This reaction also yielded red crystals which proved to be clusters of the type $M_2Mn_4(CO)_{18}$ (M = Ga or In); X-ray analysis of the corresponding digallium tetrarhenium compound reveals a planar bridged ring of two Ga and two Re atoms, as in Figure 3.44, with a pair of $Re(CO)_4$ groups forming an Re–Re bond, which measures 3.22 Å. The In–Mn bond lengths in the diindium tetramanganese compound are 2.61 (bridging) and 2.59 Å (terminal).[648] The cluster remains unsaturated and is able to accept an additional pyridine or acetone molecule at each of the tri-coordinated M atoms.

Members of the series of compounds $X_nIn[Co(CO)_4]_{3-n}$ and $X_nIn[Mn(CO)_5]_{3-n}$ (n = 0–3) are moderately air-stable solids. The In–M bonds (M = Co or Mn) are readily cleaved by dihalogens, X_2, or hydrogen halides, HX, in reactions which replace the transition-metal ligand by X. Derivatives containing both X and $M(CO)_x$ ligands undergo redistribution reactions in methanol solution to form InX_3 and $In[M(CO)_x]_3$ species. Other

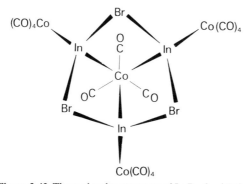

Figure 3.43 The molecular structure of $In_3Br_3Co_4(CO)_{15}$.

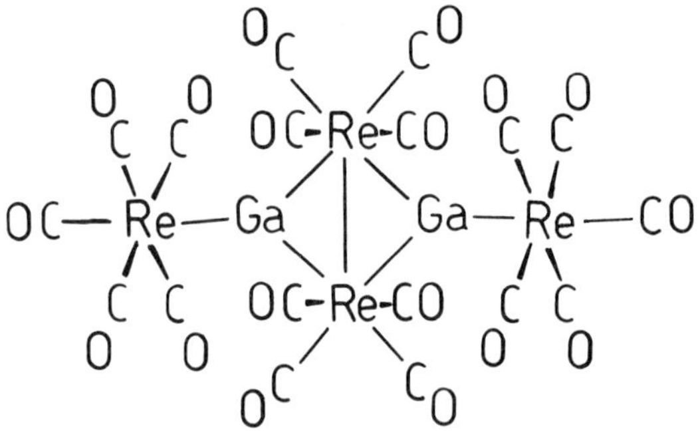

Figure 3.44 The molecular structure of $Ga_2Re_4(CO)_{18}$.

reactions include the addition of a halide ion to form complex anions of the sort mentioned earlier, or the replacement of halide by a donor solvent molecule to give a cationic species, e.g. $\{(MeCN)_2In[Mn(CO)_5]_2\}^{+}$.[461] Although most is known about the indium–cobalt and indium–manganese systems, other indium–transition-metal derivatives can be prepared by the treatment of a dinuclear transition-metal compound with an indium mono-halide in refluxing dioxan or tetrahydrofuran solvent, with the insertion of InX into the metal–metal bond.[649] The insertion of InX into a transition-element–halogen bond and the interaction of a Group 13 halide with a mercurial of the type $Hg(M'L'_x)_2$ (where $M'L'_x$ represents a transition-metal group including cyclopentadienyl, CO and other substituents) also lead to indium–transition-metal derivatives.[649,650] The list of transition metals which are known to form bonds to the heavier members of Group 13, particularly indium, now includes most members of Groups 6, 7 and 8. A novel development finds metal–metal-bonded species of this type incorporated into porphyrin complexes; there are reports of a range of Al–, Ga–, In– and Tl–transition-metal carbonyl moieties in such products.[651,652] There are other examples in which the gallium or indium entity is an organo-metallic fragment R_2M–[653] (see also chapter 6, section 6.7).

Access can be gained to thallium derivatives of the type $Tl(M'L'_x)_3$ containing Tl–Cr, Tl–Mo, Tl–W, Tl–Mn, Tl–Fe, Tl–Ru or Tl–Co bonds.[654,655] Both the trivalent and univalent states of thallium are involved in their chemistry; for example, $Tl[Co(CO)_4]_3$ dissociates:

$$Tl[Co(CO)_4]_3 \rightleftharpoons Tl[Co(CO)_4] + Co_2(CO)_8 \qquad (3.60)$$

The Tl^I derivative, $Tl[Co(CO)_4]$, can be prepared by treating thallium metal with $Co_2(CO)_8$ in benzene.[656] It is moderately air-stable, and can be used as a reagent in place of $Na[Co(CO)_4]$. Although essentially ionic in the solid state, like $Tl[Fe(CO)_3(NO)]$,[657] it forms the labile complex anion $\{Tl[Co(CO)_4]_2\}^{-}$ in solution. $Tl[Mn(CO)_5]_3$, a dark red crystalline solid, is prepared by the reaction of $TlCl_3$ with $Na[Mn(CO)_5]$ in dry thf at room temperature, or from $TlMe_3$ and $HMn(CO)_5$.[654,655] That this compound does not form adducts with pyridine or thf shows it to be a weaker Lewis acid than its indium counterpart. The Tl–Mn bonds are easily cleaved by HX, X_2 or alkyl halides, RX. The mixed-ligand derivatives, $XTl[Mn(CO)_5]_2$ and $X_2Tl[Mn(CO)_5]$, cannot be isolated, apparently because they dissociate to give TlX, $Tl[Mn(CO)_5]$ and $XMn(CO)_5$.

The solid $TlAu[Ph_2P(CH_2)S]_2$ is significant as a one-dimensional polymer containing Tl^I–Au^I bonds (with an intramolecular link of 2.959 and an inter-molecular link of 3.003 Å),[658] the shortness of which can be attributed to relativistic effects. The unusual structure apparently confers luminescent properties on this material and sets it alongside $Tl_2Pt(CN)_4$ and a complex with an Ir–Tl^I core in a class of heavy metal compounds whose detailed study is of considerable interest.

A valuable review[636] gives X-ray structural details for 18 examples of transition-metal carbonyl derivatives of gallium, indium or thallium. Among these, the pair $M[Mo(CO)_3(C_5H_5)]_3$ (M = In or Tl)[659] are notable in that the indium compound has a planar $InMo_3$ group, whereas in one polymorphic form of the thallium compound the $TlMo_3$ core is distinctly pyramidal. Here the average Tl–Mo bond length is 2.97 Å (0.10 Å less than the sum of the single-bond covalent radii), and the Tl atom lies 0.59 Å out of the plane of the three attached Mo atoms.[660] Other examples[661] include several in which Ga, In or Tl atoms are in bridging positions between transition-metal centres. Further developments have brought to light indium- and thallium-iron cluster complexes, e.g. $\{(2,2'\text{-bipyridyl})In[Fe(CO)_4]_2\}^-$, $[M_2Fe_6(CO)_{24}]^{2-}$ (M = In or Tl), $[Tl_4Fe_8(CO)_{30}]^{4-}$ and $[Tl_6Fe_{10}(CO)_{36}]^{6-}$.[662-664] The bonding in these remarkable structures has been investigated by molecular orbital calculations to evaluate factors such as electron deficiency, multicentre bonding and possible direct Tl$\cdots$Tl inter-action.[636,658,665] Yet another cluster, namely $\{In[Mn(CO)_4]_5\}^{2-}$, finds indium attached to five manganese atoms and displaying idealized penta-gonal planar coordination.[666] The In–Mn bond distances of 2.62–2.66 Å are within the expected range and the unusual bonding situation is forced on the indium atom by the steric requirements of its place at the centre of a rigid five-membered ring. Hence this is a rather special case, but even so serves to emphasise that flexibility in both coordination number and geometry is a feature of the structural chemistry of Group 13, and of indium especially (see chapter 1). It will be recalled from elsewhere in this chapter that indium provides examples of both the trigonal bypyramid and the square pyramid, which are the two more common polyhedra associated with the coordination number of five.

Some novel indium-containing clusters of a different kind have recently been synthesised; transition-metal carbonyl indium halides $(L'_xM')_nInX_{3-n}$ $(L'_xM' = Cp(CO)_2Fe$ or $Cp(CO)_3Mo)$ react with silylated chalcogens

$E(SiMe_3)_2$ (E = S, Se or Te) in tetrahydrofuran solution to form compounds $(L'_xM')_4In_4E_4$ which contain a central In_4E_4 cube.[667] Examples with each of the elements S, Se and Te in the cube have been studied by X-ray crystallography, and the following bond lengths found: In–S, 2.56; In–Se, 2.68; In–Te, 2.87; In–Fe, 2.49; In–Mo, 2.77 Å. Indium is four-coordinate with three bonds to the E atoms at the vertices of the cube and the extra position occupied by the transition-metal carbonyl group acting as a pseudohalide ligand.

The growing importance of materials that are complex in structure and diverse in the nature of their ligands and bonding is a dominant theme of the inorganic chemistry of Group 13. These final examples serve as a good illustration of this point.

References

Introduction

1. A. Haaland, *Angew. Chem., Int. Ed. Engl.*, 1989, **28**, 992.
2. K. Wade and A.J. Banister, in *Comprehensive Inorganic Chemistry*, ed. J.C. Bailar, Jr., H.J. Eméleus, R. Nyholm and A.F. Trotman-Dickenson, Vol. 1, Pergamon, Oxford, 1973, p. 993.
3. N.N. Greenwood, *Adv. Inorg. Chem. Radiochem.*, 1963, **5**, 91.
4. I.A. Sheka, I.S. Chaus and T.T. Mityureva, *The Chemistry of Gallium*, Elsevier, Amsterdam, 1966.
5. A.G. Lee, *The Chemistry of Thallium*, Elsevier, London, 1971.
6. N.N. Greenwood and A. Earnshaw, *Chemistry of the Elements*, Pergamon Press, Oxford, 1984.
7. M.J. Taylor and D.G. Tuck, in *Comprehensive Coordination Chemistry*, eds G. Wilkinson, R.D. Gillard and J.A. McCleverty, Pergamon, Oxford, 1987, Vol 3, pp. 105–182.
8. A.F. Wells, *Structural Inorganic Chemistry*, 5th edn., Clarendon Press, Oxford, 1984.

Hydrides

9. A.E. Finholt, A.C. Bond and H.I. Schlesinger, *J. Am. Chem. Soc.*, 1947, **69**, 1199.
10. K.N. Semenenko, Kh.A. Taisumov, A.P. Savchenkova and V.N. Surov, *Russ. J. Inorg. Chem.*, 1971, **16**, 1104.
11. D.L. Schmidt, C.B. Roberts and P.F. Reigler, *Inorg. Synth.*, 1973, **14**, 47.
12. H.C. Brown and N.M. Yoon, *J. Am. Chem. Soc.*, 1966, **88**, 1464.
13. F.M. Brower, N.E. Matzek, P.F. Reigler, H.W. Rinn, C.B. Roberts, D.L. Schmidt, J.A. Snover and K. Terada, *J. Am. Chem. Soc.*, 1976, **98**, 2450.
14. J.W. Turley and H.W. Rinn, *Inorg. Chem.*, 1969, **8**, 18.
15. A. Almenningen, G. Gundersen and A. Haaland, *Acta Chem. Scand.*, 1968, **22**, 328.
16. S. Cucinella, A. Mazzei and W. Marconi, *Inorg. Chim. Acta Rev.*, 1970, **4**, 51.
17. E.M. Marlett and W.S. Park, *J. Org. Chem.*, 1990, **55**, 2968.
18. S. Heřmánek, J. Fusek, O. Kříž, B. Čásenský and Z. Černý, *Z. Naturforsch.*, 1987, **42b**, 539.
·19. B. Čásenský and J. Macháček, *Inorg. Synth.*, 1978, **18**, 149.
20. A.B. Goel, E.C. Ashby and R.C. Mehrotra, *Inorg. Chim. Acta*, 1982, **62**, 161.
21. F. G.N. Cloke, C.I. Dalby, M.J. Henderson, P.B. Hitchcock, C.H.L. Kennard, R.N. Lamb and C.L. Raston, *J. Chem. Soc., Chem. Commun.*, 1990, 1394.

22. A.T.S. Wee, A.J. Murrell, N.K. Singh, D. O'Hare and J.S. Foord, *J. Chem. Soc., Chem. Commun.*, 1990, 11; F.M. Elms, R.N. Lamb, P.J. Pigram, M.G. Gardiner, B.J. Wood and C.L. Raston, *J. Chem. Soc., Chem. Commun.*, 1992, 1423.
23. W.L. Gladfelter, D.C. Boyd and K.F. Jensen, *Chem. Mater.*, 1989, **1**, 339 and refs. cited therein.
24. J.L. Atwood, F.R. Bennett, F.M. Elms, C. Jones, C.L. Raston and K.D. Robinson, *J. Am. Chem. Soc.*, 1991, **113**, 8183.
25. V.S. Mastryukov, A.V. Golubinskii and L.V. Vilkov, *J. Struct. Chem. (Engl. transl.)*, 1979, **20**, 788.
26. M. Hara, K. Domen, T. Onishi and H. Nozoye, *J. Phys. Chem.*, 1991, **95**, 6.
27. J.L. Atwood, F.R. Bennett, C. Jones, G.A. Koutsantonis, C.L. Raston and K.D. Robinson, *J. Chem. Soc., Chem. Commun.*, 1992, 541.
28. J.L. Atwood, K.D. Robinson, C. Jones and C.L. Raston, *J. Chem. Soc., Chem. Commun.*, 1991, 1697.
29. A.R. Barron and G. Wilkinson, *Polyhedron*, 1986, **5**, 1897; *J. Chem. Soc., Dalton Trans*, 1986, 287.
30. B.M. Bulychev, *Polyhedron*, 1990, **9**, 387.
31. M.L. McKee, *J. Phys. Chem.*, 1991, **95**, 6519.
32. P. Breisacher and B. Siegel, *J. Am. Chem. Soc.*, 1965, **87**, 4255, 5053.
33. J.C. Brand and B.P. Roberts, *J. Chem. Soc., Chem. Commun.*, 1984, 109.
34. R.L. Hudson and B.P. Roberts, *J. Chem. Soc., Chem. Commun.*, 1986, 1194.
35. A.J. Downs, M.J. Goode and C.R. Pulham, *J. Am. Chem. Soc.*, 1989, **111**, 1936.
36. C.R. Pulham, A.J. Downs, M.J. Goode, D.W.H. Rankin and H.E. Robertson, *J. Am. Chem. Soc.*, 1991, **113**, 5149.
37. A.T.S. Wee, A.J. Murrell, N.K. Singh, D. O'Hare and J.S. Foord, *J. Chem. Soc., Chem. Commun.*, 1990, 11.
38. D. O'Hare, J.S. Foord, T.C.M. Page and T.J. Whitaker, *J. Chem. Soc., Chem. Commun.*, 1991, 1445.
39. K.D. Dobbs, M. Trachtman, C.W. Bock and A.H. Cowley, *J. Phys. Chem.*, 1990, **94**, 5210.
40. C. Liang, R.D. Davy and H.F. Schaefer III, *Chem. Phys. Lett.*, 1989, **159**, 393.
41. K. Lammertsma and J. Leszczyński, *J. Phys. Chem.*, 1990, **94**, 2806, 5543.
42. B.J. Duke, *J. Mol. Struct. (THEOCHEM)*, 1990, **208**, 197.
43. G. Trinquier and J.-P. Malrieu, *J. Am. Chem. Soc.*, 1991, **113**, 8634.
44. M.J. van der Woerd, K. Lammertsma, B.J. Duke and H.F. Schaefer III, *J. Chem. Phys.*, 1991, **95**, 1160.
45. V. Barone and C. Minichino, *Theor. Chim. Acta*, 1989, **76**, 53.
46. K. Lammertsma and J. Leszczyński, *J. Chem. Soc., Chem. Commun.*, 1989, 1005.
47. B.J. Duke, T.P. Hamilton and H.F. Schaefer III, *Inorg. Chem.*, 1991, **30**, 4225.
48. R. Köppe, M. Tacke and H. Schnöckel, *Z. Anorg. Allg. Chem.*, 1991, **605**, 35.
49. R. Köppe and H. Schnöckel, *J. Chem. Soc., Dalton Trans.*, 1992, 3393.
50. B.J. Duke, C. Liang and H.F. Schaefer III, *J. Am. Chem. Soc.*, 1991, **113**, 2884.
51. C.W. Bock, C. Roberts, K. O'Malley, M. Trachtman and G.J. Mains, *J. Phys. Chem.*, 1992, **96**, 4859.
52. E. Wiberg and E. Amberger, *Hydrides of the Elements of Main Groups I–IV*, Elsevier, Amsterdam, 1971.
53. A.G. Avent, C. Eaborn, P.B. Hitchcock, J.D. Smith and A.C. Sullivan, *J. Chem. Soc., Chem. Commun.*, 1986, 988.
54. J.L. Atwood, D.C. Hrncir, R.D. Rogers and J.A.K. Howard, *J. Am. Chem. Soc.*, 1981, **103**, 6787.

Borohydrides

55. C.E. Housecroft and T.P. Fehlner, *Adv. Organomet. Chem.*, 1982, **21**, 57.
56. D.A. Coe and J.W. Nibler, *Spectrochim. Acta*, 1973, **29A**, 1789.
57. A.J. Downs, R.G. Egdell, A.F. Orchard and P.D.P. Thomas, *J. Chem. Soc., Dalton Trans.*, 1978, 1755.

58. P.C. Maybury and J.E. Ahnell, *Inorg. Chem.*, 1967, **6**, 1286.
59. J.W. Akitt, *Prog. NMR Spectrosc.*, 1989, **21**, 107.
60. H. Nöth and R. Rurländer, *Inorg. Chem.*, 1981, **20**, 1062.
61. P.R. Oddy and M.G.H. Wallbridge, *J. Chem. Soc., Dalton Trans.*, 1978, 572.
62. L.A. Jones and A.J. Downs, unpublished results.
63. M.T. Barlow, C.J. Dain, A.J. Downs, G.S. Laurenson and D.W.H. Rankin, *J. Chem. Soc., Dalton Trans.*, 1982, 597.
64. C.R. Pulham, P.T. Brain, A.J. Downs, D.W.H. Rankin and H.E. Robertson, *J. Chem. Soc., Chem. Commun.*, 1990, 177.
65. F.L. Himpsl, Jr., and A.C. Bond, *J. Am. Chem. Soc.*, 1981, **103**, 1098.
66. J. Borlin and D.F. Gaines, *J. Am. Chem. Soc.*, 1972, **94**, 1367.
67. N.N. Greenwood and J.A. McGinnety, *J. Chem. Soc. A*, 1966, 1090.
68. C.J. Dain, A.J. Downs and D.W.H. Rankin, *J. Chem. Soc., Dalton Trans.*, 1981, 2465.
69. D.R. Schultz and R.W. Parry, *J. Am. Chem. Soc.*, 1958, **80**, 4.
70. D.A. Atwood, J.A. Jones, A.H. Cowley, S.G. Bott and J.L. Atwood, *J. Organomet. Chem.*, 1992, **425**, C1.
71. C.R. Pulham, A.J. Downs, D.W.H. Rankin and H.E. Robertson, *J. Chem. Soc., Chem. Commun.*, 1990, 1520; *J. Chem. Soc., Dalton Trans.*, 1992, 1509.
72. C.R. Pulham and A.J. Downs, unpublished results.
73. N.N. Greenwood, B.S. Thomas and D.W. Waite, *J. Chem. Soc., Dalton Trans.*, 1975, 299.
74. N.N. Greenwood and J.A. Howard, *J. Chem. Soc., Dalton Trans.*, 1976, 177.

Fluorides

75. S.A. Polyschchuk, S.P. Kozerenko and Y.U. Gargarinsky, *J. Less-Common Met.*, 1972, **27**, 45.
76. N.A. Matwiyoff and W.E. Wageman, *Inorg. Chem.*, 1970, **9**, 1031.
77. J.W. Akitt, *Prog. NMR Spectrosc.*, 1989, **21**, 1.
78. R. Colton and P.G. Eller, *Aust. J. Chem.*, 1989, **42**, 1605.
79. D. Mootz, E.-J. Oellers and M. Wiebcke, *Acta Crystallogr.*, 1988, **C44**, 1334.
80. O. Knop, T.S. Cameron, S.P. Deraniyagala, D. Adhikesavalu and M. Falf, *Can. J. Chem.*, 1985, **63**, 516.
81. B. Gilbert and T. Materne, *Appl. Spectrosc.*, 1990, **44**, 299.
82. O. Bjørseth, O. Herstad and J.L. Holm, *Acta Chem. Scand.*, 1986, **A40**, 566.
83. A.N. Utkin, G.V. Girichev, N.I. Giricheva and S.V. Khaustov, *Zh. Strukt. Khim.*, 1986, **27**, 43.
84. S.C. Choi, R.J. Boyd and O. Knop, *Can. J. Chem.*, 1988, **66**, 2465.

Halides

85. D.M. Gruen and R.L. McBeth, *Inorg. Chem.*, 1969, **8**, 2625.
86. H. Schäfer, *Adv. Inorg. Chem. Radiochem.*, 1983, **26**, 201.
87. P.A. Mosier-Boss, R.D. Boss, C.J. Gabriel, S. Szpak, J.J. Smith and R.J. Nowak, *J. Chem. Soc., Faraday Trans. 1*, 1989, **85**, 11.
88. O.H. Han and E. Oldfield, *Inorg. Chem.*, 1990, **29**, 3666.
89. H. Nöth, R. Rurländer and P. Wolfgardt, *Z. Naturforsch.*, 1982, **37b**, 29 and refs. cited therein.
90. N.C. Means, C.M. Means, S.G. Bott and J.L. Atwood, *Inorg. Chem.*, 1987, **26**, 1466.
91. A. Bittner, D. Männig and H. Nöth, *Z. Naturforsch.*, 1986, **41b**, 587.
92. P. Pullmann, K. Hensen and J.W. Bats, *Z. Naturforsch.*, 1982, **37b**, 1312.
93. J.W. Akitt and J. Lelievre, *J. Chem. Soc., Dalton Trans.*, 1985, 591.
94. R. Kniep, P. Blees and W. Poll, *Angew. Chem., Int. Ed. Engl.*, 1982, **21**, 386.
95. J.J. Habeeb and D.G. Tuck, *Inorg. Synth.*, 1979, **19**, 257.
96. F.J.M. Gil, M.A. Salgado and J.M. Gil, *Synth. React. Inorg. Metal-Org. Chem.*, 1986, **16**, 663.

97. J.P. Kopasz, R.B. Hallock and O.T. Beachley, *Inorg. Synth.*, 1986, **24**, 87.
98. R. Kniep and P. Blees, *Angew. Chem., Int. Ed. Engl.*, 1984, **23**, 800.
99. V.M. Petrov, N.I. Giricheva, G.V. Girichev, V.A. Titov and T.P. Chusova, *Zh. Strukt. Khim.*, 1990, **31**, 46; 1991, **32**, 56.
100. V.P. Spiridonov, A.G. Gershikov, E.Z. Zasorin, N.I. Popenko, A.A. Ivanov and L.I. Ermolayeva, *High Temp. Sci.*, 1981, **14**, 285.
101. V.A. Maroni, D.M. Gruen, R.L. McBeth and E.J. Cairns, *Spectrochim. Acta*, 1970, **26A**, 418.
102. Yu.A. Buslaev, E.A. Kravččnko and L. Kolditz, *Coord. Chem. Rev.*, 1987, **82**, 53.
103. I.R. Beattie and J.R. Horder, *J. Chem. Soc. A*, 1969, 2655.
104. A. Haaland, A. Hammel, K.-G. Martinsen, J. Tremmel and H.V. Volden, *J. Chem. Soc., Dalton Trans.*, 1992, 2209.
105. H. Schnöckel, *Z. Anorg. Allg. Chem.*, 1976, **424**, 203; I.R. Beattie, H.E. Blayden, S.M. Hall, S.N. Jenny and J.S. Ogden, *J. Chem. Soc., Dalton Trans.*, 1976, 666.
106. M. Dalibart and J. Derouault, *Coord. Chem. Rev.*, 1986, **74**, 1.
107. P.L. Radloff and G.N. Papatheodorou, *J. Chem. Phys.*, 1980, **72**, 992.
108. J.-A.E. Bice, G.M. Bancroft and L.L. Coatsworth, *Inorg. Chem.*, 1986, **25**, 2181.
109. J. Weidlein, *J. Organomet. Chem.*, 1969, **17**, 213.
110. O.T. Beachley, Jr., R.B. Hallock, H.M. Zhang and J.L. Atwood, *Organometallics*, 1985, **4**, 1675.
111. D.A. Atwood, A.H. Cowley, R.A. Jones, M.A. Mardones, J.L. Atwood and S.G. Bott, *J. Coord. Chem.*, 1992, **25**, 233.
112. D.A. Atwood, A.H. Cowley and R.A. Jones, *J. Organomet. Chem.*, 1992, **430**, C29.
113. E.D. Hausen, K. Mertz, J. Weidlein and W. Schwartz, *J. Organomet. Chem.*, 1975, **93**, 291.
114. H.D. Hausen, E. Veigel and H.-J. Guder, *Z. Naturforsch.*, 1974, **29b**, 269.
115. R.W. Berg, E. Kemnitz, H.A. Hjuler, R. Fehrmann and N.J. Bjerrum, *Polyhedron*, 1985, **4**, 457.
116. J.H. von Barner, *Inorg. Chem.*, 1985, **24**, 1689.
117. D. Mascherpa-Corral and A. Potier, *J. Chim. Phys. Phys.-Chim. Biol.*, 1977, **74**, 1077.
118. W. Frank, W. Hönle and A. Simon. *Z. Naturforsch.*, 1990, **45b**, 1.
119. F. Stollmaier and U. Thewalt, *J. Organomet. Chem.*, 1981, **208**, 327.
120. M.J. Taylor, *J. Chem. Soc. A*, 1970, 2812.
121. P. Reich, D. Müller, M. Feist and G. Blumenthal, *Z. Naturforsch.*, 1990, **45b**, 344.
122. B. Gilbert, S.D. Williams and G. Mamantov, *Inorg. Chem.*, 1988, **27**, 2359.
123. R.W. Berg, H.A. Hjuler and N.J. Bjerrum, *Inorg. Chem.*, 1985, **24**, 4506.
124. R.W. Berg and T. Østvold, *Acta Chem. Scand.*, 1986, **A40**, 445.
125. M.-A. Einarsrud, E. Rytter and M. Ystenes, *Vibrational Spectroscopy*, 1990, **1**, 61.
126. D. Jentsch, P.G. Jones, E. Schwarzmann and G.M. Sheldrick, *Acta Crystallogr.*, 1983, **C39**, 1173.
127. E.E. Getty and R.S. Drago, *Inorg. Chem.*, 1990, **29**, 1186 and refs. cited therein.
128. U. Kliebisch, U. Klingebiel, D. Stalke and G.M. Sheldrick, *Angew. Chem., Int. Ed. Engl.*, 1986, **25**, 915.
129. T.A. Zawodzinski, Jr., and R. Osteryoung, *Inorg. Chem.*, 1989, **28**, 1710.
130. K. Ichikawa, T. Jin and T. Matsumoto, *J. Chem. Soc., Faraday Trans. 1*, 1989, **85**, 175.
131. J.R. Sanders, E.H. Ward and C.L. Hussey, *J. Electrochem. Soc.*, 1986, **133**, 325.
132. C.J. Dymek, Jr., J.S. Wilkes, M.-A. Einarsrud and H.A. Øye, *Polyhedron*, 1988, **7**, 1139.
133. J.M. Martell and M.J. Zaworotko, *J. Chem. Soc., Dalton Trans.*, 1991, 1495.
134. G. Seemann and K. Hensen, *Z. Naturforsch.*, 1986, **41b**, 665.
135. H.N. Borah, R.C. Boruah and J.S. Sandhu, *J. Chem. Soc., Chem. Commun.*, 1991, 154.
136. W.T. Robinson, C.J. Wilkins and Z. Zeying, *J. Chem. Soc., Dalton Trans.*, 1990, 219.
137. M.J. Taylor, P.N. Gates and P.M. Smith, *Spectrochim. Acta*, 1992, **48A**, 205.
138. A. Apblett, T. Chivers and J.F. Fait, *Inorg. Chem.*, 1990, **29**, 1643.
139. S. Pohl, *Z. Anorg. Allg. Chem.*, 1983, **498**, 15, 20.
140. See, for example, J.D. Corbett, *Prog. Inorg. Chem.*, 1976, **21**, 129.
141. R. Uson and A. Laguna, *Inorg. Synth.*, 1982, **21**, 72.
142. G. Thiele, H.W. Rotter and K. Zimmermann, *Z. Naturforsch.*, 1986, **41b**, 269.
143. M.J. Taylor, *Polyhedron*, 1990, **9**, 207.

144. D.G. Tuck, *Pure Appl. Chem.*, 1983, **55**, 1477.
145. I. Bányai and J. Glaser, *J. Am. Chem. Soc.*, 1989, **111**, 3186; 1990, **112**, 4703.
146. T. Staffel and G. Meyer, *Z. Anorg. Allg. Chem.*, 1990, **585**, 38.
147. M.B. Millikan and B.D. James, *Inorg. Chim. Acta*, 1984, **81**, 109.
148. D.F. Shriver and I. Wharf, *Inorg. Chem.*, 1969, **9**, 2167.
149. G.R. Clark, C.E.F. Rickard and M.J. Taylor, *Can. J. Chem.*, 1986, **64**, 1697.
150. X. Solans, M.C. Moron and F. Palacio, *Acta Crystallogr.*, 1988, **C44**, 965.
151. R. Burnus and G. Meyer, *Z. Anorg. Allg. Chem.*, 1991, **602**, 31.
152. R.A. Walton, *Coord. Chem. Rev.*, 1971, **6**, 1.
153. J. Glaser, P.L. Goggin, M. Sandström and V. Lutsko, *Acta Chem. Scand.*, 1982, **A36**, 55.
154. J. Glaser and U. Henriksson, *J. Am. Chem. Soc.*, 1981, **103**, 6642.
155. G. Thiele, H.W. Rotter and M. Faller, *Z. Anorg. Allg. Chem.*, 1984, **508**, 129.

Oxides and hydroxides

156. N. Ishizawa, T. Miyata, I. Minato, F. Marumo and S. Iwai, *Acta Crystallogr.*, 1980, **36B**, 228.
157. B.A. Huggins and P.D. Ellis, *J. Am. Chem. Soc.*, 1992, **114**, 2098.
158. C. Morterra, C. Emanuel, G. Cerrato and G. Magnacca, *J. Chem. Soc., Faraday Trans.*, 1992, **88**, 339.
159. J.W. Diggle, T.C. Downie and C.W. Goulding, *Chem. Rev.*, 1969, **69**, 365.
160. J.H. Kennedy, *Top. Appl. Phys.*, 1977, **21**, 105.
161. J.T. Kummer, *Prog. Solid State Chem.*, 1972, **7**, 141.
162. M.G. Barker, P.G. Gadd and S.C. Wallwork, *J. Chem. Soc., Chem. Commun.*, 1982, 516.
163. M.G. Barker, P.G. Gadd and M.J. Begley, *J. Chem. Soc., Dalton Trans.*, 1984, 1139.
164. L.B. Alemany and G.W. Kirker, *J. Am. Chem. Soc.*, 1986, **108**, 6158.
165. R. Roy, V.G. Hill and E.F. Osborn, *J. Am. Chem. Soc.*, 1952, **74**, 719.
166. S. Geller, *J. Chem. Phys.*, 1960, **33**, 676.
167. C.T. Prewitt, R.D. Shannon, D.B. Rogers and A.W. Sleight, *Inorg. Chem.*, 1969, **8**, 1985.
168. A.N. Christensen, N.C. Brock, O. von Heidenstam and A. Nilsson, *Acta Chem. Scand.*, 1967, **21**, 1046.
169. R. Roy and M.W. Shafer, *J. Phys. Chem.*, 1954, **58**, 372.
170. See, for example, D. Bourgault, C. Martin, C. Michel, M. Hervieu, J. Provost and B. Raveau, *J. Solid State Chem.*, 1989, **78**, 326 and references cited therein. A.J. Freeman, S. Massida and J. Yu, in *Chemistry of High-Temperature Superconductors II*, ACS Symposium Series 377, American Chemical Society, Washington, DC, 1988, p. 64. I.A. Kahwa, D. Miller, M. Mitchel, F.R. Fronczek, R.G. Goodrich, D.J. Williams, C.A. O'Mahoney, A.M.Z. Slawin, S.V. Ley and C.J. Groombridge, *Inorg. Chem.*, 1992, **31**, 3963 and refs. cited therein.

Hydrated cations

171. Y. Marcus, *Chem. Rev.*, 1988, **88**, 1475 and refs. cited therein.
172. H. Kanno, *J. Phys. Chem.*, 1988, **92**, 4232.
173. S.P. Best, J.K. Beattie and R.S. Armstrong, *J. Chem. Soc., Dalton Trans.*, 1984, 2611.
174. D.M. Adams and D.J. Hills, *J. Chem. Soc., Dalton Trans.*, 1978, 782.
175. N. Galešić and V.B. Jordanovska, *Acta Crystallogr.*, 1992, **C48**, 256.
176. R.M. Smith and A.E. Martell, *Critical Stability Constants*, Plenum Press, New York, 1974–1989, Vols. 1–6.
177. J.W. Akitt and J.M. Elders, *J. Chem. Soc., Faraday Trans. 1*, 1985, **81**, 1923.
178. J.J. Fitzgerald, L.E. Johnson and J.S. Frye, *J. Magn. Reson.*, 1989, **84**, 121; P.L. Brown,

R.N. Sylva, G.E. Batley and J. Ellis, *J. Chem. Soc., Dalton Trans.*, 1985, 1967; T. Hedlund, S. Sjöberg and L.-O. Öhman, *Acta Chem. Scand.*, 1987, **A41**, 197; M. Venturini and G. Berthon, *J. Chem. Soc., Dalton Trans.*, 1987, 1145.

179. B. Corain, A. Tapparo, A.A. Sheikh-Osman, G. Giorgio, P. Zatta and M. Favarato, *Coord. Chem. Rev.*, 1992, **112**, 19.
180. G. Johansson, *Acta Chem. Scand.*, 1960, **14**, 771; 1962, **16**, 403.
181. A.R. Thompson, A.C. Kunwar, H.S. Gutowsky and E. Oldfield, *J. Chem. Soc., Dalton Trans.*, 1987, 2317.
182. J.W. Akitt, J.M. Elders, X.L.R. Fontaine and A.K. Kundu, *J. Chem. Soc., Dalton Trans.*, 1989, 1889.
183. M. Henry, J.P. Jolivet and J. Livage, *Aqueous Chemistry of Metal Cations: Hydrolysis, Condensation and Complexation*, in *Struct. Bonding (Berlin)*, 1992, **77**, 153.
184. Y. Couturier, *Bull. Soc. Chim. France*, 1986, 171.
185. N. Parthasarathy, J. Buffle and W. Haerdi, *Can. J. Chem.*, 1986, **64**, 24.
186. S. Cheng and T.-C. Wang, *Inorg. Chem.*, 1989, **28**, 1283.
187. S.D. Kinrade and T.W. Swaddle, *Inorg. Chem.*, 1989, **28**, 1952.
188. L.-O. Öhman, *Inorg. Chem.*, 1989, **28**, 3629.
189. F. González, C. Pesquera, C. Blanco, I. Benito and S. Mendioroz, *Inorg. Chem.*, 1992, **31**, 727.
190. P.L. Brown, *J. Chem. Soc., Dalton Trans.*, 1989, 399.
191. S.M. Bradley, R.A. Kydd and R. Yamdagni, *J. Chem. Soc., Dalton Trans.*, 1990, 413, 2653.
192. S.M. Bradley, R.A. Kydd and R. Yamdagni, *Magn. Reson. Chem.*, 1990, **28**, 746.
193. S.M. Bradley, R.A. Kydd and C.A. Fyfe, *Inorg. Chem.*, 1992, **31**, 1181.
194. L.P. Tsiganok, A.B. Vishnikin and R.I. Maksimovskaya, *Polyhedron*, 1989, **8**, 2739.
195. J.W. Akitt and W. Gessner, *J. Chem. Soc., Dalton Trans.*, 1984, 147.
196. P.L. Brown, J. Ellis and R.N. Sylva, *J. Chem. Soc., Dalton Trans.*, 1982, 1911.
197. R. Caminiti, G. Johansson and I. Toth, *Acta Chem. Scand.*, 1986, **A40**, 435.
198. K. Wieghardt, M. Kleine-Boymann, B. Nuber and J. Weiss, *Inorg. Chem.*, 1986, **25**, 1654.
199. D.G. Tuck, *Pure Appl. Chem.*, 1983, **55**, 1477.
200. L.C.A. Thompson and R. Pacer, *J. Inorg. Nucl. Chem.*, 1963, **25**, 1041.
201. J. Glaser and G. Johansson, *Acta Chem. Scand.*, 1982, **A36**, 125.
202. J. Glaser and G. Johansson, *Acta Chem. Scand.*, 1981, **A35**, 639.
203. J. Glaser, *Acta Chem. Scand.*, 1979, **A33**, 789.
204. Q. Feng and H. Waki, *Polyhedron*, 1991, **10**, 659.
205. K. Ichikawa and T. Jin, *Chem. Lett.*, 1987, 1179.
206. G.V. Kozhevnikova and G. Keresztury, *Inorg. Chim. Acta*, 1985, **98**, 59.
207. K. Saito and A. Nagasawa, *Polyhedron*, 1990, **9**, 215 and refs. cited therein.
208. D. Hugi-Cleary, L. Helm and A.E. Merbach, *Helv. Chim. Acta*, 1985, **68**, 545; *J. Am. Chem. Soc.*, 1987, **109**, 4444.

Oxide halides

209. B. Siegel, *Inorg. Chim. Acta Rev.*, 1968, **2**, 137.
210. J.H. Holloway and D. Laycock, *Adv. Inorg. Chem. Radiochem.*, 1983, **27**, 157.
211. R. Ahlrichs, L. Zhengyan and H. Schnöckel, *Z. Anorg. Allg. Chem.*, 1984, **519**, 155.
212. H. Schnöckel and H.J. Goecke, *J. Mol. Struct.*, 1978, **50**, 281.
213. P.L. Goggin, I.J. McColm and R. Shore, *J. Chem. Soc. A*, 1966, 1004.
214. B.L. Chamberland, *Inorg. Synth.*, 1973, **14**, 123.
215. M. Vlasse, J.-C. Massies and B.L. Chamberland, *Acta Crystallogr.*, 1973, **B29**, 627.
216. H.E. Forsberg, *Acta Chem. Scand.*, 1957, **11**, 676.
217. M. Vlasse, J. Grannec and A. Portier, *Acta Crystallogr.*, 1972, **B28**, 3426.
218. P. Hagenmuller, J. Rouxel, J. David, A. Colin and B. Le Neindre, *Z. Anorg. Allg. Chem.*, 1963, **323**, 1.
219. J.R. Günter, *Z. Anorg. Allg. Chem.*, 1978, **438**, 203.

Alkoxides

220. D.C. Bradley, *Adv. Inorg. Chem. Radiochem.*, 1972, **15**, 259.
221. D.C. Bradley, R.C. Mehrotra and D.P. Gaur, *Metal Alkoxides*, Academic Press, London, 1978.
222. R.C. Mehrotra and A.K. Rai, *Polyhedron*, 1991, **10**, 1967.
223. V.A. Shreider, E.P. Turevskaya, N.I. Koslova and N.Ya. Turova, *Inorg. Chim. Acta*, 1981, **53**, L73.
224. W.S. Rees, Jr., and D.A. Moreno, *J. Chem. Soc., Chem. Commun.*, 1991, 1759.
225. J.G. Oliver and I.J. Worrall, *J. Chem. Soc. A*, 1970, 845, 1389.
226. K. Folting, W.E. Streib, K.G. Caulton, O. Poncelet and L.G. Hubert-Pfalzgraf, *Polyhedron*, 1991, **10**, 1639.
227. M.K. Dongare and A.P.B. Sinha, *Thermochim. Acta*, 1982, **57**, 37.
228. R.C. Mehrotra, *J. Non-Cryst. Solids*, 1988, **100**, 1.
229. P. Monsef-Mirzai, P.M. Watts, W.R. McWhinnie and H.W. Gibbs, *Inorg. Chim. Acta*, 1991, **188**, 205.
230. R. Reisfeld and C.K. Jørgensen, eds., *Chemistry, Spectroscopy and Applications of Sol-Gel Glasses*, in *Struct. Bonding (Berlin)*, 1992, **77**, 1.
231. F.J. Feher, T.A. Budzichowski and K.J. Weller, *J. Am. Chem. Soc.*, 1989, **111**, 7288.
232. H. Schmidbaur and M. Schmidt, *Angew. Chem., Int. Ed. Engl.*, 1962, **1**, 328.
233. R.H. Cayton, M.H. Chisholm, E.R. Davidson, V.F. DiStasi, P. Du and J.C. Huffman, *Inorg. Chem.*, 1991, **30**, 1020.
234. M.H. Chisholm, J.C. Huffman and J.L. Wesemann, *Polyhedron*, 1991, **10**, 1367.
235. S. Pasynkiewicz, *Polyhedron*, 1990, **9**, 429.
236. R.D. Rogers and J.L. Atwood, *Organometallics*, 1984, **3**, 271.
237. J.P. Oliver and R. Kumar, *Polyhedron*, 1990, **9**, 409; S. Pasynkiewicz, *Polyhedron*, 1990, **9**, 429.
238. H. Dislich, *Angew. Chem., Int. Ed. Engl.*, 1971, **10**, 363.
239. O. Yamaguchi and M. Shirai, *Polyhedron*, 1990, **9**, 367.
240. H. Funk and A. Paul, *Z. Anorg. Allg. Chem.*, 1964, **330**, 70.
241. H. Funk and A. Paul, *Z. Anorg. Allg. Chem.*, 1965, **337**, 142, 145.
242. D.C. Bradley, H. Chudzynska, D.M. Frigo, M.E. Hammond, M.B. Hursthouse and M.A. Mazid, *Polyhedron*, 1990, **9**, 719.
243. A.R. Barron, K.D. Dobbs and M.M. Francl, *J. Am. Chem. Soc.*, 1991, **113**, 39.
244. M.A. Petrie, M.M. Olmstead and P.P. Power, *J. Am. Chem. Soc.*, 1991, **113**, 8708.
245. F. Bélanger-Gariépy, K. Hoogsteen, V. Sharma and J.D. Wuest, *Inorg. Chem.*, 1991, **30**, 4140.

Oxyanion salts

246. F.B. Erim, E. Avsar and B. Basaran, *J. Coord. Chem.*, 1990, **21**, 209.
247. D.J. Gardiner, R.E. Hester and E. Mayer, *J. Mol. Struct.*, 1974, **22**, 327.
248. A.J. Carty and D.G. Tuck, *Prog. Inorg. Chem.*, 1975, **19**, 245.
249. G.V. Kozhevnikova and G. Keresztury, *Inorg. Chim. Acta*, 1985, **98**, 59.
250. R. Faggiani and I.D. Brown, *Acta Crystallogr.*, 1978, **B34**, 1675.
251. C.C. Addison and N. Logan, *Adv. Inorg. Chem. Radiochem.*, 1964, **6**, 72.
252. C. Sabelli and R.T. Ferroni, *Acta Crystallogr.*, 1978, **B34**, 2407.
253. J. Tudo, B. Jolibois, G. Laplace, G. Nowogrocki and F. Abraham, *Acta Crystallogr.*, 1979, **B35**, 1580.
254. L.S. Foster, *J. Am. Chem. Soc.*, 1939, **61**, 3122.
255. L.S. Foster, *Inorg. Synth.*, 1946, **2**, 26.
256. M. Fourati, M. Chaabouni, H.F. Ayedi, J.-L. Pascal and J. Potier, *Can. J. Chem.*, 1985, **63**, 3499.

Borates, carbonates, silicates and phosphates

257. (a) A. Delmastro, G. Gozzelino, D. Mazza, M. Vallino, G. Busca and V. Lorenzelli, *J. Chem. Soc., Faraday Trans.*, 1992, **88**, 2065; (b) P.E. Blackburn, A. Büchler and J.L. Stauffer, *J. Phys. Chem.*, 1966, **70**, 2469.
258. W. Lortz and G. Schön, *J. Chem. Soc., Dalton Trans.*, 1987, 623.
259. A. Bellaloui, D. Plee and P. Meriaudeau, *Appl. Catal.*, 1990, **63**, L7; A. Vieira and G. Poncelet, in *Pillared Layered Structures, Current Trends and Applications*, ed. I.V. Mitchell, Elsevier Applied Science, London, 1990, p. 185.
260. F. González, C. Pesquera, I. Benito and S. Mendioroz, *J. Chem. Soc., Chem. Commun.*, 1991, 587; F. González, C. Pesquera, C. Blanco, I. Benito and S. Mendioroz, *Inorg. Chem.*, 1992, **31**, 727.
261. I.V. Mitchell, ed., *Pillared Layered Structures, Current Trends and Applications*, Elsevier Applied Science, London, 1990.
262. G. Engelhardt, J. Felsche and P. Sieger, *J. Am. Chem. Soc.*, 1992, **114**, 1173.
263. R. Kniep, *Angew. Chem., Int. Ed. Engl.*, 1986, **25**, 525.
264. R. Kniep and D. Mootz, *Acta Crystallogr.*, 1973, **B29**, 2292.
265. R. Kniep, D. Müller, L. Gunze, E. Hallas and G. Ladwig, *Z. Anorg. Allg. Chem.*, 1983, **500**, 80.
266. J.H. Morris, P.G. Perkins, A.E. Rose and W.E. Smith, *Chem. Soc. Rev.*, 1977, **6**, 173.
267. S.T. Wilson, B.M. Lok, C.A. Messina, T.R. Cannan and E.M. Flanigen, *J. Am. Chem. Soc.*, 1982, **104**, 1146.
268. J.J. Pluth, J.V. Smith and J.M. Bennett, *Acta Crystallogr.*, 1986, **C42**, 283.
269. Q. Huo and R. Xu, *J. Chem. Soc., Chem. Commun.*, 1992, 168.
270. S. Hirano and P. Kim, *Bull. Chem. Soc. Jpn.*, 1989, **62**, 275.
271. R.C.L. Mooney, *Acta Crystallogr.*, 1956, **9**, 113.
272. J.B. Parise, *Acta Crystallogr.*, 1986, **C42**, 144, 670.
273. R.H. Jones, J.M. Thomas, Q. Huo, R. Xu, M.B. Hursthouse and J. Chen, *J. Chem. Soc., Chem. Commun.*, 1991, 1520.

Carboxylates

274. R.C. Mehrotra and A.K. Rai, *Polyhedron*, 1991, **10**, 1967 and refs. cited therein.
275. G.C. Hood and A.J. Ihde, *J. Am. Chem. Soc.*, 1950, **72**, 2094.
276. C.L. Edwards and R.L. Hayes, *J. Nucl. Med.*, 1969, **10**, 103.
277. M.T. Andras, S.A. Duraj, A.F. Hepp, P.E. Fanwick and M.M. Bodnar, *J. Am. Chem. Soc.*, 1992, **114**, 786.
278. J.J. Habeeb and D.G. Tuck, *J. Chem. Soc., Dalton Trans.*, 1973, 243.
279. R. Faggiani and I.D. Brown, *Acta Crystallogr.*, 1978, **34B**, 2845; 1982, **38B**, 2473.
280. J.K. Kochi and T.W. Bethea, III, *J. Org. Chem.*, 1968, **33**, 75.
281. A. McKillop and E.C. Taylor, in *Comprehensive Organometallic Chemistry*, eds. G. Wilkinson, F.G.A. Stone and E.W. Abel, Pergamon, Oxford, 1982, Vol. 7, p. 465.
282. E.C. Taylor and A. McKillop, *Acc. Chem. Res.*, 1970, **3**, 338.
283. G.T. Morgan and H.D.K. Drew, *J. Chem. Soc.*, 1921, 1058.
284. J.R. Blackborow, C.R. Eady, E.A. Körner von Gustorf, A. Scrivanti and O. Wolfbeis, *J. Organomet. Chem.*, 1976, **108**, C32.
285. J.J. Habeeb and D.G. Tuck, *Inorg. Synth.*, 1979, **19**, 257.
286. J.E. Fortman and R.E. Sievers, *Inorg. Chem.*, 1967, **6**, 2022.
287. M. Das, D.T. Howarth and J.W. Beery, *Inorg. Chim. Acta*, 1981, **49**, 17.

Chalcogenides

288. G.A. Steigmann and J. Goodyear, *Acta Crystallogr.*, 1966, **20**, 617.
289. G.R. Watkins and R. Shutt, *Inorg. Synth.*, 1946, **2**, 184.
290. M. Schulte-Kellinghaus and V. Krämer, *Acta Crystallogr.*, 1979, **B35**, 3016.

291. B. Krebs and H. Greiwing, *Acta Chem. Scand.*, 1991, **45**, 833.
292. A.G. Karipides and A.V. Cafiero, *Inorg. Synth.*, 1968, **11**, 6.
293. J. Goodyear and G.A. Steigmann, *Acta Crystallogr.*, 1963, **16**, 946.
294. L.I. Man, R.M. Imanov and S.A. Semiletov, *Soviet Phys. Crystallogr.*, 1976, **21**, 255.
295. H. Titze, *Acta Chem. Scand.*, 1981, **A35**, 763 and refs. cited therein.
296. C. Svensson and J. Albertsson, *J. Solid State Chem.*, 1983, **46**, 46.
297. W. Rehwald and G. Harbeke, *J. Phys. Chem. Solids*, 1965, **26**, 1309.
298. K.A. Yee and T.A. Albright, *J. Am. Chem. Soc.*, 1991, **113**, 6474 and refs. cited therein.
299. J.C.W. Folmer, J.A. Turner, R. Noufi and D. Cahen, *J. Electrochem. Soc.*, 1985, **132**, 1319.
300. B. Krebs, *Angew. Chem., Int. Ed. Engl.*, 1983, **22**, 113 and refs. cited therein.
301. H.-J. Deiseroth and H. Fu-Son, *Angew. Chem., Int. Ed. Engl.*, 1981, **20**, 962.
302. M.G. Kanatzidis and S. Dhingra, *Inorg. Chem.*, 1989, **28**, 2026.
303. B. Krebs, D. Voelker and K.O. Stiller, *Inorg. Chim. Acta*, 1982, **65**, L101.
304. R. Hoppe, W. Lidecke and F.-C. Frarath, *Z. Anorg. Allg. Chem.*, 1961, **309**, 49.

Chalcogenide halides

305. J. Fenner, A. Rabenau and G. Trageser, *Adv. Inorg. Chem. Radiochem.*, 1980, **23**, 3 and refs. cited therein.
306. R. Kniep and W. Welzel, *Z. Naturforsch.*, 1985, **40b**, 26.
307. L. Zhengyan, H. Janssen, R. Mattes, H. Schnöckel and B. Krebs, *Z. Anorg. Allg. Chem.*, 1984, **513**, 67.

Thiolates

308. G.G. Hoffmann, *Chem. Ber.*, 1983, **116**, 3858; 1985, **118**, 1655.
309. G.G. Hoffmann, *Z. Naturforsch.*, 1984, **39b**, 352.
310. R. Kumar, H.E. Mabrouk and D.G. Tuck, *J. Chem. Soc., Dalton Trans.*, 1988, 1045.
311. J.H. Green, R. Kumar, N. Seudeal and D.G. Tuck, *Inorg. Chem.*, 1989, **28**, 123.
312. G.G. Hoffmann, *Inorg. Chim. Acta*, 1984, **90**, L45.
313. A. Boardman, S.E. Jeffs, R.W.H. Small and I.J. Worrall, *Inorg. Chim. Acta*, 1985, **99**, L39.
314. G.G. Hoffmann and C. Burschka, *Angew. Chem., Int. Ed. Engl.*, 1985, **24**, 970.
315. (a) A. Boardman, R.W.H. Small and I.J. Worrall, *Inorg. Chim. Acta*, 1986, **120**, L23.
    (b) M.B. Power, J.W. Ziller and A.R. Barron, *Organometallics*, 1992, **11**, 2783.
316. K. Ruhlandt-Senge and P.P. Power, *Inorg. Chem.*, 1991, **30**, 2633.
317. R.K. Chadha, P.C. Hayes, H.E. Mabrouk and D.G. Tuck, *Can. J. Chem.*, 1987, **65**, 804.
318. W. Hirpo, S. Dhingra and M.G. Kanatzidis, *J. Chem. Soc., Chem. Commun.*, 1992, 557.

Nitrides, phosphides, arsenides, etc.

319. T. Østvold, E. Rytter and G.N. Papatheodoron, *Polyhedron*, 1986, **5**, 821.
320. R.C. Schoonmaker and C.E. Burton, *Inorg. Synth.*, 1963, **7**, 16.
321. S. Hayashi, K. Hayamizu and O. Yamamoto, *Bull. Chem. Soc. Jpn.*, 1987, **60**, 761.
322. W.E. White and A.H. Bushey, *Inorg. Synth.*, 1953, **4**, 23.
323. C.C. Wang, M. Zaheeruddin and L.H. Spinar, *J. Inorg. Nucl. Chem.*, 1963, **25**, 326.
324. M.D. Healy, P.E. Laibinis, P.D. Stupik and A.R. Barron, *J. Chem. Soc., Chem. Commun.*, 1989, 359.
325. R. Riedel, S. Schaible, U. Klingebiel, M. Noltemeyer and E. Werner, *Z. Anorg. Allg. Chem.*, 1991, **603**, 119.
326. A.H. Cowley and R.A. Jones, *Angew. Chem., Int. Ed. Engl.*, 1989, **28**, 1208.
327. P. Zanella, G. Rossetto, N. Brianese, F. Ossola, M. Porchia and J.O. Williams, *Chem. Mater.*, 1991, **3**, 225.
328. R. Nomura, S. Fujii, K. Kanaya and H. Matsuda, *Polyhedron*, 1990, **9**, 361.

329. R.M. Graves and G.E. Scuseria, *J. Chem. Phys.*, 1992, **96**, 3723.
330. K. Balasubramanian, *J. Phys. Chem.*, 1990, **94**, 7764.
331. C.W. Bock, M. Trachtman and G.J. Mains, *J. Phys. Chem.*, 1992, **96**, 3007 and refs. cited therein.
332. C.W. Bock, K.D. Dobbs, G.J. Mains and M. Trachtman, *J. Phys. Chem.*, 1991, **95**, 7668.
333. M.A. Al-Laham, G.W. Trucks and K. Raghavachari, *J. Chem. Phys.*, 1992, **96**, 1137.
334. M.B. Power and A.R. Barron, *Angew. Chem., Int. Ed. Engl.*, 1991, **30**, 1353.
335. M. Somer, D. Thiery, K. Peters, L. Walz, M. Hartweg, T. Popp and H.G. von Schnering, *Z. Naturforsch.*, 1991, **46b**, 789.

Amides

336. H. Jacobs, K. Jänichen, C. Hadenfeldt and R. Juza, *Z. Anorg. Allg. Chem.*, 1985, **531**, 125.
337. D.C. Bradley, *Adv. Inorg. Chem. Radiochem.*, 1972, **15**, 259.
338. H. Nöth and P. Konrad, *Z. Naturforsch.*, 1975, **30b**, 681.
339. H. Bürger, J. Cichon, U. Goetze, U. Wannagat and H.J. Wismar, *J. Organomet. Chem.*, 1971, **33**, 1.
340. M.F. Lappert, A.R. Sanger, R.C. Strivastava and P.P. Power, *Metal and Metalloid Amides*, Ellis-Horwood, Chichester, UK, 1980.
341. See, for example, K.J.L. Paciorek, J.H. Nakahara and S.R. Masuda, *Inorg. Chem.*, 1990, **29**, 4252.
342. P. Krommes and J. Lorberth, *J. Organomet. Chem.*, 1977, **131**, 415.
343. M. Cesari and S. Cucinella, in *The Chemistry of Inorganic Homo- and Hetero-Cycles*, eds. D.B. Sowerby and I. Haiduc, Academic Press, London, 1987, Vol. 1, p. 167.
344. A. McKillop, J.D. Smith and I.J. Worrall, *Organometallic Compounds of Aluminium, Gallium, Indium and Thallium*, Chapman and Hall, London, 1985.
345. M. Veith, *Adv. Organomet. Chem.*, 1990, **31**, 269.
346. M. Veith, *Chem. Rev.*, 1990, **90**, 3.
347. A.J. Downs, D. Duckworth, J.C. Machell and C.R. Pulham, *Polyhedron*, 1992, **11**, 1295.
348. K. Ouzounis, H. Riffel, H. Hess, U. Kohler and J. Weidlein, *Z. Anorg. Allg. Chem.*, 1983, **504**, 67.
349. K.M. Waggoner, M.M. Olmstead and P.P. Power, *Polyhedron*, 1990, **9**, 257.
350. G.M. Mclaughlin, G.A. Sim and J.D. Smith, *J. Chem. Soc., Dalton Trans.*, 1972, 2197.
351. J.L. Atwood and G.D. Stucky, *J. Am. Chem. Soc.*, 1970, **92**, 285.
352. L.V. Interrante, G.A. Sigel, M. Garauskas, G. Hejna and G.A. Slack, *Inorg. Chem.*, 1989, **28**, 252.
353. A. Ahmed, W. Schwartz and H. Hess, *Z. Naturforsch.*, 1978, **33b**, 43.
354. J.F. Janik, E.N. Duesler and R.T. Paine, *Inorg. Chem.*, 1987, **26**, 4341.
355. D.M. Choquette, M.J. Timm, J.L. Hobbs, M.M. Rahim, K.J. Ahmed and R.P. Planalp, *Organometallics*, 1992, **11**, 529.
356. S.J. Bryan, W. Clegg, R. Snaith, K. Wade and E.H. Wong, *J. Chem. Soc., Chem. Commun.*, 1987, 1223.
357. M.J. Zaworotko and J.L. Atwood, *Inorg. Chem.*, 1980, **19**, 268.
358. G.M. Sheldrick and W.S. Sheldrick, *J. Chem. Soc., A*, 1969, 2279.
359. T.C. Bartke, A. Haaland and D.P. Novak, *Acta Chem. Scand.*, 1975, **A29**, 273.
360. K.N. Semenenko, E.B. Lobouski and A.L. Dovsinskii, *J. Struct. Chem. (Engl. transl.)*, 1972, **13**, 696.
361. P.L. Baxter, A.J. Downs, D.W.H. Rankin and H.E. Robertson, *J. Chem. Soc., Dalton Trans.*, 1985, 807.
362. M.J. Almond, M.G.B. Drew, C.E. Jenkins and D.A. Rice, *J. Chem. Soc., Dalton Trans.*, 1992, 5.
363. H. Schmidbaur and M. Schmidt, *Angew. Chem., Int. Ed. Engl.*, 1962, **1**, 327.
364. W.R. Nutt, J.A. Anderson, J.D. Odom, M.M. Williamson and B.H. Rubin, *Inorg. Chem.*, 1985, **24**, 159.
365. J.L. Atwood, S.G. Bott, C. Jones and C.L. Raston, *Inorg. Chem.*, 1991, **30**, 4868.
366. M.H. Chisholm, V.F. DiStasi and W.E. Streib, *Polyhedron*, 1990, **9**, 253.

367. P.B. Hitchcock, H.A. Jasim, M.F. Lappert and H.D. Williams, *Polyhedron*, 1990, **9**, 245.
368. D.F. Clemens, W.S. Brey, Jr., and H.H. Sisler, *Inorg. Chem.*, 1963, **2**, 1251.
369. R. Snaith, C. Summerford, K. Wade and B.K. Wyatt, *J. Chem. Soc. A*, 1970, 2635.
370. W.S. McDonald, *Acta Crystallogr.*, 1969, **B25**, 1385.
371. D.R. Armstrong, P.G. Perkins and J.J. Stewart, *J. Chem. Soc. A*, 1971, 3674; 1973, 627.

PR_2 and allied derivatives

372. M.J. Taylor and S. Reithmiller, *J. Raman Spectrosc.*, 1984, **15**, 370.
373. A.D. Norman, D.C. Wingleth and C.A. Heil, *Inorg. Synth.*, 1974, **15**, 178.
374. J.W. Anderson and J.E. Drake, *Inorg. Nucl. Chem. Lett.*, 1969, **5**, 887.
375. A.H. Cowley and R.A. Jones, *Angew. Chem., Int. Ed. Engl.*, 1989, **28**, 1208 and refs. cited therein.
376. G. Fritz and G. Trenczek, *Z. Anorg. Allg. Chem.*, 1961, **313**, 236.
377. C.G. Pitt, A.P. Purdy, K.T. Higa and R.L. Wells, *Organometallics*, 1986, **5**, 1266.
378. O.T. Beachley, Jr., and G.E. Coates, *J. Chem. Soc.*, 1965, 3241.
379. C.G. Pitt, K.T. Higa, A.T. McPhail and R.L. Wells, *Inorg. Chem.*, 1986, **25**, 2484.
380. A.M. Arif, B.L. Benac, A.H. Cowley, R. Geerts, R.A. Jones, K.B. Kidd, J.M. Power and S.T. Schwab, *J. Chem. Soc., Chem. Commun.*, 1986, 1543.
381. E.K. Byrne, L. Parkanyi and K.H. Theopold, *Science*, 1988, **241**, 332.
382. D.E. Heaton, R.A. Jones, K.B. Kidd, A.H. Cowley and C.M. Nunn, *Polyhedron*, 1988, **7**, 1901.
383. R.L. Wells, A.T. McPhail and T.M. Speer, *Organometallics*, 1992, **11**, 960.
384. M.A. Banks, O.T. Beachley, Jr., J.D. Maloney and R.D. Rogers, *Polyhedron*, 1990, **9**, 335.
385. T. Douglas and K.H. Theopold, *Inorg. Chem.*, 1991, **30**, 594.
386. A.R. Barron, A.H. Cowley, R.A. Jones, C.M. Nunn and D.L. Westmoreland, *Polyhedron*, 1988, **7**, 77.
387. B.C. Harrison and E.H. Tompkins, *Inorg. Chem.*, 1962, **1**, 951.
388. A.H. Cowley, R.A. Jones, M.A. Mardones, J.L. Atwood and S.G. Bott, *Angew. Chem., Int. Ed. Engl.*, 1990, **29**, 1409.
389. (a) D.A. Atwood, A.H. Cowley, R.A. Jones and M.A. Mardones, *J. Am. Chem. Soc.*, 1991, **113**, 7050. M.A. Petrie, K. Ruhlandt-Senge and P.P. Power, *Inorg. Chem.*, 1992, **31**, 4038.
390. H. Hope, D.C. Pestana and P.P. Power, *Angew. Chem., Int. Ed. Engl.*, 1991, **30**, 691.
391. K.M. Waggoner, S. Parkin, D.C. Pestana, H. Hope and P.P. Power, *J. Am. Chem. Soc.*, 1991, **113**, 3597.
392. R.L. Wells, A.P. Purdy, A.T. McPhail and C.G. Pitt, *J. Chem. Soc., Chem. Commun.*, 1986, 487.

Borides, carbides and silicides

393. N.N. Greenwood, R.V. Parish and P. Thornton, *Quart. Rev.*, 1966, **20**, 441.
394. B. Post, in *Boron, Metallo-Boron Compounds and Boranes*, ed. R.M. Adams, Interscience, New York, 1964, p. 301.
395. M. Matkovich, ed., *Boron and Refractory Borides*, Springer-Verlag, Berlin, 1977; G. Will, *Acta Crystallogr.*, 1969, **B25**, 1219; A.J. Perrotta, W.D. Townes and J.A. Potenza, *Acta Crystallogr.*, 1969, **B25**, 1223.
396. P. Laveant, *Rev. Chim. Miner.*, 1965, **2**, 175.
397. E.L. Amma and G.A. Jeffrey, *J. Chem. Phys.*, 1961, **34**, 252.
398. J.H. Cox and L.M. Pidgeon, *Can. J. Chem.*, 1963, **41**, 671.
399. G.A. Jeffrey and M. Slaughter, *Acta Crystallogr.*, 1963, **16**, 177.
400. G.A. Jeffrey and V.Y. Wu, *Acta Crystallogr.*, 1966, **20**, 538.

Carboranes

401. N.S. Hosmane and J.A. Maguire, *Adv. Organomet. Chem.*, 1990, **30**, 99.
402. R.N. Grimes, in *Comprehensive Organometallic Chemistry*, eds. G. Wilkinson, F.G.A. Stone and E.W. Abel, Pergamon, Oxford, 1982, Vol. 1, p. 459.
403. J.F. Liebman, A. Greenberg and R.E. Williams, eds., *Advances in Boron and the Boranes*, VCH, Weinheim, Germany, 1988.
404. G.A. Olah, K. Wade and R.E. Williams, eds., *Electron Deficient Boron and Carbon Clusters*, Wiley, New York, 1991.
405. D.M.P. Mingos and D.J. Wales, *Introduction to Cluster Chemistry*, Prentice Hall, London, 1990.
406. C.E. Housecroft, *Boranes and Metallaboranes*, Ellis Horwood, Chichester, UK, 1990.
407. L.J. Todd, in *Metal Interactions with Boron Clusters*, ed. R.N. Grimes, Plenum Press, New York, 1982, p. 145.
408. M.A. Bandman, C.B. Knobler and M.F. Hawthorne, *Inorg. Chem.*, 1989, **28**, 1204. D.M. Schubert, M.A. Bandman, W.S. Rees, Jr., C.B. Knobler, P. Lu, W. Nam and M.F. Hawthorne, *Organometallics*, 1990, **9**, 2046.
409. B.M. Mikhailov and T.V. Potapova, *Izv. Akad. Nauk. SSSR, Ser. Khim.*, 1968, **5**, 1153.
410. M.R. Churchill and A.H. Reis, Jr., *J. Chem. Soc., Dalton Trans.*, 1972, 1314, 1317.
411. D.A.T. Young, R.J. Wiersema and M.J. Hawthorne, *J. Am. Chem. Soc.*, 1971, **93**, 5687.
412. P. Jutzi and P. Galow, *J. Organomet. Chem.*, 1987, **319**, 139.
413. M.A. Bandman, C.B. Knobler and M.F. Hawthorne, *Inorg. Chem.*, 1988, **27**, 2399.
414. J.S. Beck and L.G. Sneddon, *J. Am. Chem. Soc.*, 1988, **110**, 3467.
415. R.N. Grimes, W.J. Rademaker, M.L. Denniston, R.F. Bryan and P.T. Greene, *J. Am. Chem. Soc.*, 1972, **94**, 1865.
416. E. Canadell, O. Einstein and J. Rubio, *Organometallics*, 1984, **3**, 759.
417. N.S. Hosmane, K.J. Lu, H. Zhang, A.H. Cowley and M.A. Mardones, *Organometallics*, 1991, **10**, 392.
418. J.L. Spencer, M. Green and F.G.A. Stone, *J. Chem. Soc., Chem. Commun.*, 1972, 1178.
419. H.M. Colquhoun, T.J. Greenhough and M.G.H. Wallbridge, *J. Chem. Soc., Chem. Commun.*, 1977, 737; *Acta Crystallogr.*, 1978, **B34**, 2373.
420. P. Jutzi, *Pure Appl. Chem.*, 1990, **62**, 1035.
421. P. Jutzi, D. Wegener and M.B. Hursthouse, *Chem. Ber.*, 1991, **124**, 295.
422. M.J. Manning, C.B. Knobler, M.F. Hawthorne and Y. Do *Inorg. Chem.*, 1991, **30**, 3589.

Silyls, germanyls etc.

423. L. Rösch, *Angew. Chem., Int. Ed. Engl.*, 1977, **16**, 480.
424. L. Rösch and W. Erb, *Angew. Chem., Int. Ed. Engl.*, 1978, **17**, 604.
425. L. Rösch and H. Neumann, *Angew. Chem., Int. Ed. Engl.*, 1980, **19**, 55.
426. H. Bürger and U. Goetze, *Angew. Chem., Int. Ed. Engl.*, 1969, **8**, 140.
427. E.A.V. Ebsworth, A.G. Lee and G.M. Sheldrick, *J. Chem. Soc. A*, 1969, 1052.
428. A.G. Lee and G.M. Sheldrick, *J. Chem. Soc. A*, 1969, 1055.
429. A.G. Lee, *Spectrochim. Acta*, 1969, **25A**, 1841.
430. A.A. Arif, A.H. Cowley, T.M. Elkins and R.A. Jones, *J. Chem. Soc., Chem. Commun.*, 1986, 1776.
431. D.C. Wingleth and A.D. Norman, *Inorg. Chim. Acta*, 1986, **114**, 191.
432. C.H. van Dyke, *Prep. Inorg. Reactions*, 1971, **6**, 157.
433. E. Amberger, W. Stoeger and J. Hönigschmid, *J. Organomet. Chem.*, 1969, **18**, 77.
434. N. Duffaut, J. Dunogues, R. Calas, P. Riviere, J. Satge and A. Cazes, *J. Organomet. Chem.*, 1978, **149**, 57.
435. A.T. Weibel and J.P. Oliver, *J. Am. Chem. Soc.*, 1972, **94**, 8590.
436. A.T. Weibel and J.P. Oliver, *J. Organomet. Chem.*, 1973, **57**, 313.
437. J. Doe, S. Borkett and P.G. Harrison, *J. Organomet. Chem.*, 1973, **52**, 343.

Pseudohalides

438. S.J. Patel, D.B. Sowerby and D.G. Tuck, *J. Chem. Soc. A*, 1967, 1187.
439. S.J. Patel and D.G. Tuck, *Can. J. Chem.*, 1969, **47**, 229.
440. P.L. Goggin, I.J. McColm and R. Shore, *J. Chem. Soc. A*, 1966, 1314.
441. M.F. Lappert and H. Pyszora, *Adv. Inorg. Chem. Radiochem.*, 1966, **9**, 133.
442. K. Dehnicke and N. Krüger, *Z. Anorg. Allg. Chem.*, 1978, **444**, 71.
443. G. Wilkinson, F.G.A. Stone and E.W. Abel, eds., *Comprehensive Organometallic Chemistry*, Pergamon, Oxford, 1982, Vol. 1, Chapters 6, 7 and 8.

Low-valent derivatives

444. M. Tacke and H. Schnöckel, *Inorg. Chem.*, 1989, **28**, 2895.
445. W.N. Rowlands, A.D. Willson, P.L. Timms, B. Mire, J.H.B. Chenier, J.A. Howard and H.A. Joly, *Inorg. Chim. Acta*, 1991, **189**, 189.
446. P. Jutzi, *Adv. Organomet. Chem.*, 1986, **26**, 217; P. Jutzi, *Comments Inorg. Chem.*, 1987, **6**, 123; P. Jutzi, *Pure Appl. Chem.*, 1989, **61**, 1731.
447. C. Peppe, D.G. Tuck and L. Victoriano, *J. Chem. Soc., Dalton Trans.*, 1982, 2165.
448. M. Tacke and H. Schnöckel, unpublished results.
449. A.N. Red'kin and V.A. Smirnov, *Russ. J. Inorg. Chem.*, 1984, **29**, 1571.
450. V.A. Smirnov and A.N. Red'kin, *Russ. J. Inorg. Chem.*, 1984, **29**, 1738.
451. N. Jouini, *J. Solid State Chem.*, 1986, **63**, 439.
452. N. Jouini, *J. Solid State Chem.*, 1986, **63**, 431.
453. P. Schwerdtfeger, G.A. Heath, M. Dolg and M.A. Bennett, *J. Am. Chem. Soc.*, 1992, **114**, 7518.
454. D.A. Johnson, *Some Thermodynamic Aspects of Inorganic Chemistry*, 2nd edn., Cambridge University Press, Cambridge, 1982, p. 42.
455. D.G. Tuck, *Polyhedron*, 1990, **9**, 377.
456. M.J. Henderson, C.H.L. Kennard, C.L. Raston and G. Smith, *J. Chem. Soc., Chem. Commun.*, 1990, 1203.
457. F.G.N. Cloke, G.R. Hanson, M.J. Henderson, P.B. Hitchcock and C.L. Raston, *J. Chem. Soc., Chem. Commun.*, 1989, 1002.
458. F.G.N. Cloke, C.I. Dalby, M.J. Henderson, P.B. Hitchcock, C.H.L. Kennard, R.N. Lamb and C.L. Raston, *J. Chem. Soc., Chem. Commun.*, 1990, 1394.
459. W. Kaim and W. Matheis, *J. Chem. Soc., Chem. Commun.*, 1991, 597.
460. M.J. Taylor, in *Inorganic Reactions and Methods*, ed. A.P. Hagen, VCH Publishers, New York, 1991, Vol. 13, pp. 3–12.
461. M.J. Taylor, *Metal-to-Metal Bonded States of the Main Group Elements*, Academic Press, London, 1975, Chapter 3.
462. C. Janiak and R. Hoffmann, *J. Am. Chem. Soc.*, 1990, **112**, 5924.
463. P. Schwerdtfeger, *Inorg. Chem.*, 1991, **30**, 1660.
464. G.E. Coates and K. Wade, *Organometallic Compounds*, 3rd edn., Methuen, London, 1967, Vol. 1, p. 372.
465. D.G. Tuck, *Coord. Chem. Rev.*, 1992, **112**, 215.
466. T.A. Annan, R.K. Chadha, P. Doan, D.H. McConville, B.R. McGarvey, A. Ozarowski and D.G. Tuck, *Inorg. Chem.*, 1990, **29**, 3936.
467. D.G. Tuck, *Chem. Soc. Rev*, in press.
468. P. Paetzold, *Angew. Chem., Int. Ed. Engl.*, 1991, **30**, 544.
469. D.G. Tuck, *Can. Chem. News*, September 1990, 25.
470. L.M. Mikheeva and A.N. Grigor'ev, *Russ. J. Inorg. Chem.*, 1984, **29**, 241.
471. H. Schmidbaur, *Angew. Chem., Int. Ed. Engl.*, 1985, **24**, 893.
472. J.M. Parnis and G.A. Ozin, *J. Phys. Chem.*, 1989, **93**, 1215, 1220; Z.L. Xiao, R.H. Hauge and J.L. Margrave, *Inorg. Chem.*, 1993, **32**, 642.
473. K.J. Klabunde and Y. Tanaka, *J. Am. Chem. Soc.*, 1983, **105**, 3544.
474. J.M. Parnis and G.A. Ozin, *J. Am. Chem. Soc.*, 1986, **108**, 1699.
475. M.A. Douglas, R.H. Hauge and J.L. Margrave, *Am. Chem. Soc. Symp. Ser.*, 1982, **179**, 347.

476. M.A. Douglas, R.H. Hauge and J.L. Margrave, *J. Chem. Soc., Faraday Trans. 1*, 1983, **79**, 1533.
477. R.H. Hauge, J.W. Kauffman and J.L. Margrave, *J. Am. Chem. Soc.*, 1980, **102**, 6005.
478. J.A. Howard, H.A. Joly, P.P. Edwards, R.J. Singer and D.E. Logan, *J. Am. Chem. Soc.*, 1992, **114**, 474.
479. K.P. Huber and G. Herzberg, *Molecular Spectra and Molecular Structure. IV. Constants of Diatomic Molecules*, van Nostrand Reinhold, New York, 1979.
480. K. Lammertsma, O.F. Güner, R.M. Drewes, A.E. Reed and P.v.R. Schleyer, *Inorg. Chem.*, 1989, **28**, 313.
481. J. Leszczyński and K. Lammertsma, *J. Phys. Chem.*, 1991, **95**, 3941.
482. M. Wilson, M.B. Coolidge and G.J. Mains, *J. Phys. Chem.*, 1992, **96**, 4851.
483. H. Schumann, C. Janiak, J. Pickardt and U. Börner, *Angew. Chem., Int. Ed. Engl.*, 1987, **26**, 789; H. Schumann, C. Janiak, F. Görlitz, J. Loebel and A. Dietrich, *J. Organomet. Chem.*, 1989, **363**, 243.
484. P. Schwerdtfeger, P.D.W. Boyd, G.A. Bowmaker, H.G. Mack and H. Oberhammer, *J. Am. Chem. Soc.*, 1989, **111**, 15.
485. W. Klemm, E. Voss and K. Geiersberger, *Z. Anorg. Allg. Chem.*, 1948, **256**, 15.
486. R. Ahlrichs, L. Zhengyan and H. Schnöckel, *Z. Anorg. Allg. Chem.*, 1984, **519**, 155.
487. H. Schnöckel, *Z. Naturforsch.*, 1976, **31b**, 1291.
488. H. Schnöckel, M. Leimkühler, R. Lotz and R. Mattes, *Angew. Chem., Int. Ed. Engl.*, 1986, **25**, 921; R. Ahlrichs, M. Häser, H. Schnöckel and M. Tacke, *Chem. Phys. Lett.*, 1989, **154**, 104.
489. C. Dohmeier, R. Mattes and H. Schnöckel, *J. Chem. Soc., Chem. Commun.*, 1990, 358.
490. C. Dohmeier, C. Robl, M. Tacke and H. Schnöckel, *Angew. Chem., Int. Ed. Engl.*, 1991, **30**, 564.
491. G. Kim and K. Balasubramanian, *Chem. Phys. Lett.*, 1992, **193**, 109.
492. T.R. Brumleve, S.A. Mucklejoh and N.W. O'Brien, *J. Chem. Thermodynamics*, 1989, **21**, 1193.
493. P.J. Gardner and S.R. Preston, *Can. J. Chem.*, 1991, **69**, 1394.
494. C. Peppe and D.G. Tuck, *Can. J. Chem.*, 1984, **62**, 2798.
495. T.A. Annan, D.G. Tuck, M.A. Khan and C. Peppe, *Organometallics*, 1991, **10**, 2159.
496. K.-F. Tebbe and U. Georgy, *Acta Crystallogr.*, 1986, **C42**, 1675.
497. J. Glaser, V. Henriksson and T. Klason, *Acta Chem. Scand.*, 1986, **A40**, 344.
498. G.A. Olah, O. Farooq, S.M.F. Farnia, M.R. Bruce, F.L. Clouet, P.R. Morton, G.K.S. Prakash, R.C. Stevens, R. Bau, K. Lammertsma, S. Suzer and L. Andrews, *J. Am. Chem. Soc.*, 1988, **110**, 3231.
499. P.I. Fedorov, *Russ. J. Inorg. Chem.*, 1984, **29**, 325.
500. T. Staffel and G. Meyer, *Z. Anorg. Allg. Chem.*, 1990, **585**, 38.
501. H. Schmidbaur, R. Nowak, W. Bublak, P. Burkert, B. Huber and G. Müller, *Z. Naturforsch.*, 1987, **42b**, 553.
502. W. Hönle, A. Simon and G. Gerlach, *Z. Naturforsch.*, 1987, **42b**, 546.
503. J.C. Beamish, M. Wilkinson and I.J. Worrall, *Inorg. Chem.*, 1978, **17**, 2026. G. Gerlach, W. Hönle and A. Simon, *Z. Anorg. Allg. Chem.*, 1982, **486**, 7.
504. E. Chemouni and A. Potier, *J. Inorg. Nucl. Chem.*, 1971, **33**, 2343.
505. W. Hönle, G. Gerlach, W. Weppner and A. Simon, *J. Solid State Chem.*, 1986, **61**, 171.
506. W. Hönle, G. Miller and A. Simon, *J. Solid State Chem.*, 1988, **75**, 147.
507. K.L. Brown and D. Hall, *J. Chem. Soc., Dalton Trans.*, 1973, 1843.
508. M.A. Khan, D.G. Tuck, M.J. Taylor and D.A. Rogers, *J. Cryst. Spect. Res.*, 1986, **16**, 895.
509. J.C. Beamish, R.W.H. Small and I.J. Worrall, *Inorg. Chem.*, 1979, **18**, 220.
510. J.C. Beamish, A. Boardman, R.W.H. Small and I.J. Worrall, *Polyhedron*, 1985, **4**, 983.
511. W. Hönle and A. Simon, *Z. Naturforsch.*, 1986, **41b**, 1391.
512. H.J. Cumming, D. Hall and C.E. Wright, *Cryst. Struct. Commun.*, 1974, **3**, 107.
513. R.W.H. Small and I.J. Worrall, *Acta Crystallogr.*, 1982, **B38**, 250.
514. R.W.H. Small and I.J. Worrall, *Acta Crystallogr.*, 1982, **B38**, 86.
515. T. Okuda, H. Hamamoto, H. Ishihara and H. Negita, *Bull. Chem. Soc. Jpn.*, 1985, **58**, 2731.
516. H. Ishihara, K. Yamada and T. Okuda, *Bull. Chem. Soc. Jpn.*, 1986, **59**, 3969.

517. T. Okuda, M. Sato, H. Hamamoto, H. Ishihara, K. Yamada and S. Ichiba, *Inorg. Chem.*, 1988, **27**, 3656.
518. L.A. Woodward and M.J. Taylor, *J. Inorg. Nucl. Chem.*, 1965, **27**, 737.
519. C.A. Evans, K.H. Tan, S.P. Tapper and M.J. Taylor, *J. Chem. Soc., Dalton Trans.*, 1973, 988.
520. G. Meyer and R. Blachnik, *Z. Anorg. Allg. Chem.*, 1983, **503**, 126.
521. T. Staffel and G. Meyer, *Z. Anorg. Allg. Chem.*, 1987, **552**, 113; T. Staffel and G. Meyer, *Naturwissenschaften*, 1987, **74**, 491.
522. H.P. Beck and D. Wilhelm, *Angew. Chem., Int. Ed. Engl.*, 1991, **30**, 824.
523. G. Meyer, *Z. Anorg. Allg. Chem.*, 1981, **478**, 39.
524. T. Staffel and G. Meyer, *Z. Anorg. Allg. Chem.*, 1988, **563**, 27.
525. H.P. Beck, *Z. Naturforsch.*, 1984, **39b**, 310.
526. H.P. Beck, *Z. Naturforsch.*, 1987, **42b**, 251.
527. M.A. Khan and D.G. Tuck, *Inorg. Chim. Acta*, 1985, **97**, 73.
528. J.D. Corbett, G. Meyer and J.W. Anderegg, *Inorg. Chem.*, 1984, **23**, 2625.
529. H.P. Beck, D. Wilhelm and A. Hartl-Gunselmann, *Z. Anorg. Allg. Chem.*, 1991, **602**, 65.
530. R. Böhme, J. Rath, B. Grunwald and G. Thide, *Z. Naturforsch.*, 1980, **35b**, 1366.
531. J.C. Beamish, A. Boardman and I.J. Worrall, *Polyhedron*, 1991, **10**, 95.
532. M.J. Taylor, D.G. Tuck and L. Victoriano, *Can. J. Chem.*, 1982, **60**, 690.
533. M.A. Khan, C. Peppe and D.G. Tuck, *Can. J. Chem.*, 1984, **62**, 701.
534. I. Sinclair and I.J. Worrall, *Inorg. Nucl. Chem. Lett.*, 1981, **17**, 279.
535. M. Veith, F. Goffing, S. Becker and V. Huch, *J. Organomet. Chem.*, 1991, **406**, 105.
536. W. Uhl, M. Layh and T. Hildenbrand, *J. Organomet. Chem.*, 1989, **364**, 289; W. Uhl, M. Layh and W. Hiller, *J. Organomet. Chem.*, 1989, **368**, 139.
537. S.M. Tolmachev and N.G. Rambidi, *High Temp. Sci.*, 1973, **5**, 385; A.V. Demidov, A.G. Gershikov, E.Z. Zasorin, V.P. Spiridonov and A.A. Ivanov, *Zh. Strukt. Khim.*, 1983, **24**, 9.
538. J. Leszczyński and J.S. Kwiatkowski, *J. Phys. Chem.*, 1992, **96**, 4148.
539. A.J. Hinchcliffe and J.S. Ogden, *J. Chem. Soc., Chem. Commun.*, 1969, 1053.
540. A.J. Hinchcliffe and J.S. Ogden, *J. Phys. Chem.*, 1971, **75**, 3908; 1973, **77**, 2537.
541. D.M. Makowiecki, D.A. Lynch, Jr., and K.D. Carlson, *J. Phys. Chem.*, 1971, **75**, 1963.
542. M.A. Douglas, R.H. Hauge and J.L. Margrave, *High Temp Sci.*, 1983, **16**, 35.
543. I.V. Ovchinnikov, L.V. Serebrennikov and A.A. Mal'tsev, *Vestn. Mosk. Univ., Ser. 2: Khim.*, 1984, **25**, 157.
544. B.J. Kelsall and K.D. Carlson, *J. Phys. Chem.*, 1980, **84**, 951.
545. J.F. Hinton and K.R. Metz, *J. Soln. Chem.*, 1980, **9**, 197.
546. M. Touboul, P. Toledano, C. Idoura and M.-M. Bolze, *J. Solid State Chem.*, 1986, **61**, 354.
547. A. Simon, W. Mertin, H. Mattausch and R. Gruehn, *Angew. Chem., Int. Ed. Engl.*, 1986, **25**, 845.
548. J. Drowart, G. De Maria, R.P. Burns and M.G. Ingram, *J. Chem. Phys.*, 1960, **32**, 1366.
549. S.J. Bares, M. Haak and J.W. Nibler, *J. Chem. Phys.*, 1985, **82**, 670.
550. S.M. Sonchik, L. Andrews and K.D. Carlson, *J. Phys. Chem.*, 1983, **87**, 2004.
551. L.V. Serebrennikov and A.A. Mal'tsev, *Vestn. Mosk. Univ., Ser. 2: Khim.*, 1985, **26**, 137.
552. L.V. Serebrennikov, S.B. Osin and A.A. Mal'tsev, *J. Mol. Struct.*, 1982, **81**, 25.
553. M.J. Zehe, D.A. Lynch, Jr., B.J. Kelsall and K.D. Carlson, *J. Phys. Chem.*, 1979, **83**, 656.
554. S.M. Sonchik, L. Andrews and K.D. Carlson, *J. Phys. Chem.*, 1984, **88**, 5269.
555. A.I. Boldyrev and P.v.R. Schleyer, *J. Am. Chem. Soc.*, 1991, **113**, 9045.
556. L. Bencivenni, M. Pelino and F. Ramondo, *J. Mol. Struct. (THEOCHEM)*, 1992, **253**, 109.
557. A.V. Nemkhin and J. Almlöf, *J. Mol. Struct. (THEOCHEM)*, 1992, **253**, 101.
558. J.S. Uppal and R.H. Staley, *J. Am. Chem. Soc.*, 1982, **104**, 1229.
559. S. Harvey, M.F. Lappert, C.L. Raston, B.W. Skelton, G. Srivastava and A.H. White, *J. Chem. Soc., Chem. Commun.*, 1988, 1216.
560. H. Kunkely and A. Vogler, *Inorg. Chim. Acta*, 1991, **186**, 155.
561. H.W. Roesky, M. Scholz, M. Noltemeyer and F.T. Edelmann, *Inorg. Chem.*, 1989, **28**, 3829.

562. M. Veith and R. Rösler, *Angew. Chem., Int. Ed. Engl.*, 1982, **21**, 858; M. Veith and K. Kunze, *Angew. Chem., Int. Ed. Engl.*, 1991, **30**, 95.
563. M. Scholz, M. Noltemeyer and H.W. Roesky, *Angew. Chem., Int. Ed. Engl.*, 1989, **28**, 1383.
564. H.E. Mabrouk and D.G. Tuck, *Can. J. Chem.*, 1989, **67**, 746.
565. T.A. Annan, D.H. McConville, B.R. McGarvey, A. Ozarowski and D.G. Tuck, *Inorg. Chem.*, 1989, **28**, 1644.
566. L. Peter and B. Meyer, *Inorg. Chem.*, 1985, **24**, 307.
567. J.C. Gallucci, R.E. Gerkin and W.J. Reppart, *Acta Crystallogr.*, 1989, **C45**, 701.
568. J.G. Bergman and J.S. Wood, *Acta Crystallogr.*, 1987, **C43**, 1831; *J. Chem. Soc., Chem. Commun.*, 1976, 457.
569. K. Ozutsumi, H. Ohtaki and A. Kusumegi, *Bull. Chem. Soc. Jpn.*, 1984, **57**, 2612.
570. I.D. Brown and R. Faggiani, *Acta Crystallogr.*, 1980, **B36**, 1802.
571. J.J. Habeeb and D.G. Tuck, *J. Chem. Soc., Dalton Trans.*, 1973, 243.
572. V.S. Dmitriev, V.A. Smirnov, S.A. Malinov and L.G. Dubovitskaya, *Russ. J. Inorg. Chem.*, 1986, **31**, 1367.
573. R. Nesper and J. Curda, *Z. Naturforsch.*, 1987, **42b**, 557.
574. C. Svensson and J. Albertsson, *J. Solid State Chem.*, 1983, **46**, 46.
575. T. Wadsten, L. Arnberg and J.-E. Berg, *Acta Crystallogr.*, 1980, **B36**, 2220.
576. J.H.C. Hogg, H.H. Sutherland and D.J. Williams, *Acta Crystallogr.*, 1973, **B29**, 1590 and refs. cited therein.
577. H.-J. Deiseroth, D. Müller and H. Hahn, *Z. Anorg. Allg. Chem.*, 1985, **525**, 163.
578. G. Eulenberger, *Acta Crystallogr.*, 1986, **C42**, 528 and refs. cited therein.
579. S. Paashaus and R. Kniep, *Angew. Chem., Int. Ed. Engl.*, 1986, **25**, 752.
580. C. Geloso, H.E. Mabrouk and D.G. Tuck, *J. Chem. Soc., Dalton Trans.*, 1989, 1759.
581. J.H. Green, R. Kumar, N. Seudeal and D.G. Tuck, *Inorg. Chem.*, 1989, **28**, 123.
582. B. Krebs and A. Brömmelhaus, *Z. Anorg. Allg. Chem.*, 1991, **595**, 167.
583. A.J. Hinchcliffe, J.S. Ogden and D.D. Oswald, *J. Chem. Soc., Chem. Commun.*, 1972, 338.
584. D. McIntosh and G.A. Ozin, *J. Am. Chem. Soc.*, 1976, **98**, 3167.
585. P.H. Kasai and P.M. Jones, *J. Am. Chem. Soc.*, 1984, **106**, 8018.
586. P.H. Kasai and P.M. Jones, *J. Phys. Chem.*, 1985, **89**, 2019.
587. J.H.B. Chenier, C.A. Hampson, J.A. Howard, B. Mile and R. Sutcliffe, *J. Phys. Chem.*, 1986, **90**, 1524.
588. J.A. Howard, R. Sutcliffe, C.A. Hampson and B. Mile, *J. Phys. Chem.*, 1986, **90**, 4268.
589. J.H.B. Chenier, C.A. Hampson, J.A. Howard and B. Mile, *J. Chem. Soc., Chem. Commun.*, 1986, 730.
590. W.G. Hatton, N.P. Hacker and P.H. Kasai, *J. Phys. Chem.*, 1989, **93**, 1328.
591. O.T. Beachley, Jr., M.R. Churchill, J.C. Fettinger, J.C. Pazik and L. Victoriano, *J. Am. Chem. Soc.*, 1986, **108**, 4666.
592. O.T. Beachley, Jr., R. Blom, M.R. Churchill, K. Faegri, Jr., J.C. Fettinger, J.C. Pazik and L. Victoriano, *Organometallics*, 1989, **8**, 346.
593. H. Hoberg and S. Krause, *Angew. Chem., Int. Ed. Engl.*, 1976, **15**, 694.
594. W. Uhl, *Z. Naturforsch.*, 1988, **43b**, 1113.
595. W. Hiller, K.-W. Klinkhammer, W. Uhl and J. Wagner, *Angew. Chem., Int. Ed. Engl.*, 1991, **30**, 179.
596. K. Wade, *J. Chem. Soc., Chem. Commun.*, 1971, 792; *Adv. Inorg. Chem. Radiochem.*, 1976, **18**, 1.

Intermetallic derivatives

597. K.A. Gingerich, D.L. Cocke, H.C. Finkbeiner and C.-A. Chang, *Chem. Phys. Lett.*, 1973, **18**, 102.
598. E. Zintl, J. Goubeau and W. Dullenkopf, *Z. Phys. Chem., Abt. A*, 1931, **154**, 1.
599. W.B. Pearson, *Acta Crystallogr.*, 1964, **17**, 1.
600. H. Schäfer, B. Eisenmann and W. Müller, *Angew. Chem., Int. Ed. Engl.*, 1973, **12**, 694.
601. H.G. von Schnering, *Angew. Chem., Int. Ed. Engl.*, 1981, **20**, 33.

602. R. Nesper, *Angew. Chem., Int. Ed. Engl.*, 1991, **30**, 789; *Prog. Solid State Chem.*, 1990, **20**, 1.
603. R.B. King, *Inorg. Chim. Acta*, 1991, **181**, 217; 1992, **198–200**, 841; *Inorg. Chem.*, 1989, **28**, 2796.
604. G.S. Smith, Q. Johnson and D.H. Wood, *Acta Crystallogr.*, 1969, **B25**, 554.
605. R. Thümmel and W. Klemm, *Z. Anorg. Allg. Chem.*, 1970, **376**, 44.
606. C. Belin and R.G. Ling, *J. Solid State Chem.*, 1983, **48**, 40 and refs. cited therein.
607. H. Schäfer, *J. Solid State Chem.*, 1985, **57**, 97.
608. M. Charbonnel and C. Belin, *J. Solid State Chem.*, 1987, **67**, 210.
609. M. Charbonnel and C. Belin, *Nouv. J. Chim.*, 1984, **8**, 595.
610. C. Belin, *J. Solid State Chem.*, 1983, **50**, 225.
611. C. Belin and M. Charbonnel, *J. Solid State Chem.*, 1986, **64**, 57.
612. J.K. Burdett and E. Canadell, *J. Am. Chem. Soc.*, 1990, **112**, 7207.
613. J.K. Burdett and E. Canadell, *Inorg. Chem.*, 1991, **30**, 1991 and refs. cited therein.
614. M.L. Fornasini, *Acta Crystallogr.*, 1983, **C39**, 943.
615. S.M. Kauzlarich, M.M. Thomas, D.A. Odink and M.M. Olmstead, *J. Am. Chem. Soc.*, 1991, **113**, 7205 and refs. cited therein.
616. R.F. Gallup, C.Y. Fong and S.M. Kauzlarich, *Inorg. Chem.*, 1992, **31**, 115.
617. R.G. Ling and C. Belin, *Acta Crystallogr.*, 1982, **B38**, 1101.
618. U. Frank-Cordier, G. Cordier and H. Schäfer, *Z. Naturforsch.*, 1982, **37b**, 119, 127.
619. A.J. Bradley, *Z. Krist.*, 1935, **91**, 302.
620. S.E. Rodriguez and C.J. Pings, *J. Chem. Phys.*, 1965, **42**, 2435.
621. L. Pauling, *J. Am. Chem. Soc.*, 1947, **69**, 542.
622. S.C. Sevov and J.D. Corbett, *Inorg. Chem.*, 1992, **31**, 1895.
623. S. Larose and A.D. Pelton, *J. Phase Equilib.*, 1991, **12**, 371.
624. G. Bruzzone, *Acta Crystallogr.*, 1965, **18**, 1081.
625. S.C. Sevov and J.D. Corbett, *Inorg. Chem.*, 1991, **30**, 4875.
626. D.A. Hansen and J.F. Smith, *Acta Crystallogr.*, 1967, **22**, 836.
627. P.C. Schmidt, *Struct. Bonding (Berlin)*, 1987, **65**, 91.
628. J.D. Corbett, *Prog. Inorg. Chem.*, 1976, **21**, 129.
629. T.L.T. Birdwhistell, E.D. Stevens and C.J. O'Connor, *Inorg. Chem.*, 1990, **29**, 3894 and refs. cited therein.
630. P. Alemany, S. Alvarez and R. Hoffmann, *Inorg. Chem.*, 1990, **29**, 3070.
631. R.C. Burns and J.D. Corbett, *J. Am. Chem. Soc.*, 1981, **103**, 2627.
632. R.C. Burns and J.D. Corbett, *J. Am. Chem. Soc.*, 1982, **104**, 2804.
633. R.W. Rudolph, W.L. Wilson and R.C. Taylor, *J. Am. Chem. Soc.*, 1981, **103**, 2480.
634. E. Zintl and A. Harder, *Z. Phys. Chem., Abt. A*, 1931, **154**, 47.
635. J.D. Corbett and P.A. Edwards, *J. Am. Chem. Soc.*, 1977, **99**, 3313.
636. N.A. Compton, R.J. Errington and N.C. Norman, *Adv. Organomet. Chem.*, 1990, **31**, 91.
637. R.M. Campbell, L.M. Clarkson, W. Clegg, D.C.R. Hockless, N.L. Pickett and N.C. Norman, *Chem. Ber.*, 1992, **125**, 55.
638. W. Heiber and U. Teller, *Z. Anorg. Allg. Chem.*, 1942, **249**, 43.
639. W.R. Robinson and D.P. Schussler, *Inorg. Chem.*, 1973, **12**, 848.
640. L.M. Clarkson, K. McCrudden, N.C. Norman and L.J. Farrugia, *Polyhedron*, 1990, **9**, 2533.
641. L.M. Clarkson, L.J. Farrugia and N.C. Norman, *Acta Crystallogr.*, 1991, **C47**, 2525.
642. D.J. Patmore and W.A.G. Graham, *Inorg. Chem.*, 1966, **5**, 1586.
643. A.T.T. Hsieh and M.J. Mays, *J. Chem. Soc., Dalton Trans.*, 1972, 516.
644. Y.-J. Chen, H.D. Kaesz, Y.K. Kim, H.-J. Müller, R.S. Williams and Z. Xue, *Appl. Phys. Lett.*, 1989, **55**, 2760.
645. J.C. Vanderhooft, R.D. Ernst, F.W. Cagle, Jr., R.J. Neustadt and T.H. Cymbaluk, *Inorg. Chem.*, 1982, **21**, 1876.
646. P.D. Cradwick and D. Hall, *J. Organomet. Chem.*, 1970, **22**, 203.
647. H.-J. Haupt and F. Neumann, *Z. Anorg. Allg. Chem.*, 1972, **394**, 67.
648. H.-J. Haupt and F. Neumann, *J. Organomet. Chem.*, 1974, **74**, 185.
649. A.T.T. Hsieh and M.J. Mays, *Inorg. Nucl. Chem. Lett.*, 1971, **7**, 223.
650. J. Chatt, C. Eaborn and P.N. Kapoor, *J. Organomet. Chem.*, 1970, **23**, 109.

651. R. Guilard, A. Zrineh, A. Tabard, L. Courthaudon, B. Han, M. Ferhat and K.M. Kadish, *J. Organomet. Chem.*, 1991, **401**, 227.
652. S. Takagi, Y. Kato, H. Furuta, S. Onaka and T.K. Miyamoto, *J. Organomet. Chem.*, 1992, **429**, 287.
653. R.A. Fisher and J. Behm, *J. Organomet. Chem.*, 1991, **413**, C10.
654. J.M. Burlitch and T.W. Theyson, *J. Chem. Soc., Dalton Trans.*, 1974, 828.
655. D.P. Schussler, W.R. Robinson and W.F. Edgell, *Inorg. Chem.*, 1974, **13**, 153.
656. S.E. Petersen, W.R. Robinson and D.P. Schussler, *J. Organomet. Chem.*, 1972, **43**, C44.
657. L.M. Clarkson, W. Clegg, D.C.R. Hockless and N.C. Norman, *Acta Crystallogr.*, 1992, **C48**, 236.
658. S. Wang, J.P. Fackler, Jr., C. King and J.C. Wang, *J. Am. Chem. Soc.*, 1988, **110**, 3308 and refs. cited therein.
659. L.M. Clarkson, W. Clegg, D.C.R. Hockless, N.C. Norman and T.B. Mander, *J. Chem. Soc., Dalton Trans.*, 1991, 2229.
660. J. Rajaram and J.A. Ibers, *Inorg. Chem.*, 1973, **12**, 1313.
661. H.-J. Haupt, U. Flörke and H. Preut, *Acta Crystallogr.*, 1986, **C42**, 665.
662. J.M. Cassidy and K.H. Whitmire, *Acta Crystallogr.*, 1990, **C46**, 1781.
663. K.H. Whitmire, J.M. Cassidy, A.L. Rheingold and R.R. Ryan, *Inorg. Chem.*, 1988, **27**, 1347.
664. J.M. Cassidy and K.H. Whitmire, *Inorg. Chem.*, 1989, **28**, 1432, 1435.
665. K.H. Whitmire, R.R. Ryan, H.J. Wasserman, T.A. Albright and K.-S. Kang, *J. Am. Chem. Soc.*, 1986, **108**, 6831.
666. M. Schollenberger, B. Nuber and M.L. Ziegler, *Angew. Chem., Int. Ed. Engl.*, 1992, **31**, 350.
667. K. Merzweiler, F. Rudolph and L. Brands, *Z. Naturforsch.*, 1992, **47b**, 470.

4 Properties and uses of oxides and hydroxides
K.A. EVANS

4.1 Aluminium oxides and hydroxides

The hydroxides and oxides of aluminium are amongst the most important metal hydroxides/oxides in commercial terms. The worldwide production of aluminium hydroxide is in excess of 60 million tonnes per annum. Table 4.1 shows the recent data available from the International Primary Aluminium Institute for the production of aluminium oxide by region. A very large proportion of the aluminium hydroxide is calcined to aluminium oxide and used in electrolytic cells for the production of aluminium metal (see chapter 2). Approximately 5 million tonnes of aluminium hydroxide is directed annually to uses other than metal production. The products resulting from this aluminium hydroxide find their way into almost every sphere of human activity and in the industrialised world they touch our lives every day.

Uses of aluminium hydroxides and the oxides derived therefrom range from fire retardants to refractories, cement to paper, paint to toothpaste, abrasives to catalysts, water treatment feedstocks to ceramics. The key to the development of these seemingly disparate ranges of use has been the very wide range of properties that arise as one moves from gelatinous boehmites to crystalline hydroxides to activated alumina to α-alumina. The materials range from relatively soft compounds with high solubility in aqueous acids and alkalies, to the extremely inert and hard α-alumina. The growth of uses has at the same time been helped considerably by the large-scale manufacture of aluminium hydroxide which has kept the costs to a minimum.

Table 4.1 Alumina production in 1991 (excluding Eastern Europe countries).[a]

Area	Alumina ($\times$ 1000 tonnes as nominal 100% Al_2O_3)		
	Metallurgical uses	Other uses	Total
Africa	650	—	650
North America	5218	777	5995
Latin America	7549	238	7787
East and South Asia	1567	835	2402
Europe	4533	877	5410
Oceania	11475	228	11703
Total	30992	2955	33947

(a) Estimate in 1990 for Eastern Europe is 5 485 000 tonnes and for China is 1 700 000 tonnes 100% Al_2O_3. Statistics compiled by the International Primary Aluminium Institute, London.

4.2   Nomenclature

Aluminium forms a wide range of oxides and hydroxides; some of these are well-characterised crystalline compounds while others are ill-defined amorphous substances. Frequently the terms alumina hydrate, alumina trihydrate or ATH are used for aluminium hydroxide, but these names mistakenly imply forms of alumina with different amounts of water of crystallisation.

The crystalline forms of the aluminium hydroxides can be divided into the trihydroxide $Al(OH)_3$ (which exists in three forms: gibbsite, bayerite and nordstrandite) and the oxide hydroxide (which exists as boehmite or diaspore). All these forms are found in nature although bayerite is found only rarely. A less well documented form of aluminium oxide hydroxide is tohdite, which has the composition $5Al_2O_3 \cdot H_2O$.

Commercially the most important of the hydroxides is gibbsite, although bayerite and boehmite are also manufactured on a small industrial scale. Non-crystalline forms of the hydroxides encompass gelatinous pseudo-boehmites and aluminium hydroxides with amorphous X-ray patterns.

There are many forms of aluminium oxide (χ, η, δ, κ, θ, γ, ϱ) but α-aluminium oxide is the only thermodynamically stable form. The other forms are frequently termed 'transition' aluminas and arise during the thermal decomposition of aluminium trihydroxides and oxide hydroxides under varying conditions. β-Alumina, when first reported,[1] was thought to be another form of aluminium oxide, but subsequent investigations showed the presence of sodium oxide. The most common form of β-alumina is $Na_2O \cdot 11Al_2O_3$, although several other forms exist with differing compositions and incorporating other alkali or alkaline earth oxides (see section 4.3.5.1.11).

Several suboxides of aluminium have been reported to exist in the gas phase: AlO, Al_2O and Al_2O_2[2-4] (see also section 3.3.4).

4.3   Physical properties and methods of synthesis

A comprehensive review of the properties, structures and methods of production of the various aluminium hydroxides and oxides is given in papers by Wefers and Misra[5] and Wefers.[6]

4.3.1   Aluminium hydroxides

4.3.1.1   Gibbsite, γ-aluminium trihydroxide.   Gibbsite is distributed very widely geographically as a component of bauxites, tropical soils and clays.

It is a major constituent of tropical bauxites which are normally tertiary in age. It is sometimes referred to as hydrargillite.

It consists of close-packed sheets of hydroxyl ions with aluminium ions in two-thirds of the octahedral holes between close-packed layers, which are themselves in open packing with neighbouring pairs, the whole approximating to hexagonal symmetry (see Figure 4.1). The structure of gibbsite was proposed by Pauling in 1930 and confirmed by Megan in 1934. Gibbsite is invariably monoclinic, though a triclinic form has been reported by Saalfeld in 1960. Considerable confusion exists in the literature over the designations of the forms of the trihydroxides and transition aluminas. Gibbsite has been variously designated 'α' and γ'. Even recent publications use different designations: Wefers[6] and Slade et al.[7] use 'γ', while Misra[8] uses 'α'. In this review, the internationally accepted crystallographic form has been used. Hence, cubically packed structural forms, such as gibbsite and boehmite, are designated 'γ', while hexagonally packed forms, such as bayerite, diaspore and corundum, are designated 'α'.

Gibbsite cleaves in the (001) plane, is translucent or white depending upon its crystallite size and has a Mohs' hardness of 2½–3½. Its density is $2.42\,\mathrm{g\,cm^{-3}}$ and its standard heat of formation $-1293.2\,\mathrm{kJ\,mol^{-1}}$. It is virtually insoluble in aqueous media at neutral pH but becomes readily soluble at pHs greater than 9 or less than 5.

The Bayer process (described in chapter 2) was invented and first commercialised almost a century ago; it still accounts, however, for over

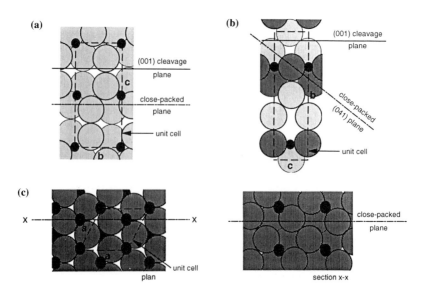

Figure 4.1 Structures of (a) gibbsite, (b) boehmite and (c) corundum. Dark shaded circles, O^{2-}; light shaded circles, OH^-; small filled circles, Al^{3+}.

95% of the 60 million tonnes of gibbsite produced on an industrial scale every year. Gibbsite produced therefrom is of high purity (>99.5%), the main impurities being Na_2O, SiO_2 and TiO_2.

The smaller plants now operating typically have an annual capacity of 100 000 tonnes; the output of these is invariably devoted to higher purity chemical or 'specials' usage. The newer plants go up to 2 million tonnes per annum in their capacity and are devoted principally to providing feedstock for calcination to alumina for use in aluminium metal production. The particle size of agglomerates produced by this route is normally in the range 30–150 μm, but crystals in the range 0.3–200 μm can be produced by controlling seed concentration, seed size, seed activity, temperature and the degree of supersaturation.

The Bayer process involves digesting the bauxite in caustic soda and then, after separating various components, gibbsite is encouraged to precipitate from the supersaturated sodium aluminate. Precipitation is normally induced by the addition of seed nuclei, but can also be effected by neutralisation with carbon dioxide. Precipitation at temperatures less than 40°C encourages the formation of new nuclei and a fine material is obtained, whereas precipitation at temperatures over 75°C encourages crystal growth, resulting in large aggregates composed of hexagonal rods and prisms.

To a limited extent, in Eastern Europe some aluminium hydroxide is still produced by the 'sinter' route which involves calcining the bauxite with sodium carbonate in the temperature range 900–1100°C to form sodium aluminate. This route is particularly advantageous when processing bauxites high in diaspore. In the United States, a combined Bayer-sinter process has been utilised to process bauxites high in silica.

4.3.1.2 Bayerite, α-aluminium trihydroxide. Bayerite exists in an approximately hexagonal close-packed structure. The lattice is built up from double layers of hydroxyl groups with the hydroxyl groups of one layer lying in the depression of the second layer. Single crystals cannot be readily synthesised and the structure has been open to some discussion; it is generally thought to be trigonal.[9] The normal growth forms are spindle or hourglass shapes. It has a density of 2.53 g cm^{-3} and a standard heat of formation of -1288.4 kJ mol^{-1}. It has been reported to be stable only in the absence of alkali ions; in their presence bayerite converts irreversibly to gibbsite.

Bayerite can be produced by a variety of methods:

1. neutralisation of sodium aluminate solution by carbon dioxide addition;
2. neutralisation of aluminium salts with aqueous ammonia at temperatures below 325 K and then ageing;
3. rehydration of ϱ-alumina;
4. addition of bayerite seed to supersaturated aluminate solutions; and

5. reaction between an amalgam of aluminium and water at room temperature.

Industrially bayerite is made by the first route, the neutralisation of the sodium aluminate from the Bayer process being achieved using flue gas containing 10–15% carbon dioxide. The resulting product contains up to 10% gibbsite, pseudo-boehmite and amorphous aluminium hydroxide.

4.3.1.3 Nordstrandite. Nordstrandite is structurally intermediate between gibbsite and bayerite and consists of double layers of hydroxyl ions with aluminium occupying two-thirds of the octahedral interstices. Two double layers are stacked with gibbsite sequences (*viz.* AB–BA–AB–BA), followed by two double layers of the bayerite sequence (*viz.* AB–AB–AB).

Currently there is no commercial production of nordstrandite. Nordstrandite of high purity can be prepared by the reaction of aluminium, aluminium hydroxide gel or hydrolysable aluminium compounds with an aqueous solution of an alkanediamine or a derivative, especially ethane-diamine and EDTA. Its existence in nature is rare but has been reported in tropical soils in West Sarawak and Guam.

4.3.2   Aluminium oxide hydroxide

4.3.2.1   Boehmite, γ-aluminium oxide hydroxide.   Boehmite has a layered structure comprising oxygen and hydroxyl ions with interstitial aluminium ions, held together by hydrogen bonding between the hydroxyl groups of adjacent layers. It has an orthorhombic habit.

Its formation is favoured by subjecting coarse crystals of gibbsite to a fast heating rate. Under these conditions the water of dehydroxylation cannot escape readily and the hydrothermal conditions thereby produced result in the formation of boehmite. Boehmite of high purity can be prepared by the hydrothermal treatment of gibbsite or aluminium metal. Under acid conditions fibrous boehmite is obtained,[10] while under alkaline conditions multiple interpenetrant twin crystals (5–15 μm) are formed.[11,12] There is a limited amount of industrial production of boehmite. European bauxites of the Tertiary and Upper Cretaceous age are especially rich in boehmite, although it is also found in Australian, Asian and African bauxites.[13] Boehmite is white, has a density of $3.01\,g\,cm^{-3}$, a Mohs' hardness of 3½–4 and a standard heat of formation of $-990.4\,kJ\,mol^{-1}$.

4.3.2.2   Diaspore, α-aluminium oxide hydroxide.   Diaspore, like boehmite, is orthorhombic. The structure is composed of AlOOH double chains which are arranged in an approximately hexagonal close-packed configuration.

It has a higher density ($3.44\,g\,cm^{-3}$) and is harder (Mohs' hardness 6½–7) than boehmite; $\Delta_f H^\ominus$ is $-998.4\,\text{kJ}\,\text{mol}^{-1}$. Diaspore is difficult to synthesise. It can be prepared hydrothermally from gibbsite or beohmite but requires higher temperature and pressure conditions than does the formation of boehmite.

Diaspore is found in bauxites from Greece, China, Romania and the C.I.S. There is no commercial production but bauxites and clays high in diaspore are used for the manufacture of refractories.

4.3.3 Aluminium oxide

4.3.3.1 Corundum (α-aluminium oxide). α-Aluminium oxide or α-alumina is composed of hexagonal close-packed layers of oxygen ions with two-thirds of the octahedral holes occupied statistically by aluminium ions (see Figure 4.1). Bragg and Bragg[14] investigated the structure, showing it to be hexagonal-rhombohedral. The solid is extremely hard with a Mohs' hardness of 9; this is surpassed by only a few other materials. It has a density of $3.98\,\text{g}\,\text{cm}^{-3}$.

Corundum is a common constituent in many igneous and metamorphic rocks. Material of gemstone quality is found in alluvial deposits in Burma, Sri Lanka, Kashmir, Australia and East Africa. The presence of metal ions gives rise to a number of precious gemstones. Traces of Cr^{3+} give ruby, of Fe^{2+}, Fe^{3+} or Ti^{4+} give blue sapphire. Synthetic rubies and both blue and white sapphires are produced in significant quantities. Lower quality corundum is a major component of emery, which is used as an abrasive; it is mined in South Africa and the Greek island Naxos.

α-Alumina is the thermodynamically most stable state of the compounds formed between aluminium and oxygen, and is the final product from thermal or dehydroxylation treatments of all the hydroxides or other oxides.

4.3.3.2 Industrial grades of aluminium oxide. Industrial grades of aluminium oxide are often divided into the following categories: smelter, activated, catalytic, calcined, low-soda, reactive, tabular, fused and high purity. These differ in their particle size and morphology, α-alumina content and impurities (especially the soda level). With the exception of very small quantities, they are produced by calcination of Bayer-derived gibbsite in a rotary kiln or 'fluid flash' calciner.

4.3.3.2.1 Smelter grade alumina. Smelter grade alumina is the name given to alumina utilised in the manufacture of aluminium metal. Historically it was manufactured using rotary kilns but is now generally produced in fluid bed or fluid flash calciners. In this process the aluminium trihydroxide is fed into a counter-current stream of hot air obtained by burning fuel oil or gas. The first effect is that of removing the free water and the next is removal of the chemically combined water; this occurs over a range of temperatures between 180 and 600°C. The dehydrated alumina is principally in the form of activated alumina and the surface area gradually decreases as the temperature rises towards 1000°C. Further calcination at temperatures >1000°C converts this to the non-adsorbent α-form. Smelter grade aluminas generally retain the physical nature of the starting aluminium trihydroxide. The conversion to the α-form is typically of the order of 25% and the specific surface area is relatively high at >50 m² g⁻¹ through the presence of transition aluminas. The nature and percentage of the transition aluminas depend on the atmosphere within the calciner, the amount of moisture present and the heating rate. The alumina most commonly used for smelting today is termed 'sandy' alumina as it has good flowing properties. The older, more traditional smelters required an alumina with a low bulk density and high α-content, and with poor flowing characteristics so that, when heaped up in the cell, it tended to stay in position. This alumina is traditionally called 'floury' alumina. Fluid flash calciners are much more efficient than rotary kilns and produce a saving of fuel in the order of 30–40%. The higher performance speciality aluminas are, however, generally manufactured in rotary or static kilns.

4.3.3.2.2 Calcined alumina.   Nearly 4 million tonnes out of the total of 40 million tonnes of alumina made annually is utilised for applications other than aluminium production. The markets for speciality calcined alumina are divided as follows (with approximate proportions):

Refractories	50%
Abrasives	20%
Whitewares and spark plugs	15%
Ceramics	10%
Others	5%

Calcined aluminas are generally manufactured in rotary kilns with the aluminium trihydroxide going through the phases described earlier. Mineralisers are frequently added to catalyse the reactions in the kiln and bring down the temperature at which the α-alumina forms. Fluorides are the most widely used type of mineraliser. Such additions have a major effect on the morphology of the crystals formed giving platelike crystals (see Figure 4.2).

Soda is the major impurity and can rise as high as 0.6 wt%; the other major impurities are SiO_2 at 0.02–0.05 wt% and Fe_2O_3 at 0.02–0.04 wt%. The α-alumina content is normally in the range 75–100% and the surface areas are consequently much lower than those for smelter grade alumina (0.5–25 m² g⁻¹). The agglomerates produced during rotary calcination are typically 40–150 μm in size; these agglomerates consist of crystals 0.5–15 μm in size. For most applications, the agglomerates are milled or micronised to

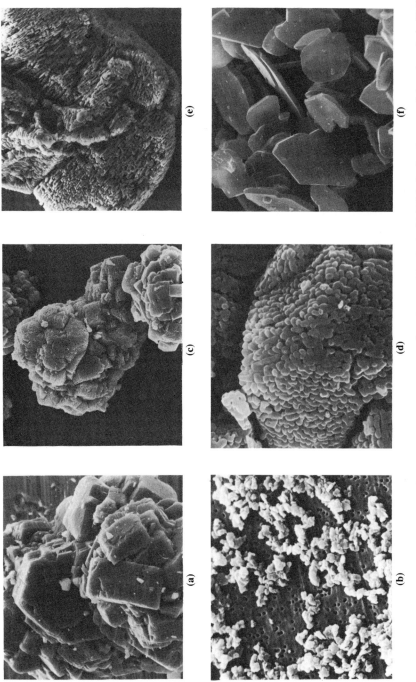

Figure 4.2 (a) Scanning electron micrograph (SEM) of coarse Bayer hydrate ($\times 1000$ magnification); (b) SEM of sub-micron precipitated grade ($\times 10000$ magnification); (c) SEM of boron-mineralised alumina ($\times 1000$ magnification); (d) SEM of boron-mineralised alumina ($\times 5000$ magnification); (e) SEM of fluoride-mineralised alumina ($\times 1000$ magnification); (f) SEM of fluoride-mineralised alumina ($\times 2000$ magnification).

release the fine crystals. On dehydroxylation, the gibbsite converts to transition aluminas with partially disordered structures. As the calcination temperature increases, the structures become more ordered until the transformation to α-alumina occurs. As demonstrated in *The Atlas of Alumina*,[12] the alumina aggregates are pseudomorphs of the original gibbsite with little change of particle size on calcination.

4.3.3.2.3 Low soda alumina. Low soda aluminas are defined generally as aluminas with a soda content of <0.1 wt%. These can be manufactured from gibbsite by a number of methods:

1. use of a low soda gibbsite prepared using modified decomposition conditions;
2. addition of chlorine gas to the kiln;
3. introduction of boron additives to the gibbsite prior to calcination (see Figure 4.2, for example);
4. washing of activated or high surface area alumina prior to final calcination; and
5. addition of compounds, such as silica sand, to the gibbsite prior to passage through the kiln; the soda reacts with the silica to form sodium silicate which is then removed by sieving.

'Medium' or 'intermediate' soda aluminas are the names given to aluminas with soda contents of 0.15–0.25 wt%.

4.3.3.2.4 Reactive alumina. The term 'reactive' aluminas is normally given to aluminas of relatively high purity and small crystal size (<1 μm) which sinter to a fully dense body at lower temperatures than low soda, medium soda or ordinary soda aluminas. Such aluminas are normally supplied after intensive ball-milling which breaks up the agglomerates produced after calcination. They are utilised to meet conditions of exceptional strength, wear resistance, temperature resistance, surface finish or chemical inertness.

4.3.3.2.5 Tabular alumina. Tabular alumina is recrystallised or sintered α-alumina, so called because its morphology consists of large (50–500 μm), flat tablet-shaped crystals of corundum. It is produced by pelletising, extruding or pressing calcined alumina into shapes, and then heating these shapes to a temperature just under their fusion point, 1700–1850°C; shaft kilns have been found to be especially effective for the production of tabular aluminas.

After calcination, the spheres or other shapes of sintered alumina can be used as they are for some applications or crushed, screened and ground to produce a wide range of sizes. As the material has been sintered, it has an especially low porosity, high density, low permeability, good chemical inertness and high refractoriness; hence it is especially good for refractory applications.

4.3.3.2.6   High purity alumina.   High purity aluminas are normally classified as those with a purity of 99.9% or higher. Purities up to approximately 99.97% can be achieved by routes starting from Bayer hydrate and using successive activations and washings, or *via* a chloride. Products of higher purity are manufactured by calcining ammonium aluminium sulfate or from aluminium metal. In the case of the route *via* ammonium alum, the necessary degree of purity is obtained by successive recrystallisations. Especially high purities can be realised from aluminium by causing the metal to react with an alcohol, purifying the aluminium alkoxide by distillation, hydrolysing it and then calcination of the product.

Applications include the manufacture of synthetic gemstones such as ruby, yttrium aluminium garnets for lasers, and sapphires for instrument windows and lasers.

4.3.4   Uses of aluminium hydroxides

Whilst the direct application of aluminium trihydroxide in paper, paint, glass, ceramic glazes, pharmaceuticals, toothpaste and as a fire retardant amounts to several hundreds of thousands of tonnes per year, the use in the manufacture of other chemicals is even larger. Largest of all is the manufacture of 'iron-free' aluminium sulfate for use in paper-making and water-treatment as a coagulant. Other very large uses are found in the manufacture of aluminium fluoride, aluminium nitrate, aluminium chloride, poly-aluminium chloride (PAC), polyaluminium silicate sulfate (PASS®, Alcan Chemicals Ltd), sodium aluminate, zeolites, catalysts and titania pigment coating. A few of the more important uses are described in detail.[15]

4.3.4.1   Fire retardancy.   Interest in the use of aluminium compounds for imparting fire resistance to materials stretches back a long way. The annals of Claudius record that in the attack on Piraeus in 83 BC, the storming towers were protected against fire by treatment with a solution of alum. The use of aluminium hydroxide as a fire retardant, however, has more recent origins, being now nearly 100 years old. In the last decade of the nineteenth century, Sir William Henry Perkin was investigating the flammability of cotton flannelette. Aluminium hydroxide was found to be the best of the 25 potential flame retardants for cellulose that he investigated. An 1896 patent covered the use of aluminium hydroxide as a fire retardant for wood fibre. The next reference to the use of aluminium hydroxide in a flame-retardant composition was in 1921 when its use, together with antimony chloride, was patented for flame-retarding rubber. There is, however, little evidence that such a mixture was adopted commercially at that time.

Several patents appeared in the mid-1950s extolling the benefits of aluminium hydroxide in polyester resin, neoprene, butyl rubber and 'epoxies' to improve the arc resistance of electrical components. The usage of aluminium hydroxide as a filler in rubbers and plastics, however, remained small throughout the 1950s and 1960s. Stimulation of the usage of aluminium hydroxide as a fire retardant was provided, however, by the work of Connolly and Thornton in 1965 which showed the improvements that could be obtained in the limiting oxygen index and reduced burning rate of polyester resins containing aluminium hydroxide.[16]

The trigger to large tonnage usage of aluminium hydroxide came a few years later after several fires in the United States where carpets were highlighted as being responsible for spreading fires. In one particular case in Ohio, 32 people died in a modern old people's home. A Senate enquiry attributed the spread of fire along a corridor to a foam-backed nylon pile carpet, the report stressing that the fatalities were due not to burns but to asphyxiation. The outcome was that legislation was introduced in the United States in 1971 covering mandatory Federal standards for the surface spread of flame on carpets and rugs when subjected to a small ignition source. Aluminium hydroxide was found to be the ideal fire retardant to meet this legislative requirement. It could be incorporated into either the foam rubber backing or the adhesive rubber anchor coat which bonds the carpet fibres. High levels of addition could be used without adversely affecting the performance of the rubber and the necessary degree of fire resistance obtained. Many thousands of tonnes of aluminium hydroxide are used for this application annually.

Fire tests for carpets exist in many countries but, in the absence of legislation requiring these fire test standards to be achieved, the growth in the market has been restricted.

During the 1970s, the use of aluminium hydroxide as a fire retardant in plastics grew from a few hundred tonnes to many tens of thousands of tonnes. The largest use was in glass-reinforced unsaturated polyester in a wide variety of applications, including electrical switch-boxes, building panels, machine housings and automotive parts. Significant markets also grew up in PVC, especially for conveyor belting, polyurethane foam, epoxies etc.

The current annual usage of aluminium hydroxide as a fire retardant worldwide is estimated at over 200 000 tonnes. The United States and Canada account for some 120 000–127 000 tonnes and Europe for 40 000–50 000 tonnes; Japan consumes about 20 000–30 000 tonnes per annum.

Combustion of a polymeric material can be divided into a number of stages: heating, decomposition of the polymer to give combustible gases, ignition, heat release to the surrounding material, and propagation. Aluminium hydroxide operates in the first two stages of the combustion process, *viz*. heating and decomposition. On heating to temperatures above

200°C, aluminium hydroxide decomposes to give alumina and water vapour:

$$2Al(OH)_3 \rightarrow Al_2O_3 + 3H_2O \tag{4.1}$$

This reaction is strongly endothermic absorbing $1.97 \, \text{kJ} \, \text{g}^{-1}$ of aluminium hydroxide. The heating of the polymeric material is thereby slowed down and the onset and rate of decomposition of the polymer are also reduced. Additionally, the water vapour that is evolved from the decomposing aluminium hydroxide dilutes any combustible gases evolved and hinders access of oxygen to the surface of the polymer, thereby further suppressing ignition.

For an endothermic material to be efficient as a fire retardant, it is important that its endothermic decomposition process occurs over the temperature range at which the polymer decomposes. The decomposition temperatures for a number of polymers are shown in Table 4.2.

Figure 4.3 shows differential thermal analyses (DTA) of two samples of aluminium hydroxide of different particle sizes. Most of the water vapour is evolved over the temperature range 200–400°C, which coincides with the decomposition temperatures of many polymers. For these samples, a heating rate of $10°C \, \text{min}^{-1}$ was used. For the coarser material, BACO FRF 5 (median particle size by Coulter Counter 65 μm) three peaks were observed, whereas for the finer material, BACO FRF 80 (median particle size 6.5 μm), the first and third peaks were considerably diminished in size. The first peak, which occurs at about 225°C, is ascribed to the conversion of the hydroxide to the oxide hydroxide:

$$Al(OH)_3 \rightarrow AlO(OH) + H_2O \tag{4.2}$$

The second peak occurs at about 310°C and corresponds to the direct conversion of the hydroxide to an activated alumina:

$$2Al(OH)_3 \rightarrow Al_2O_3(x) + 3H_2O \tag{4.3}$$

The third peak is ascribed to the conversion of the boehmite, formed at lower temperatures, to an activated alumina. This occurs at approximately 525°C.

Table 4.2 Decomposition temperatures for various polymers.

Polymer	Decomposition temperature range (°C)
Polyethylene	335–340
Polypropylene	328–410
Polyvinyl chloride	200–300
Polystyrene	300–400
Polymethyl methacrylate	170–300
Nylon 6 and 6/6	310–380
Polyvinyl acetate	213–325
Polyethylene terephthalate	280–320

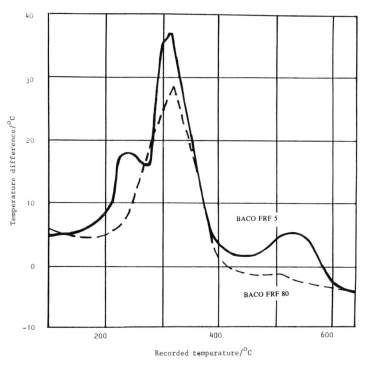

Figure 4.3 Differential thermal analysis trace of aluminium hydroxides of different median particle sizes; heating rate 10°C min^{-1}.

$$2AlO(OH) \rightarrow Al_3O_3(\gamma) + H_2O \qquad (4.4)$$

The route *via* the oxide hydroxide occurs under particular hydrothermal conditions which can exist in the centre of large crystals of the hydroxide. As the percentage of coarse material decreases, the likelihood for these conditions to occur diminishes.

It would therefore be expected that the finer the aluminium hydroxide, the more efficient it is as a flame retardant, since the percentage of water vapour being evolved over the decomposition temperature range for plastic and rubber materials is that much greater. In practice, this effect can be significant but does not manifest itself until a median particle size smaller than approximately 4 μm is reached.

The smoke-suppressing property of aluminium hydroxide is not fully understood but it is believed that the heat dissipation that occurs in the burning polymer favours cross-linking reactions over pyrolysis. This leads to the formation of a char in preference to soot particles.

The physical and chemical characteristics of aluminium hydroxide discussed earlier are mostly advantageous with respect to its use as a fire

Table 4.3 Characteristics of aluminium hydroxide which make it an effective fire retardant.

1. Endotherm occurs at an optimum temperature range for many plastics and rubbers
2. It is an effective smoke suppressant for many polymer systems
3. It does not evolve any corrosive or toxic product on decomposition
4. It presents no health hazard when handled
5. It is virtually insoluble in water, so will not leach out of a filled polymer
6. Grades with purities of 99.6% are readily obtained
7. Its electrical properties make it an ideal filler for insulators
8. It is non-volatile and so will not exude out of the polymer on ageing
9. It is relatively cheap compared to other flame retardants and to the polymer which it replaces
10. The price of the basic hydrate is likely to remain stable as it is made in very large tonnages, being an intermediate in the production of aluminium
11. It is a crystalline material and hence the absorption capacity of its surface is relatively low
12. Available in a wide variety of particle sizes (a few tenths of a μm to 100 μm)
13. It can be produced with a high level of whiteness and is unaffected by ultraviolet light
14. Mohs' hardness of 2½ so it does not cause excessive wear on equipment

retardant. These are summarised in Table 4.3; of particular note are its safe handling properties.

With such an array of benefits, one could be forgiven for asking why any other fire retardants are used. Regrettably, some of the above characteristics may make it suitable for some polymers but reduce its acceptability in others. The main limitations in the use of aluminium hydroxide are as follows: (i) it is unsuitable for polymers that are processed at high temperatures (say >200–220°C), and (ii) relatively high levels are required to impart significant levels of fire retardancy and smoke suppression; this can lead to an unacceptable loss of mechanical properties or unacceptably high process viscosities. Considerable development work has been undertaken to overcome these problem areas.[17] In order to satisfy the growing use of polymer applications and to overcome the processing and mechanical property problems associated with high aluminium hydroxide loadings, considerable effort has been devoted to tailoring materials to meet the requirements of different uses. Products in the particle size range of 100 μm to a few tenths of a micron have been developed with a range of particle size distributions in order to achieve maximum filler loading in resins at minimum viscosity.

Sub-micron grades of aluminium hydroxide are extensively used in cable-sheathing compounds. For these applications, grades of aluminium hydroxide have been specially developed with very low levels of soluble soda, such that they have the minimum adverse effect on the cable when it is immersed in water over a prolonged period. The finest grades are found to impart a reinforcing effect in several elastomer systems.[18]

The polymeric systems where aluminium hydroxide can be employed as a fire retardant/smoke suppressant are very wide ranging,[18,19] as evidenced by the summary given in Table 4.4. Aluminium hydroxide is compatible with

Table 4.4 Application areas for aluminium hydroxide as a flame retardant/ smoke suppressant.

Thermosetting polymers	
Polyester	Spray
	Hand lay-up
	SMC (sheet moulding compounds)
	DMC (dough moulding compounds)
	Foam
Epoxy	
Phenolic	
Polyurethane	
Methacrylic resins	
Thermoplastics	
PVC	Rigid
	Plasticised
Polyethylene	
Polypropylene	
Polystyrene	
Elastomers	
Natural rubber	
Styrene butadiene rubber	
Neoprene	
Nitrile rubber	
Butyl rubber	
Silicone rubber	
EPDM	Ethylene propylene diene monomer
EPR/EPM	Ethylene propylene rubber
Vamac®	Ethylene vinyl acetate
Hypalon®	Chlorosulfonated polyethylene
Vinyl acetate	Vinyl chloride copolymers
Cellulosics	
Loft insulation	
Chipboard or particle board	
Hardboard	

many other fire retardants and blends are frequently employed in order to impart the necessary degree of fire retardance/smoke suppression at minimum cost.

The use of organic and organometallic compounds has been widely investigated in order to improve the compatibility between the surface of the aluminium hydroxide and the polymer. Stearates, plasticisers, silanes, titanates, zircoaluminates and organoaluminium compounds have all been employed with varying degrees of success. The appropriate choice of substituent on the metal can result in easy processing or improved mechanical properties (especially retention of mechanical properties after immersion in water).

4.3.4.2 Toothpaste. The usage of aluminium hydroxide as an abrasive in toothpaste formulations grew dramatically in the 1970s and 1980s to an annual level of several tens of thousands of tonnes. The growth was driven by the trend towards therapeutic/orthodontic pastes and the compatibility of aluminium hydroxide with the majority of compounds which are now frequently incorporated into a toothpaste in order to bestow a therapeutic or prophylactic property.

Additives which are introduced include:

1. fluoride to minimise dental caries;
2. bacteriostats to destroy mouth bacteria, e.g. chlorohexidine;
3. zinc salts to inhibit plaque formation, e.g. zinc citrate;
4. enzyme inhibitors to prevent or reduce acid production; and
5. enzymes to accelerate the breakdown of protein, starch and lipids.

A typical toothpaste might contain *c.* 50% abrasive filler, *c.* 20% water, *c.* 25% humectant such as glycerin or sorbitol, 1–2% surfactant, e.g. sodium lauryl sulfate, 1% thickening agent, e.g. sodium carboxymethyl cellulose, 1% flavour, together with active ingredients to reduce dental caries (fluorides), minimize plaque formation (e.g. zinc citrate), or prevent gum disorders. The abrasive filler used should be sufficiently abrasive that material adhering to the teeth is removed without causing damage to the tooth enamel or any exposed dentine. A particular advantage imparted by aluminium hydroxide when incorporated into a dentrifice is that the toothpaste has very good cleansing properties but causes less damage to tooth enamel than most of the other commonly used abrasive fillers, such as calcium carbonate or dicalcium phosphate (Unilever patent, UK 1 188 353 1968). Aluminium hydroxide has a Mohs' hardness of between 2.5 and 3.5, considerably below that of tooth enamel, dentine or cementum; its abrasivity thus falls between that of chalk and dicalcium phosphate. The degree of abrasivity and cleaning properties can to some extent be controlled by the particle size and particle size distribution of the aluminium trihydroxide used in the formulation. Grades with median particle sizes of between 30 and 1 μm have been used to produce dentrifices with different characteristics. Grades of aluminium hydroxide with median particle sizes of 1 μm have the minimum abrasivity and grades of this size have been used since the 1950s to increase the viscosity of toothpastes, especially under low shear conditions. The utilisation of grades containing particles with a narrow distribution of sizes results in more cost-effective toothpastes as the higher viscosity that they impart allows higher levels of water to be incorporated.

The effect of the filler on the active ingredients is important, especially with respect to fluoride addition. The utilisation of calcium carbonate has declined sharply because it can react with the fluorides added and so obviate their effectiveness as anticaries agents. Sodium fluoride, tin(II) fluoride and sodium monofluorophosphate are all frequently used as fluoride sources in toothpastes. Tests have shown[20,21] good compatibility between aluminium hydroxide and sodium monofluorophosphate, and toothpastes containing them have been found by clinical trials to remain effective in reducing dental caries for relatively long periods. Elevated levels of soluble soda adversely influence the effectiveness of the fluoride-containing pastes. Co-milling the aluminium hydroxide with long-chain fatty carboxylic acids can improve the effective utilisation of the fluoride (US Patent 4 098 878, Colgate Palmolive 1978). Utilisation of grades of aluminium hydroxide with low surface activity also enhances the effectiveness of the fluoride,[22] as well as zinc citrate[23] and other additives. Co-milling aluminium hydroxide with acid salts and organic acids has been shown to render the toothpaste non-corrosive,[24] and to minimise fluoride loss.

4.3.4.3 Paper.

4.3.4.3 Paper. Sub-micron, normally high whiteness precipitated grades of aluminium hydroxide have been used since the early 1950s in paper manufacture, but this use is now declining. Initially aluminium hydroxide was used as a filler replacing some of the titania and substituting for calcium carbonate and China clay. Aluminium hydroxide has a lower refractive index than titania but does not adversely affect the functioning of some of the optical whiteners used in paper. It has very high retention and dispersibility characteristics during paper manufacture and improves the retention behaviour of other fillers and pigments. It produces a higher opacity than China clay and results in an improvement in printing performance.[19,25]

The use of aluminium hydroxide as a replacement for China clay in paper-coating dates back to the 1960s.[26] The aluminium hydroxide imparts high brightness, high opacity, high gloss and good ink receptivity. The high gloss is attributed to the platey morphology of the aluminium hydroxide crystals. Supercalendering makes the addition more effective as the crystal platelets are oriented parallel to the paper surface, resulting in a very smooth, glossy finish.[25]

Some use of precipitated sub-micron grades of aluminium hydroxide and an activated product made from these has taken place in carbonless copying paper. Considerable work has been undertaken in attempts to match the performance of the activated Silton clays which are used for absorption of the micro-encapsulated dyes which burst under writing pressure.

4.3.4.4 Paint. Developments have been in progress since the 1960s partially to replace the titania in paint. Williams[27] showed that a 25% replacement of the titania with a sub-micron precipitated grade of aluminium hydroxide could be made in a vinyl latex coating formulation without loss of opacity but at lower cost. The effectiveness of the aluminium hydroxide was attributed partly to the fact that it is close to the optimum size for providing the ideal spacing of titania particles for light reflection and scattering.

Some use of aluminium hydroxide as a titania substitute is made in acrylic and alkyd paint systems, but the scale is currently small. Small quantities are also used in fire-retardant paints, mastics, intumescent coatings and ablative coatings for its fire- and smoke-inhibiting characteristics.

4.3.4.5 Other uses of aluminium hydroxide

4.3.4.5.1 Sodium aluminate is manufactured by dissolving aluminium hydroxide in *c*. 50% sodium hydroxide. The pure material, $Na_2Al_2O_4$, contains equimolar proportions of Na_2O and Al_2O_3 but commercial grades contain an excess of Na_2O to prevent the spontaneous precipitation of aluminium hydroxide. Additives are sometimes incorporated to improve the stability of solutions of sodium aluminate even further.

The principal uses of sodium aluminate are as a coagulant in water purification, in paper making to improve sizing and filler retention, in the treatment of titania to improve durability, and in the manufacture of catalysts and synthetic zeolites or molecular sieves. Annual production in 1988 was estimated at 125 000 tonnes.[28]

4.3.4.5.2 Zeolites. There are over 150 types of zeolites chemically differentiated by their cation and by the Si/Al ratio in their anionic frameworks. The Si/Al ratio varies from 1, in zeolite-A, to ∞, in silicalite which is aluminium-free. Industrially, the most important zeolites are A, X and Y with Si/Al ratios of 1, 1.5 and 3, respectively. Their general formula is $M_{2/n}O \cdot Al_2O_3 \cdot xSiO_2 \cdot yH_2O$, where M is an alkali or alkaline earth metal whose valency is *n*. In addition to aluminosilicates, aluminophosphates and silicoaluminophosphates (SAPO) have been prepared with zeolite-like structures. Compositions of the more important zeolites are shown in Table 4.5.

Table 4.5 Zeolite compositions.

Type	Typical formula
Synthetic zeolites	
Zeolite-A	$Na_{12}[(AlO_2)_{12}(SiO_2)_{12}] \cdot 27H_2O$
Zeolite-X	$Na_{86}[(AlO_2)_{86}(SiO_2)_{106}] \cdot 264H_2O$
Zeolite-Y	$Na_{56}[(AlO_2)_{56}(SiO_2)_{136}] \cdot 250H_2O$
Zeolite-L	$K_9[(AlO_2)_9(SiO_2)_{27}] \cdot 22H_2O$
ZSM-5	$[Na(TPA)]_3[(AlO_2)_3(SiO_2)_{93}] \cdot 16H_2O^a$
Natural zeolites	
Chabazite	$Ca_2[(AlO_2)_4(SiO_2)_8] \cdot 13H_2O$
Mordenite	$Na_8[(AlO_2)_8(SiO_2)_{40}] \cdot 24H_2O$
Erionite	$(Ca \cdot Mg \cdot Na_2,K_2)_{4.5}[(AlO_2)_9(SiO_2)_{27}] \cdot 27H_2O$
Faujasite	$(Ca \cdot Mg \cdot Na_2,K_2)_{29.5}[(AlO_2)_{59}(SiO_2)_{133}] \cdot 235H_2O$
Clinoptilolite	$Na_6[(AlO_2)_6(SiO_2)_{30}] \cdot 24H_2O$

a TPA = tetrapropylammonium.

The basic building blocks of zeolites are SiO_4 and AlO_4 tetrahedra linked by shared oxygen atoms. Zeolite-A has a structure composed of cubo-octahedra, β cages, linked by their quadratic surfaces to produce a cage structure which has a cavity with a pore diameter of 4.1 nm. Zeolite-X and -Y produce structures with a larger pore diameter (7.4 nm). The uniform pore sizes that are produced are crucial to their use and enable them selectively to absorb or reject molecules based on their molecular size. The cations are mobile and readily undergo ion exchange. The microvoid at the centre of the structure may occupy 50% of the total volume of a crystalline zeolite. Water is readily absorbed but can be reversibly removed by heating. Zeolites can also be used to absorb carbon dioxide, thiols and hydrogen sulfide.

Zeolite-A and -X are manufactured by the reaction of sodium silicate solution with sodium aluminate, ageing of the gel produced and then filtering, washing, drying and, where appropriate, activation at c. 400°C.

Synthetic zeolite manufacture is a very large and rapidly growing market for aluminium hydroxide. Traditional uses have been gas-drying, separation processes, cracking and reforming catalysts for petroleum products, and hydrocarbon conversion. Further applications in new catalyst systems and in selective scavenging and pollution-control processes are growing rapidly. There has been a very large increase in use in recent years as a substitute for phosphate in washing powders.

4.3.4.5.3 Glasses and ceramic glazes. Some aluminium hydroxide is added directly to the glass melt during the manufacture of some glasses (see section 4.3.5.2). Aluminium hydroxide may be incorporated into glazing compositions to improve 'sparkle' and chemical resistance.

4.3.4.5.4 Pharmaceuticals. Amorphous aluminium hydroxide gel is widely used in antacid preparations.

4.3.5   Uses of aluminium oxide

4.3.5.1 Alumina in technical ceramics Alumina ceramics have outstanding resistance to heat, wear and chemicals which has led to a multiplicity of applications and their use in almost every industry. They range in composition from virtually the pure oxide with no additions to compositions containing many other constituents. These other constituents may form a glassy phase and allow components to be fabricated at lower temperatures, and sometimes with more reproducibility. Such constituents may be other oxide ceramics such as zirconia, to give materials with higher strength and toughness, or they may be platelets, fibres or whiskers, e.g. silicon carbide, to give enhanced mechanical properties. Recent years have witnessed an explosion of literature on alumina composites which are derived directly from aluminium metal. In this process — the Lanxide™ (Lanxide Inc.) process — very complex shapes can be manufactured and the oxide can be made to grow through a wide variety of ceramic materials, e.g. SiC particulates or Nicalon™ (Nippon Carlsons).[30,31]

Work on alumina ceramics dates back to the last century and a patent was filed in 1907 (German patent 220 394) covering the production of ceramic alumina. The initial use was in laboratory equipment, followed by spark plugs. Uses now extend to electronic applications, bioceramics, armour, grinding media, abrasion-resistant tiles, cutting tools, wear parts for the textile and paper-making industries etc. The properties of alumina ceramics are summarised in Table 4.6.

4.3.5.1.1 Laboratory ware. Ceramics for use in high-temperature applications, e.g. laboratory crucibles and thermocouple sheaths, are manufactured with the minimum glassy component and with an alumina content of >99%. The ceramics are produced in such a way as to give large grain sizes after sintering.[32]

4.3.5.1.2 Spark plugs. One of the most familiar examples of a high-alumina ceramic is the sparking plug insulator which typically contains 88–95% alumina. High-alumina-content insulators were developed during the Second World War to meet the needs of high compression aero-engines for which the existing mullite plugs (*c.* 55% Al_2O_3) were inadequate.[33] The high silica content of the mullite made the plugs susceptible to attack from the lead in the fuel and other additives in the lubricants. Alumina ceramics were able to meet all the necessary requirements: ability to withstand attack by tetraethyllead and the other compounds in the fuel or lubricating oils; ability to withstand temperatures from sub-zero to over 1000°C; reasonably high

Table 4.6 Properties of high alumina-containing ceramics.

- High tensile strength
- High compression strength
- Hard and abrasive
- Resistant to abrasion
- Resistant to chemical attack by a wide range of chemicals, even at elevated temperatures
- High thermal conductivity
- Resistant to thermal shock
- High degree of refractoriness
- High dielectric strength
- High electrical resistivity, even at elevated temperatures
- Transparent to radiofrequency radiation
- Low neutron capture cross-section
- Relatively inexpensive as a raw material, readily available and price not subject to violent fluctuation

thermal conductivity in order to conduct heat from the nose of the plug to the engine block and thereby avoid pre-ignition; ability to withstand potential differences of the order of 20 000 V at high temperature; and high strength. Generally low-soda aluminas ($Na_2O < 0.05\%$) are used for the manufacture of sparking plug insulators, but there is now a tendency to move to a medium-soda alumina (Na_2O level $< 0.25\%$) in a plug with a slightly lower alumina content.

4.3.5.1.3 Electronic/electrical applications. These applications include thin- and thick-film substrates for integrated circuits, electronic packages, vacuum tube envelopes, RF windows, rectifier housings, sodium vapour lamp tubes and high voltage insulators.

Initially alumina-based ceramics were used for the packaging of electronic devices in order to provide environmental protection.[34] Progressively, their use has been expanded such that the ceramic substrate becomes a functional part of the system. In 1965, IBM introduced a solid logic technology (SLT) module. The sheet of 96% alumina was metallised and resistors screened onto this substrate. Silicon device chips were then connected. Such devices have become progressively more sophisticated with more and more layers[35] and more and more chips, while becoming smaller. The relatively high thermal conductivity of the alumina enables the substrate to conduct heat away and so reduces thermal gradients. The strength of the alumina substrate also enables the devices to withstand both the thermal and mechanical stresses. Multilayer devices with 36 layers of ceramic substrate are now commonplace. The layers of alumina are made by preparing a slurry of alumina in a polymer binder (e.g. polyvinyl chloride acetate) and spreading it out as a thin film using a doctor blade. The binder is then carefully burnt out in order to control the shrinkage. Low-soda, thermally reactive aluminas are frequently utilised in these applications. As such thin films (<1 μm) are applied to the substrate, very smooth finishes (<0.1 μm roughness) are required.

High-purity aluminas (c. 99.99% Al_2O_3) find use in the manufacture of high-pressure sodium vapour lights. Normally alum-derived aluminas are utilised so that the fired polycrystalline alumina tubes are translucent allowing the transmission of $>96\%$ of the light, whilst still providing excellent resistance to the sodium vapour.

Alumina has been used in the less sophisticated area of electrical porcelain since at least 1921 when it was used to replace flint.[36] Alumina additions of up to 40% are made, together with ball clay, feldspar and China clay.

4.3.5.1.4 Bioceramics. While the quantities are small compared to other uses, the application of alumina in bioceramics is well established and presents some particularly challenging problems. The main areas are as bone substitutes in joints and dental implants. Alumina finds use in these areas because of its excellent wear characteristics and resistance to attack from body fluids. Alumina ceramics containing 99.99% Al_2O_3 are employed in order to achieve the necessary chemical inertness. Particularly exacting production conditions are required to manufacture components with small grain size ($<4\ \mu$m), high strength and close size tolerance. Chemical polishing using fluxes is sometimes employed to reduce the coefficient of friction and improve wear resistance. The use of alumina ceramic hip joints has been in practice for many years. Both alumina femoral heads (ball) and acetabulum (cup) in either alumina or ultra-high-molecular-weight poly-ethylene (UHMWPE) have been found to be far superior to the conven-tional titanium metal–UHMWPE joints, especially with respect to wear characteristics. This considerably extends the time before a patient needs a replacement joint. It has been found that the surface roughness of the metal to UHMWPE joint increases as time progresses whilst that for a ceramic system decreases.[37]

Other uses for alumina ceramics in the body include maxillary reconstruc-tions, middle ear ossicular replacements and dental implants. Dental implants have included single-crystal alumina inserts that are threaded into the alveolar bone; these can serve as anchors for bridges or they can be capped with a crown of polycrystalline alumina.[38] In addition to their high strength and high corrosion resistance, the alumina implants have the advantage that no calculus, plaque or concrements adhere to the surfaces.[39]

Some interesting developments have occurred with respect to encouraging the bonding of the alumina ceramic to bone or body tissue. It has been shown that connective tissue infiltration with initiation of bone growth to a depth of 100 μm can occur if porous alumina with a pore size of 75–100 μm is used under certain conditions.[40] Later developments have included alumina polytetrafluoroethylene composites which enable large pore size ranges (50–400 μm) to be produced and so promote rapid ingrowth of tissue.[41] Excellent bone attachment has been reported with an $MgAl_2O_4$-based ceramic composite.[42] This composite, or osteoceramic, consists of $MgAl_2O_4$ and α-$Ca_3(PO_4)_2$ in equimolar quantities. The spinel forms the skeleton and provides the strength whilst the tricalcium phosphate, which fills the skeletal pores, is biorestorable and is progressively replaced by bone.[43,44]

4.3.5.1.5 Armour applications. High-alumina ceramics have been used as armour materials since the early 1960s. The main areas where alumina ceramics have been exploited are for bullet-proof flak jackets for police and military personnel and in strategic parts of vehicles, especially helicopters. The mechanism by which ceramic materials such as alumina work as armour materials is very different from that by which metals function. Metals absorb the energy of a projectile by a plastic deformation mechanism, whilst ceramics absorb the kinetic energy of the projectile by a fracture energy mechanism. The ceramic must therefore have a high hardness, a high compressive strength and a high Young's modulus. On encountering a ceramic surface, the tip of a projectile becomes blunted so that its effective cross-section increases. This leads to a diminution in the compressive stress. A compressive shock wave occurs at the point of impact and travels through the material, resulting in star-shaped radial cracks and circumferential cracks.[45] The microstructure of the alumina ceramic plays an important role in the energy dissipation process.[46]

A backing material with a high fracture toughness is required, e.g. a high-strength steel or precipitation-hardened aluminium alloy on a fibre-reinforced plastic. There is no 'ideal' armour ceramic for all applications. The optimum choice is very dependent upon what specific ballistic threat the armour is required to resist. As the projectiles become more sophisticated, e.g. kinetic energy penetrators and shaped charges, more use has been made of ceramic matrix composites, e.g. silicon carbide whisker-filled alumina ceramics or Lanxide[TM] (Lanxide Corp., Newark) alumina/aluminium-based systems containing particulates or whiskers. Composites offer better integrity after impact and better multi-hit performance than pure ceramics.[47]

4.3.5.1.6 Ceramic tableware. Some alumina is present in ceramic tableware from the alumina occurring naturally in the kaolins, ball clays, feldspars, nepheline syenite and talcs that are used during production. Milled, calcined alumina is frequently added as a replacement for flint or quartz at a level of 10–20% to increase the fired strength of the tableware, giving better mechanical strength and thermal shock resistance.[48,49] In some cooking ware an addition of 50% Al_2O_3 is made. This is widely used for 'hotelware' intended for canteens, restaurants and hospitals where the crockery is subjected to particularly heavy use. Additional benefits that accrue from the incorporation of milled alumina include extension of the firing range, increased whiteness and fewer flaws.[50]

4.3.5.1.7 Wear-resistant components. The extreme hardness and chemical inertness of alumina ceramics make them ideal materials to solve many wear problems. These excellent wear characteristics have been attributed[51] to the absorption behaviour of the surface of the ceramics. The outer ionic layers consist of oxygen ions which possess a surface charge; water or long-chain carboxylic acids are attracted by this charge and are adsorbed onto the surface by van der Waals forces. The resulting layer of adsorbed molecules forms a protective layer that reduces wear.

One of the earliest large-scale uses of alumina ceramic components in this area was for thread guides in the textile industry. Synthetic fibres travelling at high speeds were found to be very abrasive and hardened steel, glass or glazed porcelain components were found to last for only a few days. High-alumina ceramics can be fabricated to last for up to 10 years. Additionally, the very low surface roughness that can be achieved with alumina ceramics means that no damage is caused to the thread.

Other wear components for which alumina ceramics find use are: wire-drawing pulleys for steel and copper wire fabrication; bearings for pumps; joint seals; tap-washer seals; linings for coal- or mineral-handling equipment; cyclone liners; grinding media; ball-mill linings; tines and seed drills in agricultural equipment; and the support for the wire bearing the web of paper in the fourdrinier section of paper-making equipment.

4.3.5.1.8 Cutting tools. Cutting tool inserts composed of high-alumina ceramics allow very high cutting speeds to be achieved, up to $2000\,m\,min^{-1}$. This figure is to be compared with approximately $10\,m\,min^{-1}$ for carbon steels, $50\,m\,min^{-1}$ for high-speed steels containing 25% carbide, and approximately $200\,m\,min^{-1}$ for tungsten or titanium carbide sintered materials. The requirements for an alumina ceramic to meet the stringent demands in cutting tools are a high wear-resistance; high mechanical strength at elevated temperature; and high impact strength. The key to producing a good alumina cutting tool is to maximise the impact strength which is relatively low compared with those of metals. Several routes have been developed to obtain improved impact strength:

1. fabrication of the ceramic with a small grain size;[52]
2. addition of zirconium oxide;[53]
3. addition of titanium carbide to increase the thermal conductivity and hence improve the thermal shock resistance;[53]
4. addition of silicon carbide whiskers;[54] and
5. addition of titanium carbide, titanium nitride, boron nitride or Sialon® (Lucas-Cookson).

A common way to reduce the grain size of the alumina ceramic is to add a few tenths of a percent of magnesia to inhibit exaggerated grain growth;[55,56] the target grain size is approximately $1-5\,\mu m$. High-alumina ceramic cutting tools are especially effective for cast iron and they are used with particular economic advantage for brake drums, brake discs and cylinder liners. In addition to the higher cutting speeds, high-alumina cutting tools produce a superior surface finish compared with other cutting tools and hence reduce the subsequent time and effort required for further finishing operations.

4.3.5.1.9 Zirconia-toughened alumina. As discussed previously, one characteristic of alumina ceramic materials which has limited their application in some areas has been poor fracture toughness. Claussen[57] excited great interest by incorporating zirconia, ZrO_2, into an alumina matrix to give dramatically higher toughness values and substantially higher fracture strengths.

Zirconia may exist in one of four polymorphic forms: monoclinic, tetragonal, cubic and orthorhombic; the orthorhombic form exists only under high pressure. For pure zirconia, monoclinic is the form stable at ambient temperatures. When heated to 950–1200°C, the monoclinic transforms to the tetragonal phase, which, on further heating, transforms to the cubic phase at approximately 2370°C; cooling back to ambient temperatures regenerates the monoclinic form. The transformation on cooling from the tetragonal to the monoclinic form results in a volume increase of 3–5%. The addition of certain oxides, e.g. Y_2O_3, MgO or CaO, can stabilise certain forms.[58]

If zirconia particles are incorporated into an alumina matrix, then the volume increase on cooling generates microcracks in the matrix as a result of the tetragonal → monoclinic transformation. These microcracks dissipate the energy of the propagating crack and increase considerably the toughness of the material. The size of the zirconia particles must be large enough to transform but small enough that they cause only limited microcracking. If the zirconia particles are less than a certain critical size, then the tetragonal → monoclinic transformation does not occur. Thus, if very small zirconia particles are incorporated into the alumina, cooling leaves the zirconia in the metastable tetragonal form. In the vicinity of a propagating crack tip, the stresses release the matrix constraints on the metastable tetragonal zirconia particles which, under certain conditions, will transform to the monoclinic form. The volume expansion and compressive strain thereby generated mean that additional work is required to propagate the crack further; this results in increased toughness and strength.

Typically zirconia levels of 10–15 vol.% are incorporated into the alumina matrix, but the optimum value is dictated by the zirconia particle size, the presence or otherwise of stabilisers in the zirconia, and the balance between fracture toughness and fracture strength required. Fracture toughness values twice that for conventional aluminas (approximately $10\,MPa\,m^{1/2}$) have been reported by several groups of workers.[57,58]

4.3.5.1.10 Ceramic matrix composites.   Considerable interest has been aroused by the incorporation of fibres, whiskers and platelets into alumina ceramics in order to improve specific properties, especially toughness.

The addition of 20 vol.% of silicon carbide whiskers in alumina has given a fracture toughness value of $7.8\,MPa\,m^{1/2}$ and a flexural strength of $700\,MPa$.[59] The addition of single-crystal silicon carbide fibres (diameter $0.6\,\mu m$, length 10–80 μm) raises the fracture toughness from $4.6\,MPa\,m^{1/2}$, with no addition, to $8.7\,MPa\,m^{1/2}$ (crack plane parallel to the pressing direction) and $5.4\,MPa\,m^{1/2}$ (crack plane perpendicular to the pressing direction).[60]

4.3.5.1.11 Beta-aluminas.[61]   Beta-alumina has the idealised chemical formula $Na_2O \cdot 11Al_2O_3$ with a layer structure where the sodium atoms are present in discrete layers separated by layers of aluminium and oxygen atoms forming a spinel structure. There are two main sub-groups, designated beta'- and beta''-alumina, which have different stacking sequences. Beta''-alumina has the approximate chemical formula $Na_2O \cdot 5.33Al_2O_3$; it has a structure similar to that of beta-alumina but, with a lower resistivity, is now the preferred material for use as an ionic conductor in electrochemical cells.[62] The beta''-alumina structure is stabilised by the addition of magnesium or lithium ions: final levels of magnesia in the product may be 2.5–4 mass% and lithia between 0.2 and 0.8 mass%. The presence of CaO, SiO_2 and K_2O is especially critical as they adversely affect the electrical resistivity and electrochemical behaviour of beta''-alumina. A preferred method, patented by Duncan et al.,[63] which gives beta''-aluminas with improved purity and excellent conductivity, utilises crystalline boehmite (CERA® Hydrate, Alcan Chemicals Ltd) as the alumina source. High levels of lithia (0.80% by mass) enable tubes for electrochemical cell separators to be made containing substantially 100% beta''-alumina. The addition of 5–15% zirconia to the beta''-alumina, aimed at transformation toughening of the material, inhibits the formation of large crystals and enhances the performance of the cell.[62]

Uses of beta-aluminas. The goal of a relatively light, cheap battery which could provide sufficient energy density to power electric vehicles has been a focus of researchers for decades. The development of the sodium-sulfur battery in the late 1960s was a major step in achieving that objective.[64] The battery had a much higher energy density then a lead acid battery (100–150 W h kg^{-1} compared with 40–48 W h kg^{-1}) and enabled an electric vehicle to travel twice as far for the same weight of battery.

The sodium-sulfur battery comprises molten sodium contained in a beta-alumina tube as the negative electrode and sulfur as the positive electrode. The beta-alumina keeps the sodium and sulfur separate but allows the conduction of sodium ions. The operating temperature of the battery has to be maintained at c. 350°C to keep the constituents molten. This is one of the major disadvantages as special precautions must be taken to ensure heat insulation and adequate containment in the event of an accident. Considerable opportunities exist for sodium-sulfur batteries in relatively low-cost power load-levelling systems, which would enable power plants to run at maximum efficiency. An electrically powered vehicle would require approximately 500 sodium-sulfur batteries, whereas a load-levelling storage battery would require 10 000 or more cells. Sodium-sulfur battery-powered vehicles have been running on an experimental basis in Manchester for some years.

Considerable effort has been expended in enhancing the performance of sodium-sulfur batteries.[65] A key area has been the improvement in consistency and number of charging/discharge cycles prior to failure. Recent developments with systems based on sodium-aluminium chloride show particular promise and should one day enable a relatively cheap vehicle to be produced that is free from air-pollution and conserves liquid fuels.

4.3.5.2 Alumina in glass.   The alumina content of natural glasses (mainly rhyolitic obsidians) is 12–15 wt% The substitution of silicon by aluminium in a silicate glass has a significant effect on the stability of the three-dimensional structure. Binary Al_2O_3–SiO_2 glasses are stable only up to *c.* 7 wt% Al_2O_3; above this level a liquid phase, high in alumina, separates out.[66]

Low-level additions of alumina to soda-lime-silica glasses reduce significantly the tendency for devitrification or phase separation. Alumina also reduces the liquidus temperature when added to binary silicate glasses in amounts up to 10 wt%. Moreover, it improves the durability of certain glasses, especially sodium borosilicate glasses used for scientific equipment, for pharmaceutical products or for body fluid containment where low alkali extraction rates are required. The acid resistance of Na_2O–CaO–SiO_2 glasses is improved by alumina-additions. The addition of up to 3 wt% Al_2O_3 is made to binary alkaline earth silicate and Pyrex® (Corning Inc.) type borosilicate glasses; without the alumina the borosilicate may exhibit immiscibility problems on cooling, resulting in poor durability and unpredictable properties.[67]

Alumina has a strong network-forming effect in ternary aluminosilicate glasses, thereby increasing their viscosity. The alkaline earth aluminosilicate glasses have high melting temperatures and high viscosities; this, however, allows strengthening *via* physical tempering and use at high temperature without deformation or loss of stress. The effect has been turned to advantage in Corelle® (Corning Inc.) laminated dinner ware, and a similar composition used for 'E'-glass fibres and tungsten-halogen lamp envelopes.

Calcined alumina is used in optical, lens and other special glasses requiring a high degree of purity in the raw materials, particularly with respect to impurities such as iron and chromium that can affect the colour and light transmission of the glass. High-purity aluminas are used in some photochromic glasses where control of both Al_2O_3 and impurity levels is critical.

4.3.5.3 Glass-ceramics.   Glass-ceramics have been available since the 1950s.[68] Alumina is added to control the crystallisation of glasses in order to produce a ceramic with a fine, uniform crystal size. One of the earliest materials to be developed to have practical use was cordierite, $2MgO \cdot 2Al_2O_3 \cdot 5SiO_2$; the alumina content is *c.* 35 wt%. To this magnesium aluminosilicate, titania is added as a nucleating agent. Cordierite is still widely used because of its excellent thermal shock resistance, strength, toughness and low thermal expansion. Applications include catalyst substrates and radar-transmitting missile nose-cones. Considerable interest exists in replacing alumina electronic substrates with glass-ceramics such as cordierite which have a lower dielectric constant and a lower sintering temperature than does alumina. This allows multi-layer substrates coated with Cu, Ag or Au conductor patterns to be co-fired and higher circuit densities to be obtained.

Another area of glass-ceramics rich in commercial significance is defined by the lithium aluminosilicates. There exists a wide range of compositions, between $Li_2O \cdot Al_2O_3 \cdot 4SiO_2$ and $Li_2O \cdot Al_2O_3 \cdot 10SiO_2$; all have similar structures and properties. β-Spodumene ($Li_2O \cdot Al_2O_3 \cdot 4SiO_2$) can be nucleated within a lithium aluminosilicate to produce a microcrystalline glass-ceramic with a thermal expansion coefficient close to zero. This has led to its utilisation in cooking dishes. These commercial lithium alumino-silicates have an alumina content typically of 18 wt%. The incorporation of titania or zirconia nucleating agents into lithium aluminosilicates enables β-quartz glass-ceramics to be produced with a crystallite size $<0.1\ \mu m$. This type of material is used for transparent cookware (e.g. Visions®, Corning Inc.), wood-stove windows and electric cooker tops.[67]

Glass-ceramics based on mullite, $3Al_2O_3 \cdot 2SiO_2$, with alumina contents above 35 wt%, can be produced and exhibit good high-temperature properties, excellent dielectric properties and low thermal expansion. Mullite fibres have been produced by the sol-gel process and sold under the designation Nextel® (3M Company).

Nepheline-based glass-ceramics form the basis of highly durable tableware. These materials consist in the main of $Na_2O \cdot Al_2O_3 \cdot 2SiO_2$ (nepheline) and $BaO \cdot Al_2O_3 \cdot SiO_2$ (celsian); they utilise titania as the nucleating agent and have an alumina content of c. 30 wt%.

Glass-ceramics based on fluorophlogopite ($KMg_3AlSi_3O_{10}F_2$) containing 10–20 wt% Al_2O_3 were developed in the 1970s. They can be made to have excellent dielectric properties, dimensional stability, zero porosity and the capability of being machined by ordinary metal-working tools.[69] This last feature is possible because randomly oriented, tiny flakes of mica allow only local fracturing to occur. A commercially available material, Macor® (Corning Inc.), is used for windows for microwave tube parts, precision electric insulators, seismograph bobbing and boundary retainers for the space shuttle.[67]

4.3.5.4 Refractories. Alumina is widely used in refractory products to enhance their chemical resistance and improve their thermal properties. High-alumina refractory bricks and castables are inert to most corrosive fumes and slags and will resist both reducing and oxidising atmospheres. In addition, they exhibit increased load-bearing ability, as well as resistance to spalling, thermal shock and flame impingement. Calcined, fused and tabular alumina are all used in the production of refractories for the ferrous, non-ferrous, cement, glass, chemical, petroleum, ceramics and mineral-processing industries. The usage of alumina in refractories falls into two areas which are almost equal in size: refractory bricks and monolithic refractories.

4.3.5.4.1 Refractory bricks. Alumina-silica refractories are often classified as either low-alumina ($<45\%$ Al_2O_3) or high-alumina ($>45\%$ Al_2O_3) materials. The low-alumina refractories are made from fireclays, *viz.* clays with an alumina content of $>20\%$. High-alumina refractories utilise the aluminas listed above, or bauxite or sillimanite, kyanite or andalusite. Alumina bubbles are sometimes used in refractories to improve their heat-insulating characteristics; these alumina bubbles can be made by blowing air into streams of molten alumina.

The choice of the composition of a refractory brick is dictated by its ability to withstand (i) chemical attack, (ii) erosion, (iii) thermal cycling, (iv) thermal shock and (v) transfer of heat. Whilst fully dense ceramics have many of the required properties, they do not normally have sufficient resistance to the thermal shock witnessed in furnaces. The key is therefore to combine the necessary components to give a material that has the required degree of chemical resistance but, in addition, a controlled microporosity to impart the necessary hot strength.

4.3.5.4.2 Alumina in monolithic refractories. Monolithic refractories can be sub-divided into four main types: castables, mouldables (or plastic refractories), ramming mixes and gunning mixes. They were initially introduced for the repair of furnaces but are now used for whole furnace linings. Monolithic refractories have advantages over refractory bricks when kilns or furnaces of complex shape are involved, and are frequently easier and quicker to install than refractory bricks. Monolithic refractories comprise an aggregate and a binder. The aggregate in alumina-based monolithic refractories can be any one of the following: calcined alumina, reactive alumina, fused alumina, tabular alumina, calcined bauxite, kyanite (a naturally occurring aluminosilicate from the United States, Sweden, Spain or Brazil with an Al_2O_3 content of 56–60%), sillimanite (an aluminosilicate formed under higher temperature and pressure than andalusite and found in South Africa and India), andalusite (an aluminosilicate containing 55–61% Al_2O_3 mined in Spain, South Africa, France, China and Brazil), calcined kaolin, pyrophyllite (theoretically $Al_2O_3 \cdot 4SiO_2 \cdot H_2O$ mined principally in North Carolina), mullite ($3Al_2O_3 \cdot 2SiO_2$) or calcined fireclay. Normally precalcined materials are used in order to minimise the volume change on firing. The aggregates are incorporated to a level between 50 and 80%.

There are two distinct approaches to effecting bonding of the aggregates in monolithic refractories. Depending upon the system, they are supplied either in a moist form which develops its initial strength on drying or reaction with oxygen, or in a dried mix which requires the addition of water on installation to achieve the necessary degree of hydration.

Plastic refractories that utilise clay to introduce plasticity have been commercially available since 1914. The clay provides a low level of strength on drying out and then imparts full strength on firing by ceramic bond-formation.

Phosphates, especially monoaluminium phosphate and phosphoric acid, are widely used as binders.[70] They provide excellent hot strength and abrasion resistance to the refractory. Aluminium chlorophosphate hydrate, urea phosphate, chromium aluminium phosphate and alkali meta-phosphates have all been used in this role. Aluminium monophosphate is the preferred binder as the direct use of phosphoric acid can cause bloating under certain circumstances. The ultimate heat treatment and aggregates present in the mix are particularly important in achieving the desired properties. Aluminium hydroxide chloride, alumina sols and aluminium sulfate are used as air-dried binders in plastic refractories. The aluminium hydroxide chloride and sulfate give products with good performance but evolve corrosive decomposition products on baking out.

Ramming mixes are forced into position by mechanical action and fired *in situ* in the furnace. They are similar to plastic refractories but contain less binding agent.

Castable refractories generally utilise calcium aluminate as the binder. Calcium aluminate cements with different levels of alumina (36–47%, 48–62% and 70–80% Al_2O_3) can be added to provide progressively higher hot-strength properties. In order to improve service temperatures under certain conditions, development work has focused on cement-free castables and both aluminium and ϱ-alumina[71] additions have been found to be successful.

Gunning mixes may comprise any of the other types of formulation but are generally closest to castables. Water is added to the mix immediately prior to use and it is sprayed into position by compressed air. The size of the aggregate used is carefully chosen in order to have the appropriate viscosity and flow characteristics. This is a particularly rapid method of applying a furnace lining.

The higher the alumina content in the monolithic refractory, the higher is the service temperature. Thus, the service temperature rises from 1540–1600°C for a plastic refractory containing 40–50% Al_2O_3 to 1800–1900°C for a refractory containing 90% Al_2O_3.[72] The choice of the alumina-containing aggregate or binder is, however, critically important in ensuring that the correct mineralogy is obtained in the material after installation and heating. Higher alumina contents result in refractories with higher thermal conductivity and thermal expansion.

4.3.5.4.3 Spinel. The period of the late 1980s and early 1990s has seen very significant growth in the usage of the mineral spinel, magnesium aluminate ($MgAl_2O_4$: theoretical Al_2O_3 content 71.7%). Spinel-magnesia refractories are resistant to both acidic and basic slags but have better thermal shock characteristics than do basic magnesia refractories. Spinel melts at 2135°C and has a theoretical density of $3.59\,g\,cm^{-3}$. The combination of good chemical resistance to alkalies and suitable thermal properties makes it the preferred choice in situations such as the hot zone of a cement kiln.[73,74] As cement clinker is high in calcia, it will react with aluminosilicate-based refractories. During cement-production, the clinker forms a continuous coating inside the kiln. At periodic intervals, this clinker breaks away exposing the refractory to a very high thermal shock. That the shock is too great for dolomite-based refractories led to the development of direct-bonded chrome magnesite bricks. Concerns about the disposal of chrome-based refractories has led the industry subsequently to search for alternatives. Periclase-spinel bricks provide a good solution to the problem by virtue of the following characteristics:

1. they are resistant to the alkaline environments exhibited in a cement kiln;
2. they have a lower thermal conductivity than do magnesia bricks so the kiln shell runs at a lower temperature and less heat is wasted;
3. the density is lower than that of magnesia leading to better reliability of the kiln;
4. they have much better resistance to thermal shock; and
5. they have a lower thermal expansion.

Spinels rich in either alumina or magnesia have been developed.

   Recent years have also witnessed considerable growth in the use of spinel in refractories for the Japanese steel industry. The addition of spinel to alumina improves the slag and spalling resistance of castable refractories for the ladle linings of basic oxygen furnaces (BOFs). Bricks with a chemical composition between 80% MgO and 80% Al_2O_3 can be produced to give not only a range of slag and temperature shock resistances, but also a range of creep behaviour. These materials find application in sliding gate plates, continuous casting ceramics and zone-lining in torpedo cars.

4.3.5.5. Ceramic fibres. Amorphous alumina-based ceramic fibres were developed in the 1940s. The concern about the use of asbestos led to renewed interest and they are now employed in a wide range of refractory insulation systems. They are more prone to chemical attack than is asbestos but have better thermal stability; aluminosilicate fibres can be utilised at temperatures up to 1400°C. Standard grades contain 45–50% Al_2O_3 but compositions richer in alumina are made for use under higher service temperatures. They are made from natural alumina-containing minerals or from calcined alumina, depending on the final composition required. Minor

quantities of ZrO_2, Cr_2O_3 or B_2O_3 are added to control various characteristics. The fibres are manufactured by melting the raw materials in a
de Bussey-type furnace and a stream of molten material is then allowed to
flow onto a series of rotating wheels, or the melt is subjected to a high-
velocity air-jet. High-alumina fibres can be made by utilising a solution
or sol-gel method. Saffil® (ICI Ltd) is a 95% Al_2O_3/5% SiO_2 fibre
manufactured from an aluminium oxide precursor such as basic aluminium
chloride.[75]

4.3.5.6 Polishing and grinding.   Naturally occurring alumina in the form
of corundum, either alone or when mixed with magnetite, and termed
emery, has been used as an abrasive for over two and a half thousand years.
Corundum has a Mohs' hardness of 9, a figure exceeded by only a few
materials, *viz.* diamond, boron carbide and silicon carbide. Alumina in an
appropriate form has been used for polishing a wide variety of materials:
rocks, stainless steel, non-ferrous metals, plastic spectacle lenses, plastic
aircraft windows, jewellery, or car paint finishes.

The alumina is sometimes used in an oil- or water-based slurry or it can be
mixed with oil or a wax to form a paste or block. The critical properties
which affect the performance of the alumina are its particle size, particle
shape, particle size distribution, α-alumina content and internal porosity.
For most applications, purity is not critical and 'ordinary' or 'normal' soda
products (i.e. 0.25–0.6 wt% soda) are used. Coarse aluminas (3–30 μm) are
used when a high cutting rate is required, while finer aluminas (2 μm or less)
are used when a polishing action is sought. Control of particle size and
particle size distribution is especially important for polishing plastic lenses
and aircraft windows.

Alumina is added to toothpastes specifically formulated for use by
smokers to aid the removal of nicotine deposits from teeth. Here the
avoidance of coarse particles which might cause scratching of the enamel is a
salient factor. Very fine aluminas with a narrow particle size distribution are
sometimes added to magnetic tape at a low level (e.g. 1% addition) to
prevent the build-up of iron oxide deposits on the recording/playback
mechanisms.

The production of fused alumina dates back a century to 1893 when
Wehrlein in France melted alumina-containing minerals by electrical means
to make 'electrocorundum', or artificial corundum.[76] Several hundred of
thousands of tonnes of alumina are now utilised annually in the production
of fused alumina destined for the manufacture of grinding wheels. In this
process a 'normal' soda-grade of alumina is melted in an electric arc furnace
using graphite electrodes. The melt is allowed to solidify, and the block of
alumina is then broken up into a grain of various sizes. This grain is mixed
with a resin or glass-bonding agent and formed into shapes to produce
grinding wheels. Additions of the oxides of titanium, chromium or

zirconium are sometimes incorporated into the melt to give improved performance. The addition of up to 5% TiO_2 improves the abrasive properties and up to 5% Cr_2O_3 increases the tenacity.[77] Alumina-based grains with zirconia contents in the region of 25% by weight are very effective in high-speed grinding equipment; they give longer wheel life and higher cutting rates. Particles of fused alumina of different sizes are also employed to make sanding or abrasive papers and belts.

Abrasive materials with exceptionally good performance can be manufactured by mixing colloidal aluminas with a magnesium compound and then calcining the mixture. Material with an alumina core enveloped by magnesium spinel is produced; this has especially good cutting action, giving an abrasive with a longer performance.[78]

4.4   Activated alumina

4.4.1   Introduction

As described earlier (section 4.3.3), various transition aluminas are obtained when aluminium trihydroxide or aluminium oxide hydroxide is heated at temperatures below that necessary to form α-aluminium oxide. These transition aluminas are frequently termed 'activated' or 'active' aluminas. Ulrich[79] designated the aluminas obtained by dehydroxylation of aluminium trihydroxide as γ-aluminas. This term persisted and came to be used for all the aluminas obtained by calcination at temperatures below 1000°C. The generally agreed forms of transition aluminas are γ, δ, η, θ, χ, κ, ϱ and ι. ϱ-Alumina is amorphous but the other forms have reasonably well-defined X-ray diffraction patterns. The transition phases that occur on heating aluminium trihydroxide (gibbsite) in vacuum are ϱ-alumina (100–400°C), η-alumina (270–500°C), θ-alumina (870–1150°C) and finally α-alumina. In air, two distinct routes may be followed: either *via* boehmite (180–300°C), γ-alumina (500–850°C), δ-alumina (850–1050°C) and θ-alumina to α-alumina; or *via* χ-alumina (200–500°C) and κ-alumina (900–1000°C) to α-alumina. The sequences of transformation are shown in Figure 4.4. The route followed is determined by a number of factors, in particular the particle size of the gibbsite and the hydrothermal conditions in the vicinity of the particles; other factors that have an influence on the path include the heating rate, moisture level in the feed, pressure, bed depth and soda content. Thus, for coarse gibbsite particles ($>60\ \mu$m) the route *via* boehmite is favoured as the water of dehydroxylation cannot readily escape. In the case of sub-micron gibbsite particles, however, hardly any boehmite may be produced. The amorphous form, ϱ-alumina, can be produced by heating gibbsite at temperatures above 700°C in a fast air stream or in a vacuum.[6]

in vacuum

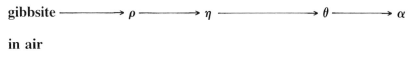

in air

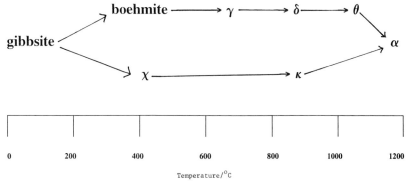

Temperature/°C

Figure 4.4 Thermal transformation sequences for gibbsite.

As for gibbsite, the dehydroxylation sequence of bayerite may follow two pathways. Coarse particles ($>100\ \mu$m), high pressures, fast heating rates and moist conditions tend to favour the route *via* boehmite, γ-alumina, δ-alumina and θ-alumina to α-alumina in the same way as with gibbsite. Fine bayerites transform under conditions of dry air and low heating rates *via* the following route: η-alumina (250–500°C) → θ-alumina (850–1150°C) → α-alumina.[80] The dehydroxylation sequence of nordstrandite is similar to that of bayerite.[81]

4.4.2 *Production of activated alumina*

Worldwide production of activated alumina is of the order of 100 000 tonnes per annum, principally for adsorption and catalysis. Historically,[82] it was produced by the dehydroxylation of gibbsite which is sometimes deposited as a scale at the bottom of the decomposition (precipitation) vessels during the Bayer process. Dehydroxylation was undertaken in a stream of air at 400°C to give a material, principally γ/η-alumina with some χ-alumina and boehmite, with a surface area of *c*. 250 m^2 g^{-1} and a broad distribution of pore sizes up to 500 nm in diameter.

Pingard[83] found that a material with a higher surface area, 250–375 m^2 g^{-1}, could be made by heating aluminium trihydroxide to temperatures between 400 and 800°C to give predominantly amorphous ϱ-alumina. Pelletisation of this material allows the making of spheres which can be rehydrated and activated to give the desired strength and pore volume. On rehydration, pseudoboehmite and bayerite are produced and these transform to

η-alumina on dehydration. Higher initial dehydration temperatures lead to very amorphous products with improved rehydration characteristics and final surface areas of $320–380 \, m^2 g^{-1}$. Activated aluminas of this type are widely used as desiccants, selective adsorbents, catalysts and catalyst supports.

An especially favoured route for the production of aluminas for catalyst manufacture is *via* alumina gels to produce pseudoboehmites. There is considerable scope in this approach for modification of the pore volume distribution of the product. A considerable body of patent literature exists covering different routes to alumina gels but the most common routes are based on the following procedures: neutralisation of aluminium sulfate or ammonium alum by ammonia; or neutralisation of sodium aluminate with carbon dioxide, sodium bicarbonate or aluminium sulfate. The mixing conditions employed to obtain the desired purity and morphological characteristics are critical. The precipitates obtained are frequently gelatinous and difficult to wash free from impurities. The filter cake may then be formed into the desired shape or spray-dried to give spheres. These materials are then activated at a temperature of 400–600°C to give products with a surface area of $300–600 \, m^2 g^{-1}$ and a pore size of approximately 4 nm, and which are mixtures of γ- and η-alumina.

Another route that is widely used for the production of activated alumina with a high surface area is *via* hydrolysis of aluminium alkoxides. The Ziegler process involves the reaction of aluminium metal with ethene and hydrogen to form triethylaluminium; controlled addition of more ethene leads to polymerisation of the ethene on each of the aluminium-ethyl arms to give the desired trialkylaluminium product which may then be partially oxidised to form the alkoxide. Hydrolysis of the alkoxide so produced results in linear alcohols and a high-purity pseudoboehmite. The main impurities are titania and traces of carbon. The product is widely used as a wash coat for automotive emission catalysts. It can be extruded to various shapes and can be peptised with various acids to give a high 'green' strength to the extruded body. The extruded material may be activated at 500–600°C to give γ-alumina with a surface area of $185–250 \, m^2 g^{-1}$.[25]

4.4.3   Uses of activated alumina

4.4.3.1   Adsorbents.   The use of activated alumina as an adsorbent goes back to the 1930s; since that time, large tonnages have been used for gas and liquid drying. Even though the surface of an activated alumina has a strong affinity for water, making it very effective as a desiccant, the process can be reversed readily at temperatures above 250°C under conditions of low relative humidity. The adsorption-desorption cycle can be repeated many hundreds of times without significant deterioration of efficacy. At a relative humidity of 50%, activated alumina can adsorb up to 20% of its own weight of water. The mechanism proposed for water-bonding assumes there to be oxide ions on the outermost surface layer and an incompletely coordinated aluminium ion in the next lower layer. This exposed cation is located in a 'hole' that is electron-deficient and acts as a Lewis acid site. The rehydroxylation of the dehydroxylated alumina surface on exposure to water vapour is accompanied by considerable heat evolution.[25]

Activated alumina can be used for removing water from a very wide range of compounds including acetylene, benzene, alkanes, alkenes and other hydrocarbons, air, ammonia, argon, chlorinated hydrocarbons, chlorine, natural gas and petroleum fuels, oxygen, sulfur dioxide and transformer oils.[84] Activated alumina can dry a gas to a water content lower than that achievable with any other commercially available desiccant.[25]

In addition to water removal, activated alumina can be used selectively to adsorb certain other chemical species from gaseous or liquid streams. Polar molecules such as fluorides or chlorides are readily adsorbed and so activated alumina is used in petroleum refining to adsorb HCl from reformed hydrogen and organic fluorides from hydrocarbons produced by the HF-alkylation process. Other important adsorption processes include the following:[85,86]

1. removal of trace chloride and iron contaminants from ethylene dichloride prior to PVC manufacture;
2. carbonyl sulfide removal from propylene prior to polypropylene manufacture;
3. removal of fluoride from waters with excessively high natural fluoride levels;
4. colour- and odour-removal from industrial effluents; and
5. removal of degradation products from transformer oils.

Chromatographic uses for activated alumina in the laboratory have been common since the 1930s but large-scale chromatography is now finding increasing application, for example in the manufacture of flavours and fragrances, vitamins, steroids, proteins and the separation of the *meta-* and *ortho-* isomers of xylene. Activated alumina has better resistance to moderately high pHs than does silica and so is the preferred choice for higher-pH systems. Activated alumina has a surface with both Lewis and Brønsted acidic and basic sites. Acidity is derived from the Al^{3+} ions and H_2O molecules coordinated to cationic sites, while basicity is due to basic hydroxide groups and O^{2-} anion vacancies.

4.4.3.2 Catalytic uses.   Activated aluminas find widespread application as both catalysts in their own right and as catalyst substrates. Some of the more significant applications are summarised below.

1. *The Claus catalyst* for the removal of hydrogen sulfide in natural gas

processing, petroleum refining and coal treatments; sulfur is obtained as a by-product.

2. *Alcohol dehydration* to give olefins or ethers.
3. *Hydrotreating* to remove oxygen, sulfur, nitrogen and metal (V and Ni) impurities from petroleum feedstocks and to increase the H/C ratio.
4. *Reforming catalysts*: Pt and Re catalysts on a γ-alumina substrate are used to raise the octane-number of petrol.
5. *Automotive exhaust catalysts* used to oxidise the hydrocarbons and carbon monoxide to water and carbon dioxide. The alumina, as a γ-alumina slurry, pseudoboehmite or boehmite sol, is coated on a cordierite or metal honeycomb. The surface is then activated by heating and platinum, palladium or rhodium coated on the alumina.

The preferred forms of activated alumina used in catalytic processes are η and γ. η-Alumina is preferred in acid-catalysed reactions because of the increased number of Lewis and Brønsted acid sites. γ-Alumina is preferred for hydrotreating applications because of its superior thermal and hydro-thermal stability. In addition to being able to control pore size, pore size distribution and pore structure, increasing attention is now devoted to modifying the alumina surface so that it is optimised for different catalyst or adsorption uses.

4.5   Oxides and hydroxides of gallium, indium and thallium

4.5.1   Gallium oxides: forms, preparation and structure

Gallium(III) oxide, Ga_2O_3, is the most stable form of gallium oxide and, like alumina, it exists in several crystalline forms depending upon the conditions of preparation. Five forms have been reported;[87] their crystal structures are listed in Table 4.7. The β-form is the most stable (melting point 1725°C) and is obtained by calcination of the hydroxide, nitrate, acetate or oxalate at temperatures above 600°C. α-Gallium(III) oxide can be obtained by heating gallium oxide hydroxide (or 'gallic diaspore'), GaO(OH), at 300–500°C, or by heating β-gallium(III) oxide at 65 kbar and 1100°C for 1 h. γ-Gallium(III)

Table 4.7 Crystal structures of gallium(III) oxides.

Form	Crystal structure	Similar compound
α-Ga_2O_3	Trigonal $R\bar{3}c$	Corundum, α-Al_2O_3
β-Ga_2O_3	Monoclinic, $C2/m$	θ-Al_2O_3
γ-Ga_2O_3	Cubic, $Fd3m$	Spinel
δ-Ga_2O_3	Cubic, $Ia3$	α-Mn_2O_3
ε-Ga_2O_3	Orthorhombic	κ-Al_2O_3

oxide is prepared by the rapid dehydration of a hydroxide gel at 400–500°C. Decomposition of gallium nitrate at approximately 250°C produces δ-gallium(III) oxide which, if briefly heated to 500°C, gives rise to ε-gallium(III) oxide.[88]

β-Gallium(III) oxide is amphoteric and is more reactive than alumina, forming salts with acids and gallate(III) derivatives with bases. It dissolves in dilute aqueous mineral acids with moderate heating. It can be reduced to gallium metal by heating to 600°C in hydrogen or carbon monoxide. Gallium(III) oxide reacts with many metal oxides on heating; it forms gallates with alkali-metal oxides, M^IGaO_2, and spinels $M^{II}Ga_2O_4$ with the oxides of magnesium, zinc, cobalt, nickel and copper.

Gallium(II) oxide, GaO, is not stable under normal conditions and has been detected only in the vapour state on the evidence of the emission spectrum.[89] Gallium(I) oxide or gallium suboxide, Ga_2O, is a dark brown powder stable in dry air at ambient temperatures. It decomposes at temperatures above 100°C to gallium metal and gallium(III) oxide. Gallium suboxide can be prepared by reducing gallium(III) oxide with gallium under vacuum at temperatures between 500 and 700°C; purification is effected by sublimination under vacuum at 500°C in the presence of gallium.[90] Gallium(I) oxide is notable for being a strong reducing agent.

4.5.2   Uses of gallium(III) oxide

Gallium(III) oxide is used in the manufacture of gadolinium gallium garnet as a substrate in bubble domain memories; these are compact devices not requiring the storage power demanded by other memory devices.

4.5.3   Gallium hydroxides: forms and preparation

Gallium(I) hydroxide, GaOH, has been detected only in flames and low-temperature matrices. The so-called 'monohydrate', GaO(OH) (equivalent to $Ga_2O_3 \cdot H_2O$) has an orthorhombic crystal structure (*Pbnm*) and is similar to diaspore;[88] it can be prepared by heating gallium metal with water in an autoclave at 200°C, by the dehydration at 100°C of gallium trihydroxide,[87] or the ageing of gallium(III) trihydroxide. Gallium(III) trihydroxide, $Ga(OH)_3$, which seems to have an amorphous structure, is prepared by the neutralisation of aqueous solutions of gallium salts or of alkaline gallates.[87]

4.5.4   Indium oxides

Indium(I) oxide, In_2O, can be prepared by heating indium(III) oxide, In_2O_3, to 700°C in a vacuum; a crystalline form has not been reported.[91] Like its gallium(II) counterpart, indium(II) oxide, InO, has been detected only in the vapour phase by spectroscopic means. Indium(III) oxide, In_2O_3, can be produced by heating the hydroxide, nitrate, carbonate or sulfate, or by burning indium in air. Indium(III) oxide decomposes at c. 2000°C; it is soluble in acid but not in alkalies. It has a cubic (Ia_3) crystal structure,[92] which transforms under high pressure to a rhombohedral modification with a corundum structure.[93] The oxide In_3O_4, which has been reported, is presumably a mixed valence compound. Peroxides of the type $In_2O_x \cdot nH_2O$ ($x = 4, 5, 6, 7, 8$ or 9) have also been described.[91]

4.5.5 Uses of indium oxides

Indium(III) oxide and mixed oxides with tin oxide, $(InSn)_2O_3$, are transparent and conductive; they find application for liquid crystal displays and demister strips on windscreens. In_2O_3 has also been used to colour glass yellow.[94]

4.5.6 Indium hydroxide

Indium(III) hydroxide, $In(OH)_3$, has a slightly distorted cubic structure, and can be prepared by the neutralisation of an aqueous solution of an indium(III) salt. It is virtually insoluble in water and low concentrations of alkali. Transparent, colloidal solutions are formed with a large excess of aqueous alkali, whilst indates are formed at very high concentrations of alkali.

4.5.7 Thallium oxides

Thallium(I) oxide, Tl_2O, has a rhombohedral structure with a density of $10.36 \, g \, cm^{-3}$. Its melting point is 596°C. It can be produced by heating thallium(I) hydroxide in a vacuum at 50°C,[95] or by heating the carbonate or hydroxide in nitrogen. It reacts with acids to give salts and with alcohols to give alkoxides. Prolonged heating ($>72 \, h$) with gallium(III) oxide or alumina at a temperature of c. 550°C gives $TlGaO_2$ or $TlAlO_2$, respectively.[91]

Thallium(III) oxide, Tl_2O_3, has a C rare earth oxide structure ($D5_3$) but above 500°C and 65 kbar it assumes a corundum structure; it has a density of $10.038 \, g \, cm^{-3}$ and a melting point of 716°C. It can be prepared by oxidation of thallium(I) nitrate with bromine or chlorine and then desiccation of the hydrated oxide thereby precipitated.[96] Thallium(III) oxide is insoluble in water but dissolves in acids to give salts.

Tetrathallium trioxide, Tl_4O_3, is a mixed oxide ($3Tl_2O \cdot Tl_2O_3$) prepared by heating Tl_2CO_3 and Tl_2O_3 at 450°C for 18 h in an inert atmosphere. It has a monoclinic (C_{2h}^2 or C_2^2) structure. The conductivity of the oxide increases when a crystal of the compound is illuminated. So-called thallium 'peroxide', TlO_2, has been prepared by the electrolysis of an aqueous solution of thallium(I) sulfate containing oxalic acid at ambient temperatures (see section 3.3.4).

4.5.8 Thallium hydroxide

Thallium(I) hydroxide, TlOH, crystallises as pale yellow needles which decompose to thallium(I) oxide, Tl_2O, on warming. Thallium(I) hydroxide can be prepared from thallium metal and ethanol in the presence of oxygen. Solutions can be prepared by the reaction of barium hydroxide with thallium sulfate, Tl_2SO_4, with the exclusion of oxygen and carbon dioxide.

4.5.9 Superconductivity of thallium-containing compounds

The revolutionary report by Bednorz and Müller, made in 1986,[97] regarding the formation of high-temperature superconducting materials of the type $(La_{5-x}Ba_x)Cu_5O_{5(3-y)}$ (where y is the oxygen vacancy concentration and x is 0.75 or 1) led to an explositon of activity in the search for other systems with superior properties. Higher values for the superconductivity transition temperature, T_c, were soon found for the large family of compounds of the type $(AO)_mM_2Ca_{n-1}Cu_nO_{2n+2}$ where A can be Tl, Pb, Bi, or a mixture thereof, $m = 1$ or 2 (only 2 when A is Bi), and M is Ba or Sr.[98-100] The highest temperature for which zero resistivity has been obtained is 122 K for $TlBa_2Ca_3Cu_4O_{11}$, $(Tl,Pb)Sr_2Ca_2Cu_3O_9$ and $Tl_2Ba_2Ca_2Cu_3O_{10}$. A list of the critical transition temperatures, T_c, for various thallium-containing systems is given in Table 4.8. Some of these superconducting ceramics, especially those having the stoichiometries $Tl_2Ba_2Ca_2Cu_3O_{10}$[101] and $Tl_2Ba_2CaCu_2O_8$,[102] have significant advantages over the more popular $YBa_2Cu_3O_7$ (T_c c.

Table 4.8 Superconductivity transition temperatures, T_c, for some compounds containing thallium.[99, 100, a]

Compound	T_c (K)
$TlBa_2Ca_3Cu_4O_{11}$	122
$(Tl,Pb)Sr_2Ca_2Cu_3O_9$	122
$Tl_2Ba_2Ca_2Cu_3O_{10}$	122
$Tl_2Ba_2Ca_3Cu_4O_{10}$	119
$TlBa_2Ca_2Cu_3O_9$	110
$Tl_2Ba_2CaCu_2O_8$	110
$TlBa_2CaCu_2O_7$	90
$(Tl,Bi)Sr_2CaCu_2O_7$	90
$(Tl,Pb)Sr_2CaCu_2O_7$	90
$Tl_2Ba_2CuO_6$	90
$(Tl,Bi)Sr_2CuO_5$	50

(a) See also section 2.5.4.

93 K),[103] such as a higher T_c and ease of preparation. On the other hand, the high toxicity of thallium (see chapter 9, for example) must be a major concern when large scale or regular usage is contemplated.

The proposed structure characterising this family of compounds consists of layers of CuO_2. The individual layers of CuO_2 are separated by Ca layers which contain no oxygen. These groups of layers are intercalated by two BaO layers and, for $m = 1$, 2..., TlO layers.[104] Determination of the structures is made difficult by the problem that many samples contain mixtures of phases.

It is difficult to envisage where this new generation of high T_c superconducting oxides will lead us. Currently the fabrication of the films required for electronic devices or the fabrication of large conductors capable of high ambient densities presents considerable problems. Once such problems have been overcome, however, there awaits us an array of devices greatly surpassing present-day technologies.

4.5.10 Superconducting components containing aluminium or gallium

Recent research at the Argonne National Laboratory and North Western University, Evanston, IL, has yielded new superconductors based on Ga, Sr, Y and oxygen in an O_2 atmosphere of 28 MPa at 900°C. These compounds give a T_c of 73 K but are the first to conduct electricity only along the planes formed by copper and oxygen atoms when they are separated by non-conducting chains.[105]

Neodymium aluminate single-crystal substrates 60 mm long and 23 mm wide have recently been produced at the National Institute for Research in Inorganic Materials in Niihara, Japan. Crystal substrates of $SrLaAlO_4$ and $CaNdAlO_4$ for high-temperature superconductors are also now commercially available.[105]

References

1. G.A. Rankin and H.E. Merwin, *J. Am. Chem. Soc.*, 1916, **38**, 568.
2. K.J.D. Mackenzie, *J. Br. Ceram. Soc.*, 1968, **5**, 183.
3. H. Yanagida and F.A. Kröger, *J. Am. Ceram. Soc.*, 1968, **51**, 700.
4. P. Ho and R.P. Burns, *High Temp. Sci.*, 1980, **12**, 31.
5. K. Wefers and C. Misra, *Oxides and Hydroxides of Aluminum*, Alcoa Technical Paper No. 19, Pittsburgh, PA, 1987.
6. K. Wefers, in *Alumina Chemicals*, ed. L.D. Hart, American Ceramic Society, Westerville, OH, 1990, p. 13.
7. R.C.T. Slade, J.C. Southern and I.M. Thompson, *J. Mater. Chem.*, 1991, **1**, 563, 875.
8. C. Misra, in *Kirk-Othmer Encyclopedia of Chemical Technology*, 4th edn., Wiley, New York, Vol. 2, 1991, p. 317.
9. F. Zigan, W. Joswig and N. Burger, *Z. Kristallogr.*, 1978, **148**, 255 and refs. cited therein.
10. J. Bugosh, *Fibrous alumina monohydrate*, US Patent 2,915,475 (1959).
11. H. Ginsberg and M. Köster, *Z. Anorg. Allg. Chem.*, 1953, **271**, 41.
12. *An Atlas of Alumina*, B.A. Chemicals Ltd., Gerrards Cross, UK, 1969.

13. I. Valeton, *Bauxites, Developments in Soil Science*, Vol. 1, Elsevier, Amsterdam, 1972.
14. W.H. Bragg and W.L. Bragg, *X-Rays and Crystal Structure*, Bell, London, 1916.
15. K.A. Evans and N. Brown, in *Speciality Inorganic Chemicals*, ed. R. Thompson, Royal Society of Chemistry, London, 1981, p. 164.
16. W.J. Connolly and A.M. Thornton, Alumina hydrate fillers in polyester systems, *Modern Plastics*, 1965, **43**, 154, 156, 202.
17. S.C. Brown, K.A. Evans and E.A. Godfrey, New developments in alumina trihydrate, in *Flame Retardants 87*, Plastics and Rubber Institute, London, 1987.
18. S.C. Brown and M.J. Herbert, New developments in ATH technology and applications, in *Flame Retardants 92*, Plastics and Rubber Institute, Elsevier Applied Science, London, 1992, p. 100.
19. L.A. Musselman, in *Alumina Chemicals*, ed. L.D. Hart, American Ceramic Society, Westerville, OH, 1990, p. 195.
20. R.J. Andlaw and G.J. Tucker, *British Dental*, 1975, **138**, 426.
21. P.M.C. James, R.J. Anderson, J.F. Beal and G. Bradnock, *Community Dent. Oral Epidemiol.*, 1977, **5**, 67.
22. K.A. Evans, A.R. Emery and K.J. Wills, *Alumina hydrate-containing toothpaste*, Eur. Patent 328,407 (1989).
23. K.A. Evans, P.L. Riley and B. Rosall, *Oral Compositions*, U.S. Patent 4,988,498 (1991).
24. A.R. Emery and J.W. Case, *Ground α-aluminium oxide trihydrate*, Brit. Patent 1,537,823 (1979).
25. C. Misra, *Industrial Alumina Chemicals*, American Chemical Society, Monograph 184, Washington, DC, 1986.
26. H.H. Murray, *Paper Coating Pigments*, Monography Series No. 30, New York, 1966.
27. J.E. Williams, Jr., *Paint Varn. Prod.*, 1967, **57**, 54.
28. W. Buchner, R. Schliebs, G. Winter and K.H. Büchel, *Industrial Inorganic Chemistry*, VCH Publishers, Weinheim, 1989, p. 247.
29. R.M. Barrer, *Hydrothermal Chemistry of Zeolites*, Academic Press, London, 1982; F.R. Ribeiro, A.E. Rodrigues, L.D. Rollmann and C. Naccache, ed., *Zeolites: Science and Technology, NATO ASI Ser., Ser. E*, Martinus Nijhoff, The Hague, Vol. 80, 1984; B. Držaj, S. Hocevar and S. Pejovnik, ed., *Zeolites*, Elsevier, Amsterdam, 1985; M.S. Spencer, *Crit. Rep. Appl. Chem.*, 1985, **12**, 64; G. Gottardi and E. Galli, *Natural Zeolites*, Springer-Verlag, New York, 1985; M.L. Occelli and H.E. Robson, ed., *Zeolite Synthesis*, ACS Symposium Series 398, American Chemical Society, Washington, DC, 1989; R.M. Barrer, *Zeolite Synthesis: an Overview*, in *NATO ASI Ser., Ser. C.* 1988, **231**, 221; J.M. Newsam, in *Solid State Chemistry: Compounds*, eds. A.K. Cheetham and P. Day, Clarendon Press, Oxford, 1992, p. 234.
30. M.S. Newkirk, A.W. Urquhart, H.R. Zwicker and E. Breval, *J. Mater. Res.*, 1986, **1**, 81.
31. M.S. Newkirk, H.D. Lesher, D.R. White, C.R. Kennedy, A.W. Urquhart and T.D. Claar, *Ceram. Eng. Sci. Proc.*, 1987, **8**, 879.
32. R. Morrell, *Handbook of Properties of Technical and Engineering Ceramics. Part 1. An Introduction for the Engineer and Designer*, H.M.S.O., London, 1985.
33. J.S. Owens, J.W. Hinton, R.H. Insley and M.E. Poland, *Am. Ceram. Soc. Bull.*, 1977, **56**, 437.
34. B. Schwartz, *J. Phys. Chem. Solids*, 1984, **45**, 1051.
35. B. Schwartz, in *Alumina Chemicals*, ed. L.D. Hart, American Ceramic Society, Westerville, OH, 1990, p. 299.
36. R. Twells, Jr., and C.C. Lin, *J. Am. Ceram. Soc.*, 1921, **4**, 195.
37. E. Dörre, H. Beutler and D. Geduldig, *Arch. Orthop. Unfall. Chir.*, 1975, **83**, 269.
38. J.W. Boretos, in *Alumina Chemicals*, ed. L.D. Hart, American Ceramic Society, Westerville, OH, 1990, p. 337.
39. E. Brinkmann, *Das Keramik – Anker Implantat*, nach E. Mutschelknauss, Zähnarztl Prax 29, 1978, p. 148.
40. S.F. Hulbert, F.A. Young, R.S. Mathews, J.J. Klawitter, C.D. Talbert and F.H. Stelling, *J. Biomed. Mater. Res.*, 1970, **4**, 433.
41. R.L. Westfall, C.A. Homsy and J.N. Kent, *J. Oral Maxillofacial Surg.*, 1982, **40**, 771.
42. G.G. Niederauer, T.D. McGee and R.K. Kudej, *Am. Ceram. Soc. Bull.*, 1991, **70**, 1010.
43. D. Cutright, S. Bhaskar, M. Brady, L. Getter and W. Posey, *Oral Surg.*, 1972, **33**, 850.

44. L.L. Hench and J. Wilson, *MRS Bull.*, 1991, **16**, 62.
45. V.D. Frechette and C.F. Cline, *Am. Ceram. Soc. Bull.*, 1970, **49**, 994.
46. S.-K. Chung, *Am. Ceram. Soc. Bull.*, 1990, **69**, 358.
47. D.J. Viechnicki, M.J. Slavin and M.I. Kliman, *Am. Ceram. Soc. Bull.*, 1991, **70**, 1035.
48. C.R. Austin, H.Z. Schofield and N.L. Haldy, *J. Am. Ceram. Soc.*, 1946, **29**, 341.
49. W.E. Blodgett, *Am. Ceram. Soc. Bull.*, 1961, **40**, 74.
50. R.J. Beals, in *Alumina Chemicals*, ed. L.D. Hart, American Ceramic Society, Westerville, OH, 1990, p. 323.
51. E. Dörre and H. Hübner, *Alumina: Processing, Properties and Applications,* Springer-Verlag, Berlin, 1984.
52. C.H. Kim, W. Roper, D.P.H. Hasselman and G.E. Kane, *Am. Ceram. Soc. Bull.*, 1975, **54**, 589.
53. E.D. Whitney, *Powder Metall. Int.*, 1978, **10**, 16, 18–21.
54. G. Fisher, in *Alumina Chemicals*, ed. L.D. Hart, American Ceramic Society, Westerville, OH, 1990, p. 353.
55. H.P. Cahoon and C.J. Christensen, *J. Am. Ceram. Soc.*, 1956, **39**, 337.
56. R.J. Brook, in *Ceramic Fabrication Processes. Treatise on Materials Science and Technology*, ed. F.F.Y. Wang and H. Herman, Academic Press, New York, Vol. 9, 1976, p. 331.
57. N. Claussen, *J. Am. Ceram. Soc.*, 1976, **59**, 49.
58. R. Stevens, *An Introduction to Zirconia*, Magnesium Elektron Ltd., Manchester, U.K., 1986.
59. P.F. Becher and G.C. Wei, *J. Am. Ceram. Soc.*, 1984, **67**, C-267.
60. G.C. Wei and P.F. Becher, *Am. Ceram. Soc. Bull.*, 1985, **64**, 298.
61. J.T. Kummer, *Prog. Solid State Chem.*, 1972, **7**, 141; J.H. Kennedy, *Top. Appl. Phys.*, 1977, **21**, 105.
62. W.T. Bakker, in *Alumina Chemicals*, ed. L.D. Hart, American Ceramic Society, Westerville, OH, 1990, p. 309.
63. J.H. Duncan, P. Barrow, A. Van Zyl and A.I. Kingon, *Making β″-alumina*, Brit. Patent 2, 175, 582 (1986).
64. J.T. Kummer and N. Weber, *Proc. 21st Annual Power Sources Conference*, 1967, p. 37.
65. J.L. Sudworth and A.R. Tilley, *The Sodium Sulfur Battery*, Chapman and Hall, London, 1985.
66. J.F. MacDowell and G.H. Beall, *J. Am. Ceram. Soc.*, 1969, **52**, 17.
67. J.F. MacDowell, in *Alumina Chemicals*, ed. L.D. Hart, American Ceramic Society, Westerville, OH, 1990, p. 365.
68. S.D. Stookey, *Ind. Eng. Chem.*, 1959, **51**, 805.
69. G.H. Beall, in *Advances in Nucleation and Crystallization of Glasses*, ed. L.L. Hench and S.W. Frieman, American Ceramic Society Special Publication, No. 5, Westerville, OH, 1972.
70. W.D. Kingery, *J. Am. Ceram. Soc.*, 1950, **33**, 239.
71. Y. Hongo, Y. Tuzuki and M. Miyawaki, *Refractory Composition*, US Patent 4,331,773 (1982).
72. L.P. Krietz and R.E. Fisher, in *Alumina Chemicals*, ed. L.D. Hart, American Ceramic Society, Westerville, OH, 1990, p. 519.
73. M. Kimura, Y. Yasuda and H. Nishio, *Interceram.*, Special Issue, 1984, p. 22.
74. J. Benbow, Cement kiln refractories, *Industrial Minerals*, January 1990, p. 37.
75. J.S. Kenworthy, M.J. Morton and M.D. Taylor, *Production of fibres*, British Patent 1,470,292 (1976).
76. I. Wehrlein, *Process for hardening aluminous materials by electric fusion*, French Patent 233,996 (1893).
77. B. Glezin, *Novosti Tekhniki*, 1937, **12**, 31.
78. M.A. Leitheiser and H.G. Sowman, *Non-fused aluminium oxide-based abrasive mineral*, US Patent 4,314,827 (1982).
79. F. Ulrich, *Norsk. Geol. Tidsskrift*, 1925, **8**, 115.
80. K. Wefers and G.M. Bell, *Oxides and hydroxides of aluminum*, Alcoa Technical Paper No. 19, Alcoa Aluminum Company of America, East St Louis, 1972.
81. U. Hauschild, *Z. Anorg. Allg. Chem.*, 1963, **324**, 15.

82. J.B. Barnitt, *Moisture-adsorbent material*, US Patent 1,868,869 (1932).
83. M.L. Pingard, French Patent 1,077,163 (1953), *Activated alumina*, US Patent 2,881,051 (1959).
84. R.D. Woosley, in *Alumina Chemicals*, ed. L.D. Hart, American Ceramic Society, Westerville, OH, 1990, p. 241.
85. H.L. Fleming and K.P. Goodboy, in *Alumina Chemicals*, ed. L.D. Hart, American Ceramic Society, Westerville, OH, 1990, p. 251.
86. H.L. Fleming, in *Alumina Chemicals*, ed. L.D. Hart, American Ceramic Society, Westerville, OH, 1990, p. 263.
87. R. Roy, V.G. Hill and E.F. Osborn, *J. Am. Chem. Soc.*, 1952, **74**, 719.
88. P. Bretèque, in *Kirk-Othmer Encyclopedia of Chemical Technology*, 3rd edn., Wiley, New York, Vol. 11, 1980, p. 604.
89. K.P. Huber and G. Herzberg, *Molecular Spectra and Molecular Structure. IV. Constants of Diatomic Molecules*, van Nostrand Reinhold, New York, 1979.
90. G. Brauer, ed., *Handbook of Preparative Inorganic Chemistry*, 2nd edn., Academic Press, New York, 1963.
91. K. Wade and A.J. Banister, *The Chemistry of Aluminium, Gallium, Indium and Thallium*, Pergamon Press, Oxford, 1975.
92. R. Roy and M.W. Shafer, *J. Phys. Chem.*, 1954, **58**, 372.
93. C.T. Prewitt, R.D. Shannon, D.B. Rogers and A.W. Sleight, *Inorg. Chem.*, 1969, **8**, 1985.
94. E.F. Milner and C.E.T. White, in *Kirk-Othmer Encyclopedia of Chemical Technology*, 3rd edn., Wiley, New York, Vol. 13, 1981, p. 207.
95. B.C. Hui, in *Kirk-Othmer Encyclopedia of Chemical Technology*, 3rd edn., Wiley, New York, Vol. 22, 1983, p. 835.
96. D. Cubicciotti and F.J. Keneshea, *J. Phys. Chem.*, 1967, **71**, 808.
97. J.G. Bednorz and K.A. Müller, *Z. Phys. B*, 1986, **64**, 189.
98. C. Michel, M. Hervieu, M.M. Borel, A. Grandin, F. Desleandes, J. Provost and B. Raveau, *Z. Phys. B*, 1987, **68**, 421.
99. A.W. Sleight, M.A. Subramanian and C.C. Torardi, *MRS Bull.*, 1989, **14**, 45.
100. I.A. Kahwa, D. Miller, M. Mitchel, F.R. Fronczek, R.G. Goodrich, D.J. Williams, C.A. O'Mahoney, A.M.Z. Slawin, S.V. Ley and C.J. Groombridge, *Inorg. Chem.*, 1992, **31**, 3963 and refs. cited therein.
101. C.C. Torardi, M.A. Subramanian, J.C. Calabrese, J. Gopalakrishnan, K.J. Morrissey, T.R. Askew, R.B. Flippen, U. Chowdhry and A.W. Sleight, *Science*, 1988, **240**, 631.
102. M.A. Subramanian, J. Calabrese, C.C. Torardi, J. Gopalakrishnan, T.R. Askew, R.B. Flippen, K.J. Morrissey, U. Chowdhry and A.W. Sleight, *Nature (London)*, 1988, **332**, 420.
103. See, for example, I.A. Kahwa and R.G. Goodrich, *J. Mater. Sci. Lett.*, 1989, **8**, 755 and refs. cited therein.
104. I.K. Schuller and J.D. Jorgensen, *MRS Bull.*, 1989, **14**, 27.
105. L.M. Sheppard, *Am. Ceram. Soc. Bull.*, 1992, **71**, 1242.

5 III–V compounds

I.R. GRANT

5.1 Introduction

The so-called III–V compounds of aluminium, gallium and indium are technologically very important as semiconductor materials. They form the material systems for solid-state optoelectronic devices and are used extensively in high-speed, high-frequency applications. Components made in these materials include infrared and visible light-emitting diodes (LEDs) and detectors, lasers and high-speed and microwave integrated circuits. Although regarded for a long time as materials potentially rivalling silicon, they perform in fact a complementary function by extending the range of applications for devices into microwave and optical communications not accessible to Group IV (14) semiconductors.

Active interest in these compounds dates from the work of Welker in the early 1950s demonstrating their semiconducting properties at the time when solid-state electronics was in its infancy. Discovery of their luminescence behaviour, including lasing action (1962), stimulated the development of light-emitting diodes and semiconductor injection lasers. By the mid-1970s, a mass consumer market had been created for LED devices based on GaAs and GaP. These remain the most important opto-electronic emitter materials in volume production.

Since the early 1980s, optical fibre telecommunications have driven the development of InP-based lasers and detectors for transmission in the optimum wavelength band. Modern telephone trunk networks, including long haul submarine cable systems, rely on InP devices. Future integrated information transmission systems, including telephony, interactive services and broadcast distribution, will require a fully fibre-based network incorporating InP technology.

GaAs has had important electronic applications since the discovery of bulk negative resistance at high electric fields and its consequent use as a source of microwave energy (the Gunn effect). The early areas of application were limited to discrete microwave devices. Transistor technology based on the metal Schottky barrier field effect transistor was developed to exploit the superior properties of the material with respect to high temperature operation, radiation 'hardness' and power consumption. This has advanced to the extent that integrated circuit manufacture is undertaken to the level of large scale integration (LSI) for digital circuits used in high-speed computing. GaAs analogue ICs are now widely used in satellite communications, including TV receiver modules.

The versatility of the III–V compounds lies in the range of properties available across the family of materials. This is greatly enhanced by the use of impurity doping. The major advantage, however, is the ability to engineer the material properties by alloying, to form substances with 2, 3 or 4 component elements within the III–V system. When these are combined in thin structures, new electronic properties can be obtained, offering tremendous flexibility in device design and the potential for novel structures and behaviour.

The suitability of a particular compound for a given application may be understood by consideration of its electronic energy band structure. This is illustrated by the case of GaAs, the most important of the III–V compounds. Figure 5.1 is a representation of the electron energy versus momentum plots for GaAs[1] and for Si[2] obtained from pseudopotential calculations.

The electron energy is measured with respect to the valence band maximum. The lowest energy conduction band states in GaAs lie at the zero momentum position, 1.4 eV above the valence band maximum at room temperature. In Si, the lowest conduction band states lie at the X point of the Brillouin zone boundary, 1.1 eV above the valence band maximum. Occupancy of the conduction band states requires an electron energy greater than the energy band gap. This may be achieved by thermal or optical excitation.

At room temperature, the equilibrium occupancy of the conduction band states, and hence the free carrier density in a pure intrinsic material, is

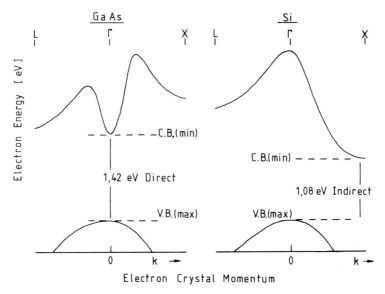

Figure 5.1 *E–k* plots for GaAs and Si.

determined by the Fermi–Dirac distribution. Equal numbers of free electrons and holes are thermally generated.

$$np = 4\left(\frac{2\pi(m_p^* m_n^*)^{1/2} kT}{h^2}\right)^3 \exp(-E_g/kT) \qquad (5.1)$$

where n is the volume density of free electrons, p is the volume density of free holes, h, k, m_n^* and m_p^* are constants, and E_g is the energy gap. Equation (5.1) is a law of mass action and shows that the electrical properties of a given material depend strongly on its energy band gap.

Optical transitions across the gap can occur by absorption of photons of minimum energy

$$E_g = h\nu \qquad (5.2)$$

where ν is the photon frequency and h is Planck's constant. In GaAs and other 'direct gap' semiconductors, such transitions occur at zero momentum, whereas in Si the transitions are indirect, involving emission or absorption of phonons to conserve momentum. Optical transitions in indirect materials are less efficient processes making them less suitable for optoelectronic applications.

The shape of the $E–k$ plot near to the conduction band minimum is approximately parabolic and determines the electron effective mass m^* in accordance with equation (5.3).[3]

$$\frac{d^2 E}{dk^2} = \frac{\hbar}{m^*} \qquad (5.3)$$

The sharp curvature of the conduction band minimum for GaAs compared with Si results in a much smaller electron effective mass. If m_0 is the free electron mass, then it emerges that

$$m^*(\text{GaAs}) = 0.068m_0$$

whereas

$$m^*(\text{Si}) = 0.97m_0$$

The consequence of this is that GaAs has a much higher electron drift velocity in low electric fields, resulting in faster operation of electronic devices.[4]

Similar considerations apply to the suitability of other III–V materials for use in particular applications, some of which are listed in Table 5.1(a). Table 5.1(b) summarises the properties of the common III–V compounds used as electronic materials.

Table 5.1 (a) Properties of III–V compounds.

Property[a]		Compound								
	AlP	AlAs	AlSb	GaN	GaP	GaAs	GaSb	InP	InAs	InSb
Band gap (eV)	2.45	2.14	1.62	3.39	2.26	1.42	0.70	1.35	0.36	0.18
Lattice constant (Å)	5.46	5.66	6.14	3.18	5.45	5.65	6.09	5.86	6.06	6.48
Density (g cm^{-3})	2.40	3.73	4.26	6.10	4.13	5.32	5.61	4.79	5.67	5.78
Optical transition (D, direct; I, indirect)	I	I	I	D	I	D	D	D	D	D
Melting temperature (°C)	>2000	1740	1080	1500	1467	1238	706	1062	943	525
Electron effective mass (m_0)	–	0.35	0.39	0.19	0.35	0.065	0.049	0.078	0.023	0.014
Electron mobility (cm^2 V^{-1} s^{-1})	80	1200	200	150	190	8.8×10^3	6×10^3	5×10^3	2.3×10^4	8.2×10^4
Hole mobility (cm^2 V^{-1} s^{-1})	–	420	550	–	120	400	800	150	260	1700

(a) All values at 300 K.

Table 5.1 (b) Some applications of III–V compound substrates.

Substrate material	Application
GaAs (n^+, p^+) (LEC, SI)[a]	IR lasers, LEDs and detectors
	Microwave and digital ICs
InP (n^+) (SI)[a]	Long wavelength IR sources and detectors for optical fibre transmission
	Opto-electronic ICs
GaP (n^+, p^+)	Visible LEDs for displays
InSb	IR photodetectors, Hall effect devices
GaSb	Long wavelength (2–5 µm) emitters
GaN	Blue/green LEDs

(a) SI, semi-insulating.

5.2   Properties of the compounds

5.2.1   Bonding and crystal structure

Preparation in single-crystal form is normally required. Crystallisation occurs in the zinc blende structure resembling that of diamond and consisting of two interpenetrating equivalent face-centred cubic lattices, one composed entirely of Group III (13) atoms, the other of Group V (15) atoms.

The electrical properties of the crystalline solid are determined by the nature of the bonding. Because of the different electronegativities of the constituents, the III–V bond is partly covalent, partly ionic in nature with an effective ionicity ranging from about 30 to 40%.[5] Strong bonding of a lattice results in a large energy band gap and a large electron effective mass, m^*, wherease weaker bonding is characterised by a smaller gap and a smaller electron effective mass and hence higher electron mobility. The bonding of atoms is dependent on their size and becomes weaker with increasing atomic number. This can be seen as a trend within a single Group of elements in the

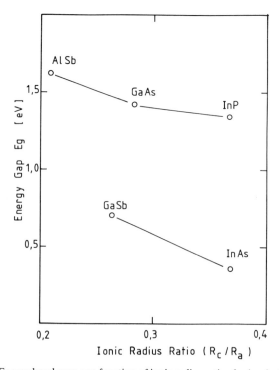

Figure 5.2 Energy band gaps as a function of ionic radius ratios for isoelectronic series.

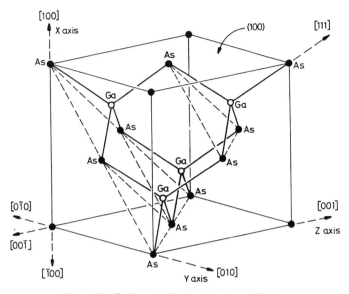

Figure 5.3 Cubic crystal lattice structure of GaAs.

Periodic Table or within the III–V compounds as the atomic number of one of the components increases. Thus, the series from GaP to GaAs to GaSb exhibits a decreasing energy gap and an increasing free electron mobility. In addition to this 'vertical' rule, a second, or 'horizontal', rule applies in which the energy gap is determined by the effective ionicity of the bond.[6] For isoelectronic compounds, i.e. those with the same total number of electrons per formula unit, both the effective ionicity and the energy gap increase as the ratio of the cation to anion radius decreases. This is illustrated in Figure 5.2 for different isoelectronic sequences. The above empirical 'rules' may be used to compare the relative electronic properties of the compounds.

Figure 5.3 is a representation of the GaAs lattice structure.[7] Using the conventional system of Miller indices, the Ga and As sublattices are translated by a distance $(-1/4, -1/4, 1/4)\, a_0$, where $a_0 = 5.654$ Å is the lattice parameter. Referring to Figure 5.3 shows that each sublattice has chemically distinct (1 1 1) lattice planes associated as closely spaced parallel pairs with a separation of 0.82 Å. The 4 : 4 coordination of the structure has a tetrahedral bonding arrangement with a Ga–As bond length of 2.45 Å resulting in strong bonding within the pairs of (1 1 1) planes, each Ga atom being bound to three As atoms in the next plane. (1 1 1) Crystal surfaces therefore consist either of Ga or As atomic planes and display different chemical reactivities. In the general case these are referred to as (1 1 1)A (Group III) and (1 1 1)B (Group V) faces.

Bonding is also a determining factor in the crystal cleavage behaviour.

(a)

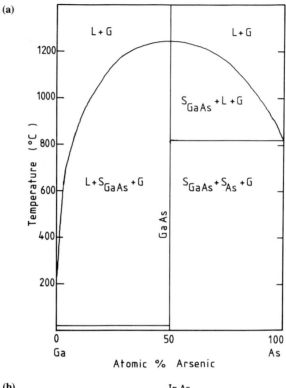

(b)

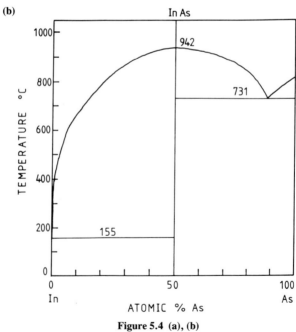

Figure 5.4 (a), (b)

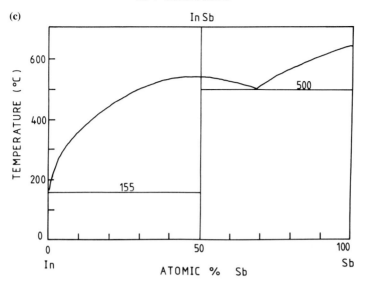

Figure 5.4 T–x phase diagrams for (a) GaAs, (b) InAs and (c) InSb.

This is of technological interest in the processing of devices on single crystal wafers, being used to define individual chips or to provide cleaved facet mirrors for laser cavities. Cleavage in the compounds AlP, AlAs, AlSb; GaP, GaAs, GaSb; InP, InAs, and InSb occurs on (0 1 1) planes as a result of bonding with an ionicity greater than 25%.[8]

5.2.2 Phase equilibria and vapour pressure

The main steps involved in the production of III–V electronic materials are:

1. synthesis of the compound;
2. bulk single crystal growth; and
3. epitaxial layer growth.

Definition and control of these processes require knowledge of the phase diagram and vapour pressure data for the material system of interest. This information is usually expressed in terms of the T–x and P–T sections of the three-dimensional phase diagram, where T is the temperature, P is the pressure and x is the composition.

Figure 5.4(a) shows the (T–x) phase diagram for GaAs with a maximum melting temperature of 1238°C at the congruent point.[9] Liquidus curves have been obtained for other compounds, namely AlSb,[10] GaP,[11] GaSb,[12] InP,[13] InAs,[14] and InSb[15] which show the same broad parabolic form around the stoichiometric composition.

The dissociation of the compound AB at high temperature proceeds by decomposition into liquid A and gaseous B with the following equilibria being established:[16]

$$AB(s) \rightleftharpoons A(g) + [1/y\, B_y(g)] \tag{5.4}$$

$$x[AB(s)] + [1 - 2x]A(l) \rightleftharpoons [(1 - x)A \cdot xB](g) \tag{5.5}$$

$$[1 - x]A(l) \rightleftharpoons A(g) \tag{5.6}$$

where x is the B fraction of the melt and B_y is the gaseous species of B ($y = 1$, 2 or 4).

Figure 5.5 shows the equilibrium vapour pressures of Ga and the various As species over the entire binary liquidus of GaAs, as determined by Arthur.[17] At the maximum melting temperature, the dissociation is dominated by As species with a total vapour pressure of around 0.9 atm.

Table 5.2 lists the melting temperatures and dissociation pressures for the important compounds. These constitute the principal factors in choosing the methods of synthesis and growth of a particular material.

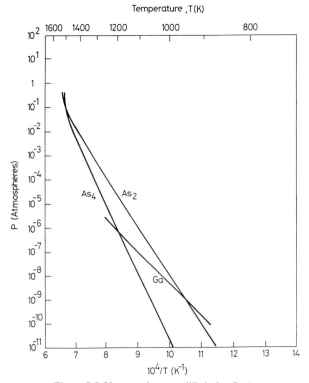

Figure 5.5 Vapour phase equilibria for GaAs.

Table 5.2   Melt dissociation of some important III–V compounds.

Compound	Melting temperature (°C)	Dissociation pressure (atm)
GaAs	1238	0.9
GaP	1467	35
InP	1062	27
InSb	525	4×10^{-8}
InAs	943	0.33
GaSb	706	1×10^{-5}

5.3   Binary compound synthesis

In almost all cases, the compounds are made by direct reaction of the constituent elements, followed by melting and crystallisation. A general requirement is the achievement of the highest possible purity, or at least a controlled impurity content, for the compound. The typical specification for the starting materials is a (6N) purity (99.9999%), and this is readily available for most of the elements. Antimony may be improved by pre-treatment with zone-refining. Gallium, indium and arsenic can be obtained in (7N) grade or better. Arsenic is used in the crystalline metallic α-form and red, rather than white, phosphorus is selected for ease of handling and purity.

5.3.1   GaAs and InAs

The preparation of large polycrystalline ingots (10 kg) of GaAs may be carried out in the apparatus shown schematically in Figure 5.6.

Gallium metal is contained in a quartz boat placed at one end of an evacuated and sealed quartz ampoule. A charge of arsenic is placed at the

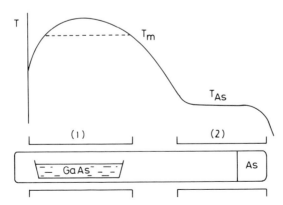

Figure 5.6 GaAs synthesis system.

other end and the whole reaction vessel is placed inside a two zone furnace as shown. The gallium is held at a temperature above the melting point of the compound while it reacts with the As vapour supplied by controlled evaporation of the solid source from the low-temperature zone. Sufficient arsenic is added to provide a balancing pressure. Solidification is achieved by controlled 'ramp-down' of the main furnace zone temperature, to give directional freezing.

Since the internal pressure remains small, the furnace equipment can be relatively simple. However, contact of the melt with the silica boat leads to the incorporation of silicon as an electrically active impurity by the reaction

$$4\text{Ga} + \text{SiO}_2(\text{s}) \rightleftharpoons 2\text{Ga}_2\text{O}(\text{g}) + \text{Si}(\text{l}) \qquad (5.7)$$

This restricts the application of this technique to material in which large concentrations of deliberately added impurities are introduced in the single crystal form. Such doped crystals may contain 10^{18} atoms cm^{-3} of Si, compared with residual pick-up from synthesis of 10^{16} atoms cm^{-3}. The potential reactivity of the melt may be countered by the use of alternative boat materials for containment. Pyrolytic boron nitride (PBN) has been used extensively but is expensive and fragile, and must be re-used many times to be economic.

InAs is synthesised in a similar way to GaAs[18] but its lower melting temperature and dissociation pressure make compounding simpler.

5.3.2   InP and GaP

The much higher dissociation pressures of these compounds lead to greater complexity in the equipment used. GaP presents the most difficult problem due to the combination of very high temperature and pressure with respect to the softening of quartz.

Horizontal zone melting methods are most commonly used.[19] Such a system is illustrated in Figure 5.7.[20] The metal is contained in a graphite tube

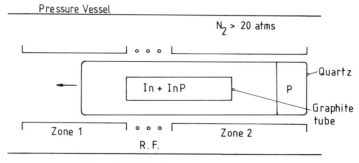

Figure 5.7 InP horizontal zone melting system.

induction-heated by a radiofrequency coil. Resistance furnaces on either side control the temperature of a solid phosphorus source at one end, and the temperature gradient for solidification at the other. The quartz ampoule is passed through the radiofrequency zone while an internal phosphorus pressure is maintained either by close measurement and control of the temperature or by monitoring the pressure difference across the ampoule.[19] The external pressure is balanced by an inert atmosphere of nitrogen at 20–30 atm in the pressure vessel chamber. The risk of explosion and release of phosphorus governs the design and ease of cleaning of the external vessel.

Growth rates between 1 mm and 40 mm per hour have been used in such equipment. The grain size, porosity and extent of metallic inclusions in the resultant self-seeded, polycrystalline ingots depend on growth rate, phosphorus pressure and zone temperature. These properties are normally determined empirically, under conditions selected to minimise explosion risks.

Use has also been made of lower-temperature, lower-pressure techniques involving growth of the III–V compound from metal-rich solutions.[21] The synthesis and solidification of InP have been carried out inside a high-pressure chamber in the form of either an autoclave[22] containing a mixture of the elements or a liquid-encapsulated Czochralski (LEC) crystal puller[23] (see section 5.3.3) in which phosphorus vapour is bubbled into an indium melt under a boric oxide encapsulant.

5.3.3 Single crystal growth

The main use of the compounds is as single crystal substrates on which device structures are formed by diffusion or ion implantation of dopant impurities, or by the epitaxial growth of thin single crystal layers on the surface. Large ingots are grown by crystallisation from stoichiometric melts using techniques which may be categorised under the following headings: (a) boat growth and (b) crystal pulling.

5.3.3.1 Boat growth. The ingot is solidified inside a container which then defines its shape and some aspects of its physical properties. The growth axis is arranged to be either horizontal or vertical. The growth is seeded using a single crystal section and crystallisation proceeds by passing the melt through a temperature gradient. Several variants are defined depending on whether the melt is moved with respect to the furnace, the furnace moved over a fixed melt, or the melting isotherm is passed along the length of the melt by ramping the furnace temperature (the gradient freeze method).

The horizontal Bridgman (HB) and gradient freeze (GF) techniques are the oldest and most widely used for the growth of doped GaAs. Figure 5.8 gives a schematic representation of a particular HB arrangement in

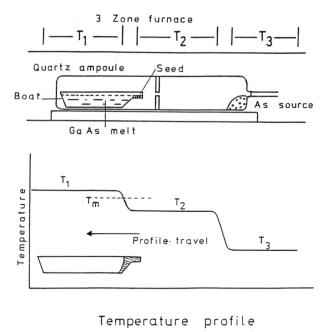

Temperature   profile

Figure 5.8 GaAs horizontal Bridgman schematic representation.

which three separate furnace zones are used to control an arsenic source, the melt temperature and the temperature gradient. The equipment can be simple since the growth is performed inside a sealed quartz ampoule and only moderate pressure and temperature are required. Large ingots can be made with low densities of crystalline dislocations reflecting the small cooling stresses. The growth axis is usually <1 1 1>, a feature due to single crystal yield considerations. This, together with the boat shape, results in non-circular wafer sections with (1 0 0) surfaces being cut from the crystal with poor efficiency in material usage. Contamination from the container results in silicon impurity pick-up (q.v), but this is insignificant for the main applications requiring heavily doped conductive substrates. The boat material is usually silica but this must be treated to avoid wetting by the melt and crystal sticking.

Although vertical configurations have been used since the earliest preparation of GaAs, these have experienced a resurgence in recent times for the growth on the <0 0 1> axis of cylindrical ingots of large diameter (up to 3 inches) and low dislocation content. There has also been considerable interest in the vertical GF method for the growth of undoped, high-purity GaAs for high resistivity substrates with low defect densities.[24]

5.3.3.2 Crystal pulling. The basis of the pulling technique was established by Czochralski[25] and is the principal method used for the production
of many materials required in the form of a single crystal substrate, most
notably silicon. The dissociation of the melt in the case of the III–V compounds adds the necessity for containment of the Group V component.
Early attempts at pulling were carried out inside sealed quartz ampoules
either with special seals or with magnetic coupling.[26] Such methods are
limited to compounds with relatively low dissociation pressures.

The most successful version of the method involves so-called 'liquid
encapsulated Czochralski growth' (LEC). This was first proposed for PbTe[27]
and then adopted for III–V compounds by Mullin *et al.*[28] The melt is
encapsulated in a molten glass inside a crucible and decomposition is
prevented by imposing an external pressure of inert gas in the chamber. This
is represented in Figure 5.9. Single crystals are seeded and pulled from the
melt through the encapsulant, while rotating both crystal and melt. Growth
rates of about 1 cm per hour are typically used.

GaAs, GaP, InP and InAs are all routinely produced in this way. GaAs
ingots up to 15 kg in weight and 150 mm in diameter are grown. Undoped
GaAs produced by this route can be made in very high purity when the
compounding is performed by direct synthesis *in situ*[29] and pyrolytic BN is
used as the crucible material.[30] All semi-insulating, high-purity GaAs for
integrated circuit substrates is grown by LEC.

The growth axis may be <0 0 1> or <1 1 1> and the crystals have circular
cross sections. Control of the shape becomes an added problem without boat

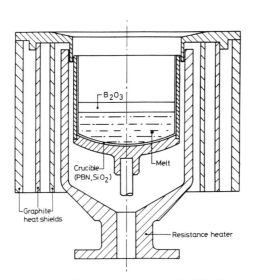

Figure 5.9 Liquid Encapsulated Czochralski (LEC) system.

containment and various types of system for automatic diameter control have been employed, mostly involving fine temperature control determined by continuous crystal weighing.[31] The single crystal yield is in general much higher than for boat growth methods. Temperature gradients tend to be very much higher in the LEC system and hence the crystals exhibit dislocation densities at least an order of magnitude greater than do those formed by well controlled boat growth.

Boric oxide, B_2O_3, is the encapsulant used. It forms a glass which flows at a temperature low enough to encapsulate the charge before decomposition occurs. It is non-volatile, transparent (allowing direct observation of the growth interface), generally inert to the melt, and wets the crucible (allowing complete encapsulation of the charge). The boric oxide also plays an important role in the control of the melt purity. It contains residual moisture up to several thousand ppm which may react with metallic impurities and with Ga to form oxides which are 'gettered' by the glass or, if volatile, transported into the growth chamber. The boric oxide must be carefully specified with respect to residual moisture, which is assessed by reference to the infrared absorption near $3580\,cm^{-1}$ associated with the O–H stretching vibration.[32]

InSb and GaSb have melting temperatures that are too low for normal boric oxide encapsulation. Although the Sb dissociation pressure is small, the melt is susceptible to oxide formation on exposure to residual air or moisture, and growth is normally carried out by conventional Czochralski pulling in an atmosphere of hydrogen. GaSb may be grown by LEC using boric oxide with flux additives to reduce the viscosity.

5.4   Influences on electrical properties

Intrinsic conduction in a semiconductor is achieved when the electrically active impurity concentration falls below the intrinsic carrier concentration. In practice, this can be achieved only in narrow gap materials, and in most cases the electrical behaviour is controlled by impurities or other defect levels.

5.4.1   Impurities

Deliberately added dopant impurities are used to modify the resistivity and conduction type (n or p) of the bulk conductor. A general principle for the III–V compounds is that Group II elements are substitutional on Group III sites to form acceptors (p-type conduction), while Group VI elements are incorporated on Group V sites as donors. Thus Zn and Cd are acceptors, while Se, S and Te are donor species. Amphoteric behaviour may be exhibited by Group IV elements such as Si, whose incorporation on Group

III or Group V sublattices of GaAs depends on the total concentration and thermal history. Sn is generally a donor while C is an acceptor in GaAs. On the other hand, Ge is an acceptor in GaAs but a donor in InP.

The binding energy for free carriers from such impurities is analogous to the hydrogen atom case. Thus, the ground state energy relative to the band states is dependent on the electron effective mass and dielectric constant for the material. Such an analysis predicts shallow states very close to the band edges for these 'hydrogenic', simple substitutional impurities, with acceptor activation energies about 50 meV and donor activation energies about 6 meV, in GaAs. The impurities remain thermally fully ionised therefore down to very low temperatures.

Transition metal elements with multiple charge states tend to form much deeper levels in the band gap and may sometimes be used to control the Fermi energy position to achieve 'intrinsic-like' high electrical resistivity.

Precise control of the dopant content during melt growth is limited by normal segregation behaviour.[33] The impurity concentration of the solid, C_s, is therefore given by

$$C_s = k \cdot C_0 (1 - g)^{k-1} \qquad (5.8)$$

where C_0 is the impurity concentration of the melt, g is the melt fraction solidified and k is the segregation coefficient which is a characteristic of a given impurity-matrix system. For $k = 1$, the impurity content is constant with crystal length. For $k < 1$, however, the solute is less soluble in the solid than in the melt and is progressively rejected into the melt, increasing its concentration as the melt is depleted. For $k > 1$, the opposite is the case and the impurity concentration falls from the first-to-freeze to last-to-freeze portions of the crystal.

A common requirement is to prepare crystals with the highest possible purity. Limitations are then imposed by inadvertent contamination from the growth system and the raw materials. In this case, segregation effects can be advantageous in the removal of impurities having $k < 1$. Table 5.3 lists the segregation coefficients of some common impurities in GaAs.[34]

Several analytical techniques are available for assessment of the chemical impurity content (see also chapter 10). Secondary ion mass spectroscopy (SIMS) has been employed for bulk analysis[35] but is especially important for its ability to provide depth profiling information in planar device structures. The usual methods applied to bulk analysis of both raw materials and compounds are spark source mass spectroscopy (SSMS) and glow discharge mass spectroscopy (GDMS).[36] The latter technique has become dominant in recent years because of its high sensitivity, with detection levels down to ppb and below. Table 5.4 shows typical GDMS analyses for an undoped GaAs sample. C, N and O are not accurately determined because of background effects, although these may be reduced by cryo-pumping.[37] Boron, isoelectronic with Ga and hence electrically inactive, is normally the

Table 5.3 Segregation co-efficients of some common impurities in GaAs.

Element	$k_{effective}$
Ag	0.1
Mg	0.3
Ca	<0.02
Zn	0.1
Al	3
In	0.1
Si	0.1
Ge	0.03
Sn	0.03
S	0.3
Te	0.3

major chemical impurity in undoped GaAs produced by LEC growth from PBN crucibles. Its concentration may vary between about 10^{18} and $10^{16} \, cm^{-3}$, depending on the encapsulant moisture content. In the high-purity undoped compounds, the residual chemical impurities originate in the reactions in the following systems:

1. melt–encapsulant, and
2. melt (including B_2O_3)–ambient gas–graphite furnace.

The reactions in (1) are governed by the oxygen partial pressure and are described by

$$2mX + nO_2 \rightleftharpoons 2X_mO_n \qquad (5.9)$$

The B_2O_3–H_2O system provides the oxygen source and carbon is a major impurity involved in the reaction processes. The equilibria for the various possible reactions are determined by the relative values of the Gibbs free energies of formation for the different oxides[38] at the melting temperature. An important mechanism for carbon incorporation in GaAs occurs by means of (2) above. Hence, CO is formed by reaction of moisture with hot graphite and subsequently dissolved in the B_2O_3 to react with the melt:

$$CO + 2Ga \rightleftharpoons Ga_2O + C \qquad (5.10)$$

$$3CO + 2Ga \rightleftharpoons Ga_2O_3 + 3C \qquad (5.11)$$

$$2B + 3CO \rightleftharpoons B_2O_3 + 3C \qquad (5.12)$$

The carbon is incorporated on the arsenic sublattice to form a shallow acceptor level.[39] Its concentration can be accurately measured by the local vibrational mode which occurs in infrared absorption (at $582 \, cm^{-1}$), and this

Table 5.4 Glow discharge mass spectrometry (GDMS) analysis of undoped GaAs.

Element	Concentration (atomic ppb)
B	180
C	8
N	38
O	33
Na	<2
Mg	<2
Al	<0.9
Si	4
P	<1
S	<2
Cl	<2
Ca	<10
Ti	<0.4
V	<0.2
Cr	<1
Mn	<0.7
Fe	2
Co	<0.4
Ni	<0.8
Cu	<2
Zn	<2
Ag	<20
Cd	<11
Sn	<2
Sb	<1
Te	<2
Hg	<2
Pb	<0.6

technique is also useful for other light element impurities, such as B and Si, or impurity complexes.[40]

Photoluminescence spectroscopy offers high sensitivity and provides identification of shallow acceptors such as C, Zn, Mn and Cu, but is non-quantitative.[41]

5.4.2 Defects

The binary compound structure provides, in addition to the impurity defects, a range of native defect types, such as vacancies, interstitials and anti-site defects, as well as complexes of native and impurity defects. In many cases, these form electrical levels in the gap and thereby contribute to the electrical conduction and recombination processes. The native point defect which has received the most attention is the EL2 centre in GaAs. This acts as a mid-gap deep donor.[42] In high-purity, undoped GaAs, EL2 is the dominant electrical level and is responsible for the semi-insulating 'intrinsic-like' electrical behaviour. It is a crystalline defect identified in electronic trap spectroscopic techniques,[43] and gives rise to optical absorption structure close to the band edge.[44] This latter effect may be used to give a quantitative estimate of EL2 in a sample of the semi-insulating material.[45] In this way it can be shown that [EL2] depends on the melt composition,[46] reaching a maximum in stoichiometric and As-rich melts. Similarly, [EL2] has been shown to increase with increasing As partial pressure in Bridgman growth. EL2 has therefore been attributed to the arsenic anti-site defect, As_{Ga}, or to a complex involving this defect.

In the highest-purity crystals, control over lattice defects and the balance of these with other impurities provides the key to the control of electrical behaviour. For example, the necessary impurity balance for undoped, semi-insulating GaAs is

$$[EL2] \quad > \quad [C_{As}] \quad > \quad N_d$$
$$\text{Deep donor} \qquad \text{Acceptor} \qquad \text{Shallow donor (Si, S etc.)}$$

where N_d is the residual donor concentration. [EL2] is dependent on the growth conditions and thermal history of the crystal. Figure 5.10 demonstrates the influence of melt composition on electrical behaviour and may be explained in terms of [EL2].

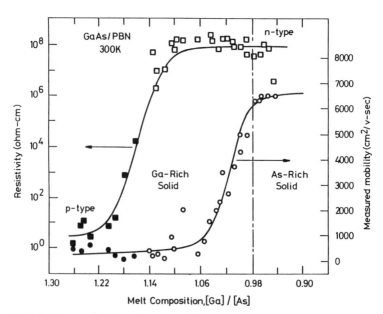

Figure 5.10 Influence of GaAs melt composition on electrical behaviour of the crystal (from Ref. 42).

The phase diagram of GaAs has a proposed 'existence range' of composition for the solid compound.[47] Knowledge of this range and the exact composition of solid ingots is desirable in order to predict the point defect content and behaviour. Numerous attempts have therefore been made to assess directly the ratio [Ga]/[As] by means of accurate coulometric titration[48] (see also chapter 10).

Heat treatments of the solid crystal are important for control of point defects, including EL2, and are a standard procedure in the production of semi-insulating GaAs.

Crystal dislocations arise from stresses imposed by differential contraction during cooling from the melt.[49] They exert an influence on bulk properties by acting as impurity- and point defect-gettering centres and sinks for micro-precipitate formation. They may thus constitute the main non-uniformity in the bulk. This influence can also be controlled to some extent by ingot and wafer heat treatments.[50]

5.5 Intermetallic alloys

The versatility of the III–V material systems lies in the ability to make alloys with properties that can be tailored to suit a given application. Band gap, lattice parameter and refractive index are important characteristics which depend on the composition of such an alloy.

The only devices fabricated directly in the bulk binary compounds are those made by ion implantation into wafer substrates, forming metal–Schottky field effect transistors (MESFETs). These devices provide the basic cells for analogue and digital integrated circuit technology in GaAs. In general, the substrate is employed as a 'foundation' for the growth of single crystal thin layer structures defining the device. These epitaxially grown layers may have different compositions and electrical properties but must all have lattice parameters closely matched to that of the substrate to avoid strain and defect formation.

Figure 5.11 is a plot of band gap energy and equivalent wavelength against lattice parameter for the III–V compounds. The variations with composition of different ternary solid solutions are indicated by the joining lines. Hence it appears, for example, that a continuous range of compositions in the Ga–Al–As system is possible without change of lattice parameter. Thus, layers of $Ga_{1-x}Al_xAs$ and GaAs are grown on GaAs substrates. Such heterostructures were used to make the first GaAs-based injection lasers, and are the basis for opto-electronic devices in the whole III–V domain. The achievement of abrupt junctions between materials with different energy band gaps and refractive indices within a continuous crystal lattice is utilised in the heterojunction laser to provide charge carrier and optical confinement

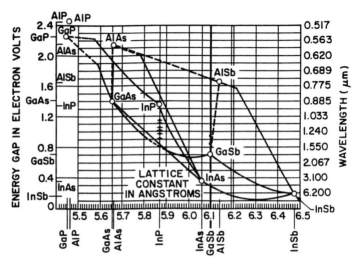

Figure 5.11 Relationship between lattice parameter, band gap, emission wavelength and alloy compositions for a range of III–V intermetallic alloys (from Ref. 55).

within the active region of the laser. This has the effect of reducing threshold current and increasing efficiency.

Electronic devices, such as the high electron mobility transistor (HEMT) based on GaAs/GaAlAs structures, also utilise the carrier confinement associated with the heterojunction to achieve enhanced performance. Periodic structures of different alloy compositions can be grown in thin layer form to produce a 'superlattice' within the basic crystal lattice. A review of heterojunction physics and device concepts is given by Morgan and Williams.[51]

The emission wavelength is a function of the composition and hence the alloy selected depends on the application. In practice, the choice of alloy is governed by the available substrate materials. Since only the binary compounds are produced in large single crystal form, the important systems are

> Ga–Al–As on GaAs: wavelength 0.7–1.0 μm
> Ga–In–As–P on InP: wavelength 1.1–1.6 μm

Infrared LEDs and lasers for applications such as remote control systems, CD players and short-distance optical signal transmission are made in the GaAs-based system. Long-haul telecommunications through optical fibres require the longer wavelengths available from InP-based alloys to match the optimum transmission characteristics of the fibre. GaSb- and InSb-based alloys are much less well developed in their materials technology, but have potential applications as sources and detectors for longer wavelengths (out to 5 μm). For the region out to 10 μm, II–VI compounds such as CdTe and (HgCd)Te are important but are much more difficult materials with respect to crystal growth and processing.

5.6 Epitaxial layer growth

Thin epitaxial layer sequences are deposited on substrates by a variety of growth techniques. The geometries involved are on the micrometre and sub-micrometre scale. Such a sequence is illustrated in Figure 5.12 where the cross-section of a GaAs-buried heterostructure laser device is represented. The operation of the device depends on the dimensions of the active region. Lower threshold current for lasing action and improved waveguiding properties are achieved for thicknesses much less than 1 μm. When the active layer thickness is of the order of 100 Å, the energy band structure across the active region approximates to a square potential well and the free carrier energies become quantised, enhancing the recombination processes. Such 'quantum well' structures exhibit greater efficiency and hence represent the trend in the requirements for growth technology for discrete and integrated opto-electronic components, where the device complexity is limited by the growth technology rather than by the photolithography, metallisation and packaging, as in the case of large-scale integrated circuits.

Epitaxial growth methods can be classified under the following general categories: (a) liquid phase epitaxy (LPE); (b) vapour phase epitaxy (VPE); and (c) molecular beam epitaxy (MBE). Consideration is now given to each of these methods.

5.6.1 Liquid phase epitaxy (LPE)

This method involves the deposition of a compound from a supersaturated solution. The usual solvents are gallium or indium metal, chosen for their low melting temperatures and vapour pressures. The mechanism of the

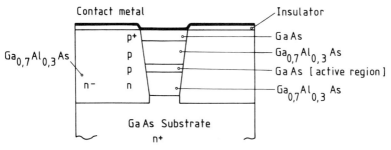

Figure 5.12 Buried heterojunction laser structure.

growth can be illustrated for the case of a saturated solution of As in Ga at a given temperature. If the temperature is reduced, the solution becomes supersaturated with As, according to the phase diagram (Figure 5.4). If a substrate of GaAs is brought into contact with the melt, equilibrium can be re-established by deposition of GaAs from solution onto the substrate. The deposit will normally be in the form of a single crystal layer having the same orientation and structure as the substrate. Growth is terminated by the physical separation of the substrate from the solution melt.

Dopant impurities can be introduced into the grown layers. Their concentration is determined by their distribution or segregation coefficients, as described previously (see section 5.4.1).

Heteroepitaxy of the important ternary and quaternary alloy compositions can also be performed by LPE. In this case, however, it is more difficult to achieve control of the compositional uniformity over the layer thickness since the components have different distribution coefficients. An added complication in the growth of alloys is created by the immiscibility of some composition ranges of solid solutions,[52] referred to as 'miscibility gaps'.

Equipment for LPE growth may be very simple, consisting of an arrangement for dipping the substrate in the solution.[53] A common form employs a graphite slider system to move the substrate between wells containing melts of different composition, thereby enabling the growth of multilayer structures.[54]

LPE is the oldest epitaxy method. It was used for the growth of the first GaAs laser structures and is still used extensively for the mass production of discrete LEDs. It is a low-cost method providing high growth rates for thick layer structures. On the other hand, its application is limited for more sophisticated thin-layer devices, where it is being superseded by the other epitaxy techniques.

5.6.2   Vapour phase epitaxy (VPE)

In this general method, gaseous species containing the Group III and Group V components are caused to react at high temperature on the substrate surface to form the epitaxial layer. Two variants exist: (i) chloride transport VPE and (ii) metal organic VPE (MOVPE).

In the chloride transport method, the final reaction is

$$GaCl(g) + 1/4As_4(g) + 1/2H_2(g) \xrightarrow{700°C} GaAs(s) + HCl(g) \qquad (5.13)$$

for the GaAs case, and

$$2InCl(g) + P_2(g) + H_2(g) \xrightarrow{650°C} 2InP(s) + 2HCl(g) \qquad (5.14)$$

for InP growth. The Group III monochloride is formed in situ by the reaction, at higher temperature, of the Group V trichloride with the Group III metal (the so-called 'trichloride' process), e.g.

$$2PCl_3(g) + 2H_2(g) + 2In(l) \xrightarrow{750°C} 2InCl(g) + P_2(g) + 4HCl(g) \quad (5.15)$$

The input components are a heated metal source, the gaseous trichloride in a purified hydrogen gas carrier stream, possible dopant gases, such as H_2S, H_2Se and SiH_4 as sources of n-type dopants, and metallic zinc, in a separate heated source, for p-type doping.

In the hydride process, the input components are a heated metal source and the Group V trihydride and HCl gas in a hydrogen carrier stream. The reactant species are formed, in the GaAs case, in the following stages:

$$Ga(l) + HCl(g) \xrightarrow{800°C} GaCl(g) + 1/2H_2(g) \quad (5.16)$$

$$AsH_3(g) \rightarrow 1/4As_4(g) + 3/2H_2(g) \quad (5.17)$$

the arsine decomposition occurring by pyrolysis in the reactor.

The growth reaction is readily reversible thermodynamically, and therefore precise control over the component partial pressures is required to ensure reproducible growth rates. This in turn needs good control over the source reactions.

Ternary and quaternary deposition can also be achieved, for example via the following process:

$$xInCl(g) + (1-x)\,GaCl(g) + 1/2y\,P_2(g) + (1-y)\,As(g) + 1/2\,H_2(g)$$
$$\rightarrow Ga_{1-x}In_xAs_{1-y}P_y(s) + HCl(g) \quad (5.18)$$

Conventional VPE systems of this type have the disadvantages of requiring a double furnace arrangement and of the problems posed by the need to control the generation of the reactant species.

In metal-organic VPE, also referred to as 'metal-organic chemical vapour deposition' (MOCVD), the reactant species are introduced directly as gaseous compounds of the III–V components, usually as a Group V hydride (AsH_3 or PH_3) and an alkyl derivative of the Group III element ($(CH_3)_3M$ or $(C_2H_5)_3M$, where $M = Ga$ or In). These are pyrolysed together at the heated substrate. Hence, an overall reaction such as

$$(CH_3)_3Ga(g) + AsH_3(g) \xrightarrow[700°C]{H_2} 3CH_4(g) + GaAs(s) \quad (5.19)$$

is brought about and epitaxial deposition occurs by this irreversible process.

The alkyl vapour is obtained typically by bubbling an H_2 carrier gas stream through the liquid compound. The reaction chamber pressure may be atmospheric or below. The full range of ternary and quaternary alloy compositions can be obtained by the addition of trimethyl or triethyl derivatives of Al (TMAl or TEAl) and/or Sb. Oxygen and water vapour must be strictly excluded from the system for the growth of Al alloys in order to eliminate the possibility of reaction with the alkyl. Other side-reactions may occur, for example between triethylindium and phosphine prior to pyrolysis to form involatile polymeric solids, or incomplete decomposition may result in the incorporation of carbon into the grown layer. The problems of instability and side-reactions may be overcome by the use of Group III and Group V alkyls combined as molecular adducts[55] (see also chapter 6), e.g.

$$(C_2H_5)_3P + (CH_3)_3In \rightarrow (C_2H_5)_3P \cdot In(CH_3)_3 \qquad (5.20)$$

Dopants can be introduced as H_2S, H_2Se, SiH_4 or $(CH_3)_4Sn$ for n-type and as dimethyl- or diethylzinc for p-type properties.

Growth in such reactors is critically dependent on gas flow control. Mass flow controllers are used for precise adjustment of the mole fractions of the reactant species, and the gas circuitry has received a great deal of attention in terms of valving and switching to achieve abrupt junctions and composition changes. The gas dynamics over the substrate determine the uniformity of the layer, and hence considerable effort has been devoted to reactor cell design. Growth rates are of the order of 0.05 μm/min for hydrogen flows of tens of litres per minute. The technique is therefore ideally suited for thin-layer structures. The extremely toxic and hazardous nature of some of the materials used imposes stringent safety considerations which add to the complexity and expense of the equipment. Nonetheless, MOVPE has become prominent in the large-scale commercial production of epitaxial structures and is superseding some of the older methods through its ability to grow more advanced device structures.

5.6.3   Molecular beam epitaxy (MBE)

In molecular beam epitaxy, crystal growth is performed in ultra-high vacuum (10^{-9} Torr) by directing thermally generated elemental or molecular beams at a heated substrate surface. Atomic or molecular fluxes are produced from separate sources in the form of effusion furnaces, with the elements held in PBN crucibles, and controlled by varying the furnace temperatures.

Growth takes place by the adsorption of Ga (or other Group III elements) on the surface, with a sticking probability of about one, followed by the adsorption of the Group V atoms to form the III–V compound. The Group V sticking coefficient is less than one; it is also temperature-dependent and is influenced by the molecular species present in the vapour, e.g. As_2, As_4 etc. Growth is stoichiometric since the reaction involves dissociative chemisorption of the Group V species by the Ga surface and the excess of Group V molecules is desorbed. The Group V flux is maintained in excess to prevent decomposition of the substrate surface at the growth temperature.

All the important alloy compositions can be grown using suitable elemental sources, and dopants may be incorporated in the same way. Metal-organic gas sources have been used in the technique of MOMBE to overcome problems experienced particularly with P-based compounds.[56] This method also has the advantage that the elemental source is external to the chamber and can thus be replaced simply, without breaking the high vacuum. The flux can be controlled and metered using techniques similar to those in MOCVD. Since the beam does not originate from a collimated source, there is no need to rotate the substrate to achieve good uniformity, as is the practice in the basic technique. This avoids the difficult technical problem of providing efficient rotating seals in the chamber. The only heating source is that for the substrate, avoiding one or more potential sources of contamination. Set against these benefits are the hazards of handling toxic and/or pyrophoric metal-organic compounds.

In the conventional arrangement, the growth rate is varied by control of the fluxes, being typically tens of Ångstroms per second. Very thin layers and abrupt junctions can be achieved by the use of shutters on the sources. The chamber environment (UHV) allows the use of a variety of diagnostic and analytical techniques to monitor *in situ* the progress of the growth. Typical attachments to MBE equipment include the facilities for reflection high energy electron diffraction (RHEED), Rutherford back-scattering (RBS), Auger electron spectroscopy (AES) and X-ray electron spectroscopy (XES). RHEED is particularly helpful for observing the two-dimensional crystal structure of the surface during growth and is normally a standard attachment. In terms of equipment, MBE is thus the most expensive of all the layer growth techniques. Because of its ability to achieve precise growth rate control, it has been used extensively to research novel device structures, especially those involving 'superlattice' and quantum well multiple thin-layer sequences.

5.7   Chemical reactivity

In normal usage of the compounds, the important considerations for reactivity are those concerned with atmospheric exposure (oxide formation) and the use of etchants for wafer processing.

5.7.1   Surface oxides

Exposure of a clean surface of a III–V sample to air or aqueous solutions will cause oxide formation by adsorption and chemical reaction. The nature of the oxide is determined by the chemistry of the individual constituents and especially that of the Group V component. Studies of the relative oxidation rates of the compounds have been carried out.[57]

The existence of native surface oxides is important for material processing and has some bearing on device performance through the influence of surface electronic states controlling the energy band structure in the near-surface region. Successful epitaxial growth requires an oxide-free substrate surface. This is achieved by thermal desorption *in situ* under a low pressure of hydrogen gas or by 'etch back' in VPE and LPE. However, the composition of the oxide, and hence its desorption characteristics, depends strongly on the conditions of formation.

The GaAs oxide layer, which can have a thickness of between about 10 and 50 Å, consists of a mixture of the oxides of Ga and As, as well as elemental As close to the oxide-semiconductor interface. The nature and composition of the oxide layer depends on wet chemical treatment, including washing in pure water for wafer cleaning, since the As oxides are soluble. InP has a greater tendency to form the phosphate $InPO_4$, in addition to the oxides of the elements.

The instability and incorporation of impurities in the oxides with storage necessitate some pre-treatment prior to epitaxy in order to normalise the surface and achieve reproducible desorption and growth. This typically involves wet chemical etching or controlled oxide formation by thermal oxidation or ozone exposure from UV irradiation in an oxygen atmosphere.[58]

5.7.2 Wet chemical etching

Etching solutions are employed at many stages of the wafer-processing cycle for the production of devices. The treatment is typically designed for one of the following purposes:

1. damage removal from sawn wafers;
2. polishing by chemo-mechanical action;
3. surface cleaning and removal of residual polish damage; and
4. definition of device structures after photolithography.

In addition, defect-selective etches may be employed for material characterisation of substrates and epitaxial layers.

With most of the etchant systems used, the chemical attack is by oxidation of the surface through the action of a strong oxidant, such as H_2O_2, and a complexing agent, such as H_2SO_4 or aqueous ammonia, to assist dissolution of the oxidation products in a diluent, normally water.

Etch mechanisms may be classified as being either (a) diffusion-limited or (b) reaction rate-limited. In the first case, a diffusion boundary layer exists at the surface under attack, and the rate of dissolution depends on the diffusion rates of the oxidant through this layer to the surface and on those governing the removal of the products from the surface. Diffusion-limited processes are characterised by a strong dependence on agitation and a weak dependence on crystallographic orientation. By contrast, reaction rate-

Table 5.5 Liquid etchants for III–V compounds.

Compound	Etchant	Volume ratio
GaAs	$H_2SO_4/H_2O_2/H_2O$	8 : 1 : 1
	$NH_4OH/H_2O_2/H_2O$	3 : 1 : 50
InP	HNO_3/HCl	1 : 1
	$H_2SO_4/H_2O_2/H_2O$	5 : 1 : 1
	Br_2/CH_3COOH	1 : 50
GaP	HNO_3/HF	1 : 1
	HNO_3/HCl	1 : 1
InSb	$HF/HNO_3/CH_3COOH$	2 : 1 : 1
InAs	HNO_3/HCl	1 : 1
AlSb	$HNO_3/HF/H_2O$	5 : 5 : 1

limited etching is strongly dependent on the chemical nature of the surface and hence is sensitive to orientation in III–V crystals. It is also sensitive to chemical and electrical inhomogeneities in the material and so may be more defect-sensitive and, as a result, less useful for polishing. Table 5.5 lists some examples of etch solutions used for the compounds.

5.7.3 Dry etching

Dry etching methods consist of sputtering by ion bombardment or chemical attack by reactive ion species, perhaps also combined with directed bombardment. Sputtering can be achieved by placing the sample in an argon plasma or by directing a collimated beam from a separate ion source. When reactive elements are introduced, either into the plasma (reactive ion etching, RIE) or into the ion beam (reactive ion beam etching, RIBE), the etch rate is enhanced. Typical species used are chlorine-based molecules such as CCl_4 and Cl_2 which are highly reactive with respect to the semiconductor, yielding volatile products.

Such dry etching methods offer much better control over etch rate, resolution and anisotropy with crystal orientation, and are important elements of the device-processing technologies for GaAs integrated circuits.

5.8 Conclusion

The last 30 years have seen a great deal of research activity involving the III–V material system and the emergence of GaAs and InP as important semiconductor materials with a wide range of applications. Many important new techniques for material-processing, such as MBE, have arisen from this, and mature technologies for large-scale integration are now established. Future demands in areas of telecommunications, such as direct satellite broadcasting and broadband optical fibre networks, will ensure an increasing importance for these compounds.

The normal considerations of extremely high purity demanded in semi-conductor applications can be expected to place yet more stringent requirements on the elements themselves and the techniques employed for their preparation and analysis.

References

1. J.S. Blakemore, *J. Appl. Phys.*, 1982, **53**, 10.
2. C. Kittel, *Introduction to Solid State Physics*, 6th edn., Wiley, New York, 1986.
3. J.P. McKelvey, *Solid State and Semiconductor Physics*, Harper International, New York, 1969, p. 266.
4. B.M. Welch, R.C. Eden and F.S. Lee, in *Gallium Arsenide: Materials, Devices and Circuits*, eds. M.J. Howes and D.V. Morgan, Wiley, Chichester, UK, 1985, p. 517.
5. J.C. Phillips, *Bonds and Bands in Semiconductors*, Academic Press, New York, 1973, Chapter 2.
6. O.G. Folberth, in *Compound Semiconductors*, Vol. 1, eds. R.K. Willardson and H.L. Goering, Reinhold, New York, 1962, p. 21.
7. S.D. Mukherjee and D.W. Woodard, in *Gallium Arsenide: Materials, Devices and Circuits*, eds. M.J. Howes and D.V. Morgan, Wiley, Chichester, UK, 1985, p. 119.
8. G.A. Wolff and J.D. Broder, *Acta Crystallogr.*, 1959, **12**, 313.
9. C.D. Thurmond, *J. Phys. Chem. Solids*, 1965, **26**, 785.
10. M. Hansen and K. Anderko, *Constitution of Binary Alloys*, 2nd edn., McGraw-Hill, New York, 1958.
11. H.C. Casey, Jr., and F.A. Trumbore, *Mater. Sci. Eng.*, 1970, **6**, 69.
12. W. Köster and B. Thoma, *Z. Metallk.*, 1955, **46**, 291.
13. K.J. Bachmann and E. Buehler, *J. Electrochem. Soc.*, 1974, **121**, 835.
14. J. van den Boomgard and K. Schol, *Philips Res. Repts.*, 1957, **12**, 127.
15. T.S. Liu and E.A. Peretti, *Trans. Am. Soc. Metals*, 1952, **44**, 539.
16. K. Weiser, in *Compound Semiconductors*, Vol 1, eds. R.K. Willardson and H.L. Goering, Reinhold, New York, 1962, p. 471.
17. J.R. Arthur, *J. Phys. Chem. Solids*, 1967, **28**, 2257.
18. J.B. Schroeder, in *Compound Semiconductors*, Vol 1, eds. R.K. Willardson and H.L. Goering, Reinhold, New York, 1962, p. 222.
19. J.E. Wardill, D.J. Dowling, R.A. Brunton, D.A.E. Crouch, J.R. Stockbridge and A.J. Thompson, *J. Cryst. Growth*, 1983, **64**, 15.
20. D. Rumsby, R.M. Ware and M. Whittaker, *Semi-Insulating III–V Materials*, ed. G.J. Rees, Shiva, Orpington, UK, 1980, p. 59.
21. G.A. Antypas, *Inst. Phys. Conf. Ser.*, 1977, **33B**, 55.
22. R.O. Savage, J.E. Anthony, T.R. Aucoin, R.L. Ross, W. Harsch and H.E. Cantwell, *Semi-insulating III–V Materials*, eds. D.C. Look and J.S. Blakemore, Shiva, Nantwich, UK, 1984.
23. J.-P. Farges, *J. Cryst. Growth*, 1982, **59**, 665.
24. W.A. Gault, E.M. Monberg and J.E. Clemans, *J. Cryst. Growth*, 1986, **74**, 491.
25. J. Czochralski, *Z. Phys. Chem.*, 1917, **92**, 219.
26. R. Gremmelmaier, *Z. Naturforsch.*, 1956, **11a**, 511.
27. E.P.A. Metz, R.C. Miller and R. Mazelsky, *J. Appl. Phys.*, 1962, **33**, 2016.
28. J.B. Mullin, R.J. Heritage, C.H. Holliday and B.W. Straughan, *J. Cryst. Growth*, 1968, **3**, **4**, 281.
29. T.R. Aucoin, M.J. Wade, R.L. Ross and R.O. Savage, *Solid State Technol.*, 1979, **22**, 59.
30. E.M. Swiggard, S.H. Lee and F.W. Von Batchelder, *Inst. Phys. Conf. Ser.*, 1976, **33B**, 23.
31. W. Bardsley, G.W. Green, C.H. Holliday, D.T.J. Hurle, G.C. Joyce, W.R. MacEwan and P.J. Tufton, *Inst. Phys. Conf. Ser.*, 1975, **24**, 355.

32. M.R. Shropshall and P.E. Skinner, in *Semi-insulating III–V Materials*, eds. D.C. Look and J.S. Blakemore, Shiva, Nantwich, UK, 1984, p. 178.
33. J.A. Burton, R.C. Prim and W.P. Slichter, *J. Chem. Phys.*, 1953, **21**, 1987.
34. S. Skalski, in *Compound Semiconductors*, Vol. 1, eds. R.K. Willardson and H.L. Goering, Reinhold, New York, 1962, p. 385.
35. J.B. Clegg, in *Semi-insulating III–V Materials*, eds. S. Makram-Ebeid and B. Tuck, Shiva, Nantwich, UK, 1982, p. 80.
36. A.P. Mykytiuk, P. Semeniuk and S. Berman, *Spectrochim. Acta Rev.*, 1990, **13**, 1.
37. J.B. Clegg, I.G. Gale and E.J. Millett, *Analyst*, 1973, **98**, 69.
38. U. Lambert and U. Wiese, *Adv. Mater.*, 1991, **3**, 429.
39. D.E. Holmes, R.T. Chen, K.R. Elliott and C.G. Kirkpatrick, *Appl. Phys. Lett.*, 1982, **40**, 46.
40. M.R. Brozel, J.B. Clegg and R.C. Newman, *J. Phys. D*, 1978, **11**, 1331.
41. B. Hamilton and A.M. Hennel, in *Properties of Gallium Arsenide*, 2nd edn., INSPEC EMIS Datareviews Series No. 2, London, 1990, Chapter 12.
42. L.B. Ta, H.M. Hobgood, A. Rohatgi and R.N. Thomas, *J. Appl. Phys.*, 1982, **53**, 5771.
43. G.M. Martin, A. Mitonneau and A. Mircea, *Electron. Lett.*, 1977, **13**, 191.
44. G.M. Martin, G. Jacob, G. Poilblaud, A. Goltzene and C. Schwab, *Inst. Phys. Conf. Ser.*, 1981, **59**, 281.
45. G.M. Martin, *Appl. Phys. Lett.*, 1981, **39**, 747.
46. D.E. Holmes, R.T. Chen, K.R. Elliott, C.G. Kirkpatrick and P.W. Yu, *IEEE Trans. Electron Devices*, 1982, **29**, 1045.
47. D.T.J. Hurle, *J. Phys. Chem. Solids*, 1979, **40**, 613.
48. K. Kurusu, Y. Suzuki and H. Takami, *J. Electrochem. Soc.*, 1989, **136**, 1450.
49. A.S. Jordan, R. Caruso and A.R. Von Neida, *Bell Syst. Tech. J.*, 1980, **59**, 593.
50. T. Inada, Y. Otoki, K. Ohata, S. Taharasako and S. Kuma, *J. Cryst. Growth*, 1989, **96**, 327.
51. D.V. Morgan and R.H. Williams, *Physics and Technology of Heterojunction Devices*, IEE series 8, 1991.
52. P. Henoc, A. Izrael, M. Quillec and H. Launois, *Appl. Phys. Lett.*, 1982, **40**, 963.
53. H. Nelson, *RCA Rev.*, 1963, **24**, 603.
54. M.B. Panish, I. Hayashi and S. Sumski, *Appl. Phys. Lett.*, 1970, **16**, 326.
55. R.H. Moss, *J. Cryst. Growth*, 1984, **68**, 78.
56. R.A. Laudise, *J. Cryst. Growth*, 1983, **65**, 3.
57. A.J. Rosenberg, *J. Phys. Chem. Solids*, 1960, **14**, 175.
58. H.L. Hartnagel, in *Properties of Gallium Arsenide*, 2nd edn., INSPEC EMIS Datareviews Series No. 2, London, 1990, p. 369.

6 Organometallic compounds: synthesis and properties

K.B. STAROWIEYSKI

6.1 Introduction and basic literature[1-12]

Organometallic compounds of aluminium, gallium, indium and thallium have been known for many years; some of them were first synthesised in the 19th century. However, since the discoveries of Ziegler's group in the 1950s, the largest share of investigations has been directed towards aluminium compounds. The importance of Ziegler's discoveries cannot be underestimated because the use of alkylaluminium compounds as co-catalysts in olefin-polymerisation revolutionised polymer chemistry at that time. Industrial methods of polymerisation were quickly developed, resulting in the production of polymers in millions of tons all over the world. For these reasons, the discovery of direct methods of synthesis of organoaluminium compounds and the synthesis of long-chain alcohols from these compounds and ethylene led to Ziegler's being awarded a Nobel Prize. It is estimated that industrial production of organoaluminium compounds is, after organotin compounds, the second biggest in the whole field of organometallics.

Since the early 1960s there has been steady progress in studies of the reactivity and structure of organothallium compounds, mainly because of a rapid expansion of organic syntheses mediated by organothallium compounds (see chapter 7). The organic chemistries of gallium and indium remained for many years the least explored fields of the Group 13 organometallics. However, the industrial application of these and organoaluminium compounds for the preparation of semiconductors and optoelectronic materials (see chapter 5) has prompted many research centres to undertake basic investigations of the synthesis, structure and reactivity not only of gallium, but also of indium and aluminium compounds. In a similar way, the ceramic industry now uses organoaluminium compounds for the synthesis of aluminium nitride. For all these reasons, the 1980s witnessed an upsurge in the number of publications devoted to organometallic compounds of the heavier Group 13 metals, although it should be stressed that the basic work on the use of organometallics in the fabrication of epitaxial layers was published by Manasevit as early as 1968.[13]

There are numerous books and reviews pertinent to the chemistry of organometallic compounds of aluminium, gallium, indium and thallium.[1-12] General considerations and laboratory procedures for derivatives of all four elements are given in a volume of the *Houben–Weyl* series published in 1970;[4] the literature is covered for the period up to the first half of 1969.

Many methods of synthesis of organo derivatives of the Group 13 metals are also presented in a book by Eisch published in 1981.[7]

The most extensive review of organoaluminium chemistry, which covers the literature up to late 1971, is presented in the book by Mole and Jeffery;[5] a more up-to-date summary is given by Eisch in *Comprehensive Organo-metallic Chemistry*, with literature coverage up to 1981.[8] The heavier elements feature in the following chapters in the same volume. The complete literature on organogallium compounds up to 1984 (in some cases even later) is collected in a volume of the *Gmelin Handbook of Inorganic Chemistry*.[11] Similarly comprehensive treatment of organoindium compounds brings the literature coverage up to early 1991. Detailed information is also to be found in Refs. 1, 2, 3 and 6.

The *Dictionary of Organometallic Compounds* gives the most extensive listing of well characterised organo compounds of the Group 13 metals with some of their physicochemical properties, as well as key references.[10] Compounds of every element are listed in structural, name and molecular formula indices, complete with the appropriate CAS registry numbers.

Very thorough annual reviews, in most cases prepared separately for each element, appeared up to 1981.[14] Since 1971, annual surveys of the literature of a more selective kind have appeared in the series *Organometallic Chemistry* published by the Royal Society of Chemistry.[15]

General descriptions of the bonding in the Group 13 organometallics can be found in the reviews of O'Neill and Wade,[8] of Jutzi,[9] and of Haaland.[12c] A special issue of *Polyhedron* comprises 30 original research papers covering timely areas of inorganic and organometallic chemistry of the Group 13 elements; these are supplemented by four short review articles on specific topics.[12a] A recent review by Alexandrov and Chikinova concerns the chemistry of peroxide derivatives of Group 13 organometallics.[12b]

6.2 Synthesis of organo derivatives

6.2.1 General considerations

Most organometallics of aluminium, gallium and indium, especially the *tris*-(alkyl) derivatives, are dramatically oxygen- and moisture-sensitive and demand for preparative work that manipulation be *in vacuo* or under an inert gas like nitrogen or argon including less than 10 ppm of oxygen-containing impurities. Any solvent or reagent and all internal surfaces of reaction apparatus must be free from any traces of active hydrogen or peroxide. The sensitivity drops as the size of the organic substituents increases and as the atom or atoms directly connected to the metal become more basic. Pyrophoric ability is associated with the vapour pressure and is greatest for the most volatile compound, *viz.* trimethylgallium.

The sensitivity to air and moisture of thallium compounds depends on the type of substituents and on the degree of association, being highest for trimethylthallium, which ignites spontaneously in air and reacts violently with water. However, many reactions of organothallium compounds can be carried out even in water as a solvent. Generally the sensitivities to air and water drop slightly in the order $Al > Ga > In$ and markedly from In to Tl.

6.2.2   Methods of synthesis

There are four basic ways to effect the preparation of organo compounds of the Group 13 metals:

1. reaction of the metal with a reagent containing the organic group;
2. oxidative addition to a subvalent metal compound, e.g. MX;
3. metathesis between a metal alkyl and a Group 13 metal compound; and
4. insertion of an organic reagent into an M–X bond (where, nearly exclusively, X = H and the reagent is an unsaturated hydrocarbon).

Under this general division of synthetic methods, there is a great variety of possibilities for obtaining any desired compound. Most of the options were described long ago and so are treated here only briefly.

The most versatile method, albeit only for very small-scale operations, is the reaction of the metal with the appropriate diorganomercury compound. Usage of the metal in a highly active form is advantageous.

$$2M + 3R_2Hg \xrightarrow[\text{time}]{\Delta} 2R_3M + 3Hg \quad (M = Al, Ga \text{ or } In) \qquad (6.1)$$

This procedure cannot, however, be used for organothallium compounds since they undergo the reverse reaction with mercury.

Electrochemical methods have been used to obtain aluminium, gallium and indium compounds. Neutral or ionic salt-like complexes of organometallic compounds of magnesium and aluminium are used in the form of melts or solutions (in alkyl halides or, better, in donor solvents) as alkylating agents. The sacrificial anode is made of the metal which is to form the organometallic compound.[16,17]

So-called 'direct synthesis' of organoaluminium compounds is the most important industrial method for the manufacture of large quantities of the products. It is a two-step reaction demanding elevated pressure (about 200 atm) and a temperature of 120°C in the first step of the process, which leads to the formation of R_2AlH. The second step, i.e. the addition of olefin, proceeds at much lower pressure, about 20 atm and a temperature of only 60°C. The overall process can be summarised in the case of triethylaluminium by the following equation:

$$2Al + 3H_2 + 6CH_2{=}CH_2 + Et_3Al \rightarrow 4Et_3Al \qquad (6.2)$$

The reaction of the metal with an alkyl halide is perhaps the method of synthesis most often used for organoaluminium compounds, especially on the laboratory scale, e.g.

$$2Al + 3RX \rightarrow R_3Al_2X_3 \tag{6.3}$$

However, the direct interaction of the metal with an alkyl halide does not always result in an organometallic product. Sometimes there is no reaction; sometimes only decomposition products are observed. An alloy or a mixture of metals, mainly with magnesium, of defined stoichiometry can help in obtaining the desired organometallic. Use of a donor solvent is also advantageous in the reaction but usually results in the formation of a stable adduct. Lower alkyl halides react rather easily with aluminium but higher ones demand donor solvents and/or alloys. Activation of aluminium by iodide, mercury, grinding with AlX_3, or addition of small amounts of organoaluminium compounds helps to initiate the reaction. For the smooth operation of the process a small amount of iodine or alkyl iodide should be permanently present in the reactor. A basic study of the reaction of an atomically clean and chemically modified (1 1 1) aluminium surface with alkyl halides has been performed. Only MeI, but not MeCl or MeBr, was adsorbed molecularly and dissociatively at the surface in the temperature range 150–500 K. It was stated that MeI decomposes completely to form Al–I and Al–CH$_3$ bonds on the surface at temperatures of 250–450 K. These surface fragments decompose at temperatures above 450 K forming chemically bonded carbon residues.[18] The investigation confirms the need for the presence of iodine or its compounds to catalyse the synthesis of organoaluminium compounds, especially when RCl is used. It is plausible that some metal impurities also catalyse the process.

The direct synthesis from gallium is claimed to proceed in a matter of weeks with low yields under special conditions, and so does not have any real practical value.[19] Alloys of different composition demand special preparation and specific conditions for the synthesis. They react with lower alkyl halides in donor solvents to give trialkylgallium complexes, it is claimed, in yields of 30[17] to 95%.[20]

To obtain free R_3Ga, a weak, involatile base must be added, e.g. diphos $(R_2PCH_2CH_2PR_2)$, triphos $(R_2PCH_2CH_2PRCH_2CH_2PR_2)$ or Ph_3P. Such compounds form complexes that are stable at room temperature and allow for the evaporation of a volatile solvent like diethyl ether. However, the Ga–P bond is so weak that the parent gallium compound can be evaporated free from any complexing agent at elevated temperatures. This is also one of the methods used to free trialkyl compounds of the metals (Al, Ga or In) from volatile uncomplexed impurities like hydrocarbons or organometallics of Si, Mg, Zn or Sn. Involatile transition metals and complexes of their compounds are left in the residues. The product obtained is very pure, being suitable for electronic applications.[21] The method results in contamination levels well below 1 ppm for the sum of all the undesired elements.[17,22,23]

Only under extreme conditions does bulk indium react with organic halides, but more active halides like C_6F_5I or alkyl bromides and iodides do give the corresponding organoindium compounds. A specially prepared In/Mg alloy reacts smoothly with alkyl bromides in diethyl ether which forms weak complexes with the products and can be eliminated, albeit with some difficulty, at elevated temperature.[24] It is proposed that thallium formed, together with trimethylthallium, in the disproportionation of MeTl reacts with MeI regenerating MeTl and forming TlI.[25]

Rieke developed a method of reduction of metal halides with alkali metals in a hydrocarbon or ethereal solvent, resulting in the formation of extremely active metal powders. Because of the reaction of the aluminium and indium powders with ethers, their reductions were conducted in xylene. The slurry thus formed then reacts readily with aryl halides (Al + PhX above 70°C and In + PhI above 80°C).[26,27]

The vapours of aluminium, gallium or indium react during codeposition with an alkyl bromide or iodide at 77 K, forming in the matrix an insertion product R–M–X.[28] Abstraction appears also to occur because the IR spectra of the matrices formed from CH_3Br and Al, Ga or In include a band attributable to the CH_3 radical. Thallium, by contrast, does not react with MeBr under these conditions. It was stated that only monatomic gallium reacts with MeBr and MeI; clustering in an inert matrix deactivates it completely.[29] This behaviour is the opposite of that shown by magnesium which is more reactive towards alkyl halides when aggregated than in the atomic form.[30] The difference is rationalised by deactivation of gallium on clustering because of the formation of strong Ga–Ga bonds (with energies of about $135\,kJ\,mol^{-1}$), as compared with the very weak Mg–Mg bonds of Mg_x clusters (with energies in the order of $8.5\,kJ\,mol^{-1}$). During the reaction of Mg_x with an alkyl halide, moreover, a relatively strong Mg–Mg bond is formed, thereby making the enthalpy of the reaction more favourable and decreasing its activation energy.[30] The aluminium and indium clusters obtained in warmed hydrocarbon matrices are reactive towards many organic halides.[31] This is probably due to the lower energy levels of empty acceptor orbitals in the case of aluminium and the much weaker M–M bonds in the case of indium ($c.\ 97\,kJ\,mol^{-1}$). Rather unexpectedly, alkyl chlorides do not react even with atomic gallium and indium, although the colouring of the matrix suggests that some interaction occurs.[29]

The synthesis of organogallium and -indium compounds can be scaled up even to the order of kilograms using a rotating cryostat with external cooling such as that produced by G.V. Planer Ltd. (Ref. 32, p. 21) or with internal cooling of the type described by Billups et al.[33] The gallium clusters are then reactive only if the vapours are codeposited with alkylaluminium compounds and an inert solvent, and only if an alkyl bromide or iodide is added to the resulting slurry. During matrix formation it was observed that a vigorous decomposition reaction takes place. Clusters obtained in such a manner include about 5% of Al together with some metal-bonded alkyl groups; both of these features are responsible in all probability for the reactivity of the gallium slurry.[29]

Compounds of the type $M(CF_3)_3$ can be obtained by a modified method involving free radicals (formed by the action of a low-temperature glow discharge on hexafluoroethane) co-deposited on a cooled, rotating drum with the corresponding metal vapours. Only *tris*(trifluoromethyl)thallium is stable at room temperature; the stability of the other drops from indium to aluminium and demands low temperatures and coordinating ligands for their survival. The reaction is accompanied by oxidative addition of C–F bonds to thallium, but as a side-reaction resulting in the formation of pentafluoroethyl-substituted compounds.[34] Trifluoromethylgallium, -indium and -thallium compounds have also been obtained in dismutation reactions of $Cd(CF_3)_2$ complexes and are stable, but only in the presence of strong donors, e.g. $(CF_3)_3In \cdot 2NCCH_3$.[35]

Another application of metal vapour synthesis leads to subvalent indium and thallium compounds. These are formed, for example, by co-deposition of the metal vapours in a matrix of cyclopentadiene. Warming up the matrix yields compounds of the type $(C_5H_5)M$ (M = In or Tl).[36]

Oxidative addition of organic iodides to subvalent metal compounds of the types MX or M_2X_4 (which can exist in the form $X_2MMX_2 \cdot 2D$ or $M^+[MX_4]^-$, where D is a donor species and X is a halogen or an organic group; see chapter 3) has been applied to the synthesis of organo derivatives of M^{III} (M = Ga, In or Tl).[37,38] The reaction is effective and fast for methyl and ethyl halides and Ga_2X_4 species (X = Cl, Br or I), whereas larger organic groups tend to undergo radical dimerisation. For indium compounds it is said that the reaction is slow but effective. The synthesis of Me_3Tl has been accomplished using equimolar amounts of TlI and MeI and bimolar amounts of MeLi as the alkylating agent. Some other triorgano compounds have also been produced by this method. However, the oxidative addition products of thallium(I) compounds often undergo reversible reductive elimination:

$$RTlX_2 \rightleftharpoons TlX + RX \qquad (6.4)$$

Metathesis is another useful laboratory method for the preparation of organo compounds of the Group 13 metals, as represented by the general equation

$$(m - n)/3MX_3 + R_{m-n}M'X_n \rightarrow (m - n)/3MR_3 + M'X_m \qquad (6.5)$$

(where $m = 1$–4 and $n = 0$–3). The alkylating agents most commonly used are organo compounds of metals from Groups 1, 2, 12 and 13 in the first and second rows of the Periodic Table and of the heavier Group 14 elements. The species MX_3 in equation (6.5) is usually a halide of a Group 13 metal. The method is commonly used for the synthesis of organoaluminium,

-gallium and -indium compounds, but it does not always work for organothallium compounds because of the facility of reduction of TlX_3. Depending
on the substituents, type of co-reagents and conditions used, the attempted
synthesis of organothallium compounds often gives poor yields because of
oxidation of the organic donor by TlX_3 or coupling of the organic substituents. The formation of reduction products like TlX or metallic thallium
can be avoided by using a thallium *tris*(carboxylate) instead of a thallium-
(III) halide. Reduction of MX_3 has also been observed in the reaction of
gallium and indium halides with organoaluminium compounds, but only as a
minority side-reaction.[24]

Thallium compounds of the type $Tl(O_2CCX_3)_3$ react with terminal
acetylenic compounds to give monoorgano-substituted thallium
compounds. When X = F, then reaction with aromatic compounds occurs
to give, depending on the ring substituents, mono- or diorganothallium
trifluoroacetate[39,40] (see also chapter 7).

6.3   Metal vapour reactions

A new approach to the chemistry of organo compounds of the Group 13
metals has been pioneered by Skell, Klabunde, Kasai, Mile and several
others. The principle of metal vapour synthesis (MVS) has been applied to
the reactions of the metal atoms M (hardly contaminated with M_2 species for
any of the four metals under discussion[41]) or their clusters with organic
reagents. Such reactions, or at least the first steps, proceed in the matrix or
in its vicinity. Many intermediates of these reactions have been observed
and characterised by spectroscopic means (see chapter 1), giving information about mechanism, bonding and structure. More general accounts of
the MVS technique and chemistry, with literature coverage up to 1980, are
given in two books.[32,42] There is no information about the commercial
application of MVS, but the methodology is now such that practical usage in
synthesis and catalysis is *ante portem*.

The reactions of aluminium, gallium or indium vapours with organic
halides have already been discussed in the preceding section in connection
with synthetic routes. Most of the other studies have focused on the interactions of aluminium (and, to a lesser degree, of gallium and indium) with
methane, ethylene, acetylene, allene, butadiene, benzene, and also carbon
monoxide. Unless stipulated otherwise, the reacting metal species were
atoms and in their electronic ground state. They were obtained mainly by
resistively heated evaporation, but also by laser ablation and pyrolysis of
triorgano derivatives.

Activation of inert C–H bonds by a metal has been recognised for some
years as a very important goal. The insertion reactions of aluminium and
gallium, but not indium, into this bond have been described by Klabunde

and Ozin and their associates.[43-45] There has been some controversy about whether aluminium demands UV photolysis or the reaction is a thermally induced one involving Al atoms in their ground state, but very different sources of metal atoms used by the Klabunde group may be responsible for the variations in reactivity. The following scheme for the reaction with methane is proposed:

$$:M^{\cdot} + CH_4 \longrightarrow \left[\begin{array}{c} \overset{\cdot}{M} \\ H\text{---}CH_3 \end{array} \right] \longrightarrow \begin{array}{c} \overset{\cdot}{M} \\ H \quad CH_3 \end{array} \tag{6.6}$$

$$(M = Al \text{ or } Ga)$$

The non-linear geometry of the CH_3–$\overset{\cdot}{M}$–H molecule is evident from its ESR spectrum. Ozin *et al.* stated that photolysis of the Al–CH_4 system with radiation having $\lambda = 368$ or $305\,nm$ results in rapid depletion of all the features associated with Al atoms ($^2D \leftarrow {}^2P$ (315, 308, 303 and 295 nm) and $^2S \leftarrow {}^2P$ (368 and 348 nm)), while giving rise to a broad absorption between 450 and 600 nm attributed to the CH_3AlH molecule. Subsequent broadband photolysis (with $\lambda > 400\,nm$) resulted in decomposition of this species and the appearance of Al and Al_2.

Very interesting species of the type $M(C_2H_4)$ ($M = Al$, Ga or In) have been formed in dilute neon matrices at about 4 K. The ESR spectrum of the green matrix containing Al–C_2H_4 or Al–C_2D_4 complexes implies bonding which is similar to that of transition metal-ethylene derivatives (**I**). The association reaction energy was calculated to be $-37 \pm 8\,kJ\,mol^{-1}$ for Ga + C_2H_4 with gallium atoms obtained by pyrolysis of trimethylgallium.[46] The $GaH_3\cdots C_2H_4$ interaction has been computed and, depending on the method of calculation, a similar enthalpy change (in the range -17 to $-43\,kJ\,mol^{-1}$) is found.[47]

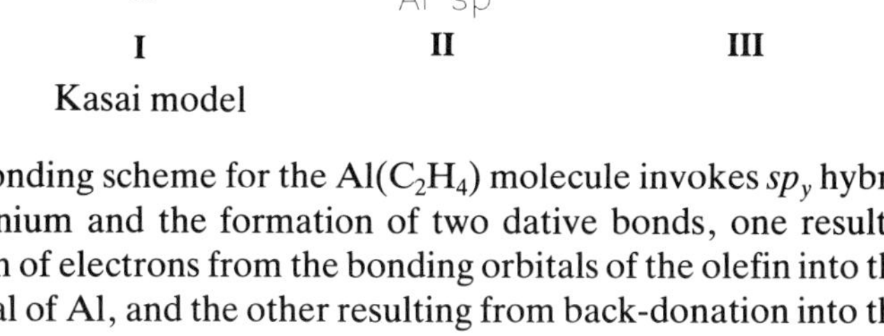

I   II   III

Kasai model

The bonding scheme for the $Al(C_2H_4)$ molecule invokes sp_y hybridisation of aluminium and the formation of two dative bonds, one resulting from migration of electrons from the bonding orbitals of the olefin into the vacant sp_y orbital of Al, and the other resulting from back-donation into the vacant antibonding orbital correlating with the semi-filled p_x orbital of the Al atom.[48,49] Co-condensation of propylene with Al atoms in adamantane or cyclohexane matrices at 77 K leads to an insertion reaction into the C–H bond forming a cyclic π-allylaluminium hydride with a folded 4-membered ring (IV).[50]

$$:Al\cdot \;\; + \;\; CH_2{=}CH{-}CH_3 \longrightarrow \;\; \cdots \longrightarrow \qquad (6.7)$$

IV

At higher temperature there seems little doubt that σ Al–C bonds are formed and that some C–C coupling is induced for ethylene as well as for propylene.[51–54] Visible or UV irradiation promotes this reaction.[49] The ESR spectrum of the matrix containing the $Al(C_2H_4)$ complex typically also includes a feature very similar to that of the aluminocyclopentene radical obtained from the interaction of butadiene with Al atoms.[55] A mechanism of cyclodimerisation *via* formation of the *bis*(ethylene)aluminium complex (V) is proposed, although a direct route not involving such a complex is feasible.[48,51] Although further reactions occur at higher temperatures, the species believed to be $Al(C_2H_4)$ (I) and the aluminocyclopentane radical (III) are observed at temperatures up to 343 K in an adamantane matrix.

V   **VI**

Calculations affecting the Al_2-ethylene complex have also been performed. Kos *et al.*[56] state that the most stable product is likely to be 1,2-dialuminoethane with an *anti* disposition of the Al(I) atoms, but So concluded recently that the structure **VI** has the lowest energy of all the six most probable configurations.[57] Of these six structures, only *anti*-1,2-dialuminoethane and **VI** are found to be thermodynamically stable and to offer appreciable barriers to dissociation. This contrasts with $1,2\text{-}Li_2C_2H_4$ which is unstable and decomposes exothermally.

It is surprising that gallium does not seem to interact with C_2F_4, even allowing for the possibility that abstraction is the preferred channel for this system. The findings indicate, however, that the binding energy is very small for $Ga\cdots C_2F_4$ compared with $Ga\cdots C_2H_4$.[46]

Mile *et al.* have investigated the interactions in the allene–Al and allene–Ga systems.[58] Ground-state aluminium and gallium atoms react rapidly with allene at 77 K, adding exclusively to the central carbon atom to give the resonance-stabilised metal-substituted allyl derivative **VII**. This

VII   VIII

suggests that the selectivity is controlled mainly by the presence of empty low-lying *p*-orbitals on the metal atom. Under the same conditions butadiene reacts with aluminium giving two major paramagnetic products.[55] The first of these is the aluminium-substituted allyl radical **VIII**; the second, formed in a cheletropic reaction, is the aluminocyclopentene radical with the Al atom bonded strongly to two sp^3 carbon atoms.

In the purple matrix formed by the co-condensation of benzene with Al atoms, there is formed an $Al \cdots C_6H_6$ complex which, on the evidence of its ESR spectrum, finds aluminium bound only through the π orbitals of one C–C bond in a manner similar to that of $Al–C_2H_4$ in **I**.[59]

Most thoroughly investigated of the low-temperature matrix reactions of aluminium is that with acetylene. A dilute neon matrix containing these species appears purple. The complexation evidently demands some activation energy because only when the Al and acetylene mix in the vapour state does the matrix give an intense ESR spectrum. In fact, the adduct was found to have not the expected π-bonded structure similar to that of **I** but a vinyl structure with *trans* hydrogen atoms.[9] The *cis* form, a metastable configuration, was observed only when the matrix was photolysed.[48,60] A related phenomenon is apparent in the structure of the molecule $[Me_2AlC\equiv CMe]_2$. Here one can see that attack of the Al atom of the first molecule on the triple bond of the second molecule results in a similar *trans* deformation of the $Al–C\equiv C–Me$ unit, but amounting to only about 20° and with only a small elongation of the $C\equiv C$ bond.[61] In either case, attack does not proceed through a π channel because *cis*-deformation of the bond system takes precedence.[62] By contrast, the $M–C\equiv C–Me$ bonds in the corresponding gallium and indium compounds, $[Me_2MC\equiv CMe]_2$ (where M = Ga or In), have a *cis* configuration.[63] It is not possible therefore to draw any clear inferences in these cases or to proceed to any obvious generalisations.

IX   X

Ab initio SCF calculations by Schaefer *et al.* lead to the conclusion that the aluminium-vinylidene complex **X** is about $50 \, kJ \, mol^{-1}$ more stable than the *cis*-vinyl isomer **IX**, but the barrier height of $37 \, kJ \, mol^{-1}$ would prevent spontaneous isomerisation of **IX** to **X** in a rare gas matrix at $4 \, K$.[64–66] The same authors discuss eight possible structures before concluding that, in energetic terms, **IX** and **X** are not only the most favourable, but also nearly degenerate.[66]

Aluminium, gallium or indium atoms react in rare gas or hydrocarbon matrices with CO molecules, even at $4 \, K$, to give $M(CO)_2$ (M = Al, Ga or In) as the main product, **XI**, containing the metal in the zero oxidation state.[67–69] The ESR and IR properties of the matrices are consistent with a bent, planar radical of C_{2v} symmetry having the unpaired electron in a molecular orbital perpendicular to the molecular plane and constructed from the metal np_x and CO $2\pi_x^*$ orbitals, while the metal lone pair electrons reside in an sp^2 orbital directed along the C_2 axis.

XI

The spectra of the compounds were observed to persist in an adamantane matrix at temperatures up to $240 \, K$. Some other, secondary products, which have also been detected, can be tentatively identified as $Al_3(CO)$ and $Ga_2(CO)$; these are formed as a result of clustering of the metal atoms during deposition. The matrices showed, however, little sign of the expected $M(CO)$ molecule, although there is some evidence indicating the presence of a small amount of linear $Ga(CO)$.

Several investigations of the interaction of Group 13 metal atoms with oxygen-containing compounds have been performed.[42] The deoxygenation yields were not found to be high in the reactions with ketones and aldehydes, and Al-induced coupling was one of the main reaction pathways. For ethers, however, deoxygenation appears to be more effective.

6.4   Physical properties and structures

The dominating factor that influences the properties of organometallics of the Group 13 metals in the oxidation state +3 — and especially of the lighter metals — is the presence of low-lying empty orbitals ready to accept electrons from one or more filled atomic orbitals (as in –Cl:) or bonding molecular orbitals (as in M–H; see (11) in Figure 6.1), to give products with structures adapted to the formation of one or more new bonds. This results typically in the formation of a 4-coordinated, sp^3-hybridised metallic centre.

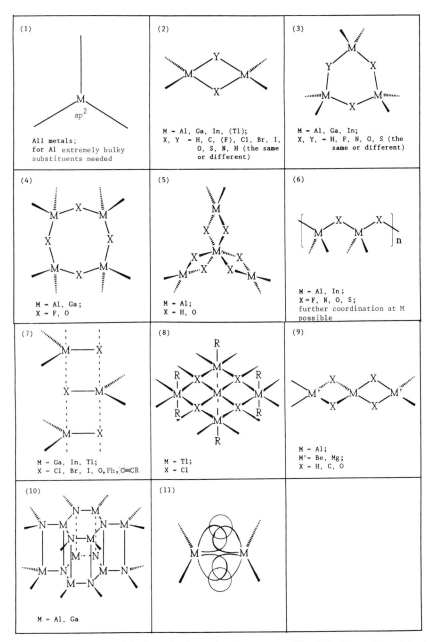

(1)

sp^2

All metals;
for Al extremely bulky
substituents needed

(2)

M - Al, Ga, In, (Tl);
X, Y - H, C, (F), Cl, Br, I,
O, S, N, H (the same
or different)

(3)

M - Al, Ga, In;
X, Y, - H, F, N, O, S (the
same or different)

(4)

M - Al, Ga;
X - F, O

(5)

M - Al;
X - H, O

(6)

M - Al, In;
X = F, N, O, S;
further coordination at M
possible

(7)

M - Ga, In, Tl;
X - Cl, Br, I, O, Ph, C≡CR

(8)

M - Tl;
X - Cl

(9)

M - Al;
M'- Be, Mg;
X - H, C, O

(10)

M - Al, Ga

(11)

Figure 6.1 Structures of the kernels of associated organometallics of the Group 13 metals.

The heavier metals of the Group, by virtue of their size and, possibly, access to empty nd-orbitals (see chapter 1), display a tendency to increase their coordination numbers to 5 and 6 and, for thallium, even to 8.[71] If an organometallic compound has several filled orbitals of similar energy, complexation will proceed to favour not only those of highest energy, but also those which result in the formation of rings. The products are mostly dimers or trimers ((2) and (3) in Figure 6.1), but there are known many other aggregates, undefined or defined, of which some of the more important are shown in Figure 6.1.

The structures of some of the organometallics can change from one set of conditions to another. For example, some cyclic associates of nitrogen, sulfur or fluoride possessing organometallic units change to long-chain linear polymers ((6) in Figure 6.1) when solidifying.[72a] An increase in the coordination number of the metal may well attend the change to a more open structure.[72b] The most commonly observed change involves the dimer-trimer system. The trimer is for many compounds the stable species at room temperature, whereas it often reassociates to a dimer at higher temperature:

$$2[R_2AlX]_3 \overset{\Delta}{\rightleftharpoons} 3[R_2AlX]_2 \qquad (6.8)$$

Aggregates with weaker bridging bonds, like those formed by carbon atoms or doubly bridging atoms with more than one donating electron pair, dissociate at higher temperature or in solution. Trimethylaluminium, which is dimeric at room temperature, is 98% dissociated to the planar monomer, $AlMe_3$, at 215°C. The bis(pentafluorophenyl)thallium compound $(C_6F_5)_2$ TlBr is polymeric in the solid state (as in (7) in Figure 6.1) but depolymerises in benzene solution to form the bromine-bridged dimer.[73]

In dimers the bridging atoms are carbon, halogens (only very seldom F, as in R_2InF where R = mesityl or CH_2Ph[74]) and other hetero-atoms (H, O, N, S, P or As) where the ring is often substituted with larger groups. In such bridges, the endocyclic angles are strained. On the other hand, trimers, which are usually folded, are free from bridge strain, but there is significant repulsion between axial substituents bigger than methyl groups. These non-bonded interactions lead, especially at elevated temperature, to re-association in the form of dimers. The only known trimer with a flat skeleton is $[Bu^t_2AlNH_2]_3$.[75] Such a structure can be rationalised on the basis of the size of the t-butyl substituents which is sufficient to fill the space above and below the ring without strong non-bonded repulsions between the substituents, but only when the Al_3N_3 skeleton is planar; there are small strains affecting only the endocyclic angles.

The association causes measurable elongation of the M–X bond which takes part in the formation of an M–X–M bridge (Figure 6.1). Mostly symmetrisation of the bridging system occurs to give bond lengths B intermediate between covalent (T) and dative (D) bond lengths (Table 6.1). In

Table 6.1 Bond lengths (in Å) in organometallics of the Group 13 metals.[a]

M–X bond	Metal											
	Al			**Ga**			**In**			**Tl**		
	T	C	B	T	C	B	T	C	B	T	C	B
M–C	(1.86); 1.91–2.00		2.03–2.18	1.94–2.03; (2.07)		2.06–2.29			2.37–2.46	2.07–2.21		
M–C≡			1.99			2.00–2.02			2.19			
M⋯C≡ / M–C in MR_4^-	2.02–2.10		2.15–2.18		2.20–2.30	2.24–2.38		2.20–2.23	2.52–2.99			
M–H	1.51–2.02		1.50–1.94	1.49–1.67		1.71–1.93	1.85–1.95		1.87			
M–F	1.63–1.69	1.78	1.81–1.82	1.78		2.44	1.99		2.59–2.64	2.08		
M–Cl	2.05–2.21		2.21–2.44	2.09–2.20		2.22–2.46	2.37–2.41		2.40–2.94	2.49		2.54–2.94
M–Br	2.21–2.33		2.33–2.42	2.25–2.35		2.35–2.38	2.54		2.64–3.10	2.62		2.74–3.07
M–I	2.58–2.59		2.58	2.40–2.58		2.60	2.64–2.75		2.84	2.81		
M–O	1.67–2.02	1.83–2.44	1.83–2.17	1.86–1.91	2.03–2.69	2.08–2.12	2.09–2.44	2.64	2.17–2.54	2.12–2.68	2.39–3.13	2.61–2.88
M–S	2.35–2.37	2.44–2.49	2.35–2.44	2.37–2.38			2.63	2.59	2.59	2.59–2.98	2.72–3.16	3.35
M–N	1.78–1.93	1.94–2.20	1.91–2.05	1.96–1.98	2.12–2.43	1.98	2.57–2.66	2.21–2.61	2.18–2.60	2.67–2.79	2.67–2.79	
M–P	2.50	2.42–2.54	2.43–2.47		2.52				2.60–2.66			
M–As				2.44–2.50	2.62	2.51–2.63						

(a) T, terminal; C, complex; B, bridging.

most cases it is found that $B = \frac{1}{2}D + \frac{1}{2}T$, but in cubane or hexameric structures ((7) or (10) in Figure 6.1) $B = \frac{1}{3}D + \frac{2}{3}T$.[12c] The best known exemptions from this rule are the alkynyl derivatives which have long M′–CM bridging bonds, a small elongation of the C–M bonds and — astonishingly — distinctly stronger bridges than do alkyl derivatives. The donor ability of the alkynyl group is sufficient to cause even Ga and In to form stable dimers in the gas phase,[61,63,76] but $R_2TlC\equiv CR$ is monomeric with a tendency to assume the form $[R_2Tl]^+[C\equiv CR]^-$.[9] Moreover, the formation of a polymer by some compounds results in different bond lengths for the M–X–M unit, as in $[(C_6F_5)_2TlBr]_n$.[73] The range of common M–X distances for terminal, bridging and donor-acceptor bonds is indicated in Table 6.1. Bond angles are omitted from the table because they depend on the form of association and on the lengths of the bridging bonds. It can be said that the endocyclic angles are strained in nearly all dimers, being equal and even less than 90° for many E–M–E units. They are mostly little above 100° for other, unstrained sp^3-hybridised aggregates. There is an evident tendency for enlargement of the terminal C–M–C angles as the atomic number of the metal increases.

The empty p or sp^3 orbitals lie deeper in energy for aluminium than for the heavier Group 13 metals, and therefore give rise to stronger bridging bonds. This results in well-defined and rather stable associated organoaluminium compounds. If one excludes most of the cases where an internal complexing agent is present, monomeric species are found only when extremely bulky groups are connected to the aluminium atom,[77] when the aluminium has access to low-energy donor orbitals, or when the external conditions are favourable.

The space between the two metal atoms in a dimer molecule can have bonding or antibonding character, the former for two-electron three-centre bonds (involving H or C bridges), the latter for bridges involving electron-rich atoms like O or N. Weak Al–Al bonding character is also observed for dimers with mixed bridges, e.g. H and N, or O and C. In aluminium compounds with the first type of bridge, the $Al\cdots Al$ distance is intermediate between the sum of the covalent and the atomic radii (see Table 6.2). But the bridged $Al\cdots Al$ bond in $[R_3Al]_2$ is much weaker than the Al–Al bond in the diatomic Al_2 molecule, despite the fact that it is not much longer. This suggests that the overlapping of aluminium orbitals in $[R_3Al]_2$ is not significant.

Enthalpies of dissociation of the dimer $[R_3M]_2$, $\Delta_{diss}H$, for the reaction

$$\underset{sp^3}{[R_3M]_2} \longrightarrow \underset{sp^2}{2R_3M} \qquad (6.9)$$

are given for Me_3M species in Table 6.2; the corresponding values for other triorgano species are as follows (in kJ mol^{-1}):[78]

Table 6.2 The sum of the atomic and covalent radii and the M–M distances in bridged dimers and M_2 molecules (all in Å), and dimerisation energies and M_2 bond energies (kJ mol^{-1}).

Metals	2 × radius		Dimers		Diatomic molecules[a]	
	Atomic	Covalent	M–M distance	Bridge energy	Distance	Energy
Al–Al	2.86	2.36	2.60 (Me)	81.5	2.47	150
			2.70 (Ph)			
Ga–Ga	2.82	2.52	– (Me)	< 10	2.44	135
In–In	3.32	2.88			3.25	97
Tl–Tl	3.42	2.98			3.41	< 87

(a) Bond lengths and energies as in chapter 1, p. 65, and Ref. 12a, p. 377.

(cyclopropyl)$_3$Ga	12	Et$_3$Al	71
(cyclopropyl)$_3$In	8	Pr$^n{}_3$Al	65
(vinyl)$_3$Ga	<21	(decyl)$_3$Al	61

The reorganisation energy for alkyl- and hydrogen-substituted compounds was originally estimated to be close to zero. More exact calculations suggest, however, that the reorganisation energy $\Delta_r H$ is equal to 41 kJ mol^{-1} for the rehybridisation of AlH$_3$ from sp^2 to sp^3;[79] the energy change for methyl-substituted compounds should be of the same order. For substituents connected to the metal through atoms possessing one or more lone pairs of electrons, e.g. halides, the value is larger for two reasons: (i) repulsion of the substituents and (ii) semi-multiple bond formation between the p orbital of the metal and the atom bearing the lone pair. The highest reorganisation energies in this group of compounds are found for the aluminium halides; the values are 132, 116 and 78 kJ mol^{-1} for AlCl$_3$, AlBr$_3$ and AlI$_3$, respectively.[80] The corresponding energies in partially alkyl-substituted compounds should be smaller.

To calculate the mean M–X bond energy, $\bar{B}(MX\text{–}M)$, in the symmetric M–X–M bridge of a dimeric species M_2X_6, one needs to use the following equation:

$$\bar{B}(MX\text{–}M) = \frac{\bar{B}(M\text{–}X_t) + \frac{1}{2}\Delta_{diss} H}{2} + \frac{1}{2}\Delta_r H \qquad (6.10)$$

Knowing the mean bond enthalpy of the terminal M–X bond, $\bar{B}(M\text{–}X_t)$, the enthalpy of dissociation of the M_2X_6 molecule, $\Delta_{diss} H$, and the reorganisation energy, $\Delta_r H$, one can evaluate $\bar{B}(MX\text{–}M)$ to assess the strength of a bridging bond. The following approximate values (in kJ mol^{-1}) have been calculated for some aluminium compounds:[6] Al$_2$F$_6$, >411; Al$_2$Cl$_6$, 289; Al$_2$Br$_6$, 251; Al$_2$I$_6$, 192; Al$_2$H$_6$, 193; and Al$_2$Me$_6$, c. 170. Organoaluminium halides and hydrides should form bridging bonds similar in strength to those of the parent halide or hydride. Calculated values

338   CHEMISTRY OF ALUMINIUM, GALLIUM, INDIUM AND THALLIUM

for alkoxy bridges are expected to be about $10 \, \text{kJ mol}^{-1}$ bigger than for Al–Cl–Al bridges. The phenoxy bridging bond is also a little stronger than the chlorine one.

The bridging bond in Ga_2Cl_6 is reasonably strong, $\bar{B}(\text{GaCl–Ga})$ being well in excess of $150 \, \text{kJ mol}^{-1}$ (a value that does not include the substantial but unknown reorganisation energy). Simultaneous endothermic splitting of bridging bonds and exothermic rehybridisation of sp^3 to sp^2 gallium result overall, however, in only a modest energy demand for the dissociation reaction. It is for this reason that gallium(III) chloride vapour includes 0.2% of the monomer $GaCl_3$ at temperatures as low as 78°C. By contrast, there is no indication of dissociation, even at higher temperatures, of dialkyl-aluminium hydride species of the type $[R_2AlH]_2$, which rehybridise with very small gain of energy.

With some knowledge of the ability to form complexes with several organic donors, one can devise sequences for the relative strengths of bridging bonds in organoaluminium compounds which can, in all probability, be transferred to other organo compounds of the Group 13 metals:[81]

$$\text{OEt} > [\text{OPh}, \text{F}] > \text{PEt}_2 > \text{H} > \text{C}{\equiv}\text{CEt} > \text{Cl} > \text{Me}$$

$$\text{PMe}_2 > \text{SMe} > \text{Cl}$$

$$\text{C}{\equiv}\text{CPh} > \text{aryl} > \text{CH}{=}\text{CRR}' > \text{Me} > \text{Bu}$$

Not all combinations of metals and donor atoms are known for the structure types (1)–(10) listed in Figure 6.1. The figure does not include, moreover, systems where metal atoms are linked through more than one atom (but see section 6.7.3); here the mode of bonding may usually be regarded as conventional. Table 6.3 offers comparisons of some of the physicochemical and structural properties of trimethyl compounds of the metals. The crystal structure of Me_3Ga is not known, although it is iso-morphous with Me_3In.[82a] The higher energy of the empty $4p$ orbital (compared with the $3p$ orbital in Me_3Al) and the lower polarity of the Ga–C bond are probably responsible for the marked weakening of the interactions between the mononuclear gallane molecules. Both Me_3In and Me_3Tl are weakly associated in the solid state, with the metal atoms in 5-coordinated environments; this may reflect the influence of vacant metal nd orbitals which can interact with orbitals of the M–C bonds under favourable conditions with the formation of weak $MCH_3 \cdots M$ bridge bonds. These same d orbitals may also be at work in mono- and dialkyl derivatives which are mostly involatile compounds with one-, two- or three-dimensional polymeric networks. Expansion of the coordination shell of the metal is most commonly found with thallium. Coordination numbers greater than 4 are found in aluminium compounds only when more electronegative atoms are bound to the metal. The influence of electronegative atoms is also evident in compounds of the heavier elements, but to a smaller degree.

Table 6.3 Some physicochemical properties of trimethyl derivatives of the Group 13 metals.

Property	Me$_3$Al	Me$_3$Ga	Me$_3$In	Me$_3$Tl
Relative molecular mass (Me$_3$M)	72.086	114.83	159.92	249.48
Pauling electronegativity of the metal	1.61	1.81	1.78	2.04
B.p. (°C)	127/1 atm 20/8 mmHg	55.7/1 atm	136/1 atm[a]	147/1 atm[b] 20/5 mmHg
M.p. (°C)	15.4	−15.9	89.0	38.5
Density (g cm^{-3}) (°C)	0.752 (20)	1.151 (15)	1.568 (19)	
Degree of association	Solution 2 / Vapour 2, 1 (≥215°C) / Solid 2	Solution 1 / Vapour 1	Solution 1 / Vapour 1 / Solid 4[c]	Solution 1 / Vapour 1 / Solid 3-dimensional network[c]
$\Delta_f H^{\ominus}$ (kJ mol^{-1})	−81	−42	+173	
Enthalpy of dissociation, $\Delta_{diss} H^{\ominus}$ (kJ mol^{-1})	81.5	<10		
M–C bond strength (kJ mol^{-1}) $\bar{B}$	281	256	162	151
D_1	328	249	197	152
D_2		148	119	
D_3		339	170	
M–C bond length (Å)	1.964 (t) 2.140 (br) (ED)[d]	1.96 (ED)[d]	2.16 (ED)[d] 2.06–2.15 (X)[d]	2.21 (ED)[d] 2.22–2.34 (X)[d]
Negative charge on C atom in sp^2 monomer	−0.160	−0.031	−0.069	−0.002
Complexes: Et$_2$O · MMe$_3$	b.p. 68°C/15 mmHg	b.p. 99°C/760 mmHg	Dissociates at 40–50°C at 0.1 mmHg	In equilibrium with Me$_3$Tl + Et$_2$O
Enthalpy of complex formation (kJ mol^{-1}): Et$_2$O · MMe$_3$	−47[e]	−37.5		Dissociates in the vapour
Me$_3$P · MMe$_3$	−88[e]	−75	−71	

(a) Boils with decomposition.
(b) Extrapolated value; may decompose explosively at temperatures >90°C; also light-sensitive.
(c) Weakly associated with five-coordinated metal atoms.
(d) ED, electron diffraction; X, X-ray diffraction; t, terminal; br, bridging.
(e) The Al–O and Al–P bonds are stronger than these figures would suggest because complex formation entails splitting of the Al(μ-Me)$_2$Al bridges.

A feature characteristic of the organometallics of the Group 13 metals is the ease with which the metal substituents undergo exchange reactions, a property which they share with other organometallic compounds. The random exchange is best observed for trialkylmetal compounds. An ideal way to investigate the process involves measurements of the ^{1}H NMR spectra at different temperatures. The exchange is believed to depend on the fast equilibrium reactions of dimerisation and dissociation, even for the heavier metal trialkyls which are monomers in the liquid phase, but there is also evidence that some exchange processes feature splitting of only *one*

$$[R_3M]_2 \rightleftharpoons 2R_3M \qquad (6.11)$$

bridging bond of the dimer, followed by rotation (as with *tris*(cyclopropyl)-aluminium[78]). With stronger bridging bonds the exchange is much slower. Even with such bonds (like those formed by the halides) the exchange process results in an equilibrium including all the possible associated metal compounds, e.g.

$$Me_6Al_2 \underset{Me_6Al_2}{\overset{Al_2I_6}{\rightleftharpoons}} Me_5Al_2I \underset{Me_6Al_2}{\overset{Al_2I_6}{\rightleftharpoons}} Me_4Al_2I_2 \underset{Me_6Al_2}{\overset{Al_2I_6}{\rightleftharpoons}} Me_3Al_2I_3 \underset{Me_6Al_2}{\overset{Al_2I_6}{\rightleftharpoons}}$$

$$Me_2Al_2I_4 \underset{Me_6Al_2}{\overset{Al_2I_6}{\rightleftharpoons}} MeAl_2I_5 \underset{Me_6Al_2}{\overset{}{\rightleftharpoons}} Al_2I_6 \qquad (6.12)$$

but the other products undergoing equilibration are hardly detectable under normal conditions. The equilibrium can be used under special conditions to achieve the separation of a particular product. From sesquimethyl-aluminium iodide, for example, it is possible to distil off trimethyl-aluminium, formed in a symmetrisation process, in high yield. The corresponding chloride, which exists as a dimer,[82b] can be separated by rectification into $[Me_2AlCl]_2$ and $[MeAlCl_2]_2$. However, there exist also compounds like $Me_2Al(C_5H_5)$ which, while undergoing exchange, do not appear to give rise to such an equilibrium.[83]

One of the main tools for investigating Group 13 organometallics is nuclear magnetic resonance. The main goals of its application are (i) the elucidation of substituent-exchange processes, (ii) the determination of thermodynamic and kinetic parameters, (iii) the monitoring of the rate and direction of reactions, (iv) the determination of structural parameters like coordination number, and (v) the assessment of properties like the dative strengths of donors in complexes, the acidities of the organometallics and the electronegativities of the metals in their compounds. The investigations have been carried out mainly with reference to the ^{1}H but latterly also the ^{13}C NMR spectra. Electronegativities of the metals have been determined from the chemical shifts of protons bonded to α-carbon atoms or from the difference between CH_3 and α-CH_2 chemical shifts in ethylmetal compounds. The following estimates of the electronegativity of aluminium have thus been made: Et_3Al 1.37, Et_2AlCl 1.52, $EtAlCl_2$ 1.61; all these values apply to aluminium in dimer molecules where it is complexed by a basic atom of the neighbouring sub-unit to give sp^3 hybridisation.

In the last decade, with the development of effective multinuclear NMR devices, more experiments have been performed using the NMR properties of the metal nuclei of this Group (see chapter 1). In particular, investigations drawing on aluminium resonances have become relatively commonplace. ^{27}Al is a nucleus with high intrinsic NMR sensitivity and a high resonance frequency, making it favourable for such studies. Its drawback is its quadrupole moment which tends to make the resonances broad, sometimes even undetectable. Nevertheless, the investigations of Akitt,[84] Benn, Rufińska and others[85] have brought many interesting results to light. Perhaps the most important finding is the correlation between ^{27}Al chemical shift and coordination geometry at aluminium; this gives significant information about the structures of many compounds. The correlation is illustrated in Figure 6.2. In any structural evaluation, however, one has to take into account the kind of aluminium substituents because some of them produce substantial changes of chemical shift. Thus, compounds with aluminium bonded to carbon resonate at high frequency (low field), whereas those with aluminium bonded to elements of Groups 15, 16 and especially 17 are found at low frequency (high field).[82,84,85]

The heavier metal nuclei have found much less use in NMR experiments involving organometallics. Certainly one can easily distinguish the oxidation state of gallium atoms on the basis of the ^{71}Ga chemical shift. As with aluminium, however, there can be drastic variations of chemical shift with change of substituent. The mixed-valence compound $Ga^+[GaX_4]^-$ complexed with benzene and its derivatives gives rise to two ^{71}Ga resonances: one of these, due to Ga^{III}, has a chemical shift depending only

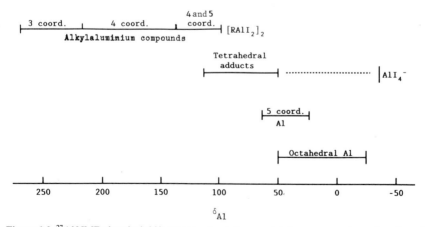

Figure 6.2 ^{27}Al NMR chemical shift range in aluminium-containing compounds as a function of the coordination number of the aluminium centre.

on the halogen X (Cl 247, Br 62), the other, due to Ga^I, a chemical shift varying from -609 for the bromo derivative complexed with mesitylene to -675 for the chloro derivative also complexed with mesitylene. The shifts for the Ga^I centres are about 100 ppm to high frequency of the values for the corresponding molten compounds $Ga^+[GaX_4]^-$; by contrast, the shifts for the Ga^{III} centres are unchanged.[86] ^{115}In and ^{205}Tl have received little attention in NMR studies of organoindium and -thallium compounds, respectively.[71,87]

6.5   Complexes with neutral bases and their reactivity

6.5.1   Synthesis and structure of complexes

It has already been noted that perhaps the most important factor influencing the properties of mononuclear organo derivatives of the Group 13 metals is the low-energy empty p orbital carried by the metal. If given the opportunity, Group 13 organometallics therefore form auto-complexes either by the use of lone-pair electrons on an M–X: group or by sharing M–X bonding electrons of a neighbouring molecule. Self-association is advantageous because it usually leads to the formation of a cyclic oligomer which can be regarded as a type of self-chelating compound. If the monomer or oligomer system contains a bonding orbital or filled atomic orbital which is not in the direct vicinity of the metal and which is high in energy (higher than that associated with the bridging atom in the case of an oligomer), then cyclic complex aggregates, e.g. **XII**, or complexes with intra- (**XIII**) or intermolecular coordination are formed.

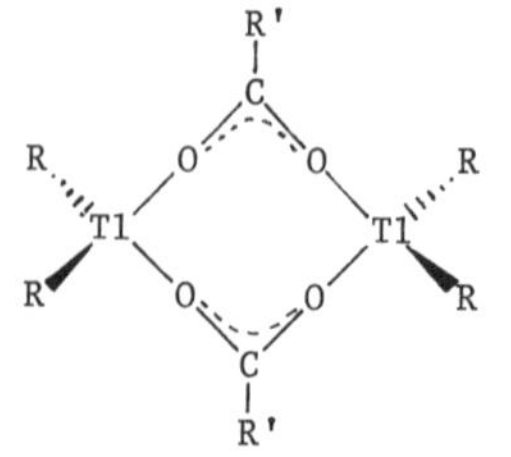

XII

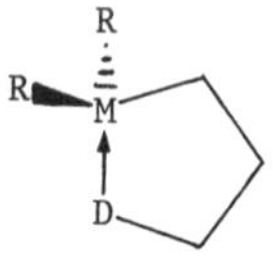

XIII

Interaction of the organometallic with a donor species leads typically to complexation. Anionic (**XIV**) or neutral (**XV**) complexes with 1:1 stoichiometry are most commonly formed. They may be stable or they may exist in equilibrium with the parent compounds. The neutral complexes with donors which are otherwise chemically unreactive are appreciably more stable with respect to thermal decomposition than are the free organometallics, especially those of the type R_3M, because of the coordinative and electronic saturation which complexation confers on the metal centre. There are obviously some complexes which are intermediate between these two

$$[R_3Al]_2 + MX \rightarrow M^+[R_3AlX]^- \tag{6.13}$$
$$\textbf{XIV}$$

$$R_3Ga + Et_2O \rightarrow Et_2O \cdot GaR_3 \tag{6.14}$$
$$\textbf{XV}$$

classes, e.g. **XVI**.[88] A neutral base often brings about the ionisation of an organometallic forming an ionic complex, but compounds of the heavier elements can exist as ion pairs, e.g. **XVII**, both in the solid state and in non-polar surroundings.[89,90] Whereas the heavier elements are prone to form

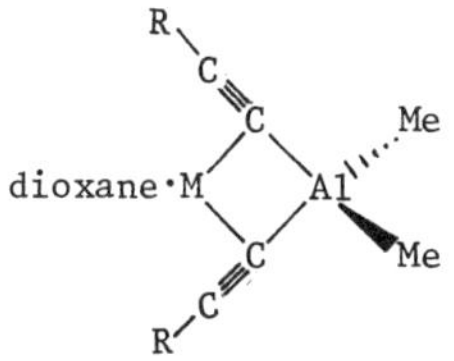

XVI 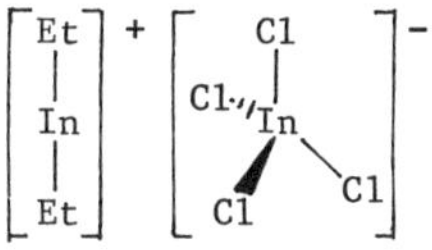   **XVII**

salts in the presence of neutral bases, aluminium compounds ionise only in the presence of strongly basic donors. For example, pyridine (py) forms $[py_2AlMe_2]^+[AlMe_4]^-$ which exists in equilibrium with the neutral $1:1$ complex $py \cdot AlMe_3$. Strong dibasic donors like $Me_2Si(N=PMe_3)_2$[91] form stable complexes with aluminium compounds, as exemplified by **XVIII** in the following reaction:

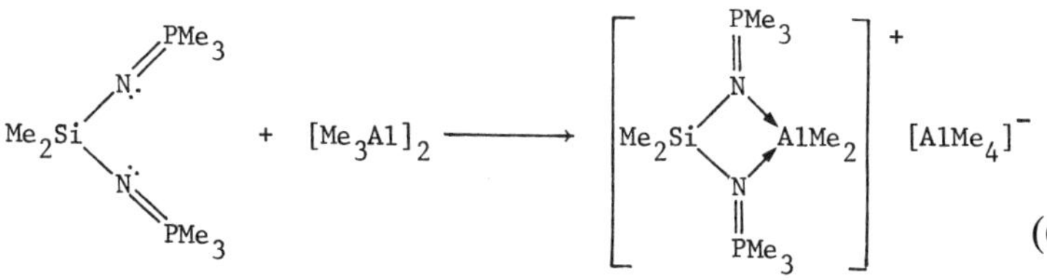

$$\tag{6.15}$$

XVIII

Although $AlCl_3$ is not an organometallic, it is noteworthy that it forms highly conducting solutions in acetonitrile. The crystals obtained from the solution are of the ionic complex $[AlCl(NCMe)_5]^{2+}[AlCl_4]_2^-$ with hexacoordinated aluminium cations.[92]

Persistent radical cation complexes (**XIX**) are formed when 4,4'-bipyridyl reacts with an organometallic of the type R_3M in the presence of sodium, but a neutral radical (**XX**) results when 4,4'-bipyridyl gives place to 2,2'-bipyridyl.[93]

The formation of a $2:1$ complex with the splitting of only one bridging

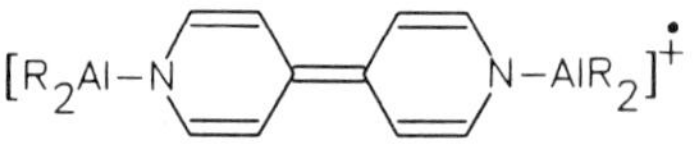

$$[R_2Al-N\!\!\overset{\displaystyle\bigcirc}{=}\!\!\overset{\displaystyle\bigcirc}{}\!\!N-AlR_2]^{\dot{+}}$$

XIX

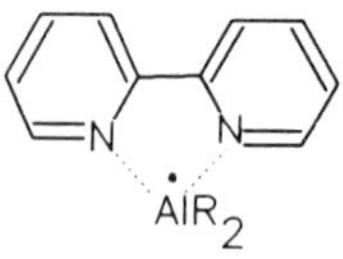

XX

bond in the dimeric organometallic molecule has been suggested as an intermediate stage of many reactions with ketones, ethers, nitriles etc.[94] There is evidence supporting the existence of such $2:1$ complexes in the cases of $Ph_2CO \cdot [Me_2AlCl]_2$[95a] and of the adducts of mesitylgallium chlorides with ether or tetrahydrofuran (**XXI**).[95b] The products formed contain only one M–O bond and retain a single chlorine bridge, but they exist in equilibrium with the $1:1$ complex and uncomplexed dimer (see equation (6.16), for example).

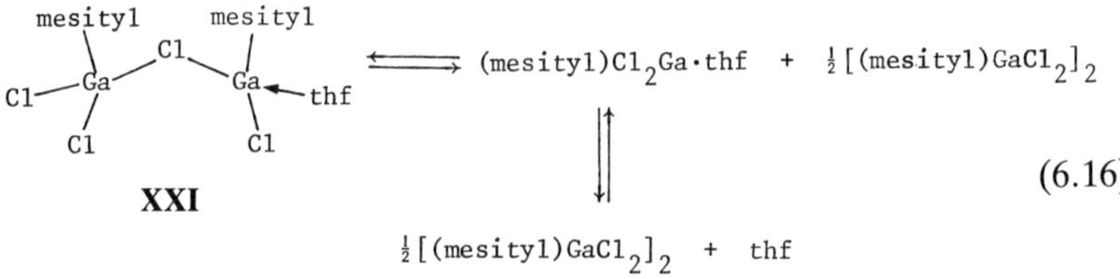

$$\text{(mesityl)}Cl_2Ga\cdot thf \;+\; \tfrac{1}{2}[\text{(mesityl)}GaCl_2]_2$$

XXI

$$\tfrac{1}{2}[\text{(mesityl)}GaCl_2]_2 \;+\; thf$$

(6.16)

Depending on the donating ability of the donor atom in the chelating chain and the chain length, the internal complex **XXII** can form an additional coordination bond, with an increase in the coordination number of the central metal atom from four to five, and the formation of a cyclic dimer **XXIV**.[96] Here the metal atoms each assume the local geometry of a deformed trigonal bipyramid.

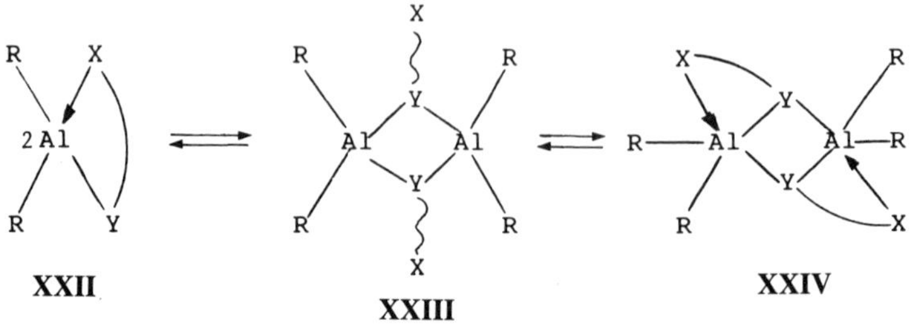

(6.17)

XXII

XXIII

XXIV

X–Y = a bidentate monoanionic ligand; X, Y can be O or N

The ability to form $1:1$ complexes decreases from aluminium to thallium, but the formation of complexes in which the metal achieves a coordination number greater than 4 is more favourable for compounds of the heavier elements. The *tris*(trifluoromethyl) compounds $(CF_3)_3M$ (M = Ga, In or Tl) add two molecules of stronger bases like acetonitrile.[35] The $1:1$ complex $Me_3In \cdot N(CH_2CH_2)_3N$ which exists in the vapour aggregates in the condensed phase to form a linear polymer incorporating penta-coordinated indium atoms. There are even known polymeric thallium compounds, e.g. (cyclopropyl)$Tl(O_2CPr^i)_2$, in which the metal is thought to be seven-coordinated.[97]

The coordinating ability of organic bases, as measured by the enthalpy of complexation, varies in the order indicated in Table 6.4. The pattern emerging for trimethylaluminium is probably typical of triorgano derivatives of all the Group 13 metals. Table 6.4 also includes some enthalpies of complexation for monomeric trimethylgallium and -indium. The *measured* enthalpy of complexation cannot be treated as a direct measure of the M–E bond strength (where E is the donor atom). In those cases where the organometallic is associated under normal conditions, it is necessary to take account of the enthalpy of dissociation in order to evaluate the strength of the M–E bond. One has also to take account of the enthalpy change $\Delta_r H$ associated with the reorganisation of MR_3 from a flat sp^2- to a pyramidal sp^3-hybridised fragment (see section 6.4). Where the complex is formed from an associated molecule, the enthalpy is hardly influenced by the reorganisation process because the product, in common with the organometallic precursor, typically features sp^3 hybridisation at the metal centre. At the same time, the coordination brings about considerable charge separation, as revealed by the bond moments (in D) determined for Et_3M complexed by dioxane, namely 3.5, 2.1, 1.3 and 0.7 for M = Al, Ga, In and Tl, respectively.[98]

In general, the stability of a complex towards dissociation depends on the acceptor ability of the organometallic compound, the electron-donating and steric properties of the base, and on the enthalpy of dissociation of any bridging bonds of the organometallic that have to be cleaved. The possibility of intramolecular complexation enhances the stability. Under favourable conditions complexing can proceed with an increase in the coordination number of the metal atom. Many such structures of dimeric complexes (like **XXIV**) or polymeric complexes (e.g. $[(cyclopropyl)Tl(O_2CPr^i)_2]_n$[97]), especially of the heavier Group 13 elements, have been confirmed by X-ray studies. The possibility of association of the complex depends on the ability of the metal atom to expand its coordination shell, but there are also known examples of complexes that associate through hydrogen-bonding, e.g. N–H$\cdots$Cl or N–H$\cdots$O interactions; such is the case, for example, in the polymeric chains of $[MeCl_2In \cdot NH_2Bu^t]_n$ or $[Me_2GaOC_2H_4NH_2]_n$.[99]

Bases with more than one donor atom are likely to form, depending on their basicity and structure, several complexes with a given organometallic. Yet maximum coordination of all the available sites is not always attained. The Me_3M compounds (M = Al, Ga, In or Tl) form with urotropine $1:1$, $2:1$ and $3:1$ complexes, but a $4:1$ product cannot be isolated.[100] The

Table 6.4 Enthalpies of complexation (kJ mol^{-1}) for monomeric Me$_3$Al, Me$_3$Ga and Me$_3$In.[a]

Me$_3$M	Me$_2$NH	Me$_3$N	NH$_3$	PhCN	thf	Me$_3$P	Et$_2$O	Me$_2$S	Me$_2$Se
Me$_3$Al	−128 (−100)	−125	−115	−106	−95	−88	−84	−70	−67
Me$_3$Ga		−88	(−77)			−75	−40[b]	c. −33	−42
Me$_3$In		−83				−71	−1[c]		

(a) Values in parentheses obtained from plots of chemical shift versus enthalpy of complexation.
(b) For Me$_2$O.
(c) For Et$_3$In (Ref. 11b, p. 70).

reaction product of 3-amino-5,6-dimethyl-1,2,4-triazine with trimethyl-aluminium with the general formula $[Me_2Al]_5[C_{11}H_{15}N_8]$ and six basic nitrogen atoms with lone pairs forms only a $1:1$ complex with Me_3Al.[101] Macrocycles with flexible rings in the *exo* conformation form mostly complexes in which all the basic atoms in the ring are coordinated, e.g. [14]aneS$_4 \cdot$ 4AlMe$_3$[102] and [Me$_4$(14)aneN$_4$] $\cdot$ 4GaMe$_3$.[103] But the macro-cycle [15]aneO$_5$ takes up only four Me$_3$Al moieties because one oxygen is forced towards the interior of the system and is sterically unavailable for complexation. For the same reason [dibenzo-18-crown-6] has 4 oxygen atoms which are unavailable for complexation.[104]

Cyclic multibasic compounds like crown ethers can form complexes with the metal in a highly coordinated environment. The dimethylthal-lium derivative $[Me_2Tl\{dicyclohexyl(16\text{-}crown\text{-}6)\}]^+[(O_2N)_3C_6H_2O]^-$ finds thallium attaining 8-fold coordination[71] including a linear C–Tl–C arrange-ment. Seven-fold coordination is assumed by aluminium in the cations formed from 15-crown-5 ether and its derivatives with organoaluminium compounds, resulting in a linear R_2Al unit located at the centre of the ring formed by 5 coplanar O atoms.[105] A very stable complex of tetraphenyl-porphyrin with Et_3Al is obtained in a reaction leading to the elimination of 2 mol of ethane at temperatures near 100°C. The resulting monoethyl-aluminium compound has the EtAl unit linked symmetrically to the four nitrogen atoms of the porphyrin ligand with the metal atom about 0.57 Å above the N_4 plane.[106]

There are also examples in which the basic atom is coordinated to two metallic centres, as with the phenoxy oxygen in $[K(dibenzo\text{-}18\text{-}crown\text{-}6)]^+$ $[Al_2Me_6OPh]^-$.[107] Potassium superoxide is bicomplexed by Me_3Al in the presence of crown ethers with the formation of a stable complex containing the anion $[Al_2Me_6O_2]^-$, **XXV**.[108] Sodium hydride in the presence of crown ethers also forms complexes incorporating bicoordinated hydrogen, e.g. **XXVI**.[109] IR measurements give reason to believe that complexes of Me_3M (M = Al or Ga) with O_2 exist at low temperatures (less than $-80°C$).[110]

XXV   **XXVI**

Many complexes are stable enough to be handled at higher temperatures or in reactive solvents without dissociation, rearrangement or elimination. The most stable towards dissociation are those with the shortest organic chains, i.e. the trimethylmetal compounds. On the other hand, complexes of the heavier metals are less stable, e.g. in the vapour phase where even a strong donor like NMe_3 coexists with Me_3Tl in equilibrium with the complex $Me_3Tl \cdot NMe_3$. Organic substituents with longer chains decrease the stability, leading to dissociation of the complexes at elevated temperatures. The trend is exemplified by the cases of $Me_3Al \cdot OEt_2$, which is stable and distils without dissociation, and $[RMe_2Si(CH_2)_n]_3Al \cdot OEt_2$ which dissociates at temperatures above 170°C (0.2 mmHg).[111] Trimethylgallium forms a stable complex with diethyl ether which can be distilled at 99°C, but the corresponding complex of the longer chain tripropylgallium loses ether at 105°C (60 mmHg; see Ref. 11a, p. 81). The more strongly hindered organometallics like Np_3Ga (where Np = neopentyl) form complexes with neither Et_2O nor thf. This contrasts with $(Me_3SiCH_2)_3Ga$ which, with slightly smaller steric contraints, coordinates thf.[112] Monoglyme complexes of R_3Al hardly dissociate. Long chain polyethers like diglymes and triglymes exist in equilibrium with their R_3Al complexes.[113] The reversible formation of adducts is of great importance for the purification of triorgano derivatives, R_3M, to obtain products of the highest electronic grade (see section 6.2[21]).

Complexes with different substituents on the metal atom are able to rearrange, especially if the free organometallic compound is present in the system, e.g.

$$2R_2R'Al \cdot OR''_2 \rightleftharpoons R_3Al \cdot OR''_2 + RR'_2Al \cdot OR''_2 \qquad (6.18)$$

In the resulting equilibrium, there is usually an excess of the species on the right-hand side of equation (6.18) as compared with what would be expected for a random distribution. Such exchange has been observed, for example, with $Me_2HAl \cdot NMe_3$.[114]

NMR investigations of exchange of the organic substituents and symmetrization involving aluminium complexes of oxygen-, sulfur- and nitrogen-bearing bases have been reviewed (Ref. 84, p. 86). Rie and Oliver have investigated such exchange for solutions containing different molar proportions of R_3Ga and donor species.[115] Faster exchange of the organic groups proceeds with donors containing Group 16 atoms. It is suggested that this is because the second lone pair orbital of the donor atom accepts a second molecule of the organometallic forming an M–E–M bond system which assists the transfer of substituents.

That multiple C–C bonds interact *via* their π orbitals with Group 13 metals in the oxidation state +3 is implied by various observations, but the effect is not strong. Dipole moment studies seem to show that Me_3M molecules in benzene solution have small but finite moments (0.6, 0.7, 1.3 and 0.5 D for M = Al, Ga, In and Tl, respectively), possibly reflecting the formation of π-type complexes.[98] The very large apparent dipole moment of $TlCl_3$ in benzene solution (3.93 D) can be interpreted in a similar manner (Ref. 6, p. 1135). The behaviours of solutions of Al_2Br_6 in methyl-substituted benzenes give grounds for believing that there are specific interactions between the halide dimer and the aromatic molecules. The possibility of a semi-dissociated dimer interacting with the aromatic compound has also been mooted. The higher degree of dissociation of organometallic dimers in aromatic hydrocarbons certainly argues in favour

$$[R_6M_2]\cdot ArH + ArH \rightleftarrows \overset{\overset{\displaystyle R}{\diagup}}{\underset{\underset{\displaystyle ArH}{\nwarrow}}{R_3M{-}R{-}M}}{-}R + ArH \rightleftarrows 2R_3M^{\delta-}\cdots ArH^{\delta+}$$

(6.19)

of such complexes. Thus, the degree of dissociation of $[Me_3Al]_2$ is 0.8% for the neat compound at 150°C, 2.31% for a hexadecane solution (mol fraction 0.01) at 100°C, but 3.48% for a mesitylene solution under similar conditions. The enthalpy of Me_3Al complexation is estimated to be $-8.8\,kJ\,mol^{-1}$ for mesitylene and $-7.6\,kJ\,mol^{-1}$ for benzene.[117] Aluminium bromide complexes somewhat more strongly.

Calculations affecting the π-complex between acetylene and AlH_3 indicate that there is indeed a defined complex with an energy slightly lower than that of the separated reagents and that its properties mark it out as an intermediate rather than a transition state.[62] Complexation of phenyl-acetylene by Et_3Al has been proposed on the evidence of NMR studies; in this case the C≡C bond and not the aromatic ring is thought to be the π donor.[118]

The energy of formation of the π-complex $C_2H_4\cdots Me_nAlCl_{3-n}$ is computed to vary from $-18\,kJ\,mol^{-1}$ for trimethylaluminium to $-58\,kJ\,mol^{-1}$ for aluminium chloride,[119] as compared with values ranging from -17 to $-43\,kJ\,mol^{-1}$ for the complex $C_2H_4\cdots GaH_3$.[47] Internal coordination of the metal by the C=C bond has been proposed for the tripenten-4-yl derivatives of aluminium and gallium, leading effectively to four-fold coordination of the metal. The interaction between the metal and double bond appears to be reasonably strong because it inhibits the formation of a dimeric structure for the aluminium compound (implying an energy in excess of $2 \times 30\,kJ\,mol^{-1}$). The interaction is weaker, however, in the gallium compound, and also in the tributen-3-yl derivative of aluminium.[120]

6.5.2 Reactivity of complexes

Nearly all the reactions of organometallic compounds of the Group 13 metals demand in the first step the formation of a complex, albeit one which cannot always be characterised. Many complexes are stable enough to be defined, others react only under drastic conditions. The reactivity can be explained on the understanding that the reactions proceed mainly when the adduct is able easily to form a transition state which undergoes rearrangement in a synchronous fashion, as with the six- or four-centred states, **XXVII** or **XXVIII**, respectively, deriving from the trialkylaluminium-ketone

$$R'_2C \overset{\delta+}{=\!=\!=} O$$
$$\delta- R \cdots\cdots AlR_2$$
$$R_2Al \text{——} R$$

XXVII

$$R'_2C \overset{\delta+}{=\!=\!=} O$$
$$\delta- R \cdots AlR_2$$

XXVIII

adducts. A wide spectrum of 1 : 1 complexes of organic carbonyl compounds is known but there is evidence too for the existence of 1:2 and 2:1 (carbonyl-MR_3) complexes. The finding that the 1:2 complexes in particular are quick to rearrange lends support to the transition state **XXVII**.[121]

Ether complexes of the type $XYZM \cdot ORR'$ (where X, Y and Z are the same or different substituents and may be alkyl groups or halogen atoms) are reasonably stable, at least those formed from organo compounds of the lighter metals. But those that do not sever the M–O linkage at higher temperatures eliminate RX (where X is a halogen) rapidly or RH slowly. On the basis of the progress of the reaction and the type of products, it has been suggested that the ether-organometallic complex is reactive, or at least more reactive, if one bridging group of the organometallic molecule is split and the ether is bonded to only one metal atom of the dimer. The system $[Me_2AlCl]_2 + PhOMe$ can be taken as an example. This rapidly eliminates 50% of the MeCl, but later the reaction slows down markedly because of the formation of $Me_2Al(\mu\text{-}OPh)(\mu\text{-}Cl)AlMe_2$.[94] Irradiation of the complex $R_3Al \cdot OPhR'$ causes the elimination not of $R'H$, as in the thermal reactions, but of PhH and PhR (a product of coupling), which are formed at similar rates.[122]

The complexes of nitriles of the type $RR'CHCN$ possessing an α-H atom react with aluminium compounds $R''AlXY$ at elevated temperatures, mainly with the elimination of $R''H$ and the formation of unstable $[RR'C\!=\!C\!=\!N\text{–}AlXY]$ which undergoes condensation. The elimination temperature is 120°C for Me_3Al but increases to 170°C for $MeAlCl_2$. The alkylation is only a side-reaction. In the presence of catalytic amounts of nickel compounds, however, the complex loses its stability and rearranges, with alkylation as the predominant pathway.[123]

A reaction of signal importance to the understanding of the processes that occur during the production of epitaxial layers of semiconductors and precursors to ceramics is the elimination of hydrocarbons from organometallic complexes, the basic part of which possesses one or more active hydrogen atoms.

$$nXYZM \cdot EHRR' \overset{T_1}{\rightarrow} [XYM\text{–}ERR']_n + nHZ \overset{T_2}{\rightarrow} [XM\text{–}ER]_n + nR'Y$$

$$\overset{T_3}{\rightarrow} [ME]_n + nRX \qquad (6.20)$$

where $E = N$, P or As; X, Y and Z = alkyl, aryl, H or halogen; R, R' = alkyl, aryl or H. Most such complexes eliminate alkane only very slowly at low temperatures; for example, $Me_3Ga \cdot AsMeH_2$ eliminates 1 mol of methane after 960 h when kept in the temperature range 0–25°C. To reach the final product, i.e. the III–V compound ME, takes three steps, only the first of which involves elimination from the original complex. Each step demands a higher temperature than the preceding one. Thus, $Me_3Al \cdot NH_3$ decomposes reasonably fast at about 50–70°C but the elimination of methane from $[Me_2Al–NH_2]_n$ takes place at a reasonable rate only when the temperature is raised to about 200°C. The two-step elimination of methane from $Me_3M \cdot NH_2Me$, carried out slowly at lower temperatures results in the formation of a hexagonal prismatic skeleton with two 'wings' (see Figure 6.1, (10)) and complying with the formula $(Me_2MNHMe)_2(MeMNMe)_6$. A planar, quasiaromatic $[AlN]_3$ ring with very short bonds (1.78 Å) is formed during the pyrolysis of $Me_3Al \cdot NH_2Ar$, where $Ar = 2,6\text{-}Pr^i_2C_6H_3$.[124]

The stability of a complex of the type $XYZM \cdot EHRR'$ depends on its composition and can be expressed by the temperature at which the reaction rate is suitable for a particular preparative objective. It is difficult to compare the literature results regarding reactivity on this basis. Strictly one needs to take account of many factors, including the catalytic influence on the reaction of the container walls, especially in the case of reactions occurring in the gas phase. Qualitatively, however, one can predict the stability of a complex in the knowledge that it decreases with the atomic number of the metal, i.e. in the order $Al > Ga > In$, and with change of the sustituents carried by the metal in the order $I > Br > Cl > Bu^i > Me > Et$. With respect to the donor part of the molecule, the stability drops in the order $N > P > As$ for gallium and indium compounds, but this order is reversed for organoaluminium species. The substituents attached to the basic atom decrease the rate of elimination in the following order: $H > Ph > Me > Bu$.[125]

6.6 Organo derivatives of the elements in lower oxidation states

Many examples of compounds of the metals in the lower oxidation states 0, $+1$ and $+2$ have been discussed in section 6.3 dealing with metal vapour reactions. Most of these are aluminium and gallium compounds; indium and thallium are more sparsely represented. The compounds are predominantly short-lived under normal conditions and their characterisation requires typically that they be maintained at low temperatures ($\lesssim 77$ K) in a solid host which is chemically inert.

Organometallics of the Group 13 metals in lower oxidation states which are stable at ambient temperatures and can be synthesised by classical methods, have been investigated more thoroughly in the last two decades.

As noted already in chapter 3, it is well established that Tl(I) compounds are the dominant feature in the inorganic chemistry of this element, both in the solid state and in aqueous solution, whereas much less is known about In(I) species. Ga(I) compounds are relatively few in number, and evidence for the existence of long-lived Al(I) species at ambient temperatures is decidedly sparse.[126]

The best known organic compounds of the elements in the oxidation state $+1$ are the η^5-bonded cyclopentadienyl compounds (see also Ref. 9, pp. 248–257) and ring-substituted derivatives like $(C_5H_4R)M$ or $(C_5R_5)M$.[127a] The less hindered cyclopentadienyl-metal derivatives of indium and thallium form in the solid phase homopolymeric zigzag chains, **XXIX**, with every metal atom sandwiched between two rings and lying on the C_5 axis of each of these rings, and at distances from them much longer than those found in

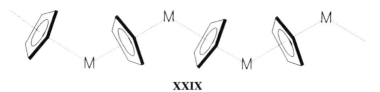

XXIX

monomeric cyclopentadienyl-metal compounds. The monomers are the only species present in the gas phase, indicating that the activation energy for $C_5H_5 \cdots M$ bond-splitting in a chain is not very high. Indeed, the corresponding gallium(I) compound, described only very recently as the product of the reaction between GaCl and either $Mg(C_5H_5)_2$ or LiC_5H_5 in a toluene/ether solution,[127b] is remarkable for its volatility. The metal-ring distances in the known polymeric and monomeric structures are, respectively, 2.68 (2.72) and 2.32 Å for $(C_5H_5)In$, and 3.19 and 2.41 Å for $(C_5H_5)Tl$. There is a small difference between the two metal-ring distances in the chain which increases as the ring takes on larger substituents. In $(C_5H_4Bu^t)In$, for example, the two distances are 2.53 and 2.85 Å; in $[C_5H_4(NC)C=C(CN)_2]Tl$ they are 3.01 and 3.06 Å.[128] There is also a slight shift of the metal atom away from its central position over a C_5 ring which is asymmetrically substituted; this shift also increases as the substituent grows in size.

Compounds with more bulky substituents in the ring may exist as oligomers rather than polymers. Thus, $(C_5Me_5)In$,[129] unlike $(C_5Me_5)Tl$ which is a linear polymer,[130] forms a hexamer in the solid phase; this consists of a near-octahedral In_6 unit with nearly equivalent In–In distances in the range 3.94–3.96 Å and with the rings placed umbrella-like, one above each In atom. The high volatility of the indium compound, with the vapour containing only the monomer $(C_5Me_5)In$, indicates relatively loose binding of the molecules within the hexamer. The corresponding aluminium derivative, prepared by the reaction of AlCl with $Mg(C_5Me_5)_2$ in a toluene/ether solution and providing the first well authenticated case of a molecular aluminium(I) compound stable under normal conditions,[126b] is a tetramer having as its kernel an Al_4 tetrahedron with Al–Al and Al–Cp distances of 2.77 and 2.03 Å, respectively (see chapter 1, p. 67). The compound $[Tl\{\mu\text{-}\eta\text{:}\eta\text{-}C_5H_3(SiMe_3)_2\text{-}1,3\}]_6$ takes the form of a *cyclic* hexamer because such a ring gives more room for the two extremely large substituents than does a zigzag chain.[131]

Even more bulky substituents may give dimeric compounds. Thus, $[C_5(CH_2Ph)_5]In$ forms a quasi-dimer through an In–In bond which is 3.63 Å long (the shortest observed to date for a cyclopentadienyl-indium derivative) and probably also through interactions between the metal and the aromatic rings of the benzyl substituents.[132] A similar dimer is formed when the same ligand is connected to Tl(I),[133] but only when the crystal growth is slow; kinetically controlled crystal growth results in a linear polymer with mononuclear units linked by secondary forces, probably involving benzyl$\cdots$Tl and dipole–dipole interactions.[134] The dimerisation energy of $[C_5(CH_2Ph)_5]M$ is not large because of the repulsive forces created by the dipole moments of the monomeric molecules. The lone pair of electrons on the metal is a dominant structural influence in cyclopentadienyl compounds of this sort, being mainly responsible for the large dipole moment (4.75 D for $(C_5H_5)In$) and the aggregating ability of the monomers.

Calculations taking in the monomeric cyclopentadienyl compounds of both indium and thallium have shown that a satisfactory description of the M–Cp bonding (Cp = cyclopentadienyl ligand) involves the interaction of ring carbon $p\pi$-orbitals with a hybrid sp and p_{xy} orbitals of the metal atom. Hence it appears too that the bonding in the system is predominantly covalent, and that the overall electron distribution on Tl, for example, is $s^{1.75}p^{0.45}d^{0.72}$, making the atom nearly neutral.[135] However, the crystalline polymers formed by both metals show much higher M–Cp bond polarities. The conductivity of tricyanovinyl-substituted cyclopentadienylthallium in dimethylformamide is two orders of magnitude higher than that of the parent compound CpTl.[128] These findings indicate strongly ionic character on the part of the tricyanovinyl compound, but the zigzag structure of the chain can be explained only by the admission of a covalent contribution to the bonding (see section 1.4.3.2). Moreover, the absence of spin-spin coupling between the ^{205}Tl and ^1H or ^{13}C nuclei confirms the considerable ionic character of CpTl and its derivatives, and particularly those with one or more strongly electron-withdrawing substituents on the C_5 ring (see Ref. 9, p. 254). For hindered derivatives with substituents having electron-donating properties and which display covalent thallium-ring interactions, this coupling is observed.

Several papers refer to the structures and applications of CpTl and its derivatives. These compounds are exceptionally useful intermediates in organometallic and organic synthesis. They are readily prepared in high yield from cyclopentadiene and thallium(I) acetate, alkoxide or hydroxide. They are easy to handle and are very reactive with respect to most metal halides and many organic substrates.[136]

Thallium acetate forms with the carborane anion $[7,8-C_2B_9H_{12}]^-$ in aqueous alkaline solution a precipitate of the compound $Tl_2[C_2B_9H_{11}]$ which is moisture- and air-stable. The product resembles in its properties CpTl. The thallium is linked to the open C_2B_3 face of the C_2B_9 cluster, the C_2B_3 ring$\cdots$Tl distance being 2.38 Å, but is not symmetrically located with respect to the cage, being shifted about 0.30 Å away from the two carbons.[137] Thallium borinates of the type $Tl[C_5H_5BR]$ containing the borabenzene moiety $[C_5H_5BR]^-$ (where R = Me or Ph) provide further examples of stable Tl(I) compounds that are only slightly air-sensitive.[138] In the complex $Tl[\eta^5-C_3Me_3B_2(H)Me]Co[\eta^5-Cp]$, the 1,3-diborole ring is centrally coordinated on one side by Co and on the other by Tl(I).[139] The Tl$\cdots$ring distance is a little longer here (by 0.08 Å) than in the carbollide compound $Tl_2[C_2B_9H_{11}]$ (see also sections 6.7.1 and 6.7.2).

The heavier Group 13 metals occur in the oxidation state +1 in salts of the type $M^+[M'X_4]^-$, where M' = Al, Ga, In or Tl, M = Ga, In or Tl, and X = Cl, Br or I (see chapter 3). These salts are able to form arene complexes with the stoichiometry M(I): arene = 1:1 and 1:2.[86] All the structures reported to date are formed from M(I) connected centrally to one or two arene molecules (depending on the bulk and mode of association) and also to the $[M'X_4]^-$ anion which links the cationic fragments into dimers, higher oligomers, or linear or three-dimensional coordination polymers (see, for example, chapter 1, p. 49). As in the cyclopentadienyl compounds, an increase in the polarisability of the aromatic ligand causes discernible strengthening of the M(I)$\cdots$ring bond, e.g. $C_6H_6 < C_6H_3Me_3 < C_6Me_6$, but only when the steric factors are favourable. Gallium(I) atoms are better acceptors than indium(I) or thallium(I) and form complexes with benzene that can be isolated, whereas indium(I) and thallium(I) do not. Attempts to complex M(I) by two aromatic rings of suitable geometry contained in the same molecule, e.g. cis-1,2-diphenylcyclopropane, have so far met with failure.[140] The only structure of a thallium(I) complex to be characterised is that of $(mesitylene)_6Tl_4(GaBr_4)_4$, where different Tl(I) sites are present, some with one and some with two coordinated aromatic rings, the Tl$\cdots$ring distance being 2.94 and 3.02 Å, respectively. In none of these complexes is there any sign of π-interactions between the aromatic rings and the M(III) centres, despite the evidence from other sources that such interactions are feasible (see section 6.4).

Little is known about the organic chemistry of the metals in the +1 oxidation state in the absence of π-type ligands. It has been suggested that compounds of the type $M'M(CH_2SiMe_3)_2$ containing M(I) (where M = Al, Ga or In) are obtained from M'H (where M' = Li, Na or K) and $(Me_3SiCH_2)_3M$ (see Refs. 8 (p. 684), 141 and 142). Reinvestigation of the systems has shown, however, that under the conditions of the experiments only $M'M(CH_2SiMe_3)_3H$ containing $M(III)$ is formed. At higher temperatures $M'M(CH_2SiMe_3)_3H$ decomposes forming products like $M'M(CH_2SiMe_3)_4$, Me_3SiCH_3, M and H_2, depending on the conditions (e.g. the nature of the solvent and the complex salt). Conspicuous by their absence, though, were products containing $M(I)$ (where $M = Al$, Ga, In or Tl).[143a] On the other hand, examples of low-valent compounds have come to light only very lately in the forms of (a) $[GaC(SiMe_3)_3]_4$ based on a Ga_4 tetrahedron with Ga–Ga distances of 2.68 Å,[143b] and (b) $K_2[Al_{12}Bu^i_{12}]$ based on an Al_{12} icosahedron with Al–Al distances of 2.68 Å.[126c]

Despite several reports of seemingly stable species containing Group 13 metals in the oxidation state $+2$, most of the compounds are diamagnetic and possess strong M–M bonds. The known paramagnetic compounds existing at room temperature are stabilised by intramolecular complexation and by multiply bonded systems incorporating donor atoms.[93,144,145] An example of such a compound is $Ga(Bu^tNCHCHNBu^t)_2$, **XXX**, obtained by MVS or other means.[144a] Reanalysis of the ESR spectrum suggests, however, that this contains not $Ga(II)$ but $Ga(III)$ bound to one singly and one doubly reduced ligand.[144b,c] Compounds of gallium and indium where the metal is similarly coordinated by nitrogen but not connected to a multiply bonded system, as in derivatives of the cyclic silazane **XXXI** of the type $[(MeSi)_2(NBu^t)_4]MCl$, dimerise immediately when reduced with the formation of M–M bonds.[146]

XXX   XXXI

Such compounds with M–M bonds are obtained usually by the action of an alkali metal or sodium naphthalide on a solution of a halide of the general type R_2MCl. They can also be synthesised by metathesis in reactions of $M_2X_4 \cdot 2D$ species (X = halogen; D = donor molecule) with the corresponding alkyllithium compound. The by-products of the latter method are the metal and the trialkylmetal compound. Coupling of metals proceeds in good yield if the substituents R are bulky and if they have a strong $+I$ effect. Although the anions $[Ga_2Me_6]^{2-}$[147] and $[Tl_2Me_6]^{2-}$[3] were the first species postulated to be M–M-bonded organometallics, the first such compound to be isolated and characterised was $Bu^i_4Al_2$.[148–150] This decomposes slowly with precipitation of aluminium, even at room temperature. The disproportionation reaction is promoted when an organic base

Table 6.5 Bond lengths and wavenumbers of the mode approximating to the M–M stretching mode in some tetraorganodimetal and related compounds.

Compound	Bond length (Å)		ν(M–M) (cm^{-1})^a
	M–M	M–C	
$[(Me_3Si)_2CH]_4Al_2$	2.66	1.98	373
$[(Me_3Si)_2CH]_4Ga_2$	2.54	1.99	337
$[Ga_2Cl_6]^{2-}$	2.40	–	237
$[Ga_2I_6]^{2-}$	2.41	–	122
$[(Me_3Si)_2CH]_4In_2$	2.83	2.19	
$(MSNB)_2In_2{}^b$	2.77	–	120–150
$In_2Br_3I \cdot 2tmeda^c$	2.78	–	

(a) This is, at best, only an approximate description of the mode; see chapter 3, p. 206.
(b) MSNB^{2-} is the ligand **XXXI**.
(c) tmeda = tetramethylethanediamine.

is added to a solution of the compound. Compounds with even bigger substituents, like the $CH(SiMe_3)_2$ group, are stable with respect to disproportionation. All the compounds with M–M bonds prepared so far are reported to be monomeric and not to form complexes, even with strong bases like tetramethylethanediamine. The steric hindrance of the substituents is probably the main factor inhibiting association; even the less hindered $[(Me_3Si)_2CH]_2AlCH_2Al[CH(SiMe_3)_2]_2$ retains its monomeric character in the solid state (Ref. 12a, p. 277). That the $C_2M–MC_2$ kernel is nearly planar has been shown by X-ray analysis.[146, 151–153] Some M–M and M–C bond lengths and Raman wavenumbers are given in Table 6.5. The results demonstrate how electronegative substituents attached to the metal lead to shortening of the M–M bond (Ref. 12a, p. 377). The M–C bonds of the $C_2M–MC_2$ units are similar in length to those in R_3M molecules; only in $R_2In–InR_2$ are they somewhat attenuated.

The final point to note is that the fragment R_2M can be treated like an organic substituent R in trialkylmetal compounds. The lack of polarity of the M–M bond is one reason for the freedom from association and complexation of compounds like $[(Me_3Si)_2CH]_4M_2$, albeit a less important one than the steric bulk of the organic substituents.

6.7   Mixed metal derivatives

The period up to about 1975 witnessed the appearance of only a few papers devoted to organometallics incorporating within a single molecule more than one kind of metal atom. The research that was carried out was concerned mainly with the catalytic activity of systems containing both Main

Group and transition metals, and with salts of the type $M'^+[MX_4]^-$. In the past two decades, however, there has been much progress in this area of research. The change has come about partly through the rapid development of methods of structural analysis, partly through the elaboration of methods of stabilising labile metal compounds by coordination involving π-donors (mainly $C_5H_5^-$ and its derivatives) and/or σ-donors (e.g. phosphines and phosphites). The research has opened up new perspectives on the applications of mixed metal compounds and, in addition, yielded a rich harvest of striking discoveries, thereby ensuring a widening interest in such studies.

The diverse types of mixed metal organometallic compounds containing a Group 13 metal call for some form of classification, but, as will soon be evident, there are many cases where the structures of the compounds defy any strict categorisation. For reasons of space, it has been necessary to omit catalytic systems and unstable multimetallic compounds from the following account. Subject to these qualifications, the compounds are grouped under the following headings:

1. compounds with direct metal–metal bonding between a Group 13 metal M and a second metal M';
2. compounds with different metals linked through one or more bridging atoms;
3. compounds with at least two different metals connected through chains of more than one heteroatom; and
4. salt-like compounds.

6.7.1   Compounds with direct M–M' bonds

In compounds with a second metal M' directly bonded to a Group 13 metal M, M' and the other groups linked to it can be treated either as a substituent in a compound of the general type MXYZ or as a donor molecule in a complex of the type $XYZM \cdot M'R_n$. In addition, more complex and extensive modes of bonding are known.

Well-authenticated examples of organo compounds in which both M and M' are Group 13 elements are largely restricted to those in which boron is one of the elements. In addition to the metallocarborane derivatives referred to previously (p. 354), *closo*-gallo- or indo-carboranes are formed by the reaction of trimethylgallium or trimethylindium with $2,3-C_2B_4H_8$ at temperatures of 180–215°C or 95–110°C, respectively.[154a] The structure of the gallium compound shows the metal to be nearly symmetrically located over the open C_2B_3 face of the carborane moiety with the Ga–CH_3 group tilted at an angle of 23° with respect to the normal to the C_2B_3 plane.[154b] The reaction of R_3M (M = Al or Ga) with $7,8-C_2B_9H_{13}$ at higher temperatures gives also a *closo* product, *viz.* $RMC_2B_9H_{11}$, this time with an icosahedral structure (Ref. 9, pp. 241–248); the properties of $EtAlC_2B_9H_{11}$ show the

AlEt group to be labile. Two molecules of PEt$_3$ attach themselves to the Al atom giving a structure with one σ B(10)–Al bond and longer range interactions of the Al with B(9) and B(11).[137b] From a *closo* compound incorporating a vertex Al–R group it is possible to prepare a *commo* sandwich compound **XXXII** with one Al atom making a common vertex between two heteronuclear icosahedral cages and the second forming an *exo*-AlR$_2$ group bonded by hydrogen bridges to two boron atoms (B(8) and B(9)) of one of the cages.[137b] The only thallium carboranes to be described are π-bonded Tl(I) compounds (*q.v.*) or Tl(I) salts of metallocarborane anions.

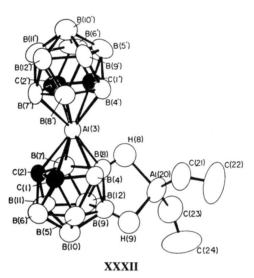

XXXII

The reaction of Me$_3$M with LiSnMe$_3$ yields, depending on the molar proportions, well-defined compounds of the type Li[(Me$_3$Sn)$_n$MMe$_{4-n}$] containing one or more direct M–Sn bonds. Me$_3$Tl gives a mixture of products, and to overcome these difficulties the corresponding thallium compounds were synthesised from Li[TlMe$_4$] and Sn$_2$Me$_6$.[155] A dimer [(Me$_3$CCH$_2$)$_2$GaTePh]$_2$, built on a Ga$_2$Te$_2$ butterfly ring, is obtained in the reaction of (Me$_3$CCH$_2$)$_2$GaCl with LiTePh.[156] There is evidence of some splitting of the rather long Ga–Te bridge bonds in solution.

Direct M–M′ bonds, where M′ is a transition metal, are a feature of the ionic complexes formed from Ph$_3$M (M = Al, Ga or In) or (C$_6$F$_5$)$_3$Tl and a carbonylate anion like [CpFe(CO)$_2$]$^-$, [CpMo(CO)$_3$]$^-$, [CpW(CO)$_3$]$^-$, [Co(CO)$_4$]$^-$ and [Mn(CO)$_5$]$^-$.[157] However, not all the complexes of this type enjoy the same mode of bonding. In compounds with less basic W atoms, like [Bu$_4$N][CpW(CO)$_3$], the carbonyl ligand is the donor centre and Ph$_3$Al forms a W–CO·Al-bonded complex. The compound obtained from CpW(CO)$_3$H and Me$_3$Al, after elimination of methane, also shows this mode of coordination, being a 12-membered cyclic dimer with the formula $[(\mu\text{-CO})(Cp)W(CO)(\mu\text{-CO})AlMe_2]_2$.[158] The bridging bonds are not very strong, splitting under the action of protonic or donor species, e.g. Et_2O. By contrast, $CpW(CO)_3H$ reacts with the weaker acid R_3Ga with the formation of W–Ga-bonded compounds. The substitution reaction with the reagents in the proportions $CpW(CO)_3H/R_3Ga = 3:1$ affords $[Cp(OC)_3W]_3Ga$.[159] It has been suggested that direct M–M′ bond-formation also occurs in similar reactions involving R_3In.

The site of complexation depends on the relation between the basicity of the central transition metal atom (which itself varies according to the substituents) and that of other electron-rich sites in the molecule. Introduction of strong Lewis bases like Ph_3P in place of CO ligands or substitution of the $\eta^5\text{-}C_5H_5$ ring with electron-donating groups enhances the basicity of the central atom and facilitates the formation of an M–M′ coordinative link. So it is that $CpMn(CO)_3$ forms mainly CO · Al bonds, but substitution of one CO ligand by Ph_3P results in the appearance of two complexes, one with CO · Al and the other with Mn–Al coordination; with time, the CO · Al-bonded complex gives way to the Mn–Al-bonded version. Further elimination of CO by phosphine gives only M–M′-bonded adducts. Certainly this is the only form known for the complexes $Cp(OC)_4Nb · AlCl_3$ and volatile $[(OC)_4Co]_2Ga(CH_2)_3NEt_2$. In the latter system, exchange of CO for Ph_3P is observed to strengthen the Ga–Co bond.[160]

There are known but few examples of compounds with cyclic structures in which Group 13 metal atoms are bridged by a second metal M′ (but see the Ga_2Te_2 ring mentioned above and the Al_2Mo_2 ring mentioned in section 6.7.3). A planar, four-membered Ga_2Fe_2 ring with a Ga–Fe distance of 2.52 Å forms the nucleus of $[(thf)(CH_2{=}CH)GaFe(CO)_4]_2$.[161] Such a structure indicates that the iron atom is in these circumstances more basic than either the vinylic carbon or the oxygen of the CO groups.

The chemistry of M–M′ bond-formation changes when the sterically hindered $(Me_3SiCH_2)_2AlPPh_2$ reacts as a soft *base* adding to $Cr(CO)_5(NMe_3)$ to form $(OC)_5Cr–PPh_2Al(CH_2SiMe_3)_2 · NMe_3$.[162] Because the Al–P and Cr–P bonds are unusually long, however, the driving force of the reaction is thought to be the formation of the strong Al–N bond.

Metalloporphyrin chlorides of the type (P)MCl (where P^{2-} is the dianion of the porphyrin and M = Al, Ga, In or Tl) react with $NaRe(CO)_5$ to give M–Re-bonded compounds.[163] The overall stability of the products $(P)MRe(CO)_5$ with a given porphyrin ring follows tne sequence Al < Ga < In ≪ Tl. The same order of stability is observed with metal carbonyl complexes $(P)MM′(L)$ where M = In or Tl and $M′(L) = Co(CO)_4$, $Mn(CO)_5$, $Cr(CO)_3Cp$, $Mo(CO)_3Cp$ or $W(CO)_3Cp$. Phenylporphyrin-indium chloride with substituted phenyl rings reacts with M′(L), where $M′(L) = Mn(CO)_5$, $Co(CO)_4$ or $Mo(CO)_3Cp$, also forming stable metal–metal-bonded complexes.[164]

Oxidative addition of the indium–carbon bond of R_3In or R_2InBr to several rhodium, iridium and platinum compounds has been investigated. One of the compounds thus obtained $(Me_3P)_3(Et)HIr–InEt_2$ has been characterised in structural terms.[165] The In–Ir bond is easily split by the addition of a soft base like a phosphine or an olefin with the regeneration of the organoindium starting material and the iridium compound complexed by the added base. In photolytic reactions of R_3In with $[CpM'(CO)_3]_2$ (where $M' = $ Mo or W), one can substitute indium-bonded alkyl groups to generate $[Cp(OC)_3M']_nInR_{3-n}$ with one or more direct In–M' bonds. The compounds with different substituents undergo symmetrization and, after several hours at room temperature, only two symmetric compounds, R_3In and $[Cp(OC)_3M']_3In$, are present. Similar procedures have resulted in the synthesis of analogous gallium and thallium compounds with molybdenum- or tungsten-bearing substituents.[166] Platinum-substituted derivatives with the general formula $(CP)(R')Pt–MR_2$, where CP is $Cy_2PCH_2CH_2PCy_2$ ($Cy = $ cyclohexyl), have been obtained in two ways: (i) by oxidative addition of R_3M to the Pt^0 compound (CP)Pt, or (ii) by protic cleavage of R_3M with the Pt^{II} compound (CP)Pt(R')H. The products, with M = Al, Ga or In, show no tendency to disproportionate. $NaCo(CO)_4$ reacts with R_2MCl (where M = Al, Ga or In) in tetrahydrofuran solution forming $(OC)_4Co–MR_2 \cdot$ thf. Compounds of this type have potential application in the production or modification of semiconductors.[167]

A cobaltocene-like compound $CpCo(C_4Ph_4AlH \cdot NEt_3)$, with one aluminole ring C_4Ph_4AlH and one Cp ring, has been described.[168a] The aluminole ring appears to be centrally coordinated to the cobalt atom and, additionally, through an Al–Co bond. The aluminium increases its coordination number to 5 by accepting an Et_3N molecule, as attested by the ^{27}Al chemical shift of $+112$ (see Figure 6.2). Similar but weaker M–M' bonding is suggested in the nickel complex $(COD)Ni(C_4Ph_4AlPh \cdot OEt_2)$, where COD = cycloocta-1,5-diene.[168b] No evidence of direct Co–B bonding is observed when the AlH fragment in the mixed cobaltocene complex is replaced by BH.[168a]

6.7.2   Compounds with bridged $M \cdots M'$ units

This is the most widely investigated class of mixed metal organometallics in which the Group 13 metals are implicated. The $M \cdots M'$ connection is established *via* one, two or three bridging atoms (**XXXIII–XXXV**, respectively). Individual bridging atoms in some compounds can even be coordinated to more than two metal atoms, as with the hydrogen atoms in the kernel of the aluminium–samarium compound **XXXVI**.[169] Known structures also include long chains with this sort of 'double' coordination of both the metal and donor atoms (and bonding as in **XXXVI**), as well as chains of cyclic dimers with the metal atoms in the spiro position (*cf.* Figure

XXXIII   **XXXIV**   **XXXV**

XXXVI

6.1, structures (7) and (9), respectively). The atoms most often found in bridging sites are hydrogen, carbon, halogen, oxygen, nitrogen or sulfur.

The reaction of $NaC_2B_8H_{11}$ with R_2AlCl in ether gives, depending on the molar proportions, either a monoalkyl-substituted *nido*-aluminocarborane or the sodium salt of the *bis*(*nido*-carboranyl)aluminate, $Na^+[Al(nido\text{-}6,9\text{-}C_2B_8H_{10})_2]^-$, with four rather long Al–C bonds (2.06 Å).[137] The former product is air- and moisture-sensitive, whereas the latter, with its steric crowding of the aluminium centre, is moderately stable to air and attacked only slowly by moisture.[170] Intermediate in the formation of *closo*-$RAlC_2B_9H_{11}$ (*q.v.*) is *exo-nido*-$R_2Al(\mu\text{-}H)_2\text{-}9,10\text{-}C_2B_9H_{11}$.[137] The reaction of trimethylaluminium with $Al(BH_4)_3$ at room temperature yields the volatile compound $Me_2Al(\mu\text{-}H)_2BH_2$.[171] This and the corresponding gallium compound can also be prepared by the reaction of Me_2MCl (M = Al or Ga) with an alkali-metal tetrahydroborate in the absence of a solvent. A similar procedure converts $MeMCl_2$ to $MeM(BH_4)_2$;[172a] the electron-diffraction pattern of the gaseous molecule has been analysed in terms of the structure $H_2B(\mu\text{-}H)_2Al(Me)(\mu\text{-}H)_2BH_2$ with a penta-coordinated Al atom. Doubly hydrogen-bridged structures are also adopted by the molecules $Me_2MB_3H_8$ (M = Al or Ga) in which the Me_2M moiety occupies the 2-position of a tetraborane(10)-like framework.[172b] Monocyclic trimetallic species $[Me_3SiCH_2(H)M]_2(\mu\text{-}H)_3Li$, with single hydrogen bridges between each pair of metal atoms, have been reported for all the Group 13 metals.[173a] An eight-membered monocyclic structure with short M–H bridging bonds is found in $[(Me_3Si)_2NAl(H)(\mu\text{-}H)Li(OEt_2O)_2]_2$.[173b]

Many compounds of the Group 13 metals, M, exist as stable aggregates incorporating other Main Group metals, M', the M and M' atoms being bridged typically to give four-membered rings. Examples are provided by compounds in which alkynyl groups bridge R_2Al and lithium (see **XVI**), beryllium or magnesium atoms;[88,174] such bonding does not appear to be effective, however, for M' = Zn. As the donor ability of the bridging atoms decreases, so it becomes more likely that M and M' will separate into distinct ionic components (see section 6.7.4), but the dividing line between bridged and ionic structures is far from clear and the properties of some compounds vary with the physical conditions in which they are found, as with $Ca[AlEt_4]_2$.[175]

Treatment of Me_3Al with $WCl_2(PMe_3)_4$ in tetramethylethanediamine yields the tungsten methylidyne complex **XXXVII**;[176a] a key feature of the

$$
\begin{array}{c}
Cl \\
\diagdown \\
\quad\quad Al \cdots X \cdots W \\
Me \diagup \quad \diagdown \quad \diagup \\
\quad\quad C \\
\quad H \diagup \quad \diagdown PMe_3
\end{array}
\qquad X = Cl \text{ or } Me
$$

XXXVII

structure is the $AlMe_2Cl/AlMeCl_2$ fragment which interacts with the methylidyne carbon $(Al-)C = 2.11$ Å) to complete four-fold coordination of the aluminium. Me_3Al also interacts with other transition metal and even lanthanide halides to give bridged bimetallic products such as $Cp_2Ti(\mu\text{-}CH_2)(\mu\text{-}Cl)AlMe_2$[176b] and $Cp_2Yb(\mu\text{-}Me)_2AlMe_2$.[176c] When films of solutions containing the compound $Bu^i_2In(SPr^n)Cu(S_2NBu^n_2)$, in which the copper and indium atoms are probably bridged through sulfur, are pyrolysed at 350°C, they give rise to layers of $CuInS_2$ suitable for photovoltaic applications. Another substrate affording conducting, transparent films, this time of indium-tin oxide, is the bimetallic species $Bu^n(EtCOO)InOSnBu^n_3$ (Ref. 12a, p. 361).

Recent research has led to the characterisation of several aluminohydride complexes in which single, double or triple hydrogen bridges secure aluminium to one or more transition metal atoms (see chapter 3, p. 120 and Ref. 12a, p. 387). Some of these carry organic groups attached to the aluminium, as with the complexes $(Me_2PhP)_2H_4Re(\mu\text{-}H)_2AlMe_2$, $(MePh_2P)_3HRe(\mu\text{-}H)_3AlMe_2$ and $Cp_2W(\mu\text{-}H)_2AlMe_3$.

The growing tendency to exploit zirconium-containing systems for homogeneous Ziegler–Natta catalysts has resulted in the preparation of many new compounds incorporating zirconium together with a Group 13 metal. In the reaction of Cp_3ZrEt with Et_3Al, for example, ethylene is eliminated to generate a complex $Cp_3Zr(\mu\text{-}H)AlEt_3$ in which the basic hydrogen of Cp_3ZrH forms a single hydrogen bridge (**XXXIII**) to the $AlEt_3$ fragment.[177] A similar singly bridged compound, but now with chlorine replacing hydrogen, is $Cp_2ZrCl(\mu\text{-}Cl)AlCl_3$ derived from the reaction of Cp_2ZrCl_2 with $AlCl_3$; this is reported to be a model for a catalyst precursor.[178] The same type of chlorine bridge is observed in $Me_3CC\equiv Ta(H)(dmpe)_2(\mu\text{-}Cl)AlMe_3$ (dmpe $= Me_2PCH_2CH_2PMe_2$).[179] These compounds are characterised by angular M′–Cl–Al bridges, with an angle between 126° and 150° at chlorine. Rather unexpected, however, is the product obtained in the reaction of Cp_2ZrMe_2 with $[Me_2AlC\equiv CR]_2$. For this has a bicyclic, planar kernel (**XXXVIII**) made up of fused ZrC_2Al and ZrC_2 rings. The

XXXVIII   **XXXIX**

geometry at the four-coordinated bridging carbon atom common to the two rings is remarkable for being almost planar.[180] An analogous compound is obtained from the reaction of Cp_2HfMe_2 with $[Me_2AlC{\equiv}CR]_2$.[181] Another type of stereochemistry hitherto unknown for carbon and involving a trigonal bipyramidal arrangement with a planar or nearly planar CH_3 group has now been identified in a $Zr-CH_3-Zr$ bridge. This forms part of a six-membered ring, ZrCZrOAlO, in the compound $(\mu\text{-Me})[Cp_2Zr(\eta^2\{C,O\}\text{-OCMe}_2)_2(\mu\text{-AlMe}_2)$ **XXXIX**.[182]

6.7.3 Compounds with different metals connected through chains of more than one heteroatom

Compounds of this class are less common, and most of the examples that have been synthesised contain a Group 13 metal together with another Main Group metal (which may itself be from Group 13). Thus, the complex formed by the tetradentate aza crown ether [14]aneN$_4$ with two molecules of Me_3Al reacts at room temperature with two molecules of Me_3Ga to afford the heterobimetallic condensation product $(MeAl)_2[14]aneN_4(GaMe_3)_2$.[183] Here the four nitrogen atoms are bridged through two Al atoms each of which is 4-coordinated. In the reaction of chloromercuriferrocene with $[Me_3Al]_2$ at 60°C, a new dimer is formed in which one methyl bridge is replaced by a CH vertex of one of the Cp rings of ferrocene.[184] However, the iron influences the aluminium atoms only slightly. Many bimetallic compounds with the metal atoms connected through the nitrogens of one or more pyrazolyl ligands and/or other bifunctional bridges have been investigated by Storr and his group[185] (see also Ref. 11, pp. 345–418). The thallium(I) salt $Tl^+[\{O(EtO)_2P\}_3CoCp]^-$ (TlL_{OEt}), with three strongly basic centres, reacts with an excess of Me_3Al reducing Tl^+ to the metal and generating the product $L_{OEt}(AlMe_2)(AlMe_3)$ where the $AlMe_2$ moiety is connected to the Co atom through two Co–P–O–Al bridges; Me_3Al is co-ordinated to the terminal P=O unit of the third of the $O(EtO)_2P$ groups linked to cobalt.[186]

The trispirocyclic dimer $[(Me_3P)_2Ni(Me_2PCH_2)_2Al(CH_2PMe_2)]_2$ dissociates partially in solution with the formation of $Ni(PMe_3)_4$ in equilibrium with a chain-like bimetallic polymer having alternating Al and Ni spiro atoms.[187] Similar spiropolymeric chains are formed in the reaction of $[(Me_2PCH_2)_3Al]_2$ with $Li(CH_2PMe_2)$.[188] Many compounds of the type $CpZr(CH_2CH_n)_m(AlEt_2)_p$ $(n,m = 1$ or 2; $p = 1-4)$ have been described; as mentioned previously, these are noteworthy mainly for their catalytic potential.[189]

Gallium linked to one or more strong π-donor groups as in *tris*(mesityl)-gallium, reacts with $M'(CO)_6$ (where $M' = Cr$ or Mo) to form π-complexes with the stoichiometry $1:1$ or $1:2$.[190] The Ga–C bonds are stabilised (but lengthened) by complexation of the mesityl rings, so that only the un-complexed mesityl groups are eliminated by HCl in a stoichiometric reaction. In the reaction of Cp_2MoH_2 with $[Me_3Al]_2$, methane-elimination is accompanied, *inter alia*, by the formal addition of two Cp_2Mo to two essentially planar Al_2Me_3 fragments giving the compound **XL**.[191] This contains an Mo_2Al_2 ring and 5- as well as 4-coordinated Al atoms. Each Mo atom is centrally coordinated to two cyclopentadienyl rings and one carbon atom of each of these (which should therefore be formulated as C_5H_4) in turn acts as a bridge linking the AlMe groups of the kernel with the two terminal $AlMe_2$ groups. The resulting molecule therefore exemplifies all three types of connection between M and M' discussed in this and the preceding sections. The thallium(I) derivative $(C_5H_4PPh_2)Tl$ reacts with $AuCl(SMe_2)$ to give a tetrametallic product having a $[(C_5H_4PPh_2)Au(Ph_2PC_5H_4)Tl]_2$ ring, with the thallium atoms centrally coordinated to the cyclopentadienyl groups at rather short distances (2.73 Å from Tl to the centroid of the C_5 ring).[192]

XL

6.7.4 Salt-like compounds

These compounds are usually composed of solvated cations in company with anions of the type $[MX_4]^-$, where X_4 represents a set of four organic or inorganic substituents; these may be all the same or they may be different. The Group 13 metal is most likely to appear in the anionic part, but can also appear in the cationic part of the salt. For example, decaborane(14) reacts with Me$_3$In forming [Me$_2$In]$^+$[B$_{10}$H$_{12}$InMe$_2$]$^-$, which behaves as an electrolyte in solution.[92,193] Many of the compounds of this class can also exist under appropriate conditions as bridged associates either exclusively or in equilibrium with the ions, e.g. Ca[AlEt$_4$]$_2$.[175]

The magnesium derivative of dihydroanthracene reacts in tetra-hydrofuran solution with R$_2$AlX (R = alkyl or H; X = H, R or R′) giving androgynous salts with ions bridged by the X groups and formulated as Mg[μ-(9,10-H$_2$-9,10-anthrylene)](μ-X)AlR$_2$.[194] Here the Mg carries a positive and the Al a negative charge. The reaction of [Me$_3$Al]$_2$ with Fe(acac)$_3$ (acac = acetylacetonate) in the presence of dmpe delivers a stable Fe^{II} complex with the formula [Fe(dmpe)$_2$(acac)]$^+$[AlMe$_4$]$^-$ (dmpe = Me$_2$PCH$_2$CH$_2$PMe$_2$).[195] The bipyridyl complex of Co^{III} reacts with a large excess of trialkylaluminium, AlR$_3$, to form an ionic complex [R$_2$Co(bipy)]$^+$ [AlR$_4$]$^-$.[196] An example of a trimetallic compound is afforded by the lithium salt of the unsaturated metallacycle incorporating a *spiro*-aluminium atom and π-complexed to two Ni(COD) groups (COD = cycloocta-1,5-diene), as in **XLI**.[197]

XLI

Bibliography and reviews

1. K. Ziegler, *Organo-Aluminium Compounds*, in *Organometallic Chemistry*, ed. H. Zeiss, ACS Monograph No. 147, Reinhold, New York, 1960, pp. 194–269.
2. A.N. Nesmeyanov and R.A. Sokolik, *The Organic Compounds of Boron, Aluminium, Gallium, Indium and Thallium*, North-Holland, Amsterdam, 1967.
3. G.E. Coates and K. Wade, *Organometallic Compounds*, Vol. 1, *Organic Compounds of Aluminium, Gallium, Indium and Thallium*, Methuen, London, 1967, pp. 295–374.
4. *Methoden der Organischen Chemie*, ed. E. Mueller, Thieme, Stuttgart, Germany, 1970, Vol. XIII/4; H. Lehmkuhl and K. Ziegler, *Aluminium Organic Compounds*, pp. 1–314; G. Bahr and P. Burba, *Gallium Organic Compounds*, pp. 315–362; *Indium Organic Compounds; Thallium Organic Compounds*, pp. 363–390.
5. T. Mole and E.A. Jeffery, *Organoaluminium Compounds*, Elsevier, New York, 1972.
6. K. Wade and A.J. Banister, *The Chemistry of Aluminium, Gallium, Indium and Thallium*, Pergamon Press, Oxford, 1975.
7. J.J. Eisch, *Organometallic Synthesis*, Vol. 2, *Nontransition Metal Compounds*, Academic Press, New York, 1981.
8. *Comprehensive Organometallic Chemistry*, eds. G. Wilkinson, F.G.A. Stone and E.W. Abel, Pergamon Press, Oxford, 1982, Vol. 1; J.J. Eisch, *Aluminium Organic Compounds*, pp. 555–682; D.G. Tuck, *Gallium and Indium Organic Compounds*, pp. 683–724; H. Kurosava, *Thallium Organic Compounds*, pp. 725–754; M.E. O'Neill and K. Wade, *Structural and Bonding Relationships Among Main Group Organometallic Compounds*, pp. 1–42.

9. P. Jutzi, *Adv. Organomet. Chem.*, 1986, **26**, 217.
10. *Dictionary of Organometallic Compounds*, Chapman and Hall, London, 1984–1989; J.D. Smith, *Aluminium*, pp. 9–120; I.J. Worrall and J.D. Smith, *Gallium*, pp. 940–962; I.J. Worrall and J.D. Smith, *Indium*, pp. 1138–1151; A. McKillop and J.D. Smith, *Thallium*, pp. 2271–2344. See also entries under each element in the 1st, 2nd, 3rd, 4th and 5th supplements.
11. (a) *Gmelin Handbook of Inorganic Chemistry*, 8th edn., *Organogallium Compounds*, Part 1, Syst. No. 36, Springer-Verlag, Berlin, 1987; (b) *Gmelin Handbook of Inorganic and Organometallic Chemistry*, 8th edn., *Organoindium Compounds 1*, Springer-Verlag, Berlin, 1991.
12. (a) *Polyhedron*, Symposium-in-print No. 10: Al, Ga, In, ed. A.R. Barron, *Polyhedron*, 1990, **9**, 149–453; (b) Yu. A. Alexandrov and N.A. Chikinova, *J. Organomet. Chem.*, 1991, **418**, 1; (c) G.H. Robinson, ed., *Coordination Chemistry of Aluminium*, WCH Publishers, New York, in press.
13. H.M. Manasevit, *J. Cryst. Growth*, 1981, **55**, 1; *Appl. Phys. Lett.*, 1968, **11**, 156.

Annual surveys

14. *Organometallic Chemistry Reviews*, eds. D. Seyferth and R.B. King, Elsevier, Amsterdam, 1963–1971. In 1972, this was merged with *J. Organomet. Chem.* as *Annual Surveys* appearing up to 1981.
15. *Chemical Society Specialist Periodical Reports: Organometallic Chemistry*, eds. E.W. Abel and F.G.A. Stone, Chemical Society, London, 1971 onwards.

References

16. A.C. Jones, N.D. Gerrard, D.J. Cole-Hamilton, A.K. Holliday and J.B. Mullin, *J. Organomet. Chem.*, 1984, **265**, 9.
17. A.C. Jones, D.J. Cole-Hamilton, A.K. Holliday and M.M. Ahmad, *J. Chem. Soc., Dalton Trans.*, 1983, 1047.
18. J.G. Chen, T.P. Beebe, Jr., J.E. Crowell and J.T. Yates, Jr., *J. Am. Chem. Soc.*, 1987, **109**, 1726.
19. M.J.S. Gynane and I.J. Worrall, *J. Organomet. Chem.*, 1972, **40**, C59.
20. V.I. Bregadze, L.M. Golubinskaya, L.G. Tonoyan, B.I. Kozyrkin and B.G. Gribov, *Dokl. Akad. Nauk SSSR*, 1973, **212**, 880.
21. D.C. Bradley, H. Chudzyńska, M.M. Factor, D.M. Frigo, M.B. Hursthouse, B. Hussain and L.M. Smith, *Polyhedron*, 1988, **7**, 1289.
22. A.H. Moore, M.D. Scott, J.I. Davis, D.C. Bradley, M.M. Factor and H. Chudzyńska, *J. Cryst. Growth*, 1986, **76**, 19.
23. A.C. Jones, *Chemtronics*, 1989, **4**, 15.
24. K.B. Starowieyski and A. Chwojnowski, unpublished results.
25. A.N. Nesmeyanov, D.A. Lemenowski and E.G. Perevalov, *Izv. Akad. Nauk SSSR, Ser. Khim.*, 1975, 1667.
26. R.D. Rieke, *Acc. Chem. Res.*, 1977, **19**, 301.
27. R.D. Rieke and L. Chao, *Synth. React. Inorg. Met.-Org. Chem.*, 1974, **4**, 101.
28. Y. Tanaka, S.C. Davies and K.J. Klabunde, *J. Am. Chem. Soc.*, 1982, **104**, 1013.
29. K.B. Starowieyski and K.J. Klabunde, *Appl. Organomet. Chem.*, 1989, **3**, 219.
30. K.J. Klabunde and A. Whetten, *J. Am. Chem. Soc.*, 1986, **108**, 6529.
31. K.J. Klabunde, T.O. Murdock, *J. Org. Chem.*, 1979, **44**, 3901.
32. J.R. Blackborow and D. Young, *Metal Vapour Synthesis in Organometallic Chemistry*, Springer-Verlag, Berlin, 1979.
33. W.E. Billups, J.P. Bell, R.H. Hauge, E.S. Kline, A.W. Moorehead, J.L. Margrave and F.B. McCormick, *Organometallics*, 1986, **5**, 1917.
34. T.R. Bierschenk, T.J. Juhlke, T.E. Bailey, and R.J. Lagow, *J. Organomet. Chem.*, 1984, **277**, 1.

35. D. Naumann, W. Strauss and W. Tyrra, *J. Organomet. Chem.*, 1991, **407**, 1.
36. G.M. Kuz'yants, *Izv. Akad. Nauk SSSR, Ser. Khim.*, 1976, 1895.
37. J.S. Poland and D.G. Tuck, *J. Organomet. Chem.*, 1972, **42**, 315.
38. W. Lind and I.J. Worrall, *J. Organomet. Chem.*, 1972, **40**, 35.
39. S. Uemura, H. Miyoshi, M. Okano and K. Ichikava, *J. Chem. Soc., Perkin Trans. 1*, 1981, 991.
40. E.C. Kooyman, J.P. Huygens, J. de Lepper, D. de Vos and J. Wolters, *Rec. Trav. Chim. Pays Bas.*, 1981, **100**, 24.
41. T.G. Dietz, M.A. Dunkan, D.E. Powers and R.E. Smalley, *J. Chem. Phys.*, 1981, **74**, 6511.
42. K.J. Klabunde, *Chemistry of Free Atoms and Particles*, Academic Press, New York, 1980.
43. K.J. Klabunde and Y. Tanaka, *J. Am. Chem. Soc.*, 1983, **105**, 3544.
44. J.M. Parnis and G.A. Ozin, *J. Am. Chem. Soc.*, 1986, **108**, 1699.
45. G.H. Jeong and K.J. Klabunde, *J. Am. Chem. Soc.*, 1986, **108**, 7103.
46. (a) S.A. Mitchell, P.A. Hackett, D.M. Rayner and M. Cantin, *J. Phys. Chem.*, 1986, **90**, 6148; (b) L. Manceron and L. Andrews, *J. Phys. Chem.*, 1990, **94**, 3513.
47. S. Oikawa, M. Tsuda, M. Morishita, M. Mashita and Y. Kuniya, *J. Cryst. Growth*, 1988, **91**, 471.
48. P.H. Kasai and D. McLeod, Jr., *J. Am. Chem. Soc.*, 1975, **97**, 5609.
49. P.H. Kasai, *J. Am. Chem. Soc.*, 1982, **104**, 1165.
50. M. Histed, J.A. Howard, H. Morris and B. Mile, *J. Am. Chem. Soc.*, 1988, **110**, 5290.
51. J.H.B. Chenier, J.A. Howard and B. Mile, *J. Am. Chem. Soc.*, 1987, **109**, 4109.
52. P.S. Skell and L.R. Wolf, *J. Am. Chem. Soc.*, 1972, **94**, 7919.
53. P.S. Skell and M.J. McGlinchey, *Angew. Chem., Int. Ed. Engl.*, 1975, **14**, 195.
54. P.S. Skell, D.L. Williams and M.J. McGlinchey, *J. Am. Chem. Soc.*, 1973, **95**, 3337.
55. J.H.B. Chenier, J.A. Howard, J.S. Tse and B. Mile, *J. Am. Chem. Soc.*, 1985, **107**, 7290.
56. A.J. Kos, E.D. Jemmis, P.v.R. Schleyer, R. Gleiter, U. Fischbach and J.A. Pople, *J. Am. Chem. Soc.*, 1981, **103**, 4996.
57. S.P. So, *J. Organomet. Chem.*, 1991, **420**, 293.
58. B. Mile, J.A. Howard and J.S. Tse, *Organometallics*, 1988, **7**, 1278.
59. P.H. Kasai and D. McLeod, Jr., *J. Am. Chem. Soc.*, 1979, **101**, 5860.
60. P.H. Kasai, D. McLeod, Jr., and T. Watanabe, *J. Am. Chem. Soc.*, 1977, **99**, 3521.
61. A. Almenningen, L. Fermholt and A. Haaland, *J. Organomet. Chem.*, 1978, **155**, 245.
62. A. Okniński and K.B. Starowieyski, *J. Mol. Struct. (THEOCHEM)*, 1986, **138**, 249.
63. T. Fjeldberg, A. Haaland, R. Seip and J. Weidlein, *Acta Chem. Scand., Ser. A*, 1981, **35**, 437.
64. P.H. Kasai, *J. Phys. Chem.*, 1982, **86**, 4092.
65. M. Trenary, M.E. Casida, B.R. Brooks and H.F. Schaefer III, *J. Am. Chem. Soc.*, 1979, **101**, 1638.
66. Y. Xie, B.F. Yates and H.F. Schaefer III, *J. Am. Chem. Soc.*, 1990, **112**, 517.
67. P.H. Kasai and P.M. Jones, *J. Am. Chem. Soc.*, 1984, **106**, 8018.
68. J.H.B. Chenier, C.A. Hampson, J.A. Howard, B. Mile and R. Sutcliffe, *J. Phys. Chem.*, 1986, **90**, 1524.
69. (a) J.A. Howard, R. Sutcliffe, C.A. Hampson and B. Mile, *J. Phys. Chem.*, 1986, **90**, 4268; (b) W.G. Hatton, N.P. Hacker and P.H. Kasai, *J. Phys. Chem.*, 1989, **93**, 1328.
70. M.V. Castano, A. Sanchez, J.S. Casas, J. Sorto, J.L. Brionso, J.H. Piniella, X. Solous, G. Germain, T. Debaerdemaeker and J. Gloser, *Organometallics*, 1988, **7**, 1897.
71. (a) J. Crowder, K. Hendrick, R. Matthews and B.L. Podejma, *J. Chem. Res. (S)*, 1983, 82, 1451. (b) Y. Kawasaki and N. Okuda, *Chem. Lett.*, 1982, 1161.
72. (a) D.A. Atwood, R.A. Jones, A.H. Cowley, J.L. Atwood and S.G. Bott, *J. Organomet. Chem.*, 1990, **394**, C6; (b) R. Kumar, V.S.J. de Mel and J.P. Oliver, *Organometallics*, 1989, **8**, 2488.
73. G.B. Deacon, P.J. Phillips, K. Henrick and M. McPartlin, *Inorg. Chim. Acta*, 1979, **35**, L335.
74. B. Neumüeller and F. Gahlmann, *J. Organomet. Chem.*, 1991, **414**, 271.
75. L.V. Interrante, G.A. Sigel, M. Garbauskas, C. Hejna and G.A. Slack, *Inorg. Chem.*, 1989, **28**, 252.
76. W. Fries, W. Schwarz, H.-D. Hausen and J. Weidlein, *J. Organomet. Chem.*, 1978, **159**, 373.

77. A.P. Shreve, R. Mulhaupt, W. Fultz, W. Robbins and S.D. Ittel, *Organometallics*, 1988, **7**, 409.
78. (a) R.D. Thomas and J.P. Oliver, *Organometallics*, 1982, **1**, 571; (b) O. Yamamoto, K. Hayamizu and M. Yanagisawa, *J. Organomet. Chem.*, 1974, **73**, 17.
79. A. Okniński and S. Pasynkiewicz, *Inorg. Chim. Acta*, 1978, **28**, L125.
80. F.A. Cotton and J.R. Leto, *J. Chem. Phys.*, 1959, **30**, 993.
81. K.B. Starowieyski, *Bridging Bonds in Organoaluminium Compounds*, Warsaw Technical University Monographs, Warszawa, Poland, 1973.
82. (a) E.M. Johnson, D. Phil. thesis, University of Oxford, 1972. (b) Z. Cerny, J. Machacek, J. Fusek, S. Hermanek, O. Kriz and B. Casensky, *J. Organomet. Chem.*, 1991, **402**, 139.
83. A.G. Lee, *J. Chem. Soc. A*, 1970, 2157.
84. J.W. Akitt, *Prog. NMR Spectrosc.*, 1989, **21**, 1.
85. (a) R. Benn and A. Rufińska, *Angew. Chem., Int. Ed. Engl.*, 1986, **25**, 861; (b) R. Benn, P. Janssen, H. Lehmkuhl, A. Rufińska, K. Angermund, P. Betz, R. Goddard and C. Krüger, *J. Organomet. Chem.*, 1991, **411**, 37; (c) J. Lewiński, J. Zachara, B. Mańk and S. Pasynkiewicz, *J. Organomet. Chem.*, in press; (d) M.D. Healy and A.R. Barron, *Angew. Chem., Int. Ed. Engl.*, 1992, **31**, 921.
86. H. Schmidbaur, *Angew. Chem., Int. Ed. Engl.*, 1985, **24**, 893 and refs. cited therein.
87. J.F. Hinton, K.R. Metz and R.W. Briggs, *Prog. NMR Spectrosc.*, 1988, **20**, 423.
88. K.B. Starowieyski, A. Chwojnowski and Z. Kuśmierek, *J. Organomet. Chem.*, 1980, **192**, 147.
89. H.D. Hausen, K. Mertz, E. Veigel and J. Weidlein, *Z. Anorg. Allg. Chem.*, 1974, **410**, 156.
90. B. Neumüller and F. Gahlmann, *J. Organomet. Chem.*, 1991, **414**, 271.
91. H. Schmidbaur, W. Wolfsberger and K. Schwirten, *Chem. Ber.*, 1969, **102**, 556.
92. J.A.K. Howard, L.E. Smart and C.J. Gilmore, *J. Chem. Soc., Chem. Commun.*, 1976, 477.
93. (a) W. Kaim, *J. Organomet. Chem.*, 1981, **215**, 325, 337; (b) W. Kaim, *J. Organomet. Chem.*, 1983, **241**, 157; (c) W. Kaim, *Z. Naturforsch.*, 1981, **36b**, 677.
94. (a) H. Reinheckel, K. Haage and D. Jahnke, *Organomet. Chem. Rev. A*, 1969, **4**, 47; (b) K.B. Starowieyski and Z. Rzepkowska, *J. Organomet. Chem.*, 1987, **322**, 309; (c) K.B. Starowieyski, A. Becalska and A. Okniński, *J. Organomet. Chem.*, 1985, **293**, 7.
95. (a) E.C. Ashby and R. Scott Smith, *J. Org. Chem.*, 1977, **42**, 425; (b) O.T. Beachley, Jr., M.R. Churchill, J.C. Pazik and J.W. Ziller, *Organometallics*, 1987, **6**, 2088.
96. (a) M.R.P. van Vliet, P. Buysingh, G. van Koten, K. Vrieze, B. Kojic-Prodic and A.L. Spek, *Organometallics*, 1985, **4**, 1701; (b) J. Lewiński, J. Zachara and B. Mańk, *J. Chem. Soc., Dalton Trans.*, in press.
97. (a) D.C. Bradley, H. Dawes, D.M. Frigo, M.B. Hursthouse and B. Hussain, *J. Organomet. Chem.*, 1987, **325**, 55; (b) F. Brady, K. Henrick and R. W. Matthews, *J. Organomet. Chem.*, 1979, **165**, 21.
98. W. Strohmeier and K. Hümpfner, *Z. Electrochem.*, 1957, **61**, 1010.
99. M. Veith and O. Recktenwald, *J. Organomet. Chem.*, 1984, **264**, 19.
100. (a) K.S. Chong, S.J. Rettig, A. Storr and J. Trotter, *Can. J. Chem.*, 1979, **57**, 586; (b) H. Krause, K. Sille, H.-D. Hausen and J. Weidlein, *J. Organomet. Chem.*, 1982, **235**, 253.
101. M.F. Self, W.T. Pennington and G.H. Robinson, *J. Coord. Chem.*, 1991, **24**, 69.
102. G.H. Robinson, H. Zhang and J.T. Atwood, *Organometallics*, 1987, **6**, 887.
103. B. Lee, W.T. Pennington, G.H. Robinson and R.D. Rogers, *J. Organomet. Chem.*, 1990, **396**, 269.
104. (a) J.L. Atwood, D.C. Hrncir, R. Shakir, M.S. Dalton, R.D. Priester and R.D. Rogers, *Organometallics*, 1982, **1**, 1021; (b) G.H. Robinson, W.E. Hunter, S.G. Bott and J.L. Atwood, *J. Organomet. Chem.*, 1987, **326**, 17.
105. (a) S.G. Bott, A. Alvanipour, S.D. Morley, D.A. Atwood, C.M. Means, A.W. Coleman and J.L. Atwood, *Angew. Chem., Int. Ed. Engl.*, 1987, **26**, 485; (b) S.G. Bott, H. Elgamal and J.L. Atwood, *J. Am. Chem. Soc.*, 1985, **107**, 1796.
106. V.L. Goedken, H. Ito, and T. Ito, *J. Chem. Soc., Chem. Commun.*, 1984, 1453.
107. M.J. Zaworotko, C.R. Kerr and J.L. Atwood, *Organometallics*, 1985, **4**, 238.
108. D.C. Hrncir, R.D. Rogers and J.L. Atwood, *J. Am. Chem. Soc.*, 1981, **103**, 4277.

109. J.L. Atwood, D.C. Hrncir, R.D. Rogers and J.A.K. Howard, *J. Am. Chem. Soc.*, 1981, **103**, 6787.
110. Yu A. Alexandrow, N.N. Vyshinskii, V.N. Kokorev, V.A. Alferov, N.V. Chikinova and G.I. Makin, *J. Organomet. Chem.*, 1987, **332**, 257.
111. G. Sonnek and H. Reinheckel, *Z. Chem.*, 1976, **16**, 64.
112. O.T. Beachley, Jr., and J.C. Pazik, *Organometallics*, 1988, **7**, 1516.
113. R.L. Shuler, R.A. De Marco and A.D. Berry, *Inorg. Chim. Acta*, 1984, **85**, 185.
114. O.T. Beachley, Jr., and J.D. Bernstein, *Inorg. Chem.*, 1973, **12**, 183.
115. J.R. Rie and J.P. Oliver, *J. Organomet. Chem.*, 1977, **133**, 147.
116. S.U. Choi, W.C. Frith and H.C. Brown, *J. Am. Chem. Soc.*, 1966, **88**, 4128.
117. M.B. Smith, *J. Organomet. Chem.*, 1974, **70**, 13 and refs. cited therein.
118. P.E.M. Allen and R.M. Lough, *J. Chem. Soc., Faraday Trans. 1*, 1973, **69**, 849.
119. J. Chey, H-S. Choe, Y-M. Chook, E. Jensen, P.R. Seida and M.M. Francl, *Organometallics*, 1990, **9**, 2430.
120. T.W. Dolzinc and J.P. Oliver, *J. Am. Chem. Soc.*, 1974, **96**, 1737.
121. A. Sporzyński and K.B. Starowieyski, *J. Organomet. Chem. Library*, 1980, **9**, 19 and refs. cited therein.
122. J. Furukawa, K. Omura, O. Yamamoto and K. Ishikawa, *J. Chem. Soc., Chem. Commun.*, 1974, 77.
123. (a) L. Bagnell, E.A. Jeffery, A. Meisters and T. Mole, *Aust. J. Chem.*, 1974, **27**, 2577; (b) W. Kuran, S. Pasynkiewicz and A. Saťek, *J. Organomet. Chem.*, 1974, **73**, 199.
124. (a) S. Amirkhalili, P.B. Smith and J. David, *J. Chem. Soc., Dalton Trans.*, 1979, 1206; (b) P.P. Power, *J. Organomet. Chem.*, 1990, **400**, 49.
125. (a) K. Gosling and R.E. Bowen, *J. Chem. Soc., Dalton Trans.*, 1974, 1961; (b) K. Haage, K.B. Starowieyski and A. Chwojnowski, *J. Organomet. Chem.*, 1979, **174**, 149; (c) Ref. 5, p. 229.
126. (a) E.P. Schram, R.E. Hall and J.D. Glore, *J. Am. Chem. Soc.*, 1969, **91**, 6643; (b) C. Dohmeier, C. Robl, M. Tacke and H. Schnöckel, *Angew. Chem., Int. Ed. Engl.*, 1991, **30**, 564; (c) W. Hiller, K.-W. Klinkhammer, W. Uhl and J. Wagner, *Angew. Chem., Int. Ed. Engl.*, 1991, **30**, 179.
127. (a) P. Jutzi, *J. Organomet. Chem.*, 1990, **400**, 1; (b) D. Loos, H. Schnöckel, J. Gauss and U. Schneider, *Angew. Chem., Int. Ed. Engl.*, 1992, **31**, 1362.
128. (a) M.B. Freeman, L.G. Sneddon and J.C. Huffman, *J. Am. Chem. Soc.*, 1977, **99**, 5194; (b) O.T. Beachley, Jr., J.F. Lees and R.D. Rogers, *J. Organomet. Chem.*, 1991, **418**, 165 and refs. cited therein.
129. O.T. Beachley, Jr., M.R. Churchill, J.C. Fettinger, J.C. Pazik and L. Victoriano, *J. Am. Chem. Soc.*, 1986, **108**, 4666.
130. H. Werner, H. Otto and H.J. Kraus, *J. Organomet. Chem.*, 1986, **315**, C57.
131. S. Harvey, C.L. Raston, B.W. Skelton, A.H. White, M.F. Lappert and G. Srivastava, *J. Organomet. Chem.*, 1987, **328**, C1.
132. H. Schumann, C. Janiak, F. Gorlitz, J. Loebel and A. Dietrich, *J. Organomet. Chem.*, 1989, **363**, 243.
133. H. Schumann, C. Janiak, J. Pickard and U. Börner, *Angew. Chem., Int. Ed. Engl.*, 1987, **26**, 790.
134. (a) H. Schumann, C. Janiak, M.A. Khan and J.J. Zuckerman, *J. Organomet. Chem.*, 1988, **354**, 7; (b) P. Schwerdtfeger, *Inorg. Chem.*, 1991, **30**, 1660.
135. C.S. Ewig, R. Osman and J.R. Van Waser, *J. Am. Chem. Soc.*, 1978, **100**, 5017.
136. (a) P. Singh, M.D. Rausch and T.E. Bitterwolf, *J. Organomet. Chem.*, 1988, **352**, 273; (b) D. Morcos and W. Tikkanen, *J. Organomet. Chem.*, 1989, **371**, 15; (c) M. Arthus, H.K. Al-Daffaee, J. Haslop, G. Kubal, M.D. Pearson, P. Thatcher and E. Curzon, *J. Chem. Soc., Dalton Trans.*, 1987, 2615.
137. (a) H.M. Colquhoun, T.J. Greenhough and M.G.H. Wallbridge, *J. Chem. Soc., Chem. Commun.*, 1977, 737; (b) D.M. Schubert, M.A. Bondman, W.S. Rees, Jr., C.B. Knobler, P. Lu, W. Nam and M.F. Hawthorne, *Organometallics*, 1990, **9**, 2046.
138. G.E. Herberich, H.J. Becker and C. Engelke, *J. Organomet. Chem.*, 1978, **153**, 265.
139. K. Stumpf, H. Pritzkow and W. Siebert, *Angew. Chem., Int. Ed. Engl.*, 1985, **24**, 71.
140. H. Schmidbaur, W. Bublak, A. Shier, G. Reber and G. Mueller, *Chem. Ber.*, 1988, **121**, 1373.

370   CHEMISTRY OF ALUMINIUM, GALLIUM, INDIUM AND THALLIUM

141. O.T. Beachley, Jr., and R.G. Simmons, *Inorg. Chem.*, 1980, **19**, 3042.
142. O.T. Beachley, Jr., C. Tessier-Youngs, R.G. Simmons and R.B. Hallock, *Inorg. Chem.*, 1982, **21**, 1970.
143. (a) R.B. Hallock, O.T. Beachley, Jr., Y. Li, W.M. Sanders, M.R. Churchill, W.E. Hunter and J.L. Atwood, *Inorg. Chem.*, 1983, **22**, 3683. (b) W. Uhl, W. Hiller, M. Layh and W. Schwarz, *Angew. Chem., Int. Ed. Engl.*, 1992, **31**, 1364.
144. (a) F.G.N. Cloke, G.R. Hanson, M.J. Henderson, P.B. Hitchcock and C.L. Raston, *J. Chem. Soc., Chem. Commun.*, 1989, 1002; (b) F.G.N. Cloke, C.I. Dalby, M.J. Henderson, P.B. Hitchcock, C.H.L. Kennard, R.N. Lamb and C.L. Raston, *J. Chem. Soc., Chem. Commun.*, 1990, 1394; (c) W. Kaim and W. Matheis, *J. Chem. Soc., Chem. Commun.*, 1991, 597.
145. U. Kynast, B.W. Skelton, A.H. White, M.J. Henderson and C.L. Raston, *J. Organomet. Chem.*, 1990, **384**, C1.
146. M. Veith, F. Goffing, S. Becker and V. Huch, *J. Organomet. Chem.*, 1991, **406**, 105.
147. C.A. Kraus and F.E. Toonder, *J. Am. Chem. Soc.*, 1933, **55**, 3547.
148. H. Hoberg and S. Krause, *Angew. Chem., Int. Ed. Engl.*, 1976, **15**, 694.
149. H. Hoberg and S. Krause, *Angew. Chem., Int. Ed. Engl.*, 1978, **17**, 949.
150. E.P. Schram and M.M. Miller, *Inorg. Chim. Acta*, 1986, **113**, 131.
151. W. Uhl, *Z. Naturforsch.*, 1988, **43b**, 1113.
152. W. Uhl, M. Layh and T. Hildenbrand, *J. Organomet. Chem.*, 1989, **364**, 289.
153. W. Uhl, M. Layh and W. Hiller, *J. Organomet. Chem.*, 1989, **368**, 139.
154. (a) L.J. Todd, in *Metal Interactions with Boron Clusters*, ed. R.N. Grimes, Plenum Press, New York, 1982, p. 148; (b) R.N. Grimes, W.J. Rademaker, M.L. Denniston, R.F. Bryan and P.T. Greene, *J. Am. Chem. Soc.*, 1972, **94**, 1865.
155. A.T. Weibel and J.P. Oliver, *J. Organomet. Chem.*, 1974, **74**, 155.
156. M.A. Banks, O.T. Beachley, Jr., H.J. Gysling and H.R. Luss, *Organometallics*, 1990, **9**, 1979.
157. R. Uson, A. Laguna, J.A. Abad and E. de Jesus, *J. Chem. Soc., Dalton Trans.*, 1983, 1127.
158. A.J. Conway, G.J. Gainsford, R.R. Schrieke and J.D. Smith, *J. Chem. Soc., Dalton Trans.*, 1975, 2499.
159. A.J. Conway, P.B. Hitchcock and J.D. Smith, *J. Chem. Soc., Dalton Trans.*, 1975, 1945.
160. (a) B.V. Lokshin, E.B. Rusach, Z.P. Valueva, A.G. Ginzburg and N.E. Kolobova, *J. Organomet. Chem.*, 1975, **102**, 535; (b) R.A. Fischer, J. Behm and T. Priermeier, *J. Organomet. Chem.*, 1992, **429**, 275.
161. J.C. Vanderhooft, R.D. Erst, F.W. Cagle, R.J. Neustadt and T.H. Cymbaluk, *Inorg. Chem.*, 1982, **21**, 1876.
162. C. Tessier-Youngs, C. Bueno, O.T. Beachley, Jr., and M.R. Churchill, *Inorg. Chem.*, 1983, **22**, 1054.
163. R. Guilard, A. Zrineh, A. Tabard, L. Courthaudon, B. Han, M. Ferhat and K.M. Kadish, *J. Organomet. Chem.*, 1991, **401**, 227.
164. (a) R. Guilard, C. Lecomte and K.M. Kadish, *Struct. Bonding (Berlin)*, 1987, **64**, 205; (b) K.M. Kadish, *Prog. Inorg. Chem.*, 1986, **34**, 435; (c) R. Guilard and K.M. Kadish, *Comments Inorg. Chem.*, 1988, **7**, 287; (d) S. Takagi, Y. Kato, H. Furuta, S. Onaka and T.K. Miyamoto, *J. Organomet. Chem.*, 1992, **429**, 287.
165. D.L. Thorn and R.L. Harlow, *J. Am. Chem. Soc.*, 1989, **111**, 2575.
166. D.L. Thorn, *J. Organomet. Chem.*, 1991, **405**, 161.
167. R.A. Fischer and J. Behm, *J. Organomet. Chem.*, 1991, **413**, C10.
168. (a) A.A. Aradi, F-E. Hong and T.B. Fehlner, *Organometallics*, 1991, **10**, 2726; (b) C. Krüger, J.C. Sekutowski, H. Hoberg and R. Krause-Göing, *J. Organomet. Chem.*, 1977, **141**, 141.
169. V.K. Belsky, Yu.K. Gun'ko, B.M. Bulychev and G.L. Soloveichik, *J. Organomet. Chem.*, 1991, **419**, 299.
170. D.M. Schubert, C.B. Knobler, W.S. Rees and M.F. Hawthorne, *Organometallics*, 1987, **6**, 201, 203.
171. M.T. Barlow, A.J. Downs, P.D.P. Thomas and D.W.H. Rankin, *J. Chem. Soc., Dalton Trans.*, 1979, 1793.

172. (a) M.T. Barlow, C.J. Dain, A.J. Downs, P.D.P. Thomas and D.W.H. Rankin, *J. Chem. Soc., Dalton Trans.*, 1980, 1374; C.R. Pulham and A.J. Downs, unpublished results; (b) C.J. Dain, A.J. Downs and D.W.H. Rankin, *J. Chem. Soc., Dalton Trans.*, 1981, 2465.
173. (a) A.G. Avent, C. Eaborn, M.N.A. El-Kheli, M.E. Molla, J.D. Smith and A.C. Sullivan, *J. Am. Chem. Soc.*, 1986, **108**, 3854; (b) A. Heine and D. Stalke, *Angew. Chem., Int. Ed. Engl.*, 1992, **31**, 854.
174. K.B. Starowieyski and A. Chwojnowski, *J. Organomet. Chem.*, 1981, **215**, 151.
175. (a) L.L. Ivanov, S. Ya. Zavizion and L.I. Zakharkin, *Synth. Inorg. Metal-Org. Chem.*, 1973, **3**, 323; (b) L.S. Bresler, Yu.I. Kul'velis and A.V. Lubnin, *Zh. Obshch. Khim.*, 1984, **54**, 1306.
176. (a) M.R. Churchill, A.L. Rheingold and J. Wasserman, *Inorg. Chem.*, 1981, **20**, 3392; (b) F.N. Tebbe, G.W. Parshall and G.S. Reddy, *J. Am. Chem. Soc.*, 1978, **100**, 3611; (c) J. Holton, M.F. Lappert, D.G.H. Ballard, R. Pearce, J.L. Atwood and W.E. Hunter, *J. Chem. Soc., Dalton Trans.*, 1979, 45.
177. H.J. Wollmer and W. Kaminsky, *Cryst. Struct. Commun.*, 1980, **9**, 985.
178. M.V. Gaudet, M.J. Zaworotko, T.S. Cameron and A. Linden, *J. Organomet. Chem.*, 1989, **367**, 267.
179. M.R. Churchill, H.J. Wasserman, H.W. Turner and R.R. Schrock, *J. Am. Chem. Soc.*, 1982, **104**, 1710.
180. G. Erker, M. Albrecht, C. Krüger and S. Werner, *Organometallics*, 1991, **10**, 3791.
181. M. Albrecht, G. Erker, M. Nolte and C. Krüger, *J. Organomet. Chem.*, 1992, **427**, C21.
182. R.W. Waymouth, K.S. Potter, W.P. Schaefer and R.H. Grubbs, *Organometallics*, 1990, **9**, 2843.
183. G.H. Robinson, W.T. Pennington, B. Lee, M.F. Self and D.C. Hrncir, *Inorg. Chem.*, 1991, **30**, 809.
184. G.D. Rogers, W.J. Cook and J.L. Atwood, *Inorg. Chem.*, 1979, **18**, 279.
185. (a) G.A. Banta, B.M. Louie, E. Onyiriuka, S.J. Rettig and A. Storr, *Can. J. Chem.*, 1986, **64**, 373; (b) D.A. Cooper, S.J. Rettig and A. Storr, *Can. J. Chem.*, 1986, **64**, 566 and refs. cited therein.
186. A. Looney, N. Cornebise, D. Miller and G. Parkin, *Inorg. Chem.*, 1992, **31**, 989.
187. H.H. Karsch and A. Appelt, *J. Organomet. Chem.*, 1986, **314**, C5.
188. H.H. Karsch, A. Appelt and G. Müller, *J. Chem. Soc., Chem. Commun.*, 1984, 1415.
189. W. Kaminsky and H.-J. Vollmer, *Liebigs Ann. Chem.*, 1975, 438.
190. O.T. Beachley, Jr., T.L. Royster, Jr., W.J. Youngs, E.A. Zarate and C.A. Tessier-Youngs, *Organometallics*, 1989, **8**, 1679.
191. R.A. Forder and K. Prout, *Acta Crystallogr.*, 1974, **B30**, 2312.
192. G.K. Anderson and N.P. Rath, *J. Organomet. Chem.*, 1991, **414**, 129.
193. N.N. Greenwood, B.S. Thomas and D.W. Waite, *J. Chem. Soc., Dalton Trans.*, 1975, 299.
194. R. Lehmkuhl, K. Mehler, R. Benn, A. Rufińska, G. Schroth and C. Krüger, *Chem. Ber.*, 1984, **117**, 389.
195. S. Komiya, M. Katch, T. Ikariya, R.H. Grubbs, T. Yamamoto and A. Yamamoto, *J. Organomet. Chem.*, 1984, **260**, 115.
196. S. Komiya, M. Bundo, T. Yamamoto and M. Yamamoto, *J. Organomet. Chem.*, 1979, **174**, 343.
197. H. Hoberg and W. Richter, *J. Organomet. Chem.*, 1980, **195**, 347.

7 Organic transformations mediated by Group 13 metal compounds

J.A. MILLER

7.1 Organoaluminium compounds

7.1.1 Introduction

The reactivity exhibited by organoaluminiums resembles some aspects of both organoboron and organolithium chemistry. There are, however, some key differences with organoaluminium chemistry that distinguish these compounds from other organometals and make them versatile tools for effecting many organic transformations. While the intrinsic nucleophilicity of organoaluminiums is, in general, well below that of organolithium or Grignard reagents, it is markedly higher than that found for organoboron derivatives. As is the case with organoboranes, the central atom in R_3Al compounds (alanes) formally possesses an empty p-orbital which leaves it with a sextet of valence electrons. On treatment with a base, typically an organolithium reagent, the coordinatively unsaturated aluminium atom is alkylated to yield the respective $Li^+AlR_4^-$ species (an alanate), in which it is now coordinatively saturated with a filled octet of electrons. Thus, monomeric organoalanes are quite Lewis acidic in nature, whereas the formal negative charge on the aluminium atom in organoalanate complexes imparts an increased nucleophilicity to its ligands. It is important to bear in mind, however, that the organic ligands of organoalanes can still function as nucleophiles. In fact, when coupled with their inherent Lewis acidity, this latent nucleophilicity can provide an added dimension to the reactivity of organoalanes compared to other reagents which act solely as either Lewis acid catalysts or nucleophilic species. This unique characteristic of organoalanes can be well exemplified by comparing their reactivity in alkyl halide coupling reactions with that of similar organolithium reagents. For example, the order of reactivity of alkyl halides towards RLi is primary > secondary > tertiary, with the hindered tertiary halides undergoing considerable β-elimination side-reactions. On the other hand, organoaluminiums cross-couple with alkyl halides with precisely the opposite order of reactivity: tertiary > secondary > primary. In contrast to the S_N2-like mechanism operating with an organolithium in this coupling reaction, complexation of R_3Al with the halide moiety, followed by transfer of the organic group from aluminium to the tertiary centre, must be a crucial sequence in this modified S_N1 process. The relative ease with which alkenyl-aluminiums are prepared from hydro- and carbometallation of acetylenes also represents a distinct advantage for these compounds as synthetic intermediates compared to the corresponding alkenyllithium and Grignard reagents. Finally, an important difference between the organic chemistry of aluminium and boron is their primary avenue of participation in carbon–carbon bond-forming reactions. While the vast majority of these reactions occur *via* migratory insertion processes for organoboranes and -borates, only recently has this pathway been recognised as a viable synthetic route for organoaluminiums. Clearly, however, most carbon–carbon bond-forming reactions involving organoaluminiums result from intermolecular transfer of an organic group.

Typically, carbon–aluminium bonds undergo facile protonation, oxidation and halogenation to produce the corresponding hydrocarbons, alcohols and alkyl halides, respectively. Moreover, organoaluminiums participate in a variety of carbon–carbon bond-forming reactions, including alkylation, carbonation, cyanation, addition to carbonyl compounds, conjugate addition to enones and epoxide ring-opening reactions. Since the general chemistry of organoaluminiums has been extensively reviewed previously,[1–14] including the pioneering contributions of Ziegler's group,[3,11,15,16] this discussion emphasises organoaluminium chemistry of current synthetic interest.

7.1.2 Transformations of alkenes and alkynes

Compared to their boron congeners, aluminium hydrides exhibit a relatively weak affinity for hydrometallation of unactivated carbon–carbon double bonds. As a result, the direct hydroalumination of alkenes has never approached the success of the corresponding hydroboration reaction in terms of being recognised as a useful synthetic procedure. Because of the elevated reaction temperatures required to effect hydroalumination of alkenes, the reaction is typically complicated by competititve dehydro-alumination (displacement) and carboalumination processes.[3–5,11,15,16] In an effort to circumvent these side-reactions, several versions of catalysed hydroalumination reactions of alkenes have been developed which allow for a facile conversion of olefins into the respective alkylaluminiums. These reactions are generally efficient processes and retain the high degree of anti-Markovnikov regioselectivity for addition of Al–H that is observed for the uncatalysed hydroalumination. Furthermore, since the alkylaluminium compounds can be directly functionalised in ways not possible for the analogous organoboranes, the catalysed hydroalumination-functionalis-ation sequences now constitute valuable synthetic procedures. For example, the combination of $LiAlH_4$ and titanium catalysts ($TiCl_4$,[17,18] supported $TiCl_4$[19] or Cp_2TiCl_2[20]) has been frequently used regioselectively to prepare tetraalkylalanates from terminal alkene substrates. These tetraalkylalanates have been converted *in situ* into alkenes,[17,18] alcohols[17,18] and alkyl acetates,[21] and have also shown an interesting chemoselective preference for alkylation of carbonyl versus ester functionality.[22] The combination of $LiAlH_4$ with UCl_3 or UCl_4 is also effective for the hydroalumination of terminal olefins.[23]

The Zr-catalysed hydroalumination of alkenes achieved by the system Bu^i_3Al-Cp_2ZrCl_2 is distinct from the above procedures since it does not utilise an aluminium hydride reagent and affords a trialkylalane, and not an alanate, reaction product.[24] The transformation proceeds readily under mild conditions and typically produces high yields of derivatised products (equation (7.1))

$$n\text{-}C_6H_{13}CH=CH_2 \xrightarrow[\substack{Cp_2ZrCl_2 \\ (10\ mol\%),\ 0°C}]{Bu^i_3Al} n\text{-}C_6H_{13}CH_2CH_2Al(Bu^i)_2 \qquad (7.1)$$

$$\xrightarrow{I_2} n\text{-}C_6H_{13}CH_2CH_2I \quad (94\%)$$

The hydroalumination of olefins may also be carried out using Cl_2AlH in the presence of catalytic amounts of certain boron compounds, such as BEt_3 or $PhB(OH)_2$.[14,25] Thus, boron-catalysed hydroalumination of 1-dodecene, followed by acylation with acetyl chloride, provides 2-tetradecanone in 82% yield.

Since it was first reported in 1956,[26-28] hydroalumination of alkynes has evolved into a standard method for the construction of stereodefined di- and trisubstituted alkenes.[8-10,13] The alkenylaluminium intermediates formed in the reaction can be functionalised with retention of double bond geometry in a number of ways not possible for the related alkenylboranes or -borates, thus adding to the synthetic utility of the process. To illustrate, reaction of a 1-alkyne with Bu^i_2AlH in a hydrocarbon solvent at 50°C furnishes an alkenylalane in which Al–H has added stereoselectively in a *cis* manner while placing the aluminium atom regioselectively at the terminal carbon (equation (7.2)).[13]

$$RC{\equiv}CH + Bu^i_2AlH \xrightarrow[50°C]{heptane} \underset{\substack{H \\ (90\%)}}{\overset{R\quad H}{\underset{}{C=C}}}_{Al(Bu^i)_2} + \underset{(6\%)}{RC{\equiv}CAl(Bu^i)_2} + \underset{(4\%)}{RCH_2CH[Al(Bu^i)_2]_2} \qquad (7.2)$$

This hydroalumination reaction is also accompanied by small amounts of metallation and diaddition products. The amount of metallation-derived alkynylalane increases on hydroalumination of relatively acidic alkynes, such as phenylacetylene,[29-33] and decreases with substituted derivatives (i.e. R = secondary and tertiary alkyl groups).[13] The extent to which these side-products are formed is also highly dependent on the choice of solvent for the hydroalumination reaction. For instance, use of ether or THF as the solvent affords high yields of the dihydroalumination product,[34,35] whereas NEt_3 leads exclusively to alkynylalanes *via* metallation.[36] Further treatment of this resultant alkynylalane with 2 equivalents of Bu^i_2AlH cleanly produces the 1,1,1-trialuminoalkane.[37,38] Hence, the products of mono-, di- or tri-hydroalumination of terminal alkynes can be selectively synthesised by merely changing the reaction solvent and $Bu_2^i AlH$: 1-alkyne stoichiometry (equation (7.3)).

$$
RC≡CH
\begin{cases}
\xrightarrow[\text{heptane}]{Bu_2^i AlH} & \underset{H}{\overset{R}{\diagdown}}C=C\underset{Al(Bu^i)_2}{\overset{H}{\diagup}} \\[2mm]
\xrightarrow[\text{THF or ether}]{2\,Bu_2^i AlH} & RCH_2CH[Al(Bu^i)_2]_2 \\[2mm]
\xrightarrow[\text{NEt}_3]{3\,Bu_2^i AlH} & RCH_2C[Al(Bu^i)_2]_3
\end{cases}
\qquad (7.3)
$$

Internal alkynes also undergo *cis*-hydroalumination readily in hydrocarbon solvents to yield the corresponding disubstituted vinyl-alanes.[13,27–29,39] Because of complexation of the alkenylalane product with unreacted $Bu_2^i AlH$, the rate of these hydroaluminations decreases significantly after 50% reaction. As a result, a stoichiometry of $2Bu_2^i AlH$: 1-RC≡CR is highly beneficial for this transformation and delivers nearly quantitative yields of (*E*)-alkenylalanes derived from *cis*-monohydro-alumination of the respective internal alkynes (equation (7.4)).[39,40]

$$
n\text{-}C_3H_7C≡CC_3H_7\text{-}n + Bu_2^i AlH \xrightarrow[50°C]{\text{hexane}} \underset{H}{\overset{n\text{-}C_3H_7}{\diagdown}}C=C\underset{Al(Bu^i)_2}{\overset{C_3H_7\text{-}n}{\diagup}} \qquad (7.4)
$$
$$
\text{(2 equiv.)}
$$

By utilising a nucleophilic aluminium hydride reagent in place of the Lewis acidic $Bu_2^i AlH$, a stereoselective *trans*-hydroalumination of internal alkynes can be accomplished. Thus, heating 3-hexyne at 125°C with $LiAlH_4$ in THF-diglyme solvent, followed by hydrolysis, affords *trans*-3-hexene in 91% yield.[41,42] Lithium diisobutylmethylaluminium hydride, prepared from reaction of $Bu_2^i AlH$ with MeLi, also hydroaluminates internal alkynes stereoselectively in a *trans* manner (equation (7.5)).[43]

$$
RC≡CR + Li[AlHMe(Bu^i)_2] \xrightarrow[100–130°C]{\text{dimethoxyethane (DME)}} \underset{H}{\overset{R}{\diagdown}}C=C\underset{R}{\overset{AlMe(Bu^i)_2Li}{\diagup}} \qquad (7.5)
$$

An advantage of these trialkylalkenylalanate reaction products over the products derived from $LiAlH_4$ is that they react readily with carbonyl compounds to form allylic alcohols, whereas the latter alkenylalanates do not. It is noteworthy that reaction of the nucleophilic aluminium hydride reagents with terminal alkynes does not generally result in hydro-alumination, but rather leads to appreciable amounts of metallation to produce the corresponding alkynylalanates.

As discussed earlier, the polarity of the solvent has a dramatic influence with regard to the nature of the products obtained from hydroalumination of terminal acetylenes. Since the monohydroalumination of 1-alkynes is best carried out in a non-coordinating medium, it is not surprising that the presence of heteroatom-containing functionality on the substrate can complicate the course of the reaction. For example, low yields of desired alkenylalanes are obtained upon hydroalumination of a propargylic alcohol with 2 equivalents of Bu_2^iAlH (equation (7.6)).[44]

$$
\underset{\substack{\text{OH}\\|}}{n\text{-}C_5H_{11}\overset{|}{C}HC{\equiv}CH} + Bu_2^iAlH \longrightarrow \underset{(25\text{–}40\%)}{\underset{H}{n\text{-}C_5H_{11}\overset{\displaystyle \overset{OAl(Bu^i)_2}{|}}{C}H}\underset{H}{\overset{}{}}\overset{H}{\underset{Al(Bu^i)_2}{C{=}C}}} \tag{7.6}
$$

(2 equiv.)

Acid-sensitive hydroxyl protecting groups, as in tetrahydropyranyl derivatives, are readily cleaved under these hydroalumination reaction conditions.[44,45] Protection of propargylic alcohols as triphenylmethyl ether derivatives, however, allows hydroalumination to proceed in reasonable yield.[46] On the other hand, blocking the same propargylic alcohol moiety as a t-butyl ether induces a net *trans*-hydroalumination of the triple bond by Bu_2^iAlH.[46,47] It should be noted, however, that t-butyl-protected alkynyl alcohols, with the exception of propargylic and homopropargylic derivatives, are well tolerated under typical hydroalumination reaction conditions to provide the expected alkenylalanes derived from *cis*-addition of Al–H.[47] Finally, the reduction of propargyl alcohols by $LiAlH_4$ to the corresponding allylic alcohols is a frequently used synthetic method.[48–53] The reaction is known to proceed *via* an alkenylalanate intermediate and affords a C–Al bond that can be protonated or halogenated. Interestingly, by varying the hydroalumination conditions, propargyl alcohols may be converted into (Z)-2-iodo-2-alken-1-ols[54] or (Z)-3-iodo-2-alken-1-ols (equation (7.7)).[55,56]

$$
RC{\equiv}CCH_2OH \quad
\begin{array}{c}
\xrightarrow[\substack{\text{2. } Bu_2^iAlH\\ \text{3. } I_2}]{\text{1. } Bu^nLi}\\[1.2em]
\xrightarrow[\text{2. } I_2]{\text{1. } LiAlH_4,\ NaOMe}
\end{array}
\quad
\begin{array}{c}
\underset{H}{\overset{R}{C}}{=}\underset{CH_2OH}{\overset{I}{C}}\\[1.5em]
\underset{I}{\overset{R}{C}}{=}\underset{CH_2OH}{\overset{H}{C}}
\end{array}
\tag{7.7}
$$

The facile synthesis of stereo- and regiodefined alkenylaluminiums *via* hydro- and carboalumination reactions provides an expedient route to a wide array of stereochemically pure olefinic derivatives. As briefly discussed in section 7.1.1, conversion of an alkenylalane into the respective -alanate *via* reaction with an alkyllithium reagent boosts its nucleophilicity and endows it with a reactivity approaching that of an alkenyllithium or Grignard reagent.[8–10,13] Scheme 7.1 illustrates the wide variety of stereospecific transformations that alkenylalanes and -alanates undergo, including halogenation,[43,53–61] alkylation,[61–65] carbonation,[13,43,58,66–70] cyanation,[71] alkylation with chloromethyl alkyl ethers,[62,69,72,73] addition to carbonyl compounds[13,73,74] (including paraformaldehyde[58,67,69,73,75,76]), conjugate addition to enones[46,77–80] and α-nitroolefins,[81,82] epoxide[83–89] and oxetane[89] ring-opening reactions, esterification *via* reaction with chloroformates,[69,90] sulfuridation,[91] and Cu(I)-induced homocoupling[92] and allylation.[93] Furthermore, alkenylaluminiums participate in Pd- or Ni-catalysed arylation,[94–96] benzylation,[97] allylation,[98–102] acylation[103,104] and alkynyl-ation[95] reactions. Alkenylalanes also serve as efficient precursors for the synthesis of stereodefined conjugated dienes *via* Pd- or Ni-catalysed cross-coupling with haloalkenes[95,96,101,105–109] and enol phosphates.[110–112] It should

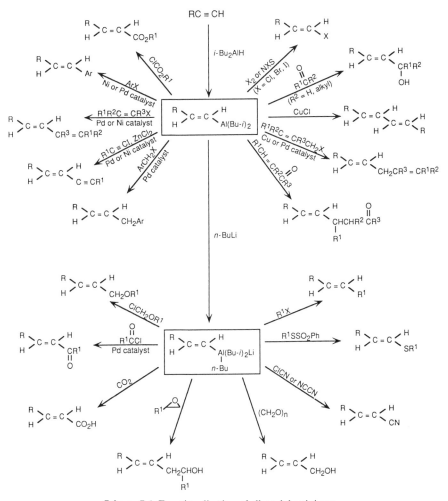

Scheme 7.1 Functionalisation of alkenylaluminiums.

be pointed out that in all of these transformations the electrophilic reagent preferentially attacks the alkenyl moiety of the organoaluminium substrate and only small amounts, if any, of isobutyl-derived by-products are typically formed.

The hydroalumination of 1-silyl-1-alkynes proceeds stereo- and regio-selectively in polar solvents, with Al–H adding to the triple bond in a *cis* manner and placing the aluminium atom solely at the silicon-bearing carbon (equation (7.8)).[13,59,63,64,113] In a non-polar medium, the hydroalumination of 1-silyl-1-alkynes also occurs with the same regioselectivity, but preferentially yields the corresponding (*E*)-alkenylalanes (equation (7.8)).[63,64,113]

$$\tag{7.8}$$

Formation of the (*E*)-α-silylalkenylalane is believed to result from isomeris-ation of the kinetically formed *cis* addition product to the thermo-dynamically preferred (*E*)-isomer.[114,115] The relative proportion of (*E*)- and (*Z*)-alkenylalanes derived from hydroalumination of an alkynyl-silane in a hydrocarbon solvent is highly dependent on the nature of the particular substrate, with compounds bearing R = *t*-butyl, phenyl or alkenyl groups affording essentially pure (*E*)-alkenylalanes.[114] In addition, alkynyl-silanes with silyl groups sterically larger than SiMe$_3$, such as SiEt$_3$, give hydroalumination product mixtures enriched in the (*E*)-alkenylalane.[64] It should also be emphasised that functionalization of these α-silylalkenyl-aluminium intermediates *via* reaction of the C–Al bond with electrophiles (E$^+$) can be readily accomplished and generally parallels the types of transformation depicted in Scheme 7.1.[13,59,63–66,70]

Despite its large steric requirement, the trialkylsilyl group exerts a powerful directive influence on the regioselectivity of alkynylsilane hydro-alumination, guiding the aluminium atom exclusively to the silicon-bearing carbon. In this regard, it is interesting to contrast the regiochemistry derived from hydroalumination of 1-silyl-1-alkynes with the corresponding *t*-butyl-substituted acetylenes. Thus, hydroalumination of 1-*t*-butyl-1-propyne with Bu$_2^i$AlH leads to a mixture of alkenylalanes in which only 15% of the alkenylalane product contains aluminium at the 1-position.[8] On the other hand, hydroalumination of 1-trimethylsilyl-1-propyne selectively yields the alkenylalane isomer possessing the aluminium atom at the carbon adjacent to the Me$_3$Si group.[70] The trimethylsilyl moiety also has an activating influence upon the acetylenic group with respect to the rate of hydroalumination. As a result, the trimethylsilylethynyl group has been utilised as an efficient means by which to achieve hydroalumination of the triple bond of tetrahydropyranyl-protected acetylenic alcohols, thereby alleviating the acetal cleavage side-reactions which are frequently encountered using alternative hydroalumination procedures.[45] Further-more, the silylacetylenic group can be selectively hydroaluminated in the presence of a dialkyl-substituted triple bond within the same molecule (equation (7.9)).[13]

$$n\text{-}C_6H_{13}C{\equiv}CCH_2C{\equiv}CSiMe_3 \xrightarrow[\text{ether}]{Bu^i_2AlH}$$

$$(7.9)$$

(92%)

Notably, the chemoselectivity between triple bonds reverses in the case of silylated 1,3-butadiynes. For example, *trans*-hydroalumination of 1-trimethylsilyl-1,3-decadiyne with $Li[(Bu^n)(Bu^i)_2AlH]$ results in the exclusive addition of Al–H to the internal triple bond (equation (7.10)).[116] The observed regioselectivity in this instance is derived from the powerful electron-withdrawing nature of the trimethylsilylethynyl group compared to the non-silicon-bearing triple bond, providing the highest electron density at the 3-position. The resultant enynylalanate intermediate in this reaction can be stereospecifically derivatised in a number of ways, including protonation,[116] bromination with BrCN,[13,61] allylation using allyl bromide,[13,61] and methylation *via* MeI/CuI (equation (7.10)).[13]

$$n\text{-}C_6H_{13}C{\equiv}CC{\equiv}CSiMe_3 \xrightarrow{Li[(Bu^n)(Bu^i)_2AlH]}$$

$$(7.10)$$

where E = H (95%), Br (90%), allyl (94%) or methyl (81%).

Chloromethylsilyl-substituted alkynylsilanes undergo a regio- and stereoselective hydroalumination in a hydrocarbon solvent to yield the (Z)-α-silylalkenylalanes.[117,118] The initially produced (Z)-α-silylalkenylalane is believed to be stabilised from geometrical isomerisation *via* intramolecular coordination of aluminium by the proximal chlorine atom. Treatment of this alkenylalane with MeLi (3 equivalents) leads to formation of the respective vinyllithium adduct of inverted geometry, which can then be elaborated *via* reaction with electrophiles such as allyl bromide (equation (7.11)).[118]

$$n\text{-}C_6H_{13}C{\equiv}CSiMe_2CH_2Cl \xrightarrow[\text{heptane}]{Bu^i_2AlH}
\underset{H}{\overset{n\text{-}C_6H_{13}}{>}}C{=}C\underset{Al(Bu^i)_2}{\overset{SiMe_2CH_2Cl}{<}}
\xrightarrow[\text{(3 equiv.)}]{MeLi}$$

(7.11)

$$\underset{H}{\overset{n\text{-}C_6H_{13}}{>}}C{=}C\underset{CH_2SiMe_3}{\overset{Li}{<}}
\xrightarrow[\text{CuBr (cat.)}]{H_2C{=}CHCH_2Br}
\underset{H}{\overset{n\text{-}C_6H_{13}}{>}}C{=}C\underset{CH_2SiMe_3}{\overset{CH_2CH{=}CH_2}{<}}$$

(84%)

Hydroalumination of 1-chloro-1-alkynes by $LiAlH_4$ proceeds stereoselectively in a *trans*-fashion, placing the aluminium atom in the 1-position (equation (7.12)).[119] The α-chloroalkenylalanate products are moderately stable at 0°C and furnish *trans*-1-chloro-1-alkenes in high yield on methanolysis:

$$n\text{-}C_6H_{13}C{\equiv}CCl \xrightarrow{LiAlH_4}
\underset{H}{\overset{n\text{-}C_6H_{13}}{>}}C{=}C\underset{Cl}{\overset{AlH_3Li}{<}}
\xrightarrow{MeOH}
\underset{H}{\overset{n\text{-}C_6H_{13}}{>}}C{=}C\underset{Cl}{\overset{H}{<}}$$

(7.12)

(80%)

While hydroalumination of a 1-chloro-1-alkyne by a lithium or sodium trialkylaluminium hydride also proceeds with the identical regio- and stereoselectivity, subsequent warming of this reaction mixture in the presence of NaOMe, followed by hydrolysis, selectively yields a *cis*-disubstituted alkene (equation (7.13)):[120]

$$n\text{-}C_6H_{13}C{\equiv}CCl \xrightarrow{NaAlHEt_3}
\underset{H}{\overset{n\text{-}C_6H_{13}}{>}}C{=}C\underset{Cl}{\overset{\overset{Et}{|}AlEt_2Na}{<}}
\xrightarrow{NaOMe}
\underset{H}{\overset{n\text{-}C_6H_{13}}{>}}C{=}C\underset{\underset{OMe}{|}AlEt_2Na}{\overset{Et}{<}}$$

$$\xrightarrow{H_3O^+}
\underset{H}{\overset{n\text{-}C_6H_{13}}{>}}C{=}C\underset{H}{\overset{Et}{<}}$$

(7.13)

(74%)

This reaction proceeds *via* migratory insertion of an alkyl group from aluminium to the adjacent vinylic carbon atom with concomitant loss of chloride. The inherent stability of the intermediate 1-chloro-1-alkenyl-alanate at 0°C enabled crossover reactions to be conducted, and these experiments clearly demonstrate the intramolecular nature of the rearrangement. The intermediate methoxyalanate in equation (7.13) also undergoes stereospecific iodination to afford comparable yields of the corresponding (E)-vinyl iodide. The wide variety of means available for functionalisation of alkenylaluminiums (Scheme 7.1), including iodination, makes this methodology very useful for the stereodefined synthesis of trisubstituted olefins as well.

Whereas migratory insertion processes are commonplace in organoboron chemistry, this type of 1,2-migration reaction has seldom been observed for related organoaluminiums.[5,10,13] Until recently, the only reported examples in which organoaluminiums appear to have undergone migratory insertion, other than that shown in equation (7.13), are the reaction of organoalanes with diazomethane to yield the homologous alanes[121–123] and the reaction of lithium chloropropargylide with $(Hex)_3Al$ to produce a mixture of the respective propargylic and allenic alanes.[8] The discovery that a facile migratory insertion process takes place during the reaction of $LiCH(Cl)SiMe_2Ph$ with R_3Al, yielding the alkylated RCH_2SiMe_2Ph, suggests that this reaction pathway could be more common for organoaluminiums than was previously believed.[124] As shown in equation (7.14), application of the reaction to an alkenylalane substrate results in selective migration of the vinyl group from aluminium to yield the respective allylic alane(s). Hydrolysis of the reaction mixture then affords the vinylsilane product in high yield.

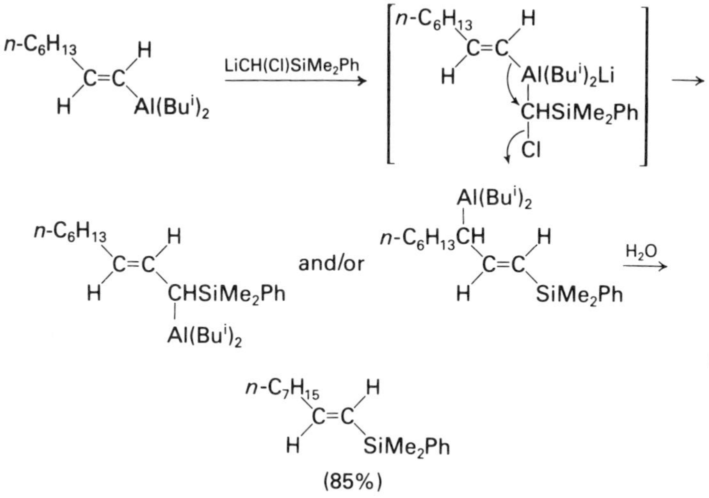

(7.14)

(85%)

Finally, lithium dihydropyranyl- and dihydrofuranylalanates, prepared from reaction of 2-lithiodihydropyran or 2-lithiodihydrofuran with R_3Al, react with both aldehydes and epoxides in the presence of $Et_2O \cdot BF_3$ by migration of an organic group from aluminium to the adjacent α-carbon with concomitant trapping of the electrophile at the β-position.[125] Thus, reaction of lithium dihydropyranyltrimethylalanate with benzaldehyde and $Et_2O \cdot BF_3$ produces the methylated tetrahydropyranyl derivative which must form by means of a migratory insertion process (equation (7.15)).

$$(7.15)$$

Because of competitive side-processes, such as the displacement reaction and polymerisation,[9–11] the carboalumination of alkenes has not evolved into a generally useful synthetic procedure. In those instances when single-stage alkene carboalumination products are realised, they are typically obtained as the corresponding vinylidene olefins which result from a reaction-terminating dehydroalumination step. Consequently, these products no longer possess carbon–aluminium bonds for further elaboration. As with the corresponding hydroalumination reaction, alkene carboalumination can be catalysed by use of titanium or zirconium compounds. For example, the reaction of 1-alkenes with 2 equivalents each of R_3Al and Cp_2TiCl_2 at room temperature affords the respective vinylidene olefins in fair to good yields.[126] These reaction conditions can be quite chemoselective, as the presence of an ester group is readily tolerated (equation (7.16)).

$$(7.16)$$

Catalytic amounts of zirconium ($Zr(OBu)_4$ or Cp_2ZrCl_2) or titanium ($TiCl_4$ or $Ti(OBu)_4$) compounds have also been used in conjunction with R_3Al or R_2AlCl to produce 1,1-disubstituted alkenes from terminal olefins.[127–129] These latter reactions may also be carried out, however, using R_2Mg or $RMgX$ reagents in place of the organoalanes with the difference that they now deliver carbometallation-derived products which bear an active carbon–metal bond.

The 1,5-dienyl moiety can be efficiently transformed into the corresponding cyclopentyl derivative via reaction with Bu_2^iAlH.[130–133] Accordingly, hydroalumination of one of the olefinic groups of 1,5-hexadiene with Bu_2^iAlH at 70°C is followed by an intramolecular carboalumination of the other alkene linkage to produce the cyclopentylmethylaluminium derivative. Hydrolysis of the reaction mixture then furnishes methylcyclopentane in nearly quantitative yield (equation (7.17)):[130,131]

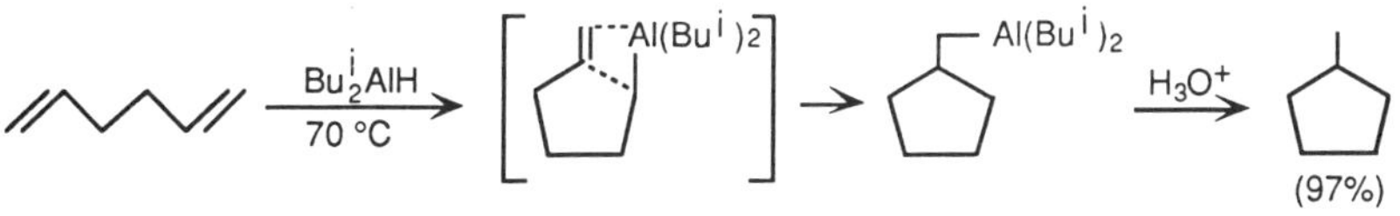

$$(7.17)$$

(97%)

Ring closure on hydroalumination of the first double bond is significantly faster than addition of Al–H to the second double bond, even at lower reaction temperatures or in the presence of an excess of Bu_2^iAlH. The reaction is also effective for the preparation of 6-member ring compounds, albeit in lower yields than those realised for the cyclopentyl derivatives.[134] In addition, α,ω-dienes can be transformed into the respective methylene-cyclopentanes or -cyclohexanes by using catalytic amounts of Bu_2^iAlH.[135] For instance, catalytic hydroalumination-tandem cyclic carboalumination of 5-methylene-1,8-nonadiene affords the bicyclic spiro derivative shown in equation (7.18) in good yield.

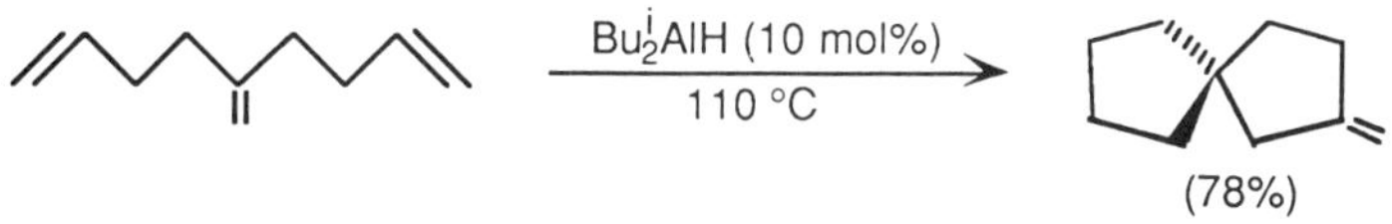

$$(7.18)$$

(78%)

The fact that the terminal alkyne group can be selectively hydro-aluminated to the respective di- and trialuminoalkanes through the proper choice of reaction solvent allows this cyclic alkene carboalumination procedure to be extended to 1-en-5-ynes as well.[136–138] Moreover, the methylcyclopentyl derivatives obtained from this enyne cyclisation reaction possess two aluminium atoms for potential elaboration. Thus, reaction of 1-hexen-5-yne with 2 equivalents of Bu_2^iAlH in refluxing ether provides methylcyclopentane in 80% yield (equation (7.19)).[136] The positions of the two aluminium atoms were clearly demonstrated by oxidation with O_2 to a mixture of the corresponding cis- and trans-(2-hydroxycyclopentyl)-methanols.

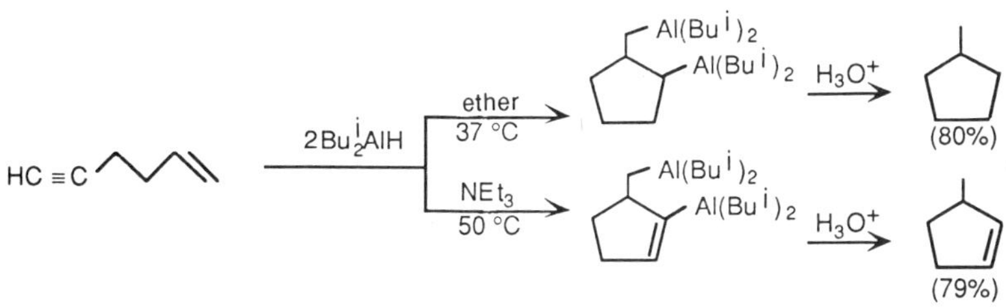

$$(7.19)$$

(80%)

(79%)

In a related reaction, hydroalumination of 1-hexen-5-yne with Bu_2^iAlH (2 equivalents) in triethylamine solvent gives 3-methylcyclopentene in good yield (equation (7.19)).[136] In this case, metallation of the triple bond initially occurs, followed by hydroalumination of the resultant alkynylalane to produce the geminal 1,1-dialuminoalkene functionality. Cyclic carbo-alumination of the remote double bond by the dialuminoalkene moiety then takes place to form the observed cyclopentenyl ring.

Controlled carboalumination of homoallylic alcohols is achieved by the reagent combination Me_3Al–$TiCl_4$ which produces the respective methylated olefins.[139–141] In this transformation, an olefinic hydrogen undergoes, in a formal sense, substitution by a methyl group to yield an olefin product of inverted geometry. The stereochemical inversion, as well as the synthetic potential of this methodology, is exemplified by the efficient stereoselective preparation of the homologated geraniol analogue depicted in equation (7.20):[141]

$$
\text{(structure)} \xrightarrow[\text{2) } H_3O^+]{\text{1) } Me_3Al,\ TiCl_4} \text{(structure)} \quad (7.20)
$$
$$
(80\%)
$$

Ethylation of homoallylic alcohols can also be accomplished using the Et_2AlCl–ROH/$TiCl_4$ reagent system, with fair to good yields of ethylated alkene products possible.[142] Interestingly, treatment of the tosylated 3-nonen-1-ol with Bu_3^iAl at room temperature quantitatively produces the alkylated cyclopropane (equation (7.21)):[143]

$$
\begin{array}{c} n\text{-}C_5H_{11} \\ H \end{array} C = C \begin{array}{c} H \\ \diagdown \diagup OTs \end{array} \xrightarrow[25\ ^\circ C]{Bu_3^i Al} \quad \begin{array}{c} Bu^i \\ n\text{-}C_5H_{11} \end{array} \triangleright\!\!\triangleleft \quad (7.21)
$$
$$
(\sim 100\%)
$$

The combination of a trialkylalane and an alkylidene iodide forms a highly efficient reagent system for the facile cyclopropanation of alkenes.[144] For instance, treatment of 1-dodecene at room temperature with equimolar amounts of Me_3Al and CH_2I_2 furnishes n-decyclopropane in virtually quantitative yield (equation (7.22)):

$$
n\text{-}C_{10}H_{21}CH = CH_2 \xrightarrow[CH_2I_2,\ 25\ ^\circ C]{Me_3Al} n\text{-}C_{10}H_{21} \triangleleft \quad (7.22)
$$
$$
(98\%)
$$

Substitution of higher trialkylalane homologues for Me_3Al does not alter the course of this cyclopropanation reaction, and use of ethylidene iodide affords the corresponding methyl cyclopropyl derivatives in comparable yield. Of the various R_3Al/CH_2I_2 combinations studied, that involving Bu_3^iAl furnishes the highest degree of selectivity in the cyclopropanation of various dienyl systems. It is noteworthy that the double bond remote from the hydroxyl group in geraniol is preferentially cyclopropanated by the

Bu^i_3Al/CH_2I_2 system. As such, the observed chemoselectivity complements that delivered by the Simmons–Smith reagent (equation (7.23)).[144]

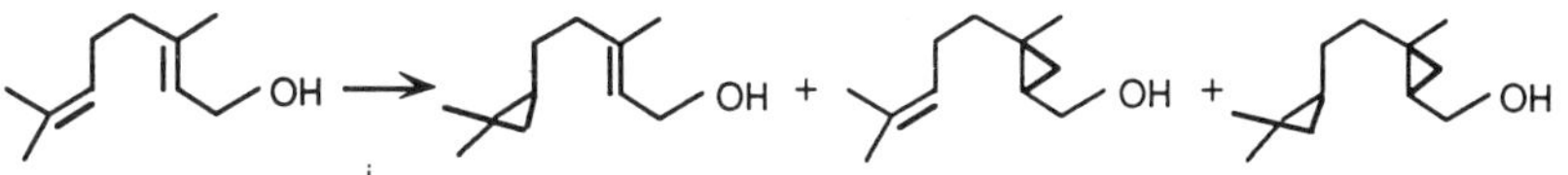

				(7.23)
Bu^i_3Al / CH_2I_2	76	1	4	
Et_2Zn / CH_2I_2	2	74	3	

Carboalumination of acetylene by a trialkylalane occurs readily in a highly stereoselective manner, affording the corresponding alkenylalane derived from a *cis*-addition of Al–R to the triple bond.[27,28] Although differing geometrically from the more commonly encountered (*E*)-alkenylalane, the reactivity of this (*Z*)-isomer closely parallels that of the (*E*)-isomer.[58,85,145] As an example, treatment of tri-*n*-octylalane with acetylene at 60°C in a hydrocarbon solvent produces the respective (*Z*)-alkenylalane which, on further reaction with paraformaldehyde in thf, furnishes (*Z*)-2-undecen-1-ol in high yield (equation (7.24)):[145]

$$HC{\equiv}CH \xrightarrow{(n\text{-}C_8H_{17})_3Al} (n\text{-}C_8H_{17})HC{=}CH(Al(C_8H_{17}\text{-}n)_2) \xrightarrow[2.\ H_3O^+]{1.\ (CH_2O)_n} (n\text{-}C_8H_{17})HC{=}CH(CH_2OH) \quad (7.24)$$

The carboalumination of acetylene has also been employed as a key step in the synthesis of (*Z*)-5-decen-2-one, a pheromone constituent of the bontebok.[146]

Carboalumination of substituted alkynes with trialkylalanes does not readily proceed in an uncatalysed fashion. For instance, heating Me_3Al with a terminal alkyne results predominantly in metallation to form the alkynylalane.[5,147] However, when this reaction is carried out in the presence of catalytic amounts of Cp_2ZrCl_2, a smooth carboalumination of the triple bond takes place under mild conditions to produce the corresponding 2,2-disubstituted alkenylalane (equation (7.25)):[80,148,149]

$$RC{\equiv}CH \xrightarrow[Cp_2ZrCl_2\ (cat.),\ 25°C]{Me_3Al} (R)(Me)C{=}CH(AlMe_2) \quad (7.25)$$

The addition of Me–Al to the alkynyl group occurs stereoselectively in a *cis*-manner, with the Al atom placed regioselectively at the terminal carbon (≥95%). At least in typical cases, the reaction has been shown mechanistically to involve a Zr-assisted direct carboalumination.[80,150,151] The C–Al bond in these 2,2-disubstituted alkenylalanes, together with that of the derived alanates, readily participates in the types of functionalisation reaction depicted in Scheme 7.1.[80,149] As a result of its highly stereo- and regioselective nature, the Zr-catalysed carboalumination of 1-alkynes with Me$_3$Al has been found to be ideally suited to the construction of terpenoids, including geraniol,[69] farnesol[76] and monocyclofarnesol.[75] Moreover, the same reaction has been utilised in the synthesis of such natural products as mokupalide,[88] brassinosteroids,[152,153] milbemycin,[154] verrucarins[155-158] and udoteatrial.[159] A variety of proximal heteroatom-containing functional groups are well tolerated under the reaction conditions to provide the carbo-alumination products of expected stereo- and regiochemistry.[80,160] In this connection, it is interesting to compare the regioselectivities derived from carbometallation of a heteroatom-containing 1-alkyne promoted by Zr and Ti compounds. As equation (7.26) illustrates, Cp$_2$ZrCl$_2$-catalysed carbo-alumination of 3-butyn-1-ol with an excess of Me$_3$Al proceeds with the typical direction of Me–Al addition,[160] whereas the analogous TiCl$_4$-catalysed reaction takes place with the opposite regioselectivity.[161]

$$
\text{HOCH}_2\text{CH}_2\text{C}\equiv\text{CH}
\quad
\begin{array}{c}
\xrightarrow[\substack{\text{Cp}_2\text{ZrCl}_2\\ \text{(cat.)}}]{\text{Me}_3\text{Al}} \xrightarrow{\text{H}_3\text{O}^+} \quad
\begin{array}{c}
\text{HOCH}_2\text{CH}_2 \quad\quad\ \text{H} \\
\diagdown\quad\ \diagup \\
\text{C=C} \\
\diagup\quad\ \ \diagdown \\
\text{Me}\quad\quad\quad \text{H} \\
(85\%)
\end{array}
\\[2em]
\xrightarrow[\text{TiCl}_4\text{ (cat.)}]{\text{Me}_3\text{Al}} \xrightarrow{\text{H}_3\text{O}^+} \quad
\begin{array}{c}
\text{HOCH}_2\text{CH}_2 \quad\quad\ \text{H} \\
\diagdown\quad\ \diagup \\
\text{C=C} \\
\diagup\quad\ \ \diagdown \\
\text{H}\quad\quad\quad \text{Me} \\
(73\%)
\end{array}
\end{array}
\qquad (7.26)
$$

Utilisation of alkynylalanes possessing β-hydrogens in the Zr-catalysed carboalumination of terminal acetylenes leads to decreased regio-selectivities and competing hydroalumination.[80,148,151] This latter side-reaction can be most conveniently controlled by use of the R$_2$AlCl/Cp$_2$ZrCl$_2$ reagent system, although regioisomeric product mixtures persist.[150,151] Disubstituted alkynes also undergo Zr-catalysed carboalumination to yield the corresponding stereodefined alkenylalanes. For example, reaction of 5-decyne with Me$_3$Al/Cp$_2$ZrCl$_2$ at 50°C, followed by hydrolysis of the reaction mixture, affords (Z)-5-methyl-5-decene in 89% yield.[80,151] Interestingly, related carboalumination reactions of silyl-substituted alkynes with R$_2$AlCl and R$_3$Al, promoted by Cp$_2$ZrCl$_2$,[162,163] Cp$_2$TiCl$_2$[162,164] or R$_2$Mg,[165] are completely regioselective transformations, placing the aluminium atom at the silicon-bearing carbon. Unfortunately, however, these reactions are generally not highly stereoselective processes.

The allylalumination of 1-alkynes is also promoted by Cp$_2$ZrCl$_2$ and provides an essentially quantitative yield of cis-monoallylated products.[166] In contrast to the high degree of regioselectivity and moderate stereoselectivity observed for allylzincation of 1-silyl-1-alkynes,[167,168] the Zr-catalysed allylalumination of 1-alkynes is >98% stereoselective but only c. 75% regioselective with respect to placement of aluminium at the terminal position of the triple bond. Notably, treatment of the regioisomeric mixture of allylalumination adducts with a deficient quantity of iodine (0.6–0.7 equivalent) results in the selective iodination of the predominant regio-isomer to furnish essentially pure 1-iodo-1,4-pentadiene derivatives. As an illustration, allylalumination of 1-trimethylsilyl-1,4-pentadiene with allyl-diisobutylalane and Cp_2ZrCl_2 at 25°C quantitatively produces a 65:35 mixture of regioisomeric alkenylalanes. Allylalumination of the trimethyl-silylethynyl moiety does not compete with reaction at the terminal triple bond. Treatment of this reaction mixture with 0.6 equivalent of iodine exclusively produces the terminally iodinated alkenyl product shown in equation (7.27) in 53% isolated yield based on the starting diyne:

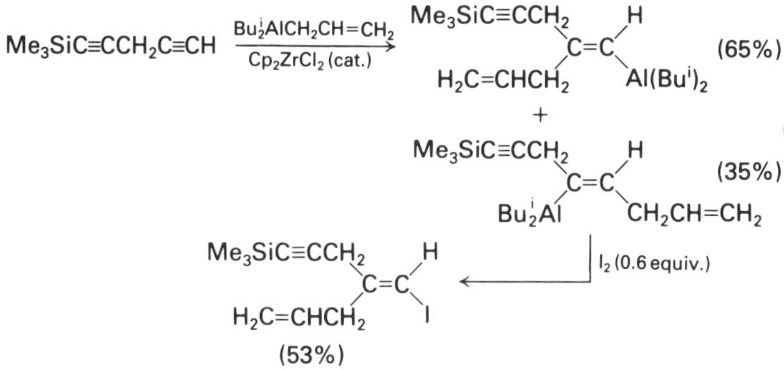

The corresponding Zr-promoted benzylalumination of 1-alkynes by tri-benzylalane also occurs readily and affords high yields of regioisomeric alkenylalane products with a similar distribution.[166]

Carboalumination of acetylenes by alkenylalanes to form the corresponding dienylalanes has been observed but, at present, is of rather limited synthetic value.[13] For example, reaction of Bu_2^iAlH with 3-hexyne in a 1:2 molar ratio cleanly yields the dienylalane derived from sequential hydroalumination and alkenyl-carboalumination (equation (7.28)):

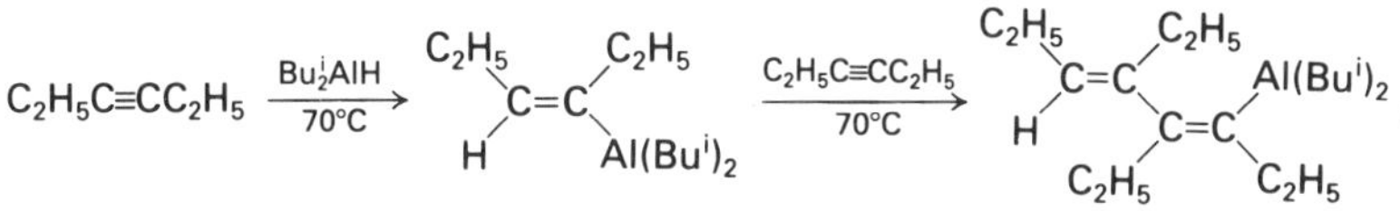

Although unoptimised, alkenyl-carboalumination of a terminal alkyne has been accomplished with the aid of Zr-catalysis. Thus, carboalumination of 1-octyne at room temperature by (E)-1-octenyldiisobutylalane in the presence of Cp_2ZrCl_2 proceeds regioselectively to produce, after proton-olysis, the respective diene product in modest yield (equation (7.29)).[151,169] Intramolecular alkenyl-carboalumination of the remote triple bond takes place upon hydroalumination of 1-trimethylsilyl-1,7-dodecadiyne with Bu_2^iAlH in hexane, and subsequent heating at 65°C, to furnish the exocyclic 7-member ring diene product (equation (7.30)).[169] A key to this trans-formation is the chemoselective hydroalumination of the silyl-substituted

$$
\begin{array}{c}
\underset{H}{\overset{n\text{-}C_6H_{13}}{>}}C=C\underset{Al(Bu^i)_2}{\overset{H}{<}} + n\text{-}C_6H_{13}C\equiv CH \xrightarrow[25°C]{Cp_2ZrCl_2} \xrightarrow{H_3O^+} \underset{H}{\overset{n\text{-}C_6H_{13}}{>}}C=C\underset{\underset{n\text{-}C_6H_{13}}{|}}{\overset{H}{<}}C=C\underset{H}{\overset{H}{<}}
\end{array} \quad (7.29)
$$

(25%)

alkynyl group in the presence of the other triple bond. Furthermore, the non-stereoselective nature of the hydroalumination of alkynylsilanes in hydrocarbon solvents, as well as the configurational instability of the resultant α-silylalkenylalane adducts,[115] provides a means for the intermediate alkenylalanes to achieve the required geometry for cyclic alkenylcarboalumination to occur.

$$
n\text{-}C_4H_9C\equiv C(CH_2)_4C\equiv CSiMe_3 \xrightarrow[\text{Hexane, 25 °C}]{Bu_2^iAlH} \xrightarrow[\text{2) } H_3O^+]{1)\ 65\ °C}
$$

(41%)

$$(7.30)$$

The reaction of an equimolar mixture of 1-octene and either 5-decyne or 1-trimethylsilyl-1-octyne with 1 equivalent of Bu_2^iAlH in hexane results in selective hydroalumination of the respective acetylene.[163] Interestingly, the opposite chemoselectivity is exhibited by the Zr-catalysed hydroalumination. Hence, treatment of a 1:1 mixture of 1-octene and 1-trimethylsilyl-1-octyne with Bu_3^iAl (1 equivalent) and Cp_2ZrCl_2 (10 mol%) results in the nearly exclusive hydroalumination of the alkene, leaving the silylacetylene virtually untouched.[163] Extending the Zr-catalysed hydroalumination to a single substrate containing both the terminal alkenyl and silylacetylenic groups causes the selective hydroalumination of the alkenyl moiety to be followed by a cyclic carboalumination of the triple bond. For example, treatment of 6-trimethylsilyl-1-hexen-5-yne with 1 equivalent of Bu_3^iAl in the presence of 10 mol% of Cp_2ZrCl_2 in 1,2-dichloroethane solvent at room temperature cleanly produces, after protonolysis, (trimethylsilylmethylene)cyclopentane in 85% yield (equation (7.31)).[163] Treatment of the same reaction mixture with iodine furnishes the corresponding vinyl iodide in 68% yield, confirming the presence of the C–Al bond in the cyclised intermediate (equation (7.31)).

$$
Me_3SiC\equiv C \diagup\!\diagdown\!\diagup\!\diagdown \xrightarrow[Cp_2ZrCl_2\ (cat.)]{Bu_3^iAl} \xrightarrow{E^+} \underset{E}{\overset{Me_3Si}{>}}C=\!\!< \bigcirc \quad (7.31)
$$

E = H (85%), I (68%)

The process can also be carried out with 7-trimethylsilylhept-1-en-6-yne to prepare (trimethylsilylmethylene)cyclohexane in comparable yield under mild reaction conditions. Extension of this reaction to 6-trimethylsilyl-4-methyl-1-hexene affords, through the lack of configurational stability of

α-silylalkenylalanes in non-polar solvents,[115] approximately equal amounts of the E and Z isomeric methylated (trimethylsilylmethylene)cyclopentanes. This isomerisation reaction can be advantageously utilised, however, to effect 'chelation-controlled' cyclic carboalumination and provide an exocyclic olefin product of high stereochemical purity. Thus, reaction of the trimethylsilyl-substituted enynyl alcohol shown in equation (7.32) with 2 equivalents of Bu^i_3Al and a catalytic amount of Cp_2ZrCl_2 at 60°C produces, after protonolysis, the (E)-methylenecyclopentanol product in 82% yield and 97% isomeric purity.[163]

The Zr-catalysed carboalumination of 1-trimethylsilyl-4-halo-1-butyne, or the corresponding tosylate, results in an unstable alkenylalane intermediate which undergoes cyclisation to afford 1-trimethylsilyl-2-methylcyclobutene in high yield (equation (7.33)):[170,171]

The transformation involves participation of the alkene π-electrons in the cyclisation step, with the stereochemistry of the intermediate α-silylalkenyl-alane demonstrated to be unimportant. The synthetic utility of this cyclobutene synthesis is exemplified by its use in the preparation of ($\pm$)-grandisol.[170,171] It is noteworthy that the reaction can be extended to the preparation of 3- and 6-membered rings as well, but is not applicable to the preparation of 5-membered ring systems. These results are in accord with both the existence of a π-cyclisation mechanism and Baldwin's cyclisation rules.[172-174] Hydroalumination of both 1-trimethylsilyl-4-bromo-1-pentyne and 1-trimethylsilyl-3-methyl-4-bromo-1-butyne with Bu^i_2AlH affords the identical product, 1-trimethylsilyl-3-methylcyclobutene, on protonolysis (equation (7.34)).[171] Therefore, the synthesis of cyclobutenes from 1-tri-methylsilyl-4-halo-1-alkynes is highly regioselective but is non-regiospecific.

$$(7.34)$$

Cyclobutene products are also derived from the reaction of trialkylalanes with homopropargyl tosylates. Thus, treatment of 4-nonyn-2-ol tosylate with Bu_3^iAl quantitatively produces a 40:60 mixture of the corresponding alkylated cyclobutene and acetylene products (equation (7.35)).[143]

$$(7.35)$$

The Zr-catalysed carboalumination of 5-iodo-1-pentynyldimethylalane is followed by cycloalkylation to furnish the respective cyclopentenylalane intermediate which, upon iodinolysis, produces 2-methyl-1-iodocyclo-pentene (equation (7.36)).[175] In contrast to the π-type cyclisation reaction depicted in equation (7.33), formation of the 5-membered ring in this case is believed to occur via a σ-cyclisation process involving the 1,1-dialumino-alkene intermediate.

$$(7.36)$$

Stereodefined exocyclic alkanes can be obtained in excellent yield from the Pd-catalysed reaction of ω-(o-iodoaryl)alkynes with organoaluminiums. For example, tandem carbopalladation/cross-coupling of 5-(o-iodophenyl)-2-pentyne with Ph_3Al in the presence of a catalytic amount of $Pd(PPh_3)_4$ results nearly exclusively in the formation of the phenylated cyclo-pentylidene product (equation (7.37)).[176] These results demonstrate that cyclic carbopalladation of the triple bond by the arylpalladium iodide inter-mediate can compete favourably relative to its direct cross-coupling with Ph_3Al.

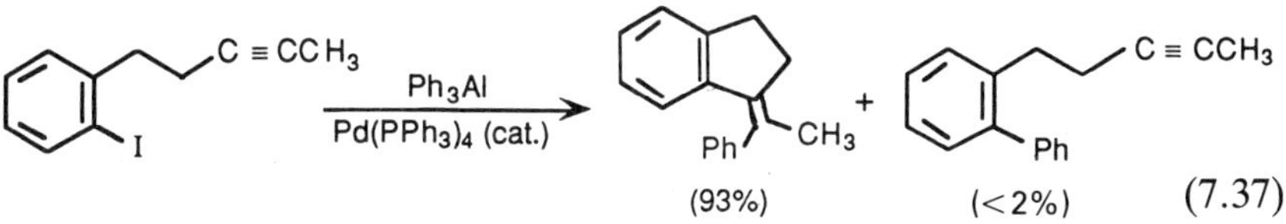

(7.37)

7.1.3 Selective reduction and alkylation of carbonyl compounds

Organoaluminium compounds readily participate in addition reactions to aldehydes, ketones, carboxylic acids and their derivatives, and nitriles.[4,5,10,11] Depending on their nature, the reaction of alkylaluminiums with carbonyl compounds can proceed along several pathways, including alkylation, reduction and enolisation. An important difference between organoalanes and the corresponding organolithium or Grignard reagents in their reaction with carbonyl compounds lies with the inherent oxygeno-philicity of alanes which can lead to initial complexation of the carbonyl oxygen.[10,14,177] These complexes may be fairly long-lived, depending on the reaction conditions and the nature of the particular alane reagent, and can result in chemistry quite different from that observed for organolithium or Grignard reagents. For example, treatment of citronellal at $-78°C$ in hexane with Me_3Al produces a stable $1:1$ complex which, on warming to room temperature, affords the acyclic $2°$ alcohol in high yield (equation (7.38)).[178] Carrying out the same reaction in 1,2-dichloroethane solvent leads to cyclisation without alkylation to produce isopregol, whereas addition of excess Me_3Al to the citronellal–Me_3Al complex in CH_2Cl_2 yields the product derived from both cyclisation and alkylation.[178]

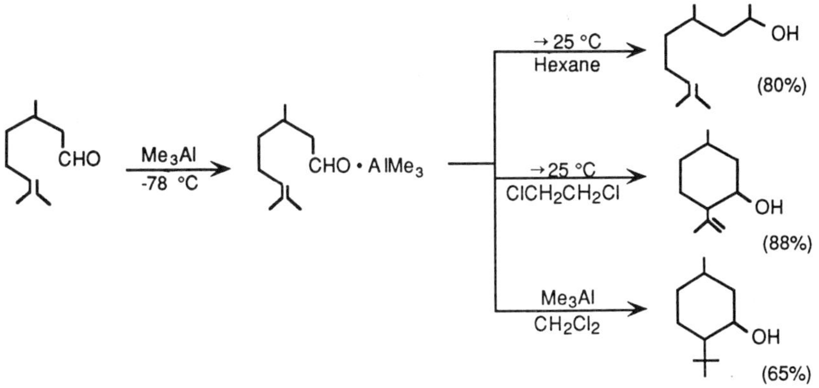

(7.38)

Excellent selectivity in carbonyl alkylation reactions can be achieved through prior coordination of the carbonyl oxygen atoms by highly hindered organoaluminium reagents. Thus, addition of a reagent such as MAD or MAT to a ketone or aldehyde, followed by reaction with an organolithium

MAD : R = Me
MAT : R = Bu^t

or Grignard reagent, furnishes a product mixture that usually differs in stereoselectivity from that derived *via* direct reaction with the nucleophile alone.[14,178,179] For instance, reaction of 3-methylcyclohexanone with Bu^{n}MgBr provides the axial and equatorial alcohols in the proportions 79 : 21, respectively, while addition of MAD to the same ketone prior to reaction with Bu^{n}MgBr results in nearly complete selectivity confined to the equatorial alcohol (equation (7.39)).[179]

$$(7.39)$$

The carbonyl oxygen is presumably coordinated by MAD in a manner placing the bulky aluminium reagent in the sterically more accessible equatorial position. Therefore, attack of the nucleophile takes place from the axial direction and, as such, the overall observed stereochemistry is typically the opposite of that derived from use of the nucleophile alone. This approach has also proved to be useful for the stereoselective alkylation of steroidal ketones.[14,177]

The bulky aluminium reagents MAD and MAT are also effective at promoting stereoselection in the alkylation of α-chiral aldehydes.[14,177,179] In fact, this procedure complements existing methodology as it generally affords a high degree of anti-Cram selectivity in the addition of nucleophiles to typical α-chiral aldehydes. For example, sequential treatment of α-phenylpropionaldehyde with MAT and MeMgI at −78°C produces a mixture of Cram and anti-Cram alcohols in 96% yield and in the proportions 7 : 93, respectively (equation (7.40)).[179] On the other hand, direct reaction of the same alcohol with MeMgI furnishes the Cram and anti-Cram products in the proportions 72 : 28.

The reduction of a specific carbonyl group in the presence of others is often a critical factor in syntheses utilising multifunctional molecules. However, while most conventional methodology entails selective reduction of the least hindered ketone group, use of MAD, in conjunction with a

	Cram		Anti-Cram	
Additive				
None	72	:	28	(64%)
MAT	7	:	93	(96%)

$$(7.40)$$

reducing agent, effects the preferential reduction of the more hindered ketone site.[14,180] Thus, reduction of an equimolar mixture of acetophenone and pivalophenone with Bu_2^iAlH alone yields 1-phenyl-1-ethanol and 2,2-dimethyl-1-phenyl-1-propanol in the proportions 2.6 : 1. Conversely, use of 1 equivalent of MAD, in combination with the same reducing agent, reverses the ratio of these alcohol products to 1 : 10. Utilisation of 2 equivalents of MAD further increases this same ratio to 1 : 16 (equation (7.41)).[180]

Additive				
None	2.6	:	1	(99%)
MAD (1 equiv.)	1	:	10	(66%)
MAD (2 equiv.)	1	:	16	(51%)

$$(7.41)$$

Similarly, the use of MAD together with Bu^tMgCl forms an efficient method for the selective reduction of cyclohexanones to the corresponding equatorial alcohols.[177,181] This procedure also provides a means for the chemoselective reduction of cyclic ketones in the presence of acyclic ketones.

The reaction of organolithium reagents with α,β-unsaturated ketones typically results in 1,2-addition to the carbonyl group to furnish the respective allylic alcohols. As an example, treatment of 6-methyl-2-cyclohexenone with MeLi at $-78°C$ produces the 1,2-adduct in 75% yield.[177,182] In contrast, complexation of the same enone with MAD, followed by treatment with MeLi under otherwise comparable reaction conditions, results exclusively in conjugate addition of the nucleophile to afford 2,5-dimethylcyclohexanone in 68% yield (equation (7.42)).[177,182] None of the allylic alcohol derived from 1,2-addition is observed under these conditions.

Without MAD	(-)	75%
With MAD	68%	0%

$$(7.42)$$

It is interesting that reaction of the same enone with a pre-mixed solution of MeLi and MAD gives almost exclusively the 1,2-addition product. This result supports the notion that initial complexation of MAD to the carbonyl oxygen, and not prior *ate* complexation of MAD with MeLi, is a key step leading to the observed 1,4-addition mode of reaction.

The conversion of aromatic aldehydes to the corresponding methyl ketones can be accomplished in a simple one-step procedure by use of the complex derived from Me_3Al, (BHT)H (2,6-di-*t*-butyl-4-methylphenol) and ether. Hence, treatment of benzaldehyde with 1.5 equivalents of $AlMe_2$(BHT)(OEt$_2$) at room temperature in toluene produces aceto-phenone in quantitative yield on hydrolysis (equation (7.43)).[183]

$$PhCHO \xrightarrow[\text{2. } H_3O^+]{\text{1. } AlMe_2(BHT)(OEt_2)} \underset{(100\%)}{\overset{O}{\underset{Ph}{\parallel}}{Me}} \qquad (7.43)$$

The conversion of easily enolisable ketones into the respective methylidene olefins utilising conventional Wittig methodology is often complicated by the fact that these reagents can function as strong bases and deprotonate the carbonyl compounds. However, use of Tebbe's reagent, $Cp_2Ti(\mu\text{-}CH_2)(\mu\text{-}Cl)AlMe_2$, provides a method for the facile methylenation of ketones which are prone to enolisation. Thus, reaction of β-tetralone with Tebbe's reagent in THF affords an excellent yield of the methylidene olefin (equation (7.44)):[184]

$$(84\%) \qquad (7.44)$$

The reactive species in this transformation appears to be the titanium methylene fragment, $Cp_2Ti=CH_2$, and it reacts chemoselectively with a ketone carbonyl in the presence of an ester group. Extension of the procedure to α,α-disubstituted ketones results in the formation of the corresponding titanium enolates.[184] Tebbe's reagent is also effective for the conversion of ester groups into the corresponding vinyl ethers.[185] A related carbonyl olefination reaction which is specific to aldehyde functionality utilises the reagent combination CH_2I_2–Zn–$AlMe_3$.[186] This procedure allows for the chemoselective methylenation of an aldehyde moiety in the presence of a ketone group (equation (7.45)).

$$(96\%) \qquad (7.45)$$

7.1.4 Rearrangements

As noted in the Introduction (section 7.1.1), organoalanes are powerful Lewis acids which can also function as alkylating agents. Consequently, alanes are distinct from other typical Lewis acid catalysts since they often serve a dual role of inducing carbocation-mediated rearrangements, as well as terminating the same reaction with the transfer of an organic group from aluminium to the substrate. An example of this unique reactivity is illustrated by the organoaluminium-promoted Beckmann rearrangement of oxime sulfonates. Thus, treatment of the oxime tosylate shown in equation (7.46) with Pr^n_3Al converts it cleanly into the respective alkylated imine, which is then reduced *in situ* with Bu^i_2AlH into pulmiliotoxin C in an overall yield of 60%:[187]

$$(7.46)$$

Beckmann fragmentation of α-alkoxycycloalkanone oxime acetates also proceeds with organic group transfer from the organoaluminium reagent to yield the corresponding functionalized ω-nitriles. For instance, reaction of α-benzyloxycyclohexanone oxime acetate with 1-hexynyldiethylalane affords the respective alkynylated nitrile in good yield (equation (7.47)).[188]

$$(7.47)$$

While the Claisen rearrangement of allyl phenyl ethers can be carried out thermally, Lewis acids are often utilised to promote this same transformation under milder reaction conditions. Interestingly, the bulky aluminium reagent methylaluminium *bis*(4-bromo-2,6-di-*t*-butylphenoxide) induces the aromatic Claisen rearrangement to furnish a product distribution different from that generally obtained either thermally or from use of typical Lewis acid reagents. As an illustration, both the nature and proportions of prenylated phenol products derived from the Claisen rearrangement of phenyl prenyl ether promoted by this aluminium reagent differ significantly from those obtained under strictly thermal conditions (equation (7.48)):[189]

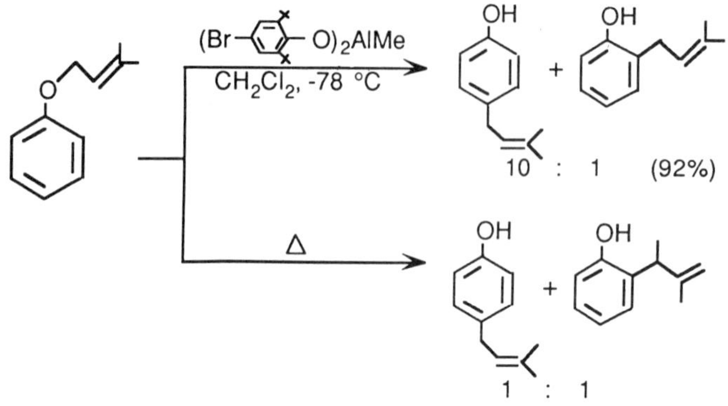

(7.48)

This same bulky organoaluminium reagent also induces Claisen rearrangement of *bis*(allyl) vinyl ethers with regiochemical control contrasting that observed for the corresponding thermal rearrangement.[190] While these aliphatic Claisen rearrangements are not promoted by typical Lewis acid catalysts ($Et_2O \cdot BF_3$, $TiCl_4$, $SnCl_4$ or $ZnBr_2$), use of a trialkylalane not only elicits the desired [3,3]-sigmatropic rearrangement, but also transfers an organic group to the carbonyl function of the resulting aldehyde as well.[191] For example, treatment of (*E*)-1-octenyldiisobutylalane at room temperature with 2-cyclohexenyl vinyl ether results in Claisen rearrangement, followed by selective transfer of the alkenyl moiety from aluminium to the newly generated aldehyde group to afford the (*E*)-allylic alcohol product (equation (7.49)):[191]

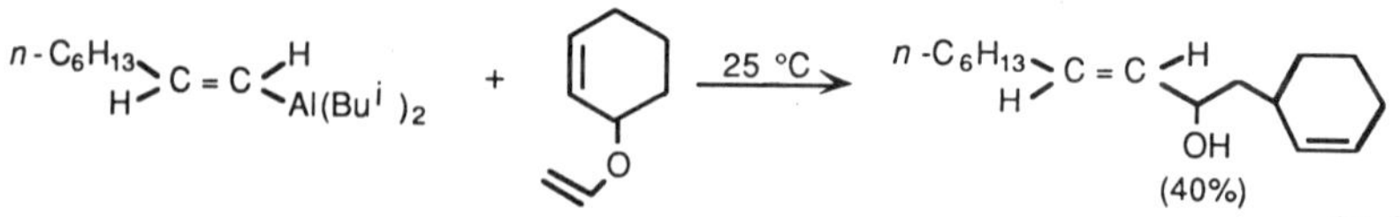

(7.49)

2-Vinyloxolane enol ethers also undergo organoaluminium-promoted Claisen rearrangement with alkylation to yield the corresponding 7-membered carbocyclic products. As shown in equation (7.50), alkynyl-alane-induced Claisen rearrangement of the cyclic enol ether furnishes the alkynylated cycloheptanol product in high yield.[192]

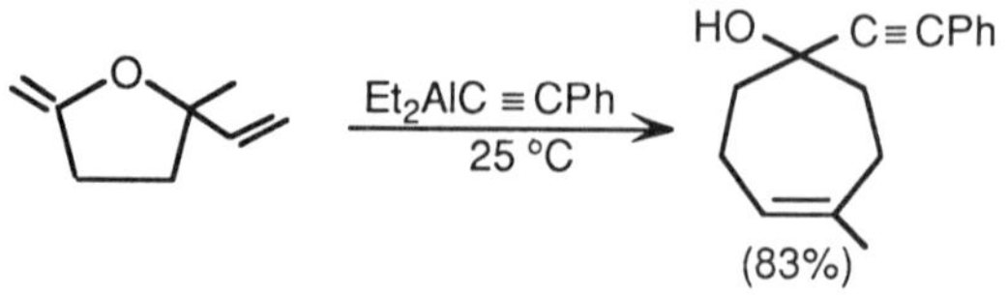

(7.50)

3-(Trimethylsilyl)methylcyclohexyl mesylates undergo cationic rearrangement upon treatment with an excess of dimethylaluminium triflate to provide hydrocarbon products derived from successive rearrangement of hydride and carbon groups.[193] The driving force behind these rearrangements is formation of the stabilised β-silyl cationic species. Thus, reaction of the bicyclic mesylate shown in equation (7.51) with Me$_2$AlOTf at low temperature yields predominantly the *trans*-fused methylenedecalin product *via* rearrangement of hydride and the Me$_3$SiCH$_2$- group.

$$(7.51)$$

The benzylic alcohol shown in equation (7.52) undergoes an interesting and synthetically useful rearrangement upon treatment with AlCl$_3$ to produce the 4-hydroxy-2,3-dimethylindene in high yield.[194] Notably, Lewis acids such as Et$_2$O · BF$_3$, SnCl$_4$ and TiCl$_4$ did not provide any of this indene product under comparable reaction conditions. This rearrangement may be viewed as proceeding *via* formation of the phenolic-stabilised benzylic carbocation followed by 1,2-methyl shift, loss of proton to yield the styrene analogue, and, finally, intramolecular Friedel–Crafts alkylation of the aromatic nucleus by the resultant allyl chloride moiety.

$$(7.52)$$

Cyclisation of γ,δ-unsaturated cycloheptenyl ketones to the corresponding mixture of *cis*- and *trans*-fused bicyclic cyclopentanones can be effected in high yield by treatment with MeAlCl$_2$ at room temperature (equation (7.53)).[195] The transformation is highly dependent on the nature of the aluminium reagent used, with MeAlCl$_2$ affording superior results to either AlCl$_3$ or EtAlCl$_2$. The reaction can be interpreted as proceeding *via* cyclisation followed by hydride and methyl group shifts.

$$(7.53)$$

An aluminium-promoted ring contraction in the cedrane series is observed upon treatment of the tricyclic allylic alcohol substrate in equation (7.54) with an excess of Bu_2^iAlH.[196] Thus, hydroalumination of the allylic alcohol moiety is followed by ring contraction to yield the respective tricyclic terminal alkene. Hydroalumination of the olefinic group in this species takes place under the action of the Bu_2^iAlH remaining in the reaction mixture and the sesquiterpene product, (8βH)-cedrane, is then liberated by protonolysis.

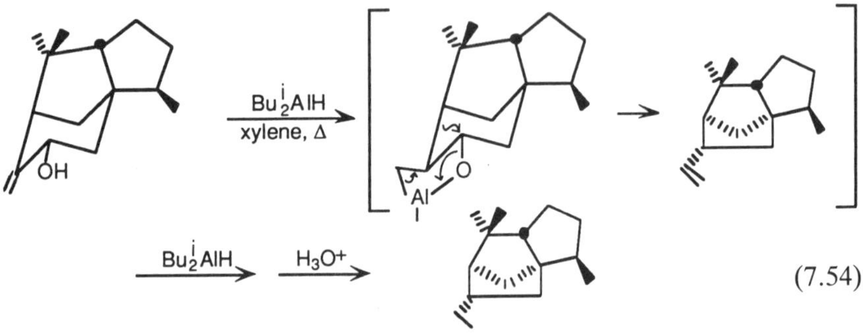

$$(7.54)$$

7.1.5 Asymmetric transformations

The concept illustrated in section 7.1.3 concerning the diastereoselective activation of carbonyl groups by MAD or MAT can be extended to utilise bulky, chiral organoaluminium reagents for the enantioselective activation of the carbonyl group. Optically pure binaphthol-based organoaluminium reagents of the types I and II have been synthesised and shown to promote various asymmetric transformations, including the induction of asymmetry in the hetero-Diels–Alder reaction.[14,177,197,198]

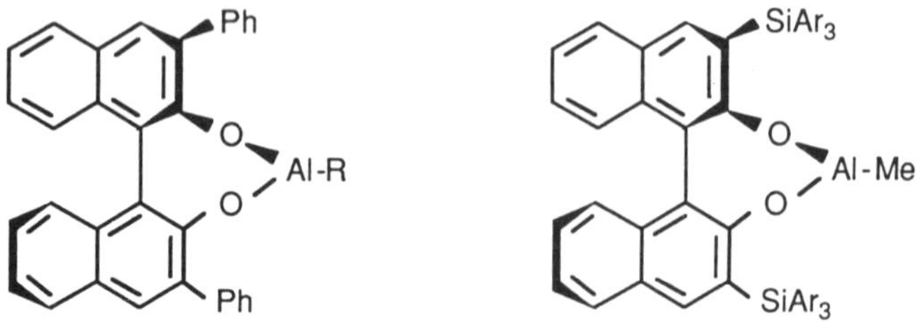

I

II

For example, catalysts of type I promote the hetero-Diels–Alder reaction between the dienyl silyl ether shown in equation (7.55) and benzaldehyde to afford, after acidic hydrolysis, the ketone product in good yield and with excellent enantioselectivity:[177]

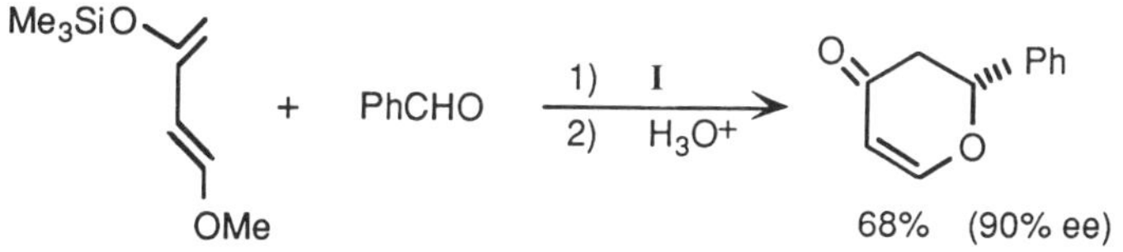

$$
\text{(7.55)}
$$

Me$_3$SiO, + PhCHO $\xrightarrow[\text{2)}\ \text{H}_3\text{O}^+]{\text{1)}\ \textbf{I}}$, OMe → (structure) 68% (90% ee)

While **I** is required in stoichiometric quantity,[14] derivatives of catalyst **II** efficiently promote the hetero-Diels–Alder reaction at levels of 5–10 mol%.[14,177,197,198] Moreover, improved yields and enantioselectivities are typically realised for **II** relative to those obtained using catalyst **I**. Thus, catalysis of the hetero-Diels–Alder reaction depicted in equation (7.56) by **II** gives, after treatment with trifluoroacetic acid, high levels of the *cis*-dihydropyrone derivative with excellent enantioselectivity for the major isomer.[197]

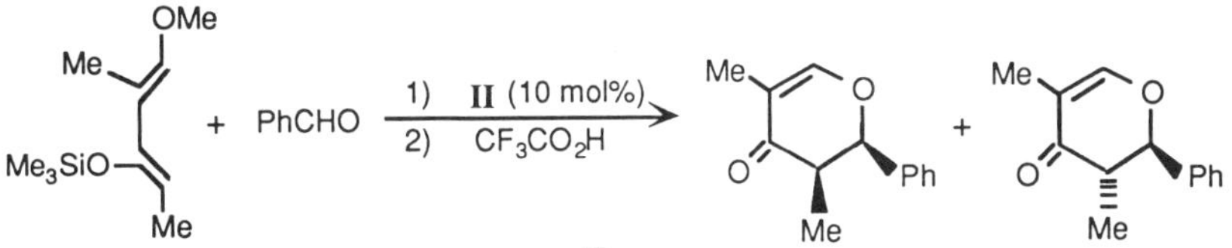

Me, OMe + PhCHO $\xrightarrow[\text{2)}\ \text{CF}_3\text{CO}_2\text{H}]{\text{1)}\ \textbf{II}\ (10\ \text{mol}\%)}$ (structures) + (7.56)

II
Ar = Ph   77% (95% ee)   7%
Ar = 3,5-xylyl   90% (97% ee)   3%

The chiral organoaluminium reagent **II** is also effective at catalysing asymmetric ene reactions of prochiral aldehydes with alkenes.[198,199] Hence, reaction of pentafluorobenzaldehyde with 2-(phenylthio)propene at −78°C using an equimolar amount of **II** (Ar = Ph) provides the corresponding homoallylic alcohol product in 90% yield and high (88% ee) enantio-selectivity.[199] Interestingly, the presence of 4-Å molecular sieves in the reaction mixture allows this transformation to be carried out with almost identical product yield and optical purity while requiring only 20 mol% of the same catalyst (equation (7.57)):

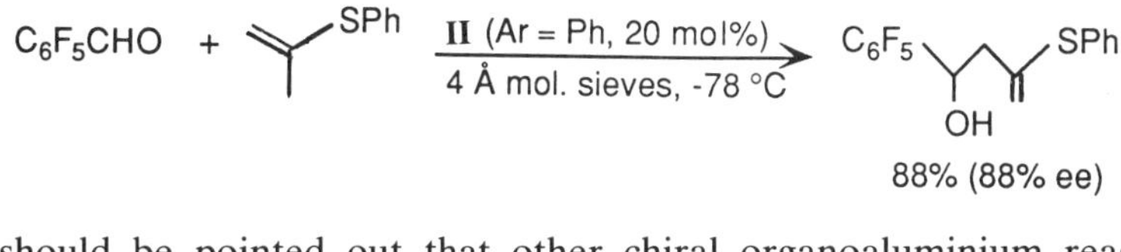

C_6F_5CHO + (structure)$\xrightarrow[\text{4 Å mol. sieves, -78 °C}]{\textbf{II}\ (\text{Ar = Ph, 20 mol}\%)}$ (structure) (7.57)

88% (88% ee)

It should be pointed out that other chiral organoaluminium reagents, including **I**, are generally ineffective at promoting this ene reaction and afford a racemic product.[199]

Chiral organoaluminium reagents which contain branching in the β-position offer an efficient means for reducing prochiral ketones in an enantioselective fashion. The absolute configuration and degree of enantiomeric purity found for the product depend strongly upon the nature of the chiral organoaluminium reducing agent. For instance, reduction of pivalophenone at room temperature with $tris[(S)$-2-methylbutyl]aluminium furnishes the respective (S)-alcohol product with an enantiomeric excess of 31% (equation (7.58)).[200] On the other hand, use of $tris(cis$-myrtanyl)-aluminium under comparable reaction conditions affords (R)-t-butyl phenyl carbinol with nearly complete enantioselectivity:[201]

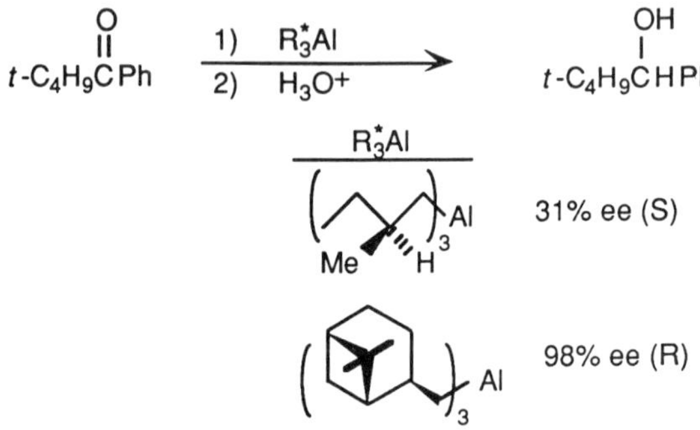

$$t\text{-}C_4H_9\overset{O}{\overset{||}{C}}Ph \quad \xrightarrow[\text{2) } H_3O^+]{\text{1) } R_3^*Al} \quad t\text{-}C_4H_9\overset{OH}{\overset{|}{C}}HPh$$

31% ee (S)   (7.58)

98% ee (R)

The manner of reduction of prochiral ketones by chiral organoaluminium reagents is complementary to that derived from the analogous organo-boranes. Thus, reduction of α-tetralone with myrtanylaluminium dichloride produces the (R)-alcohol product with excellent enantioselectivity,[202] whereas the cis-myrtanyl-9-BBN reagent delivers the alcohol of opposite configuration in slightly diminished stereoselectivity (equation (7.59)).[203]

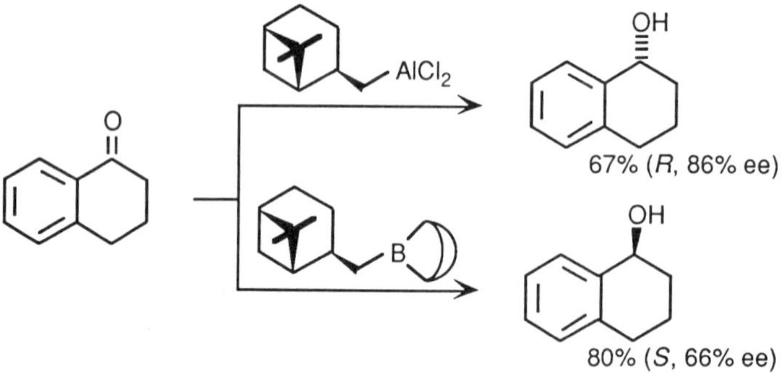

67% (R, 86% ee)

80% (S, 66% ee)   (7.59)

Asymmetric pinacol-type rearrangements are promoted by organo-aluminium reagents and can proceed with nearly complete enantio-selectivity.[204–211] As an example, the reductive pinacol-type rearrangement of chiral α-mesyloxy ketones occurs upon sequential treatment with Bu_2^iAlH and Et_3Al and yields a homoallylic alcohol product of high enantiomeric purity (equation (7.60)):[206]

77% (> 95% ee)

$$\tag{7.60}$$

Although an excess of Bu_2^iAlH is used in this transformation, it is interesting to note that 1,2-migration does not occur until Et_3Al is added to the reaction mixture. The aldehyde product of the pinacol-type rearrangement is then reduced upon formation by the residual Bu_2^iAlH present. From a mechanistic viewpoint, the retention of alkene geometry during rearrangement suggests a highly concerted nature for this step.

Organoaluminium reagents can selectively add to chiral α,β-unsaturated acetals in a 1,4- or 1,2-sense with a remarkably high degree of asymmetric induction.[14,212,213] To illustrate, reaction of the acetal shown in equation (7.61) with Me_3Al in 1,2-dichloroethane solvent affords predominantly the 1,4-addition product with an 88% excess of the (S)-enantiomer.[213] Conversely, carrying out the identical transformation in chloroform provides exclusively the 1,2-derived adduct of (R)-configuration in 85% yield and high enantiomeric purity:[213]

Solvent		
ClCH$_2$CH$_2$Cl	84% (S, 88% ee)	
CHCl$_3$	0%	85% (R, 88% ee)

$$\tag{7.61}$$

The chiral acetal starting materials utilised for this chemistry are quantitatively prepared *via* transacetalisation of the respective α,β-unsaturated aldehyde diethylacetal with N,N,N',N'-tetramethyltartaric acid diamide.[213] Furthermore, since both the (R,R)- and (S,S)-tartaric acid diamides are readily obtained in optically pure form, the method allows for the synthesis of both possible enantiomers. These 1,4- and 1,2-addition products can be readily transformed into the corresponding optically active α-substituted aldehydes, β-substituted aldehydes, α-substituted carboxylic acids and substituted allylic alcohols by the methods shown in equation (7.62).[14,213]

$$(7.62)$$

Chiral α,β-alkynyl acetals react with aluminium hydride-containing reagents such as Bu_2^iAlH or Br_2AlH to produce optically pure propargylic alcohols after removal of the chiral auxiliary.[215] For example, treatment of the chiral acetal prepared from 4-nonyn-3-one and $(-)-(2R,4R)-2,4$-pentanediol with Bu_2^iAlH in methylene chloride at $0°C$ furnishes predominantly the diastereomeric ether possessing the (R)-configuration at the propargylic position (equation (7.63)).[215] Separation of the minor diastereomeric component by column chromatography, followed by treatment of the purified major product with pyridinium chlorochromate (PCC) and potassium carbonate, then affords the optically pure (R)-4-nonyn-3-ol.

$$(7.63)$$

The alkylative cleavage of chiral acetals by certain dialkylaluminium aryloxide reagents can provide the corresponding secondary alcohols in an efficient and highly stereoselective fashion.[216] The process may be rationalised in terms of a selective coordination of the aluminium reagent to one of the acetal oxygen atoms, followed by a retentive transfer of an alkyl group from aluminium to the resulting oxocarbenium ion. Thus, reaction of dimethylaluminium pentafluorophenoxide with $(4R,6R)-4,6$-dimethyl-2-hexyl-1,3-dioxane at room temperature yields almost exclusively the corresponding alkylated acetal cleavage product derived from cationic retentive alkylation in 70% yield (equation (7.64)).[216] By comparison, a similar reaction using Me_3Al produces a 70:30 mixture of retentive and invertive alkylation products, respectively.

Me-AlL$_2$			
$Me_2AlOC_6F_5$	1	:	99 (70%)
Me_3Al	30	:	70

7.2 Organogallium and -indium compounds

7.2.1 Introduction

Until recently, organogallium and -indium compounds have been virtually ignored as reagents or intermediates in organic synthesis. While their reactivity is generally reminiscent of that of the synthetically versatile organoaluminiums, both organogallium and organoindium compounds exhibit important differences. For example, whereas aluminium and gallium form few stable compounds where the metal is not in the $+3$ oxidation state (see chapters 1, 3 and 6), cyclopentadienyl derivatives of In(I) are well known and entirely stable entities.[217] In this respect, therefore, indium is more reminiscent of thallium than its lighter congeners. Organogallium and -indium compounds also possess a decreased Lewis acidity compared to organoaluminiums, with the relative acidity being in the order Al > Ga > In. Consequently, organogallium and -indium compounds are less prone to the formation of electron-deficient, organic-bridged complexes than are the corresponding organoaluminiums. Although R_3In compounds tend to exhibit a greater reactivity towards inorganic and organic electrophiles than do similar R_3Ga derivatives, both are in general significantly less reactive than are their organoaluminium counterparts. Nonetheless, both organo-galliums and -indiums participate in protonolysis, oxidation, halogenation, carbonation and carbonyl addition reactions. As with the analogous aluminium hydride reagents, R_2GaH compounds also hydrometallate olefins and acetylenes to yield alkyl- and alkenylgallium derivatives, respectively. Since the general reaction patterns of organogalliums and -indiums have been well delineated previously,[9,217,218] the present section focuses on recently developed applications of these organometallics in organic synthesis.

7.2.2   Reactions with carbonyl compounds

The Barbier reaction of allylic halides with carbonyl compounds to produce homoallylic alcohols can be effectively promoted by gallium metal,[219] indium metal[220] or indium(I) iodide.[221] While gallium-mediated Barbier-type reactions are essentially limited to allylic iodide substrates, use of either In or InI as the promoter allows the same transformation to accommodate allylic bromides as starting materials. Notably, allylic halides substituted in the γ-position give rise exclusively to rearrangement in their reactions with carbonyl compounds when promoted by indium metal, whereas product mixtures derived from both α- and γ-coupling of the allylic organometallic are formed when using InI. Both aldehydes and ketones undergo allylation using these Ga- and In-based methodologies to produce the respective homoallylic alcohols. Moreover, α,β-unsaturated carbonyl substrates are allylated strictly with a 1,2-mode of addition. To illustrate these points, reaction of prenyl bromide with 3,3-dimethylacrylaldehyde in the presence of indium metal at room temperature affords exclusively the dienol derived from both rearrangement of the intermediate allylic indium reagent and 1,2-addition to the carbonyl compound (equation (7.65)).[220]

$$\text{(structure)} \quad + \quad \text{(structure)}\,CHO \quad \xrightarrow[\text{DMF, 25 °C}]{\text{In}} \quad \text{(structure)}\,OH \qquad (7.65)$$

$$\text{(75\%)}$$

Metallic indium[222, 223] and InI[221] also effectively promote the Reformatsky-type reaction of α-iodoesters with carbonyl compounds to furnish β-hydroxy esters. Thus, treatment of a mixture of ethyl iodoacetate and cyclohexanone with InI in THF at room temperature gives the corresponding β-hydroxy ester in high yield (equation (7.66)):[221]

$$ICH_2CO_2Et \quad + \quad \text{(structure)} \quad \xrightarrow[\text{THF, 25 °C}]{\text{InI}} \quad \text{(structure)}\begin{array}{l} CH_2CO_2Et \\ OH \end{array} \qquad (7.66)$$

$$\text{(79\%)}$$

As with the Barbier allylation reactions discussed above, α,β-unsaturated carbonyl compounds also undergo exclusive 1,2-addition with this modified Reformatsky-type procedure. Although the methodology is limited to α-iodoesters using ordinary metallic indium or InI, utilisation of activated indium metal, produced *in situ* from reduction of InCl$_3$ with potassium, allows the same reaction to be carried out successfully on both α-bromo- and α-chloroesters.[222] Finally, quinones also react, under In-mediated Reformatsky conditions, with α-iodoesters to provide an efficient route to the respective quinol esters.[224]

Allylation of quinones by preformed allylic indium sesquihalides, followed by oxidation with Ag_2O, serves as an efficient protocol for the synthesis of allylated p-benzoquinones.[225] As shown in equation (7.67), reaction of p-benzoquinone with allyl indium sesquiiodide in DMF at $-45°C$ affords the allylated product derived from 1,2-addition to the carbonyl group. Subsequent treatment of the crude reaction product with Ag_2O induces [3,3]-sigmatropic rearrangement to furnish the allylated p-benzoquinone in 91% overall yield.

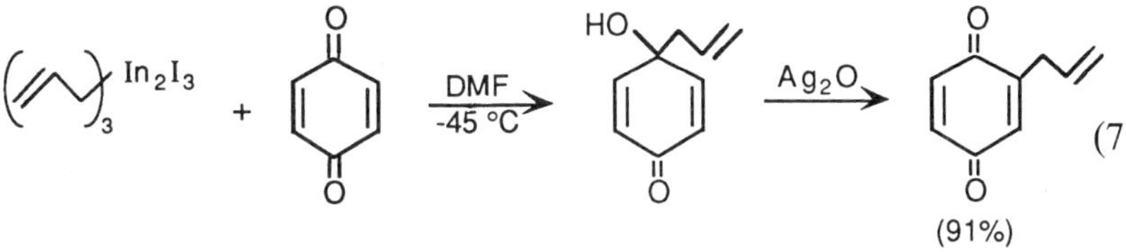

$$(7.67)$$

(91%)

The aldol condensation of α-iodoketones with aldehydes is also effectively promoted by indium metal. For instance, the indium-mediated reaction of phenacyl iodide with benzaldehyde produces a mixture of the corresponding β-hydroxy ketone and its dehydration product, chalcone (equation (7.68)).[226]

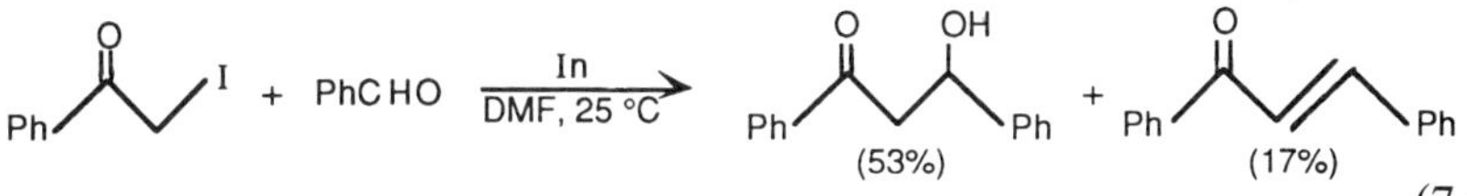

$$(7.68)$$

An interesting *gem*-diallylation of acid anhydrides is observed upon their reaction with allyl iodide in the presence of indium metal. Thus, equation (7.69) illustrates that this type of diallylation of phthalic anhydride can take place under mild reaction conditions to deliver the diallylphthalide product in high yield:[227]

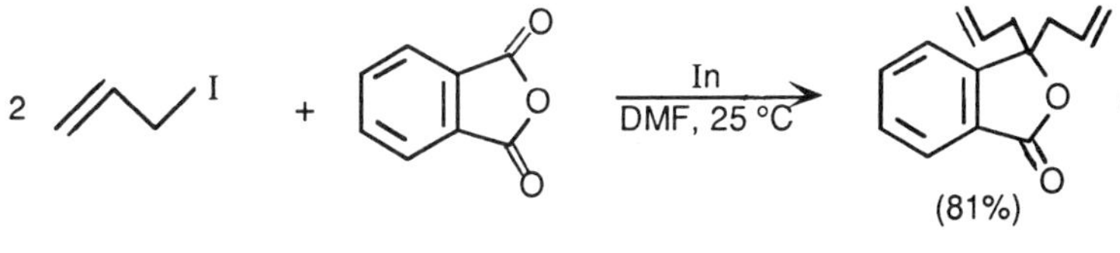

$$(7.69)$$

(81%)

Reaction of cyclic anhydrides with γ-substituted allylic halides stops at the monoallylation stage, yielding only the hydroxy lactones derived from addition of the organoindium reagent to the carbonyl moiety with allylic rearrangement. It should be pointed out that the related reaction between allylic indium sesquihalides and cyclic imides produces, among other products, diallylated lactones of the type shown in equation (7.69).[228]

Lithium tetraorganoindates, prepared by the reaction of a trialkylindium with an organolithium compound, react with cyclic enones selectively *via* conjugate addition of an organic group. As an example, the reaction of $LiIn(Bu^n)_4$ with 2-cyclohexenone at room temperature affords the 1,4-derived adduct in 76% yield (equation (7.70)).[229a] None of the respective 1,2-addition product is formed.

$$LiIn(Bu^n)_4 \quad + \quad \text{[cyclohexenone]} \quad \xrightarrow{25\ °C} \quad \text{[product]} \qquad (7.70)$$

(76%)

Interestingly, reaction of tetraorganoindates with acyclic enones leads to mixtures of 1,4- and 1,2-derived addition products. On the other hand, use of α,β-unsaturated aldehydes under the same reaction conditions results exclusively in 1,2-alkylation of the aldehyde.

The reaction of β-branched trialkylgalliums with ketones proceeds *via* reduction and enolisation pathways, with alkylation of the carbonyl group generally not found to be a competing process.[230] Significantly, reduction of carbonyl-containing compounds can also be carried out using optically active organogalliums, providing a useful procedure for the asymmetric reduction of ketones. The nature of the chiral trialkylgallium reagent has considerable influence with regard to the degree of optical activity delivered to the alcohol product. Hence, reduction of 2-methyl-4-nonyn-3-one with optically pure *tris(cis*-myrtanyl)gallium[230,231] provides for relatively high enantiomeric selectivity in the alcohol product. Conversely, reduction of the same substrate using *tris*[(S)-2-methylbutyl]gallium[230] proceeds in nearly quantitative yield but results in very little asymmetric induction (equation (7.71)).

$$i\text{-}C_3H_7\overset{O}{\overset{\|}{C}}C\equiv CC_4H_9\text{-}n \xrightarrow[50\ °C]{R_3^*Ga} \xrightarrow{H_3O^+} i\text{-}C_3H_7CHC\equiv CC_4H_9\text{-}n$$

R_3^*Ga	Yield (% ee)	
[structure]	99% (R, 9%)	(7.71)
[structure]	79% (R, 75%)	

7.2.3   Other carbon–carbon bond-forming reactions

The alkylation of allylic bromides by lithium tetraorganoindates proceeds in a regio- and stereospecific fashion to afford the corresponding olefin products. For example, the reaction of geranyl bromide at room temperature with LiInBu$_4$ cleanly furnishes the stereodefined trisubstituted olefinic derivative (equation (7.72)):[229a]

$$\text{(structure)} + \text{LiIn(Bu}^n)_4 \xrightarrow{25°C} \text{(structure)} \quad (7.72)$$

(62%)

In a similar manner, allylic indates, prepared from the reaction of an allylic indium sesquihalide with an alkyllithium reagent, react selectively with allylic bromides to yield 1,5-diene products which are specifically derived from coupling at the α-position of the allylic bromide and the γ-position of the allylic indate.[229a] The excellent regio- and stereoselectivity found for this head-to-tail coupling of allylic components makes this procedure an especially valuable synthetic tool for the controlled preparation of 1,5-dienes. In another variation and without the mediation of catalyst or solvent, cross-coupling of a trialkylindium with a chloroalkene appears also to offer considerable promise as a method of preparing olefins.[229b] For example, the reaction of tetrachloroethylene with tri-sec-butylindium delivers a high yield of 1,1,2-trichloro-3-methyl-1-pentane.

The reaction of lithium acetylides with oxiranes to produce β-hydroxy acetylenes is generally a sluggish process. Accordingly, stoichiometric amounts of Lewis acids, such as Et$_2$O · BF$_3$, are often utilised to promote this ring-opening reaction. An attractive alternative procedure for the alkynylation of oxiranes by lithium acetylides involves the use of Me$_3$Ga as a catalyst.[232,233] Thus, reaction of 1-lithio-1-octyne with propylene oxide in the presence of Me$_3$Ga (8 mol%) at room temperature provides the respective β-hydroxy acetylene in excellent yield (equation (7.73)). By comparison, the same product is formed in only 3% yield under similar reaction conditions without the aid of the Me$_3$Ga catalyst.

$$n\text{-}C_6H_{13}C\equiv CLi + \text{(oxirane)} Me \xrightarrow[\text{THF, 25 °C}]{\text{Me}_3\text{Ga (cat.)}} \xrightarrow{H_3O^+} n\text{-}C_6H_{13}C\equiv CCH_2\overset{\overset{\displaystyle OH}{|}}{C}HMe$$

(87%)

$$(7.73)$$

The isomerisation of acid-sensitive oxiranes by conventional Lewis acid promoters prior to the desired ring-opening reaction can pose serious difficulties. Therefore, this Me$_3$Ga-based procedure offers the advantages of requiring only catalytic amounts of the promoter, as well as its relatively low Lewis acidity compared to more typical Lewis acid reagents.

Cyclopropanation of electron-deficient alkenes is promoted by metallic indium under mild conditions using methylene dibromide reagents. As shown in equation (7.74), treatment of methyl vinyl ketone at room temperature with dibromomalononitrile in the presence of indium metal and LiI produces a nearly quantitative yield of the corresponding cyclo-propane derivative:[234]

$$
\text{(diagram: alkene with Me and C=O group)} + Br_2C(CN)_2 \xrightarrow[\text{DMF, 25 °C}]{\text{In, LiI}} \text{(cyclopropane product with NC, CN, Me and C=O)} \qquad (7.74)
$$

(94%)

The yield of cyclopropane product in these reactions is generally lower if LiI is omitted from the procedure. Non-activated alkenes, such as cyclohexene, fail to yield any trace of cyclopropanes under the same reaction conditions. It should be noted, however, that the reaction of Et_3In with CH_2Br_2 in cyclohexene solvent is reported to furnish the respective cyclopropanation product, norcarane.[235]

7.3   Organothallium compounds

7.3.1   Introduction

With the exception of boron, thallium is the most electronegative element of Group 13, possessing a Pauling electronegativity of 1.8, compared to the lower values of 1.5, 1.6 and 1.7 for aluminium, gallium and indium, respectively. Thus, the intrinsic nucleophilicity of organothalliums is usually less than that found for the corresponding organic derivatives of these congeners. Organothallium(III) compounds are also the least Lewis acidic of those derived from Group 13 metals and, therefore, tend to possess relatively weak acceptor properties towards donor molecules. Consequently, organothalliums do not exhibit the coordination-alkylation manner of reactivity that is so commonly displayed by the respective organo-aluminium compounds. As a result of these two factors, reactions involving the intermolecular transfer of an organic group from thallium to electron-deficient centres of organic compounds are rather uncommon. Generally, trialkylthalliums do not exhibit a marked tendency towards self-association. In this respect, therefore, they are reminiscent of the corresponding gallium and indium compounds, while quite different from the strongly associated trialkylalanes. For example, Me_3Tl, Me_3In and Me_3Ga all exist as monomeric species both in the vapour state and in solution, whereas Me_3Al is present as the dimer at normal temperatures.[217] Formation of tetra-organothallium(III)-*ate* complexes takes place upon reaction of $TlCl_3$ or TlI with an organosodium compound,[236] but the chemistry of these species has received little attention.

   Thallium can exist in two stable oxidation states besides the neutral state (see chapter 3), with inorganic salts typically being more stable with thallium in the $+1$ oxidation state and organothallium compounds usually stable only with the metal in the $+3$ state. The readily prepared and isolable organothallium(I) compound, cyclopentadienylthallium, is an exception to this generalisation.[236,237] Triorganothalliums, R_3Tl, are highly reactive towards electrophilic reagents and readily undergo protonation, oxidation or halogenation of a single ligand to afford the corresponding derivatised organic product and R_2TlX. These diorganothallium compounds are the most stable of the organothalliums and are not readily affected by exposure to moisture or molecular oxygen. On the other hand, monoalkylthallium-(III) compounds, $RTlX_2$, are highly unstable and undergo spontaneous reductive decomposition to TlX and the respective organic product RX. The monoalkenyl- and monoarylthallium(III) analogues are considerably more stable than their alkyl counterparts and, depending upon the nature of the anionic ligand, can often be isolated and characterised. The Tl–C bond in trialkylthalliums is rather weak thermodynamically, with an approximate mean bond enthalpy of $105–125 \, kJ \, mol^{-1}$.[10] On the basis of this weak bond, an organothallium might be expected to serve as an efficient source of the corresponding organic free radical. However, organothalliums usually participate in two-electron transfer processes, with the driving force presumably being the disproportionation of the highly unstable Tl(II) intermediate into a mixture of Tl(III) and Tl(I) species. Accordingly, the dominant characteristic of organic transformations which are mediated by thallium is the highly favourable reduction of Tl(III) to Tl(I) (the standard reduction potential for the aqueous couple $Tl^{3+} + 2e^- \rightarrow Tl^+$, $E^\ominus$, is $+1.25 \, V$; see chapter 1). This process, in fact, provides the basis for the majority of the synthetically useful organic chemistry that is discussed in this section. Finally, it should be pointed out that organothallium chemistry has been the subject of several reviews in the past.[9,10,217,236–242]

7.3.2 Thallation and functionalisation of aromatic compounds

The direct thallation of aromatic compounds takes place readily on treatment with thallium(III) trifluoroacetate (TTFA) in trifluoroacetic acid (TFA).[237–242] Typically, aromatic substrates which are activated towards electrophilic substitution are completely thallated within minutes at room temperature, whereas mildly deactivated arenes usually require longer reaction times or heating. The arylthallium *bis*(trifluoroacetate) products tend to crystallise directly from the reaction mixture and are easily isolated by filtration. This lack of solubility, as well as the deactivating nature of the $-Tl(TFA)_2$ substituent, aids in limiting dithallation of the aromatic substrate during the reaction. Interestingly, the reaction of particularly electron-rich aromatic substrates with TTFA does not result in thallation but leads instead to intermolecular oxidative dehydrodimerisation and formation of the corresponding biaryls in high yield.[240,241] While TTFA is the most commonly utilised reagent for the thallation of unactivated aromatic compounds, thallium(III) salts of other strong acids have also been employed as electrophilic thallation agents; these include thallium(III) nitrate, trichloroacetate, perchlorate, mixed perchlorate/acetate and trifluoromethanesulfonate (triflate). The last compound, thallium(III) triflate,[243] is an extremely powerful reagent that is capable of thallating aromatic systems which are strongly deactivated towards electrophilic substitution.

The thallation of aromatic compounds is a reversible electrophilic substitution reaction with an extremely large steric requirement. As a result, *para* substitution usually predominates under conditions of kinetic control for aromatic substrates bearing typical *ortho-para*-directing substituents. Under equilibrating conditions, however, the *meta* isomer often accumulates at the expense of the kinetically formed *ortho* and *para* isomers. For instance, the thallation of cumene at the *meta* position is increased 17-fold by carrying out the reaction at 73°C (boiling point of TFA) instead of room temperature (equation (7.75)):[244]

Aromatic compounds that contain polar substituents capable of complexing with thallium are thallated nearly exclusively in the *ortho* position. Thus, the reaction of methyl benzoate with TTFA at room temperature produces a 95 : 5 mixture of the *ortho-* and *meta*-thallated products (equation (7.76)).[244] None of the corresponding *para* isomer is produced. The formation of the *ortho* thallation product from a substrate possessing a *meta*-directing group implies that an initial complexation of the entering thallium reagent by the carboxylate functionality takes place and is then followed by intramolecular delivery of thallium to the *ortho* position. Aromatic compounds which bear

other polar substituents can also lead to predominant or exclusive *ortho* substitution, provided that a five- or six-membered ring chelation to the *ortho* position is possible. For example, *ortho* substitution is observed upon thallation of benzoic acid, phenylacetic acid, methyl phenylacetate, benzyl alcohol, anisole, β-phenylethanol and β-phenylethyl methyl ether.[240,244]

Although less direct, the preparation of arylthallium compounds can also be effected through the transmetallation of other arylmetals with thallium(III) halides or acetates. An alternative method for preparing arylthallium *bis*(trifluoroacetates), including those derived from aromatic substrates containing deactivating groups, involves thallation of arylsilanes with TTFA.[245] The reaction generally proceeds under mild conditions and affords strictly the *ipso* substitution products in good to excellent yields (equation (7.77)).

$$X = H, Me, CF_3, Ph, OMe, halogen$$

The $ArTlX_2$ products derived from thallation of aromatic compounds serve as highly efficient precursors for the preparation of a variety of functionalised benzenes. Thus, Scheme 7.2 shows that aryl iodides, bromides, chlorides, fluorides, cyanides, thiols, thiocyanates and seleno-cyanates, and nitroarenes can all be prepared from arylthalliums with the same overall regiochemical control inherent in the initial thallation process.[240,242] Furthermore, thallium-based methodology can also be utilised to transform benzenes into anilines,[238] phenols[246,247] and un-symmetrical diarylsulfones.[242,248] From a mechanistic viewpoint, treatment of the $ArTlX_2$ thallation product with a nucleophilic reagent usually leads to anionic ligand replacement to form the respective $ArTlX'_2$ derivative. As noted earlier in section 7.3.1, the stability of such a mono-arylthallium compound varies considerably depending on the nature of the group X'. Whereas spontaneous breakdown of $ArTlX'_2$ to yield ArX' and TlX' occurs only with X' = I,[240] decomposition to produce the respective substituted aromatic compound can be induced under a variety of reaction conditions, some of which are illustrated in Scheme 7.2. The use of a copper salt can serve as an efficient means for converting $ArTlX_2$ into

III

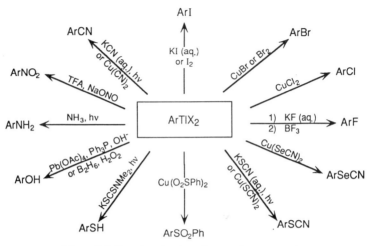

Scheme 7.2 Functionalisation of aromatic compounds.

a functionalised arene and has been proposed to involve the five-centre transition state **III**.[240] Hence, complexation of copper by an anionic ligand of the thallium compound effectively lowers the Tl^{III}/Tl^{I} reduction potential and facilitates an intramolecular nucleophilic displacement on the C–Tl bond. Lastly, it should be pointed out that many carbon–carbon bond-forming reactions involving arylthalliums are promoted by Pd complexes; these and other related transformations are discussed in section 7.3.5.

7.3.3 Oxythallation of unsaturated organic compounds

Thallium(III) salts react with certain unsaturated organic compounds to yield monoalkylthallium(III) intermediates. As stated in section 7.3.1, these $RTlX_2$ oxythallation adducts are generally unstable species and undergo rapid decomposition to provide oxidised organic products along with reduction of Tl^{III} to Tl^{I}. The product distributions are highly dependent on both the nature of the thallium salt and the solvent employed. In many cases, the decomposition of oxythallation adducts takes place via carbonium ion-like intermediates, a channel which leads to oxidative rearrangement of the carbon skeleton of the substrate. For example, the reaction of cyclohexene at room temperature with thallium(III) acetate (TTA) in acetic acid requires several days to reach completion and furnishes a mixture of glycol and acetal derivatives (equation (7.78)).[249] Conversely, oxidation of the same substrate with thallium(III) nitrate (TTN) in methanol solvent affords the product of oxidative rearrangement, cyclopentanecarboxaldehyde, in high yield within seconds at 25°C.[250] The differences with respect to reaction rate and product distribution in this example can be accounted for in terms of several factors,

$$(7.78)$$

including the relative electrophilicities of the two thallium salts, the extent to which the RTlX$_2$ oxythallation adducts tend to ionise by loss of an anionic ligand, and the ability of the acetate group to function intramolecularly as a nucleophile.

The TTN-mediated ring-contraction of cyclobutenes, cyclohexenes and cycloheptenes to the corresponding cyclic carboxaldehydes is a valuable synthetic transformation.[241] On the other hand, oxythallation of acyclic terminal and internal alkenes is more complex and usually gives mixtures of glycol and carbonyl products.[240–242] Oxidative rearrangement of styrenes using TTN/methanol does, however, constitute an efficient procedure for the preparation of arylacetaldehydes.[240,241] Moreover, the reaction of cyclic styrene derivatives, such as 1-phenylcyclohexene, with TTN in methanol can take place *via* ring-contraction to produce benzoylcycloalkanes in high yield (equation (7.79)).[250]

$$(7.79)$$

Synthetically useful ring-expansion reactions are also derived from oxy-thallation of olefins. Thus, thallium(III)-mediated oxidation of methylene-cyclobutanes and -cyclopentanes yields cyclopentanones and cyclo-hexanones, respectively.[251,252] Exocyclic styrene derivatives undergo TTN-promoted ring-expansion in methanol solvent to furnish the corresponding bicyclic ketones.[253] Changing the solvent in this same reaction to a mixture of methanol and trimethylorthoformate (TMOF) provides the analogous dimethyl ketal in nearly quantitative yield (equation (7.80)).[253]

$$(7.80)$$

Oxythallation of substituted acetylenes leads to a variety of synthetically valuable transformations, with the overall reaction pathway largely dictated by the nature of the substituents on the triple bond.[254] As an example, diaryl-substituted alkynes are oxidised by TTN (2 equivalents) in a mixture of aqueous perchloric acid and 1,2-dimethoxyethane to the corresponding benzils in high yield. Alternatively, dialkylacetylenes are converted into acyloins under essentially the same reaction conditions. The reaction of terminal alkynes with TTN under identical conditions follows yet another pathway, proceeding *via* cleavage of the triple bond to carboxylic acids and formaldehyde co-products. Accordingly, treatment of 1-octyne with TTN under acidic conditions affords heptanoic acid in 80% yield (equation (7.81)).[254] This reaction is distinct from other standard synthetic methods available for conversion of terminal alkynes into carboxylic acids[255,256] in that it gives a product derived from chain-shortening of the substrate. Similar oxidation of alkylarylacetylenes using TTN under acidic conditions

$$n\text{-}C_6H_{13}C\equiv CH \xrightarrow[\text{DME, 25°C}]{\text{TTN, HClO}_4 \text{ (aq.)}} n\text{-}C_6H_{13}CO_2H + CH_2O \qquad (7.81)$$
$$(80\%)$$

results in a complex product mixture. However, when the same substrates are treated with TTN in methanol solvent, a smooth oxidative rearrangement is observed and methyl arylacetates are obtained in high yield. For instance, treatment of 1-phenyl-1-propyne with TTN (1.1 equivalents) in methanol furnishes methyl α-methylphenylacetate in nearly quantitative yield (equation (7.82)):[254]

$$ \text{R = Me (95\%), Bu}^n \text{ (98\%), } -CH_2CHCl \text{ (95\%) or } -CH_2Ph \text{ (90\%)} $$

While the mechanism shown in equation (7.82) has been suggested for this transformation on the basis of kinetic data and the effects of aromatic ring substituents, it should be noted that attempts to isolate the vinylthallium intermediates have not been successful. This thallium-mediated oxidative rearrangement of methylarylacetylenes has proved useful as a means of preparing a number of α-methylarylacetic acid-type non-steroidal anti-inflammatory drugs, such as naproxen[240,257] and ketoprofen.[240,258]

By virtue of their rapid enolisation under the acidic conditions of oxy-thallation, ketones react with TTN at the resultant electron-rich enolic double bond. Decomposition of these oxythallation adducts leads to a number of synthetically useful transformations which are highly dependent on both the nature of the ketone substrate and the specific reaction conditions. As an illustration, oxidative rearrangement of acetophenone with TTN in acidic methanol proceeds under mild conditions and provides methyl phenylacetate in 94% yield (equation (7.83)):[259-261]

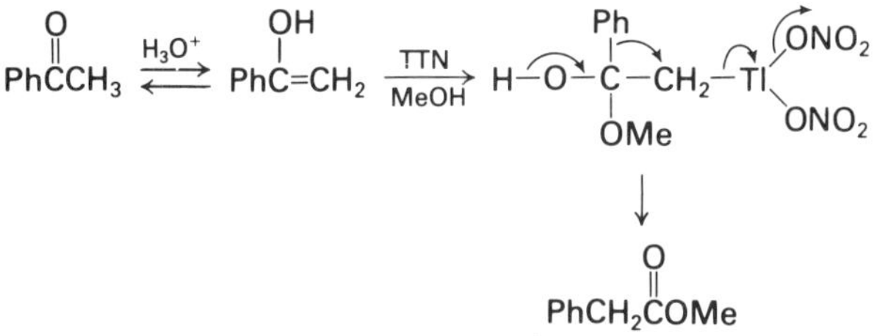

$$
\underset{\text{PhCCH}_3}{\overset{O}{\|}} \overset{H_3O^+}{\rightleftharpoons} \underset{\text{PhC=CH}_2}{\overset{OH}{|}} \xrightarrow[\text{MeOH}]{\text{TTN}} H-O-\underset{\overset{|}{\text{OMe}}}{\overset{\overset{\displaystyle Ph}{|}}{C}}-CH_2-Ti\underset{ONO_2}{\overset{ONO_2}{\big\langle}}
$$

$$\downarrow \qquad (7.83)$$

$$\underset{\text{PhCH}_2\text{COMe}}{\overset{O}{\|}}$$

The remainder of the product mixture consists of α-methoxyacetophenone (6% yield), which results from solvolysis of the oxythallation intermediate. Formation of this by-product can be completely suppressed by use of the TTN/K-10 reagent (TTN adsorbed on acidic montmorillonite K-10 clay).[262] The reaction can also be extended to the preparation of both α-substituted and α,α-disubstituted arylacetic esters. In these instances, however, use of methanol alone as the solvent leads to the formation of significant amounts of the α-methoxyketones derived from solvolysis of the oxythallation adduct. These problems are completely eliminated by several alternative procedures, including utilisation of the TTN/TMOF/methanol[263] or TTN/K-10[262] reagent systems, or by employing the enol ether of the ketone as the substrate.[264] For example, isobutyrophenone is converted into α,α-dimethylphenyl acetate in a straightforward manner using the TTN/TMOF/methanol system (equation (7.84)):[265]

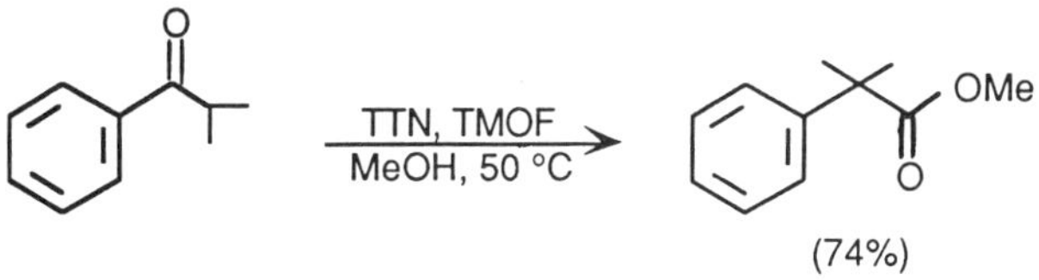

$$\xrightarrow[\text{MeOH, 50 °C}]{\text{TTN, TMOF}}$$

(7.84)

(74%)

The oxidative rearrangement of the preformed enol ether of 4-isobutyl-propiophenone shown in equation (7.85) using TTN in methanol also proceeds smoothly and furnishes the methyl ester of the commercially important anti-inflammatory agent ibuprofen:[264]

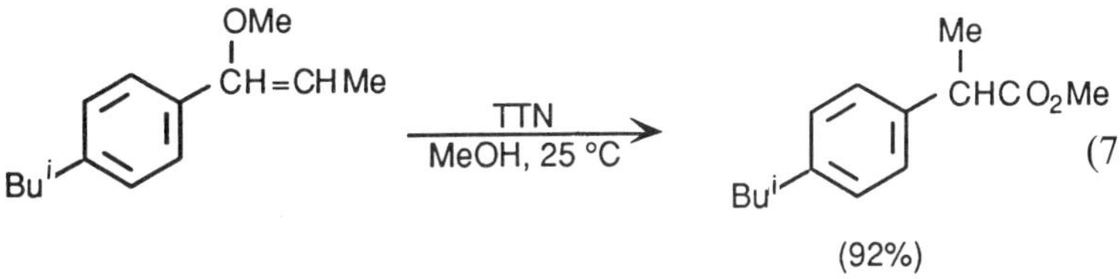

$$\xrightarrow[\text{MeOH, 25 °C}]{\text{TTN}}$$

(7.85)

(92%)

Because of the toxic nature of thallium compounds, it is of considerable importance that this same transformation is possible utilising the thallium salt in catalytic quantities. Thus, the reaction shown in equation (7.85) can also be carried out using a catalytic amount of thallium(I) acetate in the presence of manganese acetate and peroxyacetic acid oxidants to deliver the methyl ester of ibuprofen in 81% yield.[266]

While acetophenone undergoes oxidative rearrangement to methyl phenylacetate upon treatment with a stoichiometric amount of TTN in methanol (equation (7.83)), its reaction with 2 equivalents of the same thallium reagent in methanol/TMOF solvent instead produces methyl α-methoxyphenylacetate (equation (7.86)):[263]

$$\underset{\text{PhCCH}_3}{\overset{\overset{\text{O}}{\|}}{}} \xrightarrow[\text{MeOH, TMOF}]{\text{TTN (2 equiv.)}} \underset{\text{PhCHCO}_2\text{Me}}{\overset{\overset{\text{OMe}}{|}}{}} \qquad (7.86)$$
$$(70\%)$$

Methyl phenylacetate itself is not an intermediate in this conversion since it is recovered unchanged under the reaction conditions. The transformation involves two successive methoxythallation reactions, with α-methoxy-styrene and the trimethyl orthoester of phenylacetic acid established to be intermediates.

Acyclic aliphatic ketones do not undergo thallium-mediated oxidative rearrangement, but rather yield products derived from α-substitution. Hence, the reaction of aliphatic ketones with TTA in acetic acid furnishes the corresponding α-acetoxy derivatives.[241,267] Cyclic ketones, to the contrary, readily undergo oxidative rearrangement to afford ring-contraction-derived cycloalkanecarboxylic acid products in excellent yields.[240–242] To illustrate, treatment of cyclohexanone with TTN in acetic acid at 40°C produces cyclopentanecarboxylic acid in 84% yield.[268,269] As shown in equation (7.87), the ring-contraction also takes place in the expected fashion when carried out on a conjugated carbonyl group encompassed in a steroidal skeleton.[241,270]

(7.87)

7.3.4   Heterocyclisation reactions

The decomposition of the intermediate oxythallation adducts discussed in section 7.3.3 involves direct participation of the solvent as a nucleophile and/or the intramolecular migration of an organic group to the electron-deficient carbon centre adjacent to thallium. In a related fashion, substrates which possess suitably positioned heteroatom-containing functional groups may undergo intramolecular decomposition of the oxythallation inter-mediate via cyclisation and subsequent solvolysis. For example, reaction of 1-buten-4-ol and 1-penten-4-ol with aqueous thallium(III) perchlorate cleanly affords the cyclised products 3-hydroxytetrahydrofuran and 2-methyl-4-hydroxytetrahydrofuran, respectively.[271] Similar reactions of both 1-penten-5-ol and 1-hexen-6-ol, however, give predominantly the corresponding acyclic methyl ketones.[271] It should be noted, however, that there exist many examples of thallium-mediated heterocyclisations to form 6-membered rings, and the viability of this process depends on the nature of the particular substrate, as well as the reaction conditions employed.[240,242,272] The cyclisation reaction proceeds in a predictable regio- and stereospecific manner, and constitutes an efficient preparative route to substituted tetrahydrofurans and -pyrans. Thus, the cyclisation of isopulegol to the stereodefined *trans*-fused bicyclic alcohol occurs readily via reaction with TTA at room temperature (equation (7.88)).[273] In a similar fashion, neoisopulegol is converted to the corresponding alcohol with a *cis*-ring junction (equation (7.89)).[273] The cyclisation reactions illustrated in equations (7.88) and (7.89) are particularly noteworthy since they formally represent 5-*endo-trig* processes which are disfavoured by Baldwin's rules.[172-174]

$$(7.88)$$

(80%)

$$(7.89)$$

(82%)

The thallium-mediated cyclisation of 1,1-disubstituted-1-en-5-ols also furnishes tetrahydrofuran derivatives in a highly stereoselective fashion. Notably, these reactions are believed to occur *via* the favoured 6-*endo-trig* mode of cyclisation, with ring-contraction to the tetrahydrofuranyl moiety taking place during the dethallation step. For instance, the conversion of the 2-methyl-2-hexen-6-ols shown in equation (7.90) into their respective

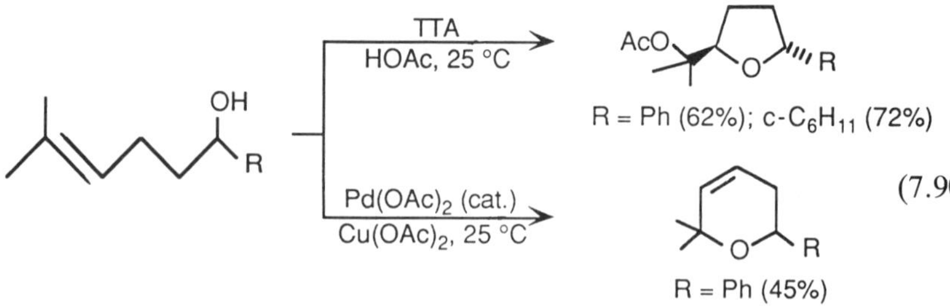

$$R = Ph (62\%); c\text{-}C_6H_{11} (72\%)$$

$$R = Ph (45\%)$$

(7.90)

tetrahydrofuran products occurs readily and in good yield upon reaction with TTA.[274,275] In contrast, the same alcohol substrate (R = Ph) undergoes cyclisation induced by palladium(II) acetate to yield the dihydropyran derivative exclusively (equation (7.90)).[276] Unlike the similar adduct derived from oxythallation, the palladated intermediate formed from cyclisation in the latter example suffers elimination to a dihydropyran rather than rearrangement to a tetrahydrofuran. Replacement of the hydroxyl group in this same substrate (R = Ph) by ether, ester, amide or carbamate functionalities suppresses the cyclisation reaction and leads to products of acetoxythallation, followed by either solvolysis of the C–Tl bond or by methyl group migration.[274] The cyclisation methodology is applicable to more complex substrates and has been utilised to prepare various building blocks useful for the synthesis of polyether antibiotics.[277,278] As an example, the regio- and stereocontrolled cyclisation of the dienol shown in equation (7.91) proceeds selectively to the respective tetrahydrofuranyl alcohol by treatment with TTA in the presence of aqueous acid:[277]

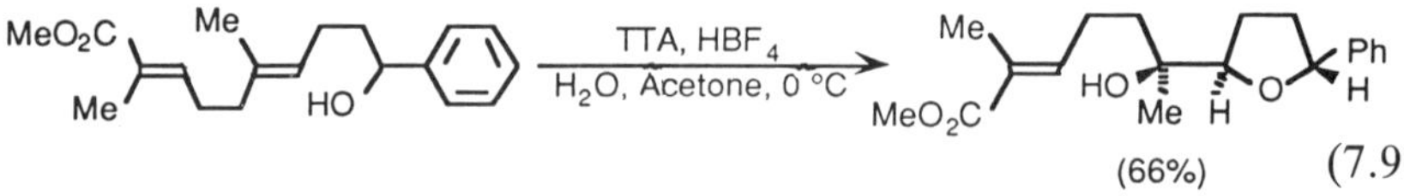

(66%)   (7.91)

The cyclisation of a 1-en-5-ol structural unit contained in a steroidal framework can also be effected by thallium reagents. Accordingly, reaction of 19-hydroxy-5α-cholest-2-ene with TTN in the presence of water produces the expected hydroxylated tetrahydrofuran in 93% yield.[279] It is worth noting that the use of an excess of the thallium reagent in the cyclisation of 1-en-5-ols can lead to more complex product mixtures. For example, treatment of the cyclopentanol derivative shown in equation (7.92) with 3 equivalents of TTA in acetic acid affords predominantly the corresponding bicyclic hemiketal.[280]

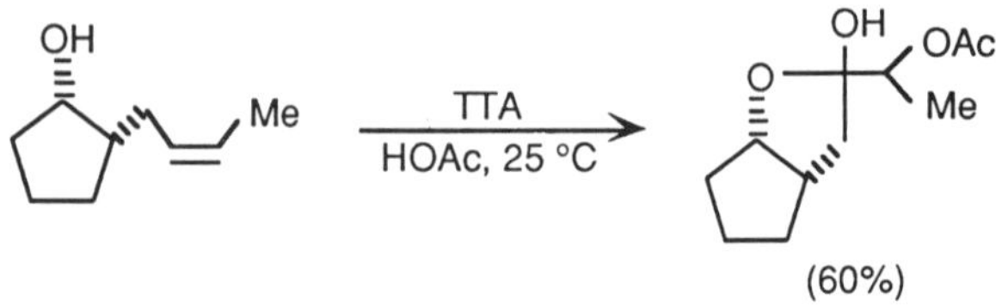

(60%)   (7.92)

Phenolic groups also participate as internal nucleophiles in thallium-promoted heterocyclisation reactions. Hence, *o*-allylphenol is smoothly converted into the acetoxymethyl dihydrobenzofuran derivative by reaction with TTA in acetic acid at 80°C (equation (7.93)).[281]

$$\text{(7.93)}$$

The related reactions of *o*-geranyl- or *o*-nerylphenol with TTFA provide a one-step generation of tricyclic systems, although the product type derived from these substrates varies from one to the other.[282] For instance, thallium(III)-promoted cyclisation of *o*-geranylphenol takes place with ring-contraction to yield cyclopentyl derivatives with a *trans*-ring junction, as shown in equation (7.94). On the other hand, *o*-nerylphenol is cyclised cleanly by treatment with TTFA to the *cis*-fused tricyclic product (equation (7.95)). In the latter case, the cyclohexenyl moiety is left intact as a result of the preferential migration of a methyl group (equation (7.95)). Thus, the olefinic geometry of these two substrates dictates the nature of the ring fusion in their respective tricyclic oxythallation intermediates. Moreover, the opposite relative stereochemistry in the intermediate is the factor determining whether ring contraction or methyl migration occurs during dethallation.

$$\text{(7.94)}$$

$$\text{(7.95)}$$

7.3.5 Carbon–carbon bond forming reactions

As described in section 7.3.2, arylthalliums are readily synthesised from the reaction of benzene derivatives with thallium(III) salts. The thallium moiety in these compounds is easily cleaved by a variety of inorganic reagents to provide useful routes to numerous types of functionalised arenes (Scheme 7.2). In addition, the arylthallium(III) compounds undergo many carbon–carbon bond-forming reactions, a property which further enhances their value as synthetic intermediates. As an example, photolysis of p-tolyl-thallium bis(trifluoroacetate) in benzene solvent affords a mixture of p-methylbiphenyl (93%) and o-methylbiphenyl (5%) in the yields indicated.[283] Alternatively, the phenylation of phenylthallium bis(trifluoro-acetate) can be achieved under non-photochemical conditions at 65°C, albeit at a much slower rate.[284] Stoichiometric[285] and catalytic[284–288] quantities of palladium salts also promote the coupling of arylthalliums to furnish the respective symmetrical biaryls. Thus, the reaction of p-tolylthallium bis(trifluoroacetate) with 1 mol% of Li_2PdCl_4 in refluxing THF produces p-methylbiphenyl in high yield (equation (7.96)).[287] Only a trace of the same coupling product was detected under similar reaction conditions in the absence of the palladium catalyst. This synthesis of biaryls

$$Me \text{—} \hspace{-0.3em} \langle \hspace{-0.3em} \bigcirc \hspace{-0.3em} \rangle \hspace{-0.3em} \text{—Tl(TFA)}_2 \quad \xrightarrow[\text{THF, 67 °C}]{Li_2PdCl_4 \text{ (cat.)}} \quad Me \text{—} \hspace{-0.3em} \langle \hspace{-0.3em} \bigcirc \hspace{-0.3em} \rangle \hspace{-0.3em} \langle \hspace{-0.3em} \bigcirc \hspace{-0.3em} \rangle \hspace{-0.3em} \text{— Me} \atop (90\%)$$

(7.96)

can be carried out in a more convenient manner by generating the aryl-thallium compound *in situ*. Hence, heating a toluene solution containing TTFA and a catalytic amount of $Pd(OAc)_2$ at 105°C provides p-methyl-biphenyl in 93% yield.[284] Such procedures are generally not applicable to highly substituted arenes or arylthalliums which bear substituents at the *ortho*-position. While the mechanism of the transformations remains unclear, it is most likely to involve a transmetallation between the arylthallium compound and the palladium salt ultimately to yield a diaryl-palladium(II) species. Reductive elimination then occurs from this Ar_2Pd intermediate to produce the biaryl product and palladium(0). The thallium(III) salt present in the reaction mixture serves to reoxidise $Pd(0)$ to $Pd(II)$, thus enabling the process to be carried out using catalytic amounts of palladium.

Palladium salts also promote the reaction of arylthalliums with activated olefins to afford the corresponding arylated alkenes.[289,290] For example, the reaction of $PhTlCl_2$ with methyl acrylate in the presence of a catalytic amount of Li_2PdCl_4 gives *trans*-methyl cinnamate in 76% yield.[289] In this case, an equivalent amount of copper(II) chloride is employed in order to reconvert $Pd(0)$ to $Pd(II)$ and allow the transformation to take place catalytically with respect to palladium. The thallation and subsequent palladium-promoted olefination of benzoic acid provides an efficient route for the synthesis of a variety of isocoumarins and 3,4-dihydroisocoumarins. Olefinic reaction partners accommodated in this process include simple alkenes, allylic halides, vinyl halides, vinyl esters, 1,2- and 1,3-dienes, and vinylcyclopropanes. Interestingly, the reaction requires only catalytic amounts of palladium when carried out on the organic halides and dienes from this group of olefin substrates. To illustrate, treatment of the *ortho*-thallated benzoic acid with 1,3-butadiene in the presence of a catalytic amount of $PdCl_2$ (10 mol%) furnishes 3-vinyl-3,4-dihydroisocoumarin in reasonable yield (equation (7.97)):[291]

$$\text{(7.97)} \quad (56\%)$$

Similarly, the sequence of thallation-olefination can be applied to p-tolyl-acetic acid, N-methylbenzamide, benzamide and acetanilide to provide an efficient entry into a number of important oxygen- and nitrogen-containing heterocyclic ring systems. As an example, the Pd-promoted reaction of the ortho-thallated acetanilide with allyl chloride produces the respective lactam derivative in 61% yield (equation (7.98)).[292] Stoichiometric amounts of palladium are required for this transformation.

$$\text{(7.98)} \quad (61\%)$$

The direct carbonylation of arylthalliums can be achieved but requires elevated temperatures and pressures. Furthermore, poor yields of carbonylation products are generally obtained under these conditions.[293] On the other hand, the same carbonylation reaction can be carried out under exceedingly mild conditions by simply employing catalytic amounts of $PdCl_2$.[294,295] Thus, $PhTl(TFA)_2$ undergoes carbonylation readily at room temperature in the presence of 10 mol% of $PdCl_2$ and 1 atm of carbon monoxide to afford methyl benzoate in 52% yield. As a result of the highly selective nature associated with the thallation of many functionalised benzenes, the sequence of thallation-carbonylation constitutes a valuable synthetic method for preparing a variety of aromatic lactones. For instance, thallation of m-methoxybenzyl alcohol with TTFA, followed by carbonylation, furnishes 5-methoxyphthalide nearly exclusively (equation (7.99)). This procedure complements the lithiation-carbonation methodology which

$$\text{(7.99)} \quad (89\%)$$

gives selectively 7-methoxyphthalide from the same starting material (equation (7.99)).[296]

Aromatic compounds are allylated in a single step by reaction with allyl-silanes, -germanes or -stannanes in the presence of TTFA.[297-299] Hence, treatment of p-xylene with allyltrimethylsilane and TTFA at room temperature provides the allylated aromatic product in high yield (equation (7.100)):

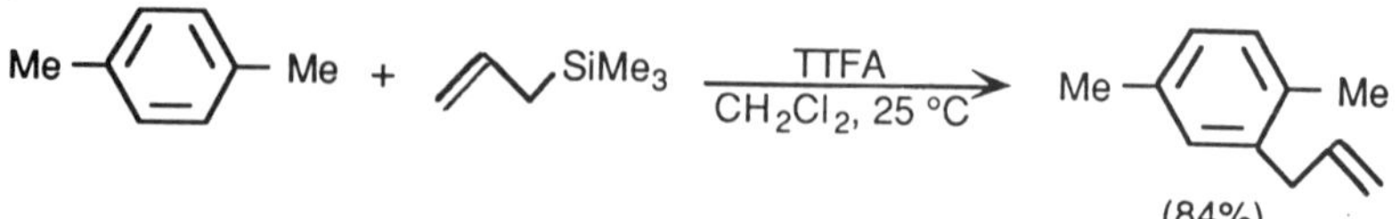

$$Me\text{—}\langle\rangle\text{—}Me \ + \ \diagup\!\!\diagdown\!\!\diagup SiMe_3 \ \xrightarrow[CH_2Cl_2,\ 25\ °C]{TTFA} \ Me\text{—}\langle\rangle\text{—}Me$$

(84%)   (7.100)

The interaction of the allylic organometallic with thallium(III) is believed to produce an allylic cation-like species which alkylates the aromatic substrate.

Triorganothalliums react stoichiometrically with acid chlorides to afford the corresponding ketones in high yield.[300] For example, treatment of Et_3Tl in ether at room temperature with decanoyl chloride results in the immediate precipitation of Et_2TlCl and produces 3-dodecanone in 91% yield. The reaction is general for both aliphatic and aromatic substrates and tolerates the presence of other functionalities such as olefin, ester and ketone groups. Activated 2° and 3° alkyl halides also couple with triorgano-thalliums to yield the respective alkylated derivatives.[301,302] Since thallium compounds are extremely toxic substances, it is significant that this alkylation reaction can also be effected using the thallium compound in catalytic amounts. Thus, the reaction of 2-chlorotetrahydropyran with lithium phenylacetylide in the presence of Me_2TlCl (10 mol%) furnishes the alkynylated product shown in equation (7.101) in nearly quantitative yield:[302]

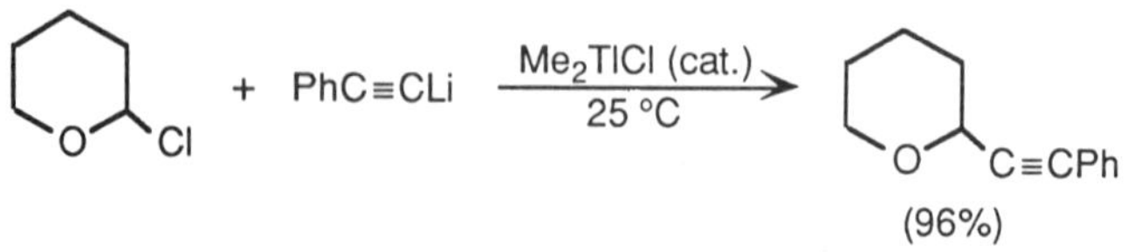

$$\langle\rangle + PhC\equiv CLi \ \xrightarrow[25\ °C]{Me_2TlCl\ (cat.)} \ \langle\rangle$$

(96%)   (7.101)

It is noteworthy that the alkynyl group is selectively transferred from thallium in this reaction; none of the corresponding product derived from competitive methyl transfer is formed. As expected, no alkynylation adduct is produced in the absence of the thallium catalyst.

References

1. R. Köster and P. Binger, *Adv. Inorg. Chem. Radiochem.*, 1965, **7**, 263.
2. H. Reinheckel, K. Hoage and D. Jahnke, *Organometal. Chem. Rev. A*, 1969, **4**, 47.

3. H. Lehmkuhl, K. Ziegler and H.G. Gellert, *Houben–Weyl, Methoden der Organischen Chemie*, 4th edn., Vol. XIII, Part 4, Thieme, Stuttgart, 1970, p. 1.

4. G. Bruno, *The Use of Aluminium Alkyls in Organic Synthesis*, Ethyl Corporation, 1970, 1973, 1977.

5. T. Mole and E.A. Jeffrey, *Organoaluminium Compounds*, Elsevier, Amsterdam, 1972.

6. K.L. Henold and J.P. Oliver, in *Organometallic Reactions*, Vol. 5, eds. E.I. Becker and M. Tsutsui, Wiley–Interscience, New York, 1975, p. 387.

7. E. Negishi, *J. Organomet. Chem. Lib.*, 1976, **1**, 93.

8. G. Zweifel, in *Aspects of Mechanism and Organometallic Chemistry*, ed. J.H. Brewster, Plenum Press, New York, 1978, p. 229.

9. G. Zweifel, in *Comprehensive Organic Chemistry*, Vol. 3, eds. D.H.R. Barton and W.D. Ollis, Pergamon Press, Oxford, 1979, p. 1013.

10. E. Negishi, *Organometallics in Organic Synthesis*, Wiley, New York, 1980.

11. J.R. Zietz, Jr., G.C. Robinson and K.L. Lindsay, in *Comprehensive Organometallic Chemistry*, eds. G. Wilkinson, F.G.A. Stone and E.W. Abel, Pergamon Press, Oxford, 1982, Vol. 7, p. 365.

12. J.J. Eisch, in *Comprehensive Organometallic Chemistry*, eds. G. Wilkinson, F.G.A. Stone and E.W. Abel, Pergamon Press, Oxford, 1982, Vol. 1, p. 555.

13. G. Zweifel and J.A. Miller, *Org. React.*, 1984, **32**, 375.

14. K. Maruoka and H. Yamamoto, *Tetrahedron*, 1988, **44**, 5001.

15. K. Ziegler, in *Organometallic Chemistry*, ed. H. Zeiss, Reinhold, New York, 1960, p. 194.

16. G. Wilke, *Coordination Polymerization*, ed. J.C.W. Chien, Academic Press, New York, 1975.

17. F. Sato, in *Fundamental Research in Homogeneous Catalysis*, ed. Y. Ishii and M. Tsutsui, Plenum Press, New York, 1978, Vol. 2, p. 81.

18. F. Sato, S. Sato, H. Kodama and M. Sato, *J. Organomet. Chem.*, 1977, **142**, 71.

19. F. Sato, H. Ishikawa and Y. Takahashi, *Tetrahedron Lett.*, 1979, 3745.

20. E.C. Ashby and S.A. Nodding, *J. Org. Chem.*, 1980, **45**, 1035.

21. F. Sato, Y. Mori and M. Sato, *Tetrahedron Lett.*, 1979, 1405.

22. G. Boireau, A. Korenova, A. Deberly and D. Abenhaim, *Tetrahedron Lett.*, 1985, **26**, 4181.

23. J.F. Lemarechal, M. Ephritikhine and G. Folcher, *J. Organomet. Chem.*, 1986, **309**, C1.

24. E. Negishi and T. Yodisha, *Tetrahedron Lett.*, 1980, **21**, 1501.

25. K. Maruoka, H. Sano, K. Shinoda, S. Nakai and H. Yamamoto, *J. Am. Chem. Soc.*, 1986, **108**, 6036.

26. G. Wilke and H. Müller, *Chem. Ber.*, 1956, **89**, 444.

27. G. Wilke and H. Müller, *Justus Liebigs Ann. Chem.*, 1958, **618**, 267.

28. G. Wilke and H. Müller, *Justus Liebigs Ann. Chem.*, 1960, **629**, 222.

29. J.J. Eisch and W.C. Kaska, *J. Am. Chem. Soc.*, 1963, **85**, 2165.

30. J.J. Eisch and W.C. Kaska, *J. Organomet. Chem.*, 1964, **2**, 184.

31. T. Mole and J.R. Surtees, *Aust. J. Chem.*, 1964, **17**, 1229.

32. J.R. Surtees, *Aust. J. Chem.*, 1965, **18**, 14.

33. V.V. Gavrilenko, B.A. Palei and L.I. Zakharkin, *Izv. Akad. Nauk SSSR, Ser. Khim.*, 1968, 910.

34. G. Zweifel and R.B. Steele, *Tetrahedron Lett.*, 1966, 6021.

35. G. Cainelli, F. Bertini, P. Grasseli and G. Zubiania, *Tetrahedron Lett.*, 1967, 1581.

36. P. Binger, *Angew. Chem., Int. Ed. Engl.*, 1963, **2**, 686.

37. G. Zweifel, G.M. Clark and R.A. Lynd, *J. Chem. Soc., Chem. Commun.*, 1971, 1593.

38. G. Wilke and W. Schneider, *Bull. Soc. Chim. Fr.*, 1963, 1462.

39. J.J. Eisch and S.G. Rhee, *J. Am. Chem. Soc.*, 1974, **96**, 7276.

40. G.M. Clark and G. Zweifel, *J. Am. Chem. Soc.*, 1971, **93**, 527.

41. L.H. Slaugh, *Tetrahedron*, 1966, **22**, 1741.

42. E.F. Magoon and L.H. Slaugh, *Tetrahedron*, 1967, **23**, 4509.

43. G. Zweifel and R.B. Steele, *J. Am. Chem. Soc.*, 1967, **89**, 5085.

44. C.J. Sih, R.G. Salomon, P. Price, R. Sood and G. Peruzzotti, *J. Am. Chem. Soc.*, 1975, **97**, 857.

45. H.P. On, W. Lewis and G. Zweifel, *Synthesis*, 1981, 999.

46. K.F. Bernady, M.B. Floyd, J.F. Poletto and M.J. Weiss, *J. Org. Chem.*, 1979, **44**, 1438.
47. A. Alexakis and J.M. Duffault, *Tetrahedron Lett.*, 1988, **29**, 6243.
48. A.B. Bates, E.R.H. Jones and M.C. Whiting, *J. Chem. Soc.*, 1954, 1854.
49. R.A. Raphael, *Acetylenic Compounds in Organic Synthesis*, Butterworth, London, 1955.
50. W.J. Borden, *J. Am. Chem. Soc.*, 1970, **92**, 4898.
51. B. Grant and C. Djerassi, *J. Org. Chem.*, 1974, **39**, 968.
52. R.E. Doolittle, *Synthesis*, 1984, 730.
53. P.A. Zoretic and R.H. Khan, *Synth. Commun.*, 1985, **15**, 367.
54. E.J. Corey, H.A. Kirst and J.A. Katzenellenbogen, *J. Am. Chem. Soc.*, 1970, **92**, 6314.
55. E.J. Corey, J.A. Katzenellenbogen and G.H. Posner, *J. Am. Chem. Soc.*, 1967, **89**, 4245.
56. E.J. Corey, J.A. Katzenellenbogen, N.W. Gilman, S.T. Roman and B.W. Erickson, *J. Am. Chem. Soc.*, 1968, **90**, 5618.
57. G. Zweifel and C.C. Whitney, *J. Am. Chem. Soc.*, 1967, **89**, 2753.
58. B.A. Palei, V.V. Gavrilenko and L.I. Zakharkin, *Izv. Akad. Nauk SSSR, Ser. Khim.*, 1969, 2760.
59. G. Zweifel and W. Lewis, *J. Org. Chem.*, 1978, **43**, 2739.
60. E. Negishi, D.E. Van Horn, A.O. King and N. Okukado, *Synthesis*, 1979, 501.
61. J.A. Miller, W. Leong and G. Zweifel, *J. Org. Chem.*, 1988, **53**, 1839.
62. S. Baba, D.E. Van Horn and E. Negishi, *Tetrahedron Lett.*, 1976, 1927.
63. J.J. Eisch and G.A. Damasevitz, *J. Org. Chem.*, 1976, **41**, 2214.
64. K. Uchida, K. Utimoto and H. Nozaki, *J. Org. Chem.*, 1976, **41**, 2215.
65. F.E. Ziegler and K. Mikami, *Tetrahedron Lett.*, 1984, **25**, 131.
66. J.J. Eisch and M.W. Foxton, *J. Organomet. Chem.*, 1963, **11**, P7.
67. G. Zweifel and R.B. Steele, *J. Am. Chem. Soc.*, 1967, **89**, 2754.
68. V.M. Bulina, L.L. Ivanov and Y.B. Pyatnova, *Zh. Org. Khim.*, 1973, **9**, 491.
69. N. Okukado and E. Negishi, *Tetrahedron Lett.*, 1978, 2347.
70. R.L. Dansheiser and H. Sard, *J. Org. Chem.*, 1980, **45**, 4810.
71. G. Zweifel, J.T. Snow and C.C. Whitney, *J. Am. Chem. Soc.*, 1968, **90**, 7139.
72. G. Zweifel and R.A. Lynd, *Synthesis*, 1976, 816.
73. A.P. Kozikowski and Y. Kitigawa, *Tetrahedron Lett.*, 1982, **23**, 2087.
74. H. Newman, *Tetrahedron Lett.*, 1971, 4571.
75. E. Negishi, A.O. King, W. Klima, W. Patterson and A. Silveira, Jr., *J. Org. Chem.*, 1980, **45**, 2526.
76. E. Negishi, L.F. Valente and M. Kobayashi, *J. Am. Chem. Soc.*, 1980, **102**, 3298.
77. K.F. Bernady and M.J. Weiss, *Tetrahedron Lett.*, 1972, 4083.
78. J. Hooz and R.B. Layton, *Can. J. Chem.*, 1973, **51**, 2098.
79. P.W. Collins, E.Z. Dajani, M.S. Bruhn, C.H. Brown, J.R. Palmer and R. Pappo, *Tetrahedron Lett.*, 1975, 4217.
80. E. Negishi, *Pure Appl. Chem.*, 1981, **53**, 2333.
81. A. Pecunioso and R. Menicagli, *Tetrahedron*, 1987, **43**, 5411.
82. A. Pecunioso and R. Menicagli, *J. Org. Chem.*, 1988, **53**, 2614.
83. S. Warwel, G. Schmitt and B. Ahlfaenger, *Synthesis*, 1975, 632.
84. E. Negishi, S. Baba and A.O. King, *J. Chem. Soc., Chem. Commun.*, 1976, 17.
85. D.B. Malpass, J.C. Watson and G.S. Yeargin, *J. Org. Chem.*, 1977, **42**, 2712.
86. K. Utimoto, K. Uchida, M. Yamaya and H. Nozaki, *Tetrahedron Lett.*, 1977, 3641.
87. M. Kobayashi, L.F. Valente, E. Negishi, W. Patterson and A. Silveira, Jr., *Synthesis*, 1980, 1034.
88. M. Kobayashi and E. Negishi, *J. Org. Chem.*, 1980, **45**, 5223.
89. A. Alexakis and D. Jachiet, *Tetrahedron*, 1989, **45**, 6197.
90. G. Zweifel and R.A. Lynd, *Synthesis*, 1976, 625.
91. A.P. Kozikowski, A. Ames and H. Wetter, *J. Organomet. Chem.*, 1979, **164**, C33.
92. G. Zweifel and R.L. Miller, *J. Am. Chem. Soc.*, 1970, **92**, 6678.
93. R.A. Lynd and G. Zweifel, *Synthesis*, 1974, 658.
94. E. Negishi and S. Baba, *J. Chem. Soc., Chem. Commun.*, 1976, 596.
95. E. Negishi, N. Okukado, A.O. King, D.E. Van Horn and B.I. Spiegel, *J. Am. Chem. Soc.*, 1978, **100**, 2254.
96. E. Negishi, T. Takahashi, S. Baba, D.E. Van Horn and N. Okukado, *J. Am. Chem. Soc.*, 1987, **109**, 2393.

97. E. Negishi, H. Matsushita and N. Okukado, *Tetrahedron Lett.*, 1981, **22**, 2715.
98. H. Matsushita and E. Negishi, *J. Am. Chem. Soc.*, 1981, **103**, 2882.
99. E. Negishi, S. Chatterjee and H. Matsushita, *Tetrahedron Lett.*, 1981, 3737.
100. H. Matsushita and E. Negishi, *J. Chem. Soc., Chem. Commun.*, 1982, 160.
101. E. Negishi, in *Current Trends in Organic Synthesis*, ed. H. Nozaki, Pergamon Press, Oxford, 1983, p. 269.
102. S. Chatterjee and E. Negishi, *J. Org. Chem.*, 1985, **50**, 3406.
103. E. Negishi, V. Bagheri, S. Chatterjee, F.-T. Luo, J.A. Miller and A.T. Stoll, *Tetrahedron Lett.*, 1983, **24**, 5181.
104. K. Wakamatsu, Y. Okuda, K. Oshima and H. Nozaki, *Bull. Chem. Soc. Jpn.*, 1985, **58**, 2425.
105. S. Baba and E. Negishi, *J. Am. Chem. Soc.*, 1976, **98**, 6729.
106. E. Negishi, *Acc. Chem. Res.*, 1982, **15**, 340.
107. E. Negishi and F.-T. Luo, *J. Org. Chem.*, 1983, **48**, 1560.
108. V. Ratovelomanana and G. Linstrumelle, *Tetrahedron Lett.*, 1984, **25**, 6001.
109. E. Negishi, T. Takahashi and S. Baba, *Org. Synth.*, 1988, **66**, 60.
110. K. Takai, K. Oshima and H. Nozaki, *Tetrahedron Lett.*, 1980, **21**, 2531.
111. M. Sato, K. Takai, K. Oshima, and H. Nozaki, *Tetrahedron Lett.*, 1981, **22**, 1609.
112. K. Takai, M. Sato, K. Oshima and H. Nozaki, *Bull. Chem. Soc. Jpn.*, 1984, **57**, 108.
113. J.J. Eisch and M.W. Foxton, *J. Org. Chem.*, 1971, **36**, 3520.
114. J.J. Eisch and S.G. Rhee, *J. Am. Chem. Soc.*, 1975, **97**, 4673.
115. E. Negishi and T. Takahashi, *J. Am. Chem. Soc.*, 1986, **108**, 3403.
116. J.A. Miller and G. Zweifel, *J. Am. Chem. Soc.*, 1983, **105**, 1383.
117. H. Shiragami, T. Kawamoto, K. Utimoto and H. Nozaki, *Tetrahedron Lett.*, 1986, **27**, 589.
118. H. Shiragami, T. Kawamoto, K. Imi, S. Matsubara, K. Utimoto and H. Nozaki, *Tetrahedron*, 1988, **44**, 4009.
119. G. Zweifel, W. Lewis and H.P. On, *J. Am. Chem. Soc.*, 1979, **101**, 5101.
120. J.A. Miller, *J. Org. Chem.*, 1989, **54**, 998.
121. H. Hoberg, *Ann. Chem.*, 1962, **656**, 1.
122. H. Hoberg, *Ann. Chem.*, 1966, **695**, 1.
123. H. Hoberg, *Angew. Chem., Int. Ed. Engl.*, 1966, **5**, 513.
124. E. Negishi and K. Akiyoshi, *J. Am. Chem. Soc.*, 1988, **110**, 646.
125. A. Alexakis, J. Hanaizi, D. Jachiet, J.F. Normant and L. Toupet, *Tetrahedron Lett.*, 1990, **31**, 1271.
126. J.J. Barber, C. Willis and G.M. Whitesides, *J. Org. Chem.*, 1979, **44**, 3604.
127. U.M. Dzhemilev and O.S. Vostrikova, *J. Organomet. Chem.*, 1985, **285**, 43.
128. U.M. Dzhemilev, O.S. Vostrikova and G.A. Tolstikov, *J. Organomet. Chem.*, 1986, **304**, 17.
129. U.M. Dzhemilev, A.G. Ibragimov, A.P. Zolotarev, A.R. Muslukov and G.A. Tolstikov, *Izv. Akad. Nauk SSSR, Ser. Khim.*, 1989, 207.
130. K. Ziegler, *Angew. Chem.*, 1956, **68**, 721.
131. G. Hata and A. Miyake, *J. Org. Chem.*, 1963, **28**, 3237.
132. R. Schimp and P. Heimbach, *Chem. Ber.*, 1970, **103**, 2122.
133. A. Stefani, *Helv. Chim. Acta*, 1974, **57**, 1346.
134. T.W. Dolzine and J.P. Oliver, *J. Organomet. Chem.*, 1974, **78**, 165.
135. P.W. Chum and S.E. Wilson, *Tetrahedron Lett.*, 1976, 1257.
136. G. Zweifel, G.M. Clark and R. Lynd, *J. Chem. Soc., Chem. Commun.*, 1971, 1593.
137. R. Rienäcker and D. Schwengers, *Liebigs Ann. Chem.*, 1977, 1633.
138. M.J. Smith and S.E. Wilson, *Tetrahedron Lett.*, 1982, **23**, 5013.
139. A.V. Youngblood, S.A. Nichols, R.A. Coleman and D.W. Thompson, *J. Organomet. Chem.*, 1978, **146**, 221.
140. K. Fujita, E. Moret and M. Schlosser, *Chem. Lett.*, 1982, 1819.
141. E. Moret and M. Schlosser, *Tetrahedron Lett.*, 1985, **26**, 4423.
142. F.W. Schultz, G.S. Ferguson and D.W. Thompson, *J. Org. Chem.*, 1984, **49**, 1736.
143. G.A. Tolstikov and U.M. Dzhemilev, *J. Organomet. Chem.*, 1985, **292**, 133.
144. K. Maruoka, Y. Fukutani and H. Yamamoto, *J. Org. Chem.*, 1985, **50**, 4412.
145. N.I. Andreeva, A.V. Kuchin and G.A. Tolstikov, *Zh. Obshch. Khim.*, 1985, **55**, 1316.

146. A.V. Kuchin, N.I. Andreeva and G.A. Tolstikov, *Zh. Org. Khim., SSSR*, 1987, **23**, 449.
147. D.E. Van Horn, L.F. Valente, M.J. Idacavage and E. Negishi, *J. Organomet. Chem.*, 1978, **156**, C20.
148. D.E. Van Horn and E. Negishi, *J. Am. Chem. Soc.*, 1978, **100**, 2252.
149. E. Negishi, *Acc. Chem. Res.*, 1987, **20**, 65.
150. T. Yoshida and E. Negishi, *J. Am. Chem. Soc.*, 1981, **103**, 4985.
151. E. Negishi, D.E. Van Horn and T. Yoshida, *J. Am. Chem. Soc.*, 1985, **107**, 6639.
152. S. Fung and J.B. Siddall, *J. Am. Chem. Soc.*, 1980, **102**, 6580.
153. K. Mori, M. Sakakibara and K. Okada, *Tetrahedron*, 1984, **40**, 1767.
154. D.R. Williams, B.A. Barner, K. Nishitani and J.G. Phillips, *J. Am. Chem. Soc.*, 1982, **104**, 4708.
155. W.R. Roush and A.P. Spada, *Tetrahedron Lett.*, 1983, **24**, 3693.
156. W.R. Roush and T.A. Blizzard, *J. Org. Chem.*, 1983, **48**, 758.
157. W.R. Roush and T.A. Blizzard, *J. Org. Chem.*, 1984, **49**, 1772.
158. W.R. Roush and T.A. Blizzard, *J. Org. Chem.*, 1984, **49**, 4332.
159. J.K. Whitesell, M. Fisher and P. Da Silva Jardine, *J. Org. Chem.*, 1983, **48**, 1556.
160. C.L. Rand, D.E. Van Horn, M.W. Moore and E. Negishi, *J. Org. Chem.*, 1981, **46**, 4093.
161. J.C. Ewing, G.S. Ferguson, D.W. Moore, F.W. Schultz and D.W. Thompson, *J. Org. Chem.*, 1985, **50**, 2124.
162. B.B. Snider and M. Karras, *J. Organomet. Chem.*, 1979, **179**, C37.
163. J.A. Miller and E. Negishi, *Isr. J. Chem.*, 1984, **24**, 76.
164. J.J. Eisch, R.J. Manfre and D.A. Komar, *J. Organomet. Chem.*, 1978, **159**, C13.
165. H. Hayami, K. Oshima and H. Nozaki, *Tetrahedron Lett.*, 1984, **25**, 4433.
166. J.A. Miller and E. Negishi, *Tetrahedron Lett.*, 1984, **25**, 5863.
167. E. Negishi and J.A. Miller, *J. Am. Chem. Soc.*, 1983, **105**, 6761.
168. G.A. Molander, *J. Org. Chem.*, 1983, **48**, 5409.
169. J.A. Miller and E. Negishi, unpublished results.
170. E. Negishi, L.D. Boardman, J.M. Tour, H. Sawada and C.L. Rand, *J. Am. Chem. Soc.*, 1983, **105**, 6344.
171. E. Negishi, L.D. Boardman, H. Sawada, V. Bagheri, A.T. Stoll, J.M. Tour and C.L. Rand, *J. Am. Chem. Soc.*, 1988, **110**, 5383.
172. J.E. Baldwin, *J. Chem. Soc., Chem. Commun.*, 1976, 734.
173. J.E. Baldwin, *J. Chem. Soc., Chem. Commun.*, 1976, 736.
174. J.E. Baldwin, *J. Chem. Soc., Chem. Commun.*, 1976, 738.
175. E. Negishi, H. Sawada, J.M. Tour and Y. Wei, *J. Org. Chem.*, 1988, **53**, 913.
176. E. Negishi, Y. Noda, F. Lamaty and E.J. Vawter, *Tetrahedron Lett.*, 1990, **31**, 4393.
177. H. Yamamoto and K. Maruoka, *Pure Appl. Chem.*, 1988, **60**, 21.
178. S. Sakane, K. Maruoka and H. Yamamoto, *J. Chem. Soc., Jpn.*, 1985, 324.
179. K. Maruoka, T. Itoh and H. Yamamoto, *J. Am. Chem. Soc.*, 1985, **107**, 4573.
180. K. Maruoka, Y. Araki and H. Yamamoto, *J. Am. Chem. Soc.*, 1988, **110**, 2650.
181. K. Maruoka, M. Sakurai and H. Yamamoto, *Tetrahedron Lett.*, 1985, **26**, 3853.
182. K. Maruoka, K. Nonoshita and H. Yamamoto, *Tetrahedron Lett.*, 1987, **28**, 5723.
183. M.B. Power and A.R. Barron, *Tetrahedron Lett.*, 1990, **31**, 323.
184. L. Clawson, S.L. Buchwald and R.H. Grubbs, *Tetrahedron Lett.*, 1984, **25**, 5733.
185. S.H. Pine, R. Zahler, D.A. Evans and R.H. Grubbs, *J. Am. Chem. Soc.*, 1980, **102**, 3270.
186. T. Okazoe, J. Hibino, K. Takai and H. Nozaki, *Tetrahedron Lett.*, 1985, **26**, 5581.
187. K. Maruoka, T. Miyazaki, M. Ando, Y. Matsumura, S. Sakane, K. Hattori and H. Yamamoto, *J. Am. Chem. Soc.*, 1983, **105**, 2831.
188. H. Fujioka, T. Yamanaka, K. Takuma, M. Miyazaki and Y. Kita, *J. Chem. Soc., Chem. Commun.*, 1991, 533.
189. K. Maruoka, J. Sato, H. Banno and H. Yamamoto, *Tetrahedron Lett.*, 1990, **31**, 377.
190. K. Maruoka, H. Banno, K. Nonoshita and H. Yamamoto, *Tetrahedron Lett.*, 1989, **30**, 1265.
191. K. Takai, I. Mori, K. Oshima and H. Nozaki, *Bull. Chem. Soc. Jpn.*, 1984, **57**, 446.
192. I. Mori, K. Takai, K. Oshima and H. Nozaki, *Tetrahedron*, 1984, **40**, 4013.
193. K. Tanino, Y. Hatanaka and I. Kuwajima, *Chem. Lett.*, 1987, 385.
194. J.A. Miller, *J. Org. Chem.*, 1987, **52**, 322.
195. B.B. Snider and C.P. Cartaya-Marin, *J. Org. Chem.*, 1984, **49**, 153.

196. A. Tenaglia, R. Faure and P. Brun, *Tetrahedron Lett.*, 1990, **31**, 4457.
197. K. Maruoka, T. Itoh, T. Shirasaka and H. Yamamoto, *J. Am. Chem. Soc.*, 1988, **110**, 310.
198. H. Yamamoto, K. Maruoka, K. Furuta and Y. Naruse, *Pure Appl. Chem.*, 1989, **61**, 419.
199. K. Maruoka, Y. Hoshino, T. Shirasaka and H. Yamamoto, *Tetrahedron Lett.*, 1988, **29**, 3967.
200. G. Giacomelli, L. Lardicci, F. Palla and A.M. Caporusso, *J. Org. Chem.*, 1984, **49**, 1725.
201. M. Falorni, L. Lardicci, C. Rosini and G. Giacomelli, *J. Org. Chem.*, 1986, **51**, 2030.
202. G. Giacomelli, L. Lardicci and F. Palla, *J. Org. Chem.*, 1984, **49**, 310.
203. M.M. Midland and J.I. McLoughlin, *J. Org. Chem.*, 1984, **49**, 4101.
204. G. Tsuchihashi, K. Tomooka and K. Suzuki, *Tetrahedron Lett.*, 1984, **25**, 4253.
205. K. Suzuki, E. Katayama and G. Tsuchihashi, *Tetrahedron Lett.*, 1984, **25**, 4997.
206. K. Suzuki, E. Katayama, T. Matsumoto and G. Tsuchihashi, *Tetrahedron Lett.*, 1984, **25**, 3715.
207. K. Suzuki, E. Katayama and G. Tsuchihashi, *Tetrahedron Lett.*, 1985, **26**, 1817.
208. K. Suzuki, K. Tomooka, M. Shimazaki and G. Tsuchihashi, *Tetrahedron Lett.*, 1985, **26**, 4781.
209. K. Suzuki, T. Ohkuma, M. Miyazaki and G. Tsuchihashi, *Tetrahedron Lett.*, 1986, **27**, 373.
210. H. Honda, E. Morita and G. Tsuchihashi, *Chem. Lett.*, 1986, 277.
211. K. Suzuki, K. Tomooka, E. Katayama, T. Matsumoto and G. Tsuchihashi, *J. Am. Chem. Soc.*, 1986, **108**, 5221.
212. Y. Fukutani, K. Maruoka and H. Tamamoto, *Tetrahedron Lett.*, 1984, **25**, 5911.
213. J. Fujiwara, Y. Fukutani, M. Hasegawa, K. Maruoka and H. Yamamoto, *J. Am. Chem. Soc.*, 1984, **106**, 5004.
214. K. Maruoka, S. Nakai, M. Sakurai and H. Yamamoto, *Synthesis*, 1986, 130.
215. K. Ishihara, A. Mori, I. Arai and H. Yamamoto, *Tetrahedron Lett.*, 1986, **26**, 983.
216. K. Ishihara, N. Hanaki and H. Yamamoto, *J. Am. Chem. Soc.*, 1991, **113**, 7074.
217. K. Wade and A.J. Banister, *The Chemistry of Aluminium, Gallium, Indium and Thallium*, Pergamon Press, Oxford, 1975.
218. H. Gilman and R.G. Jones, *J. Am. Chem. Soc.*, 1940, **62**, 2353.
219. S. Araki, H. Ito and Y. Butsugan, *Appl. Organomet. Chem.*, 1988, **2**, 475.
220. S. Araki, H. Ito and Y. Butsugan, *J. Org. Chem.*, 1988, **53**, 1831.
221. S. Araki, H. Ito, N. Katsumura and Y. Butsugan, *J. Organomet. Chem.*, 1989, **369**, 291.
222. L.-C. Chao and R.D. Rieke, *J. Org. Chem.*, 1975, **40**, 2253.
223. S. Araki, H. Ito and Y. Butsugan, *Synth. Commun.*, 1988, **18**, 453.
224. S. Araki, N. Katsumura, K. Kawasaki and Y. Butsugan, *J. Chem. Soc., Perkin Trans. 1*, 1991, 499.
225. S. Araki, N. Katsumura and Y. Butsugan, *J. Organomet. Chem.*, 1991, **415**, 7.
226. S. Araki and Y. Butsugan, *Bull. Chem. Soc. Jpn.*, 1991, **64**, 727.
227. S. Araki, N. Katsumura, H. Ito and Y. Butsugan, *Tetrahedron Lett.*, 1989, **30**, 1581.
228. S. Araki, T. Shimizu, P.S. Johar, S.-J. Jin and Y. Butsugan, *J. Org. Chem.*, 1991, **56**, 2538.
229. (a) S. Araki, T. Shimizu, S.-J. Jin and Y. Butsugan, *J. Chem. Soc., Chem. Commun.*, 1991, 824; (b) R. Nomura, S.-I. Miyazaki and H. Matsuda, *J. Am. Chem. Soc.*, 1992, **114**, 2738.
230. M. Falorni, L. Lardicci and G. Giacomelli, *Main Group Met. Chem.*, 1988, **11**, 49.
231. M. Falorni, L. Lardicci and G. Giacomelli, *Tetrahedron Lett.*, 1985, **26**, 4949.
232. K. Utimoto, C. Lambert, Y. Fukuda, H. Shiragami and H. Nozaki, *Tetrahedron Lett.*, 1984, **25**, 5423.
233. Y. Fukuda, S. Matsubara, C. Lambert, H. Shiragami, T. Nanko, K. Utimoto and H. Nozaki, *Bull. Chem. Soc. Jpn.*, 1991, **64**, 1810.
234. S. Araki and Y. Butsugan, *J. Chem. Soc., Chem. Commun.*, 1989, 1286.
235. T. Maeda, H. Tada, K. Yasuda and R. Okawara, *J. Organomet. Chem.*, 1971, **27**, 13.
236. A.G. Lee, in *Organometallic Reactions*, eds. E.I. Becker and M. Tsutsui, Wiley–Interscience, New York, 1975, Vol. 5, p. 1.
237. A. McKillop and E.C. Taylor, *Adv. Organometal. Chem.*, 1973, **11**, 147.
238. E.C. Taylor and A. McKillop, *Acc. Chem. Res.*, 1970, **3**, 338.
239. E.C. Taylor and A. McKillop, *Aldrichemica Acta*, 1970, **3**, 4.

240. A. McKillop and E.C. Taylor, in *Comprehensive Organometallic Chemistry*, eds. G. Wilkinson, F.G.A. Stone and E.W. Abel, Pergamon Press, Oxford, 1982, Vol. 7, p. 465.
241. A. McKillop and E.C. Taylor, in *Organic Synthesis by Oxidation with Metal Compounds*, eds. W.J. Mijs and C.R.H.I. de Jonge, Plenum Press, New York, 1986, p. 695.
242. S. Uemura, in *The Chemistry of the Metal-Carbon Bond*, ed. F.R. Hartley, Wiley, Chichester, U.K., 1987, p. 473.
243. G.B. Deacon and D. Tunaley, *Aust. J. Chem.*, 1979, **32**, 737.
244. E.C. Taylor, F. Kienzle, R.L. Robey, A. McKillop and J.D. Hunt, *J. Am. Chem. Soc.*, 1971, **93**, 4845.
245. H.C. Bell, J.R. Kalman, J.T. Pinhey and S. Sternhell, *Tetrahedron Lett.*, 1974, 3391.
246. S.W. Brener, G.M. Pickles, J.C. Podesta and F.G. Thorpe, *J. Chem. Soc., Chem. Commun.*, 1975, 36.
247. E.C. Taylor, H.W. Altland, R.H. Danforth, G. McGillivray and A. McKillop, *J. Am. Chem. Soc.*, 1970, **92**, 3520.
248. R.A. Hancock and S.T. Orszulik, *Tetrahedron Lett.*, 1979, 3789.
249. C.B. Anderson and S. Winstein, *J. Org. Chem.*, 1963, **28**, 605.
250. A. McKillop, J.D. Hunt, F. Kienzle, E. Bigham and E.C. Taylor, *J. Am. Chem. Soc.*, 1973, **95**, 3635.
251. P. Abley, J.E. Byrd and J. Halpern, *J. Am. Chem. Soc.*, 1973, **95**, 2591.
252. D. Farcasiu, P.v.R. Schleyer and D.B. Ledlie, *J. Org. Chem.*, 1973, **38**, 3455.
253. E.C. Taylor, C.-S. Chiang and A. McKillop, *Tetrahedron Lett.*, 1977, 1827.
254. A. McKillop, O.H. Oldenziel, B.P. Swann, E.C. Taylor and R.L. Robey, *J. Am. Chem. Soc.*, 1973, **95**, 1296.
255. G. Zweifel and S.J. Backlund, *J. Am. Chem. Soc.*, 1977, **99**, 3184.
256. S.R. Abrams, *Can. J. Chem.*, 1983, **61**, 2423.
257. Syntex Corporation, Ger. Pat. 2,322,932 (1973); *Chem. Abstr.*, 1974, **80**, 47707.
258. LEK Tovarna Farmacevtskih in Kemicnih Izdelkov, *Fr. Demande* 2,367,727 (1978); *Chem. Abstr.*, 1979, **90**, 103656.
259. A. McKillop, B.P. Swann and E.C. Taylor, *J. Am. Chem. Soc.*, 1971, **93**, 4919.
260. A. McKillop, B.P. Swann and E.C. Taylor, *J. Am. Chem. Soc.*, 1973, **95**, 3340.
261. S.D. Higgins and C.B. Thomas, *J. Chem. Soc., Perkin Trans. 1*, 1982, 235.
262. E.C. Taylor, C.-S. Chiang, A. McKillop and J.F. White, *J. Am. Chem. Soc.*, 1976, **98**, 6750.
263. E.C. Taylor, R.L. Robey, K.-T. Liu, B. Favre, H.T. Bozimo, R.A. Conley, C.-S. Chiang, A. McKillop and M.E. Ford, *J. Am. Chem. Soc.*, 1976, **98**, 3037.
264. J.A. Walker and M.D. Pillai, *Tetrahedron Lett.*, 1977, 3707.
265. J.A. Miller and R.S. Matthews, *J. Org. Chem.*, 1992, **57**, 2514.
266. J.A. Walker, US Pat. 4,135,051 (1979); *Chem. Abstr.*, 1980, **90**, 137517.
267. D.J. Rawlinson and G. Sosnovsky, *Synthesis*, 1973, 567.
268. A. McKillop, J.D. Hunt and E.C. Taylor, *J. Org. Chem.*, 1972, **37**, 3381.
269. A.J. Irwin and J.B. Jones, *J. Org. Chem.*, 1977, **42**, 2176.
270. E. Mincione, P. Barraco and M.L. Forcellese, *Gazz. Chim. Ital.*, 1980, **110**, 515,
271. J.E. Byrd and J. Halpern, *J. Am. Chem. Soc.*, 1973, **95**, 2586.
272. S. Uemura, H. Miyoshi, A. Toshimitsu and M. Okano, *Bull. Chem. Soc. Jpn.*, 1976, **49**, 3285.
273. H.M.C. Ferraz, T.J. Brocksom, A.C. Pinto, M.A. Abla and D.H.T. Zocher, *Tetrahedron Lett.*, 1986, **27**, 811.
274. J.P. Michael and M.M. Nkwelo, *Tetrahedron Lett.*, 1990, **46**, 2549.
275. J.P. Michael, P.C. Ting and P.A. Bartlett, *J. Org. Chem.*, 1985, **50**, 2416.
276. T. Hosokawa, M. Hirata, S.-l. Murahashi and A. Sonoda, *Tetrahedron Lett.*, 1976, 1821.
277. P.A. Bartlett and C. Chapuis, *J. Org. Chem.*, 1986, **51**, 2799.
278. A. Kaye, S. Neidle and C.B. Reese, *Tetrahedron Lett.*, 1988, **29**, 1841.
279. P. Kocovsky, V. Langer and A. Gogoll, *J. Chem. Soc., Chem. Commun.*, 1990, 1026.
280. V. Simonidesz, A Behr-Papp, J. Ivanics, G. Kovacs, E. Baltz-Gacs and L. Radics, *J. Chem. Soc., Perkins Trans 1*, 1980, 2572.
281. H.-J. Kabbe, *Ann. Chem.*, 1962, **656**, 204.
282. Y. Yamada, S. Nakamura, K. Iguchi and K. Hosaka, *Tetrahedron Lett.*, 1981, **22**, 1355.

283. E.C. Taylor, F. Kienzle and A. McKillop, *J. Am. Chem. Soc.*, 1970, **92**, 6088.
284. A.D. Ryabov, S.A. Deiko, A.K. Yatsimirsky and I.V. Berezin, *Tetrahedron Lett.*, 1981, **22**, 3793.
285. S. Uemura, Y. Ikeda and K. Ichikawa, *J. Chem. Soc., Chem. Commun.*, 1971, 390.
286. P.M. Henry, *J. Org. Chem.*, 1970, **35**, 3083.
287. R.A. Kjonaas and D.C. Shubert, *J. Org. Chem.*, 1983, **48**, 1924.
288. A.K. Yatsimirsky, S.A. Deiko and A.D. Ryabov, *Tetrahedron*, 1983, **39**, 2381.
289. T. Spencer and F.G. Thorpe, *J. Organomet. Chem.*, 1975, **99**, C8.
290. S. Uemura, H. Miyoshi, M. Wakasugi, M. Okano, O. Itoh, T. Izumi and K. Ichikawa, *Bull. Chem. Soc. Jpn.*, 1980, **53**, 553.
291. R.C. Larock, S. Varaprath, H.H. Lau and C.A. Fellows, *J. Am. Chem. Soc.*, 1984, **106**, 5274.
292. R.C. Larock, C.-L. Liu, H.H. Lau and S. Varaprath, *Tetrahedron Lett.*, 1984, **25**, 4459.
293. J.M. Davidson and G. Dyer, *J. Chem. Soc. A*, 1968, 1616.
294. R.C. Larock and C.A. Fellows, *J. Org. Chem.*, 1980, **45**, 363.
295. R.C. Larock and C.A. Fellows, *J. Am. Chem. Soc.*, 1982, **104**, 1900.
296. M. Uemura, S. Tokuyama and T. Sakan, *Chem. Lett.*, 1975, 1195.
297. M. Ochiai, M. Arimoto and E. Fujita, *Tetrahedron Lett.*, 1981, **22**, 4491.
298. M. Ochiai, E. Fujita, M. Arimoto and H. Yamaguchi, *Chem. Pharm. Bull.*, 1983, **31**, 86.
299. M. Ochiai, E. Fujita, M. Arimoto and H. Yamaguchi, *Chem. Pharm. Bull.*, 1982, **30**, 3994.
300. I.E. Markó and J.M. Southern, *J. Org. Chem.*, 1990, **55**, 3368.
301. I.E. Markó, J.M. Southern and M.L. Kantam, *Synlett*, 1991, 235.
302. I.E. Markó and M.L. Kantam, *Tetrahedron Lett.*, 1991, **32**, 2255.

8   The coordination and solution chemistry of aluminium, gallium, indium and thallium

D.G. TUCK

8.1   Introduction

It is obviously necessary that a monograph devoted to the chemistry of the heavier elements of Group 13 should include a review of the coordination chemistry, since it can be argued that the growth of interest in the chemistry of these metals has been driven to a considerable extent by the developing interest in their coordination compounds. In this regard, there is an interesting contrast between boron, which has extensive areas of importance other than coordination chemistry, and the heavier metallic elements dealt with in this chapter. We can begin with the simple definition that a co-ordination compound is the product of the association of a base with an acid, using the Lewis definitions of acid and base to identify molecules that can accept or donate a pair of electrons. This approach makes a somewhat arbitrary, but nevertheless useful, distinction between coordination compounds and derivatives of organometallic compounds, in which there is a substantial amount of metal–carbon bonding, and which are treated in chapter 6. There is a massive literature on the coordination compounds of the four elements in question, partly because MX_3 compounds are readily available electron-pair acceptors, and this chapter is therefore intended as a general review rather than a detailed and comprehensive compilation of all the coordination complexes which have been prepared. Several reviews of the coordination chemistry of both inorganic and organometallic compounds have appeared,[1–7] and there is no purpose in merely extending these by lists of papers, so that references are given in general to reviews, and to work that has appeared since 1984 which effectively closes the period surveyed in *Comprehensive Coordination Chemistry*.[3,4] Rather an attempt is made to review the overall trends and factors which appear to this author to have defined the coordination chemistry of these elements in the past, and at the present.

Since chemistry is an experimental science, it is not surprising that the development of the coordination chemistry of the four elements has been driven by the availability of experimental methods. For many years the chemists working in these areas had, in addition to the normal techniques of synthetic chemistry, access to those structural methods based on vibrational — and particularly infrared — spectroscopy, and on the limited use of nuclear magnetic resonance spectroscopy for the investigation of behaviour in solution. Some authors also attempted to use nuclear quadrupole resonance spectroscopy as a means of investigating bond character, but it was not until the ready availability of X-ray crystallography that this area of coordination chemistry really flourished, and the very large number of structural determinations now available in the literature has contributed greatly to our understanding. Since most of the classical pioneering workers did not have access to this technique, but rather relied on other, more 'sporting' methods of structure determination, it is only right to pay tribute at the outset to those who successfully exercised their skills in this field before the advent of modern crystallographic methods.

It is also worth noting the marked contrast here, as elsewhere, between the rate of progress in Main Group chemistry and transition-metal chemistry. In the latter field, the impressive developments of the 1950s and 1960s depended in large measure on the use of magnetochemistry and electronic spectroscopy as delicate and incisive probes of the ligand-metal interaction. These techniques have no application in Main Group chemistry, with the exception of certain studies of specific ligands, and in consequence the rapid progress made in transition-metal chemistry during the post-war development of inorganic chemistry was not immediately reflected in the study of Main Group compounds. Equally, the chemistry of the heavier Main Group elements has lacked the theoretical probes that have been applied to the structures of compounds of the lighter elements. The problems of performing meaningful theoretical molecular orbital calculations on elements with as many electrons and levels as (say) indium and thallium have for the most part resisted the attack of theoreticians. It follows that there is not much definitive information about the bonding in the coordination compounds of these four elements,[8] and that much of what can be stated in this context is essentially descriptive.

Finally, the important role of solution chemistry in the development of the coordination chemistry of these elements must be emphasised. The study of stability constants has depended to some extent on the availability of suitable metal electrodes, and since indium and, to a lesser extent, thallium have proved very suitable in this regard, there is a large number of stability constant determinations on the complexes of these elements. Some of these have been reviewed elsewhere, and it is worth pointing out in this context that as much or more is known about the aqueous solution chemistry of these elements as is known about the corresponding chemistry of many transition-metal elements. We return to this topic later, and to the relationship between solution chemistry and the solid phase.

Despite the lack of definitive experimental methods in the early investigations, there is now a very wide and instructive understanding of the coordination chemistry of aluminium, gallium, indium and thallium. In this chapter, the stress is on the different types of structures that have been reported, on the range of coordination numbers that is known, and on some of the areas where our knowledge is far from complete.

8.2   Hydride complexes

8.2.1   Derivatives of MH_3

The synthesis and properties of alane and gallane, and the absence of authenticated evidence on the formation of the indium and thallium analogues, are discussed in section 3.2.1, and the organometallic derivatives are referred to in chapter 6. The anionic species MH_4^- and MH_6^{3-}, and the neutral adducts, such as $L \cdot MH_3$ where L is a Lewis base, form the basis of this section. These compounds have played a fascinating part in the development of the chemistry of inorganic hydrides,[9,10] and the importance of their use in organic synthesis is probably equal to that of $AlCl_3$ and related Friedel–Crafts catalysts. They are of equal significance in inorganic and organometallic chemistry, since they provide a convenient entry to the synthesis of hydride derivatives of transition metals. The neutral adducts of AlH_3 and GaH_3 are readily obtained by displacing Et_2O with an amine, the etherate being the form in which the parent hydrides are most commonly prepared. An alternative route utilises a reaction such as

$$LiMH_4 + [Me_3NH]Cl \rightarrow Me_3N \cdot MH_3 + LiCl + H_2 \qquad (8.1)$$

and a third method is by reduction of (say) $Me_3N \cdot AlX_3$ with LiH. No such reactions have been reported for indium or thallium, in keeping with the absence of the parent hydride. The range of adducts known for AlH_3 encompasses both amine and ether donors; here, as in the halide system, ammonia does not form stable derivatives.

Structurally, the 1:1 adducts have the expected pseudo-tetrahedral stereochemistry at the Group 13 metal, and a sequence of stabilities[11] has been established, in the general order $N > P > O$, but with some interesting, but as yet unexplained, fine structure. Higher adducts are known for AlH_3, and $AlH_3(NMe_3)_2$ has been shown to have a trigonal bipyramidal structure,[12] but the derivative of the bidentate $N(C_2H_4)_3N$ is a polymer in which planar AlH_3 units are linked by the ligand.[13] Bridging AlH_5 units have also been identified in transition-metal complexes.[14] A significant lengthening of the Al–H bond is observed with increasing coordination number. The structure of the gallane adduct $Me_3N \cdot GaH_3$ has been shown to be pseudo-tetrahedral.[15] The compound $[H_2GaNMe_2]_2$, obtained by thermal decomposition of $H_3Ga \cdot NHMe_2$, has a Ga_2N_2 ring as its core.[16] Adducts are also known for mixed hydrido-halogeno compounds such as $AlHBr_2$, AlH_2I, etc.

Anionic derivatives of the type $LiAlH_4$, $NaGaH_4$, etc. can be prepared by the appropriate reduction process,

$$MX_3 + 4LiH \rightarrow LiMH_4 + 3LiX \qquad (8.2)$$

and for $NaAlH_4$, direct combination of the elements at high temperature and pressure is a useful commercial route. The indium analogue $LiInH_4$ is known, but $LiTlH_4$ readily decomposes, so that the stability of these compounds clearly decreases markedly down the Group. In addition to the MH_4^- derivatives, the compound Li_3AlH_6 can be prepared by the thermal decomposition of $LiAlH_4$. The predicted tetrahedral structure for an MH_4^- anion is generally found in practice, with $r(Al-H) = 1.55$ Å in $LiAlH_4$, and 1.532 Å in $NaAlH_4$.[17] Significant distortions occur in the lattice, however, as a result of interactions between the H^- ligands and Group 1 cation; in keeping with this, there are strong ion–ion interactions in non-aqueous solution, as shown by both conductance and NMR studies.[18] Compounds between AlH_4^- and transition-metal moieties have been reviewed by Bulychev.[19]

8.2.2 Borohydride derivatives

It is convenient to include a discussion of the derivatives of BH_4^- at this point, since their chemistry is interwoven with that of the hydrides of aluminium and gallium, and, to a lesser extent, of indium and thallium. The aluminium compound $Al(BH_4)_3$, one of the earliest borohydrides to be reported, can be prepared by the reaction of B_2H_6 with AlH_3, $LiAlH_4$ or Me_3Al, or from $NaBH_4 + AlCl_3$. It is an extremely volatile and hazardous liquid, which has a D_{3h} prismatic molecular structure in all phases. The molecule is fluxional in solution, and the bonding involves both two-electron $B-H_t$ and three-centre-two-electron $B-H_{br}-Al$ bonds.[20] Adducts with monodentate nitrogen donors are known, as are the mixed hydride-borohydride species $HAl(BH_4)_2$ and H_2AlBH_4,[21–23] anions such as $[HAl(BH_4)_3]^-$, etc., organo-compounds, e.g. $MeAl(BH_4)_2$, and the mixed chloro species $Cl_{3-n}Al(BH_4)_n \cdot OEt_2$ and $[ClAl(BH_4)_3]^-$. Other aluminium boranes can be formed by the thermal decomposition of $Al(BH_4)_3$, or by treating $Al(BH_4)_3$ with higher boron hydrides, and there is, in principle, an extensive coordination chemistry arising from such molecules.

The gallium compound $Ga(BH_4)_3$ has not been prepared, but the reduction of $GaCl_3$ by $LiBH_4$ produces $HGa(BH_4)_2$. In the gas phase, this molecule has a GaH_5 kernel, with one $Ga-H_t$ bond and four $Ga-H_{br}-B$ bonds.[24] The indium analogue, $In(BH_4)_3$, is known only as the tetra-hydrofuran adduct, and an unstable $In(AlH_4)_3$ has also been reported, as have compounds of the type $Tl(MH_4)_2Cl$ (M = B or Al). Not surprisingly, there is no structural information on these compounds.

8.3 Halide complexes

8.3.1 Introduction

The complexes formed with halide ligands comprise one of the most important areas of the coordination chemistry of aluminium, gallium, indium and thallium, including as it does the formation of neutral adducts, of anionic complexes and of mixed ligand species, the competitive formation of neutral and ionic species, and the factors which govern the stability of these various complexes.

This section deals firstly with the chemistry of the adducts of the trihalides with monodentate donors and their stabilities, and then examines studies of polydentate donor derivatives, after which the anionic complexes are reviewed. Some features of these discussions reappear in the later treatment of solution chemistry.

8.3.2 Neutral complexes with monodentate ligands

The study of the neutral adducts of the boron trihalides, and of other, related boron compounds, which afforded for many years one of the most fruitful fields of inorganic chemistry, was driven both by the intrinsic interest of the subject itself, and by the relevance of the behaviour and structure of these adducts, and those of the corresponding aluminium compounds, to the important topic of the mechanism of the Friedel–Crafts reaction.[25] The catalytic effect of such compounds has been discussed in detail elsewhere, and much useful information on the coordination chemistry of the Group 13 elements was garnered as a result of the extensive work in this area. The contrast between boron, whose adducts are essentially all of the 1 : 1 variety, and aluminium, gallium and indium is striking, since combining acid : base ratios of 1 : 1, 1 : 2, 1 : 3 and even (apparently) 1 : 4 have been reported for the heavier elements. It is proper here to acknowledge the pioneering work of Coates, whose studies of gallium adducts gave others the incentive to pursue both the inorganic and the organometallic chemistry of these elements. Another important reason for the investigation of the neutral adducts lies in the formation of stable products by alkane-elimination from boron adducts, leading as it does to boron–nitrogen heterocycles, a feature which prompted the investigation of the formation and decomposition of similar adducts of aluminium and gallium. Although not within the purview of the present chapter (but see chapter 3), the importance of this is encouraging structural studies should be recognised.

It would be satisfying to be able to point here towards a discussion to be found at the end of the section presenting a complete analysis of the factors which determine the coordination number of the Group 13 metal and the thermodynamic stability of these compounds. There are certainly some interesting trends which illustrate dependency upon both the metal and the ligand, but no sound theoretical basis has yet been developed, largely because of the lack of an unambiguous experimental technique for determining bond strengths in complexes.

The literature contains a number of reviews of the complexes that are formed when a Lewis base reacts with a trihalide of a Group 13 metal. These include reviews of the aluminium halides by Post and Cotz,[26] and by Dalibert and Deroualt,[27] and of the adducts of gallium, indium and thallium by Pidcock,[28] in addition to the general reviews already cited.[1-7] In the light of this existing extensive literature, and in keeping with the philosophy set out in the Introduction, there will be no attempt to give comprehensive lists of the many hundreds of adducts that have been reported; rather is it more instructive to try to identify the main themes and the important factors. There are various ways of organising the material, either by ligand or by metal, and, given that such a choice is arbitrary, the following account reviews the complexes that are formed by the trihalides in the order of increasing atomic number of the metal.

This is probably a suitable point to comment on the increasing, and deplorable, tendency to refer to AlX_3 and similar Lewis acids as being 'electron-deficient'. The original, and correct, use of this term was to identify compounds such as B_2H_6 in which there are insufficient electrons to form a classical structure based on two-centre-two-electron bonds. Aluminium trihalides and their analogues are not electron-deficient by this definition, but they are electron-pair acceptors, a property simply explained by their tendency to bond to donor molecules in a way that allows the central atom to achieve a stable electronic configuration, often described as the 'inert gas' configuration, although this is not a unique prerequisite. The two phenomena, electron-deficiency and electron-pair acceptance, are quite different, even though they have a common root in the electronic properties of specific atoms; to confuse the terminology is to refuse to acknowledge the individual importance of each of the two forms of behaviour.

8.3.2.1 Aluminium. Aluminium trifluoride has all the characteristics of an ionic compound in the crystalline state, with six-coordinate aluminium. Several hydrates are known ($AlF_3 \cdot nH_2O$; $n = 1$, 3 or 9), but there appears to be no report of other neutral adducts of AlF_3, in keeping no doubt with its ionic nature. The remaining halides are very effective Lewis acids. All are dimeric in the gas phase and in non-donor solvents, and the bromide and iodide are dimeric in the solid phase; crystalline aluminium trichloride has a layer lattice, with six-coordinate Al, changing to the dimeric Al_2Cl_6 at the melting point. These bridged dimers have been carefully studied in all three phases, with the structural details given in section 3.2.2.2. The dimerisation is clearly the result of halide cross-donation, and itself evidence of the strong acceptor character of the aluminium centre.

There has been reported a large number of adducts of AlX_3 (X = Cl, Br or I) with monodentate nitrogen donor molecules, the combining ratios being 1 : 1, 1 : 2, etc. Before examining these, it is important to emphasise that a range of compounds can be formed under different experimental conditions, here and elsewhere, and the $AlCl_3$–CH_3CN system illustrates the point nicely. The 1:1 adduct is known, and is presumably pseudo-tetrahedral, like $Me_3Al \cdot NCCH_3$,[29] but the 1:2 adduct, which one might have expected to have D_{3h} core symmetry, is in fact ionic, with $[(CH_3CN)_5 AlCl]^{2+}$, $AlCl_4^-$ and CH_3CN units in the crystal lattice.[30] In acetonitrile solution, octahedral $[Al(NCCH_3)_6]^{3+}$ cations, and the mixed species $[(CH_3CN)_5AlCl]^{2+}$, cis- and trans-$[(CH_3CN)_4AlCl]^+$, and $CH_3CN \cdot AlCl_3$ have all been detected by ^{27}Al NMR studies.[31] Even the bis-ammine has been shown to be $[trans\text{-}AlCl_2(NH_3)_4]^+[AlCl_4]^-$ in the solid state,[32] and yet mononuclear D_{3h} bis-adducts can and do exist, as in the case of $AlCl_3$-(morpholine)$_2$.[33] Perhaps the important conclusion to be drawn from such findings is that one cannot refer to 'a stable complex', or even less to 'the most stable complex', without making careful delineation of the conditions under which the system in question has been studied. Apart from the fundamental problems implicit in this caveat, it follows that derivatives of $AlCl_3$ (and other MX_3 species) with a given ligand will necessarily be discussed at different points in this chapter, in that both neutral and ionic derivatives of a given MX_3/L system are possible, indeed probable.

The simplest route to these various adducts is by direct addition, using a solution of AlX_3 in a weak donor such as Et_2O. In addition to adducts of amines, similar compounds are known with other nitrogen donors, and with phosphorus, arsenic, antimony and bismuth species, all with the metal in the +3 state. With pyridine (py), structural investigations show that $AlCl_3 \cdot py$ is a simple 1:1 molecular adduct, that $AlCl_3py_2$ is $[trans\text{-}AlCl_2py_4]^+[AlCl_4]^-$, and that $AlCl_3py_3$ is the mer-isomer of a neutral molecule with pseudo-octahedral stereochemistry.[34] As with the acetonitrile system above, different patterns of addition are clearly possible, and it is quite feasible that this is a general conclusion, applying even to those systems for which only 1:1 adducts have been reported to date.

Complexes formed with $Al \leftarrow O$ bonds follow a pattern similar to those with $Al \leftarrow N$ coordination. Ethers, carbonyl compounds, carboxylates, sulfoxides, and the like, give predominantly adducts with 1:1 stoichiometry, although some 1:2 adducts have been reported. The tetrahydrofuran[18] system illustrates that the possible range of behaviour is similar to that for the CH_3CN system. The 1:1 adduct with $AlCl_3$ or $AlBr_3$ is pseudo-tetrahedral in the solid state, but in solution there is evidence for a series of ionic species; the crystalline bis-adduct $AlCl_3 \cdot 2thf$ obtained from Et_2O solution has a trigonal-bipyramidal structure, but the ionic isomer $[AlCl_2(thf)_4]^+ [AlCl_4]^-$, which has been identified spectroscopically in CH_2Cl_2 solution, is the product obtained on crystallisation from thf/toluene mixtures.[35–37] Evidence of a different type testifying to the delicate balance between different coordination states is seen in the formation of only $X_3Al \cdot OEt_2$ for $X = Cl$, Br or I, but of both $X_3Al \cdot OMe_2$ and $AlX_3(OMe_2)_2$; given the comments above on the CH_3CN and thf systems, these may represent the tip of the iceberg of a large number of possible etherates. Inorganic oxygen donors such as $OPCl_3$, $ONCl$, SO_2, etc., have been studied. The structural results are in keeping with the discussion above, in that with $OPCl_3$, for example, both $1:1$ and $1:2$ adducts have been identified; conductivity and NMR results indicate the existence of $[AlL_6]^{3+}$ and $AlCl_4^-$ species in solution, and no doubt mixed ligand intermediates are also formed under the appropriate conditions. Similarly, with $NOCl$ the product is $[NO]^+[AlCl_4]^-$ rather than a neutral adduct.

Complexes of especial interest in terms of Friedel–Crafts reactions include those with carbonyl donors.[38] With CH_3COCl and $AlCl_3$, the $1:1$ adduct has the structure $[CH_3CO]^+[AlCl_4]^-$ in the crystal, but behaves as the neutral adduct in dichloromethane, with $Al \leftarrow O$ linking, also found in other analogues. An extensive list of relative donor strengths has been published,[39] being based on competitive ligand-displacement studies, with $COCl_2 < CH_3NO_2 < C_6H_5NO_2 < C_6H_5COCl < OMe_2 < Ph_2CO \approx OEt_2 < HCONMe_2(dmf) < MeCN < Cl^-$. Given the important effect of phase and halide ligand noted above, one can assume that this order is valid only under the experimental conditions used (see section 8.3.4).

Various adducts of AlX_3 with dialkyl sulfides are known, but they add little to what has already been stated about the overall chemistry of this important class of compounds. Extensive crystallographic studies of R_3PE adducts ($R = Ph$ or NMe_2; $E = S$ or Se) have been published.[40]

8.3.2.2 Gallium and indium.

It is convenient to take these elements together, in part because the range of coordination numbers is generally larger than for either aluminium or thallium. Interestingly, there is a deal of structural information on such compounds sometimes lacking in the case of aluminium, and this may well be due to the coincident development of crystallographic methods and an increasing interest in the chemistry of these two elements. As with aluminium, it will be impossible to make a strict separation between the different aspects of the complexes, since neutral and anionic species are frequently related in a given donor-acceptor system. Equally with aluminium, the trifluorides of gallium and indium are essentially ionic, and few neutral adducts are known; $InF_3 \cdot 3H_2O$ forms a lattice composed of In^{3+}, F^- and H_2O, and the ammonia analogue is presumably of the same type.

The remaining parent MX_3 compounds are generally halogen-bridged M_2X_6 dimers in the solid state, although $InCl_3$ is an exception to this statement, a circumstance reflected in some of the coordination chemistry of this compound. Ammoniates of varying stoichiometries have been reported, but no structural information has been obtained, in part because of their poor thermal stability. With amines, the products may be $1:1$, $1:2$ or $1:3$ adducts depending on the ligand, metal and halide involved, and these compounds have been shown by crystallographic and spectroscopic methods to have C_{3v} ($1:1$), D_{3h} ($1:2$) or distorted octahedral stereochemistry, respectively; for example, a study of $InCl_3(NMe_3)_2$ shows this molecule to be trigonal bipyramidal.[41] No comparable 1 : 3 adducts are known for the gallium trihalides, although six-coordination is not unusual in gallium(III) chemistry. Similar results have been found with nitrogen heterocyclic ligands. An interesting feature of these adducts is that thermogravimetric and related measurements have shown that loss of (e.g.) pyridine occurs under identifiable conditions, so that there is a sequence of decomposition reactions:

$$MX_3L_3 \overset{-L}{\to} MX_3L_2 \overset{-L}{\to} MX_3L \overset{-L}{\to} MX_3 \qquad (8.3)$$

Such findings imply that caution must be used in discussing the stability of adducts in anything other than a semi-quantitative fashion, as noted for the AlX_3 systems above. Many adducts with phosphorus and, to a lesser extent, arsenic donors are known for both elements. The 1 : 1 species are pseudo-tetrahedral, and the 1 : 2 species trigonal bipyramidal molecules; the much-quoted case of $InCl_3 \cdot 2PPh_3$ has apical phosphine ligands. Complexes with bidentate phosphorus ligands are reviewed below.

The complexes formed with oxygen and sulfur donors follow the pattern seen with aluminium, allowing for the ease with which higher coordination numbers can be achieved, especially for indium, and lists of such complexes have been published. The neutral 1 : 2 adducts have been shown to have trigonal bipyramidal stereochemistry (e.g. InX_3L_2, for $L = Et_2O$), but the danger of relating stoichiometry and structure too closely and uncritically is exemplified by $InI_3 \cdot 2dmso$,[42] whose solid-state structure is $[cis\text{-}InI_2(dmso)_4]^+[InI_4]^-$, but whose solution chemistry involves $[In(dmso)_6]^{3+}$ species (see section 8.6.1). Some interesting structures have been found in adducts of $InCl_3$ and $InBr_3$ with Me_3PO, $PhMe_2PO$, Ph_2MePO, Ph_3PO and dmso; both fac- and mer-isomers have been identified in InX_3L_3 molecules, along with $[InX_2L_4]^+[InX_4]^-$ for $L = Ph_3PO$ and $X = Cl$ or Br.[43] Sulfur donors form 1 : 1, 1 : 2 and 1 : 3 derivatives with trihalides of both gallium and indium, with the higher coordination numbers being more prevalent in the case of the heavier element. Adducts of InX_3 ($X = Cl$ or Br) with Me_3MS ($M = P$ or As) are trigonal bipyramidal, with apical S atoms, while the mixed species $InCl_3(SAsMe_3)_2(OH_2)$ takes the form of a distorted octahedron.[44]

8.3.2.3 Thallium.

Adduct-formation, like any other chemical reaction, can be observed only if there is no other energetically preferred competing process. In the case of thallium(III), which is strongly oxidising in both aqueous and non-aqueous media, this means that many ligands are oxidised by the metal rather than being complexed. Since both oxidation and co-ordination involve electron transfer from the donor to the thallium(III) centre, this competition is not surprising, but the end result is to preclude the existence of many of the possible analogues of those species found for the lighter elements.

Thallium(III) halides do not form ammoniates, or derivatives of amines, but pyridine adducts have been studied, and both neutral ($TlX_3 \cdot 2py$) and ionic ($[TlX_2L_4]^+[TlX_4]^-$) formulations have been proposed on the basis of their vibrational spectra. Oxygen donor adducts are few in number, but there are some interesting structural findings, with trigonal bipyramidal TlX_3O_2 kernels being identified. Because of the oxidising power of TlX_3, no neutral adducts of donors containing sulfur, phosphorus, etc., have been prepared.

8.3.3 Neutral complexes with polydentate ligands

Any discussion of the composition or structure of the complexes formed when a neutral bidentate donor interacts with an MX_3 compound immediately underlines our ignorance of the exact nature of the competitive factors which are involved. It is clear that if X is a weakly coordinating ligand, such as ClO_4^- or NO_3^-, the product is a salt of the cationic complex $[ML_3]^{3+}(X^-)_3$, and these species are discussed in section 8.6.1. The halides (X = Cl, Br or I) may also give such cations, but this is not the only type of behaviour observed. Some examples will be discussed in order to illustrate the complexity of the problem.

Two types of complexation can be readily distinguished. In one, the halide ligand is completely displaced, to give an $[ML_3]^{3+}$ cation which can be identified crystallographically or spectroscopically in the solid state, and by the molar conductivity in non-aqueous solution. The ligands most commonly used in such work are 2,2'-bipyridyl (bpy), 1,10-phenanthroline (phen) and ethanediamine (en). With aluminium and bidentate donors, a variety of structural types is possible. With ligands such as 1,10-phenanthroline and 2,2'-bipyridyl, the crystalline products are the ionic species, e.g. $[Al(bpy)_3]Cl_3$ etc., but the solution chemistry of such compounds appears to reveal the same complexity as seen above for the $AlCl_3/CH_3CN$ system, since species formulated as $[AlX_2(bpy)_2]X$ have been reported for X = Cl or Br. Examples demonstrating the complete substitution of halide include $[Al(bpy)_3]I_3$, $[Ga(bpy)_3]Br_3$, $[Ga(phen)_3]X_3$ (X = Cl, Br or I) and $[Tl(en)_3]X_3$ (X = Cl or Br), showing that such behaviour can and does occur for each of the metals in the Group. In the second common mode of complexation, the stoichiometry may be MX_3L, $MX_3L_{1.5}$ or MX_3L_2, and each of these can be formulated as an ionic compound, so that, for example, $AlX_3 \cdot bpy$ can be written $[AlX_2(bpy)_2][AlX_4]$ (X = Cl or Br). This particular example suggests that one obviously important factor is ligand size, since for X = I under comparable conditions (reaction in acetonitrile), iodide is completely removed to give $[Al(bpy)_3]^{3+}$, presumably because a six-coordinate AlN_4I_2 kernel would be unstable. Several crystal structures have confirmed the existence of such ionic structures. The $MX_3L_{1.5}$ stoichiometry appears to be common for M = In and L = bpy or phen, and here the proposed structure is, typically, $[InX_2(bpy)_2][InX_4(bpy)]$, a formulation supported by conductivity measurements, spectroscopic data, and the preparation of salts such as $[Ph_4As][InCl_4(bpy)]$ and $[InCl_2(bpy)_2]ClO_4$.

Some possible reasons for the diversity of behaviour have been discussed previously, specifically in the case of indium,[2] but here as elsewhere a completely satisfactory rationale is lacking, and there are several reasons for this. Firstly, one must accept that preparative studies, especially those involving reactions in a single non-aqueous system, can never span the range of variables which are at issue, and that with four metals, three halides and a variety of ligands, the number of possible combinations is obviously high. Secondly, the crystallisation of a given compound from solution is not necessarily evidence that this is the predominant species in solution, and there are many cases which underline this point. Thirdly, and with relevance to the last point, the stabilisation of a given structure in the crystalline state is dependent on lattice energies, which may be in opposition to a process like halide transfer in a way which is presently unpredictable. To illustrate this, we may consider the hypothetical reaction which involves the transfer of

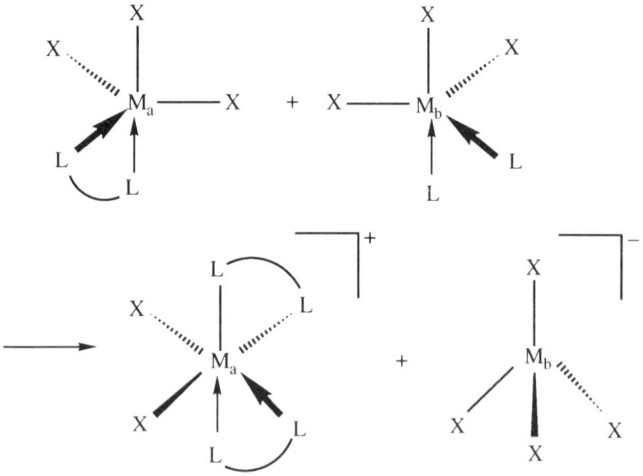

$$(8.4)$$

one bidentate ligand from M_b to M_a, and the concomitant transfer of a halide ion from M_a to M_b. The bond energy changes involved are not known, nor is the lattice energy of the product, nor are the rearrangement energies as M_a becomes six-coordinate and M_b tetrahedral in place of the initially five-coordinate molecules. Given these uncertainties, to use a gentle word, it would be foolish to attempt to identify the factors which determine the coordination state of such derivatives.

Other bidentate ligand systems which have been studied include 1,2-bis-(diphenylphosphino)ethane, which gives $GaX_3 \cdot \frac{1}{2}L$ and $InX_3 \cdot L$; none of these compounds has been structurally characterised. Compounds with monoglyme are discussed in section 8.6.4. The arsenic donor ligand *o*-phenylene*bis*(dimethylarsine) forms $InX_3 \cdot L$ for $X = Cl$ or Br and $InI_3 \cdot L_{1.5}$, and ionic structures have been proposed for these complexes; crystallographic studies would be valuable here. When two donors are present, a range of structural possibilities appears, and some interesting results have been found for $InCl_3$/bpy/H_2O adducts, and the related $InCl_3$(bpy)EtOH.[45]

The terdentate donor 2,2′,2″-terpyridyl forms mononuclear $MX_3 \cdot$ terpy for Al, Ga and In, and $X = Cl$, Br and I (not all combinations). The crystal structure of $GaCl_3 \cdot$ terpy shows a distorted octahedral $GaCl_3N_3$ kernel, with a range of M–Cl and M–N distances that are not easily rationalised, and a similar situation pertains in $InCl_3 \cdot$ terpy.[46] The aluminium compounds with $X = Cl$ or Br are isomorphous with $GaCl_3 \cdot$ terpy.

8.3.4   The stability of neutral adducts

One challenge in the study of any series of chemically related addition compounds is to identify the sequence of stability. Underlying such a question, which has received a variety of answers, is our understanding of the word 'stability' itself in this context. From the practical viewpoint of a synthetic chemist, a complex is stable if the reaction

$$D: + A \rightarrow D:A \qquad (8.5)$$

goes to completion under the conditions chosen, and this is entirely justified. The fact that the reaction of a given donor or acceptor with a number of different partners gives products in high (i.e. less than 100%) and slightly differing yields may then be taken to be due to differences in the thermo-dynamic factors which govern equation (8.5). The identification and rationalisation of these factors make a much more difficult task than that which faces the synthetic chemist, who is gratified if reaction (8.5) gives reasonably high yields, with a pragmatism which reflects the reality of synthesis, where yield and purity are generally accepted as the most important criteria.

The interpretation of such preparative work in fundamental terms is con-strained by the fact that reactions such as (8.5) generally involve both species in non-aqueous solution, or the addition of a gas (e.g. NH_3) to a solid, or other thermodynamically unattractive conditions. The process for which primary information is needed is

$$D(g) + A(g) \rightleftharpoons D:A(g) \qquad (8.6)$$

since $\Delta G^{\ominus}$ or $\Delta H^{\ominus}$ for different D and A leads to a meaningful comparison of the stabilities of DA, and hence to orders of donor or acceptor strength. Few of the complexes discussed above lend themselves to rigorous investigation by equation (8.6), so that it becomes necessary to revert to something more accessible. For example, ΔH measurements for the systems

$$D(g) + A(sol) \rightarrow D:A(sol) \qquad (8.7)$$

will yield results which allow a meaningful comparison of relative acceptor strength of A^1, A^2, etc., if it can be reasonably assumed that ΔH_{sol} is constant through the series of acceptors, and that a similar constancy applies to the set of D:A products. Given the restrictions imposed, such experiments have produced series of donor or acceptor strengths which are derived from thermochemical, as opposed to preparative, studies.

Unfortunately the conclusions of such experiments do not lead to a single order in the case of compounds of aluminium, gallium, indium and thallium. The available results have been reviewed by Greenwood and others;[47] hence the sequences listed in Table 8.1 emerge. The order in terms of all donor ligands and varied metal acceptors is

$$MX_3 > MPh_3 > MMe_3$$

for M = B, Ga, Al or In. The absence of values for thallium(III) compounds is easily understood in terms of the competing oxidation reactions discussed earlier, and one might be excused for concluding that this is perhaps the only thing that can be readily derived from Table 8.1.

Amongst the factors that contribute to the overall energy change in eq. (8.7), and hence to the results which form the basis of the table, are the following:

1. the D:A bond strength, from which it follows that variations in σ/π character from metal to metal, or donor to donor, or metal-ligand to metal-ligand, may affect the order;
2. the bond rearrangement energy associated with the change in stereochemistry at M, say, from D_{3h} to C_{3v} as MX_3 accepts a pair of electrons;
3. molecular energy level changes in the donor as donation occurs, especially in molecules with delocalised bonding systems such as py, or with d_π character as in PPh_3;

Table 8.1 Relative acceptor strengths of the Group 13 metal trihalides.[47]

Donor	Acceptor sequence
Amines, ethers	$MCl_3 > MBr_3 > MI_3$ (M = Al, Ga or In)
Phosphines	$MCl_3 > MBr_3 > MI_3$ (M = Ga)
Me_2S, Et_2S	$MCl_3 > MBr_3 > MI_3$ (M = Al)
C_4H_8S	$MI_3 > MBr_3 > MCl_3$ (M = Ga or In)
N, O donors	$AlX_3 > GaX_3$ (X = Cl or Br)
S donors	$GaX_3 > AlX_3$ (X = Cl or Br)
$CH_3COOC_2H_5$	$BCl_3 > AlCl_3 > GaCl_3 > InCl_3$
py	$AlPh_3 > GaPh_3 > BPh_3 \approx InPh_3$
py	$AlX_3 > BX_3 > GaX_3$ (X = Cl or Br)
Me_2S	$GaX_3 > AlX_3 > BX_3$ (X = Cl or Br)

4. stereochemical changes in the donor, as in PPh$_3$, AsPh$_3$, etc., on complexation;
5. conventional changes related to the electronic effect of different organic substituents on the availability of the electron pair of the donor molecule;
6. steric effects in the rearrangement terms in (2) and (4); and
7. the molecularity of the parent acceptor which may be monomeric (e.g. BX$_3$) or dimeric (e.g. In$_2$Br$_6$) under the particular experimental conditions.

The list is extensive, but certainly not exclusive, and serves to illustrate the complexity of identifying unambiguously the relative importance of the different terms in any given sequence. Such a conclusion may appear depressing after so much careful research; an alternative and more positive position is to stress the many challenges which remain in this area of Main Group chemistry.

8.3.5 Anionic complexes with halide ligands

There is a very extensive literature on the complex anions which are formed when halide ions interact with the M^{III} cation of one of the elements of this Group. The experimental results which have been reported have involved a variety of preparative techniques which are discussed below, the study of the products by vibrational spectroscopy, the use of NMR methods in non-aqueous and aqueous solution, and the application of X-ray crystallography. The last technique was applied at the level of powder diagrams before single-crystal techniques became generally available, so that in some cases there is information on the overall crystalline form and the lattice constants, but a lack of the detailed information which is normally provided by modern methods.

The preparation of anionic halide complexes can obviously be achieved simply by mixing the appropriate trihalide with a salt, typically one with a quaternary ammonium cation, in a non-aqueous solvent. This technique will certainly give products, many of which have been known for some years, and which were originally identified as double salts, but the product of such a reaction depends not only on the metal, but also on the halide and the cation in question, and occasionally on the solvent. These results point to two important factors that must be considered. One is that there are clearly equilibria in the solution phase which can be affected by both the metal and the ligand, and by the solvent; the other, equally significant, is that there are solid-state parameters, and notably the lattice energy, which will determine whether a product is stable in the solid state against dissociation into anions of lower coordination number. The interplay of the different factors cannot be predicted, or even quantitatively understood, at the present time, but an acceptance of their existence is important if the coordination chemistry of the Group 13 elements is to be properly appreciated. In particular, it should be emphasised that the species that appears in the solid state as the result of precipitation or crystallisation is not necessarily the major component of the solution from which the crystals have been obtained (cf. section 8.3.2 above, and the case of $AlCl_3$/thf in particular).

A second approach to the formation of anionic complexes is through the use of molten salts, and this is particularly relevant in view of the use of the cryolite process in the electrochemical extraction of aluminium from bauxite (see chapter 2). The AlF_6^{3-} anion is certainly present in aluminium fluoride melts, but studies have shown that four-coordinate AlX_4^- species can also be obtained in such melts,[48] and in those containing chloride or bromide. Since the factors that influence stability in molten salt media at high temperatures are clearly different from those which govern the behaviour of non-aqueous media at room temperature, it is not surprising that different and apparently conflicting results are obtained from the two preparative routes.

A final method is to heat a metal fluoride ($M'F$ or $M'F_2$) and an MF_3 compound ($M = Al$, Ga, In or Tl) together at high temperature, to cool the resultant mixture to room temperature, and to examine the phases that have been formed. It should be emphasised that the results obtained from such studies, in which powder crystallography has played an important role in identifying the products, will often be different from those obtained from room-temperature experiments. Of course, it follows that if one mixes, say, equimolar quantities of $M'F$ and MF_3, the product will necessarily have the stoichiometry $M'MF_4$; if this phase is stable as the reaction mixture cools, then a single well-defined phase will be formed, even if the compound in question cannot be obtained by other experimental methods.

The highly regular stereochemistry which is associated with these anionic complexes makes them ideal for study by vibrational spectroscopy, and the combination of far-infrared and Raman spectroscopy led to the complete identification of the structures of many of these molecular anions long before X-ray crystallographic methods made further information available. There seems little point in trying to list the results of these extensive studies, since the data are readily available, but it is quite clear that the regular tetrahedral and octahedral stereochemistries predicted by VSEPR theory are indeed the ones adopted by the anions in question. The interesting exception comes in the study of the relatively small number of five-coordinate complexes, and especially the unusual case of the anion $InCl_5^{2-}$ which has been the subject of a number of studies; there is no doubt that the structure is not the expected trigonal bipyramid, but rather a square-based pyramid. We return to this below.

Each of the elements in question has a readily accessible, magnetically active isotope, and in consequence the solution chemistry has been investigated by multinuclear NMR techniques (see section 1.3.1.1). The only drawback to these studies is that the linewidths are often significant, but fortunately the chemical shifts between one species and another are also large, and there is usually no difficulty in distinguishing the different species, provided that the symmetry of the molecule is high enough to allow observation of the NMR signal. The linewidth problem is most acute for indium, since the large magnetic moment ($I = 9/2$ for ^{115}In) means that the resonance is observed only for small, symmetrical molecules; it is also significant, to a lesser extent, for gallium. Despite these problems, there is a considerable body of information on the behaviour of the anionic halide complexes in non-aqueous media. The general conclusion is that tetrahedral and octahedral species can exist in such solutions, whether prepared by dissolving salts directly or by the extraction of the anion from an aqueous solution. The tetrahedral species in particular have lent themselves to study because of their high symmetry and, in addition to the identification of the MX_4^- anions, there have been studies of the formation of mixed halide systems, and the range of chemical shifts associated with the various ligand combinations has been measured for each of the elements.

The most convenient basis for outlining the results is to deal with fluoride derivatives separately, largely because the preparative methods used do not involve the non-aqueous room-temperature methods that have predominated in the study of the other halides. The remaining anionic complexes are then treated successively by coordination number.

8.3.5.1 Fluoride complexes. Several different stoichiometries are found in the anionic species containing aluminium and fluoride, but for the most part these arise by the solid phase interaction of AlF_6 octahedra. The parent octahedral AlF_6^{3-} unit is well characterised in salts such as Li_3AlF_6, Na_3AlF_6, etc., although even here M–F interactions (M = Li or Na) are significant. The sharing of two fluorine atoms by two AlF_6 units gives the stoichiometry AlF_5 (e.g. Tl_2AlF_5), and further sharing can lead to $Na_5Al_3F_{14}$ and $MAlF_4$ (e.g. $TlAlF_4$). At the same time, discrete AlF_4^- species have been identified in molten Li–Al–F systems, as the dissociation product of the process

$$AlF_6^{3-} \rightleftharpoons AlF_4^- + 2F^- \qquad (8.8)$$

so that the coordination number is clearly dependent upon the phase in question. A similar situation holds for gallium, since the octahedral GaF_6^{3-} anion is known, as are salts with anions formulated as GaF_5^{2-}, $Ga_3F_{14}^{5-}$, etc., which arise by the linking of GaF_6 units. High-temperature phase studies have also given rise to $MGaF_4$ (M = Na, K, Rb or Cs), while aqueous solutions yield GaF_6^{3-}, $[GaF_5(OH_2)]^{2-}$ and $[GaF_4(OH_2)_2]^-$ salts with NH_4^+, Rb^+ and Cs^+. The results for indium follow the same pattern and octahedral InF_6^{3-} has been characterised crystallographically. Thallium(III) compelexes can be obtained by high-temperature methods, but since Tl^{3+}(aq) forms only very weak complexes with fluoride ion, it is not surprising that analogous species have not been obtained from aqueous solutions.

The overall conclusion must be that all the elements of this group will form predominantly octahedral anionic complexes with fluoride by high-temperature routes, and that aqueous methods give the same, or substituted six-coordinate anions for gallium and indium. Such a pattern is not surprising, given the small size of the fluoride ligand, but there are still unanswered (or even unposed) questions about complexes of lower co-ordination number, in terms of both synthesis and structure.

8.3.5.2   Chloride, bromide and iodide complexes.   Salts of all three AlX_4^- species have been prepared by high-temperature methods, and via non-aqueous systems under more moderate conditions. Crystallographic methods have confirmed the slight distortions from regular tetrahedral symmetry identified earlier from the vibrational spectra. The ion $Al_2X_7^-$, or $[X_3Al(\mu\text{-}X)AlX_3]^-$, can be prepared in molten salt media, and from the reaction between CH_3COCl and AlX_3 (X = Cl, Br or I), and crystallo-graphic studies have confirmed the structure in which two tetrahedra are joined through a corner. There is NMR evidence for the anion $Al_3Cl_{10}^-$. By contrast with the heavier metals, aluminium does not apparently form anionic complexes with coordination numbers higher than four, except in the case of fluoride.

With *gallium*, four-coordination is again apparently the only known mode, with the exception that M_2GaBr_5 and M_3GaBr_6 have been prepared by high-temperature methods, but not structurally characterised. The GaX_4^- anions can be prepared by conventional non-aqueous routes, and can also be extracted from aqueous hydrohalic acid solution by basic organic solvents. Vibrational spectra establish the tetrahedral symmetry, which has been confirmed by X-ray measurements.

The anionic complexes of *indium* present a range of coordination numbers which some have found fascinating. High-temperature syntheses have produced compounds with the stoichiometries $MInCl_4$, M_2InCl_5 and M_3InCl_6, M_2InBr_5 and M_3InBr_6, and $MInI_4$ and M_3InI_6 for M = Na, K, Cs, etc., but reliable structural information is lacking. None of the ambient-temperature methods (reaction in non-aqueous media and extraction into basic solvents) has produced any iodo complex other than InI_4^-, a finding which emphasises the difficulties of comparing the results of ambient- and high-temperature experiments. Salts with tetralkylammonium and similar cations (M^+) were first prepared by Ekeley and Potratz,[49] who identified the various products $MInX_4$, M_2InX_5, M_3InX_6 and M_4InX_7. For X = I, as noted, only the first of these is obtained, irrespective of the size of the cation, while for X = Cl or Br, higher coordination numbers are associated with small cations, underlining the importance of lattice stabilisation in such solids. The substances M_4InX_7 were shown to be based on lattices of $4M^+$ +

$InX_6^{3-} + X^-$, with one uncoordinated halide ion, as found in some other Main Group salts. The InX_6^{3-} units, like those in M_3InX_6, are octahedral species, as shown by X-ray crystallography.

The anion $InCl_5^{2-}$ has been the subject of both X-ray and spectroscopic study,[50,51] and, while there may be arguments about the details of the molecular structure, there is no doubt that the stereochemistry is essentially that of a square-based pyramid. This is rare in Main Group chemistry, and all the more striking because the isoelectronic molecules $CdCl_5^{3-}$, $SnCl_5^-$ and (liquid) $SbCl_5$ are all trigonal bipyramids. Any speculation about the reasons for the difference can be stilled by the realisation that each of these four species exists in different lattice types, or phases, and by the knowledge that $[Ni(CN)_5]^{3-}$ can exist as both the D_{3h} and C_{4v} isomers in the same lattice, thereby establishing that the energy difference between the two states may indeed be very small. Further evidence of the difficulty in pursuing the matter quantitatively in our present state of knowledge comes from the fact that $(Et_4N)_2[InCl_5]$ recrystallises unchanged from dichloromethane, but gives $Et_4N[InCl_4]$ as the product when the solvent in question is ethanol. In keeping with this and other findings, $(Me_4N)_2InBr_5$ has been shown to be a lattice of $2Me_4N^+ + InBr_4^- + Br^-$, so that the $InCl_5^{2-}$ anion is apparently the only binary indium species with C_{4v} stereochemistry.

Despite the difficulties in preparing the parent TlX_3 compounds because of the oxidising power of *thallium(III)*, the anionic complexes are accessible through reactions in non-aqueous phases. Tetrahedral TlX_4^- anions (X = Cl, Br and I) have been obtained and characterised by X-ray and spectroscopic studies, and the octahedral $TlCl_6^{3-}$ and $TlBr_6^{3-}$ species are also well understood. The five-coordinate anions $TlCl_5^{2-}$ and $TlBr_5^{2-}$ have been identified in aqueous and non-aqueous solution by Raman spectroscopy. There is some dispute about the structure of the anion in $Cs_3Tl_2Cl_9$, which has been claimed variously to be $[Cl_3Tl(\mu\text{-}Cl)_3TlCl_3]^{3-}$, or a mixture of $TlCl_4^-$ and $TlCl_5^{2-}$ units.[52,53]

8.3.6 *The stabilities of anionic complexes*

We have already seen that several careful studies of the neutral adducts of MX_3 halides have produced sequences of donor or acceptor strength that do not allow any simple explanation in terms of the electronic and/or stereochemical factors involved, and it should therefore come as no surprise to find that the known range of anionic MX_n^{m-} complexes presents an equally difficult situation. A fundamental barrier to any quantitative analysis is that the results reviewed in section 8.3.5 are derived from preparative studies, and the nature of such work is that the absence of a given species (e.g. $InBr_5^{2-}$) may indicate its intrinsic instability in the solid state, or equally may tell us merely that the correct synthetic approach has yet to be found.

With due acceptance of these shortcomings in the available information, some generalities can be deduced from what is known. Fluoride gives the maximum coordination number of six with each +3 metal ion, so that it produces no differentiation between the metals. For chloride, the tendency to six-coordination increases with the size of the M^{3+} cation, as is the case for bromide, while iodo complexes are never more than four-coordinate. Caution should be exercised in explaining these results entirely in terms of the M–X interactions, since calculations[54] have shown that $X \cdots X$ inter-ligand repulsions are important, and in particular the singular existence of MI_4^- anions reflects the large forces that would be developed in the idealised MI_5^{2-} and MI_6^{3-} geometries. The M–X bonds are, however, clearly affected by the coordination state, since the bond distances change significantly in, say, the MCl_4^-, MCl_5^{2-} and MCl_6^{3-} anions. While all this is quantitatively sensible, there is no quantitative treatment to allow a predictive approach. It is proper to conclude this brief discussion by emphasising the solution-phase effects discussed below; crystallisation is a two-phase process, and the importance of solid-state lattice effects should not lead to any devaluation of solvation, dielectric constant, and similar factors in the liquid phase.

The range of coordination numbers found in crystalline complexes of an M^{3+}–Cl^- system can be approached in the following way. Consider the solution-phase equilibrium

$$MX_5^{2-} \rightleftarrows MX_4^- + X^- \qquad (8.9)$$

which has been identified for M = In or Tl and X = Cl, and for M = In and X = Br. The product of crystallisation in the presence of an excess of a cation M'^+ may be either $M_2'MX_5$ (i.e. $2M'^+ + InX_5^{2-}$) or $2M'^+ + InX_4^- + X^-$. A proper analysis of the thermodynamic cycle relating equation (8.9) to these lattices will involve terms such as

$$
\begin{array}{ccc}
MX_5^{2-}(sol) & \rightarrow & MX_4^-(sol) + X^-(sol) \\
\uparrow {\scriptstyle \Delta H(sol)} & \uparrow & \uparrow \\
MX_5^{2-}(c) & \rightarrow & MX_4^-(c) \quad + \quad X^-(c)
\end{array}
\qquad (8.10)
$$

and, since the solid-state stabilities can be discussed in terms of the lattice energies, one can also write

$$
\begin{array}{ccc}
M_2'MX_5(c) & \xrightarrow{\;U^{II}\;} & 2M'^+(g) + MX_5^{2-}(g) \\
\downarrow & & \uparrow \\
M'MX_4(c) + M'X(c) & \xrightarrow{\;U^{I}+U^{0}\;} & 2M'^+(g) + MX_4^-(g) + X^-(g)
\end{array}
\qquad (8.11)
$$

The only possible conclusions here are that bond energies, rearrangement energies, lattice energies and enthalpies of solution must be considered in any rigorous treatment of such systems, and that of course we do not have enough information to identify which of these, if any, is of predominant importance. The effect of cation size is generally that large cations lead to low coordination numbers in the balancing anions, thus emphasising the lattice energy factor, without leading to any quantitative enlightenment.

8.3.7 Anionic mixed halide and halide/other ligand complexes

The existence of $MX_nY_{4-n}^-$ anions in non-aqueous solution is readily established by NMR spectroscopy. For aluminium, the series with $X = I$, $Y = Cl$ or Br has been studied in CH_2Cl_2 and CH_2Br_2, and the chemical shifts of the various mixed ligand anions have been measured. These anions are also formed in molten salt media.[55] Similar results have been reported for gallium complexes in NMR and Raman investigations, and crystalline derivatives of $GaCl_3Br^-$, etc., have been prepared. There are three important differences between the aluminium and gallium systems. One is the magnitude of the change in chemical shift from MCl_4^- to MI_4^-, another is that the rate of halide-exchange is appreciably slower for the gallium complexes in non-aqueous media (but not in aqueous solution), and the third is the apparent absence of $MX_2Y_2^-$ species in the gallium system. This area has recently been reviewed.[56]

Exchange is rapid in InX_4^-/InY_4^- mixtures in non-aqueous solution, and the whole possible range of mixed species was identified for mixtures of Cl^-, Br^- and I^-. Vibrational spectra of $InX_2Y_2^-$ anions, prepared by the oxidation of InX_2^- by Y_2, have been reported, and X-ray analysis of salts of $InCl_3Br^-$ and $InBr_3Cl^-$ confirm that these anions have C_{3v} stereochemistry. Thallium has not been as well investigated in this context, although the structures of $TlCl_3X^-$ ($X = Br$ or I) and similar anions have been solved by X-ray methods.

No mixed halide complexes have been reported for five- or six-coordinate anions for any of the elements under discussion, but several six-coordinate mixed ligand anionic species have been prepared and characterised. These include MX_5L^{2-} and $MX_4L_2^-$ species, and the aquo complexes have received special attention. There is a complete set of $[InX_4(OH_2)_2]^-$ and $[InX_5(OH_2)]^{2-}$ for $X = Cl$ and Br, and the variations in bond length, etc., have been discussed.[57] Other ligands include urea and thiourea, while complexes with bidentate neutral ligands are numerous, since these $[MX_4L]^-$ species have been prepared independently, and are also invoked in discussing the structures of $MX_3L_{1.5}$ adducts, formulated as $[MX_2L_2]$ $[MX_4L]$ (see section 8.3.3).

8.4   Complexes with Group 16 donors

8.4.1   Oxygen donor complexes

The adducts formed by the trihalides and neutral oxygen donors, such as Me_2O, dmso, etc., have already been noted in section 8.3.2. The cationic complexes are best discussed in terms of the solution behaviour of the M^{III} state (section 8.6.1), and it is sufficient to record here that a large number of ML_6^{3+} species has been prepared and characterised. The thallium salt $[Tl(pyNO)_8](ClO_4)_3$ is known, but no structural information is available on what may be an eight-coordinate cationic complex (pyNO = pyridine-N-oxide).

8.4.2   Complexes with Group 15 oxyanions

Hydrated aluminium *nitrates* contain the $[Al(OH_2)_6]^{3+}$ cation in the lattice, and dehydration by heating yields a basic nitrate $Al(OH)_2NO_3 \cdot 1.5H_2O$, which also lacks Al–$NO_3$ complexation, but the reaction of N_2O_5 with $AlCl_3$ gives $Al(NO_3)_3 \cdot N_2O_5$, which has been formulated as $NO[Al(NO_3)_4]$. The analogous $Et_4N[Al(NO_3)_4]$ can be prepared by the reaction of $AlCl_3$ with Et_4NCl and N_2O_4. Salts with Rb^+ or Cs^+ are of the type $M_2[Al(NO_3)_5]$ or $M[Al(NO_3)_4]$, and crystallographic studies show that the anion in the former has one bidentate and four monodentate nitrate ligands. For gallium, the hydrated salts and the basic nitrate appear to be similar in structure to the aluminium compounds, but salts of $[Ga(NO_3)_4]^-$ are the only products in anhydrous systems; salts of $[In(NO_3)_4]^-$ have been prepared in non-aqueous media, and appear to be isostructural with the gallium species. Thallium nitrate hydrate, $Tl(NO_3)_3 \cdot 3H_2O$, has been shown to contain bidentate asymmetric nitrate groups bonded to the metal ion, and the anionic $[Tl(NO_3)_4]^-$ complex is again accessible *via* the reaction of N_2O_5 with $TlNO_3$. The distinctive coordination chemistry of these complexes is typical of the interesting and unusual ligating properties of the nitrate anion, and is in contrast to that observed with other oxyanion ligands.

Aluminium *phosphates* are of some commercial importance, and the solution- and solid-phase structures have therefore been investigated by a variety of methods. Strong complexing of aluminium occurs in aqueous phosphoric acid (see also chapter 9). The detailed model is complicated by the presence of different phosphate anions (PO_4^{3-}, HPO_4^{2-}, etc.), but the major complex is believed to be $[Al(HPO_4)_3]^{3-}$, although several other entities have been identified, including binuclear species. Mixed fluoride phosphate complexes are also reported to exist in aqueous solution. Tetrahedral AlO_4 units are present in crystalline $AlPO_4$, but in other structures the octahedral AlO_6 group predominates, so that in $Al(PO_3)_3$, for example, infinite chains of PO_4 and octahedral AlO_6 are found, and the

AlO_6 unit is also present in $NaAl_3(PO_4)_2$ and $Al(H_2PO_4)_3$. The fluoro-phosphates $M(PO_2F_2)_3$ (M = Al or Ga) presumably have similar six-coordinate units with both oxygen and fluorine ligands. The physiological properties of the Al^{3+}/2,3-diphosphoglycerate system have been reviewed.[58]

Gallium-phosphate complexes such as $Ga(HPO_4)^+$, $Ga(H_2PO_4)^{2+}$, etc., have been observed in aqueous solution, and various phosphate and pyro-phosphate solid derivatives have been prepared. The octahedral InO_6 unit is believed to be central to the structures of solid indium(III) phosphate, fluorophosphate and methylphosphonate, and similar derivatives. There have been some studies of complexing in aqueous solution, and although $InPO_4$ is insoluble in water, complexes with the $H_2PO_4^-$ ligand have been detected. In $TlPO_4$, the TlO_6 kernel is formed with oxygen atoms from six different PO_4^{3-} groups, and a similar situation pertains in the $TlO_4(OH_2)_2$ unit of $TlPO_4 \cdot 2H_2O$. The bidentate ligand $PO_2Cl_2^-$ also gives rise to TlO_6 octahedra in $Tl(O_2PCl_2)_3$.

Although there are references in the literature to derivatives of Group 13 metals with arsenate, there does not appear to have been any structural investigation of these species.

8.4.3   Complexes with Group 16 oxyanions

Aluminium-*sulfate* systems have attracted considerable interest, partly because of the industrial importance of aluminium sulfate, and partly because the alums are of such structural interest. Here, as elsewhere, ^{27}Al NMR spectroscopy has been used to advantage in the investigation of solution equilibria. Six-coordinate complexes of the type $[Al(OH_2)_{6-n}(SO_4)_n]^{(3-2n)+}$ have been identified, as have protonated derivatives. The alums $M^IM^{III}(SO_4)_2 \cdot 12H_2O$ (M^{III} = Al, Ga, etc.) are now known to involve the $[M^{III}(OH_2)_6]^{3+}$ cation,[59] but the thermal decomposition of these double salts is variously reported to generate $Al–SO_4$ complexes, which are also present in the anhydrous double sulfates $M^IM^{III}(SO_4)_2$ (M^{III} = Al or Ga). The salt $NO[Al(SO_4)_2]$ can be obtained by treating $NO[AlCl_4]$ with sulfuric acid. The Al-*selenate* systems have been investigated in less detail, but are clearly analogous to the sulfates, whereas the *tellurates* have not apparently been prepared.

Octahedral AlO_6 coordination of Al^{3+} has also been found in *tungstates*, of formulae $Al_2(WO_4)_3$ and $M^IAl(WO_4)_2$, and polytungstates and *poly-molybdates* have been investigated (e.g. $AlW_{12}O_{40}^{5-}$), and here again the gallium complexes follow similar patterns, with six-coordination being the characteristic structural feature.

Indium(III) sulfates and double salts have been known for many years, and a variety of six-coordinate InO_6 kernels has been identified in these substances. The salt $NH_4In(SO_4)_2$ is believed to involve infinite chains of $[In(SO_4)_2]^-$ units, leading to a trigonal prismatic stereochemistry, but in

$(NH_4)_3In(SO_4)_3$ one finds octahedral InO_6 units in which each oxygen belongs to a different sulfate as the repeating unit. *Trans*-$[InO_4(H_2O)_2]$ entities have been identified crystallographically in various hydrated indium sulfate lattices, and in the selenate derivative $(NH_4)[In(SO_4)_2] \cdot 4H_2O$. Weak complexing of In^{3+}(aq) by sulfate ions has been treated quantitatively. Thallium(III) sulfate, $Tl_2(SO_4)_3$, and various hydrates and double salts are known, and are presumed to contain TlO_6 units in the lattice, and this has been confirmed crystallographically in the case of $Tl_2(OH)_2(SO_4)_2 \cdot 4H_2O$. The selenate system is similar, but the unusual tellurate Tl_6TeO_{12} consists of TeO_6 octahedra so arranged as to give seven-coordinate TlO_7 entities.

With this last exception, it seems fair to summarise the overall picture as illustrating the importance of MO_6 octahedra, distorted octahedra, or trigonal prisms in the solid-state coordination chemistry of Group 16 oxyanions with aluminium, gallium, indium and thallium. The marked tendency to form M—O bonds in the crystalline state contrasts with the weak complexing ability of sulfate in aqueous solutions. Not for the first time in this field, phase appears to be a most important factor to be considered in understanding the coordination chemistry.

8.4.4   Group 14 oxyanion complexes

The basic *carbonates* of aluminium have not been structurally characterised; the parent compound is unknown and it appears that this is also the case for the heavier Group 13 elements. The basic carbonate $NH_4[Ga(OH)_2CO_3]$ has gallium coordinated by a bidentate carbonate ligand. The *silicates* of these elements, and especially of aluminium, have received extensive study, but lie outside the scope of this review; a review of the kinetic and thermodynamic behaviour of aluminosilicates under blood plasma conditions[60] underlines the increasing importance of coordination chemistry in physiological studies (see also chapter 9).

Large numbers of *carboxylates* of each element are known, and in many cases stability constant measurements have shown that complexing occurs in aqueous media to give ML^{2+}, ML_2^+ and ML_3 complexes. In the aluminium system, ^{27}Al NMR spectroscopy has confirmed these results for L = lactate and citrate, for example. Given the complexity of aqueous solution equilibria, the existence of mixed M/L/OH species is not surprising. A careful review of this area has been published.[61] Crystalline complexes of the type $MAl(O_2CR)_4$ can be prepared from solution, or by the reaction of $RAlMe_4$ with CO_2 under forcing conditions. Both bridging and monodentate carboxylate ligands have been identified; in $Me_4N[Me_3AlO_2CMe]$, the aluminium atom is four-coordinate. Both normal and basic carboxylates of gallium(III) are known, but structural results are more readily available for the organo- derivatives, such as $MeGa(O_2CMe)_2$ in which both GaO_4

and GaO_5 units are present. More is known about gallium oxalate complexes, and $Ga(ox)_2^-$ and $Ga(ox)_3^{3-}$ anions have been identified by vibrational spectroscopy and X-ray crystallography.[62] Gallium complexes with ligands such as lactate, tartrate and citrate find application in physiological studies, since radioactive gallium isotopes are important in imaging work (section 1.3.1.3).

Indium carboxylates have also been reported, but again much of the structural information relates to organometallic species. The metal is six-coordinate in both $Me_2In(O_2CCH_3)$ and $Et_2In(O_2CCH_3)$, with stacking of the molecules to produce $InC_2O_2O_2'$ coordination, while mononuclear units with $InCl_2O_2N_2$ kernels are present in $Cl_2In(O_2CPh)py_2$. Lewis acidity is also a property of $In(O_2CR)_3$, with 1:1 and 1:3 adducts being identified. The bpy and phen adducts of $In(O_2CMe)_3$ are unusual in that eight-coordination occurs, with InO_6N_2 kernels. The acceptor properties of Me_2InO_2CMe are related to the breakdown of the parent polymeric lattice, while the unusual $(Me_4N)_3In(O_2CMe)_6$ is one of a number of derivatives of $In(O_2CMe)_3$. The oxalato complexes include both $M_3[In(ox)_3]$ and $M[In(ox)_2(OH_2)_2]$, and in such complexes some interesting stereochemistry still remains to be explored. Eight-coordination has been found in $NH_4[In(ox)_2(OH_2)_2]$, with Archimedean antiprismatic geometry, while in $[In_2(ox)_3(OH_2)_4] \cdot 2H_2O$ the structure is based on pentagonal bipyramidal units with oxalate bridging to give infinite chains.

In thallium acetate, $Tl(O_2CMe)_3$, three acetates chelate to the metal, but longer Tl–O bonds link these units to give TlO_8 chains, and eight-coordination is also found in the acetate monohydrate. The change from four-coordinate aluminium to eight-coordinate thallium through five-, six- and seven-coordination illustrates the importance of the size of the metal atom, but there may be other factors involved, and this area is worth further structural investigation.

8.4.5 Hydroxide and alkoxide complexes

Aluminium hydroxide, normally obtained as a gelatinous precipitate, can be crystallised to give a lattice in which Al^{3+} ions are in octahedral sites, while the related substance AlO(OH) has a structure in which –Al–O–Al–O chains are bridged by OH groups. At least two forms of GaO(OH) are known; the metal is six-coordinate in both, but trigonal prismatic stereochemistry has been identifed in one form. Indium(III) hydroxide is also based on octahedral distribution of OH around the metal; polymeric core-linked species are formed in the early stages of the hydrolysis of In^{3+} (aq) ions when the pH rises above about 3.0. There are no reports on the structure of the thallium analogues. The role of hydroxide derivatives in the hydrolysis of the M^{3+} ions is discussed in section 8.6.2.

Compounds that can be represented as hydroxide complexes of the metals are known. The relationship between a substance such as $Cs_3[Al(OH)_6]$, in which octahedral $[Al(OH)_6]^{3-}$ units are present, and the AlO_4, AlO_5 and AlO_6 polyhedra found in polycrystalline aluminates, is a complicated one and beyond the scope of this review. A similar comment applies to the corresponding solid-state chemistry of the gallium and indium systems, and the main conclusion in the present context is that variations in coordination number and in stereochemistry are the rule rather than the exception.

Alkoxides of aluminium have been known for many years. They are readily obtained by treating $AlCl_3$ with the appropriate alcohol, or by the direct reaction of the metal plus alcohol in the presence of a suitable activating agent. As noted in section 3.2.3.4, the molecularity of these compounds is strongly dependent on the organic group in question, since dimeric, trimeric and tetrameric species are known, and bridging and terminal OR ligands have been identified, with both four- and six-coordinate aluminium in the solid state. In some cases (e.g., $Al(OPr^i)_3$) the oligomeric structure is maintained in the vapour phase. The compounds have the inherent Lewis acidity of AlX_3 species, and form adducts of the type $L \cdot Al(OR)_3$ by cleaving the oligomer. Complexes of $Al(OR)_3$ and other Main Group and transition-element alkoxides are known, and may involve either discrete $[Al(OR)_4]^-$ units, or bridging between Al and the second metal (e.g. Mg) by alkoxide to give low-melting, volatile covalent substances. Alkoxide hydrides, such as $LiAlH_2(OBu^t)_2$, mixed alkoxide halide complexes (e.g. $ROAlCl_2$), mixed alkoxides, and organometallic compounds (e.g. $[Me_2AlOMe]_3$) are amongst the mixed ligand species which are derived from the parent structure. Oligomerisation is a common feature of such derivatives, unless bulky ligands are used to suppress cross-linking.[63–65]

The gallium complexes follow a similar pattern, and have been more fully explored than those of indium. The latter element yields insoluble compounds because of the tendency of $In(OR)_3$ to form six-coordinate indium by cross-linking; the extent to which such bonding can proceed is shown by the structure[66] of the *iso*-propoxide complex, $In_5O(OPr^i)_{13}$, properly written as $In_5(\mu_5\text{-}O)(\mu_3\text{-}OPr^i)_4(\mu_2\text{-}OPr^i)_4(OPr^i)_5$. No thallium(III) compounds have been reported.

8.4.6 Complexes of β-ketoenolate, diolate and related ligands

Neutral compounds of aluminium, gallium and indium with β-ketoenolate ligands have been known for many years and in fact $Al(acac)_3$ (acac = acetylacetonate) was first reported in 1887. The classical synthesis is generally straightforward, involving reaction of the appropriate M^{3+} salt with the ligand in aqueous ammonia, but metal atom synthesis, direct electrochemical synthesis, and the reaction of Hacac with an amalgam have also been successfully used. No thallium(III) complexes have been reported, and this is again presumably the result of the strong oxidising properties of the cation. In addition to acac itself, ligands with C_6H_5, CF_3, and related substituents have been used to prepare a wide range of ML_3 complexes. The structures of $M(acac)_3$ have been established for each element by X-ray crystallography.

Structural investigations have confirmed the presence of an MO_6 distorted octahedral kernel in all these compounds, the distortion being due to the bite of the bidentate ligand. In addition to the crystal structure investigations, there has been considerable interest in the processes by which these stereochemically non-rigid molecules rearrange in non-aqueous solution (see Ref. 67 for a recent interview). Ligand-exchange can be studied by NMR methods, or by using radiotracer techniques. We should first note that Al(acac)$_3$ has been resolved into its diastereoisomers by column chromatography at low temperature, so that there is a significant barrier to interconversion under these conditions, although this resolution has not been repeated for complexes of either gallium or indium. For aluminium species, it is necessary to invoke bond-breaking to give a five-coordinate intermediate with one monodentate ligand when the ligand carries a CF_3 group; an intramolecular twist mechanism has been used to explain the interconversion of enantiomers of other complexes, for which the estimated energy barrier is $85\,kJ\,mol^{-1}$. For gallium and indium complexes, ligand exchange proceeds *via* a five-coordinate intermediate.

Organometallic derivatives of the type $R_2M(acac)$ can be prepared by the reaction of R_3M and Hacac; other mixed ligand species include $[(acac)_2Al(OH_2)_2]^+$, $[(acac)Ga(dmf)_4]^{2+}$ and $[(acac)_2Ga(dmf)_2]^+$, and $[Al(OR)_2(acac)]$,[68] and it seems likely that the preparation of other similar compounds could be achieved. The indium complexes $X_2In(acac)L$ (X = Cl, Br or I; L = phen, bpy, etc.) were obtained in an unusual reaction in which InX is oxidised in refluxing Hacac; $Cl_2In(acac)bpy$ has an $InCl_2O_2N_2$ six-coordinate kernel. Monothio-β-diketonate compounds have also been prepared and structurally characterised.[69]

Tropolonate ML_3 complexes of aluminium and indium are known, as is the anionic complex InL_4^-. Derivatives of γ-pyrone and 3-hydroxy-4-pyridone have been intensively studied by Orvig and his co-workers because of their potential application in medicinal chemistry.[70-71]

Diolate or semi-quinonate complexes of gallium and indium can be prepared by the oxidation of the metal or of the monohalide; gallium metal with dbbq ($= 3,5$-di-tert-butyl-1,2-benzoquinone) gives $Ga(dbb\dot{s}q)_3$, which has a GaO_6 kernel.[72] This compound, like the indium and thallium analogues, can also be prepared from $MX_3 + 3Na^+dbb\dot{s}q^-$. Indium monohalides and o-quinones yield semi-quinone or catecholate derivatives, depending on the oxidising power of the quinone, by one-electron transfer reactions.[73,74]

8.4.7 Sulfur, thiolate and selenolate complexes

Adducts of organic sulfides with the trihalides of the Group 13 elements have been noted in section 8.3.2. Thiolates of organoaluminium residues, such as R_2AlSR, have been synthesised and structurally characterised,[75] but the parent $Al(SR)_3$ compounds do not appear to have been prepared. A similar comment applies to gallium, but indium species are known, since $In(SPh)_3$ can be prepared by metathesis between InX_3 and $NaSPh$, by the direct oxidation of indium metal by Ph_2S_2,[76] or by direct electrochemical synthesis.[77] With Ph_2Se_2 the second method yields $In(SePh)_3$, whose structure is that of a homopolymer in which each indium is coordinated by six selenium atoms in a highly distorted octahedral arrangement.[78] Adducts of $In(SPh)_3$ can be prepared; $In(SPh)_3py_2$ has a distorted trigonal bipyramidal geometry,[78] while $[BrIn(SPh)_3]^-$ has C_{3v} symmetry in the $InBrS_3$ core.[79]

Anionic complexes of gallium $(Ga(SR)_4^-; R = Me, Et, Ph, etc.)$ have a tetrahedral GaS_4 kernel, and four-coordination is also found in the $[Ga_2S_2(SPh)_4]^{2-}$ anion, which incorporates a Ga_2S_2 ring.[80] Gallium thiolate iodide compounds are dimeric, with μ-SR ligands,[81–82] similar in structure to the $[S_2Ga(\mu\text{-}S)_2GaS_2]^{6-}$ unit in $Ba_3Ga_2S_6$.[83]

Mixed neutral thiolate (or selenolate) halide complexes of indium can be synthesised through the oxidation of InX by Ph_2S_2 (or Ph_2Se_2) to give $XIn(SPh)_2$, which is readily converted to adducts such as $XIn(SPh)_2py_2$ or $[InX_2(SPh)_2]^-$.[84] The oxidation of elemental indium by a mixture of Ph_2S_2 and elemental iodine also gives $IIn(SPh)_2$. The use of these direct methods with thallium gives thallium(I) derivatives.[73] An interesting and unusual selenium derivative is the $[In_2Se_{21}]^{4-}$ anion in which $InSe_5$ trigonal bipyramids are formed by the fusion of $InSe_5$ rings joined by an Se_5 link.[85]

8.4.8 Bidentate sulfur ligand derivatives

Two classes of ligand which have been much studied with these Main Group elements are dithiocarbamates and dithiolenes. The preparative routes in each case may involve metathesis starting with MX_3, but, in addition, direct electrochemical methods,[86] and CS_2 insertion into $M(NR_2)_3$, have been used for dithiocarbamates. The molecule $Al(S_2CNMe_2)_3$ is mononuclear, with six-coordinate aluminium, but dimers are also known, as are mixed ligand complexes (e.g. $[ClAl(S_2CNMe_2)_2]_2$) and compounds with CS_2, COS and CO_2 ligation. The structure of $Ga(S_2CNEt_2)_3$ shows almost prismatic coordination in the GaS_6 kernel. A number of RCS_2 ligands (e.g. dithiobenzoate, alkyl xanthate, etc.) form neutral InL_3 complexes, and both pseudo-octahedral and trigonal prismatic stereochemistries have been identified; an InS_6 kernel is also present in the *tris-O,O*-diethyldithiophosphate complex. The uncertainties of coordination chemistry are nicely illustrated by the compound In(dithizonate)$_3$, where two of the N–S ligands are bidentate and one is monodentate (S-bonded).

The reaction of Me$_3$Al with toluene-3,4-dithiol (H$_2$TDT) gives a number of Al–S bonded species. The TDT^{2-} anion and the related maleonitrile-dithiolate (MNT^{2-}) and 1,1-dicyanoethylene-2,2-dithiolate (i-MNT^{2-}) anions feature in indium complexes prepared typically by the reaction of InCl$_3$ with the sodium salt of the appropriate ligand. The product anions show a range of In–S coordination numbers, from four in, for example, [In(MNT)$_2$]$^-$, to six as in [In(i-MNT)$_3$]$^{3-}$. The insertion of InX (X = Cl, Br or I) into the S–S bond of 1,2-bis(trifluoromethyl)-dithiete affords a polymeric product; this in turn gives monomeric derivatives such as [InL(dmso)$_4$]$^+$ (L = (CF$_3$)$_2$C$_2$S$_2$) when treated with strong donors. The InS$_2$C$_2$ ring is common to all these compounds, and to those derived from the reaction of Me$_3$In with H$_2$TDT.

Thallium gives a series of complexes with MNT^{2-} (TlL$_2^-$, TlL$_2$X^{2-} and TlL$_3^{3-}$), but, in contrast to the indium case, TlL$_2^-$ does not appear to be an acceptor. The reaction of Tl$^+$ and TDT^{2-} gives the thallium(III) species Tl(TDT)$_2^-$, implying some interesting redox chemistry. Other TlL$_2^-$ anions where L is a bidentate sulfur donor are also known.

8.5   Group 15 derivatives

8.5.1   Complexes of neutral mono- and bidentate nitrogen donors

The formation of adducts between the trihalides MX$_3$ and nitrogen donors has already been discussed at various points in this chapter, as have the mixed ligand anionic complexes of these elements. Studies of cationic and other nitrile derivatives are conveniently dealt with in section 8.6 dealing with solution chemistry.

8.5.2   Amide and imide complexes

The simple M(NH$_2$)$_3$ compounds of the Group 13 elements have not all been prepared, but derivatives are known, as are numerous substituted amido compounds (section 3.2.5.2). The reaction of either aluminium or gallium with sodium in liquid ammonia gives NaM(NH$_2$)$_4$ and, in the case of M = Al, Na$_2$Al(NH$_2$)$_5$; both [M(NH$_2$)$_4$]$^-$ anions have almost tetrahedral symmetry. The reaction of KGa(NH$_2$)$_4$ in ammonia with NH$_4$Cl produces Ga(NH$_2$)$_3$. Amide-rich complexes like Na$_2$Al(NH$_2$)$_5$ may be based on linked M(NH$_2$)$_6$ octahedra.

Dialkylamido compounds are readily accessible, and are easier to handle than the simple amide analogues. Treatment of LiAlH$_4$ with Me$_2$NH gives LiAl(NMe$_2$)$_4$, while the reaction between the element and the amine produces $Al(NMe_2)_3$. This compound and the gallium analogue are dimeric in the solid state with a four-membered M_2N_2 ring,[87] but compounds with more bulky ligands, such as $Al[N(SiMe_3)_2]_3$, are three-coordinate and planar. A similar situation pertains in the gallium system. $In[N(SiMe_3)_2]_3$, one of the few related indium compounds, is also planar and monomeric, although in this case an adduct with Me_3PO has been reported, implying that a coordination number in excess of three can be attained.

A number of mixed ligand complexes of aluminium have been studied, with the stoichiometries $X_2Al(NR_2)$ and $XAl(NR_2)_2$ (X = H, halide or alkyl). The hydrido species can be prepared by heating a mixture of Al with, for example, Et_2NH and H_2 under pressure. Equilibration of $Al(NMe_2)_3$ with $Me_3N \cdot AlH_3$ gives H_2AlNMe_2, which is a trimer, with a six-membered Al_3N_3 ring, whereas $HAl(NMe_2)_2$ is dimeric with an Al_2N_2 ring, and this unit is also found in $[X_2AlNMe_2]_2$ (X = Cl, Br or I). Like the parent tris-(dimethylamido) compounds, the mixed ligand species show Lewis acidity, giving 1:1 adducts such as $py \cdot Al(NMe_2)_3$. There have been extensive NMR studies of the solution behaviour of these molecules. Some gallium and indium analogues have also been reported, and four-membered M_2N_2 rings are common, but not invariant, features of their stereochemistry.

Two interesting bidentate anionic nitrogen ligands merit special attention. One is the diphenyltriazene anion, dpt⁻, which gives rise to species such as $[InCl_2(dpt)_2]^-$, $InCl_2(dpt)(bpy)$, etc., and to $Tl(dpt)_3$;[88] extensive crystallo-graphic studies of the indium compounds have been published.[89] The second ligand is an anionic derivative of the neutral molecule 1,4-di-t-butyl-1,4-diazabutadiene formed as a result of the reaction of Al (or Ga) or Al plus $LiAlH_4$ with the parent compound. As described in section 6.6, the products are deeply coloured and paramagnetic, with MN_4 kernels.[90–92] They may not be the M^{II} derivatives they were thought at first to be, but they should give rise to some interesting Main Group chemistry.

Derivatives of pyrazole are more properly part of organometallic chemistry, but mention should be made of the extensive studies carried out on the use of pyrazolyl-gallate ligands in complexes of transition-metal elements.[93] The *bis*(pyrazolylborate) derivatives of gallium include $[H_2B(pz)_2]_2GaCl$ and $[H_2B(pz)_2]_3Ga$ (pz = pyrazolyl ring); the former is an unusual example of five-coordinate gallium(III).[94]

8.5.3 Azide, cyanate, thiocyanate and selenocyanate derivatives

In this section we review the information on a group of monodentate nitrogen donor ligands; those ligands which are potentially ambidentate (e.g. NCS⁻) in fact consistently bond through nitrogen in mononuclear complexes, in keeping with the comparatively 'hard' nature of the M^{3+} ions.

Aluminium *azide* can be prepared metathetically ($AlCl_3 + 3NaN_3$) or by treating $Et_2O \cdot AlH_3$ with HN_3 in ether. The homoleptic gallium and indium analogues do not appear to have been synthesised, but mixed halide azide compounds $X_2M(N_3)$ (X = Cl or Br; M = Al or Ga) and organometallic derivatives $R_2M(N_3)$ and $RM(N_3)_2$ are known (M = Al, Ga or In in various combinations), as are some anionic complexes (e.g. $[Me_3Ga(N_3)]^-$). Four-membered rings (M_2N_2) have been invoked to explain the oligomeric structures found in solution.

There are no reports of *cyanate* complexes of aluminium. For gallium, as for indium, the parent compounds have not been isolated, but adducts (e.g. $Ga(NCO)_3 \cdot bpy$; $In(NCO)_3L_3$, with L = py, dmso, etc.) have been reported; other stoichiometries are also known, including $In(NCO)_3bpy_{1.5}$, which may be $[In(NCO)_2(bpy)_2]^+[In(NCO)_4bpy]^-$. Gallium yields unstable $(Me_4N)_3Ga(NCO)_6$, and an anionic complex $[In(NCO)_4]^-$ has been prepared but may involve a coordination number higher than four, since both terminal and bridging ligands are identified by vibrational spectroscopy.

Thiocyanates, or more precisely *isothiocyanates*, are known for each of the four elements of interest. For aluminium, neutral adducts (e.g. $Et_2O \cdot Al(NCS)_3$), anionic complexes (e.g. $K_3[Al(NCS)_6]$) and organometallic species (e.g. $Me_4N[Me_2Al(NCS)_2]$) are all readily accessible by metathesis; the anion $[Al(NCS)_6]^{3-}$ is the only species detected by ^{27}Al NMR measurements in Al^{3+}/KSCN solutions in acetone or propylene carbonate.[95] Gallium compounds can be obtained by extracting $[Ga(NCS)_4]^-$ as a component of an ion-pair into basic organic solvents, and using the solution to prepare $Ga(NCS)_3L_3$ adducts. For L = en, the proposed formulation is $[Ga(en)_3](NCS)_3$. Anionic complexes with stoichiometries corresponding to $Ga(NCS)_4^-$, $Ga(NCS)_5^{2-}$ and $Ga(NCS)_6^{3-}$ have been isolated, and involve the N-bonded ligand, but no crystallographic results are available to identify the coordination number of the metal. Various routes lead to $In(NCS)_3$ in which indium is six-coordinate as a result of extensive cross-linking. With monodentate ligands, adducts of the type $In(NCS)_3L_3$ are obtained, while bpy and phen yield complexes of the type $In(NCS)_3L_{1.5}$ which are presumed to be of the form $[In(NCS)_2L_2]^+$ $[In(NCS)_4L]^-$, and en gives $[In(en)_3](NCS)_3$. Anionic complexes $[In(NCS)_5]^{2-}$ and $[In(NCS)_6]^{3-}$ have been reported, and the latter has an octahedral InN_6 kernel. The size of the cation appears to affect the stoichiometry of these salts in much the same fashion as it does for the anionic halide complexes. Neutral 1 : 1 adducts of $Tl(NCS)_3$ with phen and bpy have been prepared, although the parent compound is not known, since oxidation occurs when Tl^{3+} is mixed with NCS^-.[96]

8.5.4 Complexes with phosphorus, arsenic and antimony donors

The addition compounds formed between phosphine, or substituted phosphines, and the trihalides of aluminium, gallium, indium and thallium make up an important part of the coordination chemistry of these elements, but there is little to add here beyond what has been said earlier. Adducts of MH_3, and of the organometallic compounds MR_3, XMR_2 and X_2MR are equally significant, especially in view of the development of methods for preparing electronic materials such as InP, and are dealt with elsewhere (see chapters 5 and 6). Both mono- and bidentate phosphorus donors have been studied.

Phosphido complexes derived from organoaluminium molecules can be obtained from the reaction of R_3Al with R_2PH, and the compounds $LiAl(PR_2)_4$ (R = H, Me or SiH_3) are also known, but this area still offers room for development. No analogous derivatives of the heavier elements have been reported. An attempt to prepare $In(PPh_2)_n$ by electrochemical synthesis produced a very unstable material,[97] but a mononuclear anionic complex, $[Cl_2In(PC_4Me_4)_2]^-$, has been stabilised with the $[K(18\text{-crown-}6\text{-})]^+$ cation.[98]

8.5.5   Mixed ligand complexes including Group 15 donors

The ability of coordination chemists to design new polydentate ligands has led to an increasing number of reports of complexes in which the donation involves different donor atoms, and the only limit appears to be the ingenuity of the initial ligand synthesis. Some of this work has been initiated in order to mimic and understand the transport *in vivo* of the radioactive isotopes ^{67}Ga, ^{68}Ga and ^{111}In which are important in various diagnostic imaging techniques. Some recent work is described in Refs. 71, 99 and 100. Other studies of polydentate ligands include the structural characterisation of $Al(nta)(OH_2)_2$ (nta = the quadridentate nitrilotriacetate ligand), in which there is a distorted AlO_6 kernel;[101] the 1,4,7-triazacyclononane-triacetic acid complex of indium chloride, with seven-coordinate pentagonal bipyramidal InN_3O_3Cl coordination;[102] and the $[In(OH_2)(2,6\text{-diacetyl-pyridinesemicarbazone})]^{2+}$ cation, again a pentagonal bipyramid with $InN_3O_2O_2'$ coordination.[103] The combination of elegant synthesis and X-ray crystallography will no doubt lead to many more such structures.

8.6   Aquo complexes and solution species

8.6.1   Aquo and other cationic complexes

The compounds discussed thus far include a large number of cationic complexes of M^{3+} with a variety of mono- and bidentate neutral donors. In the absence of strongly complexing anions, salts with, say, nitrate, perchlorate, sulfate, etc., of the form $[ML_6]A_3$ or $[ML_3']A_3$, are easily prepared. For aluminium, such complex cations have been identified in non-aqueous media for L = CH_3CN, dmso, dmf, tmpa, EtOH, etc., and mixed ligand species such as $[Al(dmf)_n(dmso)_{6-n}]^{3+}$ are also known to exist in solution, as are $[Al(OH_2)_n(ORH)_{6-n}]^{3+}$ and $[Al(OMeH)_n(OEtH)_{6-n}]^{3+}$ solvates[104] (see also section 8.3.2). The salt $[Al(urea)_6](ClO_4)_3$ has been structurally characterised by X-ray crystallography and has the expected octahedral AlO_6 kernel. Bidentate ligands giving rise to cationic complexes include L = phen or bpy, again with perchlorate as the balancing anion, but the polyether 15-crown-5 has been shown to give the unusual seven-coordinate cation $[AlCl_2(crown)]^+$, with *trans* chloride ligands.[105]

Similar results have been found for gallium(III). In dmso or dmf solution, GaL_6^{3+} cations are present, and a series of $[Ga(bpy)_3]^{3+}$ and similar species has also been characterised. The solution chemistry of GaX_3 (X = Cl, Br or I) in CH_3CN again involves a series of complex equilibria, in which the cation $[Ga(NCMe)_6]^{3+}$ is certainly one of the important components. With dimethoxyethane, $GaCl_3$ yields $[cis\text{-}GaCl_2L_2][GaCl_4]$,[106] and other mixed cationic complexes with mono- and bidentate nitrogen donors have been reported. In the case of indium, NMR studies have established the presence of InL_6^{3+} cations in solutions of L = dmf, acetone, $(MeO)_3PO$, and the formation of mixed ligand species $[ML_n(OH_2)_{6-n}]^{3+}$ clearly occurs in the appropriate mixed solvent systems. Preparative work has identified InL_6^{3+} salts with nitrate, fluoroborate or perchlorate for L = MeCN, dmso, urea, dma, thiourea, etc., and the presence of $[In(dmso)_6]^{3+}$ in the perchlorate has been demonstrated crystallographically.[107] Ligand size is significant with this element, since donors such as Ph_3P, Ph_3PO or Ph_3As give $[InL_4](ClO_4)_3$ and diphos $(R_2PCH_2CH_2PR_2, L')$ yields $[InL'_2](ClO_4)_3$, so that four-coordination is possible with the appropriate conditions. Many salts are known with cations such as $[In(en)_3]^{3+}$, $[In(bpy)_3]^{3+}$ and $[In(phen)_3]^{3+}$. As with other areas of coordination chemistry, there is little definitive information on the cationic complexes of thallium(III). It should be possible to find mono- or bidentate donors which resist oxidation, and which would allow some investigation of this area.

All the above information leads one to predict the formation of $[M(OH_2)_6]^{3+}$ cations in aqueous solution, and in the crystalline lattice of suitable salts. The techniques used in studying solutions have been mainly NMR and Raman spectroscopy, but more recent work with X-ray techniques has been valuable. For aluminium, the hexa-aquo cation has been identified in acidic solutions of a number of salts, while the octahedral complex is also present in crystalline $NaAl(SO_4)_2 \cdot H_2O$ and similar lattices with $r(Al-O) = 1.878\,\text{Å}$. The $[Ga(OH_2)_6]^{3+}$ cation is the important species in aqueous solutions of nitrate or perchlorate salts, even in the presence of an excess of acetone. Various spectroscopic and dilatometric methods identify $[In(OH_2)_6]^{3+}$ in acidic aqueous media, and here, and in the case of thallium, X-ray studies of concentrated solutions confirm these conclusions. There is then a wealth of experimental evidence that $[M(OH_2)_6]^{3+}$ cations exist in aqueous solution (see also Table 3.8). There is rapid exchange between coordinated water and the bulk phase for gallium and indium (and resumably thallium), but for aluminium it is possible to distinguish the two types of water molecule, implying a much slower exchange process for this element.

8.6.2 Hydrolysis of $M^{3+}(aq)$ ions

The importance of the precipitation of gelatinous $Al(OH)_3$ in water purification chemistry is one reason for the extensive study of hydrolysis in aqueous aluminium(III) solutions (see also chapters 4 and 9), but there is also an equally practical problem that any investigation of the solution chemistry of the M(III) state must recognise the significant impact of hydrolysis in all but fairly acidic media. In strongly alkaline solution, aluminium forms $[Al(OH)_4]^-$ and the dinuclear $[Al_2O(OH)_6]^{2-}$; in media of lower pH, the mono-substituted cation $[Al(OH)(OH_2)_5]^{2+}$ is believed to be present. The ageing of aluminate solutions leads to the dinuclear $[Al_2(OH)_2(OH_2)_8]^{4+}$, and eventually to more highly polymerised species; one such, $[AlO_4Al_{12}(OH)_{24}(OH_2)_{12}]^{7+}$, has been characterised in some detail (see section 3.2.3.2). Oligomeric species have also been identified in the base hydrolysis of AlX_3,[108, 109] and the question of the Al/OH species present under physiological conditions has also been discussed.[110] The same general picture emerges for gallium, with polymeric and mononuclear complexes being identified.[111, 112] The crystal structure of a dimeric complex supports this qualitative model, which is also applicable to indium(III). Hydrolysis is important for this element in solutions with pH above about 3.0; the first step is the formation of $In-OH^{2+}$ units, which give way to polymeric species as the hydrolysis proceeds, with $In(OH)_3$ as the final product. Indium is octahedrally coordinated in this solid, as it is apparently in the hydrated salts $M_3[In(OH)_6]$ (M = Na or Rb) which can be obtained from strongly alkaline solutions (e.g. 11 M aq. NaOH).

These results, while rather patchy in distribution, give a model of successive deprotonations of $[M(OH_2)_6]^{3+}$ to $[M(OH_2)_5OH]^{2+}$ etc., with these complexes eventually undergoing oligomerisation by core-linking through M–O–M and/or M–OH–M bonding. Precipitation occurs when the polymers reach the appropriate size, and eventually $M(OH)_3$, or a hydrated derivative, is obtained. Apart from the intrinsic importance of understanding such processes, their effect on measurements of stability constants (see below) and on the formation of mixed complexes $M(OH)_nL_m$ is increasingly recognised as being significant.

8.6.3 Stability constant studies

The determination of the stability constants for the equilibria

$$M + L \rightleftarrows ML \qquad (8.12)$$

$$K_1 = \frac{[ML]}{[M][L]} \qquad (8.13)$$

or

$$M + nL \rightleftarrows ML_n \qquad (8.14)$$

$$\beta_n = \frac{[ML_n]}{[M][L]^n} \qquad (8.15)$$

is an important feature in elucidating the coordination chemistry of the metal M in solution. The K or β values can be used to obtain ΔG_n, and in conjunction with a direct determination of ΔH_n, the entropy terms can also be evaluated. The experimental measurements have been discussed by a number of authors, stressing the care which must be exercised in order to obtain meaningful results. There are several compilations of published results, and Commission V.6 of IUPAC has been responsible for the production of a series of *Critical Reviews* of specific ligand classes.[3,4] These serve to emphasise that many of the stability constants in the literature are unreliable.

This is not the place to review all the published data for complexes of aluminium, gallium, indium and thallium, but rather to make some general observations. Firstly, there is a paucity of results for aluminium, due probably to the unavailability of electrochemical methods and to the lack of a suitable radioactive tracer, since both of these are the basis of experimental techniques of wide application. In contrast, indium and, to a lesser extent, gallium have been much favoured in stability constant work because both electrochemical methods and suitable radioactive tracers are available. The results for indium have been reviewed elsewhere;[113] one important conclusion is that K_1 for the halides follows the order $F > Cl > Br > I$ (see Table 1.4), as would be expected for a hard cation. A comparable critical review of the results for thallium(III) might be useful, since an examination of the published data suggests that here the order is $I > Br > Cl > F$.

Given this situation, one cannot presently compare, in any meaningful way, the constants for, say, the 1 : 1 complexes of the four metals with a given ligand in the hope of establishing an order of complexing strength. Another reason for insisting upon a critical review before attempting any such comparison is that modern work has increasingly shown the importance of ternary mixed ligand complexes such as $M(OH)_n L_m$ in circumstances where the only species previously considered was the binary ML_m complex. Since mixed ligand complexes are commonplace in the coordination chemistry of these elements in the solid state, it would be surprising if they were not equally important in the solution phase.

8.6.4   Solution species other than $M^{3+}(aq)$

In addition to the formation of complexes by substitution of a ligand for a water molecule in $[M(OH_2)_6]^{3+}$, there is a series of reactions in which the addition of large quantities of halide ion leads to substantive changes in the coordination shell.

An example of such phenomena occurs when $AlCl_3$ is dissolved in methanol or ethanol, where the species present are the ions $[Cl_2Al(ORH)_4]^+$ and $AlCl_4^-$ (see section 8.6.1). The first of these is a member of the $MX_2L_4^+$ cation group, examples of which were discussed earlier, but the presence of the tetrahedral MX_4^- anion shows that important coordinative changes occur, beyond substitution. The strongest evidence of such processes comes from studies of the solvent extraction of gallium[56] and indium from acidic aqueous halide solutions into basic organic solvents. Raman spectroscopy shows that in the case of indium, the species extracted is InX_4^- ($X = Cl$, Br or I), but this species cannot be detected in the corresponding aqueous solution for $X = Cl$ or Br unless a hydrophilic solvent such as methanol is added, when the concentration of InX_4^- increases in the order $I > Br > Cl$.

$$[M(OH_2)_6]^{3+} \underset{H_2O}{\overset{X^-}{\rightleftharpoons}} [MX_3(OH_2)_3] \underset{H_2O}{\overset{X^-}{\rightleftharpoons}} [MX_4(OH_2)_2]^- \underset{H_2O}{\overset{X^-}{\rightleftharpoons}} [MX_5(OH_2)]^{2-}$$

$$[MX_4]^- \qquad\qquad X^- \qquad (8.16)$$

$$[MX_5]^{2-} \overset{X^-}{\rightleftharpoons} [MX_6]^{3-}$$

These results can be summarised in the complicated scheme of solution equilibria represented above; here, in addition to the successive substitution of H_2O by halide ion, dehydration reactions give rise to MX_4^-. The values of the equilibrium constants for the different equilibria will obviously depend on M and X. In terms of the information available from crystalline substances (see earlier), it is clear that each of the species in this scheme can be obtained as the mixed halide-aquo complex, or as some analogue MX_nL_{6-n} where L is an appropriate monodentate neutral ligand. It is also reasonable that the formation of MX_4^- should be especially favoured when $X = I$, since anionic iodo species with coordination numbers greater than four are not known (section 8.3.5). Finally we should emphasise that a two-phase extraction into an organic solvent or an anion exchange resin changes the normal equilibrium position in aqueous solution, and that a species present in such a solution in very low concentrations may nevertheless be the only one to be extracted into the non-aqueous phase. In general, there is a good correlation between the solution and solid-state chemistry of these four elements, once it is accepted that the stabilising factors in the two phases are very different.

8.6.5 *Redox properties*

The standard electrode potentials of aluminium have been reviewed by Perrault,[114] and those of gallium, indium and thallium by Losev.[115a] The potential relationships for the species present in aqueous acidic solution are given in Figure 8.1. The values clearly reflect the increasing oxidising power of the $+3$ ion with increasing atomic number. The reviews in question also provide a wealth of detail on the thermodynamic properties of many of the simpler compounds of these four elements.

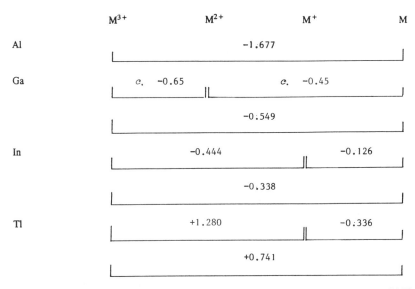

Figure 8.1 Standard reduction potentials (in volts) for the elements Al, Ga, In and Tl.[114, 115]

8.7 Lower oxidation states

Apart from the crystalline compound $[(C_5Me_5)Al]_4$ (see section 6.6) and the gaseous diatomic molecules AlX (see Table 3.12), there is no structural information on the chemistry of aluminium(I), although such species have been invoked in electrochemical processes, and in reactions of the element with CH_2X_2 (X = Cl, Br or I). Likewise there are no structurally authenticated gallium(I) compounds long-lived at ambient temperatures, with the exception of the Ga_2X_4 structures discussed below, and the η^6-aromatic complexes derived from solutions of these.[116–118] We therefore review the behaviour of inorganic indium(I) and thallium(I) complexes, and then turn to the behaviour of the elements in the $+2$ oxidation state.

8.7.1   Indium(I)

One of the difficulties in studying the coordination and other chemistry of indium(I) has been the absence of suitable starting materials, since the indium(I) halides are intractable substances. Fortunately, these compounds are soluble in mixtures of toluene and N,N,N',N'-tetramethylethane-diamine (tmed), and similar solvent systems. Other available, synthetically useful substances include cyclopentadienylindium(I) (CpIn) and its congeners, and In(dbbsq̇) (dbbsq̇$^-$ = the 3,5-di-t-butyl-1,2-semiquinonate anion).[119]

Neutral adducts of InBr and InI with ammonia disproportionate easily to $InX_3 \cdot nNH_3$, NH_3 and indium metal. Solutions of InX in toluene/tmed appear to contain $InX \cdot 3tmed$ species, but the compound precipitated on adding petroleum ether is $InX \cdot 0.5tmed$. Earlier reports of the formation of aniline and morpholine complexes of InX have not been authenticated. An adduct of phen with In(dbbsq̇) has been described, as has the mixed anionic species $[In(dbbsq̇)(Cl)(phen)]^-$. The anionic halide complexes InX_2^- and InX_3^{2-} have been produced in non-aqueous solution by a variety of simple methods, including electrochemical synthesis, metathesis of CpIn with $Et_4NX + HX$, and direct complexation of InX by X^-. These species have been used to form $[In(NCS)_2]^-$ and $[In(NCS)_3]^{2-}$ by simple metathesis in ethanol.

Complexes with bidentate monoanionic ligands can be synthesised by treating CpIn with HL (where HL = 1,3-diketone, quinoline, 2-mercapto-pentan-3-one) to give O–O, O–N, or O–S coordination, and the para-magnetic In(dbbsq̇), obtained by heating the metal with dbbq in toluene under reflux, is another related O–O species.[119] No structural information is available on any of these compounds, but it seems probable that inter-molecular cross-linking is responsible for the low solubility in the case of the first group, since the structure of the compound $[\{2,4,6-(CF_3)_3C_6H_2O\}In]_2$ demonstrates the presence of an In_2O_2 ring.[120] An important feature of the chemistry of indium(I) compounds is oxidation, so that In(dbbsq̇) reacts with I_2 to yield $I_2In(dbbsq̇)$, with oxidation at the metal being preferred to reaction at the ligand. The resultant indium(III) compounds will undergo coordination by bidentate donors to give a variety of coordination kernels involving $InO_3N_3I_2$, etc. The compound In(oxine) (oxine = quinoline-8-olate) can be oxidised by heating with Hoxine to In(oxine)$_3$; the mechanism of this interesting reaction has not been established.

The indium dihalides, which have the structures $In^I[InX_4]$ in the solid state, at least for X = Br or I (see below), have been used to form cationic indium(I) complexes with 18-crown-6 or cyclam (= 1,4,8,11-tetraaza-cyclotetradecane) of the types $[In(crown)][InX_4]$ and $[In(cyclam)][InX_4]$. Coordination of indium(I) by an excess of pyridine or γ-picoline, giving $[InL_n]^+$ and dbbsq̇$^-$, has been invoked to explain the non-aqueous solution chemistry of In(dbbsq̇).[119]

In general, significant numbers of indium(I) cationic, neutral and anionic complexes have been reported. It is unfortunate, therefore, that no significant structural information is available, and there is clearly a need for studies to fill this gap.

8.7.2 Thallium(I)

The stability of thallium(I) in aqueous solution is an important feature of the chemistry of this element, and indeed of the Group as a whole. The large size of the cation (estimated radius 1.50 Å for six-fold coordination) results in a low charge density, and hence poor complexing behaviour, so that, although numerous crystalline thallium(I) salts incorporate complexing agents in their lattices, X-ray crystallography shows that coordination to the thallium(I) is absent in many cases.

Although monodentate nitrogen donors do not coordinate to thallium(I), bidentate donors (e.g. phen and bpy) do give TlL_2^+ cations; related molecules include a volatile pyrrole derivative[121] and $[Tl(dpt)]_2$ (dpt^- = the 1,3-diphenyltriazenide anion) in which three-coordinate thallium is part of a Tl_2N_2 ring.[122] Complexes with oxygen donors include the interesting $[Tl(OMe)]_4$ which has a Tl_4O_4 cubane-like skeleton and the tetrameric $[Tl(O_2CH)]_4$.[123] The bulky ligand 2,4,6-tris(trifluoromethyl)phenoxide forms a dimer, structurally related to the indium(I) compound.[124] The Tl–O bonds of thallium(I) phenoxides undergo insertion reactions with CS_2 to give, for example, TlS_2COPh and with SO_2 to form products such as TlO_2SOPh. Thallium(I) β-diketonates appear to be monomeric in non-aqueous solution, but solid Tl(hfac) ($hfac^-$ = $CF_3COCHCOCF_3^-$) is dimeric, with chelating ligands; further intermolecular interactions give five-coordinate Tl atoms. The compound HOROTl, which is obtained by treating TlOAc with $R(OH)_2$ (= 2,2'-dihydroxy-1,1'-biphenyl), has also been shown to be a dimer, with a Tl_2O_2 core; the uncoordinated OH group is involved in intermolecular H-bonding.[125] Homopolymeric structures have also been identified by X-ray crystallography in thallium(I) carboxylates and ascorbate.

Thallium(I) dithiocarbamates have been synthesised from R_2NH, CS_2 and Tl^+ in aqueous alkali. The compound TlS_2CNPr_2 is a dimer, based on TlS_4Tl coordination, and a similar structure is found in TlS_2CNBu_2; in both cases, further cross-linking occurs to give chains of molecules. Other related compounds are the monothiocarbamates, monothioxanthates, xanthates and dithiophosphates.[4]

Many macrocyclic ligands have been used to form thallium(I) complexes, and the stability constants have been measured. In addition to [TlL]X, compounds with sandwich structures, $[TlL_2]X$, have been prepared. The comparison between indium(I) and thallium(I) suggests that in both cases the complexing power of the M^+ cation is low, and that significant M-ligand interactions require chelation by bi- or polydentate ligands.

8.7.3   Gallium(II) and indium(II) complexes

The structure of the dihalides of these elements has long been a classic problem of Main Group chemistry. Since the compounds are diamagnetic, the mononuclear MX_2 formulation is excluded, and the two dimeric structures $X_2M–MX_2$ and $M^I[M^{III}X_4]$ have both been urged. The latter has been established crystallographically and/or spectroscopically for the gallium compounds, and for indium bromide and iodide (see section 3.3.3.2). Recent investigations suggest that $InCl_2$ does not in fact exist, although other sub-chlorides have been identified, including In_2Cl_3 ($In_3^IIn^{III}Cl_6$), In_5Cl_9 ($In_3^IIn_2^{III}Cl_9$) and In_7Cl_9 ($In_6^IIn^{III}Cl_9$).[126] It may well be that the reluctance of indium(III) to form $InCl_4^-$ except in the presence of very large cations (see section 8.3.5) accounts in part for the formation of these interesting structures in preference to $In[InCl_4]$. The singularity is reflected in the overall chemistry of chloride derivatives of indium(II), which is significantly different from that of the other dihalides of gallium and indium.

The addition of various neutral donors to a solution of Ga_2X_4 ($X = Cl$, Br or I) in benzene produces complexes of the type $Ga_2X_4L_2$ which have been shown to be based on dimeric Ga–Ga units. The stereochemistry at the metal in the 1,4-dioxane derivative is pseudo-tetrahedral, with a $Ga'GaX_2N$ kernel, and a similar result is found in $Ga_2Br_4py_2$, although the conformation of the two molecules differs, with the dioxane species being eclipsed and the pyridine one staggered.[127] The bis(ether) adducts of Ga_2Br_4 have been identified in ether extracts of freshly prepared gallium solutions in aqueous hydrobromic acid, a result which seems to imply that such solutions contain some Ga^{II} species.

The corresponding indium(II) adducts have been prepared by two routes. In the first, the co-condensation of certain nitrogen or oxygen donors with the dihalide at liquid nitrogen temperatures gives on warming compounds of the type $In_2X_4L_4$ ($X = Cl$, Br or I), whose formulation as In–In bonded complexes has been derived from their Raman spectra; indium is then five-coordinate in such molecules. The alternative method involves the reaction of InX_2 ($X = Br$ or I) in benzene with the appropriate ligand; this has yielded adducts of Et_3P and tmed. The structure of $In_2Br_3I \cdot 2tmed$ has been confirmed by X-ray crystallography, which reveals an In–In bond (measuring 2.775(2) Å) between five-coordinate indium atoms ($In'InX_2N_2$). Attempts to prepare derivatives of indium dichloride by this latter method have failed, the dominant reaction being disproportionation to $InX_3 + In^0$, or to $InX_3 + InX$.

Anionic complexes of the form $[X_3M–MX_3]^{2-}$ are known for gallium and indium with all three of the heavier halogens $X = Cl$, Br or I (see section 3.3.3.2). Freshly prepared solutions of gallium in aqueous hydrohalic acids are reducing agents, and the addition of R_4N^+ cations leads to the isolation of crystalline $(R_4N)_2[Ga_2X_6]$ salts. Another preparation involves the electrochemical oxidation of anodic gallium in non-aqueous solutions containing HX and Ph_3P; salts of the type $(Ph_3PH)_2[Ga_2X_6]$ have been isolated and analysed by X-ray crystallography to identify the $[X_3Ga-GaX_3]^{2-}$ units.[128] These stable anions can be oxidised by X_2 to form GaX_4^- complexes in non-aqueous solution. Vibrational and NQR spectroscopic investigations have also been reported. The indium(II) anionic halide complexes have been prepared by the simple reaction of InX_2 with R_4NX in non-aqueous solution. The vibrational spectra are in agreement with an $[X_3In-InX_3]^{2-}$ structure; disproportionation to $InX_4^- + InX_2^-$ has been demonstrated by ^{115}In NMR spectroscopy.

A general discussion of the structural problems of gallium and indium dihalides has cast some light on these interesting compounds, and their complexes.[129] The first point to note is that, although mononuclear MX_2 compounds have not been isolated, this is not because of any thermodynamic effects, since simplistic but credible calculations establish that the M–X bond in such molecules should be reasonably strong. The important factor is surely the unpaired electron, whose reactivity leads to the formation of M_2X_4 molecules with the stability to survive in the gas phase as M–M bonded dimers; the energy of the Ga–Ga and In–In bonds is c. 135 and $100\,kJ\,mol^{-1}$, respectively. The driving force for the formation of $M[MX_4]$ structures in the solid state is provided by the substantial lattice energy (c. $500-600\,kJ\,mol^{-1}$), so that the sequence

$$2MX_2(g) \rightarrow M_2X_4(g) \rightarrow M[MX_4](c) \qquad (8.17)$$

is entirely reasonable.

The most important process in the $M_2X_4/M[MX_4]$ interconversion is intramolecular transfer of a halide ligand, which is in fact the reverse of the oxidative addition of MX to MX_3 (equation (8.18)):

$$\qquad (8.18)$$

It follows that coordinative saturation of the metal atoms, as in $X_2(L)_nM-M(L)_nX_2$ or $[X_3M-MX_3]^{2-}$, minimises or suppresses the possibility of halide transfer, and hence stabilises the M–M bond. More detailed arguments along similar lines explain the higher stability of $[Ga_2X_6]^{2-}$ anions compared with the indium analogues. The interesting challenge implicit in this argument is that mononuclear MX_2 derivatives may be accessible if appropriate ligands can be synthesised.

8.7.4   Thallium(II)

In the light of the above statement about mononuclear M^{2+} derivatives, it is interesting to note that Tl^{2+} has been identified spectroscopically in

γ-irradiated frozen aqueous solutions of Tl_2SO_4, and has also been proposed as an intermediate in the photochemical reduction of thallium(III) solutions. Subhalides have been identified, with structures such as $Tl[TlX_4]$ ($X = Cl$ or Br) and $Tl_3[TlX_6]$. An adduct has been shown crystallographically to be $[Tl(diox)][TlBr_4]$, with TlO_8 dodecahedra being formed by 1,4-dioxane-bridging. There is clearly much to be done in exploring the differences between the chemistry of thallium(II) and that of the lighter congeners, given the contrasts which obviously exist in the behaviours of the other oxidation states.

References

1. N.N. Greenwood, *Adv. Inorg. Chem. Radiochem.*, 1963, **5**, 91.
2. A.J. Carty and D.G. Tuck, *Prog. Inorg. Chem.*, 1975, **19**, 243.
3. M.J. Taylor, in *Comprehensive Coordination Chemistry*, eds, G. Wilkinson, R.D. Gillard and J.A. McCleverty, Pergamon, Oxford, 1987, Vol. 3, p. 105.
4. D.G. Tuck, in *Comprehensive Coordination Chemistry*, eds. G. Wilkinson, R.D. Gillard and J.A. McCleverty, Pergamon, Oxford, 1987, Vol. 3, p. 153.
5. J.J. Eisch, in *Comprehensive Organometallic Chemistry*, eds. G. Wilkinson, F.G.A. Stone and E.W. Abel, Pergamon, Oxford, 1982, Vol. 1, p. 555.
6. D.G. Tuck, in *Comprehensive Organometallic Chemistry*, eds. G. Wilkinson, F.G.A. Stone and E.W. Abel, Pergamon, Oxford, 1982, Vol. 1, p. 683.
7. H. Kurosawa, in *Comprehensive Organometallic Chemistry*, eds. G. Wilkinson, F.G.A. Stone and E.W. Abel, Pergamon, Oxford, 1982, Vol. 1, p. 725.
8. A. Haaland, *Angew. Chem., Int. Ed. Engl.*, 1989, **28**, 992.
9. E.C. Ashby, *Adv. Inorg. Chem. Radiochem.*, 1966, **8**, 283.
10. J.S. Pizey, *Lithium Aluminium Hydride*, Wiley, New York, 1977.
11. F.M. Brower, N.E. Matzek, P.F. Reigler, H.W. Rinn, C.B. Roberts, D.L. Schmidt, J.A. Snover and K. Terada, *J. Am. Chem. Soc.*, 1976, **98**, 2450.
12. V.S. Mastryukov, A.V. Golubinskii and L.V. Vilkov, *Zh. Strukt. Khim.*, 1979, **20**, 921.
13. E.C. Ashby, *J. Am. Chem. Soc.*, 1964, **86**, 1882.
14. A.R. Barron, M.B. Hursthouse, M. Motevalli and G. Wilkinson, *J. Chem. Soc., Chem. Commun.*, 1985, 664.
15. P.L. Baxter, A.J. Downs and D.W.H. Rankin, *J. Chem. Soc., Dalton Trans.*, 1984, 1755.
16. P.L. Baxter, A.J. Downs, D.W.H. Rankin and H.E. Robertson, *J. Chem. Soc., Dalton Trans.*, 1985, 807.
17. J.W. Lauher, D. Dougherty and P.J. Herley, *Acta Crystallogr.*, 1979, **B35**, 1454.
18. H. Nöth, R. Rurländer and P. Wolfgardt, *Z. Naturforsch.*, 1981, **36b**, 31.
19. B.M. Bulychev, *Polyhedron*, 1990, **9**, 387.
20. B.D. James and M.G.H. Wallbridge, *Prog. Inorg. Chem.*, 1970, **11**, 99.
21. N. Davies and M.G.H. Wallbridge, *J. Chem. Soc., Dalton Trans.*, 1972, 1421.
22. P.R. Oddy and M.G.H. Wallbridge, *J. Chem. Soc., Dalton Trans.*, 1978, 572.
23. H. Nöth and R. Rurländer, *Inorg. Chem.*, 1981, **20**, 1062.
24. M.T. Barlow, C.J. Dain, A.J. Downs, G.S. Laurenson and D.W.H. Rankin, *J. Chem. Soc., Dalton Trans.*, 1982, 597.
25. G.A. Olah, ed., *Friedel–Crafts and Related Reactions*, Interscience, New York, 1963, Vols. 1–4.
26. E.W. Post and J.C. Kotz, *MTP International Review of Science, Inorganic Chemistry Series 2*, ed. M.F. Lappert, Butterworth, London, 1975, Vol. 1, p. 219.
27. M. Dalibart and J. Derouault, *Coord. Chem. Rev.*, 1986, **74**, 1.
28. A. Pidcock, *MTP International Review of Science, Inorganic Chemistry Series 2*, ed. M.F. Lappert, Butterworth, London, 1975, Vol. 1, p. 281.
29. J.L. Atwood, S.K. Seale and D.H. Roberts, *J. Organomet. Chem.*, 1973, **51**, 105.

30. I.R. Beattie, P.J. Jones, J.A.K. Howard, L.E. Smart, C.J. Gilmore and J.W. Akitt, *J. Chem. Soc., Dalton Trans.*, 1979, 528.
31. F.W. Wehrli and R. Hoerdt, *J. Magn. Reson.*, 1981, **42**, 334; F.W. Wehrli and S. Wehrli, *ibid.*, 1981, **44**, 197.
32. H. Jacobs and B. Nöcker, *Z. Anorg. Allg. Chem.*, 1992, **614**, 25.
33. G. Müller and C. Krüger, *Acta Crystallogr.*, 1984, **C40**, 628.
34. P. Pullmann, K. Hensen and J.W. Bats, *Z. Naturforsch*, 1982, **37b**, 1312.
35. J. Derouault, P. Granger and M.T. Forel, *Inorg. Chem.*, 1977, **16**, 3214.
36. A.H. Cowley, M.C. Cushner, R.E. Davis and P.E. Riley, *Inorg. Chem.*, 1981, **20**, 1179.
37. N.C. Means, C.M. Means, S.G. Bott and J.L. Atwood, *Inorg. Chem.*, 1987, **26**, 1466.
38. A.V. Orlinkov, I.S. Akhrem and M.E. Vol'pin, *Russ. Chem. Rev.*, 1991, **60**, 524.
39. D.E.H. Jones and J.L. Wood, *J. Chem. Soc. A*, 1971, 3132, 3135.
40. N. Burford, B.W. Royan, R.E.v.H. Spence and R.D. Rogers, *J. Chem. Soc., Dalton Trans.*, 1990, 2111.
41. R. Karia, G.R. Willey and M.G.B. Drew, *Acta Crystallogr.*, 1986, **C42**, 558.
42. F.W.B. Einstein and D.G. Tuck, *J. Chem. Soc., Chem. Commun.*, 1970, 1182.
43. W.T. Robinson, C.J. Wilkins and Z. Zeying, *J. Chem. Soc., Dalton Trans.*, 1990, 219.
44. W.T. Robinson, C.J. Wilkins and Z. Zeying, *J. Chem. Soc., Dalton Trans.*, 1988, 2187.
45. M.A. Malyarick, S.P. Petrosyants and I.B. Ilyuhin, *Russ. J. Inorg. Chem.*, in press.
46. G. Beran, K. Dymock, H.A. Patel, A.J. Carty and P.M. Boorman, *Inorg. Chem.*, 1972, **11**, 896.
47. N.N. Greenwood and A. Earnshaw, *Chemistry of the Elements*, Pergamon, Oxford, 1984, p. 269 and refs. cited therein.
48. J.-J. Videau, J. Portier and B. Piriou, *Rev. Chim. Miner.*, 1979, **16**, 393.
49. J.B. Ekeley and H.A. Potratz, *J. Am. Chem. Soc.*, 1936, **58**, 907.
50. D.S. Brown, F.W. B. Einstein and D.G. Tuck, *Inorg. Chem.*, 1969, **8**, 14.
51. G. Joy, A.P. Gaughan, Jr., I. Wharf, D.F. Shriver and J.P. Dougherty, *Inorg. Chem.*, 1975, **14**, 1795.
52. I.R. Beattie, T.R. Gilson and G.A. Ozin, *J. Chem. Soc. A*, 1968, 2765.
53. T.J. Bastow, B.D. James and M.B. Millikan, *J. Solid State Chem.*, 1983, **49**, 388.
54. C.S. Lin and D.G. Tuck, unpublished results.
55. R.W. Berg, E. Kemnitz, H.A. Hjuler, R. Fehrmann and N.J. Bjerrum, *Polyhedron*, 1985, **4**, 457.
56. M.J. Taylor, *Polyhedron*, 1990, **9**, 207.
57. G.R. Clark, C.E.F. Rickard and M.J. Taylor, *Can. J. Chem.*, 1986, **64**, 1697.
58. I. Sóvágó, T. Kiss and R.B. Martin, *Polyhedron*, 1990, **9**, 189.
59. J.K. Beattie, S.P. Best, B.W. Skelton and A.H. White, *J. Chem. Soc., Dalton Trans.*, 1981, 2105.
60. R.B. Martin, *Polyhedron*, 1990, **9**, 193.
61. A.E. Martell, R.J. Motekaitis and R.M. Smith, *Polyhedron*, 1990, **9**, 171.
62. N. Bulc, L. Golič and J. Šiftar, *Acta Crystallogr.*, 1984, **C40**, 1829.
63. M.B. Power and A.R. Barron, *Polyhedron*, 1990, **9**, 233.
64. B. Cetinkaya, P.B. Hitchcock, H.A. Jasim, M.F. Lappert and H.D. Williams, *Polyhedron*, 1990, **9**, 239.
65. M.H. Chisholm, V.F. DiStasi and W.E. Streib, *Polyhedron*, 1990, **9**, 253.
66. D.C. Bradley, H. Chudzynska, D.M. Frigo, M.E. Hammond, M.B. Hursthouse and M.A. Mazid, *Polyhedron*, 1990, **9**, 719.
67. K. Saito and A. Nagasawa, *Polyhedron*, 1990, **9**, 215.
68. M.F. Garbauskas, J.H. Wengrovius, R.C. Going and J.S. Kasper, *Acta Crystallogr.*, 1984, **C40**, 1536.
69. C. Sreelatha, V.D. Gupta, C.K. Narula and H. Nöth, *J. Chem. Soc., Dalton Trans.*, 1985, 2623.
70. D.J. Clevette, D.M. Lyster, W.O. Nelson, T. Rihela, G.A. Webb and C. Orvig, *Inorg. Chem.*, 1990, **29**, 667.
71. D.J. Clevette and C. Orvig, *Polyhedron*, 1990, **9**, 151,
72. A. Ozarowski, B.R. McGarvey, A. El-Hadad, Z. Tian, D.G. Tuck, D.J. Krovich and G.C. DeFotis, *Inorg. Chem.*, in press.
73. T.A. Annan and D.G. Tuck, *Can. J. Chem.*, 1989, **67**, 1807.

74. T.A. Annan, R.K. Chadha, P. Doan, D.H. McConville, B.R. McGarvey, A. Ozarowski and D.G. Tuck, *Inorg. Chem.*, 1990, **29**, 3936.
75. J.P. Oliver and R. Kumar, *Polyhedron*, 1990, **9**, 409.
76. R. Kumar, H.E. Mabrouk and D.G. Tuck, *J. Chem. Soc., Dalton Trans.*, 1988, 1045.
77. J.H. Green, R. Kumar, N. Seudeal and D.G. Tuck, *Inorg. Chem.*, 1989, **28**, 123.
78. T.A. Annan, R. Kumar, H.E. Mabrouk, D.G. Tuck and R.K. Chadra, *Polyhedron*, 1989, **8**, 865.
79. R.K. Chadha, P.C. Hayes, H.E. Mabrouk and D.G. Tuck, *Can. J. Chem.*, 1987, **65**, 804.
80. L.E. Maelia and S.A. Koch, *Inorg. Chem.*, 1986, **25**, 1896.
81. G.G. Hoffmann and C. Burschka, *Angew. Chem., Int. Ed. Engl.*, 1985, **24**, 970.
82. A. Boardman, S.E. Jeffs, R.W.H. Small and I.J. Worrall, *Inorg. Chim. Acta*, 1985, **99**, L39.
83. B. Eisenmann, M. Jakowski and H. Schäfer, *Rev. Chim. Miner.*, 1984, **21**, 12.
84. C. Peppe and D.G. Tuck, *Can. J. Chem.*, 1984, **62**, 2798.
85. M.G. Kanatzidis and S. Dhingra, *Inorg. Chem.*, 1989, **28**, 2024.
86. C. Geloso, R. Kumar, J.R. Lopez-Grado and D.G. Tuck, *Can. J. Chem.*, 1987, **65**, 928.
87. K.M. Waggoner, M.M. Olmstead and P.P. Power, *Polyhedron*, 1990, **9**, 257.
88. D.St.C. Black, V.V. Davis, G.B. Deacon and R.J. Schultze, *Inorg. Chim. Acta*, 1979, **37**, L528.
89. J.T. Leman, H.A. Roman and A.R. Barron, *J. Chem. Soc., Dalton Trans.*, 1992, 2183.
90. F.G.N. Cloke, G.R. Hanson, M.J. Henderson, P.B. Hitchcock and C.L. Raston, *J. Chem. Soc., Chem. Commun.*, 1989, 1002.
91. M.J. Henderson, C.H.L. Kennard, C.L. Raston and G. Smith, *J. Chem. Soc., Chem. Commun.*, 1990, 1203.
92. F.G.N. Cloke, C.I. Dalby, M.J. Henderson, P.B. Hitchcock, C.H.L. Kennard, R.N. Lamb and C.L. Raston, *J. Chem. Soc., Chem. Commun.*, 1990, 1394.
93. B.M. Louie, S.J. Rettig, A. Storr and J. Trotter, *Can. J. Chem.*, 1985, **63**, 503, 703, 2261, 3019.
94. D.L. Reger, S.J. Knox and L. Lebioda, *Inorg. Chem.*, 1989, **28**, 3092.
95. N. Komatsu, M. Yokoi and E. Kubota, *Bull. Chem. Soc. Jpn.*, 1988, **61**, 3746.
96. J. Blixt, R.K. Dubey and J. Glaser, *Inorg. Chem.*, 1991, **30**, 2824.
97. T.A. Annan, R. Kumar and D.G. Tuck, *J. Chem. Soc., Dalton Trans.*, 1991, 11.
98. T. Douglas, K.H. Theopold, B.S. Haggerty and A.L. Rheingold, *Polyhedron*, 1990, **9**, 329.
99. G.E. Jackson, *Polyhedron*, 1990, **9**, 163.
100. D.A. Moore, P.E. Fanwick and M.J. Welch, *Inorg. Chem.*, 1990, **29**, 672.
101. G.C. Valle, G.G. Bombi, B. Corain, M. Favarato and P. Zatta, *J. Chem. Soc., Dalton Trans.*, 1989, 1513.
102. A.S. Craig, I.M. Helps, D. Parker, H. Adams, N.A. Bailey, M.G. Williams, J.M.A. Smith and G. Ferguson, *Polyhedron*, 1989, **8**, 2481.
103. J. Davis and G.J. Palenik, *Inorg. Chim. Acta*, 1985, **99**, L51.
104. Yu. A. Buslaev and S.P. Petrosyants, *Polyhedron*, 1984, **3**, 265.
105. S.G. Bott, H. Elgamal and J.L. Atwood, *J. Am. Chem. Soc.*, 1985, **107**, 1796.
106. S. Böck, H. Nöth and A. Wietelmann, *Z. Naturforsch.*, 1990, **45b**, 979.
107. J.M. Harrowfield, B.W. Skelton and A.H. White, *Aust. J. Chem.*, 1990, **43**, 759.
108. J.W. Akitt and W. Gessner, *J. Chem. Soc., Dalton Trans.*, 1984, 147.
109. J.W. Akitt, J.M. Elders, X.L.R. Fontaine and A.K. Kundu, *J. Chem. Soc., Dalton Trans.*, 1989, 1889.
110. M. Venturini and G. Berthon, *J. Chem. Soc., Dalton Trans.*, 1987, 1145.
111. I. Tóth, L. Zékány and E. Brücher, *Polyhedron*, 1985, **4**, 279.
112. S.M. Bradley, R.A. Kydd and R. Yamdagni, *J. Chem. Soc., Dalton Trans.*, 1990, 413, 2653.
113. D.G. Tuck, *Pure Appl. Chem.*, 1983, **55**, 1477.
114. G.P. Perrault, in *Standard Potentials in Aqueous Solutions*, eds. A.J. Bard, R. Parsons and J. Jordan, Marcel Dekker, New York, 1985, p. 566.
115. (a) V.V. Losev, in *Standard Potentials in Aqueous Solutions*, eds. A.J. Bard, R. Parsons and J. Jordan, Marcel Dekker, New York, 1985, p. 237; (b) S.G. Bratsch, *J. Phys. Chem. Ref. Data*, 1989, **18**, 1.

116. H. Schmidbaur, U. Thewalt and T. Zafiropoulos, *Angew. Chem., Int. Ed. Engl.*, 1984, **23**, 76.
117. H. Schmidbaur, *Angew. Chem., Int. Ed. Engl.*, 1985, **24**, 893.
118. H. Schmidbaur, R. Nowak, B. Huber and G. Müller, *Polyhedron*, 1990, **9**, 283.
119. T.A. Annan, D.H. McConville, B.R. McGarvey, A. Ozarowski and D.G. Tuck, *Inorg. Chem.*, 1989, **28**, 1644.
120. M. Scholz, M. Noltemeyer and H.W. Roesky, *Angew. Chem., Int. Ed. Engl.*, 1989, **28**, 1383.
121. E. Ciliberto, S. Di Bella, A. Gulino and I.L. Fragalà, *Inorg. Chem.*, 1992, **31**, 1641.
122. J. Beck and J. Strähle, *Z. Naturforsch*, 1986, **41b**, 1381.
123. K. Ozutsumi, H. Ohtaki and A. Kusumegi, *Bull. Chem. Soc. Jpn.*, 1984, **57**, 2612.
124. H.W. Roesky, M. Scholz, M. Noltemeyer and F.T. Edelmann, *Inorg. Chem.*, 1989, **28**, 3829.
125. T.A. Annan, J.E. Kickham, S.J. Loeb, L. Taricani and D.G. Tuck, unpublished results.
126. H.P. Beck and D. Wilhelm, *Angew. Chem., Int. Ed. Engl.*, 1991, **30**, 824.
127. J.C. Beamish, A. Boardman, R.W. H. Small and I.J. Worrall, *Polyhedron*, 1985, **4**, 983.
128. M.A. Khan, D.G. Tuck, M.J. Taylor and D.A. Rogers, *J. Crystallogr. Spectrosc. Res.*, 1986, **16**, 895.
129. D.G. Tuck, *Polyhedron*, 1990, **9**, 377.

9 The elements in the environment
R.B. MARTIN

9.1 Introduction

The quickest way to grasp the likely chemistry of ions engaged in mainly ionic bonding relates their ionic radii with those of other ions of known chemistries. Table 9.1 compares the effective ionic radii of the four tripositive ions of this discourse and the unipositive Tl^+ with alkali, alkaline earth and other selected metal ions.[1] For a given ion, the ionic radius increases with coordination number since the greater number of bonds weakens the strength of any one bond.

Size similarity ranks higher than charge identity in permitting metal ion substitutions in both mineralogy and biology. Thus Na^+ and Ca^{2+} of similar size but different charges often interchange. For example, the plagioclase feldspars form a series of six aluminium silicate mixtures in a common triclinic lattice from $NaAlSi_3O_8$ (albite) to $CaAl_2Si_2O_8$ (anorthite). Neither Mg^{2+} nor Sr^{2+} replaces Ca^{2+} in this series. The charge compensation required in minerals is seldom critical when substitution occurs in a protein or other biological ligands.

Table 9.1 Effective ionic radii in Å.[a]

Ion	Coordination number				
	4	5	6	8	12
Be^{2+}	0.27		0.45		
Al^{3+}	0.39	0.48	0.54		
Ga^{3+}	0.47	0.55	0.62		
Fe^{3+}	0.49	0.58	0.65	0.78	
Mg^{2+}	0.57	0.66	0.72	0.89	
Zn^{2+}	0.60	0.68	0.74	0.90	
Li^+	0.59		0.76	0.92	
In^{3+}	0.62		0.80	0.92	
Lu^{3+}			0.86	0.98	
Tl^{3+}	0.75		0.89	0.98	
Cd^{2+}	0.78	0.87	0.95	1.10	1.31
Ca^{2+}			1.00	1.12	1.34
Na^+	0.99	1.00	1.02	1.18	1.39
Sr^{2+}			1.18	1.26	1.44
Ba^{2+}			1.35	1.42	1.61
K^+	1.37		1.38	1.51	1.64
Tl^+			1.50	1.59	1.70
Rb^+			1.52	1.61	1.72

(a) From Ref. 1.

On the basis of the radii, Al^{3+}, though quite small, is closest in size to Fe^{3+} and Mg^{2+}. Ca^{2+} is much larger, and in its favoured eight-fold coordination exhibits a radius of 1.12 Å, yielding a volume nine times greater than that of Al^{3+}. In the mixed crystal $Ca_3Al_2(OH)_{12}$, each Al^{3+} is surrounded octahedrally by six hydroxide ions and each Ca^{2+} by a cubic array of eight hydroxide ions. Each metal ion thus adopts its own favoured coordination number. The Al–O distances are 1.92 Å and the average of the Ca–O distances is 2.50 Å.[2] The difference of 0.58 Å agrees exactly with the difference in radii between six-coordinate Al^{3+} and eight-coordinate Ca^{2+}, as given in Table 9.1. The Al^{3+} and Ca^{2+} sites are distinctly different; one metal ion cannot substitute for the other. For these and other reasons, we have argued that in biological systems Al^{3+} will mainly be competitive with Mg^{2+}, rather than Ca^{2+}.[3,4] Both Al^{3+} and Mg^{2+} favour oxygen donor ligands, especially phosphate groups.[5] Al^{3+} is 10^7 times more effective than Mg^{2+} in promoting the polymerisation of tubulin to microtubules.[6] Wherever there is a process involving Mg^{2+}, there is an opportunity for interference by Al^{3+}.

9.2 Geochemistry

Owing to their strongly reducing properties, none of the metals is found free in nature. The standard reduction potentials for the aqueous process $M^{3+} + 3e^- \rightarrow M(s)$ are -1.68, -0.55, -0.34 and $+0.74$ V for Al, Ga, In and Tl, respectively. Additionally, for $Tl^{3+} + 2e^- \rightarrow Tl^+$, we have $E^{\ominus} = +1.25$ V and for $Tl^+ + e^- \rightarrow Tl(s)$, $E^{\ominus} = -0.34$ V. Thus, unless the environment is highly oxidising, thallium should appear as Tl^+. In 1827, Wöhler liberated metallic Al by reducing $AlCl_3$ with potassium metal (see chapter 1).

9.2.1 Abundances

Table 9.2 lists the abundances of the four elements at several sites. The cosmic abundance is given relative to 10^6 atoms of silicon.[7] The earth's continental crust[7] and sea-water[8] abundances are on a mass basis, ppm for the crust and ppb for sea-water. Although they are not usually compared in this way, doing so allows a direct numerical comparison of abundances in the crust with those of solutes in sea-water. Thus the ratio of crustal to sea-water solute abundances is $10^5 : 1$, $500 : 1$, $1000 : 1$ and $50 : 1$ for Al, Ga, In and Tl, respectively, with Al the most and Tl the least depleted in sea-water. Aluminium comprises 8% of the earth's crust but less than 1 ppm of the solutes in sea-water. None of the four elements comes close to saturation in sea-water. The last column of Table 9.2 lists the major species of each element occurring in the oceans at pH 8.1.[9]

Table 9.2   Abundances of the Group 13 metals.

Metal	Cosmic per 10^6 Si[a]	Earth's crust[a] (ppm)	Sea-water[b] (ppb)	Ocean species, pH 8.1[c]
Al	85 000	81 300	<1	$[Al(OH)_4]^-$
Ga	38	15	0.03	$[Ga(OH)_4]^-$
In	0.2	0.1	0.0001	$In(OH)_3(aq)$
Tl	0.2	0.5	0.01	Tl^+, $TlCl(aq)$

(a) Ref. 7.
(b) Ref. 8 (except Al).
(c) Ref. 9.

All four elements are enriched in the surface of the earth compared to the earth as a whole. Aluminium is a lithophile (found with silicates) while indium and thallium are mainly chalcophiles (found as sulfides), and the intermediate gallium appears associated with both environments.

A brief synopsis of the geochemistry of each element follows. For a more detailed presentation element by element, see the contributions by many authors in the comprehensive handbook edited by Wedepohl.[10]

9.2.2   Aluminium

Aluminium is the most abundant metal on the earth and moon, comprising 8.1% of the earth's crust by mass and, owing to its lightness, a lesser 6.4% on an atom basis. Igneous rocks contain feldspars and micas, which on weathering yield clay minerals such as kaolinite ($Al_2(OH)_4Si_2O_5$) and hydroxides. Weathering and leaching out of the silica produces the main ore bauxite, commercial deposits of which are about half Al_2O_3 (used as the basis for estimation). Of the Al_2O_3, most is in the form of gibbsite, $Al(OH)_3$, and boehmite, $AlO(OH)$. The relative stabilities of gibbsite and several aluminosilicates have been described under a variety of conditions[11,12] (see chapter 4 for structural information). Cryolite, Na_3AlF_6, still used in the electrolytic production of metallic Al, was found in significant amounts in Greenland, but this source has been depleted, and it is now made synthetically.

Among the aluminium-containing minerals are various gemstones: these are derived from corundum, Al_2O_3, with a hardness second to that of diamond, as with ruby (trace of Cr^{3+}) and sapphire (trace of Fe^{2+} or Fe^{3+}); from the basic copper phosphate, as with turquoise; and from the silicates, as with topaz, beryl (including emerald and aquamarine) and many garnets.

The large Allende meteorite fell in northern Mexico in 1969. Its abundances of 19 condensable elements are the same within a factor of two as solar abundances. The carbonaceous chondrite contains large chondrules, up to 25 mm in diameter that are deficient in alkalis and rich in calcium and aluminium. These large chondrules exhibit an unusual composition and mineralogy, including an aluminium- and titanium-rich pyroxene. Such chondrules evidently condensed at high temperatures and may sample the first condensate from the formation of the solar system 4600 million years ago.[13]

9.2.3   Gallium

In one of his historic predictions, Mendeléev described the properties of the then unknown element gallium in 1869, 6 years before its discovery by the spectroscope, the means also predicted by Mendeléev. (As noted in chapter 1, indium and thallium were also discovered spectroscopically.) Mendeléev even wrote to the discoverer of gallium suggesting that he re-determine the element's density claimed originally to be $4.7 \, g \, cm^{-3}$, rather than the predicted and correct value of $5.90 \, g \, cm^{-3}$.[14]

Gallium occurs in all kinds of rocks, its proportion decreasing as the basicity of the rock increases. As expected from its comparable size (Table 9.1), Ga^{3+} is found as a replacement for Fe^{3+} and with Al^{3+} in bauxite, but also as a sulfide with its neighbours in the Periodic Table, zinc and germanium. The highest gallium concentration of up to 1% occurs in the rare sulfide mineral germanite. However, commercial production relies on recovery of the less than 0.01% found in bauxite, which is processed in large quantities for its aluminium (see chapter 2).

9.2.4   Indium

This widely dispersed chalcophile is found with the similarly sized zinc (Table 9.1) in sulfide minerals and is obtained as a by-product of zinc processing (see chapter 2). In the ocean, Cl^- species are not competitive with soluble $In(OH)_3$.[9]

9.2.5   Thallium

This dispersed element occurs with lead in galena, PbS, where Tl^+ and Bi^{3+} (or Sb^{3+}) replace two Pb^{2+}. Tl^+ replaces the similarly sized K^+ and Rb^+ (Table 9.1) occurring in some feldspars and micas. Thallium usually appears in the 1+ oxidation state, but appears as 3+ in the rare mineral avicennite, Tl_2O_3, found in central Asia. If Tl(III) were to appear in the ocean, it would be present as $Tl(OH)_3(aq)$.[9] Contrary to the principle of hard and soft acids and bases, Tl(III) is softer than Tl(I).[15]

9.3   Environmental aspects

Of the four elements under discussion, the three heaviest occur only in trace amounts and are not of general environmental concern. Therefore, this section focuses on aluminium, the most common metal in the earth's crust. Despite its abundance in the crust, so little aluminium occurs in the planet's waters that their Al(III) content has not been reliably measured until recently. The ocean concentration of less than 1 μg Al per litre, considerably smaller than that allowed by clay sediments, may result from accumulation of aluminium (and silicon) by diatoms.[16–18] Most natural waters contain insignificant amounts of aluminium, except for those in some volcanic regions and for alum springs. Any freed Al^{3+} is usually deposited in sediment as a hydroxide. With the advent of acid rain, metal ions such as aluminium, mercury and lead escape from mineral deposits and more frequently appear in fresh waters. Acid rain serves as the key that springs the lock for metal ion release. The Al(III) concentration increases sharply in clear water lakes at pH < 6 where greater than micromolar amounts (i.e. > 27 μg Al per litre) may occur.[19,20] Micromolar concentrations of Al(III) are more damaging to fish than increased acidity. Until recently, nature has not had to cope with a high Al(III) activity. Modern society has set in motion an epic experiment to test the ability of living creatures — including humans — to cope with this change.

Al(III)-rich soils retard plant growth and limit crop production in many places.[21,22] In a model early study, Al(III) was identified as the agent responsible for damage from acid soil to barley growth.[23] Addition of even acidic phosphates increased growth; Al(III) sequestration was more beneficial than acidification was harmful.

Plants such as tea that accumulate Al(III) do so in acidic soils and evidently detoxify the Al(III) by storing a chelated version in cell vacuoles of older leaves separate from the more metabolically active parts of the plant. Tea plants have been found with as much as 3% Al(III) in older leaves but only 0.01% in younger ones—a 300-fold difference.[24] Typical tea infusions contain about 50 times as much Al(III) as do infusions from coffee.[25] Addition of milk to tea should immobilise Al(III) as an insoluble phosphate, while addition of lemon will strongly complex the Al(III) in a dangerous citrate complex (section 9.4.2).[3,26–28]

Grass tetany is an acute muscle spasmodic condition of ruminant animals (usually females stressed by lactation) grazing on lush green pastures in the early spring and resulting in convulsive death unless treated. It is accompanied by low Mg^{2+} concentrations in growing grass and blood serum, but this deficiency alone does not produce the condition. In 1971 it was suggested that grass tetany is due to high Al(III) and low Mg(II) in grass forage, and a phosphate fertiliser with a Mg(II) supplement was recommended as a cure.[29] Like some other diseases with a claimed linkage to

Al(III), this one has proved controversial and the question remains unresolved.[30]

The aqueous chemistry of aluminium is relatively simple. The aquated Al^{3+} cation exchanges its coordinated water molecules 10^5–10^6 times faster than does the aquated Cr^{3+} cation (while being kinetically inert compared with other biologically important metal ions), the hydroxide is much more soluble than that of Fe^{3+}, and the metal exhibits only one oxidation state in biological systems, i.e. Al(III). There is, then, no oxidation-reduction chemistry to Al(III) in biology. The chemistry of aqueous Al^{3+} is made difficult, however, by hydrolytic equilibria in aqueous solutions, to which subject we turn in the next section.

9.3.1 Al^{3+} hydrolysis

Whatever ligands may be present, understanding the state of Al(III) in any aqueous system demands awareness of the species that Al(III) forms with the components of water at different pH values. Here we use Al(III) as a generic term for the 3 + ion when a specific form is not indicated. In solutions more acid than pH < 5, Al(III) exists as the octahedral hexahydrate, $[Al(OH_2)_6]^{3+}$, usually abbreviated as Al^{3+}. As the solution becomes less acidic, $[Al(OH_2)_6]^{3+}$ undergoes successive deprotonations to yield $[Al(OH)]^{2+}$, $[Al(OH)_2]^+$, and soluble $Al(OH)_3$, with a decreasing and variable number of water molecules.[3,31] Neutral solutions give an $Al(OH)_3$ precipitate that redissolves, owing to formation of tetrahedral aluminate, $[Al(OH)_4]^-$, the primary soluble Al(III) species at pH > 6.2.

The four successive deprotonations from $[Al(OH_2)_6]^{3+}$ to yield $[Al(OH)_4]^-$ squeeze into an unusually narrow pH range of less than one log unit, with pK_a values of 5.5, 5.8, 6.0 and 6.2.[32] In contrast, the corresponding four normal deprotonations from $[Fe(OH_2)_6]^{3+}$ span 6.6 log units with pK_a values of 2.7, 3.8, 6.6 and 9.3. The narrow span for Al^{3+} is explained by the cooperative nature of the successive deprotonations due to a concomitant decrease in coordination number.[31]

The upper half of Figure 9.1 shows the distribution of $[Al(OH_2)_6]^{3+}$ and mononuclear hydrolysed species based on the four successive pK_a values given above. Thus, only two species dominate over the entire pH range, the octahedral hexahydrate $[Al(OH_2)_6]^{3+}$ at pH < 5.5, and the tetrahedral $[Al(OH)_4]^-$ at pH > 6.2, while there is a mixture of hydrolysed species and coordination numbers in the range 5.5 < pH < 6.2.[3,31] These equilibria must be considered in all solutions containing Al(III). If, in addition, other ligands are incapable of holding Al(III) in solution, it becomes necessary to include the solubility equilibrium.

The lower half of Figure 9.1 applies to solutions saturated with amorphous $Al(OH)_3$. The dashed straight line of slope 3 in the lower half of the figure gives the molar concentration of the hexahydrated metal ion, $[Al^{3+}]$. The

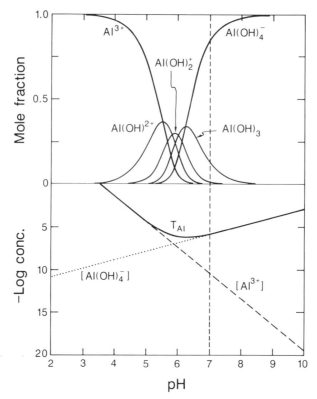

Figure 9.1 Al^{3+} hydrolysis. Upper half: mole fraction of soluble species as a function of pH. Lower half: for saturated solutions of amorphous $Al(OH)_3$, the negative logarithm of molar concentration of the free ion, $[Al^{3+}]$, straight dashed line; and the sum over all species present, T_{Al}, curved solid line. The straight dotted line represents the concentration of tetrahedral aluminate ion, $[Al(OH)_4^-]$. From Ref. 40.

solid curve represents the *total* concentration of this ion and all mononuclear hydrolysed forms, T_{Al}, with the distribution shown in the upper half of the figure. From the lower part of the figure we learn that large amounts of $Al(OH)_3$ dissolve in acidic stomachs.

Figure 9.1 also shows as a dotted line the molar concentration of aluminate, $[Al(OH)_4^-]$, the main species at pH > 6.2. At any pH, the ordinate distance between the straight lines for $[Al(OH)_4^-]$ and $[Al^{3+}]$ gives the logarithm of their concentration ratio. Thus at pH 7.0 the molar ratio $[Al(OH)_4^-]/[Al^{3+}]$ is given by $10^{4.5} = 3 \times 10^4 : 1$. Therefore, $[Al(OH)_4]^-$ (aluminate) should be the starting point for thinking about aluminium in sea-water and in most biological fluids.

What happens when an $AlCl_3$ solution is administered at a local concentration of 0.01 M Al(III) to a tissue at pH 7? Ascension of the dashed pH 7 line in Figure 9.1 indicates that the permissible free Al^{3+} concentration is only $10^{-10.3}$ M and that the concentration of all soluble forms is $10^{-5.7}$ M = 2 μM. Unless the rest of the added Al(III) has been complexed by other ligands, it will form insoluble $Al(OH)_3$. From the upper part of the figure we see that of the soluble forms 85% is $[Al(OH)_4]^-$ and 14% $Al(OH)_3$. Of the Al(III) administered at pH 7 only 2 μM appears in soluble forms, most of which is $[Al(OH)_4]^-$ since the bulk of the added Al(III) precipitates or coordinates to nearby ligands. To keep Al(III) in solution, it may be administered as a complex.

Given time and appropriate concentrations and pH, polynuclear hydroxo complexes may form. A dihydroxo bridged dimer is not a significant species in aqueous solutions.[31] Several investigators[33,34] support the formation of a soluble Al-13 polymer of composition $[AlO_4Al_{12}(OH)_{24}(OH_2)_{12}]^{7+}$, for which a crystal structure determination reveals an interesting Keggin ion structure with a unique central tetrahedral Al^{3+} surrounded by 12 octahedral Al^{3+} moieties.[35] This complex arises in diverse circumstances: it occurs in antiperspirants,[36] may be more phytotoxic than monomeric Al(III),[37,38] and is the pillaring agent in smectites.[39]

9.3.2 pAl = −log [Al³⁺]

In addition to the hydrolysis features already considered, the amount of free, aqueous Al^{3+} in solution depends upon several variables: the ligands present, the stability constants of the complexes they form with Al^{3+}, and the total Al(III) to total ligand molar ratio. For ligands with protons competing with the metal ion for binding sites in the pH range of interest, the pH is also a variable. Thus, instead of simple association of metal ion and basic ligand, $Al^{3+} + L \rightarrow AlL$, the relevant reaction may become displacement of a proton from the acidic ligand by the metal ion, $Al^{3+} + HL \rightarrow AlL + H^+$. For ligands containing phenolate, catecholate or amino groups, the amount of free, aqueous Al^{3+} in neutral solutions becomes pH-dependent. It follows that listed stability constants overstate effective binding strengths of these ligands and need to be lowered to reflect the competition of the proton with the metal ion for basic binding sites. The most practical method to allow for proton–metal ion competition at a ligand is to calculate *conditional* stability constants, K_c, applicable to a single pH.[3,26,40] Conditional stability constants may also allow for deprotonation of metal ion-coordinated water that yields, in some cases, Al^{3+} complexes which become more stable with increasing pH, as with citrate, nitrilotriacetate and EDTA.

Results from quantitative evaluation of conditional stability constants are revealingly expressed as the negative logarithm of the free Al^{3+} concentration, $-\log_{10}[Al^{3+}] = pAl$. A pAl value also includes the ligand concentration. If we let the molar concentration ratio of total Al(III) to total ligand be represented by R, then, for $R < 1$, we have

$$pAl = \log K_c - \log[R/(1 - R)] \simeq \log K_c - \log R \qquad (9.1)$$

where the last equality applies when $R < 1/10$. Table 9.3 lists conditional stability constants and pAl values at the intracellular pH 6.6 and plasma pH 7.4 for several systems with 1 μM total Al(III) under the conditions indicated in the table.[40,41] Weak Al^{3+} binders appear at the top and strong binders at the bottom of the table. The increasing pAl values as one goes down the table indicate, in keeping with the response to pH, decreasing free Al^{3+} concentrations. Thus, since 0.1 mM citrate lies lower in Table 9.3 than does 1 mM ATP^{4-}, we predict that citrate will withdraw Al^{3+} from ATP^{4-}, and experimentally citrate has been used for this purpose.[42] Despite high normal stability constants, as indicated in the second column, salicylate[43] and catecholamines[44] bind Al^{3+} relatively weakly because competition from the proton in the strongly basic ligands leads to the low conditional stability constants listed in the third and fifth columns. The catecholamines furnish models for Al^{3+} binding to catecholate-like functions in humic acids, which are also modelled by catechol[45] and gallic acid.[46] Kaolinite is the least soluble aluminium silicate.[11]

Table 9.3 Negative logarithm (base 10) of free Al^{3+} concentration, pAl.[a]

Complex or ligand[b]	$\log_{10} K_s$	pH 6.6		pH 7.4	
		$\log_{10} K_{6.6}$	pAl	$\log_{10} K_{7.4}$	pAl
DNA	<5.6	<5.6	<7.3	<5.6	<7.3
Salicylate, 0.2 mM	12.9, 10.6	6.3, 4.0	9.1	7.1, 4.8	10.7
Amorphous $Al(OH)_3$	Insoluble		9.1		11.5
Al^{3+} to $[Al(OH)_4]^-$	Ref. 31		9.1		12.1
Catecholamines	15.6, 13.0	7.4, 4.8	9.7	9.0, 6.4	12.8
Kaolinite[e]	Insoluble		10.2		12.6
$AlPO_4$	Insoluble		11.7[c]		12.1[d]
$[(HO)_2AlO_3POH]^-$	Ref. 41		11.7[c]		12.9[d]
Nitrilotriacetate (NTA)	11.1	10.0	11.7	11.6	13.3
2,3-DPG, 3 mM	12.5	11.6	12.2	12.2	13.1
ATP, 1 mM	7.9, 4.6	8.9	12.3	9.8	13.0
Citrate, 0.1 mM; Ca^{2+}, 3 mM	Ref. 40	10.3	12.3	11.7	13.7
Citrate, 0.1 mM	8.1	11.3	13.3	12.7	14.7
Transferrin	Ref. 40			13.6, 12.8	15.3
F^-, 5 mM with OH^-	Ref. 48		14.9		15.1
EDTA	16.2	13.1	14.8	14.7	16.4
Deferoxamine	24.1	16.8	18.4	19.2	20.8

(a) 1 μM total Al^{III} except for insoluble salts. Equilibria in addition to those related to listed $\log_{10} K_c$ values often required to calculate pAl.
(b) 50 μM ligand unless otherwise noted.
(c) 10 mM total phosphate.
(d) 2 mM total phosphate.
(e) $Al_2(OH)_4Si_2O_5$ with 5 μM $Si(OH)_4$, typical of plasma.

9.4   Biological aspects

For the reasons listed in section 9.5.1, by far the greatest amount of work has been performed with aluminium and this section deals exclusively with that element. The most likely Al^{3+} binding sites are oxygen atoms, especially if they are negatively charged. Phosphate, catecholate and carboxylate groups are the strongest Al^{3+} binders. Even when part of a potential chelate ring, sulfhydryl groups do not bind Al^{3+}.[47] Amines do not bind Al^{3+} strongly except as part of multidentate ligand systems such as nitrilotriacetate (NTA) and EDTA. The nitrogenous bases of DNA and RNA do not bind Al^{3+}.[3,26] Al^{3+} binds fluoride (Table 9.3), and at the 1 ppm level, at which fluoride is added to acidic drinking water, most Al(III) appears as $[AlF_2]^+$ and neutral AlF_3.[48] In mixed complexes of ADP and F^-, the ternary complex appears with the frequency expected statistically on the basis of the stabilities of the binary complexes.[49]

9.4.1   Phosphate binding

In the human body, extracellular fluids contain up to 2 mM total phosphate at pH 7.4 and intracellular fluids about 10 mM total phosphate at pH 6.6. Al^{3+} forms an insoluble salt with phosphate often designated as $AlPO_4$ or sometimes as $AlPO_4 \cdot 2H_2O$, corresponding to the composition of the mineral variscite. For intracellular fluids at pH 6.6 containing 10 mM total phosphate, we find, $-\log[Al^{3+}] = pAl = 11.7$, while for extracellular fluids at pH 7.4 containing 2 mM total phosphate, $pAl = 12.1$.[40] This pair of values, which appears in the seventh row of Table 9.3, indicates that the maximum free Al^{3+} concentration is extremely low.

It may be remarked that the hydrogen atoms have been located in the crystal structure of the metavariscite form of $AlPO_4 \cdot 2H_2O$.[50] While we might have expected $(HO)_2Al(O_2PO_2H_2)$, the structure shows intact water molecules coordinated in *cis* positions to hexacoordinate Al^{3+}. The remaining four coordination positions share a phosphate oxygen from four different phosphate groups. Each phosphate oxygen shares a vertex with a tetrahedral phosphorus and an approximately octahedral Al^{3+}. A proton magnetic resonance study of the solid also indicates bound water molecules.[51] Since the sharing of phosphate oxygens at common vertices ceases when metavariscite dissolves, however, the structure of the mineral and the state of its water molecules are irrelevant to aqueous solutions.

Al^{3+} will frequently form soluble complexes with phosphate groups in biological systems. Stability constants recently determined for Al^{3+} binding to adenosine-5'-nucleotides[52] and 2,3-diphosphoglycerate (DPG)[53] appear in logarithmic form in Table 9.3. The largest stability constants are found for ADP and ATP where chelation occurs. For comparison, the stability constant for Mg^{2+} binding to ATP and other nucleoside triphosphates is

$\log K_1 = 4.3$,[54] implying that binding is 4000 times weaker than for Al^{3+}. Thus, 0.2 μM Al^{3+} competes with 1 mM Mg^{2+} for ATP.

Many suppose that Al^{3+} binds to DNA in the cell nucleus. However, Al^{3+} binding to DNA is so weak that a quantitative study was limited to pHs up to 5.5 before the onset of metal ion hydrolysis and precipitation.[55] For Al^{3+}, therefore, DNA cannot compete with ATP, $HOPO_3^{2-}$ and other ligands in Table 9.3. With very weakly basic phosphates with pK_a c. 1, DNA serves merely as a polyelectrolyte interacting with Al^{3+} weakly and non-specifically. Residing at the very top of Table 9.3 with the lowest pAl values or highest allowed free Al^{3+} concentrations, DNA loses Al^{3+} to all other entries in the table. We deduce that Al^{3+} binding to DNA is so weak under intracellular conditions that it fails by several orders of magnitude to compete with either metal ion hydrolysis or insolubility of even an amorphous $Al(OH)_3$. No matter what other ligands are present, this competition with aqueous solvent components remains. Therefore, we conclude that the observation of aluminium with nuclear chromatin is due to its coordination not to DNA but to other ligands.

What ligands might bind Al^{3+} in the cell, especially in the nuclear chromatin region? ATP and ADP are comparably strong Al^{3+} binders. A crucial Al^{3+} binding site in chromatin promises to be phosphorylated proteins, perhaps phosphorylated histones. Phosphorylation and dephos-phorylation reactions normally accompany cellular processes. The phosphate groups of any phosphorylated protein provide the requisite basicity, and, in conjunction with juxtaposed carboxylate or other phosphate groups, become strong Al^{3+} binding sites. Abnormally phosphorylated proteins have been found in the brains of patients suffering from Alzheimer's disease.[56,57] The addition of Al^{3+} to purified cytoskeletal proteins from a bovine brain source induces selective aggregation of highly phosphorylated proteins.[58] High Al(III) contents have been found associated with increased linker histones in the nuclear region of Alzheimer's diseased brains.[59] Ternary Al^{3+} complexes have received little study, although Al^{3+} has been used as a tanning or crosslinking reagent. Very possibly Al^{3+} crosslinks proteins, and proteins and nucleic acids in the nucleus.

9.4.2   Plasma carriers of aluminium

Citrate exists mainly in the form of the tricarboxylate anion at pH > 6, and at 0.1 mM in the blood plasma, it is the leading small molecule binding Al^{3+}.[26,27] In neutral solutions, the main species is $HOAlLH_{-1}^{2-}$, followed by $AlLH_{-1}^{-}$ with $pK_a \simeq 6.5$ for the loss of a proton from metal ion-bound water.[3] Even though much of the citrate in plasma occurs as a Ca^{2+} complex, Al^{3+} easily displaces Ca^{2+} from citrate. With due consideration of alkaline earth cations in the plasma, there is a molar ratio of almost $10^8 : 1$ for citrate-bound Al^{3+} to unbound Al^{3+}.[40] There is ample proof that citrate facilitates incorporation of Al(III) into mammals. Al(III) levels were elevated in both the brain and bones of rats fed a diet containing aluminium citrate or even just citrate.[60-62] Evidently citrate alone chelates trace Al(III) in the diet. Dosing lambs with aluminium citrate promotes absorption of Al(III) and alters the balance of other minerals.[63,64] Increased levels of serum Al(III) were found in patients with chronic renal failure who were taking an Al(III)-containing phosphate binder with citrate.[65] A rapidly fatal encephalopathy in human patients with chronic renal failure has been attributed to concomitant ingestion of $Al(OH)_3$ and citrate.[66,67] Moreover, healthy adults taking $Al(OH)_3$-based antacids along with citric acid, citrate salts or citrus fruits showed substantial increases in Al(III) levels of blood[68,69] and urine.[70-72] The amount of citrate present should always be considered as a variable in Al(III) ingestion studies.

Transferrin is the main protein carrier of Al^{3+} in the plasma. Displacement of the 10^9 times more strongly binding Fe^{3+} is unnecessary because plasma transferrin is about 50 μM in unoccupied sites. Recent experiments confirm the conclusion based on stability constants[26,73] that citrate is the low molecular weight and transferrin the high molecular weight carrier of Al^{3+} in rat serum.[74]

9.5 Toxicology

Despite the penchant of nutritionists to find, at one time or another, essentiality for almost every element, there is no conclusive evidence that any of the four Group 13 metals is essential for any form of life. Indeed, rather than essentiality, toxicological discussions are dominated by the long held realisation that thallium is dangerously poisonous and the growing awareness of the potential toxicity of aluminium.

9.5.1 Aluminium

Coupled with the increasing availability of Al(III) is its identification as a causative or associative agent in several kinds of human disorders.[75,76] Al(III) is the acknowledged causative agent in conditions arising from long-term haemodialysis, vitamin D-resistant osteomalacia, iron-adequate microcytic anaemia and dialysis dementia. A recent study of 10^4 long-term haemodialysis patients finds that a serum Al(III) concentration of greater than 1.5 μM increases the mortality risk, and 40% of the patients exceed this level.[77] In normal individuals, the serum Al(III) level is less than 0.4 μM. Similar conditions have been observed in children with renal failure with prescribed high intakes of Al(III) to combine with excess phosphate of hyperphosphatemia. Upon acute ingestion, even normal adults may accumulate Al(III) in bones.[78]

Al(III) is a cause of a high frequency of amytrophic lateral sclerosis and Parkinsonism dementia among the natives of southern Guam, the Kii peninsula of Japan and western New Guinea. In these areas, the soils are high in Al(III) and low in Mg^{2+} and Ca^{2+}. Introduction of non-native diets has reduced the incidence of these ailments, which are associated with an accumulation of Al^{3+} in brain neurons that have undergone neurofibrillary degeneration into tangles, but without plaque formation.[79,80]

As noted previously, elevated levels of Al(III) have been observed in the brains of victims suffering from Alzheimer's disease. About two cases of senile dementia out of three are of the Alzheimer's type. Less than 5% of the cases are familial. In the United States, Alzheimer's disease afflicts about 1 person in 30 in the age group 65–74, about 1 in 6 in the age group 75–84, and about 1 in 3 of those aged 85 or more. Women are at greater risk. The disease ranks as one of the most frequent killers of the elderly, with 10^5 deaths annually in the United States. Alzheimer's disease is a progressive senile dementia characterised not by minor failures such as the sufferer's forgetting where his or her glasses were put, but rather by complete loss of input from a memory bank so that the sufferer forgets that he or she wears glasses at all. There is no cure. Autopsies of victims of the disease reveal deposition in the brain of fibrillar amyloid proteins as paired helical filaments that contain high levels of Al(III),[81,82] and of senile plaques that have been reported to contain aluminosilicate.[83] Whether Al(III) causes the disease or possesses an affinity for the abnormal tangles remains uncertain.[84,85] Alzheimer's disease has been associated with the content of Al(III) in drinking water,[86] even though the levels are still low compared to dietary intake.

There are four categories of exposure of healthy adults to aluminium. For an individual, wide variations in intake occur in each category; some typical values appear in Table 9.4. Natural sources contribute only about 5 mg per day and probably even less for non-tea drinkers. High aluminium levels occur naturally in only a few foods. Food additives contribute 5–100 mg per day, the richest agents being alum baking powder products and Al(III)-containing emulsifiers added to processed cheeses. Except in instances of long stewing of acidic or highly salted foods, aluminium cooking utensils, especially if not new, introduce little aluminium into the diet.[87] As indicated by the last category in Table 9.4, the greatest potential sources of intake are Al(III)-containing antacids and buffered aspirins.[88] Most ingested aluminium is poorly absorbed by the intestine and is eliminated by the body. The last entry in Table 9.4, vaccines, provides an example of solutions that bypass the intestinal barrier. Solutions given intravenously to patients may result in significant quantities of aluminium being injected into the blood and may be especially critical in infants.[89] Intakes by individuals with uremic problems,[90,91] industrial exposures, etc., are not covered in this list. The

Table 9.4 Aluminium intake by healthy adults.

Natural sources, about 2–5 mg per day
- Tea, typical 0.1% with up to 1% in dry tea, but poorly extracted in steeping, c. 1 mg per cup
- Drinking water, average is 0.02 mg Al per litre
- Herbs, many high in Al but little consumed
- Spinach

Food additives, up to about 20 mg per day
- Alum baking powder products, cornbread
- Emulsifier in American cheeses
- In cream substitute
- Aluminium silicates as anticaking agents
- $Al_2(SO_4)_3$ as anticoagulant for turbid drinking water

Al containers (new) in stewing of acidic fruits, spaghetti sauce, sauerkraut (<0.1 mg per day)

Pharmaceuticals, if used, to 4000 mg per day
- Antacids, 4000 mg per day for heavy users
- Buffered aspirin, 500 mg per day for arthritics
- Antiperspirants
- Vaccines (intramuscular injection bypasses intestinal barrier)

strong Al^{3+} binder deferoxamine (bottom of Table 9.3) has been used to remove Al(III) from aluminium-intoxicated patients.[92]

9.5.2 Gallium

Environmental exposure of humans to gallium appears negligible, with less than 1 ppb in human tissue. Like Al^{3+}, Ga^{3+} is absorbed but little through the gastrointestinal tract. Transferrin is the main plasma carrier of any Ga^{3+}. Radionuclides of gallium have been used as tracers and anti-tumour agents (see chapter 1). The main industrial exposure to gallium is from gallium arsenide used in electronics (see chapter 5), but the main reason for any toxicity associated with this compound is due to the arsenic, not the gallium. Neither levels of tolerance nor methods of detoxification have been reported for gallium. Environmental gallium appears to pose no threat to human health.[93]

9.5.3 Indium

Environmental exposure of the general population to indium should be negligible. Radionuclides have been used as diagnostic agents (see chapter 1). Tolerance levels have been established for inhalation of indium compounds at industrial sites. No methods of detoxification have been reported, but environmental indium appears to pose no threat to human health.[94]

9.5.4   Thallium

Once used as medicinal agents and rodenticides, because of their high toxicity, thallium salts are no longer available in the United States and other countries for these purposes. Other uses for thallium salts continue, especially in their increasing demand as an ingredient of fibreglass in communication systems. The sulfides in coal carry thallium, and coal-burning is the major source of thallium release into the environment. A typical coal contains 1 ppm of thallium, but some coals run much higher. Except in the vicinity of the coal-burning plants, the thallium release is probably not a hazard to humans. Worldwide, thallotoxicosis from ingestion of thallium-based rodenticides remains one of the most common acute toxic diseases caused by metals.[95]

Both Tl(I) and Tl(III) salts are readily absorbed by the gastrointestinal tract and the skin. Excretion is slow with a half-life of nearly 1 month. Thallium concentrates in the brain and testes. The lethal dose is less than 1 g of a thallium compound in a single ingestion. Thallotoxicosis involves the nervous system, skin, hair-loss, and the cardiovascular system. Tl^+ can substitute for the similarly sized K^+ cation (Table 9.1) and interfere in K^+-dependent processes. Recovery from thallotoxicosis takes months and may be incomplete as nervous system damage may be irreversible.[95] Thallium may be the most toxic non-radioactive metal.

As this brief summary indicates, thallium compounds are dangerously and cumulatively toxic even at low levels. They should be handled with great care and respect, never being allowed to touch the skin or to be ingested.

References

1. R.D. Shannon, *Acta Crystallogr.*, 1976, **A32**, 751.
2. R. Weiss and D. Grandjean, *Acta Chem. Scand.*, 1964, **17**, 1329.
3. R.B. Martin, *Metal Ions Biol. Syst.* 1988, **24**, 1.
4. T.L. Macdonald and R.B. Martin, *Trends Biochem. Sci.*, 1988, **13**, 15.
5. R.B. Martin, *Metal Ions Biol. Syst.*, 1990, **26**, 1.
6. T.L. Macdonald, W.G. Humphreys and R.B. Martin, *Science*, 1987, **236**, 183.
7. B. Mason and C.B. Moore, *Principles of Geochemistry*, 4th edn., Wiley, New York, 1982.
8. P. Henderson, *Inorganic Geochemistry*, Pergamon Press, Oxford, 1982.
9. C.F. Baes and R.E. Mesmer, *The Hydrolysis of Cations*, Wiley-Interscience, New York, 1976.
10. K.H. Wedepohl, ed., *Handbook of Geochemistry*, Springer-Verlag, Berlin, 1969–1978.
11. R.B. Martin, *Polyhedron*, 1990, **9**, 193.
12. W. Stumm and J.J. Morgan, *Aquatic Chemistry*, Wiley–Interscience, New York, 1970.
13. B. Mason, *Acc. Chem. Res.*, 1975, **8**, 217.
14. N.N. Greenwood and A. Earnshaw, *Chemistry of the Elements*, Pergamon Press, Oxford, 1984.
15. S. Ahrland, I. Grenthe, L. Johansson and B. Noren, *Acta Chem. Scand.*, 1963, **17**, 1567.
16. D. Hydes, *Nature (London)*, 1977, **268**, 136.
17. D. Hydes, *Science*, 1979, **205**, 1260.
18. M. Stoffyn, *Science*, 1979, **203**, 651.

19. S.E. Jorgensen and A. Jensen, *Metal Ions Biol. Syst.*, 1984, **18**, 61.
20. C.T. Driscoll and W.D. Schecher, *Metal Ions Biol. Syst.*, 1988, **24**, 59.
21. G.J. Taylor, *Metal Ions Biol. Syst.*, 1988, **24**, 123.
22. C.D. Foy, R.L. Chaney and M.C. White, *Annu. Rev. Plant Physiol.*, 1978, **29**, 511.
23. B.L. Hartwell and F.R. Pember, *Soil Sci.*, 1918, **6**, 259.
24. H. Matsumoto, F. Hirasawa, S. Morimura and E. Takahasi, *Plant Cell Physiol.*, 1976, **17**, 627.
25. K.R. Koch, M.A.B. Pougnet, S. deVilliers and F. Monteagudo, *Nature (London)*, 1988, **333**, 122.
26. R.B. Martin, *Clin. Chem.*, 1986, **32**, 1797.
27. R.B. Martin, *J. Inorg. Biochem.*, 1986, **28**, 181.
28. R.B. Martin, in *Aluminum and Renal Failure*, eds. M.E. deBroe and J.W. Coburn, Kluwer, Dordrecht, 1990, pp. 7–26.
29. E.J. Dennis, *Fert. Solutions*, 1971, **15**, 44.
30. J.P. Fontenot, V.G. Allen, G.E. Bunce and J.P. Goff, *J. Anim. Sci.*, 1989, **67**, 3445.
31. R.B. Martin, *J. Inorg. Biochem.*, 1991, **44**, 141.
32. These values relate to various conditions (including ionic strength): see L.-O. Öhman, *Inorg. Chem.*, 1988, **27**, 2565.
33. E. Marklund and L.-O. Öhman, *Acta Chem. Scand.*, 1990, **44**, 228.
34. P. Bertsch, *Soil Sci. Soc. Am. J.*, 1987, **51**, 825.
35. G. Johansson, *Acta Chem. Scand.*, 1960, **14**, 771.
36. D.L. Teagarden, J.F. Kozlowski, J.L. White and S.L. Hem, *J. Pharm. Sci.*, 1981, **70**, 758.
37. D. Hunter and D.S. Ross, *Science*, 1991, **251**, 1056.
38. T.B. Kinraide, *Plant and Soil*, 1991, **134**, 167.
39. D. Plee, F. Borg, L. Gatineau and J.J. Fripiat, *J. Am. Chem. Soc.*, 1985, **107**, 2362.
40. R.B. Martin, in *Aluminium in Chemistry, Biology, and Medicine*, eds. M. Nicolini, P.F. Zatta and B. Corain, Cortina International-Verona, Raven Press, New York, 1991, pp. 3–20.
41. R.B. Martin, in *Aluminium in Biology and Medicine*, Ciba Foundation Symposium 169, Wiley, Chichester, UK, 1992, pp. 5–25.
42. F.C. Womack and S.P. Colowick, *Proc. Natl. Acad. Sci.*, 1979, **76**, 5080.
43. L.-O. Öhman and S. Sjöberg, *Acta Chem. Scand.*, 1983, **A37**, 875.
44. T. Kiss, I. Sovago and R.B. Martin, *J. Am. Chem. Soc.*, 1989, **111**, 3611.
45. L.-O. Öhman and S. Sjöberg, *Polyhedron*, 1983, **2**, 1329.
46. L.-O. Öhman and S. Sjöberg, *Acta Chem. Scand.*, 1981, **A35**, 201.
47. I. Tóth, L. Zékány and E. Brücher, *Polyhedron*, 1984, **3**, 871.
48. R.B. Martin, *Biochem. Biophys. Res. Commun.*, 1988, **155**, 1194.
49. D.J. Nelson and R.B. Martin, *J. Inorg. Biochem.*, 1991, **43**, 37.
50. R. Kniep and D. Mootz, *Acta Crystallogr.*, 1973, **B29**, 2292.
51. C. Doremieux-Morin, M. Krahe and F. d'Yvoire, *Bull. Soc. Chim. France*, 1973, 409.
52. T. Kiss, I. Sovago and R.B. Martin, *Inorg. Chem.*, 1991, **30**, 2130.
53. I. Sovago, T. Kiss and R.B. Martin, *Polyhedron*, 1990, **9**, 189.
54. H. Sigel, R. Tribolet, R. Malini-Balakrishnan and R.B. Martin, *Inorg. Chem.*, 1987, **26**, 2149.
55. D. Dyrssen, C. Haraldsson, E. Nyberg and M. Wedborg, *J. Inorg. Biochem.*, 1987, **29**, 67.
56. N.H. Sternberger, L.A. Sternberger and J. Ulrich, *Proc. Natl. Acad. Sci.*, 1985, **82**, 4274.
57. I. Grundke-Iqbal, K. Iqbal, Y. Tung, M. Quinlan, H.M. Wisniewski and L.I. Binder, *Proc. Natl. Acad. Sci.*, 1986, **83**, 4913.
58. J. Diaz-Nido and J. Avila, *Neurosci. Lett.*, 1990, **110**, 221.
59. W.J. Lukiw, L. Krishnan, L. Wong, T.P.A. Kruck, C. Bergeron and D.R.C. McLachlan, *Neurobiology of Aging*, 1991, **13**, 115.
60. P. Slanina, Y. Falkeborn, W. Frech and A. Cedergren, *Food Chem. Toxicol.*, 1984, **22**, 391.
61. P. Slanina, W. Frech, A. Bernhardson, A. Cedergren and P. Mattsson, *Acta Pharmacol. Toxicol.*, 1985, **56**, 331.
62. J.L. Domingo, M. Gomez, J.M. Llobet and J. Corbella, *Kidney Int.*, 1991, **39**, 598.
63. V.G. Allen, J.P. Fontenot and S.H. Rahnema, *J. Anim. Sci.*, 1990, **68**, 2496.
64. V.G. Allen, J.P. Fontenot and S.H. Rahnema, *J. Anim. Sci.*, 1991, **69**, 792.

65. C.D. Hewitt, C.L. Poole, F.B. Westervelt, J. Savory and M.R. Wills, *Lancet*, 1988, **ii**, 849.
66. B.B. Kirschbaum and A.C. Schoolwerth, *Am. J. Medical Sci.*, 1989, **297**, 9.
67. A.A. Bakir, D.O. Hryhorczuk, E. Berman and G. Dunea, *Trans. Am. Soc. Artif. Intern. Organs*, 1986, **32**, 171.
68. P. Slanina, W. Frech, L. Ekstrom, L. Loof, S. Slorach and A. Cedergren, *Clin. Chem.*, 1986, **32**, 539.
69. R. Weberg and A. Berstad, *Europ. J. Clin. Invest.*, 1986, **16**, 428.
70. A.A. Bakir, D.O. Hryhorczuk, S. Ahmed, S.M. Hessl, P.S. Levy, R. Spengler and G. Dunea, *Clin. Nephrol.*, 1989, **31**, 40.
71. J.A. Walker, R.A. Sherman and R.P. Cody, *Arch. Intern. Med.*, 1990, **150**, 2037.
72. J.W. Coburn, M.G. Mischel, W.G. Goodman and I.B. Salusky, *Am. J. Kidney Dis.*, 1991, **17**, 708.
73. R.B. Martin, J. Savory, S. Brown, R.L. Bertholf and M. Wills, *Clin. Chem.*, 1987, **33**, 405.
74. M.F. vanGinkel, G.B. vanderVoet, H.G. vanEijk and F.A. deWolff, *J. Clin. Chem. Clin. Biochem.*, 1990, **28**, 459.
75. M.R. Wills and J. Savory, *Lancet*, 1983, **ii**, 29.
76. C.D. Hewitt, J. Savory and M.R. Wills, *Clin. Lab. Med.*, 1990, **10**, 403.
77. J.A. Chazan, N.L. Lew and E.G. Lowrie, *Arch. Intern. Med.*, 1991, **151**, 319.
78. J.R. Eastwood, G.E. Levin, M. Pazianas, A.P. Taylor, J. Denton and A.J. Freemont, *Lancet*, 1990, **ii**, 462.
79. D. Perl, D.C. Gajdusek, R.M. Garruto, R.T. Yanagihara and C.J. Gibbs, *Science*, 1982, **217**, 1053.
80. R.M. Garruto, *Neurotoxicology*, 1991, **12**, 347.
81. D.R. Crapper, S.S. Krishnan and A.J. Dalton, *Science*, 1973, **180**, 511.
82. D.P. Perl and A.R. Brody, *Science*, 1980, **208**, 297.
83. J.M. Candy, J. Klinowski, R.H. Perry, E.K. Perry, A. Fairbairn, A.E. Oakley, T.A. Carpenter, J.R. Atack, G. Blessed and J.A. Edwardson, *Lancet*, 1986, **i**, 354.
84. D.R.C. McLachlan, W.J. Lukiw and T.P.A. Kruck, *Can. J. Neurol. Sci.*, 1989, **16**, 490.
85. D.R.C. McLachlan, T.P. Kruck, W.J. Lukiw and S.S. Krishnan, *Can. Med. Assoc. J.*, 1991, **145**, 793.
86. C.N. Martyn, C. Osmond, J.A. Edwardson, D.J. Barker, E.D. Harris and R.F. Lacey, *Lancet*, 1989, **i**, 59.
87. J.L. Greger, *Food Technol.*, 1985, **39**, 73.
88. A. Lione, *Pharmacol. Ther.*, 1985, **29**, 255.
89. G.L. Klein, *Nutr. Rev.*, 1991, **49**, 74.
90. M.R. Wills and J. Savory, *Crit. Rev. Clin. Lab. Sci.*, 1989, **27**, 59.
91. M.E. deBroe and J.W. Coburn, eds., *Aluminum and Renal Failure*, Kluwer, Dordrecht, 1990.
92. R.L. Bertholf, M.R. Wills and J. Savory, in *Handbook on Toxicity of Inorganic Compounds*, eds. H.G. Seiler and H. Sigel, Marcel Dekker, New York, 1988, pp. 55–64.
93. R.L. Hayes, in *Handbook on Toxicity of Inorganic Compounds*, eds. H.G. Seiler and H. Sigel, Marcel Dekker, New York, 1988, pp. 297–300.
94. R.L. Hayes, in *Handbook on Toxicity of Inorganic Compounds*, eds. H.G. Seiler and H. Sigel, Marcel Dekker, New York, 1988, pp. 323–326.
95. L. Manzo and E. Sabbioni, in *Handbook on Toxicity of Inorganic Compounds*, eds. H.G. Seiler and H. Sigel, Marcel Dekker, New York, 1988, pp. 677–688.

10 Analytical methods

H. ONISHI

This chapter presents a brief review of classical (time-honoured) methods of separation and analysis and of recent advances in analytical technology. The literature cited is necessarily limited. It seems that analytical chemistry is going increasingly outside the realm of purely chemical methods with a growing dependence on physical methods of analysis. In the following account, photometric (spectrophotometric) methods should be understood to include molecular fluorescence or fluorimetric methods.

10.1 Aluminium

Samples or materials in which aluminium is determined are extremely diverse. For example, in *Analytical Chemistry, Application Reviews* for 1991, aluminium is found specifically in steel and related materials,[1] in clinical samples[2] and in water analysis.[3] Aluminium is ubiquitous because of both natural and artificial sources. In trace analysis, therefore, prevention or control of contamination throughout the entire course of analysis (including sampling) is of vital importance.

Speciation or identification of species of aluminium, especially in aqueous media, has attracted much attention. Discussion of the results obtained by plural analytical techniques, e.g. chromatography and nuclear magnetic resonance, will give a clue to speciation. This is a matter of considerable relevance to the environmental chemistry of aluminium (see chapter 9).

Decomposition of silicate samples is brought about by fusion with sodium carbonate. Aluminium oxide is fused with potassium hydrogen sulfate, Na_2CO_3–H_3BO_3 or Na_2CO_3–$Na_2B_4O_7$.[4] Aluminium oxide dissolves in hot hydrochloric acid under pressure.[5] Aluminium metal dissolves in hydrochloric acid or sodium hydroxide solution.

Organic samples can be decomposed by wet oxidation or dry oxidation methods, but wet oxidation is generally preferable (compare Ref. 6). A microwave oven has been used for decomposition of biological materials.

10.1.1 Separations

A detailed account of separation procedures is given in Ref. 6. Of special importance is the separation of aluminium from iron.

Precipitation of aluminium hydroxide with aqueous ammonia is not selective. In the separation of iron(III) from aluminium by precipitation with sodium hydroxide, there is a danger of co-precipitation of aluminium.

A useful method for eliminating many metals at one time involves electrolysis with a mercury cathode in dilute (0.05–0.1 M) sulfuric acid solution. The use of mercury may be undesirable, however.

Aluminium is not adsorbed on a strong-base anion-exchange resin from about 9 M hydrochloric acid solution, but uranium(VI), vanadium(V), chromium(VI), molybdenum, tungsten, iron(III), cobalt, copper, zinc, cadmium, gallium, tin(IV,II), antimony(V,III), etc. are so adsorbed.

Iron(III) reacts with cupferron in 2.5–3 M sulfuric acid and the cupferrate is extracted into chloroform; aluminium is not extracted in this way. Iron(III) can be extracted with diisopropyl ether or 4-methyl-2-pentanone (isobutyl methyl ketone) from about 6 M hydrochloric acid and can thus be separated from aluminium.

10.1.1.1 Chromatography.   Separation and determination of aluminium by ion chromatography—high-performance ion-exchange chromatography —have been studied by many workers. The aluminium eluted is detected by conductivity,[7] colour and fluorescence reactions (e.g. with Tiron[8]) and chemiluminescence.[9]

Separation and determination of aluminium and other metals by high-performance liquid chromatography (HPLC) of their chelate derivatives has attracted the attention of many workers. The chelating reagents include 8-quinolinol,[10] β-diketones, hydrazone, hydroxamic acid and monoazo reagents; spectrophotometric measurements usually afford the means of detection and estimation.

Aluminium has been determined by gas chromatography of its β-diketonate derivatives. For example, aluminium in sea-water can be determined with 1,1,1-trifluoro-2,4-pentanedione.[11]

10.1.2 Methods of determination

Earlier reviews of this subject are to be found in Refs. 12–14. For a more up-to-date account, the reader is directed to Ref. 15.

10.1.2.1 Comparison of methods.   Many papers have been published on the comparison of the different methods available for the determination of aluminium. Table 10.1 summarises the samples and methods that have been so evaluated. For details of the conclusions reached, the original articles should be consulted.

10.1.2.2 Gravimetric methods.   Aluminium hydroxide is precipitated with aqueous ammonia at pH about 7, and the precipitate is filtered, ignited and weighed as Al_2O_3.

Aluminium is also precipitated with 8-quinolinol (8-hydroxyquinoline) at

Table 10.1 Comparison of analytical methods used for the determination of aluminium.

Sample	Methods	Reference
Water	Atomic absorption, photometric and fluorimetric	16
Silicates	Atomic emission,[a] atomic absorption and X-ray fluorescence	17
Molecular sieves	Gravimetric, titrimetric and photometric	18
Zeolites	Atomic absorption, neutron activation, proton inelastic scattering etc.	19
Extracts of soils	Titrimetric, photometric, atomic absorption and atomic emission[a]	20
	Titrimetric, photometric and atomic absorption	21
	Different photometric methods and atomic absorption	22
Soil and clay	HPLC and atomic absorption	23
Biological materials	Neutron activation and atomic absorption	24
Plant material	Atomic absorption, atomic emission[a] and photometric	25
Human lungs	Atomic emission[a] and neutron activation	26

(a) Atomic emission mostly refers to inductively coupled plasma atomic emission spectrometry.

pH 5.0–5.5, and the precipitate is filtered, dried at 130°C and weighed as $Al(C_9H_6ON)_3$.

10.1.2.3 Volumetric (titrimetric) methods. The stability constant (as $\log_{10} K$) of the complex formed by aluminium(III) with the ethylene-diaminetetraacetate anion (EDTA) is 16.5. Aluminium is usually determined by back-titration with a standard metal solution rather than by direct titration with EDTA. Many indicators have been proposed for back-titration. Furthermore, after back-titration, fluoride ion is added to liberate EDTA from the aluminium complex, and the EDTA is itself back-titrated.[27–29]

An outline of a representative procedure is as follows. About $100\,cm^3$ of a solution containing 1–10 mg of Al is adjusted to pH about 3 with acetic acid of aqueous ammonia. A measured excess of a standard (0.01 M) solution of the disodium salt of EDTA is added and the solution is boiled for 2 min. After cooling, the pH of the solution is adjusted to 5–6 with hexamethylene-tetramine crystals. Using Xylenol Orange as the indicator, the excess of EDTA is titrated with a standard (0.01 M) zinc solution.

In addition, salts of the following reagents have been proposed as titrants: 1,2-cyclohexanediaminetetraacetic acid (1,2-diaminocyclohexanetetra-acetic acid), diethylenetriaminepentaacetic acid[30] and N-(2-hydroxyethyl)-ethylenediaminetriacetic acid.[31]

Another titrimetric method (oxidation-reduction titration) makes use of the precipitation of the aluminium 8-quinolinate complex, treatment of the precipitate with dilute hydrochloric acid to liberate 8-quinolinol and bromination of the 8-quinolinol.

10.1.2.4 Electrochemical methods. Aluminium can be titrated or determined with fluoride (usually sodium fluoride) using a fluoride ion-selective electrode.[32–35]

Determination of aluminium by polarography or voltammetry of its complexes with organic reagents (e.g. monoazo reagents) has been studied. The methods have involved, for example, differential pulse polarography[36] and stripping voltammetry.[37,38]

10.1.2.5 Photometric (spectrophotometric) methods. Many papers have appeared since the publication of an earlier general review.[6] It seems that features of recent developments are flow-injection analysis and fluorimetric determination with various reagents, e.g. morin, Schiff bases, hydrazones, Lumogallion (a monoazo reagent) and 8-hydroxyquinoline-5-sulfonic acid.

Although many reagents or methods are available, more selective reagents or methods would be desirable. Aluminon, Chrome Azurol S and Pyrocatechol (Catechol) Violet are common reagents for the determination of aluminium in aqueous solutions. In the use of these reagents, the pH of the sample solution must be closely controlled. 8-Quinolinol is used for the extraction-photometric and fluorimetric determination of aluminium.

An outline of the *Aluminon method* is as follows. The slightly acidic solution containing 2–20 μg of Al is transferred to a 50-cm^3 volumetric flask. Exactly 2 cm^3 of 5 M hydrochloric acid, 1.0 cm^3 of mercaptoacetic acid (1 part to 19 of water), 3 cm^3 of 1% starch solution and 5.00 cm^3 of 3.5 M ammonium acetate are added. The solution is diluted to about 45 cm^3, and 2.00 cm^3 of 0.3% Aluminon solution are added. The solution is heated in a boiling water bath for 10 min. After cooling, the solution is diluted to the mark, and the absorbance is measured at 525 nm. The sensitivity index, or Sandell sensitivity, corresponding to 0.001 absorbance is 0.002 μg cm^{-2} Al (530 nm).

8-Quinolinol method. The pH of the slightly acidic solution containing 10–50 μg of Al is adjusted to about 5 with ammonium acetate and aqueous ammonia. The solution is diluted to about 50 cm^3 in a separating funnel. Exactly 10 cm^3 of a 1% solution of 8-quinolinol in chloroform is added and the funnel is shaken for 3 min. The organic phase is separated and treated with anhydrous sodium sulfate. The absorbance is measured at 390 nm. If a small amount of iron(III) is present, the absorbance is also measured at 470 nm, and aluminium and iron can thus be determined simultaneously. The sensitivity index for aluminium is 0.004 μg cm^{-2} Al (390 nm).

10.1.2.6 Atomic spectroscopy. At present, atomic spectroscopy, whether in absorption or emission, is very useful for the determination of small amounts or traces of aluminium (and other metals also) in various matrices. Both atomic absorption spectrometry (AAS) and inductively coupled plasma atomic emission spectrometry (ICP AES) usually require that solid samples be brought into solution. Comparisons of AAS and ICP AES for the determination of aluminium in bones[39] and biological fluids[40] have been made (see also Table 10.1).

Atomic absorption spectrometry (AAS). Determination of aluminium with a dinitrogen oxide (N_2O)–acetylene flame and a wavelength of 309.3 or 396.2 nm is widely used, although the sensitivity is rather low.

Determination of aluminium with a graphite furnace (electrothermal atomisation) and the above wavelengths also finds wide application, especially for the analysis of biological materials,[41] including serum.[42] Chloride and perchloric acid interfere with the determination of aluminium, and use of matrix modifiers, such as magnesium nitrate and phosphoric acid, is recommended.

Atomic emission spectrometry (AES). AES using an ICP source and typical wavelengths of 396.152, 394.401, 309.278 and 308.215 nm are exploited for the determination of aluminium in a variety of samples, including geological, ferrous, non-ferrous and biological materials. Details of the ICP emission spectra of aluminium, gallium, indium and thallium have been presented by Anderson and Parsons.[43] Determination of aluminium by DC Ar plasma AES has also been studied. Other AES techniques use flame, graphite furnace, arc and spark sources.

10.1.2.7 X-Ray fluorescence (XRF). XRF using a K_α line of aluminium at 0.8340 nm is applied to the determination of more than traces (minor and major constituents) of aluminium in rocks (silicates), ores, steels and other refractory materials. Solid samples are suitable for such measurements. Recent papers describe the use of XRF spectrometry for the analysis of magnesite, dolomite and related materials (after fusion with $Li_2B_4O_7$–$LiBO_2$)[44] and coal powders.[45]

10.1.2.8 Nuclear magnetic resonance (NMR). ^{27}Al NMR measurements permit the determination of aluminium (^{27}Al $I = 5/2$, natural abundance = 100%, see also chapter 1) or give information about species in water,[46-48] zeolites,[19,49] tea digests and infusions[50] and plants.[51]

10.1.2.9 Mass spectrometry (MS). Trace analysis by ICP MS is being studied (e.g. for the determination of aluminium and thallium in biological materials[52]).

10.1.2.10 Neutron activation analysis. Aluminium can be determined by neutron activation; instrumental neutron activation analysis is suitable (see, for example, Table 10.1). With thermal neutrons from a nuclear reactor or a

^{252}Cf source, the reaction ^{27}Al(n,γ)^{28}Al (half-life 2.25 min, γ 1.778 MeV) is used; rocks, coal and biological materials have been analysed. With fast neutrons from a 14-MeV neutron generator or a ^{241}Am–Be source, reference is made to the product of the reaction ^{27}Al(n,p)^{27}Mg (half-life 9.45 min, γ 0.844 MeV); rocks, bauxite and coal have been analysed.

10.1.3   Analysis of aluminium and aluminium alloys

References 13 and 14, which date from 1966 and 1967, respectively, afford general reviews of this subject. The latest review[53] appeared in 1979.

Analysis of aluminium and aluminium alloys, i.e. the determination of minor and trace constituents of these matrices, has been accomplished by chemical and atomic spectroscopic methods. The spectroscopic methods involve atomic absorption (mainly by injection into a flame) and atomic emission (mainly with spark and arc sources).

Methods for the characterisation of high-purity aluminium have been reviewed.[54] The methods include activation analysis, glow-discharge mass spectrometry and wet chemical multi-step procedures. Use of glow-discharge mass spectrometry has been explored more fully in recent years.[55,56] Secondary-ion mass spectrometry (SIMS) has been employed for the determination of uranium and thorium.[55] Trace impurities in high-purity aluminium have been determined by ICP AES after distillation of aluminium as triethylaluminium under reduced pressure[57] (a procedure likely to call for significant safety precautions).

10.2   Gallium

Silicate rocks can be decomposed with sulfuric and hydrofluoric acids. Gallium oxide is fused with potassium hydrogen sulfate. Gallium metal dissolves in hydrochloric acid or sodium hydroxide solution.

10.2.1   Separations

This subject has been reviewed at some length.[6] Separation of gallium from iron and aluminium is especially important.

For the separation of small amounts of gallium, liquid-liquid extraction is the most suitable procedure. Gallium is extracted as $HGaCl_4$, aq. with diethyl ether from 5.5–6 M hydrochloric acid or with diisopropyl ether from 6.5–8 M hydrochloric acid and can thus be separated from aluminium and certain other elements. Under these conditions iron(III), gold(III) and thallium(III) are also extracted, but if a suitable reducing agent, e.g. titanium(III) chloride, is added, such extraction is inhibited. Gallium is recovered by evaporating the organic phase to dryness on a water bath.

Gallium is also extracted from hydrochloric acid solution with crown ethers.[58,59] In another development, the separation of thallium(III), indium and gallium by p-octylaniline extraction has been proposed.[60] More recently still there have been studies of the adsorption of small amounts of gallium and indium on a phosphoramidate chelating fibre.[61]

For the separation of aluminium, indium, gallium and thallium(III) on a strong-base anion-exchange resin, aluminium and indium are eluted with 7 M hydrochloric acid, gallium with 1 M hydrochloric acid and thallium with 4 M perchloric acid. The same mixture can also be separated by cation-exchange chromatography on AG 50W-X4 resin.[62] Elsewhere ion chromatography has been described as the basis of a method for the separation and determination of gallium, indium and thallium(III).[63]

Separation and determination of gallium and other metals by high-performance liquid chromatography (HPLC) of suitable chelate derivatives have been studied by several workers. The chelating reagents include 8-quinolinol (8-hydroxyquinoline)[64] and 8-hydroxyquinoline-5-sulfonic acid.[65]

10.2.2 Methods of determination

Earlier reviews of this subject are to be found in Refs. 12, 66 and 67.

10.2.2.1 Gravimetric methods. Gallium hydroxide is precipitated from dilute sulfuric acid solution with aqueous ammonia at pH about 7, and the precipitate is filtered, ignited and weighed as Ga_2O_3. Gallium may also be determined as its 8-quinolinate complex.

10.2.2.2 Volumetric (titrimetric) methods. The stability constant (as $log_{10} K$) of the complex formed by gallium(III) with the ethylenediamine-tetraacetate anion (EDTA) is 21.0. Gallium can be determined by direct titration with EDTA at pH close to 3. Copper–1-(2-pyridylazo)-2-naphthol (Cu–PAN), gallocyanine and Pyrocatechol (Catechol) Violet are typical indicators.

An outline of a representative procedure is as follows. The pH of the sample solution is adjusted to 3–3.5 with an acetate buffer solution, a few drops of Cu-PAN indicator solution are added, and the solution is heated to boiling and titrated with a standard solution (e.g. 0.01 M) of the disodium salt of EDTA.

Determination of gallium, indium and thallium(III) by successive titrations with EDTA using Semi-xylenol Orange as the indicator has been proposed.[68]

10.2.2.3 Electrochemical methods. Complexometric or EDTA titrations using electrochemical — typically coulometric — techniques have been described for the high-precision determination of gallium and indium (see also section 10.2.3). In another variation on the electrochemical theme, stripping voltammetry in the absence[69] or presence[70] of a chelating reagent is noteworthy for its high sensitivity.

10.2.2.4 Photometric (spectrophotometric) methods.   Many papers have appeared since the publication of the earlier general review.[6] The procedures they describe may be summarised as follows: (i) sensitive colour reactions with phenylfluorone and related compounds in the presence of cationic surfactants; (ii) fluorimetric determination with hydrazones, semicarbazones and thiocarbohydrazones; and (iii) spectrophotometric and fluorimetric determination with heterocyclic azo reagents (PAN and others), aryl monoazo reagents and aryl *bis*azo reagents.

Methods using an ion associate of $GaCl_4^-$ and Rhodamine B (cation) and the 8-quinolinate derivative of gallium are especially useful. The products lend themselves to measurements of fluorescence intensities as well as absorbances.

An outline of a Rhodamine B method[71] is as follows. About $7\,cm^3$ of a $0.1\,M$ hydrochloric acid solution containing 1–6 μg of Ga is transferred to a separating funnel and $0.3\,cm^3$ of 20% titanium(III) chloride solution, $3\,cm^3$ of 0.3% Rhodamine B solution (in $8\,M$ hydrochloric acid) and about $2\,g$ of sodium chloride are added. The solution is shaken with $10.0\,cm^3$ of benzene for 10 min. The separated organic phase is centrifuged and transferred to a dry cell without rinsing. The absorbance is measured at 565 nm. The sensitivity index corresponding to 0.001 absorbance is $0.0008\,\mu g\,cm^{-2}$ Ga.

10.2.2.5 Atomic spectroscopy.   Atomic spectroscopic methods are very useful. Atomic absorption spectrometry (AAS) has been compared with atomic emission spectrometry (AES) as a means of determining gallium in tumour-affected tissues.[72]

AAS.   Gallium has been determined in various matrices. An air–acetylene or dinitrogen oxide–acetylene flame or a graphite furnace is used for the analysis of samples contained in dilute nitric acid solution.[73] In either case a wavelength of 287.4 nm is widely used. In graphite furnace AAS, matrix modifiers such as nickel nitrate[74] are often added.

AES.   With a DC arc or an inductively coupled plasma (ICP) source, emission measurements have widely been made at a wavelength of 294.364 nm. The relevant ICP emission spectrum has been characterised in detail.[43] Flame emission spectrometry (with a dinitrogen oxide–acetylene flame and a wavelength of 417.2 nm) has also been employed.

10.2.2.6 X-Ray fluorescence (XRF).   XRF using a K_α line of gallium has been applied to the analysis of waste water[75] and silicate rocks.[76]

Determination of gallium in biological tissues has also been effected by an electron probe microanalyser.[77]

10.2.2.7 Mass spectrometry (MS). Less than 1 ppm of Ga in aluminium can be determined by isotope dilution ICP MS.[78] Determination of impurities, including gallium, in indium phosphide has also been achieved by secondary-ion mass spectrometry (SIMS).[79]

10.2.2.8 Neutron activation analysis. The analysis is based on the reaction $^{71}Ga(n, \gamma)^{72}Ga$ (half-life 14.10 h, γ 0.834 MeV). Recent applications include the determination of gallium in rocks, ores and biological materials;[80] separations are usually made prior to activity measurements.

10.2.3 Analysis of gallium and gallium arsenide

10.2.3.1 Gallium. Isotopic variations in commercial high-purity gallium have been measured.[81] Coulometric complexometric back-titration with electrolytically generated cadmium(II) permits the determination of gallium and indium with a relative standard deviation of 0.01%.[82] In addition, a multi-technique approach to the trace characterisation of high-purity gallium has been discussed.[83] Gallium has been removed by volatilisation of gallium(III) chloride before determination of trace elements by AAS.[84]

10.2.3.2 Gallium arsenide.[85] The coulometric method mentioned above can be used to determine the stoichiometric composition of gallium arsenide.[86] For the determination of impurities, it is necessary to appeal to sensitive methods such as flameless AAS, ICP AES, ICP MS and activation analysis. To these methods may be added SIMS and glow-discharge mass spectrometry.

The characterisation of gallium arsenide samples may well entail surface analysis introducing problems which are more amenable, *inter alia*, to particle beam analysis.[87] For example, techniques used to interrogate the surfaces of such samples include, in addition to X-ray microprobe measurements and Auger electron spectroscopy, SIMS, particle-induced X-ray emission (PIXE) and Rutherford back-scattering spectrometry.

10.3 Indium

Methods of separation and determination of indium are less selective than those of gallium and thallium, and the determination of indium in real samples is not necessarily simple.

Indium metal dissolves in hot mineral acids; indium alloys dissolve in mineral acids or aqua regia. Minerals containing indium can be decomposed first with hydrochloric acid, then with nitric acid or with aqua regia. Indium oxide is fused with potassium hydrogen sulfate.

10.3.1 Separations

A general review of this subject is to be found elsewhere.[6] The main problem arises from the lack of any outstanding separation method for indium. On the other hand, liquid-liquid extraction and ion-exchange methods are relatively effective.

Indium can be extracted as $HInBr_4$, aq. from 4–5 M hydrobromic acid into diethyl ether or diisopropyl ether. However, iron(III), gold, gallium, thallium(III), tin(II,IV) and antimony(V) are also extracted under these conditions. Indium is also extracted from 0.5–2 M hydroiodic acid or from 1.5 M potassium iodide/0.75 M sulfuric acid solutions into diethyl ether. Beryllium, iron(II), aluminium and gallium are not extracted, but cadmium, tin(II) and thallium(I,III) follow the indium into the organic layer.

Small amounts of indium are extracted from a basic solution containing potassium cyanide with a chloroform solution of dithizone. Thallium(I), tin(II), lead and bismuth are also extracted. The extractive separation of indium from other Group 13 elements has already been alluded to in section 10.2.1.

Indium can be separated from iron(III), zinc, aluminium and other metals by adsorption on the cation-exchange resin AG 50W-X8 from a solution composed of 0.2 M nitric acid and 30% acetone; elution is brought about with a solution composed of 0.5 M hydrochloric acid and 30% acetone.[88] Other ion-exchange separations have been mentioned in section 10.2.1. As noted in this section, studies have also been carried out to determine the feasibility of separating and determining indium by ion chromatography and high-performance liquid chromatography.

10.3.2 Methods of determination

Earlier reviews of this subject feature in Refs. 12, 66 and 89.

10.3.2.1 Gravimetric methods. Indium hydroxide is precipitated from a solution containing ammonium nitrate with aqueous ammonia. The precipitate is filtered, ignited and weighed as In_2O_3. Indium may also be determined as the 8-quinolinate complex.

10.3.2.2 Volumetric (titrimetric) methods. Of the various methods which involve (i) titration with $K_4[Fe(CN)_6]$, (ii) oxidation-reduction titration of the indium 8-quinolinate complex and (iii) titration with the disodium salt of ethylenediaminetetraacetic acid (EDTA), the last is recommended.

The stability constant (as $\log_{10} K$) of the complex formed by indium(III) with the EDTA anion is 24.9. Indium can be determined by direct titration with EDTA at pH 2–4. 1-(2-Pyridylazo)-2-naphthol (PAN), Cu–PAN and Xylenol Orange (XO) are among the agents which function as indicators. When Cu–PAN is used, indium can be titrated at pH 2.5–3 (in an acetate medium), as in the case of gallium (section 10.2.2.2). When XO is used, the pH of the sample solution is adjusted to 3 with acetate, the solution is heated to 50–60°C, XO is added and the solution is titrated with a standard solution of the disodium salt of EDTA.

Information about the successive determination of gallium, indium and thallium has already been given in section 10.2.2.2.

10.3.2.3 Electrochemical methods. Various electrochemical methods are available for indium. Many papers have been published on the polarography or voltammetry of indium; supporting electrolytes include hydrochloric acid, potassium chloride (as in the studies of Kurotu[90]) and hydrobromic acid. Polarography in the presence of organic complexing reagents has also been investigated.

Anodic stripping voltammetry of indium has been studied by many workers. In this way, for example, Locatelli et al.[91,92] have simultaneously determined indium and cadmium. Potentiometric stripping analysis has also been explored.

Indium(I) in $In_2O_3 \cdot xIn_2O$ has been determined by a potentiometric oxidation-reduction titration.[93] The use of coulometric complexometric techniques has been mentioned previously (see section 10.2.2.3).

10.3.2.4 Photometric (spectrophotometric) methods. Numerous papers have appeared since the publication of the earlier general review.[6] The methods described may be classified as follows: (i) sensitive colour reactions with phenylfluorone and related compounds in the presence of cationic surfactants; (ii) absorptiometric and fluorimetric determination with Schiff bases; (iii) use of heterocyclic azo reagents, including 4-(2-pyridylazo)-resorcinol and the condensation product of poly(vinyl alcohol) with 7-(4-formylphenylazo)-8-quinolinol;[94] and (iv) use of ion-association reagents.

Nevertheless, indium still lacks an outstanding photometric reagent. Methods using 8-quinolinol and dithizone are relatively good. The former involves chloroform extraction at pH 3.5 and the sensitivity index corresponding to 0.001 absorbance is 0.017 $\mu g\,cm^{-2}$ In at 400 nm. The dithizone method involves either carbon tetrachloride extraction at pH 5.5 (with a sensitivity index of 0.0017 $\mu g\,cm^{-2}$ In at 510 nm) or chloroform extraction at pH 9 (with a sensitivity index of 0.0019 $\mu g\,cm^{-2}$ In at 510 nm).

An outline of the 8-quinolinol method is as follows. The pH of the sample solution containing 10–100 μg of In is adjusted to about 3.5 and the solution is transferred to a separating funnel. A buffer solution (pH 3.5; nitric acid–potassium hydrogen phthalate–ammonia) is added, together with 20.0 cm^3 of a 0.5% solution of 8-quinolinol in chloroform, and the mixture is shaken. The separated organic phase is filtered through a dry filter paper and its absorbance is measured at 400 nm. Instead of absorbance, the fluorescence intensity of the organic phase may be measured.

10.3.2.5 Atomic spectroscopy. Atomic spectroscopic methods lend themselves well to the determination of indium; atomic absorption spectrometry (AAS), in particular, has found widespread applications. Comparisons of flame AAS and flame emission spectrometry have been made.[95,96]

AAS.[97] Procedures typically involve the injection of a dilute nitric acid solution into an air–acetylene or dinitrogen oxide–acetylene flame or into a graphite furnace. In either case a wavelength of 303.9 nm is widely used. Graphite furnace AAS commonly entails the addition of a matrix modifier such as lanthanum nitrate, palladium nitrate or a mixture of nickel nitrate, aluminium nitrate and the diammonium salt of EDTA.[98] A recent paper[99] describes the determination of 20–80 ng g^{-1} of In in silicate rocks.

AES. With a DC arc or an inductively coupled plasma (ICP) source, emission measurements are commonly made at wavelengths of 303.936 and 451.132 nm. The relevant ICP emission spectrum has been characterised in some detail.[43] Flame emission spectrometry (with an air–acetylene or dinitrogen oxide–acetylene flame and a wavelength of 451.1 nm) offers an alternative approach.

10.3.2.6 Neutron activation analysis. Indium can be determined with high sensitivity by reference to the following nuclides formed by neutron bombardment: ^{114m}In (half-life 49.51 days, γ 0.190 MeV)[100] and ^{116m}In (half-life 54.1 min, γ 1.097, 1.293 MeV).[101]

10.4 Thallium

The analytical chemistry of thallium has been the subject of two relatively recent reviews.[102,103] The commonly occurring oxidation states of thallium are +1 and +3 and consequently analytical methods (separation and determination) for thallium are many, and some of these methods are selective. However, all operations need to take into account the toxicity of thallium and its compounds.

Thallium metal dissolves in nitric acid or sulfuric acid but only with difficulty in hydrochloric acid; thallium alloys dissolve in mineral acids or aqua regia. Thallium minerals are decomposed with sulfuric acid and potassium sulfate, pyrite and sphalerite with hydrochloric acid and nitric acid and silicate rocks with hydrofluoric acid and nitric (or sulfuric) acid.

Organic samples can be decomposed by wet oxidation, e.g. with nitric acid and sulfuric acid, but dry oxidation should be avoided because thallium may be lost by volatilisation.

10.4.1 Separations

A general account of how thallium may be separated from other elements is given in Ref. 104.

Thallium(I) is separated by precipitation with hydrogen sulfide, potassium chromate, potassium iodide or thionalide. Small amounts of thallium(III) are co-precipitated with iron(III) hydroxide. Small amounts of thallium(I) are extracted from a basic solution (pH about 10) containing cyanide and citrate with a chloroform solution of dithizone; indium, tin(II), lead and bismuth follow suit.

Thallium(III) is extracted as $HTlCl_4$, aq. from 1–6 M hydrochloric acid into diethyl ether. Molybdenum, iron(III), gold, gallium and tin are also extracted. Alternatively thallium(III) is extracted from 1–6 M hydrobromic acid into diethyl ether or diisopropyl ether. From 1 M hydrobromic acid gold is also extracted, but iron(III) and gallium are not; thallium(I) follows thallium(III) from 1.1–3.2 M hydrobromic acid into diethyl ether.

The extractive separation of thallium(III), indium and gallium has already been noted (see section 10.2.1). Thus anion-exchange separation of aluminium, gallium, indium and thallium(III) and ion chromatography of gallium, indium and thallium(III) mixtures have both been investigated.

10.4.2 Methods of determination

General reviews of this subject up to 1973 appear elsewhere.[12,66,105]

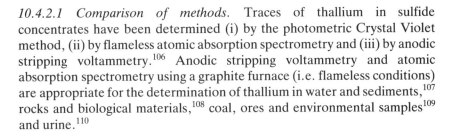

10.4.2.1 Comparison of methods. Traces of thallium in sulfide concentrates have been determined (i) by the photometric Crystal Violet method, (ii) by flameless atomic absorption spectrometry and (iii) by anodic stripping voltammetry.[106] Anodic stripping voltammetry and atomic absorption spectrometry using a graphite furnace (i.e. flameless conditions) are appropriate for the determination of thallium in water and sediments,[107] rocks and biological materials,[108] coal, ores and environmental samples[109] and urine.[110]

10.4.2.2 Gravimetric methods. There follows an outline of the thallium(I) chromate method. The acidic solution containing thallium(I) is neutralised with aqueous ammonia and then made basic. The solution is heated on a water bath and potassium chromate is added. After 12 h or more, the precipitate is filtered through a sintered-glass filter crucible, washed, dried at 120–130°C and weighed as Tl_2CrO_4. Conditions for the precipitation of thallium(I) chromate from 'homogeneous' solution have been described.[111]

In another method thallium(I) is precipitated with potassium iodide and weighed as TlI.

10.4.2.3 Volumetric (titrimetric) methods.   These may involve either oxidation-reduction or complexometric titrations.

To bring about the oxidation of thallium(I) to thallium(III) requires the use of an oxidising agent like $KBrO_3$, KIO_3, $Ce(SO_4)_2$ or $KMnO_4$. When potassium bromate is used, for example, the hydrochloric acid concentration of the sample solution is adjusted to about 2 M and Methyl Orange is added as the indicator; the solution is then heated to 50–60°C and titrated with standard potassium bromate solution. Studies have been carried out to devise ways of reducing any thallium(III) in the sample to thallium(I) prior to the titration with bromate.[112–114]

The reduction of thallium(III) to thallium(I) can be brought about by the addition of potassium iodide, and the liberated iodine is titrated with sodium thiosulfate solution.

Complexometric methods involve ethylenediaminetetraacetic acid (EDTA). The stability constants (as $\log_{10} K$) of the complexes formed by the EDTA anion with thallium(I) and thallium(III) are 6.41 and 35.3, respectively. Thallium(III) can be determined by direct titration with the disodium salt of EDTA at pH 1–8. 1-(2-Pyridylazo)-2-naphthol, 4-(2-pyridylazo)-resorcinol, Xylenol Orange (XO) and Methylthymol Blue are representative of the indicators that may be used.

An outline of a representative procedure is as follows. Thallium(I) may be separated by precipitation as its iodide; the precipitate is separated and decomposed by heating with aqua regia (resulting in the oxidation of thallium(I) to thallium(III)). The solution is diluted and adjusted to pH 4–5 with ammonium acetate, XO is added and the solution is titrated at 60–80°C with a standard solution of the disodium salt of EDTA.

In recent work,[115] after the usual back-titration with zinc sulfate with XO as the indicator, 4-amino-5-mercapto-3-propyl-1,2,4-triazole has been added and the EDTA released titrated with zinc sulfate solution.

Section 10.2.2.2 has referred already to the successive determination of gallium, indium and thallium.

10.4.2.4 Electrochemical methods.   Thallium(I)   gives   well-defined polarographic waves in various supporting electrolytes. Anodic stripping voltammetry is highly suitable for the determination of very low concentrations of thallium. The disodium salt of EDTA (which also acts as a supporting electrolyte) and surfactants have been used to mask interfering elements. In addition to the applications mentioned in section 10.4.2.1, thallium has been determined in the following materials with or without separation: Al,[116] Bi,[117] Cd salts,[118] Ni and Ni-base alloys,[92,119,120] Pb,[121] Pb salts[122] and wine.[123]

The determination of thallium by potentiometric stripping analysis[124] has also been studied.[125] In relation to other electrochemical methods, moreover, many ion-selective electrodes for thallium(I) and thallium(III) have been prepared.

10.4.2.5 Photometric (spectrophotometric) methods. A general account of this subject has been published recently.[104] Three main methods have been favoured for the determination of thallium; these involve (i) the use of ion associates of $TlCl_4^-$ or $TlBr_4^-$ and basic (cationic) dyes such as Rhodamine B; (ii) the formation of a chelate derivative of thallium(I) with dithizone; and (iii) oxidation of iodide to iodine with thallium(III).

The Rhodamine B method. The following is an outline of a typical procedure. To the sample solution containing 1–10 μg of Tl^I or Tl^{III}, 0.50 cm^3 of 3 M sulfuric acid is added and the solution is evaporated until fumes of sulfuric acid or sulfur trioxide appear. Hydrochloric acid (1 part to 9 of water, 5 cm^3) and 1.0 cm^3 of saturated bromine water are added and the solution is heated to near the boiling point. Heating is stopped when the colour due to bromine has disappeared. The solution is diluted to 10.0 cm^3 with hydrochloric acid (1 part to 9 of water). The solution is transferred to a separating funnel as completely as possible. One cm^3 of 0.2% Rhodamine B solution is added and the solution shaken with 10.0 cm^3 of benzene. The separated organic phase is centrifuged and transferred to a dry cell without rinsing. The absorbance is measured at 560 nm. The sensitivity index corresponding to 0.001 absorbance is 0.0021 μg cm^{-2} Tl. Iron, gold, mercury and antimony interfere.

As little as 0.5 μg of Tl can be determined fluorimetrically with Rhodamine B.[126] A recent paper[127] also describes the use of Pyronine G, a basic dye, for the photometric determination of thallium(III) as the ion-pair between $TlCl_4^-$ and the Pyronine G cation.

10.4.2.6 Atomic spectroscopy. As with the other Group 13 metals, atomic spectroscopic methods, and especially atomic absorption spectrometry (AAS), have proved highly effective for the determination of thallium (compare section 10.4.2.1). Comparisons of atomic spectroscopic methods have included the following: analysis of minerals and coal by AAS and inductively coupled plasma atomic emission spectrometry (ICP AES),[128] steel-manufacturing materials by flame AES, ICP AES and AAS,[129] high-temperature superconductors[130] and biological materials[131] by flame AAS and flame AES, and urine by flame AAS and emission spectrography.[132]

AAS.[133]   Thallium in rocks[134] and urine[135] has been determined by either flame or flameless (graphite furnace) AAS after extractive separations. For AAS a dilute sulfuric or nitric acid solution is injected into an air–acetylene flame or a graphite furnace. In either case, measurements are generally made at a wavelength of 276.8 nm. Flame AAS has found applications in the analysis of not only geological and biological materials, but also water and zinc and cadmium. In AAS work with a graphite furnace chloride interferes seriously and matrix modifiers, e.g. palladium nitrate and magnesium nitrate[136] and nickel nitrate and the tetraammonium salt of EDTA,[137] often need to be added. This approach has also found numerous applications, including the analysis of water, rocks, soil, coal and fly ash, nickel-base and other non-ferrous alloys and biological materials.

AES.   As mentioned above, flame AES (with a dinitrogen oxide–acetylene or an air–acetylene flame and a wavelength of 535.0 nm) has been employed for the determination of thallium. With an ICP source, wavelengths of 351.924, 276.787 and 535.046 nm are amongst those commonly used. Reference 43 includes details of the ICP emission spectrum.

Other methods.   Laser-excited atomic fluorescence spectrometry has been used for the determination of extremely small amounts of thallium in biological materials[138] and air.[139] Laser-enhanced ionisation spectrometry affords another promising method for determining traces of thallium.[140]

10.4.2.7 X-Ray fluorescence (XRF).   Thallium in plant materials has been determined by XRF.[141,142]

10.4.2.8 Mass spectrometry (MS).   Isotope dilution MS has been applied by many workers to the determination of traces of thallium in various matrices, e.g. reagents, water, rocks, coal and biological materials. Another variation involves ICP MS which has been exploited for the trace analysis of geological materials, nickel-base alloys and biological materials.[52] In addition and more specifically, dimethylthallium ions have been determined by secondary-ion mass spectrometry (SIMS).[143]

10.4.2.9 Neutron activation analysis.   The method is based on the measurement of the β radiation (0.763 MeV) of ^{204}Tl (half-life 3.78 years) produced by neutron bombardment of the sample.[144]

References

1. T.R. Dulski, *Anal. Chem.*, 1991, **63**, 65R.
2. C.J. Menendez-Botet and M.K. Schwartz, *Anal. Chem.*, 1991, **63**, 194R.
3. P. MacCarthy, R.W. Klusman, S.W. Cowling and J.A. Rice, *Anal. Chem.*, 1991, **63**, 301R.
4. British Standards Institution, *British Standard*, BS 4140: Part 4: 1986.
5. British Standards Institution, *British Standard*, BS 4140: Part 13: 1986.
6. H. Onishi, *Photometric Determination of Traces of Metals, Part IIA: Individual Metals, Aluminum to Lithium*, 4th edn., Wiley, New York, 1986.
7. N.E. Fortier and J.S. Fritz, *Talanta*, 1985, **32**, 1047.
8. J.R. Dean, *Analyst*, 1989, **114**, 165.
9. P. Jones, T. Williams and L. Ebdon, *Anal. Chim. Acta*, 1990, **237**, 291.
10. Y. Nagaosa, H. Kawabe and A.M. Bond, *Anal. Chem.*, 1991, **63**, 28.
11. C.I. Measures and J.M. Edmond, *Anal. Chem.*, 1989, **61**, 544.
12. W.F. Hillebrand, G.E.F. Lundell, H.A. Bright and J.I. Hoffman, *Applied Inorganic Analysis*, 2nd edn., Wiley, New York, 1953.
13. G.H. Farrah and M.L. Moss, *Aluminum*, in *Treatise on Analytical Chemistry*, eds. I.M. Kolthoff and P.J. Elving, Wiley, New York, 1966, Part II, Vol. 4.
14. G.P. Koch, E. Morgan and C.L. Hilton, *Aluminium*, in *Encyclopedia of Industrial Chemical Analysis*, eds. F.D. Snell and C.L. Hilton, Wiley, New York, 1967, Vol. 5.
15. G.H. Jeffery, J. Bassett, J. Mendham and R.C. Denney, *Vogel's Textbook of Quantitative Chemical Analysis*, 5th edn., Longman Scientific & Technical, Harlow, UK, 1989.
16. R. Playle, J. Gleed, R. Jonasson and J.R. Kramer, *Anal. Chim. Acta*, 1982, **134**, 369.
17. I.A. Voinovitch, J.P. Degre, J. Louvrier and N. Musikas, *Analusis*, 1984, **12**, 214.
18. I. Sarghie, C. Simion and D. Bilba, *Rev. Chim. (Bucharest)*, 1986, **37**, 244.
19. D.R. Corbin, B.F. Burgess, Jr., A.J. Vega and R.D. Farlee, *Anal. Chem.*, 1987, **59**, 2722.
20. R.C. Bruce and D.J. Lyons, *Commun. Soil Sci. Plant Anal.*, 1984, **15**, 15.
21. A. Mosquera and F. Mombiela, *Commun. Soil Sci. Plant Anal.*, 1986, **17**, 97.
22. E.J. Willoughby, *Commun. Soil Sci. Plant Anal.*, 1986, **17**, 667.
23. M. Meaney, M. Connor, C. Breen and M.R. Smyth, *J. Chromatogr.*, 1988, **449**, 241.
24. B. Kratochvil, N. Motkovsky, M.J.M. Duke and D. Ng, *Can. J. Chem.*, 1987, **65**, 1047.
25. F.C. Thornton, M. Schaedle and D.J. Raynal, *Commun. Soil Sci. Plant Anal.*, 1985, **16**, 931.
26. A. Alimonti, E. Coni, S. Caroli, E. Sabbioni, G.E. Nicolaou and R. Pietra, *J. Anal. Atom. Spectrom.*, 1989, **4**, 577.
27. S. Banerjee and R.K. Dutta, *Analyst*, 1976, **101**, 516.
28. O.P. Bhargava, *Talanta*, 1979, **26**, 146.
29. S. Dasgupta, B.C. Sinha and N.S. Rawat, *Analyst*, 1984, **109**, 39.
30. V.N. Tikhonov and O.A. Mikhailova, *Zh. Anal. Khim.*, 1983, **38**, 1982.
31. N. Zhou, Y. Gu, Z. Lu and W. Chen, *Talanta*, 1985, **32**, 1119.
32. A. Hulanicki, R. Lewandowski, A. Lewenstam, M. Chmurska and H. Matuszak, *Mikrochim. Acta*, 1985, **III**, 253.
33. A. Campiglio, *Mikrochim. Acta*, 1986, **III**, 425.
34. L.N. Moskvin, A.E. Zeimal and A.E. Mogilevskaya, *Zh. Anal. Khim.*, 1990, **45**, 990.
35. N. Radic and M. Bralic, *Analyst*, 1990, **115**, 737.
36. K.E. Johnson, A.K. Brichta and K.-L. Holter, *Can. J. Chem.*, 1988, **66**, 139.
37. J. Wang, P.A.M. Farias and J.S. Mahmoud, *Anal. Chim. Acta*, 1985, **172**, 57.
38. C.M.G. van den Berg, K. Murphy and J.P. Riley, *Anal. Chim. Acta*, 1986, **188**, 177.
39. R. Giordano, S. Costantini, I. Vernillo, B. Casetta and F. Aldrighetti, *Microchem. J.*, 1984, **30**, 435.
40. A. Sanz-Medel, R. Rodriguez Roza, R. Gonzalez Alonso, A. Noval Vallina and J. Cannata, *J. Anal. Atom. Spectrom.*, 1987, **2**, 177.
41. A. Cedergren and W. Frech, *Pure Appl. Chem.*, 1987, **59**, 221.
42. W. Slavin, *J. Anal. Atom. Spectrom.*, 1986, **1**, 281.
43. T.A. Anderson and M.L. Parsons, *Appl. Spectrosc.*, 1984, **38**, 625.
44. M.H. Jones and B.W. Wilson, *Analyst*, 1991, **116**, 449.
45. B.C. Pearce, J.W.F. Hill and I. Kerry, *Analyst*, 1990, **115**, 1397.

46. P.M. Bertsch, R.J. Barnhisel, G.W. Thomas, W.J. Layton and S.L. Smith, *Anal. Chem.*, 1986, **58**, 2583.
47. P.M. Bertsch and M.A. Anderson, *Anal. Chem.*, 1989, **61**, 535.
48. C. Changui, W.F.E. Stone and L. Vielvoye, *Analyst*, 1990, **115**, 1177.
49. P.P. Man and J. Klinowski, *J. Chem. Soc., Chem. Commun.*, 1988, 1291.
50. K.R. Koch, *Analyst*, 1990, **115**, 823.
51. T. Nagata, M. Hayatsu and N. Kosuge, *Anal. Sci.*, 1991, **7**, 213.
52. J.K. Friel, C.S. Skinner, S.E. Jackson and H.P. Longerich, *Analyst*, 1990, **115**, 269.
53. H.J. Seim and R.C. Calkins, *Anal. Chem.*, 1979, **51**, 170R.
54. G. Kudermann, *Fresenius' Z. Anal. Chem.*, 1988, **331**, 697.
55. J.L. Genna, *J. Cryst. Growth*, 1988, **89**, 62.
56. L.F. Vassamillet. *J. Anal. Atom. Spectrom.*, 1989, **4**, 451.
57. I. Izumi, *Bunseki Kagaku*, 1989, **38**, 341.
58. H. Koshima and H. Onishi, *Analyst*, 1986, **111**, 1261.
59. H. Koshima and H. Onishi, *Analyst*, 1987, **112**, 335.
60. S.R. Kuchekar and M.B. Chavan, *Talanta*, 1988, **35**, 357.
61. X. Luo, Z. Su, X. Chang, G. Zhan and X. Chao, *Analyst*, 1991, **116**, 965.
62. F.W.E. Strelow and T.N. van der Walt, *Talanta*, 1987, **34**, 895.
63. D. Yan, J. Zhang and G. Schwedt, *Fresenius' Z. Anal. Chem.*, 1988, **331**, 601.
64. C. Baiocchi, G. Saini, P. Bertolo, G.P. Cartoni and G. Pettiti, *Analyst*, 1988, **113**, 805.
65. Y. Shijo, A. Saitoh and K. Suzuki, *Chem. Lett.*, 1989, 181.
66. H. Onishi, *Gallium, Indium and Thallium*, in *Treatise on Analytical Chemistry,* eds. I.M. Kolthoff and P.J. Elving, Wiley, New York, 1962, Part II, Vol. 2.
67. H. Onishi, *Gallium*, in *Encyclopedia of Industrial Chemical Analysis*, eds. F.D. Snell and L.S. Ettre, Wiley, New York, 1971, Vol. 13.
68. M.A. E.-H. Hafez, A.M.A. Abdallah and T.M.A. E.-F. Wahdan, *Analyst*, 1991, **116**, 663.
69. R. Udisti and G. Piccardi, *Fresenius' Z. Anal. Chem.*, 1988, **331**, 35.
70. J. Wang and J.M. Zadeii, *Anal. Chim. Acta*, 1986, **185**, 229.
71. Y. Hasegawa, T. Inagake, Y. Karasawa and A. Fujita, *Talanta*, 1983, **30**, 721.
72. S. Caroli, A. Alimonti, P.D. Femmine and S.K. Shukla, *Anal. Chim. Acta*, 1982, **136**, 225.
73. L.N. Sukhoveeva, B. Ya. Spivakov, A.V. Karyakin and Yu.A. Zolotov, *Zh. Anal. Khim.*, 1979, **34**, 693.
74. X. Shan, Z. Yuan and Z. Ni, *Anal. Chem.*, 1985, **57**, 857.
75. A.I. Zabreva, L.M. Philippova, N.N. Andreeva and N.V. Ivanova, *Anal. Chim. Acta*, 1987, **195**, 357.
76. K. Matsumoto and K. Fuwa, *Anal. Chem.*, 1979, **51**, 2355.
77. K. Nakamura and H. Orii, *Anal. Chem.*, 1980, **52**, 532.
78. A. Makishima, I. Inamoto and K. Chiba, *Appl. Spectrosc.*, 1990, **44**, 91.
79. T. Tanaka, Y. Homma and S. Kurosawa, *Anal. Chem.*, 1988, **60**, 58.
80. O. Stulzaft, B. Maziere and S. Ly, *J. Radioanal. Chem.*, 1980, **55**, 291.
81. J.W. Gramlich and L.A. Machlan, *Anal. Chem.*, 1985, **57**, 1788.
82. T. Tanaka, *Anal. Sci.*, 1989, **5**, 171.
83. S. Gangadharan, S. Natarajan, J. Arunachalam, S. Jaikumar and S.V. Burangey, *J. Res. Natl. Bur. Stand. (U.S.)*, 1988, **93**, 400.
84. W.-D. Hu, C. Vandecasteele, G. Wauters and R. Dams, *Anal. Chim. Acta*, 1989, **226**, 193.
85. S. Kurosawa, *Bunseki*, 1991, 588.
86. T. Tanaka, K. Watakabe, K. Kurooka and T. Yoshimori, *Bunseki Kagaku*, 1989, **38**, 724.
87. S. Kawase, *Bunseki*, 1987, 310.
88. F.W.E. Strelow, C.H.S.W. Weinert and T.N. van der Walt, *Talanta*, 1974, **21**, 1183.
89. J.M. Ramaradhya, R.C. Bell, S. Brownlow and C.J. Mitchell, *Indium*, in *Encyclopedia of Industrial Chemical Analysis*, eds. F.D. Snell and L.S. Ettre, Wiley, New York, 1971, Vol. 14.
90. T. Kurotu, *Anal. Chim. Acta*, 1990, **233**, 325.
91. C. Locatelli, F. Fagioli, C. Bighi and T. Garai, *Talanta*, 1986, **33**, 243.
92. C. Locatelli, F. Fagioli, T. Garai, C. Bighi and R. Vecchietti, *Anal. Chim. Acta*, 1988, **204**, 189.

93. I.L. Nesterova, A.V. Moev, M.A. Voronova and P.V. Kovtunenko, *Zh. Anal. Khim.*, 1984, **39**, 47.
94. S.-C. Liang and E. Zang, *Fresenius' Z. Anal. Chem.*, 1989, **334**, 511.
95. M.E. Britske and I.V. Slabodenyuk, *Zh. Anal. Khim.*, 1982, **37**, 1417.
96. J.L. Bernal, Ma.J. del Nozal, L. Deban and A.J. Aller, *Talanta*, 1982, **29**, 1113.
97. B. Ya. Spivakov, L.N. Sukhoveeva, K. Dittrich, A.V. Karyakin and Yu. A. Zolotov, *Zh. Anal. Khim.*, 1979, **34**, 1947.
98. K. Matsusaki, *Bunseki Kagaku*, 1990, **39**, 823.
99. R. Kuroda, T. Wada, T. Soma, N. Itsubo and K. Oguma, *Analyst*, 1990, **115**, 1535.
100. M. Ebihara, A. Nemoto and H. Akaiwa, *Bunseki Kagaku*, 1987, **36**, 836.
101. E. Taskaev and D. Apostolov, *J. Radioanal. Chem.*, 1979, **49**, 127.
102. M. Sager, *Spurenanalytik des Thalliums*, in *Analytische Chemie für die Praxis*, eds. H. Hulpke, H. Hartkamp and G. Tölg, Georg Thieme Verlag, Stuttgart, 1986.
103. B. Griepink, M. Sager and G. Tölg, *Pure Appl. Chem.*, 1988, **60**, 1425.
104. H. Onishi, *Photometric Determination of Traces of Metals, Part IIB: Individual Metals, Magnesium to Zirconium*, 4th edn., Wiley, New York, 1989.
105. F.A. Lowenheim, *Thallium*, in *Encyclopedia of Industrial Chemical Analysis*, eds. F.D. Snell and L.S. Ettre, Wiley, New York, 1973, Vol. 18.
106. E.A. Jones and A.F. Lee, *Rep. Natl. Inst. Metall. (S. Africa)*, 1980, No. 2036.
107. J.P. Riley and S.A. Siddiqui, *Anal. Chim. Acta*, 1986, **181**, 117.
108. I. Liem, G. Kaiser, M. Sager and G. Tölg, *Anal. Chim. Acta*, 1984, **158**, 179.
109. H. Gorbauch, H.H. Rump, G. Alter and C.H. Schmitt-Henco, *Fresenius' Z. Anal. Chem.*, 1984, **317**, 236.
110. G.J.H. Bessems, L.W. Westerhuis and H. Baadenhuijsen, *Ann. Clin. Biochem.*, 1983, **20**, 321.
111. K.N. Upadhyaya, *Analyst*, 1978, **103**, 766.
112. S.R. Sagi, G.S.P. Raju, K.A. Rao and M.S.P. Rao, *Talanta*, 1982, **29**, 413.
113. S.N. Dindi and N.V.V.S.N.M. Sarma, *Talanta*, 1985, **32**, 1161.
114. A.R.M. Rao, M.S.P. Rao, K.V. Ramana and S.R. Sagi, *Talanta*, 1989, **36**, 686.
115. N. Shetty A., R.V. Gadag and M.R. Gajendragad, *Talanta*, 1988, **35**, 721.
116. M.M. Palrecha, A.V. Kulkarni and R.G. Dhaneshwar, *Analyst*, 1986, **111**, 375.
117. A. Ciszewski, *Talanta*, 1985, **32**, 1051.
118. Z. Lukaszewski, M.K. Pawlak and A. Ciszewski, *Talanta*, 1980, **27**, 181.
119. L.G. Petrova, V.I. Ignatov and E. Ya. Neiman, *Zh. Anal. Khim.*, 1982, **37**, 22.
120. Y. Zhang, *Fenxi Huaxue*, 1982, **10**, 726.
121. I.B. Berengard and B. Ya. Kaplan, *Zavod. Lab.*, 1984, **50**(4), 8.
122. A. Ciszewski and Z. Lukaszewski, *Talanta*, 1983, **30**, 873.
123. H. Eschnauer, V. Gemmer-Colos and R. Neeb, *Z. Lebensm.-Unters. -Forsch.*, 1984, **178**, 453.
124. D. Jagner and A. Graneli, *Anal. Chim. Acta*, 1976, **83**, 19.
125. J.K. Christensen, L. Kryger and N. Pind, *Anal. Chim. Acta*, 1982, **141**, 131.
126. H. Koshima and H. Onishi, *Analyst*, 1989, **114**, 615.
127. K.K. Namboothiri, N. Balasubramanian and T.V. Ramakrishna, *Talanta*, 1991, **38**, 945.
128. H. Berndt, J. Messerschmidt, F. Alt and D. Sommer, *Fresenius' Z. Anal. Chem.*, 1981, **306**, 385.
129. K.-H. Sauer and S. Eckhard, *Mikrochim. Acta, Suppl.*, 1981, 87.
130. E.S. Zolotovitskaya, V.G. Potapova and T.V. Druzenko, *Zh. Anal. Khim.*, 1990, **45**, 1213.
131. A.S. Curry, J.F. Read and A.R. Knott, *Analyst*, 1969, **94**, 744.
132. C.D. Wall, *Clin. Chim. Acta*, 1977, **76**, 259.
133. M.S. Leloux, Nguyen Phu Lich and J.-R Claude, *Atom Spectrosc.*, 1987, **8**, 71.
134. R. Keil, *Fresenius' Z. Anal. Chem.*, 1981, **309**, 181.
135. J. Flanjak and A.E. Hodda, *Anal. Chim. Acta*, 1988, **207**, 283.
136. B. Welz, G. Schlemmer and J.R. Mudakavi, *Anal. Chem.*, 1988, **60**, 2567.
137. K. Matsusaki, *Bunseki Kagaku*, 1991, **40**, 413.
138. J.P. Dougherty, J.A. Costello and R.G. Michel, *Anal. Chem.*, 1988, **60**, 336.
139. Z. Liang, G.-T. Wei, R.L. Irwin, A.P. Walton, R.G. Michel and J. Sneddon, *Anal. Chem.*, 1990, **62**, 1452.
140. N. Omenetto, T. Berthoud, P. Cavalli and G. Rossi, *Anal. Chem.*, 1985, **57**, 1256.

141. H. Rethfeld, *Fresenius' Z. Anal. Chem.*, 1980, **301**, 308.
142. H.F. Haas and V. Krivan, *Fresenius' Z. Anal. Chem.*, 1985, **322**, 261.
143. K. Günther and F. Umland, *Fresenius' Z. Anal. Chem.*, 1989, **333**, 6.
144. R.K. Itawi and Z.R. Turel, *J. Radioanal. Nucl. Chem.*, 1987, **115**, 141.

Index

Most of the entries in the following index appear under the headings of the individual Group 13 metals. Listings are then given for each element, firstly under the heading 'general', and then under separate headings based on the formal oxidation state of the metal, e.g. gallium, oxidation states < +1 (including the element), gallium(I), gallium(II), gallium, mixed valence derivatives, and gallium(III). Under a given heading there may be additional subdivisions to accommodate important classes of compounds, e.g. complexes, organogallium(III) compounds, and Group V compounds (of gallium). Bold lettering is used to highlight some of the major entries.